EMERGING TECHNOLOGIES IN NON-DESTRUCTIVE TESTING VI

PROCEEDINGS OF THE 6TH INTERNATIONAL CONFERENCE ON EMERGING TECHNOLOGIES
IN NONDESTRUCTIVE TESTING (ETNDT6), BRUSSELS, BELGIUM, 27–29 MAY 2015

# Emerging Technologies in Non-Destructive Testing VI

*Editors*

D.G. Aggelis & D. Van Hemelrijck
*Department of Mechanics of Materials and Constructions (MeMC),
Faculty of Engineering, Vrije Universiteit Brussel, Elsene, Belgium*

S. Vanlanduit
*Department of Mechanical Engineering (MECH), Faculty of Engineering,
Vrije Universiteit Brussel, Brussels, Belgium*

A. Anastasopoulos
*Mistras Group Hellas, Athens, Greece*

T.P. Philippidis
*Department of Mechanical Engineering and Aeronautics,
University of Patras, Patras, Greece*

CRC Press is an imprint of the
Taylor & Francis Group, an **informa** business

A BALKEMA BOOK

*CRC Press/Balkema is an imprint of the Taylor & Francis Group, an informa business*

© 2016 Taylor & Francis Group, London, UK

Typeset by V Publishing Solutions Pvt Ltd., Chennai, India
Printed and bound in Great Britain by CPI Group (UK) Ltd, Croydon, CR0 4YY

All rights reserved. No part of this publication or the information contained herein may be reproduced, stored in a retrieval system, or transmitted in any form or by any means, electronic, mechanical, by photocopying, recording or otherwise, without written prior permission from the publisher.

Although all care is taken to ensure integrity and the quality of this publication and the information herein, no responsibility is assumed by the publishers nor the author for any damage to the property or persons as a result of operation or use of this publication and/or the information contained herein.

Published by: CRC Press/Balkema
P.O. Box 11320, 2301 EH Leiden, The Netherlands
e-mail: Pub.NL@taylorandfrancis.com
www.crcpress.com – www.taylorandfrancis.com

ISBN: 978-1-138-02884-5 (Hbk)
ISBN: 978-1-135-64754-8 (eBook PDF)

*Emerging Technologies in Non-Destructive Testing VI – Aggelis et al. (Eds)*
*© 2016 Taylor & Francis Group, London, ISBN 978-1-138-02884-5*

# Table of contents

| | |
|---|---|
| Preface | xi |
| Organization | xiii |

## *Plenary papers*

| | |
|---|---|
| Basics and applications of NDE based on elastodynamics toward infra-dock for concrete structures<br>*M. Ohtsu* | 3 |
| Resonant defects: A new approach to highly-sensitive defect-selective ultrasonic imaging<br>*I. Solodov* | 13 |

## *Infra-asset assessment with innovative NDT*

| | |
|---|---|
| Experimental verification of a Rayleigh-wave based technique for detecting the depth<br>of deteriorated concrete<br>*K.C. Chang, T. Shiotani & S.B. Tamrakar* | 25 |
| Evaluation of grouting conditions for PC structures with wide-band ultrasounds<br>*M. Hashinoki, M. Hara, T. Kinoshita & K.C. Chang* | 33 |
| Two-dimensional AE-Tomography based on ray-trace technique for anisotropic materials<br>*Y. Kobayashi & T. Shiotani* | 41 |
| Characterization of inclined surface crack in steel reinforced-concrete<br>by multichannel R-wave measurements<br>*F.W. Lee, H.K. Chai & K.S. Lim* | 49 |
| Elastic wave tomography with arbitrary locations of impact<br>*K. Matsumoto, S. Momoki, T. Shiotani & Y. Kobayashi* | 55 |
| An ultrasonic method utilizing anchors to inspect steel-plate bonded RC decks<br>*N. Ogura, H. Yatsumoto, K.C. Chang & T. Shiotani* | 61 |
| Strategic maintenance philosophy for infra-asset with innovative NDT<br>*T. Shiotani* | 69 |
| Use of X-ray CT and AE for detection of cracked concrete properties<br>*T. Suzuki, T. Morii & T. Shiotani* | 79 |
| Ultrasonic method for predicting residual tension of ground anchors<br>*S.B. Tamrakar, T. Shiotani, K.C. Chang & Y. Fujiwara* | 85 |
| Use of Digital Image Correlation (DIC) and Acoustic Emission (AE) to characterise<br>the structural behaviour of hybrid composite-concrete beams<br>*S. Verbruggen, S. De Sutter, S.N. Iliopoulos, D.G. Aggelis & T. Tysmans* | 93 |

## *Elastic waves*

| | |
|---|---|
| An ACA/BEM for solving wave propagation problems in non-homogeneous materials<br>*T. Gortsas, I. Diakides, D. Polyzos & S.T. Tsinopoulos* | 103 |
| Acousto-ultrasonic Structural Health Monitoring of aerospace composite materials<br>*M. Gresil, A. Muller & C. Soutis* | 109 |

v

Guided wave tomography based inspection of CFRP plates using a probabilistic reconstruction algorithm 117
J. Hettler, M. Tabatabaeipour, S. Delrue & K. Van Den Abeele

Modelling the dispersive behavior of fresh and hardened concrete specimens through non-local lattice models 125
S.N. Iliopoulos, D.G. Aggelis & D. Polyzos

The Ultrasonic Polar Scan: Past, present and future 133
M. Kersemans, A. Martens, S. Delrue, K. Van Den Abeele, L. Pyl, F. Zastavnik, H. Sol, J. Degrieck & W. Van Paepegem

Time-Of-Flight recorded Pulsed Ultrasonic Polar Scan for elasticity characterization of composites 141
A. Martens, M. Kersemans, J. Degrieck, W. Van Paepegem, S. Delrue & K. Van Den Abeele

Ultrasonic testing of adhesively bonded joints in glass panels 147
B. Mojškerc, T. Kek & J. Grum

Guided ultrasonic waves for the NDT of immersed plates 153
P. Rizzo, E. Pistone & A. Bagheri

*Nonlinear elastic waves*

Nonlinear acoustics and acoustic emission methods to monitor damage in mesoscopic elastic materials 161
M. Bentahar, R. El Guerjouma, C. Mechri, Y. Baccouche, S. Idjimarene, A. Novak, S. Toumi, V. Tournat, J.-H. Thomas & M. Scalerandi

A simulation study of Local Defect Resonances (LDR) 167
S. Delrue & K. Van Den Abeele

Highly Nonlinear Solitary Waves for the NDT of slender beams 173
P. Rizzo, A. Bagheri & E. La Malfa Ribolla

Nonlinear ultrasonic inspection of friction stir welds 179
M. Tabatabaeipour, J. Hettler, S. Delrue & K. Van Den Abeele

*Acoustic emission, damage and lifetime prediction of composites*

Fracture mechanism of CFRP-strengthened RC beam identified by AE-SiGMA 185
N. Alver, H.M. Tanarslan, Ö.Y. Sülün & E. Ercan

Acoustic emission of fiber reinforced concrete under double-punching indirect tensile loading 191
D. Choumanidis, P. Nomikos, A. Sofianos, E. Komninou, P. Oikonomou & E. Badogiannis

Acoustic emission monitoring of high temperature process vessels & reactors during cool down 197
D. Papasalouros, K. Bollas, D. Kourousis & A. Anastasopoulos

*NDT in aerospace*

Thermal strains in heated Fiber Metal Laminates 205
B. Müller, J. Sinke, A.G. Anisimov & R.M. Groves

Lamb wave dispersion time-domain study using a combined signal processing approach 213
P. Ochôa, R.M. Groves & R. Benedictus

Fuse-like devices replacing linear sensors—working examples of percolation sensors in operational airliners and chemical installations 219
H. Pfeiffer, I. Pitropakis, M. Wevers, H. Sekler & M. Schoonacker

*Applications*

Non-Destructive Testing techniques to evaluate the healing efficiency of self-healing concrete at lab-scale 227
E. Gruyaert, J. Feiteira, N. De Belie, F. Malm, M. Nahm, C.U. Grosse, E. Tziviloglou, E. Schlangen & E. Tsangouri

Assessment of grouted connection in monopile wind turbine foundations
using combined non-destructive techniques 235
*A.N. Iliopoulos, D. Van Hemelrijck, D.G. Aggelis & C. Devriendt*

Detection of damaged tool in injection molding process with acoustic emission 241
*T. Kek, D. Kusić & J. Grum*

An experimental investigation of Electromagnetic Acoustic Transducers applied
to high temperature plates for potential use in solar thermal industry 247
*M. Kogia, L. Cheng, A. Mohimi, V. Kappatos, T.-H. Gan, W. Balachandran & C. Selcuk*

Monitoring of self-healing activation by means of Acoustic Emission and Digital
Image Correlation 255
*P. Minnebo, D.G. Aggelis & D. Van Hemelrijck*

Finite element model updating using thermographic measurements: Comparison between
materials with high and low thermal conductivity 261
*J. Peeters, B. Ribbens, J.J.J. Dirckx & G. Steenackers*

Portable automated Radio-Frequency scanner for Non-Destructive Testing
of Carbon-Fibre-Reinforced Polymer composites 267
*B. Salski, W. Gwarek, P. Kopyt, P. Theodorakeas, I. Hatziioannidis, M. Koui,
A.Y.B. Chong, S.M. Tan, V. Kappatos, C. Selcuk & T.-H. Gan*

Detection of the interface between two metals by DC current
stimulated thermography 275
*N.J. Siakavellas & J. Sarris*

Water-Absorption-Measurement instrument for masonry façades 281
*M. Stelzmann, U. Möller & R. Plagge*

Assessment of eSHM system combining different NDT methods 287
*M. Strantza, D.G. Aggelis, D. De Baere, P. Guillaume &
D. Van Hemelrijck*

Non-destructive evaluation of an infusion process using capacitive
sensing technique 293
*Y. Yang, T. Vervust, F. Bossuyt, J. Vanfleteren, G. Chiesura, G. Luyckx,
J. Degrieck & M. Kaufmann*

In-situ testing using combined NDT methods for the technical evaluation
of existing bridge 299
*N.V. Zoidis, E.N. Tatsis, E.D. Manikas & T.E. Matikas*

*Biological applications*

Kinect sensor for 3D modelling of logs in small sawmills 309
*J. Antikainen & D. Xiaolei*

A noninvasive approach for the assessment of dental implants stability 315
*P. Rizzo, E. La Malfa Ribolla & A. Di Cara*

Monitoring techniques for nanocrystalline stabilized zirconia from some
medical prosthesis 319
*A. Savin, M.L. Craus, V. Turchenko, A. Bruma, S. Malo & T.E. Konstantinova*

Raman spectroscopy: An emerging technique for minimally-invasive clinical testing 327
*M.Z. Vardaki, P.S. Papaspyridakou, G.L. Givalos, C.G. Kontoyannis & M.G. Orkoula*

*Combination of NDT techniques*

Concrete compressive strength estimation by means of combined NDT 335
*G. Concu, B. De Nicolo, N. Trulli & M. Valdés*

Development of a Condition Monitoring system for tidal stream generators rotating
components combining Acoustic Emission and vibration analysis 341
*J.L.F. Chacon, V. Kappatos, C. Selcuk, A. Romero, J. Jimenez, S. Soua & T.-H. Gan*

Detection of incipient SCC damage in primary loop piping using various novel
NDE technologies 347
*B.K. Jackson, J.L.W. Warwick, W. Li & J.J. Wall*

Comparative study of NDT inspection methods in carbon fiber composite laminates 353
*G. Steenackers, J. Peeters, G. Arroud & S. Wille*

### Embedded sensors

Investigations on the structural integrity and functional capability of embedded
piezoelectric modules 361
*S. Geller, A. Winkler & M. Gude*

Healing performance monitoring using embedded piezoelectric transducers in concrete
structures 367
*G. Karaiskos, E. Tsangouri, D.G. Aggelis, D. Van Hemelrijck & A. Deraemaeker*

Fibre-reinforced composites with embedded piezoelectric sensor-actuator-arrays 375
*A. Winkler, M. Dannemann, E. Starke, K. Holeczek & N. Modler*

### Application of NDT/SHM techniques to cultural heritage

Comparative surface damage determination at a Jewish grave, found in front
of the central building of Aristotle University of Thessaloniki, using two different
mobile ultrasonic velocity equipments 383
*B. Christaras, A. Moropoulou, M. Chatziangelou, L. Dimitraki & K. Devlioti*

A new visual-based diagnostic protocol for cultural heritage exploiting
the MPEG-7 standard 389
*A. Doulamis, A. Kioussi & A. Moropoulou*

Comprehensive energy diagnosis methodology integrating non destructive testing 397
*M.A. García-Fuentes, J.L. Hernández, A. Meiss & C. Colla*

Non-contact contemporary techniques for the geometric recording of Cultural Heritage 405
*A. Georgopoulos*

Numerical methods for the interpretation and exploitation of AE monitoring results 413
*S. Invernizzi, G. Lacidogna & A. Carpinteri*

Digital cultural heritage—A challenge for the chemical engineering: Contextualizing
materials in a holistic framework 421
*M. Ioannides, E. Alexakis, C.M. Coughenour, M.L. Vincent, M.F. Gutierrez,*
*V.M. Lopez-Menchero Bendicho, D. Fritsch, A. Moropoulou, V. Rajcic & R. Zarnic*

Raman spectroscopy. A non-destructive tool for on-site chemical analysis
of artifacts and monuments 427
*E.Th. Kamilari, S.N. Kouvaritaki, M.G. Orkoula & C.G. Kontoyannis*

Evaluation of seismic risk in regional areas by AE monitoring of historical buildings 433
*G. Lacidogna, P. Cutugno, F. Accornero, S. Invernizzi & A. Carpinteri*

Non-destructive evaluation of historic natural stone masonry with GPR 441
*F. Lehmann & M. Krüger*

Application of the mortar static penetration test to historical buildings 449
*D. Liberatore, L. Sorrentino, L. Frezza, N. Masini, M. Sileo & V. Racina*

Non destructive testing to perform service of the evaluation of conservation works 457
*R. Manganelli Del Fà, C. Riminesi, S. Rescic, P. Tiano & A. Sansonetti*

Tube-jack and sonic testing for the evaluation of the state of stress in historical masonry 465
*E.C. Manning, L.F. Ramos, P.B. Lourenço & F.M. Fernandes*

Integration of EFD, MRM and IRT for moisture mapping on historic masonry:
Study cases in northern Italy 471
*R. Olmi, C. Riminesi & E. Rosina*

Pre-diagnostic prompt investigation of a historic Bell Tower by visual inspection
and microwave remote sensing 477
*A. Saisi, C. Gentile & L. Valsasnini*

On site investigation and continuous dynamic monitoring of a historic tower in Mantua, Italy 485
*A. Saisi, M. Guidobaldi & C. Gentile*

Focus on soluble salts transport phenomena: The study cases of Leonardo monochrome
at Sala delle Asse (Milan) 493
*A. Sansonetti, M. Realini, S. Erba & E. Rosina*

Novel applications of micro-destructive cutting techniques in cultural heritage 499
*M. Theodoridou & I. Ioannou*

Integrated ND methodologies for the evaluation of the adhesion of frescoes
on stone masonry walls 505
*M.R. Valluzzi, G. Salemi, R. Deiana, E. Faresin, G. Giacomello, M. Giaretton,
M. Panizza, M. Pasetto, S. Calò, M. Battistella & A. Frestazzi*

Detection and localization of debonding damage in composite-masonry strengthening
systems with the Acoustic Emission technique 511
*E. Verstrynge, K. Van Balen, M. Wevers, B. Ghiassi & D.V. Oliveira*

NDTs in the monitoring and preservation of historical architectural surfaces 519
*L. Falchi, E. Zendri & G. Driussi*

*Optical sensors for structural health monitoring*

The novel potential for embedded strain measurements offered by micro-structured
optical fiber Bragg gratings 529
*T. Geernaert, S. Sulejmani, C. Sonnenfeld, H. Thienpont, F. Berghmans, G. Luyckx,
J. Degrieck & D. Van Hemelrijck*

Combining embedded Fibre Bragg Grating sensors and modal analysis techniques
to monitor fatigue induced propagating delaminations in composite laminates 537
*A. Lamberti, G. Chiesura, B. De Pauw & S. Vanlanduit*

SMARTFIBER: Miniaturized optical-fiber sensor based health monitoring system 543
*N. Lammens, G. Luyckx, E. Voet, W. Van Paepegem & J. Degrieck*

*Electromagnetics and X-rays*

Micro-CT as a well-established technique to investigate the internal damage state
of a composite laminate subjected to fatigue 551
*G. Chiesura, G. Luyckx, E. Voet, W. Van Paepegem, J. Degrieck, M.N. Boone,
J. Dhaene & L. Van Hoorebeke*

Damage detection and classification in composite structure after water-jet cutting
using computed tomography and wavelet analysis 557
*A. Katunin*

Post Weld Heat Treatment surface residual stress measurements using X-ray diffraction 563
*S. Kumar, M.J. Tan, B.S. Wong & N. Weeks*

Enhancement of spatial resolution using metamaterial sensor in NonDestructive Evaluation 569
*A. Savin, N. Iftimie, R. Steigmann & A. Bruma*

Author index 577

*Emerging Technologies in Non-Destructive Testing VI – Aggelis et al. (Eds)*
*© 2016 Taylor & Francis Group, London, ISBN 978-1-138-02884-5*

# Preface

The 6th International Conference on Emerging Technologies in Nondestructive Testing (ETNDT6) took place in Brussels, Belgium, May 27–29, 2015 (www.etndt6.be). It was organized by the Department of Mechanics of Materials and Constructions (MEMC) of the Vrije Universiteit Brussel (VUB).

The conference gathered world-wide leading experts and covered all fields of *Non-Destructive Testing (NDT)* in materials and structures. This is a subject of great significance, with great social impact as it is strongly related to detection of damage in engineering structures, such as buildings, bridges, aircrafts, pressure vessels, monuments etc. using non-invasive techniques (among others ultrasound, X-rays, Radar, neutrons, thermography, vibrations, acoustic emission). Health monitoring of structures and components is in close connection to the developments in the field of nondestructive testing. Information on the structural condition is important for planning maintenance and overall management of structures. Fast, reliable and low cost methodologies for evaluation are certainly in demand in the life cycle engineering of the structures.

The ETNDT6 conference aimed to provide the latest developments in these techniques, contribute to the exchange of ideas among scientists and industry and disseminate significant knowledge. The conference attracted more than one hundred participants from 22 countries showing that the area of NDT receives continuously more attention due to its importance in safety and engineering. The conference featured the following excellent plenary talks:

- "Neutron techniques for NDT", by Prof. Helena Van Swygenhoven-Moens from Paul Scherrer Institute & EPFL, Switzerland
- "Basics and Applications of NDE based on Elastodynamics towards Infra-Dock for Concrete Structures" by Prof. Masayasu Ohtsu from Kumamoto University, Japan
- "Resonant Defects: A New Approach to Highly-Sensitive Defect-Selective Imaging" by Igor Solodov from University of Stuttgart, Germany.

Substantial keynote presentations were contributed by A. Moropoulou, A. Saisi, N. Lammens, H. Pfeiffer, M. Bentahar, E. Gruyaert, N. Godin, T. Shiotani, M. Ioannides, T. Suzuki, P. Rizzo, M. Gresil, G. Chiesura, T. Kek.

The present volume is a collection of the submitted papers grouped in different sections according to the session they were presented. We believe that it represents the current trends in nearly all fields of NDT with cutting edge applications.

We would like to thank all the contributors who submitted and presented their studies during ETNDT6, as well as the scientific committee, the members of which assisted in the review process. We would also like to thank all people (personnel and volunteers of MEMC) who contributed to the success of ETNDT6!

*Emerging Technologies in Non-Destructive Testing VI – Aggelis et al. (Eds)*
*© 2016 Taylor & Francis Group, London, ISBN 978-1-138-02884-5*

# Organization

## THE ORGANIZING COMMITTEE

Aggelis D.G., *Vrije Universiteit Brussel, Belgium*
Van Hemelrijck D., *Vrije Universiteit Brussel, Belgium*
Vanlanduit S., *Vrije Universiteit Brussel, Belgium*
Anastasopoulos A., *Mistras Group Hellas, Greece*
Philippidis T.P., *University of Patras, Greece*

## ETNDT6 SCIENTIFIC COMMITTEE

Adams R.D., *University of Bristol, UK*
Alver N., *Ege University, Turkey*
Breysse D., *Université de Bordeaux, France*
Busse G., *University of Stuttgart, Germany*
Chai H.-K., *Malaya University, Malaysia*
Chiang C., *Chaoyang University of Technology, Taiwan*
Deraemaeker A., *ULB, Belgium*
Dobmann G., *Fraunhofer Institut (IZFP), Saarbrücken, Germany*
Forde M., *Edinburgh University, UK*
Glorieux C., *University of Leuven, Belgium*
Godin N., *INSA Lyon, France*
Gresil M, *University of Manchester, UK*
Grosse C.U., *Technical University of Munich, Germany*
Grum J., *University of Ljubljana, Slovenia*
Gusev V., *University of Le Mans, France*
Karaiskos Gr., *VUB, Belgium*
Kazys R., *Kaunas University of Technology, Lithuania*
Kobayashi Y., *Nihhon University, Japan*
Lacidogna G., *Politecnico di Torino, Italy*
Mandelis A., *University of Toronto, Canada*
Matikas T.E., *University of Ioannina, Greece*
Matsuda H., *Tobishima, Japan*
Moropoulou A., *NTUA, Greece*
Ogin S., *University of Surrey, UK*
Ohtsu M., *Kumamoto University, Japan*
Ono K., *UCLA, USA*
Polyzos D., *University of Patras, Greece*
Popovics J., *University of Illinois at Urbana-Champaign, USA*
Rizzo P., *University of Pittsburgh, USA*
Ruiz G., *Universidad de Castilla-La Mancha, Spain*
Saisi A., *Politecnico di Milano, Italy*

Schulte K., *Technical University of Hamburg, Germany*
Shiotani T., *Kyoto University, Japan*
Shiwa M., *NIMS, Tsukuba, Japan*
Solodov I., *University of Stuttgart, Germany*
Soutis C., *University of Manchester, UK*
Van Breugel K., *TU Delft, The Netherlands*
Van Paepegem W., *Gent University, Belgium*
Vantomme J., *Royal Military Academy, Belgium*
Wevers M., *University of Leuven, Belgium*

*Plenary papers*

*Emerging Technologies in Non-Destructive Testing VI – Aggelis et al. (Eds)*
*© 2016 Taylor & Francis Group, London, ISBN 978-1-138-02884-5*

# Basics and applications of NDE based on elastodynamics toward infra-dock for concrete structures

M. Ohtsu
*Graduate School of Science and Technology, Kumamoto University, Kumamoto, Japan*

ABSTRACT: In order to promote the sustainable society, the long service-life of infrastructure is to be an evolutional target. Aging and disastrous damages in concrete structures have updated the critical demand for maintenance. Being stimulated by the dock system for ships, the medical checkup system has been developed for prognosis as "ningen (human in Japanese) dock". Similarly, to extend the service life of the concrete structure, we have made a feasibility study to establish an "infra-dock" system in the Japan Concrete Institute. Thus, on-site NDE (nondestructive evaluation) techniques are to be developed as constituents of diagnosis and prognosis procedures in the infra-dock. In this respect, basics of the elastic-wave methods are clarified, including Ultrasonic Testing (UT), Impact-Echo (IE), and Acoustic Emission (AE). Promising results on applications to concrete structures are discussed.

## 1 INTRODUCTION

Aging and disastrous damages in concrete structures have been reported as the critical issue worldwide as necessary for continuing maintenance and repair of the structures in service. To this end, a variety of techniques and evaluation methods for diagnosis and prognosis are under development. Accordingly, on-site inspection techniques of concrete and masonry structures by NonDestructive Testing (NDT) have been comprehensively surveyed in RILEM technical committee, "TC 239-MCM: on-site measurement of concrete and masonry structures by visualized NDT". Now, the state-of-the art report on applications of NonDestructive Evaluation (NDE) to infrastructures is in preparation.

With respect to human body, the medical checkup system for prognosis is developed as "the ningen dock" in Japan. In order to extend the service life of the concrete structures, we have proposed "the infra-dock", so-called as referring to "the ningen dock", and the technical committee "JCI-TC-125FS: establishment of infra-dock for concrete structures" has been set up in the Japan Concrete Institute (JCI). As basic constituents of the infra-dock, inspection techniques for on-site measurements should be synthesized as the prognosis procedure. Consequently, RILEM-TC 239-MCM and are in collaboration with JCI-TC-125FS (ohtsu, 2014). Based on these committee researches, it is readily found that the elastic-wave methods shall play a key role for NonDestructive Evaluation (NDE) of concrete structures.

Consequently, theoretical bases of the elastic-wave methods are clarified and reviewed. These include Ultrasonic Testing (UT), Impact-Echo (IE), and Acoustic Emission (AE). Then, successful results on applications to NDE in concrete structures are discussed, toward establishment of infra-dock.

## 2 ELASTIC-WAVE METHOD

### 2.1 Basics

The governing equation of elastodynamics (Grosse & Ohtsu, 2008) is known as Navier's equation, from which P-wave velocity, $v_p$, is derived,

$$v_p = \sqrt{\frac{(1-\nu)E}{\rho(1+\nu)(1-2\nu)}}, \tag{1}$$

where $\nu$ is Poissons's ratio, $\rho$ is the density and $E$ is the modulus of elasticity of the propagating medium. From Navier's equation, the integral representation of the solution, $u_k(x,t)$, in the elastodynamic field is obtained as,

$$u_k(x,t) = \int_S [G_{ki}(x,y,t) * t_i(y,t) \\ -T_{ki}(x,y,y) * u_i(y,t)] \, dS. \tag{2}$$

Here $G_{ki}$ means Green's functions and $T_{ki}$ are tractions derived from Green's functions. $t_i$ is the traction and $u_i$ is the displacement on the surface $S$. In the generalized theory of Acoustic Emission (AE) (Ohtsu & Ono, 1984), Eq. 2 is separated into two equations of the traction (force) at $y, f_i(t)$, and the dislocation (displacement), $b_i(y,t) = u_i^+ - u_i^-$ as,

$$u_k(x,t) = G_{ki}(x,y,t) * f_i(t). \qquad (3)$$

$$u_k(x,t) = \int_S T_{ki}(x,y,t) * b_i(y,t) dS. \qquad (4)$$

Eq. 3 is the theoretical representation of ultrasonic testing (UT) and impact-echo (IT). Since the function $G_{ki}(x,y,t)$ contains the delta function $\delta(t - \frac{|x-y|}{v_p})$ in the time, the first motion of $u_k(x,t)$ at $x$ due to the force $f_i(t)$ is observed at $t = |x-y|/v_p$. In UT measurement, determination of the arrival time, $t = |x-y|/v_p$, is the key issue.

In the case of IE, elastic waves are normally detected at the surface. In elastodynamics, the problem, where a transient displacement due to a pulse is detected at the surface in a half space, is called Lamb's problem. This is because the problem was first solved by H. Lamb (1904). Thus, Green's functions in a half space are called Lamb's solutions. The computational code has been already published in the literature (Ohtsu & Ono, 1984). Lamb's problem presents the basic concept of IE. In order to consider spectral components of elastic waves, Eq. 3 is transformed in the frequency domain,

$$U(f) = G(f)F(f), \qquad (5)$$

where $U(f)$, $G(f)$ and $F(f)$ are Fourier transforms of $u(t), G(t)$ and $f(t)$ in Eq. 3. In the case that the function $f(t)$ is the delta function (impulse without duration time), $F(f)$ is referred to as constant. Since $f(t)$ normally has the duration time as a contact time of the impact, $F(f)$ has a corner frequency, $f_c$, at which the amplitudes decrease from a plateau. In the case of steel-ball drop with diameter $D$, it is approximated as $f_c = 1.25/t_c$, where $t_c$ is the contact time obtained from the simplified equation, $t_c = 0.0043D$ (Sansalone & Streett, 1997).

Theoretical treatment of AE is derived from Eq. 4, considering the crack surface $F$,

$$\begin{aligned} u_k(x,t) &= \int_F T_{ki}(x,y,t) * b_i(y,t) dF \\ &= G_{kp,q}(x,y,t) C_{pqij} l_i n_j \\ &\times \int_F b(y) dF * S(t) \\ &= G_{kp,q}(x,y,t) C_{pqij} l_i n_j \Delta V * S(t), \qquad (6) \end{aligned}$$

where crack vector $b(y,t)$ is decomposed into crack motion $b(y)$, the unit crack vector $l$ and the source-time function $S(t)$. $G_{kp,q}$ are the spatial derivatives of $G_{kp}$, $C_{pqij}$ are the elastic constants and $n_j$ is the crack-normal vector. Defining the crack volume $\Delta V$, the moment tensor $M_{pq}$ is derived as,

$$M_{pq} = G_{kp,q} C_{pqij} l_i n_j \Delta V \qquad (7)$$

The components of the moment tensor consist of the crack-motion vector, $l_i$, the normal vector to the crack surface, $n_j$, and the crack volume $\Delta V$. Eventually, crack motion is kinematically modeled by components of the moment tensor as illustrated in Figure 1. Since the moment tensor is a symmetric tensor of the 2nd rank, the number of independent components is six. Thus, 6-channel system is, at least, necessary for determination of the components.

### 2.2 UT for crack-depth measurement

As realized in Eq. 1, P-wave velocity is associated with the modulus of elasticity and the density. Consequently, quality of concrete is readily estimated from the velocity. For example, one code to estimate the compressive strength is specified (JCMS-III B5704, 2003). Concerning defects in concrete, UT has been applied to estimation the depth of a surface crack (BS 1881, 1986). One well-known procedure is called the $T_0$-$T_c$ method. As Illustrated in Figure 2, two cases of travel timers are measured without and with a surface crack of interval distance $L$. The depth $d$ is estimated from,

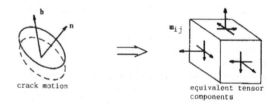

Figure 1. Crack motion and equivalent tensor components.

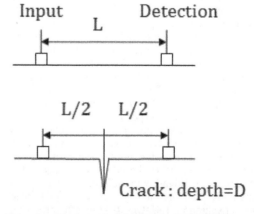

Figure 2. Crack depth estimation by UT (BS1181).

$$d = \frac{L}{2}\sqrt{(\frac{T_c}{T_0})^2 - 1} \qquad (8)$$

The accuracy of the procedure is dependent on the precise estimation of arrival times. Conventionally, a threshold reading is employed. A first-motion detection in SiGMA analysis discussed later could be useful for the estimation.

### 2.3 SIBIE analysis of IE

The basic concept of IE method is illustrated in Figure 3. Elastic waves are generated by a short duration mechanical impact, and the paths of the elastic waves are illustrated and a typical frequency spectrum is given. In the case of the void, peak frequencies appear at $f_t$ and $f_{void}$, which are calculated as,

$$f_t = v_p / 2T, \qquad (9)$$

$$f_{void} = v_p / 2d, \qquad (10)$$

where $T$ is the plate thickness and $d$ is the depth of a defect.

Since it is not easy to identify the resonance peak frequency in the frequency spectrum, SIBIE (Stack Imaging of spectral amplitudes Based on Impact Echo) procedure has been developed (Ohtsu & Watanabe, 2002). This is a post-processing technique for IE data. In the procedure, first, a cross-section of concrete is divided into square elements as shown in Figure 4.

Resonance frequencies due to reflections at each element are calculated from,

$$f_R = v_p / R \quad \text{and} \quad f_{r2} = v_p / r_2 \qquad (11)$$

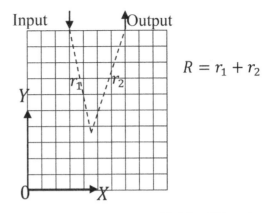

Figure 4. SIBIE imaging model (Ohtsu&Watanabe, 2002).

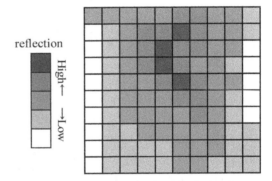

Figure 5. Example of SIBIE result.

Spectral amplitudes corresponding to these two resonance frequencies in the frequency spectrum are summed up at each element as shown in Figure 5. Thus, reflection intensity at each element is estimated as stack image.

### 2.4 SiGMA analysis of AE

On the basis of the far-filed term of P-wave, a simplified procedure was developed (Ohtsu, 1991), which is suitable for a PC-based processor and robust in computation. The procedure is now implemented as a SiGMA (Simplified Green's functions for Moment tensor Analysis) code. From Eqs. 6 and 7, the amplitude of the first motion, $A(x)$, at the detection point $x$ is simply represented as,

$$A(x) = Cs \frac{Re(t,r)}{R} r_p r_q M_{pq}, \qquad (12)$$

$Cs$ is the magnitude of the sensor response including material constants. $Re(t,r)$ is the reflection

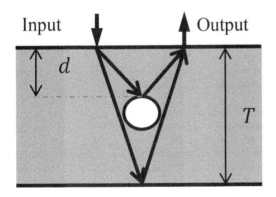

Figure 3. Configuration for impact-echo.

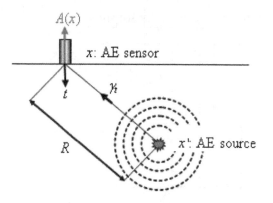

Figure 6. Crack nucleation and AE detection.

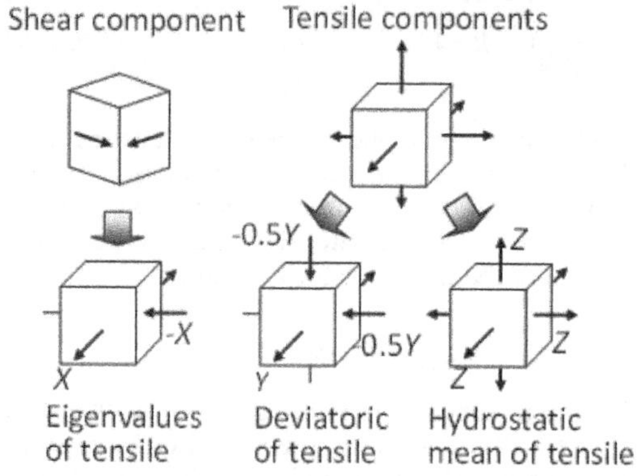

Figure 8. Unified decomposition of eigenvalues.

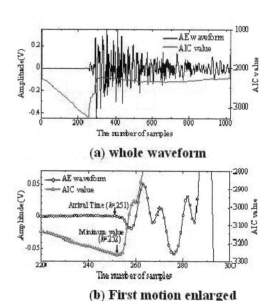

Figure 7. Recorded AE waveform (Ohno & Ohtsu, 2010).

coefficient at the surface of detection. $R$ is the distance between the source $y$ and the detection point $x$, of which direction cosine is $r = (r_1, r_2, r_3)$. These factors are illustrated in Figure 6.

In SiGMA analysis, two parameters of the arrival time (P1) and the amplitude of the first motion (P2) are to be determined from AE waveform as shown in Figure 7. Focusing on the intervals before and after the onset of AE signal, an AutoRegressive (AR) technique has been developed (Ohno & Ohtsu, 2010) as illustrated in the figure.

To identify source kinematics, the eigenvalue analysis of the moment tensor was introduced.

The eigenvalues of the moment tensor are represented by the combination of the shear crack and the tensile crack as shown in Figure 8.

Eventually, the relative ratios $X$, $Y$ and $Z$ are obtained as,

$$\frac{E_1}{E_1} = 1.0 = X + Y + Z$$
$$\frac{E_2}{E_1} = 0 - Y/2 + Z \qquad (13)$$
$$\frac{E_3}{E_1} = -X - Y/2 + Z$$

Here, $E_1$ is the maximum eigenvalue, $E_2$ is the intermediate eigenvalue and $E_3$ is the minimum eigenvalue. In the present SiGMA code, AE sources with shear ratios $X$ less than 40% are classified as tensile cracks. When the ratio $X$ is larger than 60%, AE source is classified as a tensile crack. In the case of the ratios between 40% and 60%, the cracks are referred to as mixed-mode. After determining the crack type, the direction of crack motion is derived from the eigenvectors. In the eigenvalue analysis, three eigenvectors $e_1$, $e_2$ and $e_3$,

$$e_1 = l + n$$
$$e_2 = l \times n \qquad (14)$$
$$e_3 = l - n$$

are determined. Here, vector $l$ and $n$ are interchangeable. In order to visualize kinematical information of AE sources, crack models classified as tensile, mixed-mode and shear are given in Figure 9. Here, an arrow vector indicates the crack motion vector $l$, and the circular plate corresponds to a crack surface which is perpendicular to the crack normal vector $n$.

Figure 9. Visualized models of tensile, mix-mode and shear cracks.

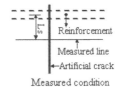

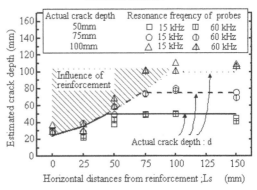

Figure 10. The effect of rebar on crack-depth evaluation (JCMS-III).

## 3 APPLICATIONS OF ELASTIC-WAVE METHODS

### 3.1 Evaluation of crack depth by UT in reinforced concrete

When UT measurement is applied to evaluate the crack depth, several factors which could lead to erroneous results are noted. The presences of water and reinforcing steel-bar (rebar) are, in particular, to be taken into careful consideration.

In the one case that a surface crack is saturated with water, the depth might be underestimated. This is because a reflected wave at water surface contaminates a diffracted wave from the crack tip. In the other case that rebar crosses a crack, a reference distance $d_r$ is calculated from,

$$d_r = \sqrt{L_c^2 + L_s^2}, \quad (15)$$

where $L_c$ is the cover-thickness from rebar and $L_s$ is the horizontal distance from the rebar to the crack as shown in Figure 10. If the depth $d$ determined by Eq. 8 is shorter than $d_r$, the presence of rebar is negligible, without the effect of sensor selections. Otherwise, the crack depth is underestimated, as shown in the figure, due the effect of rebar. This could happen if the crack is nucleated deeper than the cover-thickness and rebar is located close to the crack (JCMS-III B5705, 2003).

For determination of the depths of surface cracks, an application of SIBIE is also under development for on-site measurement (Ohtsu et al., 2008).

### 3.2 Detection of ungrouted tendon ducts by SIBIE

1. *Calibration of impact*
In the case where Lamb's solution is available (Ohtsu & Ono, 1984), Green's function in Eq. 3 is known. Consequently, impact $f(t)$ and frequency spectrum $F(f)$ are rationally recovered if the detecting system has a flat response. In a concrete block illustrated in Figure 11, elastic waves are detected

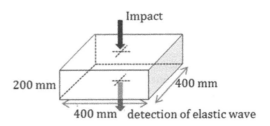

Figure 11. Concrete block for calibration.

due to impact, by the laser-interferometer, which is calibrated of a flat response up to 50 kHz. The configuration of the impactor and the detection presents Lamb's problem of buried pulse.

In the proposed procedure, a detected wave by the laser-interferometer is transformed by Fast Fourier Transform (FFT) and then $U(f)$ in Eq. 5 is obtained. After calculating $G(f)$, $F(f)$ is derived from $U(f)/G(f)$. In order to check an applicability of this procedure, a steel-ball of 20 mm diameter was dropped from 100 mm height. Then, $F(f)$ was calculated, and further transformed into the time domain by inverse FFT. According to the Sansalone (Sansalone & Streett, 1997), the impact due to a steel-ball drop is known to be approximated by a sine function of the contact time $t_c$. So, function $f(t)$ is compared with a sine function in Figure 12. A reasonable agreement is confirmed as the contact time $t_c = 108.3$ μs obtained from $0.0043D$.

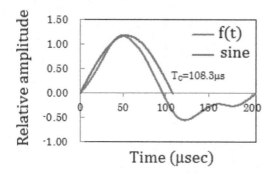

Figure 12. Recovered function of the impact due to steel-ball drop.

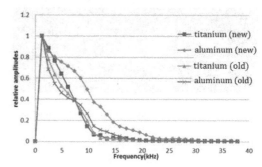

Figure 14. Frequency responses of the impactor devices.

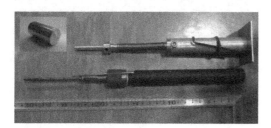

Figure 13. Impactor devices.

Although Sansalone proposed to employ the steel-ball drop as the impactor, it is not easy to apply on-site measurement. In addition, the higher frequency is in demand for shallow defects, the smaller power could be driven by the steel ball of small diameter. Therefore, a device for the impactor is under development. Previously, a bullet was shot with compressed air (Ohtsu et al. 2008). In order to make the device light-weight and easy-use for on-site measurement, an impactor driven with a spring is recently developed. So far, two types of the impactors are made as shown in Figure 13. A bullet in the figure is attached to the rod-head inside the cylinder. An old device on the bottom is made of steel cylinder, while a new device on the top is of aluminum cylinder to decrease the weight. Two bullets (heads) are prepared. One is made of titanium and the other is of aluminum.

Frequency responses $F(f)$ are estimated from Eq. 5, and normalized as shown in Figure 14. It is found that a new impactor with titanium head has the wide frequency up to 25 kHz.

2. *Experiments of ungrouted tendon ducts*

A model specimen for experiments is illustrated in Figure 15. Three ducts for post-tensioning in prestressed concrete are embedded. By employing the new impactor with titanium head, impact tests were conducted. As shown in Figure 15 (b), three ducts are ungrouted, partially grouted and fully grouted.

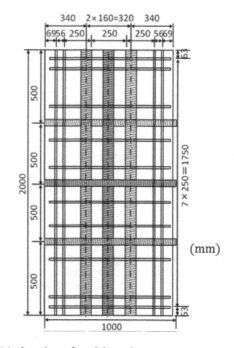

(a) plan view of model specimen.

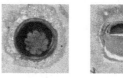  

(b) cross-section of ducts.

Figure 15. Configuration of model specimen and ducts.

Results of SIBIE analysis are shown in Figure 16. The impact was driven from the left in Figure 15 (b). In order to visualize results in the analysis, the upper-bound frequency is set to 7 kHz for a shal-

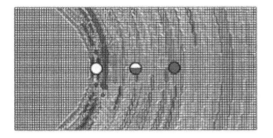

(a) 7 kHz low-pass

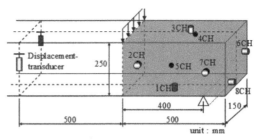

Figure 17. AE sensor array in RC beam tested.

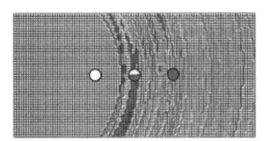

(b) 5 kHz low-pass

Figure 16. Results of SIBIE analysis.

low duct and 5 kHz for a deeper duct from the left. It is clearly found that both ungrouted and partially grouted duct are visually detected.

### 3.3 Mechanisms of shear failure in reinforced concrete by SiGMA

1. *Reinforced concrete beam*

A Reinforced Concrete (RC) beam of dimensions 150 mm × 250 mm × 2000 mm was tested. The compressive strength of concrete at 28-day standard curing was 29.7 MPa. P-wave velocity was 4230 m/s. Two reinforcement bars (13 mm in diameter) were installed with cover thickness 46.5 mm from bottom. In order to focus on the diagonal-shear failure, eight AE sensors were attached using electron wax on the surface of shear span as illustrated in Figure 17.

AE signals were detected by 150 kHz-resonant AE sensors, and sampling frequency for recording waveforms was 1 MHz by 8-channnel system. AE waves were pre-amplified by 40 dB and amplified by 20 dB at main system. The threshold level was set to 42 dB.

Based on AE activity, a loading process is divided into three stages. AE events are determined from all eight AE sensors. More than 700 AE events were recorded within the Event Definition

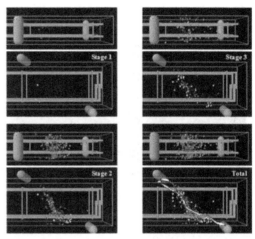

Figure 18. Results of SIGMA analysis in shear failure (Ohno & Ohtsu, 2010).

Time (EDT) of 120 μs. Results of the SiGMA analysis are shown in Figure 18. The locations of 370 AE events are determined. It is clearly realized that most of micro-cracks are located around the cracks of the diagonal-shear failure as inserted in the graph. Thus, SiGMA analysis is applicable to visually identify internal mechanisms of concrete failure.

### 3.4 Mechanisms of rebar corrosion in reinforced concrete by SiGMA

1. *Corrosion process of steel*

According to a phenomenological model of reinforcement corrosion in marine environments (Melchers & Li, 2006), a typical corrosion loss is illustrated as shown in Figure 19 (a). At phase 1, the corrosion is initiated. The rate of the corrosion process is controlled by the rate of transport of oxygen. As the corrosion products build up on the corroding surface of rebar, the flow of oxygen is eventually inhibited and the rate of the corrosion

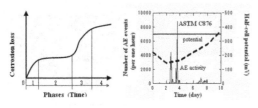

(a) Phenomenological model    (b) AE activities in corrosion

Figure 19. Phenomenological mode and AE activities in corrosion process (Ohtsu & Tomoda, 2008).

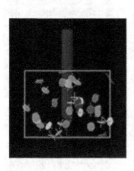 

(a) SiGMA analysis at 42 days elapsed    (b) Micrograph at 56 days

Figure 20. Results of SiGMA analysis in corrosion (Kawasaki et al., 2012).

loss decreases at phase 2. The corrosion process involves further corrosion loss as phases 3 and 4 due to anaerobic corrosion. Thus, at two stages of phase 1 and phase 3, the corrosion process is activated. We conducted an acceleration corrosion test of rebar corrosion, and found that two active generations of AE hits as shown in Figure 19 (b) (Ohtsu & Tomoda, 2008). Active AE generations are reasonably identified, corresponding the two phases. It is clarified that the 1st activity corresponds to rust nucleation on rebar surface and the 2nd generation results from concrete cracking due to expansion of corrosion products.

SiGMA analysis was applied to the 2nd activities of AE in the accelerated corrosion test (Kawasaki et al., 2012). Results of SiGMA analysis at 42 days elapsed are shown in Figure 20 (a) on a cross-section. Mostly events are located surrounding rebar. The events which are located at −78.75° orientation are mostly classified into tensile (opening) cracks. Additionally, shear cracks (sliding) are observed at 78.75° orientation, after tensile cracks observed.

Results of the stereo-microscope of cross-section at 56 days are shown in Figure 20 (b). Three microcracks were observed at the cross section toward the bottom of RC beam. Orientation of these cracks is about −78.75°. Thus, it is confirmed that tensile cracks are observed at −78.75° orientation.

Thus, the corrosion process in reinforced concrete is kinematically to be identified by SiGMA analysis.

## 4 TOWARD ESTABLISHMENT OF INFRA-DOCK

In the ningen-dock, during the course of one or two days, clients undergo such inspections as blood test, urine and fecal exams., X-ray, ultrasonography, and so forth along with a consultation with a doctor. The ningen-dock for human-body is illustrated in Figure 21, along with a conceptual illustration of the infra-dock at the bottom. Here, a percussion treatment is referred to as comparable to UT. An internal inspection by a see-through technique (X-ray CT) is compared with SIGMA and SIBIE. Thus, the elastic-wave methods could readily contribute inspection procedures in the infra-dock.

The technical committee JCI-TC-125FS is now working to establish the infra-dock programs for civil structures and buildings. The programs could present such courses as half-day, one-day, and two-day, which are to be performed by updated on-site inspections. So far, these courses for the infra-dock are under development, including an optional course.

In addition, RILEM TC-239 MCM is working on visualized inspection techniques for on-site measurement. Consequently, on-site measurement by the elastic-wave methods under development should be synthesized as constituents of prognosis programs in the infra-dock.

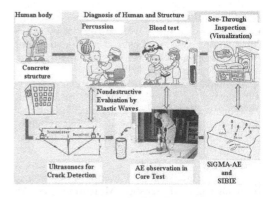

Figure 21. Concept of Infra-dock.

## 5 CONCLUDING REMARKS

The technical committee JCI-TC-125FS is to be completed in 2015. So, we are planning to pro pose the infra-dock system as a bench-mark in the world. As a future plan, a seminar for the certified engineers and the preparation for a new organizing institute are to be proposed for establishing the infra-dock.

The technical committee RILEM TC-239 MCM is going to publish the state-of-the-art report on visualized NDT for concrete and masonry structures in 2015. The report is planned to contain Computer Tomography (CT), radar technique, infra-red thermography, cross-sectional visualization of defects by SIBIE, 3D-visualization of crack kinematics by SiGMA, hybrid NDE for rebar corrosion, and on-site damage evaluation by AE and CT. Thus, advanced NDT techniques would be summarized and should be practically implemented. It is expected that the achievement could constitute the infra-dock programs and courses.

## ACKNOWLEDGEMENT

The author wishes sincerely to thank all the members of the both technical committee JCI-TC-125FS and RILEM technical committee TC-239 MCM. With their good and valuable contributions, this paper is compiled as a plenary lecture.

## REFERENCES

BS 1881:Part 203 (1986), Recommendation for measurement of the velocity of ultrasonic pulses in concrete, British Standards, London.
Grosse, U.C. & Ohtsu, M. Eds.(2008), Acoustic emission testing, Springer.
CMS-III B5704 (2003), Test method for compressive strength of concrete by ultrasonic waves, Federation of Construction Materials Industries, Japan.
JCMS-III B5705 (2003), Measuring method for estimating surface-crack depth in concrete by ultrasonic-wave velocity, Federation of Construction Materials Industries, Japan.
Kawasaki, Y. et al.(2012), Fracture mechanisms of corrosion-induced cracks in reinforced concrete by BEM and AE-SiGMA, Proc. 14th Int. Conf. on Structural Faults and Repair.
Lamb, H.(1904), On the propagation of tremors over the surface of an elastic solid, *Philo. Trans. Roy. Soc.*, A203, 1–42.
Melchers, R.E. & Li, C.Q.(2006), Phenomenological modeling of reinforcement corrosion in marine environments, *ACI Materials Journal*, 103(1), 25–32.
Ohno, K. & Ohtsu, M.(2010), Crack classification in concrete based on AE, *Construction and Building Materials*, 24(12), 2339–2346.
Ohtsu, M. & Ono, K.(1984), A generalized theory of AE and Green's functions in a half space, *Journal of AE*, 3(1), 124–133.
Ohtsu, M.(1991), Simplified moment tensor analysis and unified decomposition of AE sources, *J. Geophys. Res.*, 96(B4), 1187–1189.
Ohtsu, M. & Watanabe, T.(2002), Stack imaging of spectral amplitudes based on impact-echo for flaw detection, *NDT&E International*, 35, 189–196.
Ohtsu, M. et al.(2008), Elastic-Wave Methods for Crack Detection and Damage Evaluation in Concrete, Proc. 12th Int. Conf. on Structural Faults and Repair.
Ohtsu, M. & Tomoda, Y.(2008), Phenomenological model of corrosion process in reinforced concrete identified by AE, *ACI Materials Journal*, 105(2), 194–199.
Ohtsu, M.(2014), Toward establishment of infra-dock for concrete structures, Proc. 15th Int. Conf. on Structural Faults and Repair.
Sansalone, M.J. & Streett, W.B.(1997), Impact-echo, Bull-brier Press, New York.

*Emerging Technologies in Non-Destructive Testing VI – Aggelis et al. (Eds)*
*© 2016 Taylor & Francis Group, London, ISBN 978-1-138-02884-5*

# Resonant defects: A new approach to highly-sensitive defect-selective ultrasonic imaging

I. Solodov

*Institute for Polymer Technology (IKT-ZfP), University of Stuttgart, Stuttgart, Germany*

ABSTRACT: In this paper, a consistent way to enhance acoustic, optical and thermal defect responses is suggested by using selective ultrasonic activation of defects based on the concept of Local Defect Resonance (LDR). A straightforward phenomenology and the finite element simulation are developed to evaluate various order LDR frequencies. The LDR provides a selective excitation of a defect that results in a high local vibration amplitude and enhancement of the defect ultrasonic, thermal and optical responses readily measurable even for a few mW of acoustic input. This proposes LDR application as an efficient and highly-sensitive mode for ultrasonic, thermosonic and optical defect-selective imaging in NDT.

## 1 INTRODUCTION

Over last decades ultrasonic imaging has become a mature technology with numerous applications in medical diagnostics, underwater exploration, geophysics, and nondestructive testing. The instrumentation for ultrasonic imaging ranges from high-frequency sonars and ultrasonographs to flaw detectors and phased arrays. Despite the variety of apparatus developed, they all use the same fundamental principle: reflection/scattering of ultrasound from an object or a defect in a medium or material.

The efficiency of ultrasonic reflection/scattering from a defect depends on its acoustic impedance, geometry and orientation. An important role of ultrasonic frequency of a probing wave comes to a simple rule of thumb: the higher the frequency, the better lateral resolution is. However, in some materials (e.g. composites) the high frequency ultrasound is not applicable due to severe damping. The orientation of defects is also difficult to change; as a result a surface breaking crack is basically 'invisible' for ultrasonic wave incident normal to the surface. In many cases, therefore, a low efficiency of ultrasonic reflection/scattering has to be accepted and taken as given since no changes can be introduced in the above mentioned factors.

These shortcomings can be avoided or facilitated by applying other derivative effects in ultrasound-defect encounter. They include e.g. nonlinear, thermal, acousto-optic, etc. responses also applied for NDT and ultrasonic imaging. These secondary effects are normally relatively inefficient so that the corresponding NDT techniques require an elevated acoustic power and stand out from conventional ultrasonic NDT counterparts for their specific instrumentation particularly adapted to high-power ultrasonics.

The efficiency of ultrasonic wave-defect interaction relevant to detection and imaging can be quantified by the amplitude of the defect vibration for a given amplitude of a probing wave. A natural way to increase the vibration amplitude is to drive the specimen at one of its natural frequencies. This approach is used in various ultrasonic NDT techniques with an obvious drawback of „missing" the defect due to the presence of nodal lines in a standing wave pattern.

In this paper, a consistent way to enhance acoustic, nonlinear, optical and thermal defect responses is suggested by using selective ultrasonic activation of defects based on the concept of Local Defect Resonance (LDR) (Solodov et al., 2011). Unlike the resonance of the whole specimen, the LDR provides an efficient energy pumping from the wave directly to the defect. It is experimentally shown that the frequency—and spatially-selective ultrasonic activation of defects via the concept of LDR is the way to optimize efficiency and sensitivity of defect imaging in laser vibrometry, thermosonics, ultrasonic shearography and nonlinear NDT.

## 2 CONCEPT AND EXPERIMENTAL EVIDENCE FOR LDR

The concept of LDR is based on the fact that inclusion of a defect leads to a local decrease in stiffness for a certain mass of the material in this area, which should manifest in a particular characteristic frequency of the defect. The LDR fundamental frequency can be introduced as a natural frequency of the defect with an effective rigidity $K_{eff}$ and mass $M_{eff}$:

$$f_0 = \sqrt{K_{eff}/M_{eff}}/2\pi.$$

To derive the expressions for $K_{eff}$ and $M_{eff}$ one could evaluate potential and kinetic vibration energy of the defect. This approach applied to a circular Flat-Bottomed Hole (FBH) (radius $a$, thickness $h$) yields (Solodov et al., 2013):

$$K_{eff} = 192\pi D/a^2; M_{eff} = 1.8\,m, \qquad (1)$$

where $D = Eh^3/12(1-\nu^2)$ is the bending stiffness and $m$ is the mass of the plate in the bottom of the defect.

Equations (1) are then combined to yield the LDR frequency of the circular FBH:

$$f_0 \approx \frac{1.6h}{a^2}\sqrt{\frac{E}{12\rho(1-\nu^2)}} \qquad (2)$$

The problem in practical use of the analytical approach is concerned with the boundary conditions for the defect edges, which were assumed to be clamped in deriving (2). Instead, the finite element simulation by using the software Comsol Multiphysics (physics package "structural dynamics," "eigenfrequency analysis") was found to be suitable for analyzing the vibration characteristics of structures with defects and estimation of the LDR frequencies. Figure 1 (top) illustrates the vibration pattern at frequency 10.4 kHz, which is readily identified as a fundamental LDR of a circular FBH in a PMMA plate, followed by the higher-order LDR at higher driving frequency of 23.25 kHz (Fig. 1, bottom).

A direct way to experimentally reveal LDR is to measure an individual contribution of each point of the specimen in its overall frequency response. For this purpose an ultrasonic excitation by a wide-band piezoelectric transducer is combined with a laser vibrometer scan of the specimen surface. It enables to probe and indicate all possible resonances in every point of the specimen. The origin of each maximum is then verified by imaging the vibration pattern in the specimen at the corresponding frequency. Figure 2 shows an example of LDR vibration patterns measured for a square insert (25 × 25 mm$^2$) in a Carbon Fiber-Reinforced Polymer (CFRP) at different frequencies. A strong enhancement (about 20 dB) of the vibration amplitude observed locally in the defect area is identified as a fundamental defect resonance (Fig. 2a). Besides the fundamental LDR, zoom-in

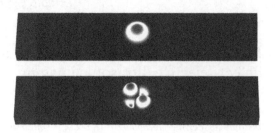

Figure 1. FEM simulation of vibration patterns of a PMMA plate (thickness 3 mm) with a circular FBH (radius 1 cm, depth 2 mm): fundamental LDR (10.4 kHz) (top), and higher-order LDR (23.25 kHz) (bottom).

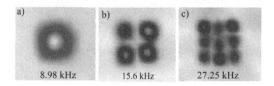

Figure 2. Fundamental (a) and higher-order LDR (b, c) for a square inset in a CFRP plate.

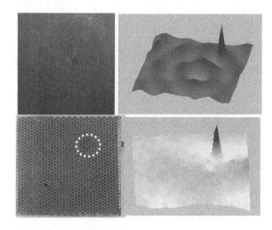

Figure 3. Fundamental LDR (48.5 kHz) in heat damaged CFRP area (top) and delamination in honeycomb glass fibre-reinforced (GFRP) structure (bottom, LDR 14.94 kHz).

scan of the vibration field inside the defect area in a wider frequency range reveals the higher-order LDR with multiple nodal lines in the vibration patterns (Fig. 2b, c). Such a methodology was successfully applied to a search of the LDR in a variety of materials. The two examples presented in Figure 3, illustrate a clear evidence of LDR in kHz-frequency range for damage in composites. Similar LDR with local resonance "amplification" of the vibration amplitude as high as ~20–40 dB were measured for other types of realistic defects in various materials.

## 3 LDR FOR EFFICIENT NONLINEAR NDT

The bottleneck problem of nonlinear NDT is a low efficiency of conversion from fundamental frequency to nonlinear frequency components. Since LDR is as an efficient resonant "amplifier" of the local vibrations, one would expect it to contribute appreciably to defect nonlinearity. An extremely high resonant nonlinearity is demonstrated in Figure 4 for a delamination in GFRP plate: Multiple Higher Harmonic (HH) generation is observed in the defect area while driving the specimen at LDR frequency even at moderate input voltage. A crucial role of the driving frequency match to LDR for nonlinearity increase is illustrated in Figure 5 for a crack in a UniDirectional (UD-) CFRP rod. As the driving frequency matches the LDR frequency (19.5 kHz), a strong enhancement of the HH amplitudes generated locally in the defect area is observed.

A high quality factor of LDR can also be used as an "amplifier" in the frequency mixing nonlinear NDT. This method is based on the nonlinear interaction of ultrasonic waves of different frequencies $(f_1, f_2)$ that results in a combination frequency output: $f_\pm = f_1 \pm f_2$.

An application of LDR as the "frequency mixing amplifier" for NDE and imaging of realistic defects is illustrated then in Figures 6–7 for an impact-induced damage (area ~5 × 5 mm$^2$) in CFRP plate (280 × 40 × 1 mm$^3$). A linear LDR frequency response of the impact demonstrates a well-defined double-maxima peak around 110 kHz (Fig. 6). In the experiment, the two interacting flexural waves were excited in a continuous wave mode by using the piezo-transducers attached to the opposite edges of the plate. One of the frequencies was fixed at $f_1 = 77.5$ kHz while the other was swept from $f_2 = 28.5$ to 37.5 kHz to provide the sum frequency variation around the LDR frequency of the defect. Figure 7 shows the velocity amplitude

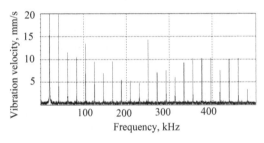

Figure 4. HH spectrum for delamination in glass fiber reinforced (GFRP) specimen driven at LDR frequency 20900 Hz. Input voltage is 7 V.

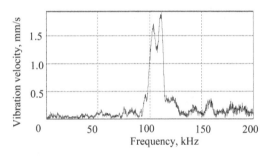

Figure 6. LDR frequency response for an impact induced damage in a CFRP plate.

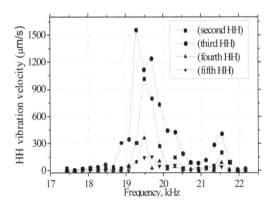

Figure 5. Higher harmonic LDR frequency responses of a crack in UD-CFRP.

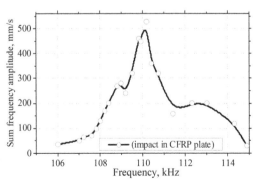

Figure 7. LDR induced amplification at the sum frequency vibration for impact damage in a CFRP plate.

at sum frequency as a function of $f_+$ measured by changing $f_2$ in the frequency range indicated above. The impact of LDR is clearly seen by comparing the data with those in Figure 6: more than 20 dB increase in the output is observed when the combination frequency matches the frequency of LDR.

According to the above, at moderate input signals the LDR enhances appreciably the nonlinearity of defects via local "amplification" of vibrations. It raises substantially the efficiency of "conventional" nonlinear effects, like HH generation and wave mixing. However, this is not the only dynamic scenario of nonlinear phenomena for resonant defects. At higher level of excitation, a combined effect of LDR and nonlinearity can result in anomalously efficient frequency conversion into higher harmonics and subharmonics via superharmonic and subharmonic resonances.

For the superharmonic resonance, the input frequency is taken as $\approx \omega_0/n$ and converted into $\omega_0$ drive via the nth-order nonlinearity of the defect.

A direct proof of superharmonic resonances in defects is demonstrated for the third-order resonance in impact damaged CFRP specimen with LDR around 5140 Hz in Figure 8. One-third of the LDR frequency (1714 Hz) was therefore selected for the excitation and the input voltage was increased up to 80 V. The spectrum (Fig. 8) of the defect vibrations measured in the defect area beyond the threshold illustrates the dominance of the third harmonic: the third harmonic amplitude is ~25 dB higher than fundamental frequency component.

Manifestation of parametric effects (e.g. resonant subharmonic vibrations) is due to the amplitude-dependent shift (modulation) of LDR frequency induced by the driving signal. Unlike conventional (linear) resonance, the parametric resonances provide an exponential growth of the vibration amplitudes in time even in the presence of damping. Such instability develops as soon as the frequency modulation index (input signal amplitude) exceeds a certain threshold determined by the energy dissipated in the system.

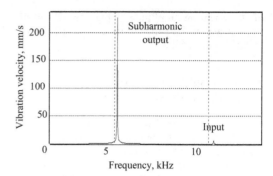

Figure 9. Subharmonic spectrum for LDR of impact damage in CFRP beyond threshold input voltage 45 V.

To observe the resonant growth of subharmonic the excitation frequency was changed to the second harmonic (10280 Hz) of the fundamental LDR for the impact damage in CFRP. The threshold for the resonance was found to be ≈ 45 V. Beyond the threshold, the subharmonic component increases dramatically and prevails in the vibration (velocity) spectrum: $V_{\omega/2}/V_\omega \approx 30$ dB at 10280 Hz input (Fig. 9). The input frequency range for both subharmonic and superharmonic resonances was measured to be within ~100–200 Hz that corresponds to a high Q factor of the LDR for this defect.

### 3.1 Nonlinear LDR imaging of defects

A local "amplification" of vibrations is a basis of sensitive frequency-selective imaging even in a linear LDR case. The benefit of the linear LDR imaging is demonstrated in Figure 10, where it is used for visualization of two small square artificial delaminations (8 × 8 mm² and 12 × 12 mm²) with LDR frequencies 91130 Hz and 71250 Hz, correspondingly, in a CFRP plate (300 × 300 × 3 mm³). The excitation at corresponding LDR frequencies results in imaging of the defects separately (Fig. 10a, b) while the excitation within the frequency range of both LDR brings an image of the both defects (Fig. 10c).

Under the LDR condition, new nonlinear frequency components are generated efficiently and highly localized in the defect area that provides a background for the high-contrast defect-selective imaging. The benefit of the higher harmonic LDR imaging is illustrated in Figure 11. A substantial improvement of the image quality (10 × 20 mm² delamination in GFRP plate) is clearly seen by comparing the fundamental and the second harmonic images.

Other examples of the resonance nonlinear imaging are given in Figures 12–13. The LDR

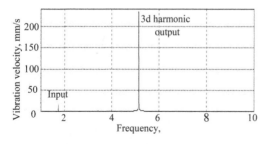

Figure 8. Spectrum of the third-order superharmonic LDR in impact damaged CFRP plate.

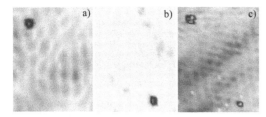

Figure 10. Frequency-selective defect imaging of delaminations in CFRP plate in a linear LDR mode: excitation frequency 71250 Hz (a); 91130 Hz (b); wideband excitation 45–100 kHz (c).

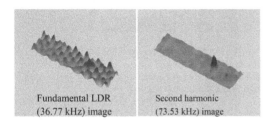

Figure 11. Linear (left) and HH (right) LDR imaging of a delamination in GFRP specimen.

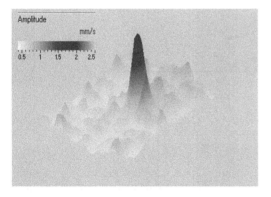

Figure 12. Sum-frequency image of the impact-induced damage (~5 × 5mm²) in a CFRP plate.

contribution to the sum-frequency signal makes it localized in the damage area and enables to be used for mixing frequency imaging with reasonable signal-to-noise level (~15 dB, Fig. 12). This image of the impact in a CFRP plate was obtained by mixing two flexural waves of frequencies 77 kHz and 30 kHz via the combination frequency resonance (LDR frequency of the defect 107 kHz). The benefit of the subharmonic LDR is illustrated in Figure 13 for the impact damage in CFRP specimen.

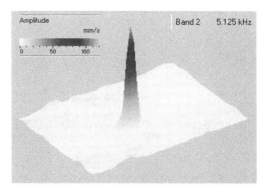

Figure 13. Subharmonic LDR imaging of impact damage in a CFRP plate: Input 10250 Hz; output 5125 Hz.

## 4 LDR FOR THERMOSONIC IMAGING

Thermosonic response of a defect is proportional to an acoustic energy delivered, i.e. to the square of its vibration amplitude. To provide a measurable temperature response, thermosonics traditionally relies on high-power ultrasonic welding equipment, which includes kW-power supply (at fixed frequencies 20 or 40 kHz) and piezo-stack converters combined with ultrasonic boosters and horns (Mignogna et al., 1981). The test specimen is pressed against the horn that results in unstable ultrasonic response and highly non-reproducible measurements. The reason for this "specificity" is concerned with a low efficiency of ultrasound-heat conversion that is usually taken for granted without an effort to be optimized.

To make ultrasonic thermography compatible with conventional ultrasonic equipment would be a step on the way to extend its applicability in non-destructive inspection. To this end, an obvious task is to find out a feasibility of ultrasonic thermography in the mW-acoustic power range typical for commercial ultrasonic applications.

The LDR-induced amplification of local vibrations is, therefore, beneficial for enhancement of acousto-thermal conversion and would enable to reduce acoustic power required for thermosonic imaging. Figure 14 illustrates the benefit of LDR thermosonic imaging of the square insert in CFRP specimen. A crucial role of LDR is readily seen: At fundamental LDR frequency (8980 Hz) the temperature response (~0.25 K) is by more than an order of magnitude higher than that outside the LDR (8000 Hz).

The laser vibrometry measurements of this defect (shown in Figure 2a, b) reveal both the fundamental (8980 Hz) and the higher-order (15600 Hz) LDR with substantially different vibration patterns. The thermal images taken at

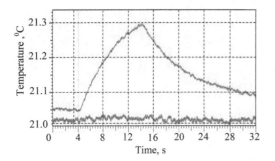

Figure 14. Temperature response of a rectangular inset in CFRP plate at LDR frequency (8980 Hz, upper curve) and outside resonance (lower curve, 8000 Hz).

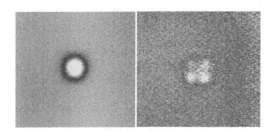

Figure 15. Thermosonic images of a rectangular insert in CFRP plate at fundamental LDR frequency (8980 Hz, left) and at higher-order LDR (15600 Hz, right).

Figure 16. LDR imaging of a square insert (25 × 25 mm$^2$) in CFRP (left) in a wideband excitation mode: vibrometry (center) and thermosonic (right) images.

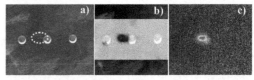

Figure 17. Laser vibrometry (b) and LDR thermography (c) LDR imaging of fatigue crack between the rivet holes (dotted area in zoomed optical image (a)) in aluminum aviation component.

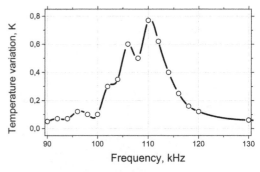

Figure 18. Thermal LDR frequency response of an impact induced damage in a CFRP plate.

the corresponding excitation frequencies (Fig. 15) demonstrate the importance of the higher-order resonances for visualization of the defect shape: while the fundamental LDR visualizes the center part, the higher-order LDR are responsible for imaging of the border areas of the defect.

This feature is beneficial for the quality of thermosonic imaging: the higher-order resonant thermal patterns visualize the contours of defects and thus applicable for recognizing a real defect size and shape. To image an entire defect area various order thermal images can apparently be superimposed. The impact of wideband excitation on resonant imaging of defects is illustrated in Figure 16 for both laser vibrometry and thermosonics. The excitation of the square insert with fundamental LDR of 8980 Hz by using a periodic sweep mode (1–200 kHz bandwidth) results in a superposition of various order resonant patterns (both thermal and vibration) and reproduces an entire shape of the defect.

Figure 17 illustrates the application of LDR thermosonics to an aluminum aviation component: (1.8 × 180 × 300 mm$^3$) plate with a fatigue crack between the rivet holes (zoomed optical image Figure 16a). The identification of LDR of cracks in metals is complicated due to high mechanical quality factors of the materials and various vibration modes of the cracked defects. One of the LDRs of the fatigue crack measured by laser vibrometry at 11600 Hz is shown in Figure 17b; the thermosonic image obtained at this frequency (Fig. 17c) demonstrates the applicability of the LDR methodology to thermosonic imaging in metals.

To illustrate the LDR benefit to enhancement of thermosonics efficiency the defect temperature response is measured as a function of driving ultrasonic frequency. The example in Figure 18 for point-like impact damage in a CFRP plate demonstrates substantial temperature increase (~15 times) at LDR frequency ~110 kHz in accord with the defect acoustic response shown in Figure 6.

To quantify the contribution of LDR to thermosonics output the temperature response of this defect is measured for calibrated ultrasonic power obtained with laser vibrometry. The data in Figure 19, a confirms highly efficient LDR thermosonics: $\Delta T = 1.4$ K for ~60 mW input acoustic power. Such temperature variation provides reliable imaging in mW power range with a high temperature contrast in lateral direction (Fig. 19b).

The results shown above imply that a strong increase in the defect temperature rise (thermal output signal) at LDR frequency enhances the Signal-to-Noise Ratio (SNR) of thermosonic imaging. On the other hand, an increase of the SNR is also known to occur in the lock-in mode primarily due to diminishing the noise level (Busse et al., 1992). By introducing the benefit of LDR in the lock-in approach a resonance thermosonic mode operating at unusually low excitation levels can be projected. To this end, following the general lock-in concept the amplitude of ultrasonic excitation at the LDR frequency was modulated sinusoidally at the lock-in frequency (between 0.01 Hz and 1 Hz). A temperature image sequence of the surface was recorded with the IR-camera and a discrete Fourier transformation at the lock-in-frequency was applied to compress the data into a pair of amplitude and phase images.

The enhancement in sensitivity and the SNR of LDR lock-in imaging are readily seen from Figure 20, where the amplitude lock-in (a) and LDR temperature (b) thermosonic images of a circular Flat-Bottomed Hole (FBH) in PMMA plate are shown. To have the SNR >1 directly in the temperature image (without lock-in) (Fig. 20b), the input power had to be increased by up to ~2 mW (to generate $\Delta T$ ~ 100 mK). On the contrary, the LDR lock-in image in Fig. 20a) was taken when the input was reduced to anomalously low power of ~200 µW. The background for such an extraordinary performance is a combined action of the lock-in (reduction SNR) and the high thermosonic quality factor (efficient heat generation) in the LDR mode.

Figure 21 illustrates the enhancement in sensitivity of thermosonics by combining LDR and lock-in

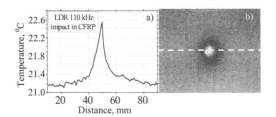

Figure 19. LDR thermosonic imaging of ~(5 × 5 mm²) impact damage area in a CFRP plate (b); quantified temperature contrast of the image along the dotted line (a).

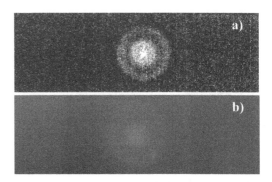

Figure 20. LDR thermosonic imaging of FBH in PMMA plate at LDR frequency 7670 Hz: (a) - amplitude lock-in (lock-in frequency 0.05 Hz) image (acoustic input ~200 µW); (b) – temperature image at input power ~ 2 mW.

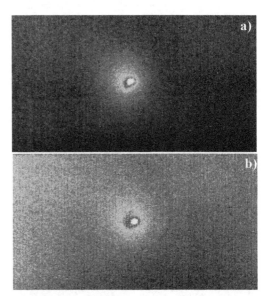

Figure 21. LDR thermosonic imaging of an impact (~5 × 5 mm²) in CFRP plate: amplitude lock-in image (a) at ~1 mW input acoustic power; (b) – temperature image at ~16 mW input power.

for an impact damage in a CFRP plate: the amplitude lock-in image (a) corresponds to ~1 mW input power while a similar contrast of the temperature image (b) requires ~16 mW of acoustic power.

## 5 LDR SHEAROGRAPHY

A conventional out-of-plane optical sensor head (Menner et al., 2010) with continuous-wave laser was used for time-averaged shearographic imaging.

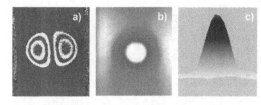

Figure 22. Sherographic (a) and laser vibrometer images (b (top view), c (side view)) of a FBH vibration in PMMA plate at fundamental LDR (7672 Hz).

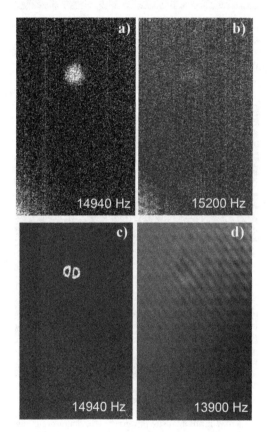

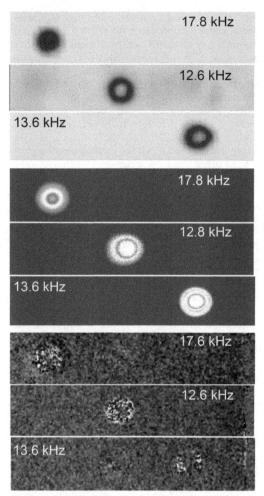

Figure 24. LDR laser vibrometer (top), thermosonic (center) and shearographic (bottom) images of three circular FBH in PMMA specimen: the frequencies of maximum output are indicated on the images.

Figure 23. Impact of frequency mismatch on LDR thermosonic (a, b) and shearographic (c, d) imaging of an adhesion lack area in honeycomb structure (Fig. 3): the excitation frequencies are indicated on the images.

In Figure 22, the laser vibrometer and shearography images of a FBH vibration in PMMA specimen obtained at the fundamental LDR (7672 Hz) are compared. The 'butterfly' fringe pattern (Fig. 22a)) clearly demonstrates that shearography is sensitive to the gradient of out-of-plane LDR displacement ("bell-like" function in Fig. 22c) in the shearing (horizontal) direction. The increase in out-of-plane vibration amplitude due to LDR provides a strong increase in the corresponding strain and thus enhancement of sensitivity in LDR shearography mode.

The effect of LDR shearography is demonstrated in Figure 23 for aluminium honeycomb structure (thickness 16 mm) with several inclusions of resin and water in GFRP liners ($0.5 \times 100 \times 100$ mm$^2$). The laser vibrometry scan of the front side shown in Figure 3 indicates fundamental LDR of one of the defects at 14940 Hz. The ultrasonic excitation for LDR shearography was carried out by conventional piezo-ceramic transducers (Conrad Elektronik GmbH) attached to the rear side of the structure. The results of both shearographic and thermosonic imaging (Fig. 23) confirm substantial enhancement of the sensitivity of detection at the LDR frequency: the images practically disappear even at a minor frequency mismatch from the LDR frequency (Fig. 23).

To trace frequency selectivity of LDR shearographic imaging (discussed above for laser vibrometry) we used a set of three circular FBH in a PMMA plate. For the plate thickness of 25.5 mm, the depths of the holes changed from 22 to 24 mm thus defining different LDR frequencies and driving excitation frequencies, respectively. The images of the defects obtained by the laser vibrometry, thermosonics and shearography at corresponding LDR frequencies are compared in Figure 24. They confirm maximum sensitivity at the LDR frequencies; a minor variation of the frequencies is caused by different clamping conditions of the specimens.

## 6 CONCLUSIONS

In summary, the frequency match between the excitation ultrasonic frequency and the frequency of LDR results in substantial "amplification" of the defect vibration amplitude. It leads to enhancement of sensitivity and efficiency in ultrasonic NDT and imaging of defects via nonlinearity, laser vibrometry, thermosonics and shearography. A strong frequency selectivity of LDR results in opportunity of detecting a certain defect among a multitude of others by using all above mentioned NDT methods. An improvement of image quality of an entire defect area is obtained in the higher-order LDR and wide-band ultrasonic excitation modes. The LDR NDT requires much lower acoustic power to activate the defects that makes it possible to avoid high-power instrumentation and use conventional ultrasonic equipment instead.

## ACKNOWLEDGEMENT

The author acknowledges support of this study in the framework of ALAMSA project funded from the European Union's Seventh Framework Programme for research, technological development and demonstration under grant agreement no. 314768.

## REFERENCES

Busse, G., Wu, D., Karpen, W. 1992. Thermal wave imaging with phase sensitive modulated themography, *Journal of Applied Physics* 71 (8): 3962–3965.

Menner, P., Gerhard, H., Busse, G. 2010. Remote defect visualization with thermal phase angle shearography, *AIP Conference Proceedings* 1211 (1): 2068–2072.

Solodov, I., Bai, J., Bekgulyan, S., Busse, G. 2011. A local defect resonance to enhance acoustic wave-defect interaction in ultrasonic nondestructive testing, *Applied Physics Letters* 99: 211911.

Solodov, I., Bai, J., Busse, G. 2013. Resonant ultrasonic spectroscopy of defects: Case study of flat-bottomed holes, *Journal of Applied Physics* 113: 223512.

*Infra-asset assessment with innovative NDT*

*Emerging Technologies in Non-Destructive Testing VI – Aggelis et al. (Eds)*
*© 2016 Taylor & Francis Group, London, ISBN 978-1-138-02884-5*

# Experimental verification of a Rayleigh-wave based technique for detecting the depth of deteriorated concrete

K.C. Chang, T. Shiotani & S.B. Tamrakar
*Department of Civil and Earth Resources Engineering, Kyoto University, Kyoto, Japan*

ABSTRACT: Concrete surfaces often deteriorate due to various factors, such as precipitation, freeze, fire, chemical reactions, loadings, etc. To detect the depth of those deteriorated concretes, Spectral Analysis of Surface Wave (SASW) was introduced, which utilizes the dispersion characteristics of Rayleigh waves (R-waves) propagating in multi-layered solids. The SASW method was conventionally performed in Fourier-based frequency domain, while a Wavelet-based time-frequency domain analysis could be an alternative although less studies were focused on it. This study aims to preliminarily investigate the feasibility of the Wavelet-based SASW method with a laboratory-scale specimen. It is verified that, firstly, dispersion phenomenon of R-waves is observed in most cases considered in this study, and secondly, from the dispersion characteristics, the depth of the deteriorated layer (modelled by mortar) can be estimated with acceptably accuracy in certain cases, although the optimal conditions are still under investigation. In addition, several factors affecting the estimation accuracy are discussed, including the size of steel sphere for generating R-waves and the receiver location.

## 1 INTRODUCTION

Concrete surfaces often deteriorate due to various factors, such as precipitation, freeze, fire, chemical reactions, loadings, etc. The surface deterioration generally presents in terms of decrease in material stiffness and strength, breaking cracks, distributed micro-cracks, and so on. From the concrete surface, the presence of certain deterioration terms can be visibly inspected, e.g. cracks extended to a visible degree, fire-damaged surfaces or so, but to evaluate the depth of those deterioration layer requires further inspection techniques. How to detect the depth of those deteriorated concretes, in a fast and reliable manner, is thus always of engineers' great interest.

For this purpose, Spectral Analysis on Surface Wave (SASW) was proposed (Cho &Lin 2001, Goueygou et al. 2004, Kim et al. 2006, Aggelis & Shiotani 2008, Chekroun et al 2009), which employs the dispersion characteristics of Rayleigh waves (R-waves) that propagate in multi-layered elastic solids. R-waves are sensitive to the change of mechanical properties of surface layers because most of their energies are transported near surfaces. It is supposed that the elastic waves would travel with different velocities in intact and deteriorated concretes, generally faster in the former and slower in the latter. It follows that the Rayleigh waves of different wave lengths (as well as penetration depths) would disperse, and the dispersion characteristics would be dependent on layer depths.

Based on this concept, the SASW method allows a surface impulse impact that excites a wave group covering a variety of wave lengths, and then analyzes the dispersion characteristics of the waves recorded by a pair of sensors with a certain spacing at the same surface. Empirically, the impulse impact is performed by a mechanical impact, namely hammering with a steel sphere. The dispersion characteristics is conventionally analyzed in Fourier-based frequency domain, while a Wavelet-based time-frequency domain analysis could be an alternative (Kim & Park 2002) for tackling various noises in some field conditions. The later technique was illustrated in some numerical studies (e.g. Tsai et al. 2014), but not yet fully verified in practice so far.

This study attempts to preliminarily verify the feasibility of the Wavelet-based SASW method by a laboratory-scale cuboid specimen. The main body of the specimen was casted with normal concrete; the top surface was covered by mortar to model the deteriorated concrete. A couple of sensors were mounted at the mortar surface with equal spacing, to record surface responses subjected to a steel-sphere impact. A pair of the recorded responses are then taken into analysis to examine whether the present technique is practically feasible or not. In addition to the feasibility, several key factors are also studied, including the spacing of the sensors and the size of the impact steel sphere.

## 2 DESCRIPTIONS OF EXPERIMENT

The specimen (see Fig. 1) was a cuboid of 40 cm in width, 60 cm in length and 26 cm in depth. The main body of 20 cm deep was casted with normal concrete, with a design strength $\sigma_c = 21 \sim 24$ N/mm². The top and bottom surfaces was covered by mortar with a w/c ratio up to 90% and design strength $\sigma_m = 15$ N/mm² to model deteriorated concrete. The depth of the surface layers is 40 and 20 mm, respectively. No rebar is embedded in the specimen for simplification and only the 40 mm side was studied in the current feasibility study.

Although only two sensors is required to perform depth evaluation with the present method, herein eight sensors were mounted on the target surface, for investigating the spacing effect. The eight sensors (denoted by P1 to P8, see Fig. 2) were equally spaced with a distance of 50 mm. High sensitivity insulated accelerometers of TEAC 707IS model were employed; their resonance frequency was 30,000 Hz and ±3dB nominal frequency response was 3–12,000 Hz. The acceleration data were recorded by a portable data acquisition system of Keyence NR-600 model, with a 100 kHz sampling rate for all channels.

The mechanical impact was performed by hammering the test surface with a steel sphere. The impact location was 50 mm apart from P8 along the- axis of sensor arrays. Two sizes of steel sphere were employed, each with 5 and 15 mm in diameter $D$.

Prior to the experiment, the velocity of P-waves in the normal concrete and mortar were measured, each with an independent cuboid specimen of 10 × 10 × 40 cm in size. Using conventional impact-echo method (Sansalone & Streett 1997), the P-wave velocity was measured as 3,880 m/s for the normal concrete and as 2,840 m/s (about 30% less than the former) for the mortar. Under a reasonable assumption that R-wave velocity is 0.55 times of the P-wave speed, the R-wave velocities could be calculated as 2,134 m/s for the normal concrete and 1,562 m/s for the mortar. It should be noted that the P-wave velocities measured in the pre-test herein merely served as references for the main tests afterwards, in acknowledging the fact that in future field tests they would be unavailable due to the absence of independent specimens.

## 3 ANALYSIS ALGORITHMS AND ILLUSTRATIONS

Let us take a typical case, $D = 5$ mm, to illustrate the operational algorithm.

### 3.1 Impact responses and their wavelet transforms

Firstly the waveforms of each sensor subjected to the impact were recorded, as shown in Figure 3.

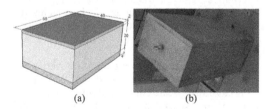

Figure 1. Test specimen: (a) design drawing with dimensions (in cm); (b) photo.

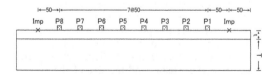

Figure 2. Sensor allocation (unit: mm).

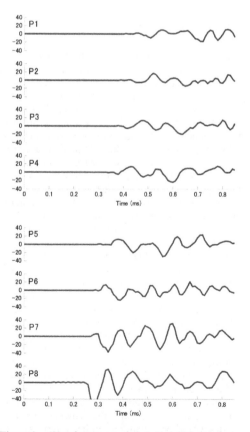

Figure 3. Waveforms recorded at all eight locations.

It should be noted that 0 ms herein indicates the starting instant of the measuring interval rather than the true impact instant; it is true for other waveforms afterwards without further indication. From the delay in arriving times, the propagation of the surface wave from P8 to P1 can be obviously observed.

To realize the dispersion characteristics, time-frequency analysis was performed. Herein, Wavelet Transform was employed as the time-frequency analysis tool. The Wavelet spectrum, as an outcome of Wavelet Transform, of the waveforms for all locations are shown in Figure 4. Those spectra can be interpreted as follows.

(1) P8 might serve as the reference point because it was the point closest to the impact point. From its Wavelet spectrum, it is observed that most of the wave components of different frequencies arrived approximately at the same time, indicating that the spacing between the impact point and reference point (P8) was short enough so that the dispersion had not significantly occurred. Also, the major frequency components covered 7 to 14 kHz, indicating that the impact of steel sphere of $D = 5$ mm might excite waves majorly covering frequencies within this range. Therefore, this frequency band was taken into further analysis in this study; it is also true for the analyses afterwards without further indication.

(2) In some spectra, dispersion phenomena can be observed. For example, in P1 spectrum, it is observed that the waves of lower frequencies, specifically lower than 12 kHz, propagated faster and arrived P1 earlier, and that the waves of higher frequencies, mostly higher than 12 kHz, propagated slower and arrived P1 later. It can be interpreted that, since high-frequency waves propagate with smaller wave lengths that penetrate shallower the multi-layered solids, they would propagate with almost identical and smaller speed if they travel completely within the surface deteriorated layer, i.e. with penetrating wave lengths too small to reach the underlying intact layer; on the other hand, since low-frequency waves propagate with larger wave lengths that penetrate deeper the multi-layered solids, they would propagate faster if travelling with wave lengths penetrating deep enough to the underlying intact layer. Such a dependence of propagating velocity and wave length (dependent on frequency as well) is known as dispersion phenomenon. It can also be observed in some other spectra like P2, P5, P6 and P7 spectra, although not be observed in P3 and P4 spectra probably due to some difficulties in picking up correct peaks with respect to certain frequencies.

### 3.2 Dispersion curves and deterioration depth evaluation

As previously stated, the R-wave dispersion phenomenon would be dependent on the depth of the deteriorated concrete layer. It is supposed that the depth of the deteriorated concrete can be evaluated if the R-wave dispersion characteristics is made known. The dispersion characteristics employed herein is a dispersion curve in velocity-wavelength plane, which expresses the relationship between the propagating velocity $C_R$ and wave length $\lambda_R$ of R-waves. Through time-frequency analysis, this curve can be derived using simply two vibration responses recorded at two locations along the R-wave propagating path.

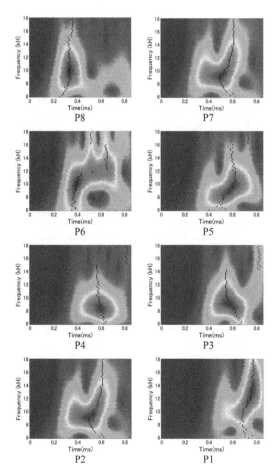

Figure 4. Wavelet spectrum of the waveforms recorded at all eight locations, impact nearby P8, $D = 5$ mm (red, yellow, green and blue counters indicate large to small amplitudes; black dots indicate the peak with respect to each frequency).

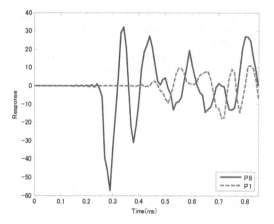

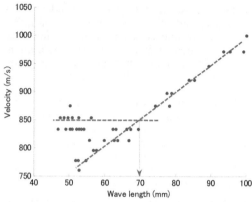

Figure 5. Waveforms recorded at P8 and P1.

Figure 6. Dispersion curve, Case I8-D5-P8P1.

Taking the vibration responses recorded by P8 and P1 sensors for example (see Fig. 5), the R-wave dispersion curve can be obtained as follows. For simplicity, this case is denoted as Case I8-D5-P8P1, where the term I8 denotes the impact point nearby P8, D5 the diameter $D = 5$ mm, and P8P1 the receivers P8 and P1. For a certain frequency, $f_i$, the arriving time of its largest peak to P8 and P1, say $t_{8,i}$ and $t_{1,i}$, can be read from their Wavelet spectra respectively. In knowing the distance between P8 and P1, i.e. $d = 350$ mm in this case, the propagating velocity $C_{R,i}$ of the wave of frequency $f_i$ can be calculated as

$$C_{R,i} = d/\Delta t_i = d/(t_{1,i} - t_{8,i}) \quad (1)$$

where $\Delta t_i$ is defined as the difference of arriving times between P8 and P1.

Next, the wave length $\lambda_{R,i}$ corresponding to such wave can be calculated as

$$\lambda_{R,i} = C_{R,i}/f_i \quad (2)$$

Calculating the propagating velocity and corresponding wave length with respect to a range of frequency of interest, e.g. 7 to 14 kHz in this case, and plotting their relationships in the speed-wavelength plane, a dispersion curve can be obtained, as shown in Figure 6. As expected, the wave of the largest wave length propagates the fastest (with the largest speed); the propagating speed decreases as the wave length decreases from the maximum to around 70 mm, and almost remains constant when the wave length is smaller than 70 mm. The wavelength corresponding to the turning point would be related to the depth of the surface layer. It was numerically proposed to be two times of the layer depth (Tsai et al. 2014); in other words, the layer depth could be identified as one half of the wavelength corresponding to the turning point. By this way, the depth of the mortar layer was estimated as 35 mm, which is acceptably close to the true depth, 40 mm.

It might be noted that a bifurcation appeared around 70 mm. The reason of the bifurcation is still under investigation, probably being the imperfect cast of the mortar layer.

4 RESULTS AND DISCUSSIONS

For other cases, the depth of the mortar layer could be estimated through constructing dispersion curves in the similar way. Let us consider the case with the same impact point, i.e. P8 end, and the same hammer size, i.e. $D = 5$ mm, but with different receiver spacing distances $d$. The propagating speed $C_R$ vs. wave length $\lambda_R$ relationships constructed with P8-P2 receivers (Case I8-D5-P8P2, $d = 300$ mm), P8-P4 receivers (Case I8-D5-P8P4, $d = 200$ mm), and P8-P5 receivers (Case I8-D5-P8P5, $d = 150$ mm) are plotted in Figures 7, 8, and 9, respectively.

The dispersion characteristics can be observed for Cases I8-D5-P8P2 and I8-D5-P8P5, but the depth of mortar layer estimated by the dispersion curves are quite deviated: $120/2 = 60$ mm for Case I8-D5-P8P2 and $54/2 = 27$ mm for Case I8-D5-P8P5, neither of which is close to the true depth. As for Case I8-D5-P8P4, no dispersion characteristics is observed in the $C_R - \lambda_R$ plot. It follows that the estimation of depth of mortar layer is failed. Moreover, the other cases with smaller receiver spacing distances are not considered herein, e.g. P8-P6 receivers with $d = 100$ mm or P8-P7 receivers with $d = 50$ mm, for the reasons that dispersion phenomenon may not be obvious within a

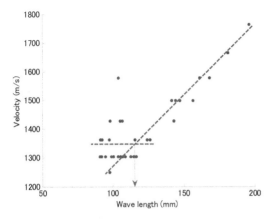

Figure 7. Dispersion curve, Case I8-D5-P8P2.

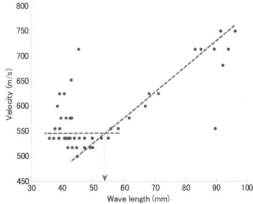

Figure 9. Dispersion curve, Case I8-D5-P8P5.

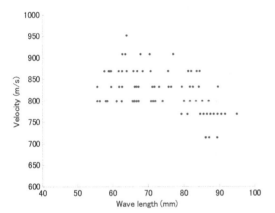

Figure 8. Dispersion curve, Case I8-D5-P8P4.

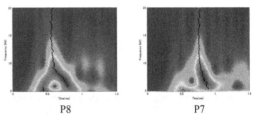

Figure 10. Wavelet spectrum of the waveforms recorded at P8 and P7, impact nearby P8, $D = 15$ mm.

short propagation distance and that too short the receiver spacing is ineffective in practice.

Next, let us consider a larger steel sphere, with diameter $D = 15$ mm. Sensor allocation remained the same and impact point remained at P8 end. In this case, it is firstly observed that P8 was not proper to serve as a reference point anymore, presenting a Wavelet spectrum that is not easy to identify correct peaks with respect to certain frequencies (see Fig. 10). The disqualification of P8 being a reference point probably due to the larger impact excited by the larger steel sphere. Contrarily, P7 was more proper to serve as a reference point, presenting a Wavelet spectrum that is easier to identify correct peaks with respect to most frequencies of interest. For this reason, P7 is taken as reference in this case.

The overall results analyzed for the cases with impact nearby P8 and steel sphere diameter $D = 5$ and 15 mm are summarized in Table 1. It is observed that, firstly, the depth of the mortar layer can be estimated from the dispersion characteristics in five out of eight cases, but fails in the other three cases. Among those five successful cases, Cases I8-D5-P8P1 and I8-D15-P7P3 yielded acceptable estimated depth that is close to the true depth, 40 mm, but Case I8-D5-P8P5 yielded much smaller while Cases I8-D5-P8P2 and I8-D15-P7P1 yielded much larger estimated depth than the true depth.

It seems that there was no strong correlation between the successful estimation of the depth of mortar layer and the receiver location (as well as the receiver spacing distance). Large variations in the estimated depth were obtained and they had little correlation to the receiver location and spacing distance. The potential factors could be discussed as follows.

(1) *Imperfect casting of the mortar layer*. Several portions of the mortar layer were found to delaminate from the concrete core. The delamination might prevent the long-wavelength R-waves to penetrate into the underlying concrete layer and accordingly make the measured waves dominated by R-waves of short wave lengths and by P-waves either propagated along

Table 1. Estimated depth $t_e$ of mortar layer.

| Case ID | $D$ (mm) | Receiver #1 | #2 | $d$ (mm) | $t_e$ (mm) |
|---|---|---|---|---|---|
| I8-D5-P8P1 | 5 | P8 | P1 | 350 | 35 |
| I8-D5-P8P2 | 5 | P8 | P2 | 300 | 60 |
| I8-D5-P8P3 | 5 | P8 | P3 | 250 | NA |
| I8-D5-P8P4 | 5 | P8 | P4 | 200 | NA |
| I8-D5-P8P5 | 5 | P8 | P5 | 150 | 27 |
| Mean | | | | | 41 |
| I7-D15-P7P1 | 15 | P7 | P1 | 300 | 65 |
| I7-D15-P7P2 | 15 | P7 | P2 | 250 | NA |
| I7-D15-P7P3 | 15 | P7 | P3 | 200 | 40 |
| I7-D15-P7P4 | 15 | P7 | P4 | 150 | NA |
| Mean | | | | | 53 |

Note. $D$: diameter of steel sphere;
$d$: distance between Receiver #1 and #2;
NA: not available to be estimated.

surfaces or reflected from delaminated interfaces. As a result, the dispersion phenomenon did not occur in those delamination portions.

(2) *Surface and interface cracks.* Several obvious cracks were visually inspected on the surface of the mortar layer, and it is believed that there would be more cracks forming beneath the surface or at the interfaces between mortar and concrete layers. Those cracks might also decrease the R-wave propagating speeds and cause the R-waves to disperse. It then made the resultant dispersion curves so complicated that the turning point was hardly identified or that the turning point occurred at where corresponded to other dispersion source rather than the interface between mortar and concrete. It would be encouraging if the turning point could be identified for the latter condition, which indicates that the present method could be applicable to estimate other deteriorated conditions. However, so far no enough evidences support this statement and thus further studies are required.

The size of steel sphere for generating mechanical impacts seems to have slight effect on the accuracy of estimated depth, even though only two sizes, $D$ = 5 and 15 mm, were tested in this study. Taking average of the depths estimated with those two sizes respectively (see Table 1, 41 mm for $D = 5$ mm case and 53 mm for $D = 15$ mm case), one could indicate that the steel sphere of 5 mm in diameter was more appropriate to estimate the mortar layer

depth of 40 mm herein than that of 15 mm was. Probable reason could be that the steel sphere of $D$ = 5 mm excited R-waves of wave lengths closer to the target depth to be estimated.

In addition to the receiver location and the size of steel sphere, there are several issues could be discussed as follows. First issue is that the R-wave speed $C_R$ presented in the $C_R - \lambda_R$ plot are smaller than that calculated in the pre-test. For example, the $C_R$ presented in Figure 6 ranges from 750 to 1100 m/s, which is smaller than that calculated in the pre-test, i.e. 1562 m/s. Potential reason might be the same as those discussed previously: the delamination of surface mortar layer and the surface and interface cracks would decrease the R-wave propagating speed. The same observation could apply to the other plots, e.g. Figures 6–9.

Another issue, important but still open, is the robust identification of the turning point in the dispersion curve. In this preliminary study, the turning points were identified empirically and roughly by eye inspections. Although the identification method of turning points is out of the scope of this study, it could be briefly discussed as follows. A turning point is highly dependent to the numerical fitting of two curves: one is the line with constant slope representing the proportional relationship between propagating speed and wave length; the other one is a horizontal line, or arguably a line with a slope different from the former one. How to select proper points for fitting those two lines are critical but still open. It would be more difficult nearby the turning points, where the data generally distributes in a highly deviated manner. For some cases, the turning point would be replaced by a bifurcation points, which makes the identification more complicated. The theoretical background of the appearance of bifurcation is still under investigation.

## 5 CONCLUDING REMARKS

This study focuses on a potentially powerful NDT technique for estimating the depth of surface deteriorated concrete. The present technique employs the dispersion characteristics of R-waves propagating in multi-layered solids, simply requiring two sensors installed on the target surface. Through this preliminary experimental study on a laboratory-scale specimen, the following concluding remarks can be drawn.

Firstly, dispersion phenomenon, i.e. the dependence of propagating speed on wave length, of R-waves is observed in most cases considered in this study.

Secondly, from the dispersion characteristics, or more specifically the dispersion curve in the propagating speed vs. wavelength plane, the depth of the deteriorated layer (modelled by mortar) can be estimated with acceptably accuracy in certain cases, although the optimal condition is still under investigation.

Thirdly, the size of steel sphere for generating mechanical impacts seems to have slight effect on the accuracy of estimated depth. The steel sphere of 5 mm in diameter is more appropriate to estimate the mortar layer depth of 40 mm herein than that of 15 mm.

Fourthly, there is no strong correlation between the successful estimation of the depth of mortar layer and the receiver location (as well as the receiver spacing distance) probably due to imperfect casting of the mortar layer and to surface and interface cracks.

Although the above observations may apply to the cases similar to this study only, they provide precious references for further practical applications to real structures. To make the technique applicable to general structures, more laboratory and field experiments are requiring, which are the main topics of the ongoing studies.

## REFERENCES

Aggelis, D.G. & Shiotani, T. (2008). Surface wave dispersion in cement-based media: Inclusion size effect. *NDT&E International* 41, 319–325.

Chekroun, M., Le Marrec, L., Abraham, O., Durand, O. & Villain, G. (2009). Analysis of coherent surface wave dispersion and attenuation for non-destructive testing of concrete, *Ultrasonics* 49, 743–751.

Cho, Y.S. & Lin, F.B. (2001). Spectral analysis of surface wave response of multi-layer thin cement mortar slab structures with finite thickness, *NDT&E International* 34, 115–122.

Goueygou, M., Piwakowski, B., Fnine, A., Kaczmarek, M., Buyle-Bodin, F. (2004). NDE of two-layered mortar samples using high-frequency Rayleigh waves, *Ultrasonics* 42, 889–895.

Kim, D.S., Seo, W.S., Lee, K.M. (2006). IE–SASW method for nondestructive evaluation of concrete structure, *NDT&E International* 39, 143–154.

Sansalone, M. & Streett, W.B. (1997). *Impact-echo: nondestructive evaluation of concrete and masonry*, Bullbrier Press, New York.

Tsai, W.H., Liu, Y.T., Lin, Y.C. and Lin, Y.F. (2014). Detecting the depth of concrete deterioration using Rayleigh wave dispersion based on time-frequency analysis, *Proceedings of the 11th European Conference on Non-Destructive Testing*, 6–10 Oct., Prague.

*Emerging Technologies in Non-Destructive Testing VI – Aggelis et al. (Eds)*
*© 2016 Taylor & Francis Group, London, ISBN 978-1-138-02884-5*

# Evaluation of grouting conditions for PC structures with wide-band ultrasounds

**M. Hashinoki**
*West Nippon Expressway Company Limited, Osaka, Japan*

**M. Hara**
*Nippon P.S Company Limited, Fukui, Japan*

**T. Kinoshita**
*H&B System Company Limited, Tokyo, Japan*

**K.C. Chang**
*Graduate School of Engineering, Kyoto University, Nishikyo, Kyoto, Japan*

ABSTRACT: Grouting plays an important role in Pre-stressed Concrete (PC) structures, for keeping pre-stressed tendons working consistently with concrete bodies. However, many PC grouting defects were reported, arising due to material or construction problems and eventually resulting in fatal damage, such as corrosion and even fracture of PC tendons. To tackle this problem, a non-destructive testing method that employs ultrasounds of a wide frequency band ranging from 5 to 150 kHz was presented to evaluate the grouting conditions. This paper provides an overview of the present method and reports its practical applications to the evaluation of grouting conditions for a number of real PC structures.

## 1 INTRODUCTION

Nowadays, grouting of tendons in Pre-stressed Concrete (PC) structures can be well performed as the advances in grouting materials, construction methods and quality controls[1]. However, grouting in past times were not as well performed as it is. Some cases were reported that ducts were insufficiently grouted, resulting in corrosion or breakage of pre-stressed tendons, decrease of load-resisting capacity and durability[2]. Those poor grouting were mainly due to (and usually varied with respect to) materials of poor performance (e.g. causing bleeding), structural problems (e.g. too small spaces in ducts), poor workmanships (e.g. workers with limited knowledge of PC).

To prevent the damage caused by insufficient grouting, many researches and surveys has been devoted to evaluating grouting qualities by Non-Destructive Testing (NDT) methods. For one example, elastic-wave methods and electromagnetic-wave methods were proposed, based on the straightforward idea that the voids in the ducts might be easily detected by those waves that travel in different speeds in the air and solids. However, several difficulties arise, e.g. rebars surrounding pre-stressed tendons, ducts confining grout, etc.,

making the abovementioned methods less successful. To resolve those difficulties, more advanced methods are required. To name some, multipath array radar method[3], impact-echo method[4], X-ray method[5], and wide-band ultrasound method are regarded as appropriate methods currently, although it is noted that each method has its specific applicability conditions like the depth of concrete covers, material of ducts, etc.

This paper focuses on a wide-band ultrasound method that employs ultrasounds with a wider frequency band than those used in conventional ultrasound methods, and presents its applicability to the evaluation of grouting conditions of existing PC bridges following the introduction of its theory and instruments.

## 2 EVALUATION OF GROUTING CONDITIONS BY A WIDE-BAND ULTRASOUND METHOD

### 2.1 *Theory*

Evaluation of grouting conditions by the wide-band ultrasound method is performed by mounting probes on the concrete surface right above the target duct, and then transmitting ultrasounds of a wide band of frequencies via the transmitter probe and receiving the ultrasounds reflected from the duct via the receiver probe. The grouting conditions can be evaluated by observing the difference in the characteristics of transmitted and reflected waves. A wave propagation manner can be illustrated as in Figure 1. Ultrasounds may reflect from the interface between two different medium. It implies that, if voids are present, approximately total reflection may occur and the reflected waves are of large amplitudes. On the other hand, if voids are absent, reflection rates as well as amplitudes of reflected waves are smaller. Furthermore, in knowing that the waves reflected from ducts are majorly of high frequencies, it can be expected that majorly high-frequency waves are received in insufficient grouting cases and lower-frequency waves in full grouting cases.

Conventional ultrasound methods employ transmitting waves of mono frequency (see Figure 2), by which target reflected waves (reflected from ducts) are usually confused with waves (confusing waves) reflected from coarse aggregates and rebars, and thus not easily to be identified. The present wide-band ultrasound method employs ultrasounds of a frequency band as wide as 5 to 150 kHz and extract the information of ducts by filtering out confusing waves (see Figure 3).

### 2.2 Instruments

Instruments used in this study is shown in Photo 1, and their specifications are listed in Table 1. The transmitter and receiver probes are 76 mm in

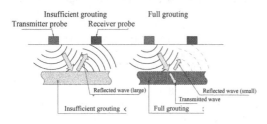

Figure 1. Illustration of ultrasound propagation.

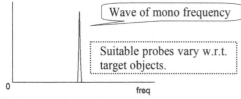

Figure 2. Frequency band employed in conventional ultrasound methods.

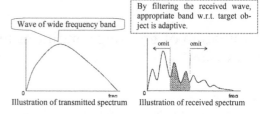

Figure 3. Frequency band employed in the wide-band ultrasonic wave method.

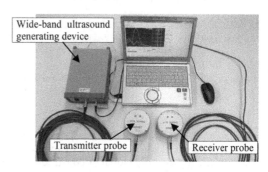

Photo 1. Instruments for wide-band ultrasonic wave method.

Table 1. Specifications of instruments.

| | |
|---|---|
| Frequency band | 2.5~1,000 kHz |
| Input voltage | 1~500 V |
| Gain | 20~60 dB |
| No. of samples | 4,000 samples (optional 16,000 samples) |
| Sampling frequency | 78 kHz~10 MHz |
| Power supply | AC100V |
| External dimensions | W200 mm × D230 mm × H72 mm, 2.8 kg |
| Probe (transmitter, receiver) | φ76 probe, φ95 mm × H60 mm |

diameter, which is large enough for a target with deep concrete cover.

### 2.3 Evaluation methods

#### 2.3.1 Multipoint measurement (confusing-waves processing)

When transmitting ultrasounds into concrete, besides the waves reflected from ducts (target waves), waves reflected from the opposite surface, rebars, etc., are also received. In addition, pre-stressed tendons are generally surrounded by rebars, indicating that the waves reflected from ducts might be smaller than those from rebars. As a result, to extract the target waves from the

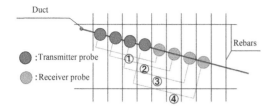

Figure 4.  Multipoint measurement approach.

Photo 2.  Multipoint measurement approach: a field measurement case.

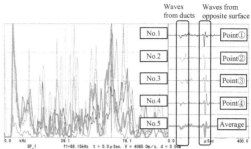

Figure 5.  An example of averaging waveforms.

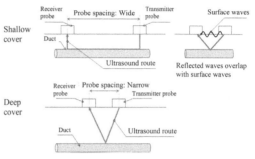

Figure 6.  Probe spacing and ultrasonic wave propagation path.

superposed waves are difficult and accordingly to evaluate grouting conditions is difficult. To solve this difficulty, multipoint measurements were conducted to amplify the waves reflected from ducts. This approach can be illustrated as in Figure 4 and a field measurement case is given in Photo 2.

In multipoint measurement, the transmitter and receiver probes were slid along the target duct and several rounds, say herein four rounds, of measurement were performed at different locations. By doing so, the relative location between the probes and ducts remained identical but that between the probes and rebars varied. Therefore, averaging the measured waveforms of all rounds might amplify the waves reflected from ducts and cancel the other confusing waves. Afterwards, the averaged waveforms were taken for further analysis.

An example of averaging waveforms is shown in Figure 5, where the waves labelled No. 1 to 4 are waveforms measured from different points respectively and their averaged waveform is labelled No. 5. Various confusing waves are observed in respective waveforms but they are cancelled and become less dominant, so that the waves reflected from ducts and opposite surface stand out.

### 2.3.2  Probe spacing

The ultrasounds that access receiver probes consist of not only aforementioned reflected waves but also surface waves that propagate on the concrete surface. In cases of deep concrete cover, the arriving time difference between surface waves and reflected waves is large, but in cases of shallow cover, it decreases and those waves might be confused. To address this issue, a probe-spacing strategy were employed as shown in Figure 6. In the case of shallow cover, the probe spacing was tuned wider so that the surface waves and target reflected waves could be distinguished. Generally, in the case of a cover smaller 140 mm, probe spacing was tuned to 500 mm; otherwise, it was tuned to 200 mm.

### 2.4  Analysis method

#### 2.4.1  Waveform processing

As mentioned above, the averaged waveforms obtained from multipoint measurements are taken for further analysis. An outline of the waveform analysis is shown in Figure 7. Figure 7(a) gives the time series of an example averaged waveform along with window functions. Window functions are functions used for spectral analysis, being shifted continuously along time axis to extract (or to window) waveforms of interest. In the figure, the first vertical line indicates the starting point (the instant when the target wave is received); the second vertical line indicates the ending point (the instant when the shift of window function ends). The window

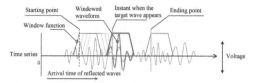

(a) Time series of an example waveform along with windows.

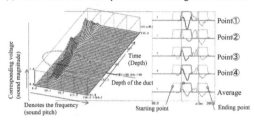

(b) Time-frequency analysis.

Figure 7. Outline of the waveform analysis.

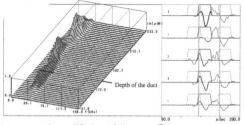

(a) An example qualified as "full grout (○)"

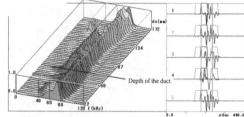

(b) An example qualified as "insufficient grout (×)"

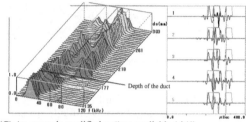

(C) An example qualified as "not available (△)"

Figure 8. Example outcomes of the wide-band ultrasonic wave method.

function in bold is located at the instant when the target wave appears; the waveform within the window is the zoom-in of the windowed waveform. The right-hand side of Figure 7(b) shows the time series of received waveforms: those labelled No. 1 to 4 were measured from different locations respectively and that labelled No. 5 is the averaged waveform. The left-hand side of Figure 7(b) shows the time-frequency spectrum of the averaged waveform, where X-axis denotes the frequency (sound pitch), Y-axis the corresponding voltage (sound magnitude), Z-axis the time (depth from concrete surface). Herein, the spectral magnitudes were normalized by the maximum, i.e. they are expressed in a ratio to the maximum.

### 2.4.2 Evaluation method

It is known that, in insufficient grouting cases, peaks of time-frequency spectrum present in higher frequency band than those in full grouting cases. Employing this knowledge, grouting conditions can be evaluated.

An example of evaluating different grouting conditions is given as in Figure 8. The time-frequency spectrum in Figure 8(a) presents large peaks in a low frequency band (lower than 40 kHz) nearly at the depth of the duct, so this case was qualified as "full grouting (○)". In comparison to Figure 8(a), the spectrum in Figure 8(b) presents large peaks in a higher frequency band (about 70 kHz), so this case was qualified as "insufficient grouting (×)". As for the spectrum in Figure 8(c), although it presents peaks in low frequency band (lower than 40 kHz) nearly at the depth of the duct, it also presents peaks in higher frequency band (about 70 kHz) at certain locations deeper than the duct. Such a case like this was qualified as "not available (△)".

Herein, the threshold between (○) and (×) were considered as 90% full grout in the duct. Moreover, the "not available (△)" qualification implies that some other factors which prevent a reliable estimation like deteriorated concrete surface or effect of rebars should be considered; final decisions are preferably made together with the outcomes of other methods like drilling tests (described later). It should be noted that the frequency band taken to evaluate grouting conditions as in Figure 8 varied with respect to several factors such as the strength of concrete, depth of concrete cover, diameter of duct, etc., and therefore would be determined for each target bridge.

### 2.5 Applicability conditions

The applicability conditions for evaluating PC grouting conditions by the present wide-band ultrasound method are listed in Table 2; it should be remarked that they were practically obtained from and verified in numerous pre-tests on laboratory-scaled specimens. Those conditions indicate the

Table 2. Applicability conditions.

| Items | Conditions |
|---|---|
| Type of duct | Steel duct, PE duct |
| Diameter of duct | φ38 mm or more |
| Depth of concrete cover | 250 mm or less |
| Spacing of cables | 110 mm or more |
| Spacing of rebars | 125 mm or more |

practical ranges within which the present method could work properly and accurately. Beyond the given ranges, since the present method might work less accurately, it is suggested to perform multiple runs of test at different locations and to interpret the outcomes complementarily.

Moreover, several preferable environmental conditions at the measurement points for the present method are listed as follows.

- No crack, honeycomb, or delamination.
- Not severely deteriorated concrete.
- Not too rough concrete surface.
- Constant concrete cover.

## 3 OUTLINE OF A PC GROUTING CONDITION INVESTIGATING SYSTEM

The procedures used for evaluating PC grouting conditions are shown in Figure 9.

### 3.1 Marking PC tendon locations referring to design drawings

This step is to mark the locations of target PC tendons referring to design drawings, for the following purposes: (1) To prevent missing of any location detection by Ground Penetration Radars (GPR) later; (2) To confirm the tendon number when locating tendons later.

### 3.2 Locating rebars and PC tendons by GPR

It possibly happens that the PC tendons in field are not located in the expected positions referred by the design drawings due to some construction reasons like a conflict with other members or construction errors. The precise localization of PC tendons is an important factor to the accurate evaluation of PC grouting conditions by the present method. For this reason, GPR were employed to locate rebars and PC tendons (see photo 3).

The GPR in use were StructureScan by GSSI company, as shown in Photo 4. StructureScan outperforms other conventional GPR products in little multiple reflected waves so that dense

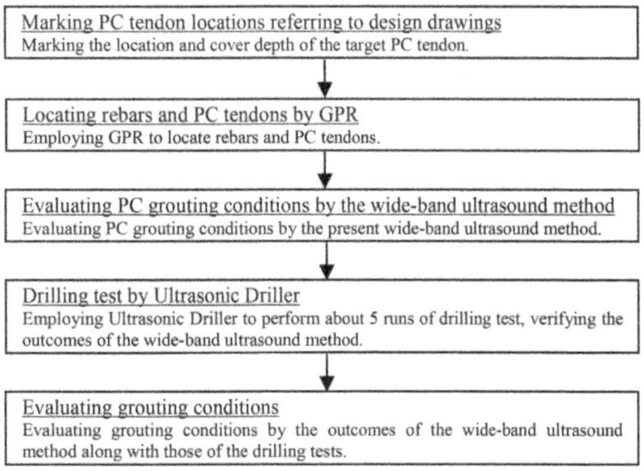

Figure 9. PC grouting condition investigating procedures.

Photo 3. Locating PC cables.

Photo 4. StructureScan.

allocated rebars and dual allocated rebars can be quickly detected. One example outcome yielded by StructureScan is shown in Figure 10, where the solid lines locate rebars and dashed lines locate

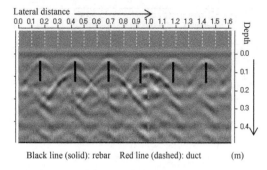

Figure 10. An example outcome.

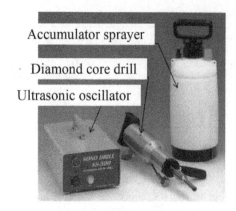

Photo 6. Ultrasonic Driller.

Photo 5. Evaluating PC grouting conditions.

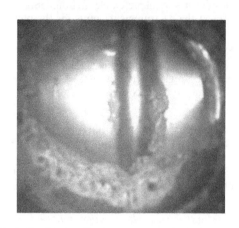

Photo 7. Duct surface.

ducts. It is observed that the ducts could be clearly detected even located beneath rebars or between rebars.

### 3.3 Evaluating PC grouting conditions by the wide-band ultrasonic wave method

This step is to evaluate PC grouting conditions by the present wide-band ultrasound method (see Photo 5).

### 3.4 Drilling tests by Ultrasonic Driller

By wide-band ultrasound method, the grouting conditions were evaluated by the frequency characteristics of waves reflected from ducts. The threshold (frequency) between full and insufficient grouting may vary with respect to the strength of concrete, depth of concrete cover, diameter of duct, etc. To verify the outcomes of the present method and to improve its accuracy, drilling tests were performed by Ultrasonic Driller. In addition, they can be used to confirm the deterioration levels of PC grouting and tendons as well as the presence of water.

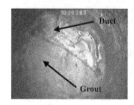 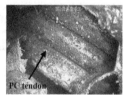

(a) Full grouting    (b) Insufficient grouting

Photo 8. An example of drilling test results.

An Ultrasonic Driller is shown in Photo 6 and a duct surface layer after a drilling is shown in Photo 7. Ultrasonic Driller is a drilling device that applies ultrasonic vibrations of a certain frequency to drilling blades; the applied frequency is tuned suitable for cutting concrete but unsuitable for steel, so that the rebars and PC tendons could be protected from damage by the drilling device. An example of the drilling test results is shown in Photo 8.

Table 3. Summary of the investigation results.

| PC tendon Type | No. | Investigation data | | Drilling results | | No. consist. | Ratio (%) |
|---|---|---|---|---|---|---|---|
| | | WUT | results | F | I | | |
| Steel strands | 156 | ○ | 69 | 69 | 0 | 69 | 100.0 |
| | | △ | 78 | 75 | 3 | – | – |
| | | × | 9 | 9 | 0 | 0 | 0.0 |
| Steel cables | 44 | ○ | 22 | 22 | 0 | 22 | 100.0 |
| | | △ | 4 | 4 | 0 | – | – |
| | | × | 18 | 4 | 14 | 14 | 77.8 |
| Steel rods | 118 | ○ | 43 | 43 | 0 | 43 | 100.0 |
| | | △ | 64 | 46 | 18 | – | – |
| | | × | 11 | 3 | 8 | 8 | 72.7 |
| Summary | 318 | ○ | 134 | 134 | 0 | 134 | 100.0 |
| | | △ | 146 | 125 | 21 | – | – |
| | | × | 38 | 16 | 22 | 22 | 57.9 |

Note. F: full grouting; I: insufficient grouting; No. consist.: number of consistent cases; Ratio: ratio of consistent cases.

## 4 INVESTIGATION RESULTS ON EXISTING PC BRIDGES

A series of investigation of PC grouting conditions was performed on main cables of 47 PC bridges. The investigation results are summarized in Table 3, where the outcomes of the wide-band ultrasound method and drilling tests can be mutually compared. Herein, although "△" cases were excluded from the consistency evaluation, most of them were identified as "full grouting" in the drilling tests.

It can be observed from these results that, by the present method, although the "full grouting" cases were highly consistent with the drilling results, the "insufficient grouting" cases were not as consistent as the former ones were. It is probably due to the confusion with the high-frequency waves reflected from rebars or so; even so, it is at the conservation side regarding to PC health evaluation.

## 5 CONCLUDING REMARKS

### 5.1 Achievements from real structure investigations

The consistency of the testing outcome and actual status is a critical issue in applying NDT methods to real concrete structures. The accuracy of new testing methods is likely to be ambiguous because of few chances to verify those methods on real structures as well as of difficulties in confirming actual status.

A PC grouting condition investigating system that integrates the wide-band ultrasound method with Ultrasonic Driller was proposed. With this system, its reliability and the consistency of the testing outcomes and actual status were confirmed. In addition, in the cases where few testing samples were available, the incorporation of minor destructive testing could be helpful in evaluating the current status of grouting conditions and tendon health.

### 5.2 Open issues

Several open issues are listed as follows

1. Currently it is difficult to evaluate grouting ratios in practice by the current method. In experimental levels, it has been shown possible to evaluate the grouting ratios of around 50%. Through a great amount of further studies, making it practical is very important.
2. Since couplants are employed when installing probes, remaining couplants were found on the concrete surface. Also, the current probes are unsuitable to very rough concrete surface, it is considered necessary to develop some dry probes that are more suitable to rough surfaces.

Once the above issues can be addressed properly, more efficient maintenance strategies and more reliable PC bridges can be expected.

## REFERENCES

1) Ikeda, S., Tezuka, M., Niitani, K. and Hosono, H. (2013). Revision of Guidelines for Design and Construction of Grouting for Prestressed Concrete Structures. *Journal of prestressed concrete, Japan, 55* (3), 74–83.
2) Tamakoshi, T., Hiraga, K. and Kimura, Y. (2012). Countermeasure to the corrosion damage of PC tendons. *Civil engineering journal, 54* (5), 50–51.
3) Morishima, H., Abe, H., Aoki, K. and Hara, M. (2005). Investigation for grouting conditions of inner cables by electromagnetic radar. *Journal of Japan Prestressed Concrete Engineering Association, 47* (3), 71–78.
4) Kamada, T. (2003). Non-destructive testing in maintenance of PC structures. *Journal of Japan Prestressed Concrete Engineering As-sociation, 45* (1), 51–58.
5) Fujii, M. and Miyagawa, T. (1989). Non-destructive testing for grouting conditions of PC structures. *Journal of Materials, Concrete Structures and Pavements, JSCE,* No.402/V-10, 15–25.

*Emerging Technologies in Non-Destructive Testing VI – Aggelis et al. (Eds)*
*© 2016 Taylor & Francis Group, London, ISBN 978-1-138-02884-5*

# Two-dimensional AE-Tomography based on ray-trace technique for anisotropic materials

### Y. Kobayashi
*Department of Civil Engineering, College of Sciece and Technology, Nihon University, Tokyo, Japan*
*C-PIER, Kyoto University, Kyoto, Japan*

### T. Shiotani
*Graduate School of Engineering, Kyoto University, Kyoto, Japan*

ABSTRACT: The authors introduce a new algorithm of two-dimensional AE-Tomography for anisotropic materials in this paper. This algorithm is implemented on the basis of ray-trace technique that is proposed by the authors to consider the refraction and diffraction of the elastic wave on heterogeneous and anisotropic elastic wave velocity structure. AE-Tomography is a technique that intends to estimate source locations of AE events and elastic wave velocity structure simultaneously by using arrival times at receivers, and the authors have proposed algorithms of AE-Tomography for two—and three-dimensional problems to evaluate integrity of structures. Although the proposed technique presumes that the materials of the structures are isotropic, it is intended to apply these techniques for the anisotropic materials e.g. FRP since use of the FRP is spreading various fields of engineering as tough materials, especially in the field of aerospace engineering, and its evaluation of integrity is significant to keep the serviceability. The authors already proposed techniques of ray-trace and source location on the anisotropic materials for two-dimensional problems as preliminary studies. In these techniques, the anisotropy of the materials is specified by setting the elastic wave velocities in various directions and linearly interpolating the specified elastic wave velocities between the directions, and the anisotropy is considered to estimate the ray-paths and the source locations. The new algorithm is structured on the basis of these techniques, and the elastic wave velocity structure is reconstructed by considering the anisotropy as well. The introduced algorithm is verified by executing a series of numerical investigations that assume some patterns of the anisotropy of the elastic wave velocity structure, and features of the algorithm are discussed on the basis of the results of the investigations.

## 1 INTRODUCTION

The authors have studied on the tomographic techniques for the evaluation of integrity of concrete structures in recent. Among the studies, elastic wave velocity tomography based on a technique for subsurface exploration (Sassa, Ashida, Kozawa, & Yamada 1984) was applied for two and three-dimensional problems and used for the evaluation of large scale structures to investigate its integrity (Momoki, Shiotani, Chai, Aggelis, & Kobayashi 2013). However, since elastic wave velociy tomography identifies elastic wave velocity distribution from first travel times of elastic waves between sources and receivers, source locations and emission times of the elastic waves must be known, and this consequently makes observation costs high because a receiver must be re-installed in the vicinity of a point of the excitation. For overcomming this difficulty, AE-Tomography was proposed by

Schurbert (Schubert 2004)(Schubert 2006). AE-Tomography is a technique that identifies elastic wave velocity distribution only from arrival times of elastic waves at receivers. This technique reduces the observation costs since the source information is innecessary; however, its algorithm assumes that the ray-path can be approximated as straight lines to simplify its computational procedure, and it would degrade the identified elastic wave velocity distribution because the ray-path is not straight on heterogeneous elastic wave velocity distribution in general. In recent, the authors proposed a new algorithm of AE-Tomography on the basis of ray-trace technique for two and three-dimensional problems(Kobayashi & Shiotani 2012)(Kobayashi, Shiotani, & Oda 2014a). The algorithm of AE-Tomography is supposed to apply for structures of isotropic bodies because materials are handled as isotropic materials in almost cases in the field of civil engineering. However, use of composit materials, e.g. FRP, are now starting in civil engineering, and its monitoring and maintainance are supposed to be significant works to keep its serviceability and safety during its service time. Therefore, the authors have studied on tomographic techniques on anisotropic bodies to achieve the objectives. As preliminary studies, the authors proposed a ray-trace technique and source location technique that considers the anisotropy anisotropy(Kobayashi, Shiotani, & Oda 2014b) (Kobayashi & Shiotani 2014). In this paper, a new algorithm of AE-Tomography and the source location technique that consider the anisotropy on the basis of the preliminary studies is proposed, and the algorithm is verified by performing a series of numerical investigations.

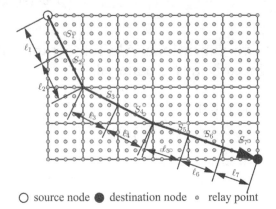

○ source node ● destination node ○ relay point

Figure 1. Representation of ray-path on two-dimesional mesh.

## 2 ELASTIC WAVE VELOCITY TOMOGRAPHY ON ANISOTROPIC MATERIALS

In this section, a new algorithm of elastic wave velocity for anisotropic materials is introduced since AE-Tomography is structured on the basis of the elastic wave velocity tomography. Elastic wave velocity tomogrpahy has been used as a technique to identify elastic wave velocity distribution on cross section or volume of interest. In Elastic wave velocity tomography, the elastic wave velocity distributions are reconstructed as an identification problem based on following observation equation.

$$\Delta T = f(V(r)) \quad (1)$$

in which, $\Delta T$ is a vector of first travel times of elastic waves between sources and receivers, $V(r)$ is elastic wave velocity at $r$, and $f$ is a nonlinear function that computes the first travel times on the given informations. Generally, the first travel times $\Delta T$ are computed as differences between arrival times of the elastic waves at receivers and emission times of the elastic waves at source locatons as follows.

$$\Delta T = A - O \quad (2)$$

in which, $A$ is a vector of arrival times of the elastic waves at receivers, $O$ is a vector of emission times of the elastic waves at source locations. Hence, the elastic wave velocity tomography is actually an identification problems that reconstructs the elastic wave velocity distribution from the first travel times $\Delta T$ and the function $f$. The function $f$ are constructed on the basis of ray-trace technique in the techniques that the authors have proposed. In case of two-dimensional problems, a cross section of interest is meshed as well as finite element analysis and ray-path is approximated as polylines as illustrated in Figure 1. On the approximation, relay points are installed on the mesh to raise the accuracy of ray-path representation in the presented technique (Kobayashi 2013). If it is assumed that elastic wave velocity is constant in each cells and materials of a cross section is isotropic, a travel time between a source and a receiver is shown as follows.

$$A - O = \Delta T = \sum_{i=1}^{n} S_i \ell_i \quad (3)$$

in which, $S_i$ is slowness that is a reciprocal of elastic wave velocity of cell $i$, and $\ell_i$ is a length of a ray-path between the source and the receiver on cell $i$. Thus, travel times on any ray-paths between two points on the cross section can be computed by using Equation 3. Because a ray-path that gives minimum travel time is required for the elastic wave velocity tomography, the ray-path is sought from all of the potential ray-path between the two points. This is the shortest path problem, and Dijkstra's algorithm is adopted for the seeking in the present technique. Consequently, the observation equations are given as

$$\begin{Bmatrix} \Delta T_1 \\ \Delta T_2 \\ \vdots \\ \Delta T_n \end{Bmatrix} = \begin{bmatrix} \ell_{00} & \ell_{01} & \cdots & \ell_{0m} \\ \ell_{10} & \ell_{11} & \cdots & \ell_{1m} \\ \vdots & \vdots & \cdots & \vdots \\ \ell_{n0} & \ell_{n1} & \cdots & \ell_{nm} \end{bmatrix} \begin{Bmatrix} S_0 \\ S_1 \\ \vdots \\ S_m \end{Bmatrix} \quad (4)$$

in which $\Delta T_i$ is a travel time of ray-path $i$, $\ell_{ij}$ is a length of the ray-path $i$ on cell $j$. On the basis of the observation equations, the slowness vector is identified by an identification technique. In this study, SIRT(Simultaneous Iterative Reconstruction

Technique) is adopted as the identification technique.

To apply the elastic wave velocity tomography for anisotropic bodies, the anisotropy of materials is implemented in the elastic wave velocity tomography by defining the the slowness $S_i$ as a function of an anlge of a ray-path. The function $S(\theta)$ is set to the individual cells on the mesh, and each of the function is represented by a slowness profile that is shown in Figure 2. In this profile, the directivity of the slowness is set by specifying slowness at specified angles to a reference axis that is defined on each cells. It should be noted that the slowness between the specified angles are given by lineary interpolating the slownesses at the adjacent specified angles as shown in Figure 2. The function is actually described as follows in this study.

$$S(\theta) = Sm\, a(\theta) \qquad (5)$$

in which, $Sm$ is the minimum slowness on the profile, i.e. this corresponds to maximum velocity in the profile, and $a(\theta)$ is a coefficient that is represented as $S(\theta)/Sm$. Consequently, the observation equations 4 are rewritten as follows to consider the anisotropy.

$$\{\Delta T\} = [\ell]\{Sm\} \qquad (6)$$

where

$$\{\Delta T\} = \begin{Bmatrix} \Delta T_1 \\ \Delta T_2 \\ \vdots \\ \Delta T_n \end{Bmatrix} \qquad (7)$$

$$[\ell] = \begin{bmatrix} \ell_{00} a_{00} & \ell_{01} a_{01} & \cdots & \ell_{0m} a_{0m} \\ \ell_{10} a_{10} & \ell_{11} a_{11} & \cdots & \ell_{1m} a_{1m} \\ \vdots & \vdots & \cdots & \vdots \\ \ell_{n0} a_{n0} & \ell_{n1} a_{n1} & \cdots & \ell_{nm} a_{nm} \end{bmatrix} \qquad (8)$$

$$\{Sm\} = \begin{Bmatrix} Sm_0 \\ Sm_1 \\ \vdots \\ Sm_m \end{Bmatrix} \qquad (9)$$

in which, $a_{ij}$ is the coefficient of the ray-path $i$ in cell $j$. By applying the identification technique to Equation 6, the vector $Sm$ is identified and elastic wave velocity distribution is visualize by using the vector. It is noteworthy that this observation equations assume that the directivities of the slowness, actually the ratios of the slowness to the minimum slowness at specified angles, does not change even if the $Sm$ is changed to simplify its computational procedure.

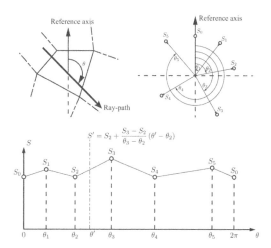

Figure 2. Slowness profile.

## 3 AE-TOMOGRAPHY ON ANISOTROPIC MATERIALS

AE-Tomography is a technique that uses AE as signals, i.e. this technique uses only arrival time of AE at receivers, to identify elastic wave velocity distribution on the basis of the elastic velocity tomography. As introduced in the previous section, elastic wave velocity tomography reconstructs elastic wave velocity distribution from first travel times that are computed from emission times and arrival times of elastic waves and ray-paths that are detemined by ray-trace technique between source locations of the elastic waves and locations of receivers. However, AE is inappropriate as the signals for the identification since it is impossible to compute the travel times and determine the ray-path bacause emission times and source locations of AE are generally unknown, Therefore, in AE-Tomography, the source locations and the emission times are identified prior to the reconstruction of the elastic wave velocity distribution by using the AE source location technique as shown in Figure 3. The source locations have been identified by assuming that elastic wave velocity distribution is homogeneous and ray-paths can be approximated as straight lines between the source location and receivers in the conventional AE source location techniques to simplify its computational procedure. However, the assumptions are violated if detorioration of the cross section is localized and severe since the elastic wave velocity locally decline in this case, and consequently, it leads to degrade the accuracy of the identified AE source locations. Hence, the authors proposed a source location technique based on the ray-trace technique to avoid the difficulty (Kobayashi, Shiotani, & Oda 2014b). And

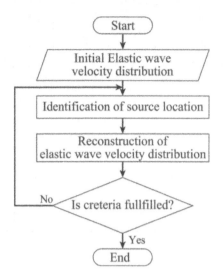

Figure 3. Conceptual flow diagram of AE-Tomography.

further, the anisotropy was implemented in the source location technique to extend its applicability for AE-Tomography on anisotropic materials (Kobayashi & Shiotani 2014). In this technique, firstly potential emission times $P_{ij}$ of AE are computed at individual nodal points and relay points on the mesh as follows by executing the ray-trace from all of receivers on the mesh.

$$P_{ij} = T_i - \delta T_{ij} \qquad (10)$$

in which, $i$ is a receiver number, $j$ is a serial number of the nodal points, and relay points and $\delta T_{ij}$ is a travel time from the receiver $i$ to the point $j$ that is computed by the ray-trace technique. It should be noted that the anisotropy is automatically considered in the source location technique because the anisotropy is already implemented in the ray-trace technique. Therefore, each of the nodal points and relay points has $n$ potential emission times in which $n$ is number of receivers. The potential emission times are identical at the source location if the elastic wave velocity distribution and shape of the ray-path are exactly represented. However, the point generally does not exist since the representation is normally insufficient due to limitations of meshing of the cross section. Thus, a point that gives minimum variance of the potential emission times is choosen as the source location, and average of the potential emission times at the source location is used as the emission time. The first travel time and the ray-path can be determined by using the estimated source locations and emission times, and the observation equation of Equation 6 can be formulated by using the results. Finally, the elastic wave velocity distribution with the anisotropy is identified by applying identification technique for the observation equations.

## 4 VERIFICATION OF ELASTIC WAVE VELOCITY TOMOGRAPHY FOR ANISOTROPIC MATERIALS

In this section, the elastic wave velocity tomography technique for anisotropic materials is verified by performing a series of numerical investigations because AE-Tomography is implemented on the basis of elastic wave velocity tomography, and its accuracy is degraded if the result of elastic wave velocity tomography is insufficiently accurate. Figure 4 shows a cross section that is used for the verification. The cross section is a square of 10 m of its edges, and it consists of two parts that are colored in white and gray in Figure 4. In this model, ratio of vertical and horizontal elastic wave velocity is 10 vs 1 on the entire cross section, and vertical elastic wave velocity of the white area is set to 4000 m/s to simulate area of integrity, and vertical elastic wave velocicy of the gray area is set to 3000 m/s to simulate damaged area as well. The velocity on an angle between the vertical and horizontal axes are computed by linear interpolation as shown in Figure 2. On this cross section, 200 AE source locations are randomly generated, and travel times of the AE are computed by executing the ray-trace form the sources to the locations of receivers. It should be noted that the anisotropy is considered on the computation of the travel times. The receivers are installed at apaxes of the cross section, and consequently, 4 receivers are used in this case. It should be noted that the source locations and the emission times, i.e. the emission times are actually zero in this case, of the artificial AE are known. Thus, we virtually used the travel time as the first travel time from the source to receivers, and the source locations that are used to compute the travel times are used as the real source locations. On these condition, the first travel times and the source locations are used as observation data for elastic wave velocity tomography, and the accuracy of the identified elastic wave velocity distribution by elastic wave velocity tomography is discussed. As the first case, the elastic wave velocity distribution is identified with the anisotropy. As the initial elastic wave velocity distribution for elastic wave velocity tomography, homogeneous elastic wave velocity distribution of 4000 m/s with the same anisotropy with the original model is given. Thus, the vertical elastic wave velocity is 4000 m/s, and the horizontal elastic wave velocity is 4000 m/s on the initial model. On this model, the elastic wave velocity distribution is updated

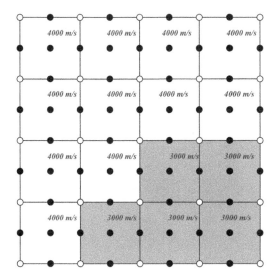

Figure 4. Mesh for verification and original elastic wave velocity distribution.

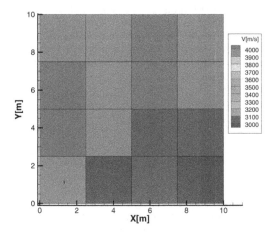

Figure 5. Identified elastic wave velocity distribution with consideration of the anisotropy.

30 times by using the first travel times as observation data on the basis of the presented algorithm of elastic wave velocity tomography. Figure 5 shows the identified elastic wave velocity distribution. It should be noted that this is a contour of the vertical elastic wave velocity distribution. Thus, horizontal velocity is 10% of the vertical velocity since the algorithm does not consider change of the anisotropy in its computational procedure. According to Figure 5, the contour shows the same tendency with the original elastic wave velocity distribution, and further, its value is also consistent with the original one. In this case, number of the ray-path is 800 due to 200 sources and 4 receivers,

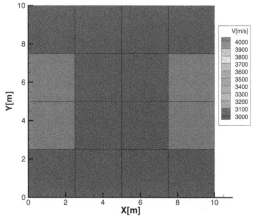

Figure 6. Identified elastic wave velocity distribution without consideration of the anisotropy.

and number of variables that are identified is 16. Thus, the number of observation equations are 50 times of the number of variables, and it implies that the observation equations sufficiently exist to identify the elastic wave velocity distribution accurately. As the second case, the elastic wave velocity distribution is identified without any consideration of the anisotropy as well as the case with consideration of the anisotropy by using a conventional algorithm of elastic wave velocity tomography. Figure 6 shows a contour of the resultant elastic wave velocity distribution. It should be noted that the elastic velocity on the contour does not have any anisotropy in this case. According to Figure 6, it is shown that the resultant elastic wave velocity distribution is not consistent with the original elastic wave velocity distribution. This reveals importance of the consideration of the anisotropy for the identification of elastic wave velocity distribution on anisotropic materials. Consequently, it is confirmed that the proposed algorithm for elastic wave velocity tomography correctly considers the anisotropy of the materials, and identifies the elastic wave velocity distribution.

## 5 VERIFICATION OF AE-TOMOGRAPHY FOR ANISOTROPIC MATERIALS

In this section, the proposed algorithm of AE-Tomography based on the introduced algorithm of Elastic wave velocity tomography is verified by performing a series of numerical investigations. For this verification of AE-Tomography, the same model ( cross section, receiver installation, AE source locations, and travel times from the source to receivers ) with the verification of elastic wave velocity tomography is used. However, since only

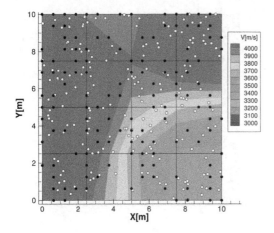
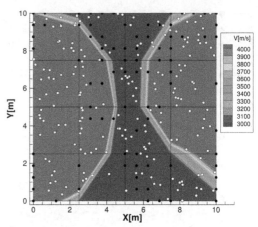

Figure 7. Identified elastic wave velocity distribution by AE-Tomography with consideration of the anisotropy (White circles: real source locations, Black circles: identified source locations).

Figure 8. Identified elastic wave velocity distribution by AE-Tomography without consideration of the anisotropy (White circles: real source locations, Black circles: identified source locations).

arrival times of AE at the receivers are necessary for AE-Tomography, the travel times are used as the arrival times of AE at receivers and the source locations and emission times are not given as computational conditions in this verification. On this condition, the elastic wave velocity distribution is identified with the proposed algorithm of AE-Tomography that considers the anisotropy and the conventional algorithm of AE-Tomography that does not consider the anisotropy. Figure 7 shows the resultant elastic wave velocity distribution that is identified by considering the anisotropy. According to Figure 7, the resultant elastic wave velocity distribution is qualitatively consistent with the original elastic wave velocity distribution. The accuracy is lower than the result of elastic wave velocity tomography because AE-Tomography identifies the elastic wave velocity distribution with less number of boundary conditions, i.e. source locations and emitted times of elastic waves. As introduced in previous section, AE-Tomography estimates the source locations and the emisstion times of AE, and then, the observation equations are structured by using the "estimated information". Thus, it is natural that the accuracy of the resultant elastic wave velocity disribution is lower than the ones by the elastic wave velocity tomography. However, the result qualitatively catches the tendency of the original elastic wave velocity distribution, and this fact reveals that AE-Tomography enables to identify the damaged area as a tool by considering the anisotropy in its algorithm. Figure 8 shows the resultant elastic wave velocity distribution that is identified without consideration of the anisotropy. The resultant elastic wave velocity distribution in Figure 8 is significantly different from the original elastic wave velocity distribution, and this result also reveals the importance of the consideration of the anisotropy in the algorithm of AE-Tomography to evaluate integrity of anisotropic materials.

6 CONCLUSIONS

In this study, a new algorithm of AE-Tomography for anisotropic materials is introduced as well as the elastic wave velocity tomography for anisotropic material as a preliminary study for implementing the AE-Tomography. The introduced techniques are verified by performing a series of numerical investigations, and following conclusions are drawn.

1. The identified elastic wave velocity distribution by the proposed algorithm of elastic wave velocity tomography was consistent with the original elastic wave velocity distribution. This result reveal that the proposed algorithm successfully considers the anisotropy.
2. The identified elastic wave velocity distribution by the conventional elastic wave velocity tomography was not consistent with the original elastic wave velocity distribution since the anisotropy is not considered.
3. The identified elastic wave velocity distribution by the introduced algorithm for AE-Tomography on the anisotropic materials is qualitatively consistent with the original elastic wave velocity distribution.

4. The identified elastic wave velocity distribution by the conventional AE-Tomogrpahy was not consistent with the original elastic wave velocity distribution as well as the conventional elastic wave velocity tomography.

## REFERENCES

Kobayashi, Y. (2013). Mesh-independent ray-trace algorithm for concrete structures. *Constructions and Building Materials 48*, 1309–1317.

Kobayashi, Y. & T. Shiotani (2012). Seismic tomography with estimation of source location for concrete structures. In *Structural Faults and Repair 2012*.

Kobayashi, Y. & T. Shiotani (2014). Two-dimensional source location technique on anisotropic medium on the basis of raytracing. In *The 22nd International Acoustic Emission Symposium*.

Kobayashi, Y., T. Shiotani, & K. Oda (2014a). Three-dimensional ae-tomography with accurate source location technique. In *Structural Faults and Repair 2014*.

Kobayashi, Y., T. Shiotani, & K. Oda (2014b). Two-dimensional ray-trace technique for anisotropic materials. In *The 6th Asia and Pacific Young Researchers and Graduates Symposium*.

Momoki, S., T. Shiotani, H.K. Chai, D.G. Aggelis, & Y. Kobayashi (2013). Large-scale evaluation of concrete repair by three-dimensional elastic-wave based visualization technique. *Structural Health Monitoring 12*(3), 241–252.

Sassa, K., Y. Ashida, T. Kozawa, &M. Yamada (1984). Improvement in the accuracy of seismic tomography by use of an effective ray-tracing algorithm. In *MIJ/IMM Joint Symposium Volume Papers*, pp. 129–136.

Schubert, F. (2004). Basic principles of acoustic emission tomography. *Journal of Acoustic Emission 22*, 147–158.

Schubert, F. (2006). Tomography technique of acoustic emission monitoring. In *ECNDT2006*.

*Emerging Technologies in Non-Destructive Testing VI – Aggelis et al. (Eds)*
*© 2016 Taylor & Francis Group, London, ISBN 978-1-138-02884-5*

# Characterization of inclined surface crack in steel reinforced-concrete by multichannel R-wave measurements

F.W. Lee, H.K. Chai & K.S. Lim
*University of Malaya, Kuala Lumpur, Malaysia*

ABSTRACT: A non-destructive methodology for evaluating concrete surface breaking crack is developed using multichannel measurements of surface Rayleigh waves (R-waves). Numerical simulations were carried out to understand and analyse propagation behaviour of R-waves in steel-reinforced concrete containing a surface crack with varying depths and degree of inclinations. The change in wave parameters, namely velocity and amplitude were examined and quantified to obtain correlations with the depth and inclination of crack. Experimental measurement on a concrete block specimen was then conducted to validate the effectiveness of the methodology, particularly in cases of crack where the generated R-wave wavelength was greater than the crack depth. Experimental assessment results are found to be in decent agreement with the analytical ones, indicating feasibility of the proposed method to acquire elastic wave data and characterize concrete surface crack by analyzing the change of R-waves parameters.

*Keywords*: steel reinforced-concrete; surface crack depth; inclination; surface Rayleigh waves; velocity; amplitude; excitation frequency

## 1 INTRODUCTION

The formation of surface cracks on concrete structures may be due to single or combined effect of overloading, drying shrinkage, temperature variations, chemical attack, weathering, differential settlement and other degradation processes [1]. Some Non-Destructive Testing (NDT) methods have been developed and detect and evaluate cracking in concrete, such as infrared thermography, ground penetrating radar, acoustic impact method, ultrasonic pulse-velocity method and pulse-echo method [2]. Majority of these methods are based on principles of elastic wave propagation. Most recently, surface Rayleigh wave (R-wave) has been studied for utilization in assessment of surface cracks of civil infrastructures [3].

Some unique features of R-waves are advantageous for concrete crack assessments, such as low attenuation and high energy contents that facilitate sensitive detection in inhomogeneous concrete medium [4]. Furthermore, R-waves are dispersive and the propagation depth into a medium is a function of the wavelength. These characteristics are potentially useful for characterization of surface cracks.

The aim of this study is to refine the R-waves based method for assessing surface crack in concrete, by examining the fundamental relationships between wave propagation behaviour and crack characteristics. Numerical simulations were carried out for the investigation. The relationships between R-waves parameters and crack characteristics were later validated by experimental measurement to examine the feasibility of proposing an in-situ method of concrete surface crack assessment using multi-channel instrumentation for R-wave measurement.

## 2 WAVE MOTION SIMULATION

The simulation work was conducted using commercial software [5] that solves two dimensional (2D) elastic wave propagation problems by temporal acoustic interrogations based on the finite difference method. Figure 1 illustrates the model of simulation, which is essentially a steel reinforced concrete section of 500 mm (width) × 300 mm (depth), resembling that of a bridge deck. The model was configured to have longitudinal wave velocities of approximately 4300 m/s and 6099 m/s, for concrete and steel reinforcement, respectively. The left, right and bottom sides of model were set as "infinite boundary" to avoid occurrence wave reflections. In the simulation, three sensors were placed on left and right side of the crack on top surface of concrete, respectively with spacing of 40 mm. Excitation of waves was made from a point source on the top surface of concrete.

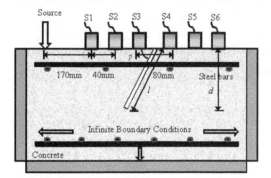

Figure 1. Wave motion simulation model.

Simultaneous recording by all the six sensors was triggered by sensor S1 located 170 mm from the excitation point. The simulations were conducted with one excitation that produced a full-cycle elastic wave. From the simulation, the influence of crack on waveform distortion, attenuation and pulse velocity was examined.

## 3 EXTRACTION OF R-WAVES

To facilitate separation and extraction of R-waves from the bulk waves in recorded signals for further analysis, the concept of Matched Filtering of Center of Energy (MFCE) was adopted. As example, in a measurement array consists of multiple sensors, detected signal at sensor $i$ for a particular time can be identified as $S_i(t)$. To locate the center of energy of the received signals, signal processing procedure has been performed based on the followings:

$$t_c = \frac{\int_{-\infty}^{\infty} t |S_i(t)|^2 dt}{\int_{-\infty}^{\infty} |S_i(t)|^2 dt} \quad (1)$$

where $t_c$ is defined as the occurrence time of center of energy for R-waves in the signal period. The power of two in the expression is to discriminate the wave components (P—and R-waves) in the waveform and help locate the position of the center of energy for R-wave. After acquiring $t_c$, windowing on the selected portion of signal was performed to eliminate the undesired tail or echo of the signal.

$$w(t) = \begin{cases} 0, |t| > \sigma \\ 1, |t| \le \sigma \end{cases} \quad (2)$$

The windowed signal:

$$S_{w,i}(t) = w(t - t_c) S_i(t) \quad (3)$$

Lastly, matched filtering on each processed signal $[S_{w,i}(t)]$ was performed with the processed signal acquired from the first sensor $[S_{w,1}(t)]$. Hence:

$$C(t) = \int_{-\infty}^{\infty} S_{w,i}(\tau) . S_{w,1}(t - \tau) d\tau, \quad (4)$$
$$i = 1 \, to \, 6$$

$$t_{peak,i} = C(t)_{max.} \quad (5)$$

Matched filtering helps eliminate the noise in the signal and $t_{peak}$ marks the position where the matching of $S_{w,1}(t)$ is optimum. While $t_{peak} = 0$ is the result of autocorrelation and it is used as the reference to locate the time position for the other signals.

## 4 WAVE SIMULATION RESULTS

Examples of simulated waveforms results of homogenous concrete model obtained from 10 kHz and 150 kHz excitation frequencies are given in Figure 2, for cases of sound concrete and 150 mm-deep crack inclining at 30° and 90° from the top concrete surface. A strong Rayleigh burst could be clearly observed after the initial longitudinal wave arrival, especially for the higher frequency excitations in the sound concrete model. Wave distortion was caused by the presence of crack, as manifested by waveforms collected from sensors S4, S5 and S6. A marked decrease in amplitude compared to those obtained from the same locations on the sound concrete model was found. Referring to the findings of simulation other crack cases, in general the depth of crack was found to have greatly affected recorded amplitude by sensors on the opposite side of crack.

It is known that the wavelength of R-wave is a function of its penetration depth and no R-waves should exist at the other side of a crack if the penetration depth is much lower than the crack depth. However, as have been demonstrated by the wave motion simulation work in this study, it seems possible for R-waves to be "reformed" after being totally blocked by the crack. For cases with high excitation frequency (low penetration depth) and a deep crack, a significant portion of wave energy was found to propagate along the crack side in downward direction, before being impinged at the crack bottom to be emitted at its tip to all possible directions. Some wave energy has reached the top concrete surface and as result of interactions of wave particle movements with the surface boundary conditions, R-waves may have been formed again to propagate and be detected by the three sensors on the opposite side of crack.

In some previous studies, it is found that the crack depth is practically divided by the major

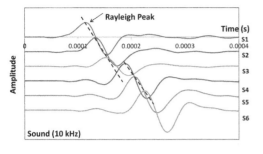

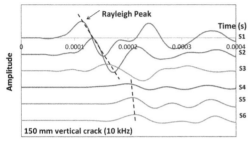

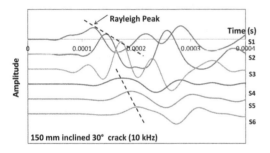

Figure 2. Simulated time domain data for sound concrete and models with 150 mm crack inclining at 30° and 90° (vertical crack).

wavelength of R-waves to give a parameter that provides flexibility and wider coverage in comparison and evaluation [6–8]. In this study, the depth-to-wavelength ratio is plotted against the R-wave velocity index as in Figs. 4. The velocity index, $VI$ for each propagation was calculated using the following equation:

$$VI = \frac{\sum_{j=2}^{6} V_{c,1\sim j}}{\sum_{j=2}^{6} V_{s,1\sim j}} \qquad (6)$$

where $\sum_{j=2}^{6} V_{C,1\sim j}$ and $\sum_{j=2}^{6} V_{S,1\sim j}$ are summation of R-wave velocities from sensor S1 to the other respective sensors, for model with crack and the sound model, respectively. It is discernible that a velocity index of 1.0 indicates that the propagation of R-waves has not been distorted by crack. To assess the influence of crack on the amplitude of R-wave, amplitude index, $AI$ was computed using the following:

$$AI = \frac{\left(\sum_{i=4}^{6} A_{C,i} / \sum_{j=1}^{3} A_{C,j}\right)}{\left(\sum_{i=4}^{6} A_{S,i} / \sum_{j=1}^{3} A_{S,j}\right)} \qquad (7)$$

where $A_c$ is R-wave amplitude recorded by sensor number $i$ in model with crack, while $A_S$ is that recorded from simulating the sound concrete model.

Figure 3 shows examples of velocity and amplitude indices plotted against the crack depth to R-wave wavelength ratio ($d/\lambda$), respectively for models with crack inclining at 30°. It is found that the velocity index decreases as $d/\lambda$ increases, in logarithmic regressions for crack cases of 30° and in linear regressions for the vertical one, respectively. Also, a different trend was found in which the velocity index decreases in polynomial regressions for cases of crack inclining more than 90°. On the other hand, it was noted that amplitude index decreases with increase of $d/\lambda$ in polynomial trend. In addition, the amplitude index becomes lower as the frequency increases due to the tendency of higher frequency components easily to lose their energy through absorption, scattering and distortion by the crack. In general, it is found that the amplitude index seemed to decrease with regards to the degree of inclination from 30° to 150°. It is suggestive of the phenomenon that more energy has been blocked from arriving at the other side of crack as the crack inclination angle increases.

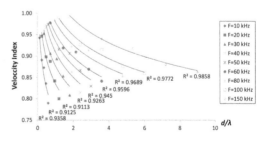

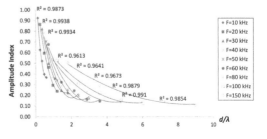

Figure 3. Velocity and amplitude indices plotted against $d/\lambda$ for different R-wave excitation frequencies.

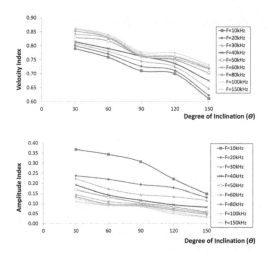

Figure 4. Velocity and amplitude indices plotted against degree of inclination for crack (constant vertical depth of 150 mm).

The attenuation of amplitude manifested to be a more suitable parameter for crack characterization than velocity index since the discrepancy between the homogenous and cracked models was greater and more noticeable than it was for velocity.

To further investigate the effect of crack inclination on R-wave propagation, velocity and amplitude indices are plotted against the crack inclination degree as exemplified in Figure 4. The vertical depth of crack was made constant at 150 mm regardless of degree of inclination. In general, both the velocity and amplitude indices decrease with increase of inclination angle from 30° to 150°, justifying the fact that large amount of energy has been blocked due to the increase in effective vertical depth as result of crack inclination angle increase. It could be confirmed as well that the drop of velocity with increase of crack inclination was due to the face that R-waves, in particular those with effective penetration depth lesser than the crack vertical depth, have consumed longer time to propagate to the opposite side of crack.

## 5 EXPERIMENTAL EXAMINATION

Experimental measurements were conducted on steel reinforced concrete block specimens of 300 mm × 300 mm × 500 mm prepared. Figure 5 shows the experimental set-up. One of the specimens was the control with no defect, while the other two comprised of one artificial crack inclining at 30° and 90° (the vertical crack), respectively as measured against the horizontal plane. The crack inclining at 30° would become 150° if measured from the other side of the specimen.

The artificial crack was instigated by suspending a polystyrene foam board in the concrete mould before casting. The polystyrene foam boards were prepared in triangular shape so as to provide different "crack depths" along the transverse direction of specimen. Six accelerometers with frequency range of 0.005 kHz to 60 kHz (PCB Inc.) were placed on the top surface of specimen in an arrangement similar to that adopted for simulation work. Elastic waves were excited by dropping steel balls of different sizes to the concrete surface at location similar to that examined in the simulation work, which was 170 mm away from the trigger sensor. The impact point formed a straight line with the locations of accelerometers. The steel balls have multiple sizes to give a variety of excitation frequencies. A waveform acquisition system (NI PXIe-4492 by National Instruments Co.) was used control measurement and record waveform data.

Examples of time domain data obtained from measuring the control specimen as well as specimens with vertical crack depths of 125 mm and 25 mm are given in Figure 6. Identification and extraction of R-waves were performed for computation of the velocity and amplitude index. The key procedures involved in crack characterization using the measured data are shown in Figure 7. The results of estimation are exemplified as in Figures 8 and 9, which are compared with the actual depth and crack inclination angle. Considering the experimental

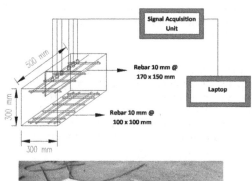

Figure 5. Experimental set-up for wave measurement.

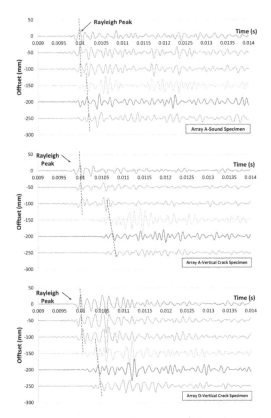

Figure 6. Time domain data measured from the sound concrete specimen and that with vertical crack depths of 125 mm and 25 mm.

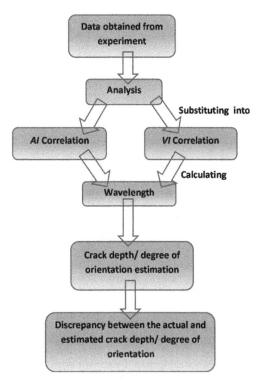

Figure 7. Proposed procedures for concrete surface crack characterization.

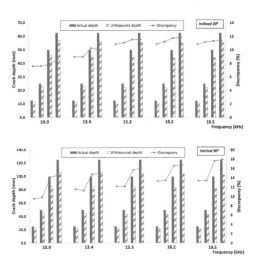

Figure 8. Comparison of actual crack depth with estimated results for 30° degree inclined and vertical cracks.

findings from all cases, it shows that when crack depths are greater than the R-wave wavelength, the corresponding discrepancy values between the actual and experiment measured crack are smaller. Besides, underestimation seems to be consistent, with maximum error of approximately 18%.

In estimating crack inclination angle by both velocity and amplitude indices, the measured data consistently underestimate. The maximum error was found to be approximately 12%. The errors as obtained from both estimating the depth and angle of inclination were considered acceptable considering the many factors that have potentially contributed to this, such as inherent porous nature and minor cracking of concrete surface layer on which accelerometers were mounted and slight unevenness of the concrete surface due to protruding aggregates. Besides, the attenuation of waves in concrete that was not actually "homogenous" should not be overlooked, even in the absence of crack as local variations in properties is common for concrete. The complex reflections and convergence of waves at boundaries of specimens, on the other hand, are deemed worth investigating further for clarification of R-wave propagation behavior.

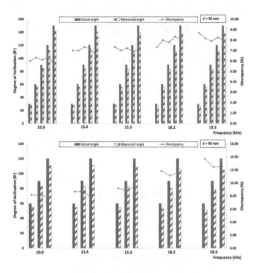

Figure 9. Comparison of actual crack inclination angle with estimated results for 30 mm and 90 mm deep cracks.

## 6 CONCLUSIONS

The velocity and amplitude indices of R-waves were computed from information obtained by simulating time domain waveform data. Correlations between these R-wave parameters characteristics of concrete surface crack, namely depth and inclination angle. The accuracy of correlations was examined through experiment measurements of concrete specimens induced with artificial crack. The overall results exhibited satisfactory agreements between the wave parameters and crack characteristics. The amplitude index seemed to be a more reliable parameter for estimating crack depth and its inclination angle, especially when the wavelength is greater than the crack depth. The proposed measurement methodology of R-waves can possibly be refined, in terms of robustness and reliability for in-situ assessment purposes.

## ACKNOWLEDGEMENT

This work was supported by the High Impact Research Grant, Ministry of Education Malaysia (UM.C/625/1/HIR/MOHE/ENG/54) and University of Malaya Research Grant Scheme (RP004B-13 AET). The authors are also grateful for the help rendered by MDC Precast Industries Sdn Bhd (501872-A) during experiment preparation.

## REFERENCES

[1] Aggelis, D.G. & Shiotani, T. (2007). Repair evaluation of concrete cracks using surface and through-transmission wave measurements. *Cement and Concrete Composite*, 29(9), 700–711.
[2] Mori, K., Spagnoli, A., Murakami, Y., Kondo, G. & Torigoe, I. (2002). A new non-contacting non-destructive testing method for defect detection in concrete. *NDT & E International*, 35(6), 399–406.
[3] Kee, S.H., & Gucunski, N. (2014). Characterizing a Surface-Breaking Crack in Concrete Bridge Decks Using Surface Wave Measurements. In *Transportation Research Board 93rd Annual Meeting* (No. 14–3944).
[4] Zewer, A., Polak, M.A. and Santamaria, J.C. (2005). Detection of surface breaking cracks in concrete members using Rayleigh waves. *Journal of Environmental & Engineering Geophysics*, 10(3), 295–306.
[5] Wave2000, Cyber-Logic, Inc., New York, http://www.cyberlogic.org.
[6] Doyle, P.A. and Scala, C.M. (1978). Crack depth measurement by ultrasonics: a review. *Ultrasonics*, 16(4), 164–170.
[7] Arias, I. and Achenbach, J.D. (2004). A model for the ultrasonic detection of surface-breaking cracks by the scanning laser source technique. *Wave Motion*, 39(1), 61–75.
[8] Chai, H.K., Momoki, S., Aggelis, D.G., & Shiotani, T. (2010). Characterization of Deep Surface-Opening Cracks in Concrete: Feasibility of Impact-Generated Rayleigh-Waves. *ACI Materials Journal*, 107(3).

*Emerging Technologies in Non-Destructive Testing VI – Aggelis et al. (Eds)*
*© 2016 Taylor & Francis Group, London, ISBN 978-1-138-02884-5*

# Elastic wave tomography with arbitrary locations of impact

K. Matsumoto & S. Momoki
*Tobishima Corporation, Chiba, Japan*

T. Shiotani
*Kyoko University, Kyoko, Japan*

Y. Kobayashi
*Nihon University, Tokyo, Japan*

ABSTRACT: Establishing a method for health evaluation of infrastructure is an urgent issue. The authors have been studying on practical implementation of elastic wave tomography that is capable of making comprehensive health evaluation of civil structures. The elastic wave tomography integrates numerous scanning line data. For collecting scanning line data, installing sensors at sending and receiving points is essential. It was therefore necessary to allocate and reallocate sensors numerous times. In this study, a new tomography measurement method is developed, which requires no sensors at wave generation points. Properties of wave made at random locations on the measured target surface by the inspector are estimated based on the receiving data at specific points. A large amount of scanning line data can be collected by allocating a minimum number of sensors. Then, efficient tomography measurement of elastic waves is possible, enabling the use of hammering tests that are conducted in various types of structures.

## 1 INTRODUCTION

In view of the aging of infrastructure systems and its resultant harm to third parties, inspection of infrastructure has been intensified in recent years. Visual inspection and hammering tests, which are basic components of inspection, enable inspection only at the surface and within the external layers. Internal evaluation of infrastructure is considered to require the application of various nondestructive evaluation techniques. Such techniques have, however, been actually applied only in limited infrastructure systems because of cost and manpower restrictions.

The authors have studied to put in practical use the elastic wave tomography capable of making a wide-area health evaluation of infrastructure including its interior [1]. As a recent research result, elastic wave tomography was established that was capable of evaluating internal health with sensors installed only on one side of the structure [2] and it was made possible to apply the tomography to all kinds of infrastructure systems. Health evaluation using elastic wave tomography with sensors on one side of the structure is greatly affected by the amount of data between scanning lines within the measured area. More detailed inspection requires the specification of numerous wave generation points to increase data volume. The number of sen-

sors to be installed for collecting wave generation point data also inevitably increases. Much work and time are therefore required. Local applicability of the evaluation method for detailed inspection remains a problem. In this study, in an attempt to solve the problem, a new measurement method is developed. Impacts generated from the hammering test are specified as transmitting signals and the wave generation point data is estimated from receiving data, which enables the collection of numerous impact signals with minimum number of sensors installed. This facilitates simultaneous and comprehensive inspections of internal healthiness using elastic wave tomography with sensors installed on one side of the structure while conducting hammering tests.

## 2 OUTLINE OF TOMOGRAPHY WITH SENSORS INSTALLED ON ONE SIDE

The wave excited by hitting the surface of a structure with a steel ball is classified into several components such as pressure wave (P wave) and transverse wave (S wave) according to the direction of vibration or progress of the wave. The component used in ordinary elastic wave tomography is the propagation velocity of pressure wave. Focus was placed on the surface waves (Rayleigh wave and Lamb wave), which propagate on the surface, among elastic waves. In this study, Rayleigh wave was mainly examined. As shown in Figure 1, a surface wave propagates near the surface while vibrating ellipsoidally. The range of the ellipsoidal vibration is identical to the length of the surface wave. Phase of the surface wave varies according to the depth, which is equivalent to the wavelength. Figure 2 shows that in the case where defects such as cracks and voids exist in the range of vibration of surface wave, which is the sum of the scanning line and depth, phase changes in the defective section of surface wave due to reflection or scattering even though there is no change in the pressure wave propagation velocity on the surface. It was believed that by properly analyzing the change in wave properties, internal health evaluation with sensors installed only on one side of the structure can be realized. As an index indicating the change of surface wave, the phase velocity of surface wave was obtained using equation (1) [3] and applied to elastic wave tomography.

$$V_{ph} = \lambda f = \left(\frac{2\pi \Delta x}{\Delta \varphi}\right) f \qquad (1)$$

where, $\lambda$ is wave length, $\pi$ is the circular constant, $V_{ph}$ is the phase velocity (m/sec), $\Delta x$ is the distance between scanning lines (m), and $\Delta \varphi$ is the phase difference between frequencies $f$ of sending and receiving waves $f$ (Hz) (radian).

Distributions of phase velocities of surface wave were obtained in concrete specimens with pseudo defects (Styrofoam plates) (Figure 3) using the elastic wave tomography with sensors only one side of the structure (Figure 4). The distributions of phase velocities of surface waves were obtained using elastic wave tomography with sensors only one side of the structure (Figure 4). The data on the wave generation point (time of generation and waveform) was obtained by hitting near sixteen acceleration sensors, and the receiving point data (receiving time and waveform) was collected form the remaining fifteen sensors. Then, analysis was made based on the phase velocity obtained from the 240 scanning lines (16 wave generation points x 15 receiving points). Wavelengths were estimated from the predominant frequency of the surface

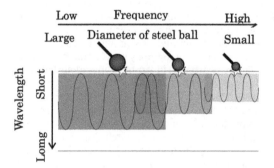

Figure 1. Directions of surface wave vibration and progress.

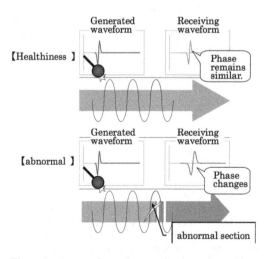

Figure 2. Image of soundness evaluation using surface wave.

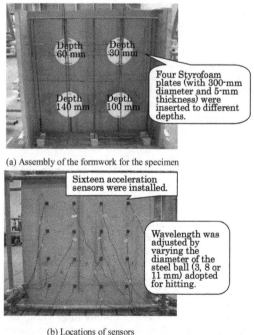

Figure 3. Outline of concrete specimen.

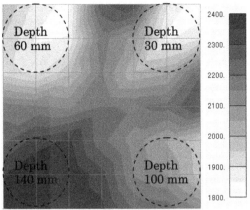

(a) diameter of the steel ball is 3 mm
(Wave Length:110 mm)

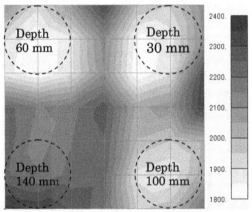

(b) diameter of the steel ball is 8 mm
(Wave Length:180 mm)

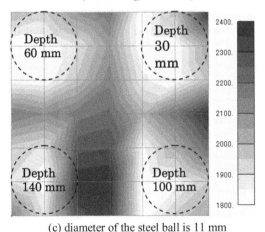

(c) diameter of the steel ball is 11 mm
(Wave Length:240 mm)

Figure 4. Distributions of phase velocities of surface wave using elastic wave tomography with sensors only one side of the structure.

wave in the generated waveform. The pseudo defects with a depth shorter than the estimated wavelength were detected as those in an area of low phase velocity. It was thus verified that internal health could be evaluated by installing sensors only on one side of the structure. More details about the methodology and the results can be seen in [2].

## 3 DEVELOPMENT OF MEASUREMENT TECHNOLOGY USING ARBITRARY HITTING SIGNALS

Conventional method required the installation of sensors at the impact point for collecting wave generation point data. Reducing the number of sensors to save labor and time resulted in reduced amount of data collected (left side of Figure 5) and low-resolution images. Acquiring images of higher resolution involved the increase of sensors and required more labor and time.

The method developed in this study (right side of Figure 5) enables the collection of arbitrary impact signals without increasing the number of sensors installed by estimating the wave generation point data from the receiving point data. The wave generation point data required in the elastic wave tomography with sensors on one side is wave generation location, time and waveform. Methods of estimating each parameter from receiving point data are described below.

### 3.1 Estimation of wave generation location and time

The wave generation location and time are estimated from the receiving location and the first arrival time for receiving waveform by making back analysis. This is a general practice in the Acoustic Emission (AE) method, a nondestructive evaluation technique [4]. Conventional AE methods estimate the wave generation location and time based

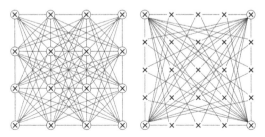

Figure 5. Images of conventional (left) and developed (right) techniques ( × : wave generation point, o: receiving point).

on the assumption that the study area is homogeneous (propagation velocity is constant). The wave generation location and time estimated based on the assumption, therefore, are highly likely to differ from the true value in heterogeneous structures deteriorated with time.

The developed method estimates the wave generation location and time using the AE tomography method. The explanation of detailed algorithm of the AE tomography method is omitted in this paper. Please refer to reference 5. When the wave generation location and time are estimated by the AE tomography method, the propagation velocity structure in the study area is also estimated. Then, highly accurate estimation is possible in reflecting the heterogeneity of the measured area, indicated by the propagation velocity structure.

Figure 6 shows the estimates of wave generation locations obtained by conventional methods using simulation and the AE tomography method. The receiving data at 49 wave generation points (x) and four points at the four corners of the area was assumed and the wave generation location and time were estimated by back analysis. As show in Figure 6 (a), waves diffract around defective sections in the study area and then its arrival is delayed. Conventional methods do not have such phenomena (Figure 6 (b)). The wave generation location therefore is not in agreement with the true location. The AE tomography method evidently estimates the wave generation location more accurately than conventional methods (Figure 6 (c)).

3.2 *Estimation of generated waveform*

The AE tomography method enables highly accurate estimation of the wave generation location and time. The elastic wave tomography with sensors on one side also requires the generated waveform. This section describes the method for estimating generated waveform from receiving waveform.

The waveform with zero amplitude before a certain point in time is referred to as the causal function [6]. The amplitude and phase of the Fourier spectrum of this function are correlated to each other. The phase can be uniquely determined from the amplitude using the Hilbert transform [7]. The waveform determined by the process is referred to as the minimum phase shift function [8]. If the Fourier spectrum of the generated waveform can be estimated, the generated waveform can be obtained as a minimum phase shift function. Accordingly, Figure 7 shows the flow of steps for estimating the generated waveform, using the developed technology.

First, Fourier spectrum of receiving waveform is obtained for an impact signal. Then, an amplitude distance attenuation model is developed using the distance between the wave generation location estimated by the AE tomography method and each receiving location. The amplitude is estimated at the wave generation point for each frequency by regression analysis. The phase is obtained for the amplitude at the wave generation point by the Hilbert transform. The signal generated from the amplitude and phase is subjected to inverse Fourier transform. Then, the generated waveform is estimated as a minimum phase shift function. Figure 8 shows the waveform obtained using the sensors installed near the impact point and the minimum phase shift function estimated based on the receiving waveform using the developed method. The figure shows that the generated waveform obtained was similar to the generated waveform used in conventional elastic wave tomography

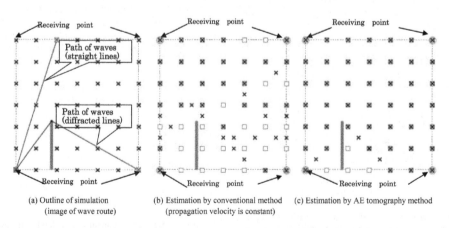

(a) Outline of simulation (image of wave route)

(b) Estimation by conventional method (propagation velocity is constant)

(c) Estimation by AE tomography method

Figure 6. Wave generation locations (x) estimated by conventional method and AE tomography method (square: true wave generation location).

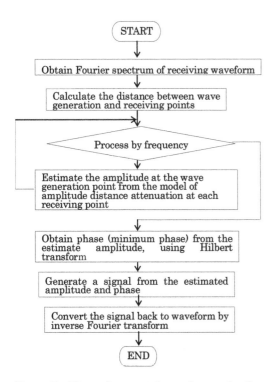

Figure 7. Flow of generated waveform estimation steps.

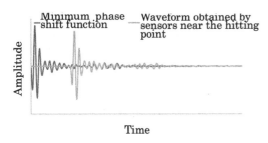

Figure 8. Minimum phase shift function and waveform obtained by sensors near the hitting point.

with sensors on one side (waveform obtained using the sensors installed near the hitting point). The minimum phase shift function exhibits a waveform starting at zero in time history. Shifting the function to the wave generation time that is estimated by the AE tomography method enables estimation as a generated waveform for an arbitrary hitting signal.

Adopting the above method enables the estimation of wave generation location, time and generated waveform without installing sensors near the hitting point, and the application of numerous arbitrary hitting signals as in hammering tests to elastic wave tomography using a minimum number of sensors on one side of the structure.

## 4 VERIFICATION FROM ACTUAL STRUCTURE MEASUREMENT

The validity of the developed technology was verified using an actual structure. Verification was made in a 1.5-by-2.4 m area of an abutment (Figure 9). A crack is found along a diagonal line in the study area. The goal of verification was to check whether the crack could be detected as a deflection or not. Accelerometers used in the developed technology were located at four edges of the study area (Figure 9). The area was hit randomly by a steel-ball hammer. For comparison, analysis was also made by the elastic wave tomography with 16 sensors attached on one side at 0.5-m vertical and 0.8-m horizontal intervals (Figure 10).

Figures 11 and 12 show phase velocity distributions obtained by the developed method and by conventional elastic wave tomography with sensors on one side, respectively. Preparation for verification and measurement required approximately 15 minutes with the developed method, while one hour was required for the conventional method. In verification, it was found that the AE tomography estimated 107 wave generation points. The number of scanning lines was 428 (107 x four receiving points), nearly double the data volume obtained by conventional technology (240 scanning lines). It was therefore possible to increase the number of elements according to the data volume for enhanced resolution. Both figures show a low-velocity area along the crack. The velocity distribution obtained using the developed technology providing higher resolution show the distribution of cracks more clearly. The developed method is applicable without using numerous arbitrary impact signals as in hammering tests and without requiring numerous sensors. It was shown in this study that the method

Figure 9. Locations of sensors (Wave generation by arbitrary hitting. Four sensors were installed.)

Figure 10. Locations of sensors (Conventional technology. Sixteen sensors were installed.)

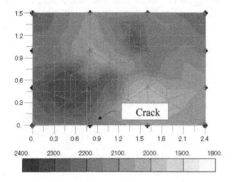

Figure 11. Distribution of phase velocities based on arbitrary hitting signals.

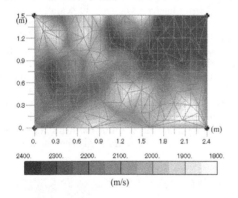

Figure 12. Distribution of phase velocities based on conventional technology.

saves labor cost and time involved in health evaluation and enables the provision of high-resolution results based on larger data volume.

## 5 CONCLUSION

In this study, a new measurement method was developed that uses numerous arbitrary impact signals as in normal hammering tests. The objective is to save labor involved in the elastic wave tomography using sensors only on one side and to improve accuracy. The following conclusions could be obtained.

1. The wave generation location and time, which are required as wave generation point data, can be accurately estimated by adopting the AE tomography method.
2. The generated waveform can be estimated by the minimum phase shift function adopting regression analysis using a distance attenuation model and the Hilbert transform.
3. As described in (1) and (2) above, numerous arbitrary impact signals can be used without installing many sensors. Thus, labor cost and time required for implementation can be saved. As a result of verification in an actual structure, it was found that high-resolution results can be obtained based on the data on numerous impact signals.

## REFERENCES

[1] Momoki, S., H.K. Chai, T. Shiotani, Y. Kobayashi and T. Miyanaga, Damage evaluation of concrete structures by three-dimensional seismic tomography, Journal of structural engineering, Vol. 57 A, pp. 959–966, 2011.
[2] Chai, H.K., D.G. Aggelis, S. Momoki, Y. Kobayashi and T. Shiotani, Single-side access tomography for evaluating interior defect of concrete, Construction and Building Materials, Vol. 24, pp. 2411–2418, 2010.
[3] W. Sachse, Y.-H. Pao, On the determination of phase and group velocities of dispersive waves in solids, J. Appl. Phys. Vol. 49, No. 8, pp. 4320–4327, 1978.
[4] Ishida, T., Science of rock destruction sound, Kinmiraisya, pp. 80–86, 1999.
[5] Momoki, S., Y. Kobayashi, T. Shiotani, Development of AE tomography for asset monitoring of infrastructures -1: algorithmic construction -, 2013 National conference on acoustic emission, pp. 57–60, 2013.
[6] Papoulis, A., Applied Fourier integral for enginerring, ohmusya, pp. 251–267, 1967.
[7] Papoulis, A., Fourier integral and its applications, McGraw-Hill, pp. 192–217, 1962.
[8] Tatsumi, Y., T. Sato, Multiple Event Analysis of 1979 Imperial Valley Earthquake Using Causality Concept of Earthquake Motion, Journal of JSCE, No. 380, pp. 475–484, 1987.

# An ultrasonic method utilizing anchors to inspect steel-plate bonded RC decks

N. Ogura
*Engineering Department, CORE Institute of Technology Corporation, Tokyo, Japan*

H. Yatsumoto
*Engineering Department, Hanshin Expressway Company Limited, Osaka, Japan*

K.C. Chang & T. Shiotani
*Graduate School of Engineering, Kyoto University, Nishikyo, Kyoto, Japan*

ABSTRACT: Road bridge deck slabs mostly made of concrete have been strengthened by steel plate bonding method in consideration of extended use of the bridges in future. Steel plates are bonded to the deck bottom surfaces to improve load carrying performance and durability. As over 30 years have passed since the technique was first applied, debonding of steel plates is often found by hammer impact test with abnormal noise suggesting poor bonding. However, damage in deck slab concrete strengthened with steel plates cannot be directly visually observed because of the steel plates covering the bottom surfaces. Although presence and extent of debonding can be estimated by hammer impact test, detection of damage in concrete remains unestablished. The steel plate bonding method uses temporary anchors which penetrate the plates into the concrete. The authors developed an inspection method using the temporary anchors as probes to directly detect damage in the concrete.

## 1 INTRODUCTION

Deck slabs of road bridges which are mostly made of concrete have been affected by fatigue damage under repeated loads of vehicles. Many of the bridges in Japan were built during the period of rapid economic growth after the WWII, and fatigue damage is frequently found on them, especially in decks which were constructed before 1970. Many field surveys and researches have been made to investigate the fatigue problem.

One of the strengthening methods developed as a result of such investigation is the steel plate bonding method. Steel plates are bonded to the deck bottom surfaces by using anchor bolts, resin or other means for the improved load carrying performance. This technique has been popularly employed on expressways located in Osaka and other urban areas. However, over 30 years have passed since the technique was first applied, and the number of vehicles and vehicle loads have increased with time to exceed the initially estimated design levels. The steel plates are often found debonding during inspection using an impact test hammer as locations with abnormal noise specific to poor bonding.

The administrators of expressways started to strengthen the inspection for enhanced damage detection. However, damage in concrete of deck slabs strengthened with steel plates cannot be directly visually observed because of the steel plates covering the bottom surfaces. Therefore, although presence and extent of debonding can be estimated from the change in sound during ordinary inspection by hammer impact test, detection of damage in concrete still remains unestablished.

The steel plate bonding method uses temporary anchors which penetrate through the plates into the concrete. The authors developed an inspection method using the temporary anchors as probes to

Figure 1. A site strengthened by steel plate bonding method.

directly detect damage in the concrete. This paper reports the result of effectiveness test for the proposed inspection method using artificial defect specimens.

## 2 BASIC EXPERIMENT USING ARTIFICIAL SPECIMENS

### 2.1 Outline of the experiment

In order to examine various factors in using temporary anchors for probes, the authors prepared control specimens having a design strength of 24 N/mm² and defect specimens having simulated deterioration only in the surface area. They were rectangular specimens of 300 × 300 × 2200 mm. Figure 2 shows their schematic diagrams.

The defective part in the defect specimens was created by placing poorly proportioned concrete to a depth of 20 mm from the surface.

Other materials used in the specimens include SS400 steel plates with a thickness of 4.5 mm which are commonly used for steel plate bonding in the field, M12 concrete anchors for temporary anchors, and epoxy resin for steel plate bonding. Steel plate debonding was reproduced by dividing each specimen into two parts at the center and leaving the gap between the steel plate and concrete surface of the right half ungrouted to create a poor bond zone. The gap in the left half was grouted fully with epoxy resin to make a full bond zone.

Table 1 shows measurement results of compressive strength and static modulus of elasticity of the concrete used.

Holes were drilled to a depth of 60 mm as specified, and temporary anchors were spaced at 400 mm as commonly practiced in the field.

### 2.2 Measurement method

Impact elastic wave method, which is a type of elastic wave method, was selected in consideration of ease of operation and practical applicability in the field. An impact was applied to the surface of a test subject, and the generated elastic wave is captured by a receiver. Because of the physical impact, the waves used for measurement consist mainly of low frequency components below the ultrasonic range (20 kHz and above). Time, frequency and phase information of the received waves is analyzed to obtain thickness of the member or determine presence or absence of internal defects or

Table 1. Physical property test results of the specimens.

|  | Compressive strength N/mm² |  | Static elastic modulus kN/mm² |  |
|---|---|---|---|---|
|  | Measured value | Average | Measured value | Average |
| Control specimens | 33.9 33.4 33.9 | 33.7 | 28.7 28.3 28.9 | 28.6 |
| Defect specimens | 22.3 23.9 23.8 | 23.3 | 25.4 25.0 25.1 | 25.2 |

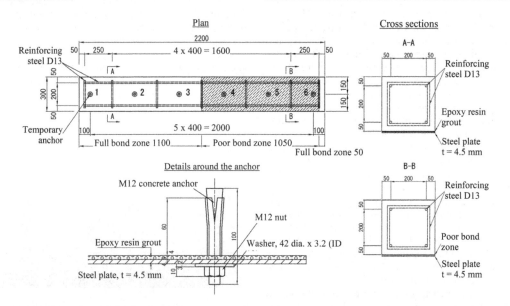

Figure 2. Schematic diagrams of the specimens.

voids. Figure 3 shows a schematic diagram of the measurement.

One end of an anchor was finished smooth and hit with a steel ball of 15 mm in diameter. Two Acoustic Emission (AE) sensors were used. One was placed near the anchor to be hit to obtain the input signal, and the other was placed on one end of the other anchor to record the waves.

### 2.3 Examination

Measurement by impact elastic wave method was carried out on both control and defect specimens before and after steel plate bonding, using the anchors to record waveforms at propagation distances of 400 to 1200 mm. The recorded data was sorted by distance, propagation velocity was calculated, and frequency spectrum distribution and prominent frequency were determined from the frequency analysis by fast Fourier transformation. Using some data selected from these analyses, evaluation focused on time domain was also made through wavelet transformation.

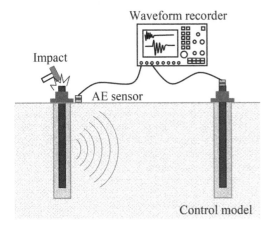

Figure 3.  Schematic diagram of measurement.

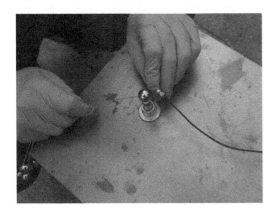

Figure 4.  Impact applied after steel plate bonding.

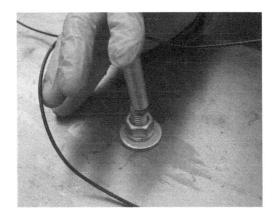

Figure 5.  Receiving sensor installed.

## 3 EXPERIMENT RESULTS AND DISCUSSION

### 3.1 Propagation velocity in the control and defect specimens before steel plate bonding

Figure 6 shows propagation velocities measured in the control and defect specimens before steel plate bonding, by distance between the sensors. Sensor-to-sensor distances were 400, 800 and 1200 mm. Propagation velocity in the control specimens was around 4300 m/s. Propagation velocity in the defect specimens was about 4000 to 4100 m/s, decreasing by about 5% as compared with the control specimens.

The difference in propagation velocity between the control and defect specimens was likely due to the defect part simulated in the concrete surface, suggesting that sensitivity was not affected by the use of temporary anchors for probes. However, variation in propagation velocity was slightly larger at a propagation distance of 400 mm which was shorter than 800 or 1200 mm in both control and defect specimens. The reason for this was likely that degree of contact or magnitude of impact of the steel ball was expressed locally due to the short propagation distance. This suggests that more accurate measurement of propagation velocity in concrete would be obtained at reasonably larger distances.

### 3.2 Frequency distribution in the control and defect specimens before steel plate bonding

Figures 7 and 8 show the analysis results for prominent frequency and spectral centroid, respectively, in frequency distribution of the received waveforms

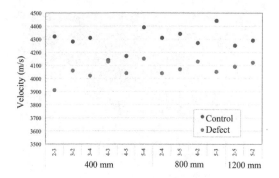

Figure 6. Propagation velocities in the control and defect specimens before steel plate bonding.

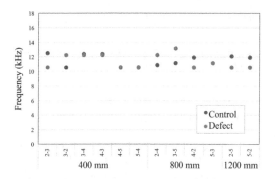

Figure 7. Prominent frequency at the receiving sensor.

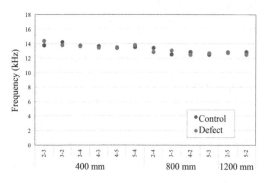

Figure 8. Spectral centroid at the receiving sensor.

in the control and defect specimens before steel plate bonding, by distance between sensors.

Spectral centroid is an analysis technique focused on elastic wave energy as explained by Kuzume et al. (2013) which can express attenuations in the elastic wave in low and high frequency components quantitatively. The spectral centroid used here is an average weighted according to the frequency spectrum obtained by propagation waveform analysis.

The spectral centroid was obtained by weighting the average with frequency components between 1 kHz and 50 kHz (slice width: 150 Hz) so that

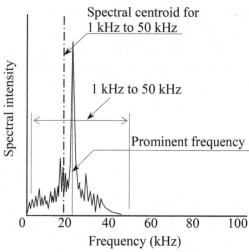

Figure 9. Schematic diagram of the spectral centroid.

influence of noise or minor vibration of the specimens would be reduced (Equation 1, Fig. 9).

$$\text{Spectral centroid (kHz)} = \frac{\sum Ei \cdot Fi}{\sum} \quad (1)$$

where, $Ei$: magnitude of the element; and $Fi$: frequency.

As shown in Figure 7, prominent frequency fell in the range from 10 to 14 kHz for both control and defect specimens. Unlike in propagation velocity, no significant relationship with properties of the concrete was found.

Calculated spectral centroid in Figure 8 mostly fell in the range from 12 to 14 kHz, showing smaller variation by sensor-to-sensor distance as compared with prominent frequency. However, the longer the distance between the sensors, the smaller the spectral centroid was. These were likely due to distance attenuation. Propagation energy attenuation was more significant at longer distances because the impact was applied in the same direction as the anchors and also because the signals were received through the medium of the anchors. Consequently, this method was considered applicable to a relative comparison at a same distance but not appropriate for a comparison at different distances.

3.3 *Propagation velocity in the control and defect specimens after steel plate bonding*

Figure 10 shows propagation velocities measured in the control and defect specimens after steel plate bonding, by distance between the sensors. Figure 11 shows the results for Measurement line 2–4 as an example of measured waveform.

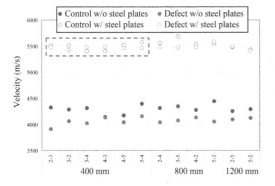

Figure 10. Propagation velocity of the control and defect specimens after steel plate bonding.

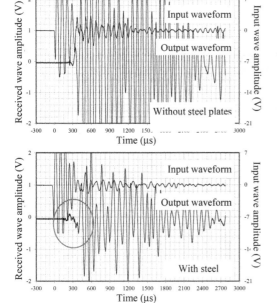

Figure 11. Initial waveforms before and after steel plate bonding.

The significant difference found in propagation velocity between the control and defect specimens before was not found generally after steel plate bonding, with velocity values in both types being about 5500 m/s. Although impact was applied to the end of the anchor, it was likely that the elastic wave also traveled through the steel plate and reached the receiving sensor before the arrival of the signals through the concrete.

The waveforms recorded at around the arrival of the primary wave exhibited obvious difference between before and after steel plate bonding. This indicates the importance of how to exclude the influence of these primary waves from evaluation.

### 3.4 Influence of epoxy resin and frequency distribution in the control and defect specimens after steel plate bonding

In order to identify how the presence of epoxy resin used in steel plate bonding could influence propagation velocity and frequency distribution, a comparison was made in each of the control and defect specimens between the following three categories: Measurement lines 2-3 and 3-2 in the full bond zone; Measurement lines 4-5 and 5-4 in the poor bond zone; and Measurement lines 3-4 and 4-3 crossing the border between the full and poor bond zones. The measurement results for these are boxed with a dotted line in Figure 10.

As shown in the diagram, no significant difference was found in propagation velocity between the three categories. This suggests that influence of the propagation of elastic waves through steel plates is likely to be much larger than that of the presence of resin.

Figure 12 shows elastic wave frequency distributions measured near the impact point of Measurement line 2-3 in the full bond zone and that of Measurement line 5-4 in the poor bond zone. When the impact point was in the poor bond zone, a significant shift appeared from the high frequency side to the low frequency side. The reason for this was likely to be a bending resonance phenomenon. When evaluated based on a parameter of spectral centroid, the frequency was almost above 15 kHz in the full bond zone and was below that in the poor bond

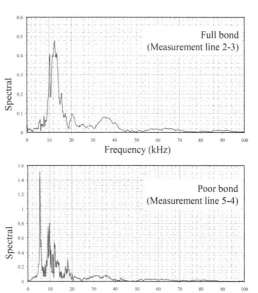

Figure 12. Frequency distributions in input waveforms after steel plate bonding.

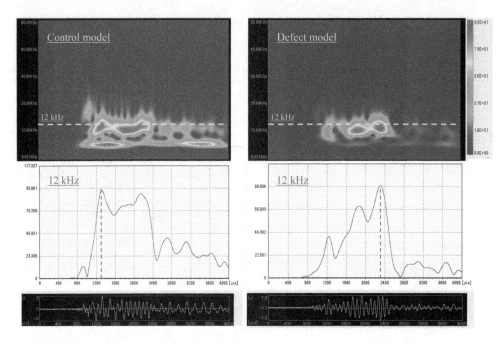

Figure 13. Wavelet analysis results for the control and defect specimens after steel plate bonding.

zone. Consequently, it was considered that frequency analysis would provide a certain level of quantitative evaluation on debonding of the steel plates.

## 3.5 Wavelet analysis

The basic experiments described above provided the following findings: (1) soundness/deterioration of concrete can be readily determined from propagation velocity when steel plates are not present; (2) when temporary anchors are used for measurement, distance attenuation has a major influence in the frequency-based evaluations due to dispersion and reduction of input energy; and (3) although the influence of the presence of epoxy resin does not appear in the propagation velocity which is more influenced by elastic waves propagating through steel plates, frequency analysis on the elastic waves at the receiving sensor provides quantitative evaluation.

These findings suggested that more valid difference would be obtained by time domain evaluation for determination of internal deterioration of steel-bonded concrete. The authors examined wavelet analysis technique for that purpose.

Figure 13 shows the analysis results for Measurement line 5-2 (sensor-to-sensor distance: 1200 mm) in the control and defect specimens after steel plate bonding. A focus was placed on a frequency of 12 kHz which was found to be prominent at both of the transmitting and receiving sensors. Difference was found between the control and defect

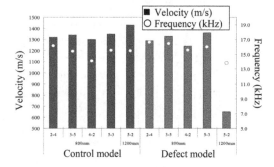

Figure 14. Relationship between spectral centroid and propagation velocity.

specimens in time until the focused frequency reached the dominant level.

In order to establish a method to determine soundness of steel-bonded concrete using temporary anchors, the authors experimentally tried a biaxial evaluation using the spectral centroid at the receiving sensor and the velocity for the time until the focused frequency reached the dominant level as shown in Figure 14.

It was found that velocity difference to be used for soundness/deterioration determination of the steel-bonded concrete appeared more distinct in evaluation focused on the prominent frequency band than in the evaluation based on propagation of primary waves. The results also suggested that steel plate debonding by epoxy resin could be detected by evaluation using spectral centroid at the receiving sensor.

## 4 CONCLUSIONS

The current study revealed that internal soundness of the concrete strengthened by steel plate bonding could be determined successfully by impact elastic wave method, with an elastic wave generated at the end of a temporary anchor.

Further research will continue, including analytical investigation on the frequency band to be focused on at the elastic wave input on the temporary anchors as well as propagation path in the part embedded in the concrete. Experiments using specimens will also be carried out to validate the evaluation index.

## ACKNOWLEDGEMENT

This study was implemented in collaboration with the Laboratory of Innovative Techniques for Infrastructures, Kyoto University. Their great assistance and support to this research are gratefully acknowledged.

## REFERENCE

Kuzume, K., Manabe, H., Ogura, N., Yamamoto, T. & Miyagawa, T. 2013. Experimental study on inspection for internal deterioration of concrete affected by alkali-silica reaction using ultrasonic tomography method. *The 13th JSMS Symposium on Concrete Structure Scenarios*: 181–188.

# Strategic maintenance philosophy for infra-asset with innovative NDT

T. Shiotani
*Department of Civil and Earth Resources Engineering, Kyoto University, Kyoto, Japan*

ABSTRACT: Among developed countries it become a social issue of most interest that how we can cope with the huge numbers of existent or ageing infrastructures under the limitation of shrinking infra-investment. In the paper, a conventional way of maintenance philosophy has been reviewed. And the ideal strategic maintenance philosophy upon the progress of failure the multi-scale of damage is discussed, the corresponding technique to the objective damage scale is introduced and some cutting edge inspection technologies as well as must-study issues on this theme for the future are discussed.

## 1 INTRODUCTION

Among developed countries it become a social issue of most interest that how can society cope with the huge numbers of existent or ageing infrastructures under the limitation of shrinking infra-investment. As those had been constructed during the rapid economy growth e.g., 1960s in Japan, the life span had been set in about 50 years assuming that those can be readily replaced over the life span. However, increase of aged people against work force due to the low birth rate makes the taxation shortage, leading to the replacement difficult.

As for the philosophy of maintenance program for the infrastructures there are two concepts: one is to respond when the damage becomes remarkable and other is to measure even in early damage. The latter is regarded as the ideal treatment for the infrastructures in terms of life cycle cost.

Let remind the current situation of infra-maintenance management. Figure 1 shows a typical decay curve of infrastructure showing the performance versus time. When implementing the management program, the minimum level of service or performance shall be set in advance. The level might be determined based on the importance of the structure in consideration both of direct and indirect impact when it has been failed. In advanced cases, external diseconomies including environmental load substances are considered. When the integrity

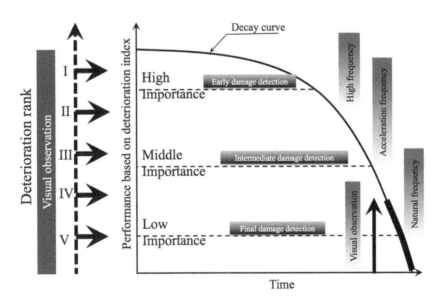

Figure 1. Current performance evaluation of infrastructures based on deterioration rank.

and estimation of remaining time of infrastructures are discussed, the present performance of the structure is generally evaluated by the visual survey for the first, then the resultant deterioration rank will be associated with the performance. Specifically the rank denoting from I to V in the chart will be corresponding to the performance level in Fig. 1. As can be understood, the rank has been determined based on the observation namely naked eyes, the commencing level of the deterioration with this is limited to be the damage as emerged on the surface. Correspondingly, the deterioration rank estimating early phase by the observation shall be corrected to be the latter phase of the deterioration as shown in the right side up-oriented arrow in the chart. In short even in the case that the early treatment was done by visual observation, which could be regarded as the late countermeasure against the whole deterioration process. Thus reliable NDT to evaluate early damage shall be studied and of high demand. Apart from the evaluation of deterioration, one shall define the term, deterioration or damage in advance. The deterioration appeared to be represented by the quality of concrete e.g., poor mechanical property and small elastic moduli. Accordingly one tries to deterioration-layered specimen as to add poor one on the rich one; however, from the view point of deterioration the both can be regarded as intact, and therefore plastic plates and styrene foam sphere balls are well included in the concrete to simulate the deterioration. These actually resemble air-voids namely cracks which has a low acoustic impedance, and intensive studies have been implemented to relate the damage and e.g., elastic waves; however no clear interpretation between them has yet to been established.

As shown above, there are so many issues to solve in order to realize substantial strategic maintenance for infra-asset. In this paper several fundamental issues when applying NDT, especially with elastic wave techniques, to in-situ infrastructures are introduced followed by several cutting-edge methods.

## 2 ELASTIC WAVE TECHNIQUES

As shown in Fig. 2, when study concrete materials objective scale shall correspond to the scale of the objectives. When evaluating the concrete particle e.g., the resolution of millimeters scale shall be necessarily measured. As the elastic wave approaches, the wavelength, being effective to the resolution, has been though that it should be smaller than the resolution i.e., in concrete a couple mega hertz will be equivalent to this millimeters scale resolution. Notation of frequency shown in the right side of Fig. 1 exhibits such corresponding frequencies to the deterioration scales. As in the chart ultrasonic measurement with high frequency shall be conducted for the early damage detection whereas seismometer will be appropriate for the eventual failure phase. It is noted however, as found in other ultrasonic applications, the wavelength with larger than resolution is not always negligible, rather effective to estimate tiny defect (Kawai et al., 2009). Accordingly it would be important to study the substantial relation between the scale of objective and wavelength/ frequency in concrete while clarification of objective scale corresponding to the maintenance level, presumably determined by infra-owner, is other crucial issue to set the strategic maintenance.

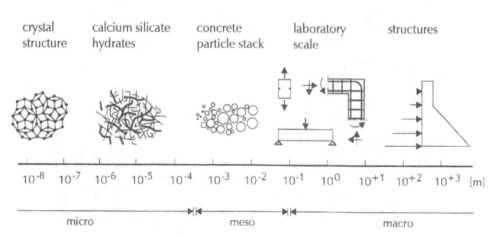

Figure 2. Various scale of observation that must be considered when studying materials and structures (Van Mier 1997).

## 3 DAMAGE/ DETERIORATION

As for on going damage, local damage can be visualized by AE sources detected with a passive approach. The damage has also been quantified based on the AE activity obtained during loading and unloading processes. The combination of RTRI or Load and Calm ratios has been employed to quantify the damage as show in Fig. 3 (Shiotani et al., 2009). As shown in the chart, the symbol is successfully shifted to the upper left side with increase of damage, wheel-load repetitions in this case.

While as for existent damage, as elastic wave generation, namely AE occurrence could not be expected during no-load conditions so that active approach with elastic waves, exciting and detecting waves through the media have been crucially utilized. In this case, the damage can be visualized by the elastic wave parameters such as velocity and attenuation of amplitude as show in Fig. 4 exhibiting velocity distributions. The velocity, however, must be recognized as one of the non-sensitive damage indices among other elastic wave parameters as shown in Fig. 5 (Shiotani & Aggelis, 2009). In addition these parameters are influenced by measurement conditions i.e., these can not be an intrinsic parameter being quantitative to the damage e.g., the velocity is a parameter showing dispersion of which the values change depending on frequency (Aggelis 2008), and amplitude can be readily decayed with propagation distance. Accordingly damage indices which are not influenced by these experimental conditions shall be studied. Currently new damage index has been proposed by the author using attenuation coefficient, Q-value, in elastic waves (Shiotani et al. 2012).

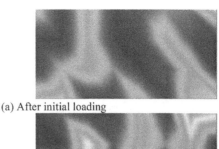

(a) After initial loading

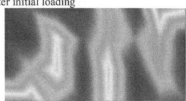

(b) 10,000-passage

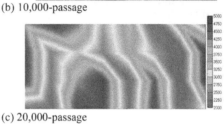

(c) 20,000-passage

Figure 4. Tomogram in a specific section. Legend shows the velocity in m/s.

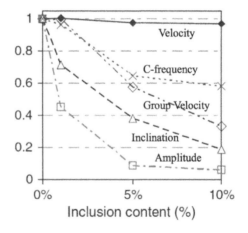

Figure 5. Normalized elastic wave parameters in terms for artificial damage contents.

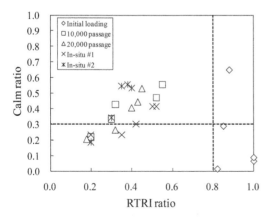

Figure 3. Calm and RTRI ratios in three incremental damage while wheel loading test of RC deck.

## 4 DAMAGE INDEX

In AE monitoring, the waveforms are obtained as convolution of functions of source, propagation media, sensor and acquisition system in the time domain, and those frequency responses can be formulated by a simple multiple equation in the frequency domain:

$$X(f) = U(f)T(f)D(f)S(f) \quad (1)$$

here $X(f)$, $S(f)$, $D(f)$, $T(f)$, $U(f)$ are Fourier transforms of detected AE waveforms: AE source, propagation media, sensor; and acquisition system, respectively. $D(f)$ in Eq. 1 is the member of media quantifying the damage; however, as $S(f)$ is not readily obtained in AE technique, an approach to identify $D(f)$ irrespective to source time function is crucial. In AE application, plural numbers of sensors are employed to locate the AE source, and therefore the comparison of waveforms detected among different sensors for an AE source could suffice this requirement as in Eq. 2 assuming the frequency responses of all the sensors employed are compatible.

$$\frac{x_2(f)}{x_1(f)} = \frac{T_2(f)}{T_1(f)} \cdot \frac{D_2(f)}{D_1(f)} \cong \frac{D_2(f)}{D_1(f)} \quad (2)$$

$$D(f) = \exp\left(-\frac{\pi f}{VQ}d\right) \quad (3)$$

On the other hand, when $D(f)$ is dependent on propagation media attenuation, Eq. 3 can be defined as well, where $f$ is a frequency (Hz), $V$ is a P-wave velocity (m/s) and $Q$ is a normalized value demonstrating attenuation rate.

By combining Eq. 2 and Eq. 3, Eq. 4 is obtained.

$$\frac{X_i(f)}{X_1(f)} \cong \frac{D_i(f)}{D_1(f)} = \exp\left(-\frac{\pi f}{VQ}d\right)$$
$$\Delta d_i = d_i - d_1 (d_i \geq d), (i = 2, \ldots, n) \quad (4)$$

Equation 4 shows that a function of frequency response can be expressed by an exponential function dependent on the difference of distance $\Delta d$ and frequency $f$.

As for the fundamental study using Q-value, standard mortar was used as test specimens (Shiotani et al., 2014). Prism-shaped specimens of 150 × 150 mm with different heights of about 100, 200, 300, 400 and 500 mm to simulate a variety of propagation distances, are used. In order to reproduce damage, two types of spherical styrene balls (6 mm and 3 mm in diameter) were employed as a false cavity, and four kinds of the mixture rate of 0%, 1%, 5% and 10% by the volume were prepared. As shown two types of different dia. ball is used in this study because it has been reported that the attenuation characteristic of elastic wave was also dependent on the size of inclusions even in the same mixture rate (Aggelis & Shiotani 2008)

The Q-values of all cases are summarized as shown in Fig. 6. For both cases of φ 6 mm and φ 3 mm, the Q-value becomes smaller with increase

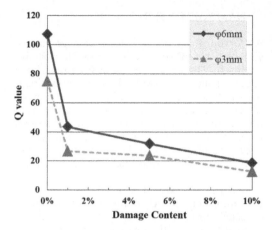

Figure 6. Q-value as a function of damage contents.

of damage content. Comparing the result of φ 6 mm and φ 3 mm, the Q-values of φ 3 mm show slightly smaller values than those of φ 6 mm. In the fact that smaller scale inclusions influences more on attenuation characteristics than that of larger in the same volume of inclusions accords quite well with the past study (Aggelis & Shiotani, 2008). However, it is noted that two results of 0%, exhibiting no artificial inclusions, namely no damage, show different Q-values. This might be attributed to the heterogeneity of the specimen. These might be attributed to existent initial air volume different when casting, a slight difference of sensors' setup condition or propagation paths. However, it can be resulted that the Q-value uniquely represents the degree of damage as a normalized parameter, and this is the very concept to quantify the damage of the materials with the frequency response ratio using AE waveforms.

According to the continuum damage mechanics, the state of damage is represented by the scalar damage parameter, $\Omega$ (Loland, 1989) as expressed by Eq. 5, where $E$ is a Young's modulus of a damage material, and $E^*$ is that of an intact material.

$$\Omega = 1 - \frac{E}{E^*} \quad (5)$$

As show in Eq. 5, the damage parameter, $\Omega$, depends on the ratio of Young's modulus of the intact state to that in the damage state, and this can be one of the quantitative damage evaluation indices. However, the Young's modulus in an intact state is not readily obtained, and therefore the method to estimate the $E^*$ has been intensively studied (Suzuki et al., 2007). Hereafter, it will be verified that if this damage parameter, $\Omega$ can be associated with the Q-value, by examining the relation between the Q-value and the $\Omega$ determined in this experiment.

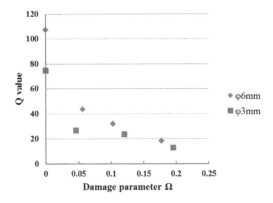

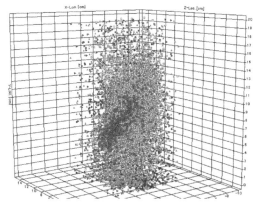

Figure 7. $Q$-value as a function of damage parameter $\Omega$.

Figure 8. Cumulative AE sources derived through triaxial compression test of tuff sample.

Subjected to uniaxial compression, both for reference and simulated damage specimens, Youngs' moduli were obtained, and $\Omega$ is calculated accordingly based on Eq. 5. The resultant relation of $Q$-values and $\Omega$ is shown as in Fig. 7. The $Q$-values decrease remarkably when the $\Omega$ value increases from 0 to 0.05 corresponding to the initial damage/deterioration, implying that $Q$-value has a potential to provide useful information to the early damage stage. Also as the good correlation of $Q$-value and $\Omega$ is obtained, the damage or the deterioration of the materials can be quantified with $Q$-values, which can be then converted to the damage parameter $\Omega$.

## 5 VISUALIZATION OF DAMAGE

As shown in Fig. 8, damage evolution can be visualized when plotting accumulated AE sources through the whole failure process. In in-situ structures, however, as there is few cases installing AE sensors from default condition, cumulative damage interpretation with AE activity cannot be performed, and therefore tomographic approaches using through the thickness or reflected elastic waves have been employed. In conventional elastic wave tomography, the excitations shall be made at designated locations with accurate time record of excitation, leading to time-consuming measurement and requiring a large numbers of sensors. For the bulky 3D structures of which no access allowed to the real area, 3D visualization of damage cannot be implemented. As a solution for this difficult configuration, the authors' research group has proposed new approach namely 'acoustic emission tomography' (hereafter referred to as AET). In AET, source locations and evaluation of velocity distributions are simultaneously implemented with the following procedure. Hereafter, the source

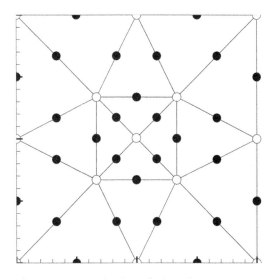

Figure 9. Conventional set of relay points.

location algorithms are demonstrated, leaving detailed tomographic procedure to the other literature (Kobayashi 2012).

The source location technique is based on raytrace algorithm (Schubert, 2004). This algorithm is characterized by installation of relay points in each cell as illustrated in Fig. 9. As the ray-paths are formed by segments among nodal points in conventional ray-trace algorithm, its resolution depends on the mesh characteristics, implying that high accuracy source location requires fine mesh. This leads to increment of the number of degrees of freedom since slowness, which is a reciprocal of velocity, shall be defined in each cell, and consequently makes the identification procedure more complicated. In this ray-trace algorithm, the relay points between nodes are proposed and a ray-path

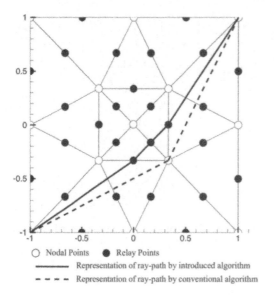

O Nodal Points   ● Relay Points
—— Representation of ray-path by introduced algorithm
- - - - Representation of ray-path by conventional algorithm

Figure 10. Revised ray path in consideration of proposed relay points.

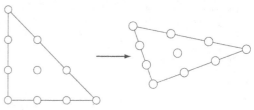

Figure 11. Mapping to the global coordinate of set relay points.

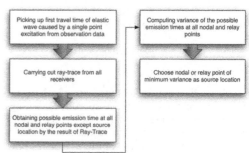

Figure 12. Procedure to estimate the source locations.

is formed by segments among nodal and relay points as shown in Fig. 10.

The resolution of ray-trace is increased without increment of the degrees of freedom by this approach. Besides as role of the relay points is relaying the signals, the relay points shall be distributed uniformly on the surface. However, it is difficult due to the heterogeneous shape of cross section of concrete structures. To solve this problem, the relay points are installed by using isoparametric mapping that is used in the ray-trace algorithm. Since the each cell is mapped into isosceles right triangle, the relay points can be uniformly installed in the mapped cell as shown in Fig. 11. This algorithm does not give exactly uniform distribution of relay points if the shape of the cell is skewed; however, the distribution is improved by avoiding use of strongly skewed cells. The source location is estimated by using this ray-trace algorithm.

The procedure of the estimation of source location is briefly described as in Fig. 12. As for the first step to estimate the source location, the ray-trace is carried out for a receiver as illustrated in Fig. 13. This procedure calculates travel times $t_{ij}$ from a receiver $i$ to all nodal and relay points that are numbered as $j$. Since first travel time $T_i$ at receiver $i$ is observed, the possible emission time of the signal $E_{ij}$ is computed by Eq. 6 at a nodal or relay point $j$.

$$E_{ij} = T_i - t_{ij} \quad (6)$$

The step is applied for all receivers, and then variance of the $E_{ij}$ is computed as follow.

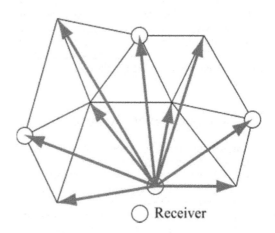

Figure 13. Radiation of waveform form a receiver.

$$\sigma_j = \frac{\sum_i (E_{ij} - m_j)^2}{N} \quad (7)$$

in which

$$m_j = \frac{\sum_i E_{ij}}{N} \quad (8)$$

where N is number of receivers. For the estimation of the source location, the variance $\sigma_j$ is evaluated. If the slowness distribution is exactly identical to real slowness distribution, $\sigma_j$ must be equal to zero at the source location and $m_j$ must be the emission time. Due to the discretization error of slowness distribution and insufficient resolution of ray-trace, generally $\sigma_j$ is not zero even at the source location while the identification procedure of seismic tomography. However, it is predicted that $\sigma_j$ gets minimum at the source location. Hence, in this paper, the source location is determined as a nodal or relay point of minimum variance $\sigma_{jmin}$. Additionally $m_j$ is used as possible emission time. It is noted that the accuracy of the estimation of source location is controlled by the density of nodal and relay points because the source location is assigned to a nodal or relay point in the proposed algorithm. Furthermore, by applying this technique to the iterative procedure of identification of wave velocity structure, the source locations are updated in every iterative step, improving the accuracy of the source location. This approach can be applied for not only AE signals, but also signals that are generated by any excitation point.

Seismic tomography requires source location, emission time and travel time to the receiver. However, the signals having neither emission time nor source location can be used for seismic tomography. It is noteworthy that the source location and emission time can be estimated from travel times to the receivers under the wave velocity distribution determined by the method introduced in previous section. Based on these facts, a procedure of seismic tomography with estimation of source location is introduced. Figure 14 illustrates the procedure of seismic tomography with estimation of source location. In the seismic tomography with estimation of source location, the first step is to estimate the source location and emission time. If the observed travel times can be separated into groups that are respectively associated with individual excitation points, the estimation of source location and emission time are carried out for each observed travel time group. The second step is applying the ray-trace to all estimated source locations. In this step, the ray-trace is carried out for the all of estimated source locations, and the travel time among the estimated source locations to the other nodal or relay points are figured out. Finally by adding the computed travel time to the estimated emission time, the theoretical travel times at receivers are given by the following equation.

$$T_i' = m_j + t_{ij} \quad (9)$$

In the third step, the slowness distribution is updated to eliminate the difference of the theoretical and observed travel time by identification technique.

In order to evolve this 2D AET practically, 3D AET has been proposed (Shiotani et al., 2014). Firstly the element in the 3D AE tomography is expressed in three dimensions different from the one of 2D AE tomography as shown in Fig. 6, where the most different point between 2D AET and 3D AET is the ray-trace technique in the algorism. In the developed ray-trace technique, the waves in 3D AET are expressed by Eqs. 10 and 11, which are expanded to three dimensions different from Eq. 12 used in 2D AET.

$$a_1 x + b_1 y + c_1 z + d_1 = 0 \quad (10)$$
$$a_2 x + b_2 y + c_2 z + d_2 = 0 \quad (11)$$
$$ax + by + c = 0 \quad (12)$$

For the sake of this developed ray-trace technique, it can be possible to verify the AE source location and the deterioration of elastic wave velocity in 3 dimensions. Here the applicability of 3D AET is demonstrated with rock material.

Cylindrical tuff specimens were sampled at the site of rock slope failure monitoring in Hokkaido, Japan. The specimen has 100 mm in diameter and 200 mm in height. The array of eight AE sensors (60 kHz resonance, R6, PAC) for the AE tomography is shown as in Fig. 15. Four strain gauges were installed onto the central surface of the specimen. Figure 16a shows the photos with regards to the triaxial compression test and b. the rock material. The triaxial compression tests were conducted under consolidated drained condition for generating AE activity in the specimen. The confining pressure was set at 294 kPa and the loading rate was kept constant by 1 kN/min. Finally, shear fracture was observed in the specimen as shown in Fig. 16d causing 907 AE events.

Figure 17 shows the accumulated numbers of AE events and the transverse strain with elapsing time. As shown, the accumulated number of AE events is increasing along with increase of transverse strain. Focusing on the relation between time and accumulated number of AE events, the state of fracture progress is divided into three stages. The stage from step-0 to step-1 is regarded as the progress stage of the microscopic failure. Then the stage from step-1 to step-2 is considered as of a steady stage of mesoscopic failure and the

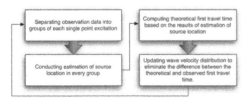

Figure 14. Procedure of seismic tomography with estimation of source location.

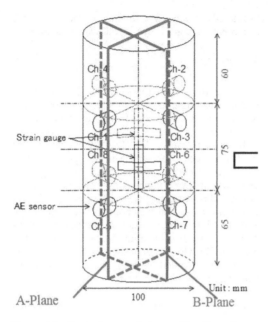

Figure 15. Revised ray path in consideration of proposed relay points.

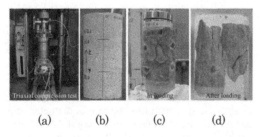

Figure 16. Photos of a. testing apparatus, b. tuff specimen, c. specimen under loading, and d. one after the test from left.

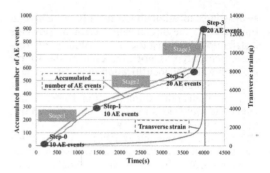

Figure 17. Cumulative numbers of AE events and transversal strain over the experimental time.

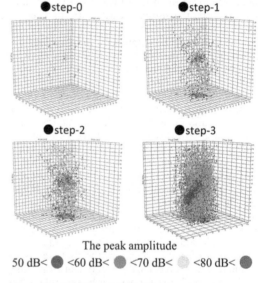

Figure 18. Evolution of 3D AE sources.

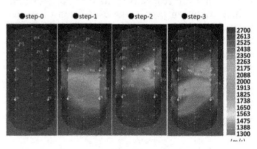

Figure 19. Velocity distributions by 3D AET.

stage from step-2 to step-3 is one to be the final macroscopic failure. Then in order to verify the change of velocity structure in each stage, 3D AE tomography analysis was conducted for four steps. The AE tomography can reflect the whole of the past-fracture/ damage phenomenon only with the most latest AE data, so that AE tomography analysis was conducted with 10 or 20 AE events data obtained around the end of the specific stage. Figure 18 shows the AE source locations in the specimen in each step, where the sources are divided with their peak amplitude as exhibited in the legend. It is obvious that the number of AE sources is increasing in accordance with the load application. Although the precise comparison shall been further examined, the plane comprised of red points in step-3 might be identical to the shear failure plane observed visually after the test. The results of 3D AE tomography are shown in Fig. 19. The decrease areas of elastic wave velocity couldn't be confirmed in step-0, implying no remarkable damage progress in this step. In the following step-1 and step-2, the low elastic wave velocity was confirmed to develop in the central area of the specimen. In the step-3, the large scale of damage evolution was apparent as the low elastic wave velocity was confirmed from the upper right area to the lower left area in the specimen, which is regarded as the final step of the failure. It is noted that this final damage evolution was in good accordance with the shear plane confirmed by both of visual observation and the result of AE source location in 3 dimensions.

## 6 CONCLUSIONS

In this paper several fundamental issues when applying NDT, especially with elastic wave techniques, to in-situ infrastructures were introduced followed by several cutting-edge methods. Still many must-study issues exist to apply NDT for in-situ infrastructures. While no one doubt that proactive maintenance will be the most ideal strategy for important infrastructures, we shall recognize that even definition of deterioration/ damage is still an issue to study. As for elastic wave approaches passive NDT such as one using primary AE activity is difficult to be employed, and rather than that secondary AE activity detecting emission due to friction among existent damages would be the principal way. Tomographic approaches e.g., AET would be the main cast to solve these facts.

## REFERENCES

Aggelis, G.D. 2008. Stress wave scattering: Friend or enemy of non destructive testing of concrete?," *Journal of Solid Mechanics and Materials Engineering*, Vol. 2, No. 4.

Kawai, K., Shiotani, T., Ohtsu, H., Tanaka, H. & Kawagoe, H. 2010. Estimation of axial stress n ground anchors by means of indirect ultrasonic technique. Engineering Technics Press, *Structural Fault & Repair 2010*, 158, CD-ROM.

Kobayashi, Y. 2012. Mesh-independent ray-trace algorithm for concrete structures, *Emerging Technologies in Non-Destructive Testing* V, Taylor & Francis Group, 103–109.

Loland, K.E. 1989. Continuous damage model for load—response estimation of concrete," *Cement and Concrete Research*, Vol. 10, 395–402.

Schubert, F. 2004. Basic principles of acoustic emission tomography, *Journal of Acoustic Emission*, Vol. 22, 147–158.

Shiotani, T. & Aggelis, G.D. 2009. Wave propagation in cementitious material containing artificial distributed damage, *Materials and Structures*, 42, pp 377–384.

Shiotani, T. & Takada, Y. 2012. Damage assessment of civil engineering materials by means of transfer function of AE waveforms, *Abstracts of AEWG54*, S6–2.

Shiotani, T., Ohtsu, H., Momoki, S., H.K. Chai, Onishi, H. & Kamada, T. 2012. Damage evaluation for concrete bridge deck by means of stress wave techniques, ASCE, *Journal of Bridge Engineering*, Vol. 17(6), 847–856.

Shiotani, T., Osawa, S., Kobayashi, Y. & Momoki, S. 2014. Application of 3D AE tomography for triaxial tests of rocky specimens, *31st conference of the European Working Group on Acoustic Emission* (EWGAE), CD-ROM.

Shiotani, T., Takada, Y., Ishizuka, K. & Momoki, S. 2014. Engineering Technics Press, *Structural Fault & Repair 2014*, CD-ROM.

Suzuki, T., Ohtsu, M. & Shigeishi, M. 2007. Relative damage evaluation of concrete in a road bridge by AE rate-process analysis, *Materials and Structures*, Vol. 22, 30–38.

Van Mier, G.M., 1997. *Fracture process of concrete*, CRC Press.

*Emerging Technologies in Non-Destructive Testing VI – Aggelis et al. (Eds)*
*© 2016 Taylor & Francis Group, London, ISBN 978-1-138-02884-5*

# Use of X-ray CT and AE for detection of cracked concrete properties

Tetsuya Suzuki & Toshihiro Morii
*Niigata University, Niigata, Japan*

Tomoki Shiotani
*Kyoto University, Kyoto, Japan*

ABSTRACT: The X ray CT test is one of the most useful method for detection and visualization of crack damage of deteriorated concrete. In recent years, damage evaluation of in-situ concrete structures is now in urgent demand. In this concern, quantitative damage estimation of concrete is proposed to be performed, applying Acoustic Emission (AE) measurement in a uniaxial compression test with X-ray CT test. Generating behavior of AE events in the core test is quantitatively analyzed, based on the rate process theory, because notable discrepancy of AE activity is observed between damaged concrete and undamaged concrete. The damage is quantitatively defined by a scalar damage parameter in damage mechanics. Correlating AE rate with the damage parameter, quantitative estimation of damage is proposed in terms of the rate '$\beta$'. Concrete core samples were taken from reinforced concrete columns of an existing canal which is strongly affected by freeze and thawed process. Prior to the compression test, distribution of micro-cracks in a concrete-core sample was inspected by helical X-ray Computed Tomography (CT), which scans at one-millimeter intervals. Thus, the results suggest that the damage of concrete could be quantitatively evaluated by comparison of AE damage parameter and X-ray CT images. A relation between AE rate and the damage parameters is correlated, and the damage of concrete is quantitatively estimated using AE and X-ray CT.

## 1 INTRODUCTION

The durability of concrete structure could decrease drastically due to the effects of environmental effects or earthquakes. Recently, the Great East-Japan Earthquake hit Tohoku area on March 11, 2011 (Kazama and Noda, 2012). As a result, damage evaluation techniques for diagnostic inspection are in great demand in concrete engineering. The degree of damage in concrete structures is, in most cases, evaluated from the decrease trend of concrete mechanical properties, such as strength. For effective damage estimation of concrete, it is necessary to evaluate not only the mechanical properties but also the degree of damage (i.e. crack development in concrete).

By the authors, quantitative damage evaluation of concrete is proposed by applying Acoustic Emission (AE) (Ohtsu and Suzuki, 2004; Suzuki et al., 2007; Suzuki and Ohtsu, 2011) method and X-ray CT test (Suzuki et al., 2010; Suzuki and Ohtsu, 2014) in the core test. The AE method is one of the most useful method for detection of material damage (Gross and Ohtsu, 2008). Therefore, to inspect existing structures for maintenance, AE techniques draw a great attention. This is because crack nucleation and extension are readily detected and monitored. In this respect, the measurement

of AE activity in the compression test of core samples was proposed.

In this study, damage estimation of concrete-core samples are investigated applying X ray CT and AE. Test samples were taken from reinforced concrete of an existing canal wall, which has been subjected to the influence of freeze and thawed process. Crack distribution in core concrete was inspected with helical CT scans, which were performed at one-millimeter intervals. After helical CT scan, concrete damage was evaluated by AE in compression test. Thus, the decreases in physical properties due to cold environment are evaluated by the CT values, mechanical properties and AE parameters.

## 2 EXPERIMENTS

### 2.1 Damage identification of concrete by X-ray CT test with spatial analysis parameters

The cracked core samples were inspected with helical CT scans at the Medical Center, Niigata University. The helical CT scan was undertaken at one-millimeter intervals before the compression test. The measurement conditions are shown in Table 1. The example of X-ray CT image is shown in Fig. 1. The output images were visualized in gray scale where air appears as a dark area and the densest parts in the image appear as white. The exact positioning was ensured using a laser positioning device. Samples were scanned constantly at 0.5 mm pitch overlapping. A total of 200 to 400 2D-images were obtained from each specimen depending on the specimen length. These 2D images can be assembled to provide 3D representation of core specimens. Detected image data is analyzed by spatial statistics, such as the concentration index $I_\delta$.

### 2.2 Freeze-thawed samples

Cylindrical samples of 5 cm in diameter and about 10 cm in height were composed of taken from the freeze-thawed damaged structures (Suzuki et al., 2010). The mechanical properties of testing samples are shown in Table 2. Test samples are classified into three types by cracking conditions (Type A~C). The heavy cracked core-sample is named "Type A". The little cracked core-samples are named "Type B". The normal samples are named "Type C".

### 2.3 Compression test with AE Monitoring

A uniaxial compression of testing samples was conducted as shown in Fig. 2. Silicon grease was pasted on the top and the bottom of the specimen, and a Teflon sheet was inserted to reduce AE events generated by friction. The SAMOS-AE system (manufactured by PAC) was employed as a measuring device (Fig. 2(a)). AE hits were counted by using an AE sensor R-15. The frequency range was from 60 kHz to 1MHz. To count the number of AE hits, the threshold level was set to 60dB with a 40dB gain in a pre-amplifier and 20dB gain in a main amplifier. For event counting, the dead time was set as 2ms. AE measurement was conducted with the measurement of axial and lateral strains using DICM (Fig. 2(b)).

Table 1. Setting used for helical CT scan.

| | |
|---|---|
| Helical Pitch | 15.0 |
| Slice Thickness | 0.5 mm |
| Speed | 7.5 mm/rotation |
| Exposure | 120 kW and 300 mA |
| Recon Matrix | 512 × 512 |
| Field of View | 100–200 mm |

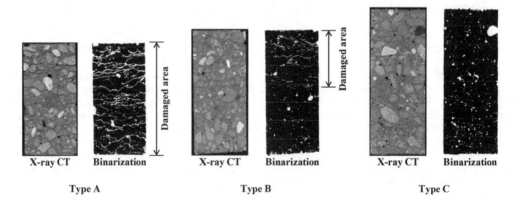

Figure 1. Characteristics of concrete damage in testing samples.

Table 2. Physical properties of testing core samples.

| | Compressive strength (N/mm²) | Maximum strain (μ) | Tangent modulus of elasticity (GPa) | AE parameter $\beta$ ($\beta < 0.0$: Damage) | Sample size |
|---|---|---|---|---|---|
| Type A Heavy Cracked Concrete | 7.0 [–] | 3,000 [–] | 5.9 [–] | –0.011 [–] | 1 |
| Type B Little Cracked Concrete | 10.4 [3.8~20.0] | 1,742 [800~3,365] | 11.2 [2.9~25.1] | –0.005 [–0.009~–0.0003] | 3 |
| Type C Non-Cracked Concrete | 27.9 [27.6~28.2] | 1,250 [1,050~1,450] | 35.8 [27.7~43.9] | +0.015 [+0.014~+0.016] | 2 |

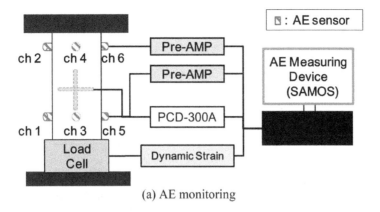

(a) AE monitoring

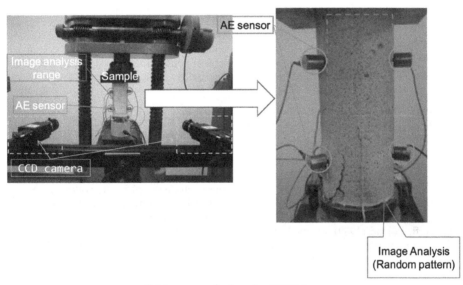

(b) Image analysis using DICM

Figure 2. Test setup for AE and DICM monitoring in compression test.

## 2.4 Analytical procedure for damage evaluation of concrete using acoustic emission

AE activity of a concrete core under compression is associated with the rate process theory was introduced (Ohtsu and Suzuki, 2004). AE behavior of a concrete sample under compression is associated with the generation of micro cracks. These cracks tend to gradually accumulate until final failure. Since this process could be referred to as stochastic, the following equation of the rate process is introduced to formulate the number of AE events, $dN$, due to the increment of strain from $\varepsilon$ to $\varepsilon + d\varepsilon$,

$$f(\varepsilon)d\varepsilon = \frac{dN}{N}, \qquad (1)$$

where $N$ is the total number of AE events and $f(\varepsilon)$ is the probability function of AE at strain level $\varepsilon\%$. For $f(\varepsilon)$ in Eq.1, the following exponential function is assumed,

$$f(\varepsilon) = \alpha \cdot \exp(\beta\varepsilon), \qquad (2)$$

where $\alpha$ and $\beta$ are empirical constants. Here, the value '$\beta$' is named the rate (Fig. 3). The probability varies in particular at low strain level, depending on whether rate '$\beta$' is positive or negative. If rate '$\beta$' is negative, the probability of AE events is high at low strain level. This indicates that the testing concrete may be damaged. If the rate is positive, probability is low at low strain level and the concrete is in stable condition. Therefore, it is

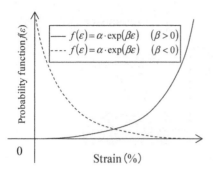
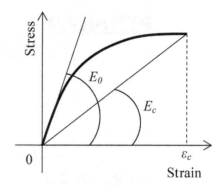

Figure 3. Two possible relations of probability function $f(\varepsilon)$.

Figure 4. Stress-Strain relation and determination of Young's modulus.

possible to quantitatively evaluate the damage in a concrete using AE under uniaxial compression by AE generation behavior.

### 2.5 Quantification of cracks in concrete by damage mechanics with X-ray CT data

Damage parameter $\Omega$ in continuum damage mechanics is defined as a relative change in the modulus of elasticity, as follows,

$$\Omega = 1 - \frac{E}{E^*}, \quad (3)$$

where $E$ is the modulus of elasticity and $E^*$ is the modulus of concrete which is assumed to be intact and undamaged. Loland (1989) assumed that the relationship between damage parameter $\Omega$ and strain $\varepsilon$ under uniaxial compression is expressed,

$$\Omega = \Omega_0 + A_0 \varepsilon^\lambda, \quad (4)$$

where $\Omega_0$ is the initial damage at the onset of the uniaxial compression test, and $A_0$ and $\lambda$ are empirical constants of the concrete. The following equation is derived from Eqs. 3 and 4,

$$\sigma = (E_0 - E^* A_0 \varepsilon^\lambda)\varepsilon. \quad (5)$$

In this study, the damage of concrete is evaluated by damage parameter "$\lambda$". The equation of $\lambda$ is expressed (Fig. 4),

$$\lambda = \frac{E_c}{E_0 - E_c}. \quad (6)$$

The CT number obtained in Hounsfield Units (HU) represents the mean X-ray absorption associated with each area on the CT image. The CT numbers vary according to the material properties, generally adjusted to 0.0 for water and to −1,000 for air. The detected X-ray CT images are analyzed by the concentration index $I_s$ which is defined as spatial statistics parameter for quantitative evaluation of concentration level of physical values, such as concrete damage (cracks).

## 3 RESULTS AND DISCUSSION

### 3.1 Physical properties of testing samples

Mechanical properties are summarized in Table 2, with the maximum, and the minimum values of all specimens. 6 samples were collected from concrete walls which is strongly influenced by freezing and thawing process.

The compressive strength is 7.0 N/mm² in the heavy cracked condition (Type A), while that of the non-cracked condition is 27.9 N/mm² as the average (max: 27.6 N/mm², min: 28.2 N/mm²). Thus, the decrease in the mechanical properties is clearly observed in Type A. On the other hand in the little cracked condition (Type B), the compressive strength is 10.4 N/mm² as the average (Type A < Type B < Type C).

Characteristics of X-ray CT data is shown in Fig .5. The shape of the probability density function $f(x)$ for each sample type is confirmed different trend.

AE generating behavior of each specimen is evaluated by the damage parameter $\beta$. Type A and Type B samples is analyzed the negative '$\beta$' value in AE rate-process analysis. These samples are strongly damaged. Compared to the average value of Type A and Type C. The obtained values in Type C is $+1.5 \times 10^{-2}$, and $-1.1 \times 10^{-2}$ of Type A. The rate '$\beta$' is negative; the probability of AE activity is high at a low stress level in compression test. It is indicating that the sampling structure is heavy damaged. Therefore, these results of mechanical properties and AE parameter $f(\varepsilon)$ suggest that the sampling structure is developed local damages.

## 3.2 Estimation of concrete damage using AE and X-ray CT

The X-ray CT data is analyzed by the concentration index $I_\delta$ which is defined as the spatial statistics parameter. Calculated $I_\delta$ is quantitatively evaluated by relation between $I_\delta$ and analytical area of X-ray CT image which is named the $c$ value (Fig. 6).

The $c$ values are compared with average void perimeter in Fig. 7. Thus, it can be clearly separated in each sample type by using relation between $c$ value and the void perimeter. Our recent studies, the concentration of crack damage in concrete was positively correlated with decrease trend of CT value (Suzuki et al., 2010). Thus, the results of Type A and Type B are plotted in low $c$ value part ($c < 100$), it is considered that these samples have been fairly damaged. In Type A sample, AE generating behavior in core test showed the negative '$\beta$' value in AE rate-process analysis (Fig. 8).

Thus, in this study, we try to be estimation of concrete damage level using comparison X-ray CT parameter '$c$' and AE index '$\beta$'.

## 3.3 Comparison of c value and $\beta$, $V_p$

The accumulation of crack damage in testing samples is positively correlated with decrease trend of '$c$' and '$\beta$'. In $\beta < 0.0$ condition, AE generation behavior is high at low strain level. This results indicates that the testing concrete may be damaged.

The P wave velocity is detected same trend of AE index $\beta$ in core test. The standard of P wave velocity in concrete is defined as 4,000 m/s (JSNDI, 2000). The decrease trend of P wave velocity is correlated with inner damage (JSCE-TC326, 2004). In $\beta = 0.0$ condition, P wave velocity is estimated about 3,296 m/s (< 4 km/s, 82.4% = 3,296/4,000). Therefore, these results detect that negative value of $\beta$ with the increase in damage. And, AE index in core test is estimated by non-destructive P wave monitoring.

## 4 CONCLUSION

For quantitative damage evaluation of concrete, AE and X-ray CT methods were applied to the experiments of core samples. The crack distributions of concrete-core were inspected with X-ray CT. The damage of concrete due to crack progressive conditions was evaluated by AE parameter in compression tests. AE generation behavior in core test is closely associated with the damage, which can be quantitatively evaluated by $c$ value calculated from the concentration index $I_\delta$ of X-ray CT data. Thus,

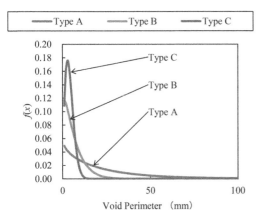

Figure 5. Characteristics of crack and void in testing concrete using the probability function $f(x)$ of X-ray CT data.

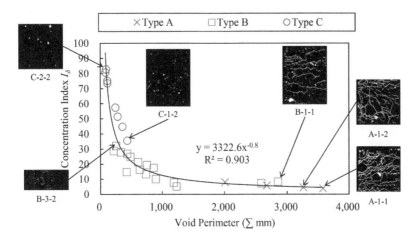

Figure 6. Comparison of void perimeter and the concentration index $I_\delta$.

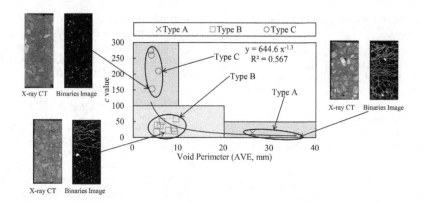

Figure 7. Comparison of void perimeter and $c$ value from the concentration index $I_\delta$.

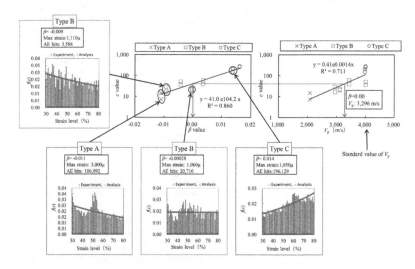

Figure 8. Comparison of X-ray CT parameter '$c$' and damage parameter ('$\beta$', '$V_p$').

the damage of concrete is quantitatively estimated as damage parameters by AE and X-ray CT.

REFERENCES

C.U. Gross, M. Ohtsu Edit. 2008. Acoustic Emission Testing, Springer; 2008: 3–10.
JSCE TC326 Edit. 2004. Nondestructive inspection of concrete by elastic wave method, JSCE: 31–37.
JSNDI Edit.1994. Non-Destructive Testing of Concrete Structures: 114.
Kazama, M. and Noda, T. 2012. Damage Statics (Summary of the 2011 off the Pacific Coast of Tohoku Earthquake damage), *Soils and Foundations*, 52(5): 780–792.
Loland, K.E. 1989. Continuous damage model for load—response estimation of concrete, *Cement and Concrete Research*, 10: 395–402.
Ohtsu, M. and Suzuki, T. 2004. Quantitative Damage Evaluation of Concrete Core based on AE Rate-Process Analysis, *Journal of Acoustic Emission*, 22: 30–38.
Suzuki, T., Shigeishi, M. and Ohtsu, M. 2007. Relative Damage Evaluation of Concrete in a Road Bridge by AE Rate—Process Analysis, *Materials and Structures*, 40(2): 221–227.
Suzuki, T., Ogata, H., Takada, R., Aoki, M. and Ohtsu, M. 2010. Use of Acoustic Emission and X-Ray Computed Tomography for Damage Evaluation of Freeze-Thawed Concrete, *Construction and Building Materials*, 24: 2347–2352.
Suzuki, T., Naka, T., Aoki M. and Ohtsu, M. 2010. Development of the DeCAT System for Damage Estimation of Concrete, *the 13th International Conference Structural Faults & Repair - 2010*.
Suzuki, T. and Ohtsu, M. 2011. Damage Evaluation of Core Concrete by AE, *Concrete Research Letters*, 2(3): 275–279.
Suzuki, T. and Ohtsu, M. 2014. Use of Acoustic Emission for Damage Evaluation of Concrete Structure Hit by the Great East Japan Earthquake, *Construction and Building Materials*, 67: 186–191.

*Emerging Technologies in Non-Destructive Testing VI – Aggelis et al. (Eds)*
*© 2016 Taylor & Francis Group, London, ISBN 978-1-138-02884-5*

# Ultrasonic method for predicting residual tension of ground anchors

S.B. Tamrakar, T. Shiotani & K.C. Chang
*Kyoto University, Kyoto, Japan*

Y. Fujiwara
*West Nippon Expressway Company Limited, Osaka, Japan*

ABSTRACT: Good maintenance of ground anchors installed along the slopes and the cuts of express-ways is very important to keep those slopes and cuts stable and safe. In this research, a cheap, quick and nondestructive method to evaluate residual tension of ground anchors is introduced. Ultrasonic waves reflected back from the upper and lower surfaces of the bearing plate are recorded in the waveform and the maximum amplitudes of each of those reflected waves are compared. While comparing the average maximum amplitude of each reflected wave with corresponding tensile load, a good linear increasing trend is observed in case of 1st reflected wave from the lower surface of the bearing plate (WRL1). Comparison is made between all the representative average maximum amplitudes of WRL1 of both laboratory and field experiments with corresponding tensile stress (or tension). A generalized equation is then recommended which can be used for predicting residual tension of the anchors.

## 1 INTRODUCTION

Ground anchors, hereafter "anchors" are generally used for the stabilization of slopes and cuts. In Japan, for the stabilization of such slopes and cuts of expressways, 12,000 anchors have been installed till date from 1950s (Takemoto et al. 2010). Among them, lots of anchors installed are with less protection measures against deterioration and ageing. Performance of those old anchors are decreasing and hence slope failures are occurring remarkably (Littlejohn & Mothersille 2008). Accidents due to such failures threatens the traffic safety as well as life and property of human beings. Such failure accidents against deterioration can be possibly avoided if a proper maintenance of those anchors is possible.

A good maintenance is only possible if information of external and internal conditions of the anchor is known. But no such method exists which can obtain both of those information directly and jointly. In addition, as there is no accessibility in the interior of the anchor structure, the assessment should be conducted through the area located in the surface only, i.e. the anchor head (Kawai et al. 2009). Therefore, residual tension measurement of anchor without disturbing it is the only way to know the deterioration of anchor. But till now, there is no such established rational method with which residual tension measurement is possible.

In general, visual inspection is done which only gives the information of external condition of

anchor, especially around the anchor head. So, this is not a reliable method to check the soundness of whole anchor. Another method that has been used is X-ray imaging method (Mitchell 1991). This method gives the insight of the physical condition of the anchor. But it cannot measure the tension of the anchor. This test is time consuming, expensive and requires a special care against radiation.

Magnetostriction method has been introduced where magnetic permeability is measured (Akutagawa et al. 2008). But as magnetic permeability greatly depends on the slight difference in the shape of the anchor head as well as the properties of the materials, application of this method for various types of anchors is difficult.

Another test, known as "lift-off test" directly estimates the residual tension of anchors. In this method, hydraulic jack is set up at the anchor head and it is pulled out against the bearing plate. Then, by plotting a graph between hydraulic jack pulling load and displacement of anchor head, residual tension is evaluated. This test is very expensive and requires lots of time in handling and set up of large size apparatus. So, testing of only few anchors out of many anchors in a long time interval is possible with this apparatus which is not sufficient for stability analysis of whole area. Moreover, during the test, anchor head (anchor tendon) is pulled out from its original position by few millimeter.

For the weak and deteriorated anchors, it might jeopardize the safety of the anchor itself (Littlejohn & Mothersille 2007). Use of comparatively smaller and lighter weight, SAAM (Sustainable Asset Anchor Maintenance) jack has been developed for lift-off test (Sakai et al. 2008). However, as more than 12000 anchors have been already installed, a simplification in device only for lift-off test does not suffice the demands of ready-to-apply test so that time and cost can be reduced.

A method based on ultrasonic elastic wave has been used for monitoring the stress level in the stranded wires (Chaki & Bourse 2009). But this technique does not produce reliable results for anchors installed deep into the ground where the access is limited to one single side. Therefore, it is necessary to develop a method which is fast and easy and can do safe inspection without damaging the original structural condition of the anchors. In addition, it would be better if measurement is possible to conduct on or from the visible portion of the anchor, i.e., anchor head.

Recently, possible application of ultrasonic wave technique for estimating residual tension of anchor has been thought of. Laboratory tests with ultrasonic wave transmission method have been carried out with nut-fix type and wedge-fix type anchors for the measurement of tensile loads (Kleitsa et al. 2010 & Kawai et al. 2010). Accordingly, transmission method is applicable to nut-fix type anchors only. But as there is a large disparity in transmitted energy during loading and unloading stages, its field application is yet to be considered. This method cannot be applied to wedge fix type anchors as it requires to set up the sensor at specific position which is very difficult.

Researches on ultrasonic wave reflection method have been carried with different types of anchors in the laboratory (Iwamoto et al. 2012). Cumulative as well as individual energy of reflected waves from the upper surface of bearing plate are then compared with the tensile load in loading and unloading steps. Accordingly, no hysteresis is observed between the unloading and second loading steps. This indicates the possible application of this method in the field anchors. But no relationship between the energy of reflected wave and the tensile load has been shown. Also, reflection wave from the lower surface of the bearing plate has not been considered and compared.

The main aim of this research is to provide a generalized correlation between residual tension and reflected waves for wedge-fix type anchors using ultrasonic wave reflection method. For this, different laboratory and field tests are carried out. Properties of reflected waves from the upper and the lower surfaces of bearing plate are then compared to determine the most appropriate reflection wave for the analysis.

Finally, a simple correlation between residual tensile load and the determined reflection wave is obtained so that it can be used to predict the residual tension of wedge-fix type anchors of the field.

## 2 GROUNT ANCHORS AND ITS TYPES

General layout of the anchor is shown in Figure 1. In Figure 2a, outline of the top portion of the anchor is shown. Top portion of the anchor can be divided into anchor head, bearing plate, concrete plate and anchor tendon.

Anchor tendon portion can be either a single steel rod or a strand of steel wires. Depending on the type of tendon, the shape of anchor head varies.

For a steel rod tendon, hexagonal shaped anchor head is used (see Fig. 3a). This type of anchor is known as nut-fix type anchor. Most of the anchors installed in the past are of this type and they were installed without proper remediation against water permeation. So, they undergo corrosion-induced deterioration.

Another type of anchor is known as wedge-fix type anchor (see Fig. 3b). It consists of number of steel stranded wires which are fixed to the round shaped anchor head by wedges. Most of the recent anchors are of this type and they are strong against corrosion. Depending upon the size and number of stranded steel wires present, this type of anchor is marked as E5-3, E5-4, E5-12, etc. Here, in this research, wedge-fix type anchors are only considered.

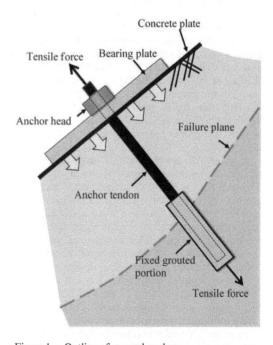

Figure 1. Outline of ground anchor.

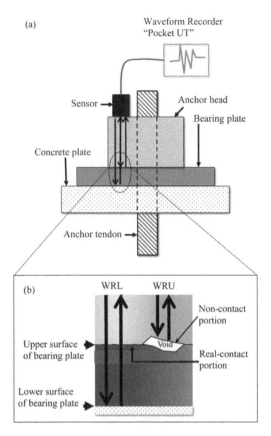

Figure 2. (a) Outline external portion of anchor and (b) Close up view of contact boundary surfaces.

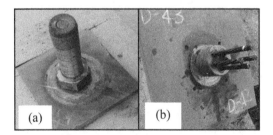

Figure 3. (a) Nut-fix type and (b) Wedge-fix type anchor.

## 3 ULTRASONIC WAVE TECHNIQUE

### 3.1 Ultrasonic wave technique

Ultrasonic wave technique is a non-destructive technique which is generally used to measure the internal defect of the materials. In this technique, at first ultrasonic wave is emitted on the surface of the media and then the properties of either transmitted or reflected ultrasonic waves are studied.

In this research, ultrasonic wave reflection method is explained and applied.

### 3.2 Ultrasonic wave reflection method

In this method, at first an ultrasonic wave is emitted at the top (upper) surface of the anchor head via a transducer (sensor) placed on the surface of the anchor head in such a way that the emitted incident wave passes downward towards the bearing plate and concrete plate along the direction parallel to the axis of the anchor (see Fig. 2b). Incident waves on their way downward, when comes in contact with boundary surfaces are either reflected or transmitted. There are two boundary contact surfaces considered in this research; the upper and the lower surfaces of the bearing plate. Let the waves reflecting back from the boundary contact surface between the anchor head and the bearing plate be WRU and those reflecting back from the boundary contact surface between the bearing plate and the concrete plate be WRL. Depending upon the path each reflected wave travels, they are indicated as WRU1, WRU2, WRU3, WRL1, WRL2, WRL3, etc. Initial incident wave and corresponding reflected waves from different boundary surfaces are measured in the waveform as shown in Figure 4.

Ratios of reflection and transmission waves with respect to incident wave at the contacting boundary surfaces depend on the acoustic impedances of those contacting surfaces (materials) and tightness between them. The larger the axial load, the higher the tightness between contacting surfaces. Therefore, depending upon the axial load, the property (amplitude or energy) of the reflected wave changes. This can be explained by tribology (Matsuoka 2003). Contacting surfaces between two solid materials can be divided into two portions; (i) real-contact portion if solids touch each other directly and (ii) non-contact (void) portion if solids do not touch each other directly.

Real-contact portion increases with the increase in axial tension and vice versa. Schematic diagram of contact portions for the boundary contact surfaces between the lower surface of anchor head and the upper surface of the bearing plate is shown in Figure 2b. In real-contact portion, two contacting surfaces have similar acoustic impedances and they are under higher tension. Here, all the incident waves are transmitted downward. In contrary, in non-contact portion as there exist empty or void portion between two contacting surfaces, all the incident waves are reflected back because the acoustic impedance of air (void) is very small. In general, both real-contact and non-contact portions exist between two contacting surfaces. But their ratio depends on the axial tension of the anchor.

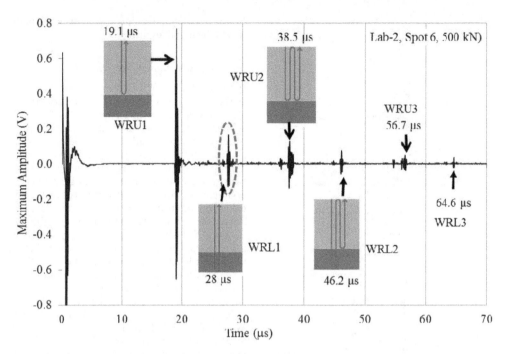

Figure 4. Waveform showing different reflected waves (Lab-3, 500 kN).

## 4 ULTRASONIC WAVE TECHNIQUE

### 4.1 *Ultrasonic wave technique*

In this research, Pocket UT system, here after "Pocket UT" is used for ultrasound wave generation, data acquisition and analysis. This standalone data acquisition system is portable and ideal for on-site inspection. It includes an internal ultrasonic pulser (or receiver) which generates (or receives) ultrasonic waves. This system is generally used for assessing the presence of cracks, flaws, corrosion/erosion, delamination and internal anomalies in variety of structures, materials and surfaces. Here, in this research, it is used to generate and receive the waves in the waveform so that the analysis of required reflected wave can be done. Along with Pocket UT, transducer (5 MHz), connecting cables and couplant are also used.

### 4.2 *Outline of the experiment*

In this research, only wedge-fix type anchors are considered. Both laboratory and field experiments conducted are shown in Tables 1–2. Laboratory experiments are divided into three groups; (i) Lab-1, (ii) Lab-2 and (iii) Lab-3, depending upon the type of anchor head used. Lab-2 group consists of anchor heads having two different dimensions. Lab-3 is a special type of anchor where cylindrical

Table 1. Anchor type and measurement points.

| Name | Anchor type (Wedge-fix) | Number of measuring points per anchor load |
|---|---|---|
| Lab-1 | E5-4 | 16 |
| Lab-2 | E5-7 | 12 |
| Lab-3 | Cylindrical steel | 8 |
| F-Kumamoto | E5-3 | 6 |
| F-Yamaguchi | E5-3 | 12 |
| F-Fukuchi | E5-4 | 16 |
| F-Shinihama | E5-5 | 10 |
| F-Ootoyo | E5-7 | 18 |

Table 2. Total numbers, loading range and contact area of anchors.

| Name | Load range (kN) (Total numbers) | Calculated contact area (cm$^2$) |
|---|---|---|
| Lab-1 | 0~200 (total 5) | 32.4 |
| Lab-2 | 20~500 (total 46) | 94.94 & 56.52 |
| Lab-3 | 0~500 (total 11) | 17.5 |
| F-Kumamoto | 0~250 (total 6) | 25 |
| F-Yamaguchi | 0~231 (total 11) | 25 |
| F-Fukuchi | 0~277 (total 11) | 27.6 |
| F-Shinihama | 0~293 (total 14) | 32.4 |
| F-Ootoyo | 348~557 (total 5) | 46.37 |

steel anchor head is used. Tension in the laboratory anchors is created by providing compression load on the anchor tendon. Loading range and total number of loading steps for each laboratory experiment is shown in Table 2. Loading and unloading processes are followed in laboratory tests so that effect of hysteresis on the measured values can be checked. In case of field experiments, five field experiments are conducted; F-Kumamoto, F-Yamamoto, F-Fukuchi, F-Shinihama, and F-Ootoyo. Types of anchor head, total number of anchors tested, and the residual tensile load range of each field experiment are shown in Tables 1–2. In case of field experiments, residual tensile load of each anchor is measured in advance by lift-off tests.

### 4.3  *Set up and measurement*

Anchor head upper surface is used for setting up of the transducer. But before setting up the transducer, cleanliness of the anchor head surface must be confirmed. Laboratory experiments are generally performed with new anchor heads. So, the surfaces of anchor heads are comparatively clean. But field experiments consist of old anchor head with caps. In the case of field experiments, anchor head caps must be removed at first and any grease or oil present on the anchor head surface must be wiped off. In addition, if rust is present, it should be removed by using either sand paper or grinder. Once the anchor head upper surface is cleaned, then mark the measuring points on the surface of anchor head. At least marking of six measurement points are recommended for one anchor.

During marking, uneven parts are avoided and markings are made near to the outer circumference of the anchor head. Marking at the outer circumference is suggested as contact ratio over this place is good. These are done in order to increase the accuracy in the measurement. After marking, transducer is connected with the Pocket UT. Apply some couplants on contacting surface of the sensor. Couplants make smooth contact between the transducer and the anchor head. Place the sensor on the marked positions.

An excitation frequency of 5 MHz is generated via Pocket UT on the sensor. Pressure on the transducer is applied by hand and the waveform obtained is checked. Beyond certain pressure, there is no change in the maximum amplitude. Once the waveform showing the maximum amplitude is confirmed, then that waveform is saved. This saved waveform data is later used for further analysis. As mentioned earlier, one anchor head must have at least six measurement points and hence six measurement data. Therefore, in order to obtain a representative maximum amplitude of one anchor

head, averaging of maximum amplitudes of each measurement point must be done. This representative maximum amplitude is then used for further comparison.

In order to change the axial tensile load of each anchor into the axial tensile stress, one must know the contact area between the anchor head and the bearing plate. For this, diameters of anchor head and bearing plate hole must be measured. Measurement of these diameters are easy in case of laboratory experiments as anchor head and bearing plate are free to move. But it is difficult to measure the diameter of bearing plate hole in case of field experiment. For this, ultrasonic wave reflection method technique can be used. After calculating the areas of both anchor head and bearing plate hole, then contact area between them can be obtained by deducting the area of bearing plate hole from the anchor head area. Calculated contact area for each experiment is shown in Table 2.

## 5  RESULTS AND DISCUSSIONS

Waveform obtained for 500 kN load in one of the measurement points of Lab-3 experiment is shown in Figure 4. It shows the relationship between the arrival time and the maximum amplitude for each reflected wave. Knowing the arrival time of each reflected wave from the waveform and taking the wave propagation velocity for anchor head and bearing plate as 5800 m/s, travel path for corresponding waves can be determined and accordingly, those reflected waves can be marked as WRU1, WRU2, WRU3, WRL1, WRL2 and WRL3. Then for each such wave, maximum amplitude can be measured from the waveform. For one anchor, there will be at least six such maximum amplitude values. Hence averaging of all those measured maximum amplitudes for one anchor should be done so that one representative average maximum amplitude with corresponding tensile load of that anchor can be obtained.

With the increasing loads, magnitudes of reflected waves from upper surface of the bearing plate showed decreasing tendency. With the increase on the load, real contact portion is increased. This causes more waves to transmit from the anchor head towards bearing plate and hence, less reflection. In relation to the waveform mentioned in Figure 4, relationship between the average maximum amplitudes and the corresponding tensile loads is shown in Figures 5–6.

Figure 5 represents the reflected waves from the upper surface of the bearing plate. Here, with the increase in applied tensile loads, there is decreasing trend in average maximum amplitude. WRU1 shows almost same value in all the loads.

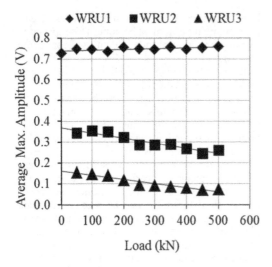

Figure 5. Comparison of WRU1, WRU2 and WRU3.

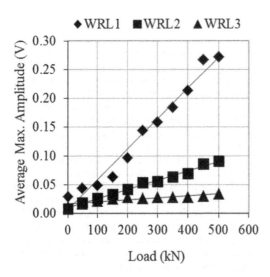

Figure 6. Comparison of WRL1, WRL2 and WRL3.

At higher loads, WRU2 and WRU3 show very less difference in their average maximum amplitudes. Figure 6 represents the reflected waves from the lower surface of the bearing plate. It shows increasing trend in average maximum amplitudes with increasing tensile loads. Comparing the average maximum amplitude values of all the reflected waves with corresponding tensile loads, it can be observed that the 1st reflected wave from the lower surface of the bearing plate (WRL1) shows the best linear relationship showing wider range in amplitude values and tensile loads. Hence, in all the experiments, the average maximum amplitude value of WRL1 with corresponding tensile load is considered as a representative parameter for further analysis.

From the external diameter of anchor head and area of bearing plate hole, the contact area between the anchor head and the bearing plate can be obtained. See Table 2 for the calculated contact areas for each experiment. Axial tensile load of each anchor is then divided by corresponding contact area to obtain axial tensile stress of that anchor. Comparison between this axial tensile stress and average maximum amplitude of WRL1 for all the laboratory and field experiments performed is shown in Figure 7. As shown in the figure, a generalized relationship between the average maximum amplitude and the axial tensile stress can be obtained. This relationship can be used for the possible prediction of residual tension of deteriorated anchors (wedge-fix type) in the field in future.

Comparison between this axial tensile stress and average maximum amplitude of WRL1 for all the laboratory and field experiments performed is shown in Figure 7. As shown in Figure 7, a generalized relationship between the average maximum amplitude and the axial stress can be obtained.

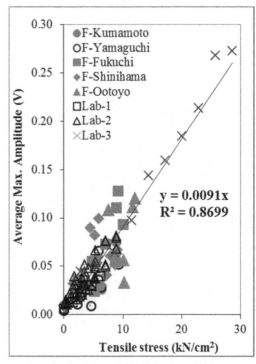

Figure 7. Comparison of average maximum amplitude and tensile stress.

# 6 CONCLUSIONS

This research is focused on wedge-fix type anchors using ultrasonic wave reflection method. By comparing the results obtained from the laboratory and the field experiments, following conclusions can be made.

a. Magnitudes of the reflected waves from upper surface of bearing plate decrease with increasing loads. While the magnitudes of the reflected waves from lower surface of the bearing plate increases with increasing loads.
b. Average maximum amplitude of the 1st reflected wave from the lower surface of the bearing plate (WRL1) is recommended to consider as a representative value of the anchor for the analysis as it gives good linear relationship between the average maximum amplitude and the tensile load.
c. Generalized relationship between the average maximum amplitudes of WRL1 and the axial tensile stress can be obtained. This relationship can be used to predict the residual tension of anchors.

## ACKNOWLEDGEMENT

This is a collaborative research funded by West Nippon Expressway Company Limited (W-NEXCO), Japan. Authors kindly acknowledge their assistance. Authors are thankful to Kyoto University graduates, Mr. T. Nakai & Mr. K. Iwamoto for their contribution in conducting experiments.

## REFERENCES

Akutagawa, S., Arimura, Y., Nakori, E., Kasurai, S., Baba, S. & Mori, S. 2008. New evaluation method of PS-anchor force by using magnetic sensor and its application in a large underground powerhouse cavern. *Journal of construction management and engineering* 64(4): 413–430.

Chaki, S. & Bourse, G. 2009. Guided ultrasonic waves for no-destructive monitoring of the stress levels in pre-stressed steel stands. *Ultrasonics 2009* 49: 162–171.

Iwamoto, K., Shiotani, T., Ohtsu, H., Maeda, Y., Hamasaki, S. & Yanagisako, S. 2012. Axial stress evaluation of ground anchors with reflection waves of ultrasonic. *Structural Fault & Repair 2012, 14th Int'l Conference and Exhibition, Edinburgh, UK, 3–5 July 2012*.

Kawai, K., Shiotani, T., Ohtsu, H., Yanase, T. & Tanaka H. 2009. Estimation of axial stress in ground anchors by means to indirect ultrasonic technique. *Proceedings of EIT-JSCE, Geotechnical infrastructure management*: CD–9.

Kawai, K., Shiotani, T., Ohtsu, H., Tanaka, H. & Kawagoe, H. 2010. Estimation of axial stress in ground anchors by means of indirect ultrasonic technique. *Engineering Technics Press, Structural Fault & Repair 2010, Exhibition showguide*: 158.

Kleitsa, D., Kawai, K., Shiotani, T. & Aggelis, D. 2010. Assessment of metal strand write pre-stress in anchor head by ultrasonic. *NDT&E International* 43: 547–554.

Littlejohn, S. & Mothersille, D. 2007. Maintenance testing and service behavior monitoring of permanent ground anchorages. *Institution of civil engineers, International Conference on Ground Anchorages in Service*.

Littlejohn, S. & Mothersille, D. 2008. Maintenance and monitoring of anchorages: guideline. *Proceedings of the Institute of Civil Engineers, Geotechnical Engineering* 56(12): 42–45.

Matsuoka, H. 2003. Physics of friction (in Japanese). *Surface science* 24(6): 328–333.

Mitchell, T.M. 1991. Radioactive/nuclear methods. In V.M. Malhotra & N.J. Carino (eds), *CRC Handbook on Non Destructive Testing of Concrete*: 227–252. Florida: CRC Press.

Sakai, T., Tsunekawa, Y., Fukuda & Nagano, M. 2008. A way of pulling test for ground anchor using SAAM jack (In Japanese). *Monthly Magazine of Japanese Geotechnical Society* 56(12) 611: 42–45.

Takemoto, M., Fujiwara, Y. & Yokota, S. 2010. Ground anchorage integrity evaluation of the lift-off test (In Japanese): *Annual Conference of Japanese Geotechnical Society* 45(2) 940: 1897–1880.

# Use of Digital Image Correlation (DIC) and Acoustic Emission (AE) to characterise the structural behaviour of hybrid composite-concrete beams

S. Verbruggen, S. De Sutter, S.N. Iliopoulos, D.G. Aggelis & T. Tysmans
*Department of Mechanics of Materials and Constructions, Vrije Universiteit Brussel, Brussels, Belgium*

ABSTRACT: The use of composites such as TRCs (Textile Reinforced Cements) and FRPs (Fibre Reinforced Polymers) enable the development of lightweight civil structures. A lightweight solution for floor renovation is a hybrid composite-concrete structure: prefabricated beams (TRC-CFRP reinforced hollow boxes with concrete on top) which support sandwich panels and a finishing concrete layer to create a monolithic hybrid floor. As the hybrid beams are the main structural element of this floor system, their loadbearing and failure behaviour should be fully understood. In order to examine the optimal design of the structures in terms of load bearing capacity, the beams are tested in four point bending while the amount of CFRP reinforcement and concrete thickness are varied. The Digital Image Correlation (DIC) and Acoustic Emission (AE) measuring techniques are applied in a complimentary fashion to monitor the bending and failure behaviour of the full scale hybrid beams. Development of surface strain fields and monitoring of the exact cracking patterns in relation to the applied load are performed by DIC. Furthermore, AE contributes in defining the load at the onset of serious cracking activity and characterizing of the contribution of the different fracture modes that may vary from cracking in the concrete layer, debonding in the interphase between concrete and the TRC hollow box or delamination between the successive layers of TRC and CFRP.

## 1 INTRODUCTION

High performant composite materials have enabled the use of composite materials in structural application. Their high strength-to-weight ratio and good corrosion properties offer possibilities to eliminate the drawbacks of current Reinforced Concrete (RC) elements. Different authors have investigated mixed composite-concrete designs to replace the standard RC cross-sections (El-Hatcha et al. 2012, De Sutter et al. 2014).

Most hybrid designs combine the compressive strength of concrete with the stiffness of Fibre Reinforced Polymers (FRPs). These FRPs however lack the necessary fire resistance for many structural applications. A solution to this problem is offered by fire safe and low-cost Textile Reinforced Cements (TRC). Researchers at the Vrije Universiteit Brussel (VUB) developed an Inorganic Phosphate Cement (IPC) (Wu et al. 1997), which can be combined with dens glass fibre textiles, resulting in a high performant TRC (GFR.IPC) with fibre volume fractions of up to 25%. TRC's and concrete are combined in a new hybrid design.

### 1.1 Hybrid design

The authors propose a composite-concrete design composed of a hollow GFR.IPC box with a Carbon FRP (CFRP) strip (TRADECC 2007) inside and concrete on top (Figure 1). The experimentally determined material properties are given in Table 1. Two dimensional parameters ($h_{co}$: height of concrete; $h_{ca}$: thickness of CFRP) are varied within the proposed concept to test the feasibility of the hybrid beams (Table 2). Of every type two beams are tested

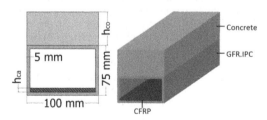

Figure 1. Geometry and dimensions of the proposed hybrid beam design. Two parameters are varied in the experiments: the concrete height and the thickness of the CFRP reinforcement.

Table 1. Material properties. Values determined according to *(CEN 2000), **(Cuypers 2002) and ***(CEN 2004).

| Material | Tensile strength [MPa] | Compressive strength [MPa] | Young modulus [GPa] |
|---|---|---|---|
| Concrete | n/a | 52.4* | 36.2*** |
| GFR.IPC | 50.5 | 80** | 10.1–3.8 |
| CFRP | 2375 | n/a | 137 |

Table 2. Dimensions of the tested beams.

| Beam's name | 'NC-1/2' | 'C-1/2' |
|---|---|---|
| $h_{co}$ | 25 | 50 |
| $h_{ca}$ | 0 | 3 |

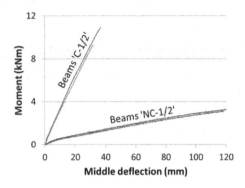

Figure 2. The non-carbon reinforced beams 'NC' clearly exhibit a less stiff behaviour.

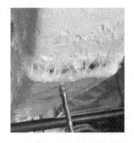 

Figure 3. Tensile failure of beam 'NC-1' (left) and the ILS-failure of beam 'C-1' (right).

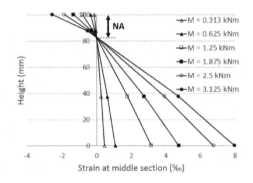

Figure 4. The strain profile of beam 'NC-1' at the midspan for increasing load steps.

under four-point bending (500 mm between the point loads). The first beam type contains no carbon reinforcement (NC-1/2), while the other beam type is reinforced with 3 mm thick CFRP strips (C-1/2).

To fully understand and characterise the structural behaviour of the hybrid beams under the applied loads, Non-Destructive Techniques (NDT) such as Linear Variable Differential Transformers (LVDTs), strain gauges, Digital Image Correlation (DIC) and Acoustic Emission (AE) are applied. In this paper, the authors analyse the possibilities of these techniques to accurately measure the structural response of the beams. Displacements, strains and failure modes are determined and combined from the different measuring techniques to understand the overall mechanical behaviour.

## 2 MECHANICAL BEHAVIOUR

Moment-deflection curves (Figure 2) are plotted based on load measurements from a load cell and displacement measurements by a LVDT placed in the middle of the beam's span. These curves indicate the significant difference in stiffness between beams 'NC-1/2' and 'C-1/2'.

The absence of carbon reinforcement for beams 'NC-1/2' leads to too high stresses at the bottom of the GFR.IPC box, causing failure in tension due to fibre pull-out for both beams (Figure 3 left). The presence of sufficient CFRP reinforcement, which is the case for beams 'C-1/2', results in a theoretical concrete crushing failure. Experimentally, both carbon reinforced beams exhibit Interlaminar Shear Failure (ILS) at the bottom of the box. This may have limited a further monolithical action between the CFRP strip on the one hand and the concrete-box on the other hand, resulting in a limited further load increase (Figure 3 right).

Four strain gauges, located at mid-span and along the height of the cross-section (top, middle concrete, middle GFR.IPC box, bottom) allow us to draw the strain profile at different load steps. All strain profiles exhibit a linear variation. Despite the tensile failure of beams the non-carbon reinforced beams, the tensile strain only reaches 8.0 ‰ for NC-1 (Figure 4), which is lower than the failure strain obtained with material tests (11 ‰). The ultimate concrete strain in the reinforced beam types also does not approach the theoretical failure

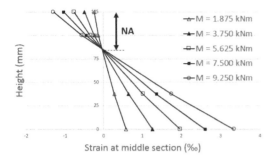

Figure 5. The strain profile of beam 'C-1' at the midspan for increasing load steps.

Figure 6. Two 3D DIC systems are used to monitor the behaviour of one half of the beam's span.

strain of 3.5 ‰ for beam C-1 (Figure 5) due to the premature ILS failure.

Both strain profiles indicate that the Neutral Axis (NA) is situated close to the interface between concrete and GFR.IPC (NC-1 '25 mm concrete': NA at 17 mm; C-1 '50 mm concrete: NA at 40 mm).

## 3 DIGITAL IMAGE CORRELATION

Digital Image Correlation (DIC) allows the determination of the displacement—and deformation field of a specimen's surface under any type of loading condition by optical, non-contacting measurements. Out of grey scale pictures taken of a speckle pattern (pattern of black spots on a white background, applied on the specimen's surface) during testing by two charged coupled device cameras, a 3D image can be created. The measurements are based on the comparison of a reference image, generally in the unloaded stage, with the images taken at the different load steps. (Sutton et al. 2009).

The DIC technique is a valuable and promising tool to identify and monitor the overall behaviour and failure mechanisms of (hybrid) concrete beams. Therefore do many recent studies already apply a similar image processing technique to monitor concrete structures in general (Ferrier et al. 2003, Avril et al. 2004, Contamine et al. 2013, Verbruggen et al. 2014).

In this specific test set-up one pair of DIC cameras follows an approximately 1 m wide area on the side surface of the beam from the support onwards (DIC 1 in Figure 6). Another pair of cameras is focused on an approximately 0.4 m wide area starting from the middle of the beam (DIC 2 in Figure 6).

### 3.1 Longitudinal strains

Detailed information on the overall loadbearing behaviour of the hybrid beams can be obtained by studying the evolution of the longitudinal strain field. Plotting this strain along the beams' length at consecutive load steps up to the maximum load

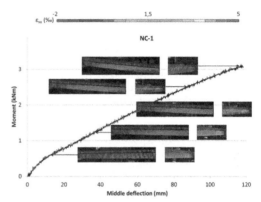

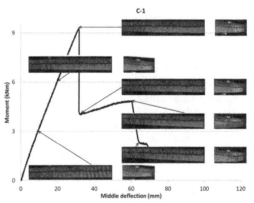

Figure 7. The longitudinal strains gradually increase up to the maximum load. During consecutive load drops, a varying amount of strain is released.

for beams 'NC-1' and 'C-1' (Figure 7), indicates a gradual increase in tensile strain in the GFR.IPC box and compressive strain in the concrete. This smooth and gradual increase corresponds to the continuous evolution of the experimental moment-deflection curves up to their maximum load.

For beam C-1, the drops in the experimental moment-deflection curve correspond to sudden reductions in these strain fields (Figure 7 after reaching the maximum load). However, for the first load drop (from 9.3 kNm to 4.1 kNm) this strain reduction remains limited (absolute difference in the order of magnitude 0.01%) on the visual surface of the monitored beam half. The fall back load (4.1 kNm) and the maximum load before the second load drop (4.8 kNm) are comparable to the maximum load of beams NC-1/2 (3.2 kNm), without internal CFRP reinforcement. Considering this observation and the fact that the CFRP reinforcement is the only non-visible structural component for the DIC systems, the first load drop must be caused by a loss of composite action between the GFR.IPC box and the CFRP reinforcement. This confirms the observations discussed in "2 Mechanical behaviour", where interlaminar shear debonding is observed within the internal layers of the GFR.IPC box, preventing the further contribution of the CFRP to the overall structural behaviour. The slightly higher fall back load of beam C-1 compared to the failure load of beams NC-1/2 is due to the only partial loss of composite action, which will not occur along the full length of the beam.

The second drop (from 4.8 kNm to 2.6 kNm) on the other hand corresponds to the tensile failure of the GFR.IPC box in the centre of the beam's span, similar to the failure of beams NC-1/2. This results in a strain reduction within the GFR.IPC box (absolute difference in the order of magnitude 1%) and an increase in tensile strain in the concrete volume and thus a sudden increase the concrete crack opening.

3.2 *Horizontal displacements*

The evolution and especially the continuity of the horizontal displacement along the beam's length or height yields valuable information on the cracking and failure behaviour of the hybrid beams. To study this continuity, horizontal and vertical lines, each consisting of 2000 data points, are extracted from the DIC results.

A visualisation of the crack development in the concrete compression layer of beam C-1 can be obtained by plotting the horizontal displacement (at a height of 15 mm above the GFR.IPC—concrete interface) versus the longitudinal location in the beam for different load steps (Figure 8). Considering the higher bending moment in the central beam area, only the results of DIC2 are represented in Figure 8. The vertical jumps, indicated by numbers 1 to 8, correspond to a sudden increase in displacement and thus a crack in the concrete. In this paper a crack is defined as a difference in horizontal displacement larger than 0.02 mm over a horizontal interval of 5 mm, which is not adjacent to another crack interval.

The horizontal displacement curves clearly indicate a continuous evolution until the second load drop. This indicates the absence of concrete cracks up to beam failure, which is a direct consequence of the position of the neutral axis around the GFR.

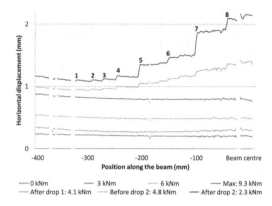

Figure 8. No concrete cracks are present up to the maximum load. However, 8 cracks develop between the first and second load drop.

IPC box—concrete interface (and thus the absence of severe tensile stresses in the concrete volume), as discussed in "2 Mechanical behaviour". As bending cracks often introduce debonding between different material layers (CEB 2006), the absence of these cracks prevents a brittle and undesired premature beam failure. However, 8 cracks start developing between the first and second load drop and their width seriously increases after the tensile failure of the GFR.IPC box, especially for crack number 7, which is located right above the failure location of the GFR.IPC box.

Plotting the horizontal displacement over a vertical line allows the verification of the composite action between the concrete top layer and the GFR.IPC box (Figure 9). When the composite action and thus the connection between the GFR.IPC box and the concrete is lost, both elements can deform separately and will exhibit a different (horizontal) displacement pattern. The verification of this connection is an important tool because the analytical predictions, based on Eurocode 2 for concrete calculation (CEN 2004), assume a perfect bond and the validity of the Bernouilli criterion.

Plotting the horizontal displacement over the beam's height at six different locations along the beam's length (Figure 9) for an unloaded specimen, half of the maximum load and the maximum load, results for both beam types in a linear and continuous evolution for all load steps. This indicates the retention of the connection between the GFR.IPC box and the concrete top layer up to beam failure and thus the correctness of a perfect bond assumption within the Bernouilli criterion.

Additionally these horizontal displacements show the decreasing cross section rotation towards the beam's centre. An increasing beam curvature is indicated by the increasing ratio of the difference

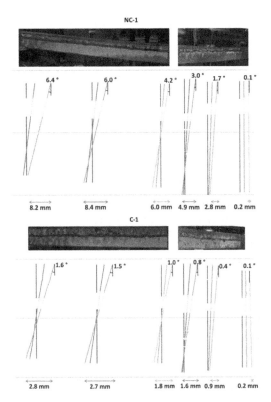

Figure 10. Four AE sensors mounted on the concrete layer of the composite beams by means of magnetic clamps.

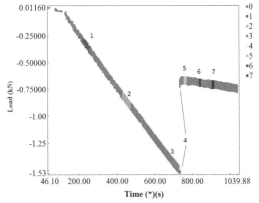

Figure 9. The continuous evolution of the horizontal displacement over the beam's height proves the assumption of a perfect bond between the concrete and the GFR.IPC box and the Bernouilli criterion.

Figure 11. Indicative selection of AE classes during loading of a CFRP reinforced beam. Class 4 is exactly at the load drop.

in cross section rotation over the distance between the extracted lines.

## 4 ACOUSTIC EMISSION RESULTS

Monitoring by AE sensors allows recording of all transient waves emitted by the damage propagation events. On top of that, the different mechanisms induce different crack tip motion resulting in quite different AE characteristics. This enables passive monitoring and characterization of the dominant fracture mode (Ohtsu 2010). In the present case, four resonant sensors (R15 Mistras Holdings) were applied along the length of the beams on the concrete layer, covering the central 500 mm, as shown in Fig. 10. The signals were pre-amplified by 40 dB and the threshold was defined at 35 dB. Some of the major features of the received waveforms are (among others) the "average frequency" AF, in kHz which is the number of threshold crossings over the duration and the "rise time" RT, which is the time between the onset of the waveform and its maximum peak.

The total AE activity accumulated to the order of more than 50000 hits for each beam. In this case the trends of AE parameters are of particular interest. As already mentioned above the reinforcement by CFRP played an important role in the fracture behavior; the two reference beams fractured by tensile cracks at the bottom of the IPC box while the CFRP reinforced beams induced additional delaminations in between the layers of the IPC apart from the tensile stress cracking. In order to check the transition of AE indices, the total population of each experiment was preliminary divided in 7 classes. Class #1 corresponded to the AE recorded during low load at the start of the experiment, while #7 was at the end, after the major damage has occurred. An indicative separation of classes for one beam is shown in Fig. 11.

The AE activity recorded at the different windows is plotted in terms of AF vs. RT axes and shown in Fig. 12.

It is obvious that the activity of the reference beam (NC-1) does not exhibit strong shifts throughout the experiment. The average frequency of the emissions stays approximately constant at 130 kHz

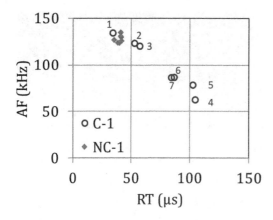

Figure 12. Average values of AF vs. RT for CFRP reinforced (C-1) and reference (NC-1) composite beams, for different loading stages (numbers 1–7).

and the rise time of the signals at 40 μs from the initial stages of loading until the end, denoting a similar way of fracturing, or else that the dominant fracture mechanism (in this case cracking at the bottom of the IPC box) does not seriously change. On the other hand it is interesting to note the transition of the CFRP reinforced beam (C-1). While at the initial stages the activity starts at similar values with the reference beam (see point 1 in Fig. 12), there is a gradual shift to higher RT and lower AF resulting in a large leap for stage 4, when the average RT is more than 100 μs and the AF drops to 62 kHz. This shift denotes the debonding shear action that has its peak during the macroscopic load drop moments (Aggelis 2011) and has been reported in other cases of debonding of IPC layers (Blom et al. 2014). For the subsequent stages the values remain much higher than those of the initial stage. The above results indicate that in controlled laboratory conditions it is possible to realize the fracturing trends by AE monitoring. However, in any case, care should be taken to the geometry of the specimens and the location of sensors relatively to the zone sustaining most of the damage. In case the experimental layout changes, although the trends may still exist, the values will not be directly comparable. In this case in all four beams, geometry and test setup was identical enabling the comparison between the different types of beams. It is mentioned that the other beams of the same types exhibited exactly the same trend, but are not presented or discussed due to size limitations.

## 5 CONCLUSION

Mechanical tests prove the feasibility of the proposed composite-concrete beams. A tensile composite failure was observed for the non-carbon reinforced beams, while ILS caused the premature failure of the carbon reinforced beams.

Using DIC as an NDT monitoring technique yields additional information on the structural behaviour of the beams. Strain releases during load drops confirmed the ILS failure of the carbon reinforced beams. Horizontal displacements, and especially their discontinuities, indicate the absence of concrete cracks during the load increase and proved the monolithical action of the cross-section and thus the validity of the Bernouilli criterion.

The complimentary use of AE enables to check the transition between fracture mechanisms. Specifically, for the CFRP reinforced beams that induced debonding in the IPC layers of the box, a strong shift to longer AE signals and lower frequencies was recorded, while on the other hand the reference beams follow a remarkably similar AE pattern throughout the whole experiment.

## ACKNOWLEDGEMENTS

Research partially funded by the Brussels Capital Region through the Innoviris Strategic Platform Brussels Retrofit XL for the first two authors and Fonds Wetenschappelijk Onderzoek-Vlaanderen (FWO) for funding the research of the third author through a PhD scholarship.

The authors gratefully acknowledge the cooperation with the company TRADECC, trough the delivery of the epoxy glue and CFRP.

## REFERENCES

Aggelis D.G. 2011. Classification of cracking mode in concrete by acoustic emission parameters. Mechanics Research Communications; Vol. 38; pp. 153–157.

Avril S., Ferrier E., Vautrin A., Hamelin P. & Surrel Y. 2004. A full-field optical method for the experimental analysis of reinforced concrete beams repaired with composites. Compos part A-appl s; Vol. 35; pp. 873–884.

Blom J., El Kadi M., Wastiels J., Aggelis D.G. 2014. Bending fracture of textile reinforced cement laminates monitored by acoustic emission: Influence of aspect ratio. Construction and Building Materials 70 (2014) 370–378.

CEB-FIP 2006. fib bulletin 35; Retrofitting of concrete structures by externally bonded FRPs with emphasis on seismic applications. Lausanne, Switzerland; ISBN 2-88394-075-4.

CEN (Comité Européen de Normalisation). 2000. Concrete—Part 1: Specification, performance, production and con-formit. EN 206–1.

CEN (Comité Européen de Normalisation) 2004. Eurocode 2: Design of concrete structures—Part 1-1: General rules and rules for buildings. ENV 1992-1-1.

Contamine R., Si Larbi A. & Hamelin P. 2013. Identifying the contributing mechanisms of textile reinforced concrete (TRC) in the case of shear repairing damaged and reinforced concrete beams. Eng struct, Vol. 46, pp. 447–458.

Cuypers H. 2002. Analysis and Design of Sandwich Panels with Brittle Matrix Composite Faces for Building Applications. Doctoral thesis. Vrije Universiteit Brussel, Faculty of Engineering, Brussels, Belgium.

De Sutter, S. & Remy, O. & Tysmans, T. & Wastiels, J. 2014. Development and experimental validation of a lightweight stay-in-place composite formwork for concrete beams. Construction and building materials. Vol. 63, pp. 33–39. http://dx.doi.org/10.1016/j.conbuildmat.2014.03.032.

El-Hacha, R. & Chen, D. 2012. Behaviour of hybrid FRP-UHPC beams subjected to static flexural loading. Composites: Part B; Vol. 43, pp. 582–593.

Ferrier E., Avril S., Hamelin P. & Vautrin A. 2003. Mechanical behaviour of beams reinforced by externally bonded CFRP sheets. Mater struct; Vol. 36, pp. 522–529.

Ohtsu M. 2010. Recommendations of RILEM Technical Committee 212-ACD: acoustic emission and related NDE techniques for crack detection and damage evaluation in concrete: 3. Test method for classification of active cracks in concrete structures by acoustic emission. Mater Struct; 43(9), pp. 1187–9.

Sutton M.A., Orteu J.J. & Schreier H.W. 2009. Image correlation for shape, motion and deformation measurements. Basic concepts, theory and applications. Springer Science+Business Media; New York, USA; ISBN 978-0-387-78746-6.

TRADECC. 2007. PC® CARBOCOMP. http://www.frp.co.il/uploadimages/12.pdf. July 2007.

Verbruggen S., Tysmans T. & Wastiels, J. 2014. TRC or CFRP strengthening for reinforced concrete beams: An experimental study of the cracking behaviour. Eng Struct, Vol. 77, pp. 49–56.

Wu, X. & Gu, J. 1997. inventors. Inorganic Resin Compositions, Their Preparation And Use Thereof. *European patent EP 0 861 216 B1*. 29 May 1997.

*Elastic waves*

*Emerging Technologies in Non-Destructive Testing VI – Aggelis et al. (Eds)*
*© 2016 Taylor & Francis Group, London, ISBN 978-1-138-02884-5*

# An ACA/BEM for solving wave propagation problems in non-homogeneous materials

T. Gortsas, I. Diakides & D. Polyzos
*Department of Mechanical Engineering and Aeronautics, University of Patras, Patras, Greece*

S.T. Tsinopoulos
*Department of Mechanical Engineering, Technical Research Institute, Patras, Greece*

ABSTRACT: The present work reports a study on the propagation of elastic waves in unidirectional fiber reinforced composite materials. The fiber composite is modeled as a slab consisting of a large number of randomly or uniformly distributed elastic fibers embedded in an elastic matrix medium. That slab is also embedded in an infinitely extended matrix material, where a longitudinal plane wave propagates and impinges upon the considered composite material. The goal of the present work is twofold: first to solve numerically the just described large-scale scattering problem and second to compare the obtained results with those taken by the solution of the same problem with the composite material being homogenized according to an Iterative Effective Medium Approximation (IEMA) [J. Acoust. Soc. Am. 116, 3443–3452 (2004)]. To this goal an advanced Boundary Element Method (BEM) effectively combined with an Adaptive Cross Approximation (ACA) algorithm is employed. For very high number of inclusions, the proposed ACA/BEM technique solves the considered large-scale scattering problem with memory requirements of almost linear complexity.

## 1 INTRODUCTION

The propagation of a plane wave in particulate and fiber composites is always characterized by dispersion and attenuation due to its multiple scattering by the embedded in-homogeneities. Thus, even in the case where the constituents of the composite are non-dispersive and non-attenuative materials, any elastic wave propagating in the main body of the composite undergoes both dispersion and attenuation. A methodology of estimating the dispersive and attenuative properties of a composite elastic medium is that of Iterative Effective Medium Approximation (IEMA) proposed by Tsinopoulos et al. (2000), Verbis et al. (2001) and Aggelis et al. (2004). The IEMA makes use of the single inclusion self-consistent condition of Kim et al. (1995) and assumes that the effective stiffness of the composite is the same with the corresponding static one and evaluates iteratively the effective and frequency dependent dynamic density of the composite. The complex value of the effective dynamic density and the static effective stiffness of the composite determine, eventually, the wave speed and the attenuation coefficient of the plane wave propagating through the composite material.

The goal of the present work is to check numerically the validity of IEMA by comparing the results of two main problems. The first problem considers a slab of a large number of randomly distributed elastic fibers embedded in an infinitely extended elastic matrix medium where a longitudinal plane wave propagates and impinges perpendicularly upon the fibers. In the second problem a homogeneous material with frequency dependent properties evaluated by the IEMA replaces the slab of fibers. In both problems the elastic energy scattered in the forward direction is evaluated by the Boundary Element Method (BEM) and the obtained results are compared to each other.

The Boundary Element Method (BEM) is a very well known and robust numerical tool successfully used for the solution of wave scattering problems. Two remarkable advantages it offers as compared to other numerical methods is the reduction of the dimensionality of the problem by one and its high solution accuracy. Despite the advantages the brutal application of BEM to large-scale problems, like that of scattering by a large number of fibers, suffers from very time consuming computations and high demands for computer memory capacity. Both problems come from the generation of the non-symmetric coefficient matrix $[\mathbf{A}]$ and the solution of the final system of algebraic equation $[\mathbf{A}] \cdot \{\mathbf{x}\} = \{\mathbf{b}\}$. More precisely, the fully populated matrices produced by BEM require $O(N^2)$

operations for its buildup and O($N^3$) operations for the solution of the final matrix system through Gaussian elimination or typical LU-decomposition solvers. The use of iterative solvers decreases the operation requirements from O($N^3$) to O(MxN$^2$), with M being the number of iterations, but still remains inefficient for large scale problems. To the same conclusion we reach when parallel computing methods are exploited for the solution of the problem.

An alternative approach to accelerate BEM is the recently proposed Adaptive Cross Approximation Algorithm (ACA) along with hierarchical matrices (Bebendorf and Rjasanow (2003), Borm et al. (2003), Brunner et al. (2010)). The acceleration here, is achieved because only a small number of elements of the collocation matrix [**A**] are calculated, while the rest ones are approximated via the already calculated values. In the present work an advanced ACA/BEM efficiently combined with iterative solvers, proposed by Gortsas et al. (2015)), is employed for the solution of the aforementioned scattering problems. That ACA/BEM technique is explained in brief in the next section while the IEMA is illustrated in the section after next. Finally, the obtained results are reported and discussed in the forth section.

## 2 SOLUTION OF SCATTERING PROBLEM VIA ACA/BEM

Consider $q$ identical, circular elastic fibers of radius $a$, density $\rho_f$, Young modulus $E_f$ and Poisson ratio $v_f$ embedded in an infinitely extended elastic matrix with material properties $\rho_m$, $E_m$ and $v_m$ respectively. A harmonic longitudinal plane wave propagating in $\hat{k}$ direction with frequency $\omega$ impinges upon the cloud of randomly or uniformly distributed elastic fibers (Fig. 1).

The just described two-dimensional (2D) elastostodynamic problem admits an integral representation of the form:

$$\frac{1}{2}\mathbf{u}^{(m)}(\mathbf{x}) + \int_S \tilde{\mathbf{t}}^*(\mathbf{x},\mathbf{y},\omega)\cdot\mathbf{u}^{(m)}(\mathbf{y})dS_y$$
$$= \int_S \tilde{\mathbf{u}}^*(\mathbf{x},\mathbf{y},\omega)\cdot\mathbf{t}^{(m)}(\mathbf{y})dS_y + \mathbf{u}^{inc}(\mathbf{x}) \quad (1)$$

$$\frac{1}{2}\mathbf{u}^{(f)}(\mathbf{x}) + \int_{S_i} \tilde{\mathbf{t}}^*(\mathbf{x},\mathbf{y},\omega)\cdot\mathbf{u}^{(f)}(\mathbf{y})dS_y$$
$$= \int_{S_i} \tilde{\mathbf{u}}^*(\mathbf{x},\mathbf{y},\omega)\cdot\mathbf{t}^{(f)}(\mathbf{y})dS_y, \quad i=1,2,...,q \quad (2)$$

where $S_i$ is the boundary of each fiber, $S = S_1 + \cdots + S_q$, $\mathbf{x}$ and $\mathbf{y}$ are points at $S_i$, $\mathbf{u}$ and $\mathbf{t}$ are the displacement and traction vector, respectively,

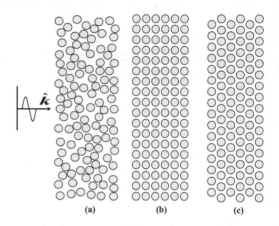

Figure 1. A longitudinal plane wave propagating in $\hat{k}$ direction impinges upon the $q$ (a) randomly, (b) square and (c) hexagonal distributed elastic fibers.

the superscripts ($f$) and ($m$) indicate fiber and matrix, respectively and $\tilde{\mathbf{u}}^*(\mathbf{x},\mathbf{y},\omega)$ and $\tilde{\mathbf{t}}^*(\mathbf{x},\mathbf{y},\omega)$ are the 2D free space frequency domain elastodynamic fundamental displacement and traction tensor, respectively, explicitly given in Polyzos et al. (1998).

According to a conventional BEM formulation, the boundaries $S_i$ are discretized into three-noded quadratic line elements with the general rule of using four elements per wave length of the incident wave. Collocating Eqs. (1) and (2) at all nodes and applying the continuity boundary conditions valid for displacements ant tractions at all fiber surfaces, one obtains a system of algebraic equations written in matrix form as

$$[\mathbf{H}]\cdot\{\mathbf{u}\} = [\mathbf{G}]\cdot\{\mathbf{t}\} + \{\mathbf{u}^{inc}\} \quad (3)$$

where the matrices [**G**] and [**H**] contain evaluated integrals having kernels corresponding to fundamental displacement and traction, respectively, the components of vectors {**u**} and {**t**} comprises all the nodal displacements and tractions, respectively and the vector {**u**$^{inc}$} contains all the known values of the incident wave at each node of the discretized surfaces.

Rearranging Eq. (3) and keeping only the known vectors at the right hand side, one obtains the final linear system of algebraic equations

$$[\mathbf{A}]\cdot\{\mathbf{x}\} = \{\mathbf{u}^{inc}\} \quad (4)$$

As it has been already mentioned in the introduction, in conventional BEM, [**A**] is a full populated matrix requiring O($N^2$) operations for its buildup and O($N^3$) operations for the solution of

Eq. (4) via Gaussian elimination or typical LU-decomposition solvers, which is prohibitive for solving realistic problems where the degrees of freedom $N$ are of the order of hundreds of thousands. In order to overcome the conventional BEM memory limitations and solve the above described problem for a large number of fibers, a hierarchical ACA accelerated BEM is proposed. More precisely, the matrices [H] and [G] appearing in Eq. (3), are represented hierarchically using a block tree structure. By means of simple geometric considerations the blocks, which correspond to large distances between source and collocation points, are characterized as far field blocks (or admissible) and compressed using low rank matrices found by an ACA algorithm (Bebendorf and Rjasanow (2003), Borm et al. (2003), Brunner et al. (2010)). The rest blocks of the tree, which are dominated by the singular behavior of the fundamental displacement and traction kernels, are characterized as near field blocks (or non-admissible) and are fully calculated as in conventional BEM. Furthermore, a significant reduction of the solution time of the problem is accomplished by utilizing the iterative solver GMRES for the solution of Eq. (4). According to that solver, the matrix [A] is never formed explicitly, saving significant amount of memory which corresponds to the zero values appearing in [A] due to the fact that [G] and [H] are uncoupled between each other. Thus, the GMRES multiplications are performed directly in Eq. (3) and a block left diagonal preconditioner is used to accelerate the convergence. The block dimensions are chosen to be approximately equal to the number of nodes that each fiber is discretized into and the block's inversion is performed using the LU decomposition algorithm. More details on the aforementioned ACA/BEM technique can be found in Gortsas et al. (2015).

## 3 THE ITERATIVE EFFECTIVE MEDIUM APPROXIMATION (IEMA)

In this section the main steps of IEMA are illustrated. When a plane wave propagates in a nonhomogeneous medium, it can be considered as a sum of a mean wave travelling in the medium with the dynamic effective properties of the composite and fluctuating waves derived from the multiple scattering of the mean wave. The basic idea of IEMA is that the fluctuating waves should be vanished at any direction within the effective medium. That hypothesis is reflected by the relation (Kim et al. (1995)):

$$n_1 g_d^{(1)}(\hat{\boldsymbol{d}};\hat{\boldsymbol{k}},\hat{\boldsymbol{k}}) + n_2 g_d^{(2)}(\hat{\boldsymbol{d}};\hat{\boldsymbol{k}},\hat{\boldsymbol{k}}) = 0 \quad (5)$$

where $n_1$, $n_2$ represent the volume concentration of fibers and matrix medium, respectively, with $n_1 + n_2 = 1$, $\hat{\boldsymbol{k}}$ is the direction of wave propagation, $\hat{\boldsymbol{d}}$ is the polarization vector of incident wave and $g_d^{(1)}$, $g_d^{(2)}$ are the forward scattering amplitudes derived from the solution of two single-scatterer problems, i.e. the scattering of a longitudinal (d = P) or transverse (d = S) wave by an inclusion and matrix fiber, respectively, embedded in a material having the effective properties of the composite.

The mean wave is both dispersive and attenuated and has a complex wavenumber $k_d^{\,eff}(\omega)$ defined as:

$$k_d^{\,eff}(\omega) = \frac{\omega}{C_d^{\,eff}(\omega)} + i\alpha_d^{\,eff}(\omega), \quad (6)$$

where $C_d^{\,eff}(\omega)$, $\alpha_d^{\,eff}(\omega)$ stand for the effective and frequency dependent phase velocity and attenuation coefficient, respectively of a mean wave propagating with circular frequency $\omega$. In order to determine both $C_d^{\,eff}(\omega)$, $\alpha_d^{\,eff}(\omega)$, the IEMA proposes the following iterative procedure.

First, the composite medium is replaced by an elastic homogeneous and isotropic medium with material properties $E^{\,eff}$ and $\nu^{\,eff}$, calculated by using the static mixture model of Christensen (1990).

The effective density of the nonhomogeneous medium has been taken equal to

$$(\rho^{\,eff})_{step1} = n_1\rho_1 + n_2\rho_2 \quad (7)$$

In the first step of the iteration procedure, the real effective wave number $(k_d^{\,eff})_{step1}$ of the mean wave can be calculated by the well-known equations valid for longitudinal and shear waves, respectively:

$$(k_p^{\,eff})_{step1} = \omega\sqrt{\frac{(\rho^{\,eff})_{step1}}{E^{\,eff}}\frac{(1+\nu^{\,eff})(1-2\nu^{\,eff})}{(1-\nu^{\,eff})}}$$

$$(k_s^{\,eff})_{step1} = \omega\sqrt{\frac{(\rho^{\,eff})_{step1}}{E^{\,eff}}2(1+\nu^{\,eff})} \quad (8)$$

Next, utilizing $E^{\,eff}$, $\nu^{\,eff}$, $(k_d^{\,eff})_{step1}$, the forward scattering amplitudes $g_d^{(1)}$, $g_d^{(2)}$ are evaluated and the validity of the self-consistent relation (5) is checked. If it is not true, then we proceed to the second step where the dispersion relation proposed by Foldy (1945) is exploited for the evaluation of wavenumbers of second step, i.e.

$$(k_d^{\,eff})_{step2} = (k_d^{\,eff})_{step1} + \frac{3n_1 g_d^{step1}(\hat{\boldsymbol{d}};\hat{\boldsymbol{k}},\hat{\boldsymbol{k}})}{a^2(k_d^{\,eff})_{step1}} \quad (9)$$

with $a$ being the radius of the fiber and

$$g_d^{step1} = [n_1 g_d^{(1)} + n_2 g_d^{(2)}]^{step1} \quad (10)$$

The procedure is continued until the self-consistent Eq. (5) to be satisfied. If $l$ is the final step, the phase velocity $C_d^{eff}(\omega)$ and the attenuation coefficient $\alpha_d^{eff}(\omega)$ are evaluated from the relation

$$\left(k_d^{eff}(\omega)\right)_{step l} = \frac{\omega}{C_d^{eff}(\omega)} + i\alpha_d^{eff}(\omega), \quad (11)$$

More details can be found in Aggelis et al. (2004) and Verbis et al. (2001).

## 4 NUMERICAL RESULTS

In Fig. 2 the magnitude of radial scattering amplitudes in the forward direction, for three different fiber arrangements (Fig. 1) are presented for a propagating longitudinal wave, with respect to the range of non-dimensional frequencies $ka = 0.1 - 2$. The size of the virtual control volume containing the fibers for all three cases is $180 \times 540 \, \mu m$. The volume fraction is 35% for the square and random fiber arrangement and 32% for the hexagonal arrangement. The total number of the considered identical fibers is 108 for the square and random arrangements and 99 for the hexagonal, while their radius is $a = 10 \, \mu m$. The material properties used are presented in Table 1.

Similarly, in Fig. 3 the magnitude of the vertical scattering amplitudes in the forward direction, for a propagating shear wave, are presented for the three different arrangements and the same non-dimensional frequency range.

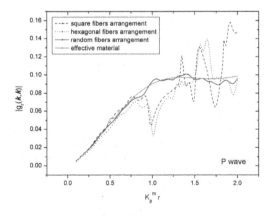

Figure 2. Radial scattering amplitude in the forward direction for a propagating longitudinal wave, calculated for the square, random and hexagonal fiber arrangements, as well as the effective material provided by IEMA.

Table 1. Material properties for the fibers and the matrix.

| Material | ρ (kg/m³) | E (GPa) | ν |
|---|---|---|---|
| Aluminum AA520 (matrix) | 2600 | 66 | 0.31 |
| Alumina Al₂O₃ (fibers) | 3700 | 360 | 0.25 |

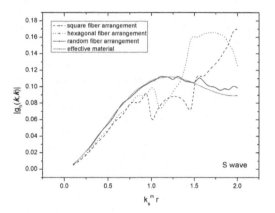

Figure 3. Horizontal scattering amplitude in the forward direction for a propagating longitudinal wave, calculated for the square, random and hexagonal fiber arrangements, as well as the effective material provided by IEMA.

In both cases, the results obtained for the three different fiber arrangements are compared to the corresponding ones taken when fibers and surrounding material is replaced by a homogeneous slab with frequency dependent properties provided by IEMA for the same frequency range.

Observing Figs 2 and 3 one can say that the results obtained for the effective material provided by IEMA are in very good agreement with the corresponding ones taken by the random arrangement of fibers at all the considered frequencies. For low frequencies ($ka < 1$) the same conclusion is valid for the periodic arrangements of fibers. For dimensionless frequencies $ka > 1$ there are significant differences between the periodic and the random arrangement of fibers. In the case of periodic arrangement of fibers there are certain frequencies where a sudden decrease of the value of the scattering amplitude is observed. For those frequencies part of the scattered energy is either diverted in different directions than the forward one or it is trapped between the scatterers as standing waves. Moreover there are also frequencies where the scattered energy in the forward direction is amplified due to the multiple scattering effect.

Finally, in order to examine the prediction capability of IEMA in other directions than the forward

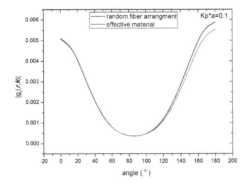

Figure 4. Magnitude of the radial scattering amplitude for different angles and frequency k*a = 0.1.

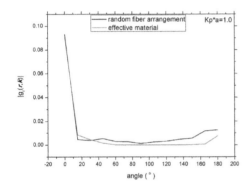

Figure 5. Magnitude of the radial scattering amplitude for different angles and frequency k*a = 1.0.

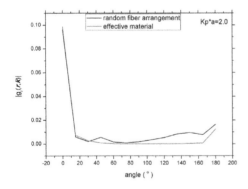

Figure 6. Magnitude of the radial scattering amplitude for different angles and frequency k*a = 2.0.

one, the magnitude of the radial scattering amplitude is calculated for a longitudinal incidence in randomly distributed fibers and for three different frequencies $k\alpha = 0.1, 1, 2$ at the angle range 0°–180°, with the angles 0° and 180° representing the forward and backward direction, respectively. The obtained results are depicted in Figs. 4, 5 and 6 and as it is observed IEMA predicts pretty well the scattered energy by the composite with the random arrangement of fibers at all directions, while provides almost identical results for the angle range 0°–20°.

## 5 CONCLUSIONS

The propagation of longitudinal and shear elastic waves in unidirectional fiber reinforced composite slabs has been studied for three different arrangements of fibers (random, square and hexagonal). The corresponding problems have been solved with the aid of an advanced ACA/BEM and the obtained results have been compared to the corresponding ones taken after the homogenization of the composite through the IEMA. The main conclusions are: (a) for propagation in the forward direction and for low frequencies IEMA works well for all the arrangements of fibers, (b) for high frequencies IEMA supports effectively the wave propagation only in the composite with the random distribution of the fibers and (c) IEMA predicts pretty well the scattered energy by the composite with the random arrangement of fibers at almost all directions.

## ACKNOWLEDGEMENT

This research has been co-financed by the European Union (European Social Fund—ESF) and Greek national funds through the Operational Program "Education and Lifelong Learning" of the National Strategic Framework (NSRF)—Research Funding Program: ARCHIMEDES III.

## REFERENCES

Aggelis D.G., Tsinopoulos S.V., Polyzos D. 2004. An iterative effective medium approximation for wave dispersion and attenuation predictions in particulate composites, suspensions and emulsions. J. Acoust. Soc. Am. 9: 3443–3452.

Bebendorf M., Rjasanow S. 2003. Adaptive low-rank approximation of collocation matrices. Computing 70:1–24.

Borm S., Grasedyck L., Hackbusch W. 2003. Introduction to hierarchical matrices with applications. Eng Anal Boundary Elem 27:405–422.

Brunner D., Junge M., Rapp P., Bebendorf M., Gaul L. 2010. Comparison of the Fast Multipole Method with Hierarchical Matrices for the Helmholtz-BEM. CMES Computer Modelling of engineering structures 58:131–158.

Christensen R.M. 1990. A critical evaluation for a class of micro-mechanics models. J. Mech. Phys. Solids 38:379–404.

Foldy L.L. 1945. The multiple scattering of waves. Phys. Rev. 67:107–119.

Gortsas T., Tsinopoulos S.V., Polyzos D. 2015. A study on the microstructural effects of unidirectional fiber composite plates via advanced hierarchical ACA and accelerated BEM formulations. Submitted for publication.

Kim J.Y., Ih J.G., Lee B.H. 1995. Dispersion of elastic waves in random particulate composites. J. Acoust. Soc. Am. 97: 1380–1388.

Polyzos D., Tsinopoulos S. V., Beskos D.E. 1998. Static and Dynamic Boundary Element Analysis in Incompressible Linear Elasticity. European Journal of Mechanics A/Solids, 17(3):515–536.

Tsinopoulos S.V., Verbis J.T., Polyzos D.2000. An Iterative Effective Medium Approximation for Wave Dispersion and Attenuation Predictions in Particulate Composites. Advance Composites Letters, Vol.9:193–200.

Verbis J.T., Kattis S.E., Tsinopoulos S.V., Polyzos D. 2001. Wave dispersion and attenuation in fiber composites. Computational Mechanics 27:244–252.

*Emerging Technologies in Non-Destructive Testing VI – Aggelis et al. (Eds)*
*© 2016 Taylor & Francis Group, London, ISBN 978-1-138-02884-5*

# Acousto-ultrasonic Structural Health Monitoring of aerospace composite materials

## M. Gresil
*School of Materials, University of Manchester, Manchester, UK*

## A. Muller & C. Soutis
*Aerospace Research Institute, University of Manchester, Manchester, UK*

ABSTRACT: Piezoelectric transducers have a long history of applications in Nondestructive Evaluation (NDE) of material and structure integrity owing to their ability of transforming mechanical energy to electrical energy and vice versa. From an acoustic point of view, there is no difference between Structural Health Monitoring (SHM) and conventional NDE since both rely on the same physics in the sense that in either case acoustic waves are generated and then detected. SHM was 'born' from the conjunction of several techniques and has a common basis with NDE. In fact, several NDE techniques can be converted into SHM techniques, by integrating sensors and actuators inside the monitored structure. For instance, traditional ultrasonic testing can be easily converted into an acousto-ultrasonic SHM system, using embedded or surface-mounted Piezoelectric Wafer Active Sensors (PWAS). These sensors should be affordable, lightweight, and unobtrusive such as to not impose cost and weight penalty on the structure and to not interfere with the structural strength and airworthiness. Other damage measuring methods based on large area measurements have been used in SHM development for verification and validation of damage and/or for understating the proposed SHM approach; however, they do not seem appropriate for permanent installation onto the monitored structure and will not be discussed under the heading of 'SHM sensors'. This paper presents and discusses an overview of ultrasonic SHM techniques for composite materials. After a brief introduction, it presents the PWAS-based SHM principle, which is followed by a discussion of the passive and active ultrasonic SHM techniques. The paper identifies advantages and disadvantages of these in-situ NDE methods and guidelines for future work on heterogeneous, anisotropic materials, like aerospace composite polymer.

## 1 INTRODUCTION

Structural Health Monitoring (SHM) was born from the conjunction of several techniques and has a common basis with Non-Destructive Evaluation (NDE). In fact, several NDE techniques can be converted into SHM techniques, by integrating sensors and actuators inside the monitored structure. For instance, traditional ultrasonic testing can be easily converted into an acousto-ultrasonic SHM system, using embedded or surface-mounted Piezoelectric Wafer Active Sensors (PWAS). These sensors should be affordable, lightweight, and unobtrusive such as to not impose cost and weight penalty on the structure and to not interfere with the structural strength and airworthiness.

Other damage measuring methods based on large area measurements (ultrasonic C-scans, thermography, etc.) have been used in SHM development for verification and validation of damage and/or for understating the proposed SHM approach; however, they do not seem appropriate

for permanent installation onto the monitored structure and will not be discussed in this paper.

This paper presents a brief overview of acousto-ultrasonic SHM techniques and discusses the PWAS-based SHM principle of aerospace composite materials. It follows with a discussion of the passive and the active acousto-ultrasonic SHM techniques. The paper ends with a conclusion and guidelines for future work.

### 1.1 *Background*

The end of the twentieth century has brought new problematic to the aeronautical industry and research. On the one hand, the air transport has given rise to an exponential growth of passenger in air traffic, resulting in an increased demand. On the other hand, a global objective such as the reduction of the impact on environment has to be considered. Therefore, major modern challenges for aeronautical research are to reconcile the improvement of aircraft efficiency, the economical aspect (ie: manufacturing, fuel, and maintenance cost) and the extension of operational lifetime without compromising the aircraft safety and reliability (ACARE, 2001; Pfeiffer & Wevers, 2007). In consideration of those challenges, the European Commission has published in 2001 a report titled "European Aeronautics: a vision for 2020" describing the research objectives and expected advancement of the aeronautical research for 2020 (ACARE, 2001).

Over the past 30 years composite materials have been increasingly used in aircraft structures (Diamanti & Soutis, 2010; Grondel, Assaad, Delebarre, & Moulin, 2004; Staszewski, Mahzan, & Traynor, 2009). Recent designed aircraft such as the Airbus A380 introduced in 2007 and the Boeing B787 introduced in 2011 have their structures made of 25% and 50% of composite materials respectively (Staszewski et al., 2009). This growing interest for composite materials in aviation over aluminium alloys is mainly due to their excellent specific properties such as high stiffness and strength for weight ratio (especially in case of polymer matrix composite). The particular properties of composite materials offer new design perspective. For instance their anisotropic properties can be tailored to design requirements and improve the aerodynamic while using lighter components. This, on a larger scale, can impact aircraft efficiency and therefore, contribute to the achievement of current research objectives.

### 1.2 Common damages in aircraft composite structure

Many researchers agreed that impact damage is one of the most frequent damage encountered by aircraft structure (Diamanti & Soutis, 2010; Grondel et al., 2004; Haase, Thomson, Bishop, & Isambert, 2013; Staszewski et al., 2009). A major concern is that some impact damages in composite materials do not provide visible evidence, therefore may not be detected during regular surface inspection. Those particular damages are known as BVID (Barely Visible Impact Damage) and concern principally two type of impact damages: the low velocity (and low energy) impact damages, which are associated with a velocity ranging between 4 to 8 m.s$^{-1}$ and an impact energy up to 50 J; and the high energy low velocity impact damages (Haase et al., 2013). Most of the incidents generating BVID occur during ground operations, common examples include: bird strikes, runway debris, tool drops during maintenance, or collision with ground vehicles (Grondel et al., 2004; Haase et al., 2013). In addition to BVID, Grondel (2004) identified debondings are another common type of hidden damages difficult to detect with visual inspection.

Those damages or defects often initiate further damages, such as matrix crack growth, delamination, or fibres breakage. Therefore, they are the starting point to reduction of the composites strength and degradation of its structural integrity (Grondel et al., 2004; Staszewski et al., 2009). Failure to detect damage before it expands and reach a critical size may lead to failure of the component. A late damage detection can result in extra cost due for instance to replacement but also to aircraft service interruption. Moreover, the cost of repair and maintenance represent a quarter of the expense for commercial aircraft in-service (Giurgiutiu, 2008). For instance, in 2000 the cost by the Airports Council International (ACI) related to aircraft collision with ground vehicle was estimated to \$3 billion (Haase et al., 2013). In the worst-case scenario such situation may lead to catastrophic failure of the aircraft and jeopardize users safety. In consequence, the detection of those damages as well as other flaws or defect is crucial. On the one hand to allow detection of the damaged area before it reaches a critical size, hence to ensure structure integrity and overall aircraft safety, on the other hand to provide sufficient time for action, such as by monitoring or preventing further growth of the damage (Staszewski et al., 2009).

Therefore, current research focuses on the development or the improvement of NDE for damages detection (especially BVID or other hidden defects) in composite structure. In particular Structural Health Monitoring (SHM) systems have gained growing interest for the development of smart or self-sensing composite structures, leading to a growing amount of literature in field (Diamanti & Soutis, 2010; Giurgiutiu, 2008; Gresil, Soutis, & Giurgiutiu, 2014; Grondel et al., 2004; Ihn & Chang, 2008; Staszewski et al., 2009).

### 1.3 Piezoelectric wafer active sensors

Piezoelectric wafer active sensors (Giurgiutiu, 2008) (PWAS) are small, lightweight, and relatively low-cost sensors based on the piezoelectric principle that couples the electrical and mechanical variables in the material (mechanical strain, $S_{ij}$, mechanical stress, $T_{kl}$, electrical field, $E_k$, and electrical displacement $D_j$) in the form:

$$S_{ij} = s_{ijkl}^E T_{kl} + d_{kij} E_k$$
$$D_j = d_{jkl} T_{kl} + \mathcal{E}_{jk}^T E_k \tag{1}$$

where $s_{ijkl}^E$ is the mechanical compliance of the material measured at zero electric field ($E = 0$), $\mathcal{E}_{jk}^T$ is the dielectric permittivity measured at zero mechanical stress ($T = 0$), and $d_{kij}$ represents the piezoelectric coupling effect. The direct piezoelectric effect converts the stress applied to the sensor into electric charge. Similarly, the converse piezoelectric effect produces strain when a voltage is applied to the sensor.

At ultrasonic frequencies (k-MHz range), PWAS can sense and excite guided Lamb waves traveling long distances along the thin-wall shell structures of aircraft and space vehicles. PWAS are made of thin inexpensive piezo-ceramic wafers electrically poled in the thickness direction and provided with top and bottom electrodes. PWAS can be bonded to the structure with strain-gage installation methodology. They have also been experimentally inserted between the layers of a composite laminate, but this option has raised some structural integrity issues that are still being examined. PWAS have been extensively used for SHM demonstrations because they convert directly electric energy into elastic energy and vice-versa and thus require very simple instrumentation: effective measurements of composite impact waves and guided-waves transmission/reception have been achieved with experimental setups consisting of no more than a signal generator, a digitising oscilloscope, and a PC (Giurgiutiu, 2008).

As shown in Figure 1, PWAS transducers can serve several purposes (Giurgiutiu, 2008): (a) high-bandwidth strain sensors; (b) high-bandwidth wave exciters and receivers; (c) resonators; (d) embedded modal sensors with the electromechanical (E/M) impedance method. By application types, PWAS transducers can be used for (i) **active sensing of far-field damage** using pulse-echo, pitch-catch, and phased-array methods, (ii) **active sensing of near-field damage** using high-frequency E/M impedance method and thickness mode, and (iii) **passive sensing of damage-generating events** through detection of low-velocity impacts and acoustic emission at the tip of advancing cracks. By using Lamb waves in a thin-wall structure, one can detect structural anomaly, i.e., cracks, corrosions, delaminations, and other damage. Because of the physical, mechanical, and piezoelectric properties of PWAS transducers, they act as both transmitters and receivers of Lamb waves travelling through the structure. Figure 1a illustrates the **pitch-catch method**. An electric signal applied at the transmitter PWAS generates, through piezoelectric transduction, elastic waves that travel into the structure and are captured at the receiver PWAS. Figure 1b illustrates the **pulse-echo** method. In this case, the same PWAS transducer acts as both transmitter and receiver. Figure 1c illustrates the use of PWAS transducers in **thickness mode**. The thickness mode is usually excited at much higher frequencies than the guided wave modes discussed in the previous two paragraphs. For example, the thickness mode for a 0.2-mm PWAS is excited at around 11 MHz, whereas the guided wave modes are excited at tens and hundreds of kHz. When operating in thickness mode, the PWAS transducer can act as a thickness gage. Figure 1d illustrates the detection of **impacts and acoustic emission (AE) events**. In this case, the PWAS transducer is operating as a passive receiver of the elastic waves generated by drop weight impacts or by AE events. Figure 1e illustrates the electromechanical impedance spectrum measured in the k-MHz range. When a structure is excited with sustained harmonic excitation of a given frequency, the waves travelling in the structure undergo multiple boundary reflections and settle down in a standing wave pattern know as vibration. Structural vibration is characterised by resonance frequencies at which the structural response goes through peak values. A natural extension of the PWAS pulse-echo method is the development of a PWAS phased array (Figure 1f), which is able to scan a large area from a single location. The PWAS phased arrays utilise the phase array principles to create an interrogating beam of guided waves that travel in a thin-wall structure and can sweep a large area from a single location.

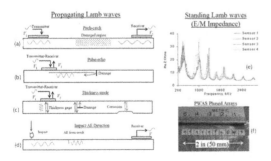

Figure 1. Use of Piezoelectric Wafer Active Sensors (PWAS) for damage detection with propagating and standing guided waves in thin-wall structures (Giurgiutiu, 2008).

## 2 PASSIVE SHM

This section aims to describe the use of the Acoustic Emission (AE) technique to contribute to the general problem of SHM and, more generally, to the prediction of the remaining lifetime of industrial materials and structures. AE is primarily used to study the physical parameters and the damage mechanisms of a material, but it is also used as an on-line NDT. The AE phenomenon is based on the release of energy in the form of transitory elastic waves within a material having dynamic deformation processes. The waves, of various types and frequencies, propagate in the material and undergo possible modifications before reaching the surface of the studied sample. A typical source of an AE wave within a material is the appearance of a crack from a defect when the material is put under constraint, or when a pre-existing crack grows. This technique makes it possible to detect in real time the existence of evolutionary defects.

The piezo-based AE sensors are relatively well established in conventional ultrasonic NDE; however, these conventional AE sensors are not quite appropriate for deploying in large numbers on a flight structure due to both cost and size. The SHM sensors (both PWAS and Fibre Bragg Grating (FBG)) have also been shown capable of AE monitoring: Several authors (Koh, Chiu, Rajic, & Galea, 2003; Martin, Hudd, Wells, Tunnicliffe, & Das-Gupta, 2001; Sung, Oh, Kim, & Hong, 2000) used PWAS, whereas others (Jong-In, Hyung-Joon, Chun-Gon, & Chang-Sun, 2005; Perez, Cui, & Udd, 2001) used FBG sensors for AE emission monitoring. Existing AE monitoring methodology for signal capturing and interpretation (noise filtering, AE events counting algorithms, etc.) can also be used with SHM sensors. PWAS and conventional R15I transducer produced very similar signals (Gresil, Yu, Shen, & Giurgiutiu, 2013). Moreover, using the discrete and the continuous wavelet transform, the AE signal energy is not uniformly distributed between the symmetric and anti-symmetric mode using wavelet transform signal based processing (Gresil et al., 2013).

New concepts, such as distributed sensors and bio-mimetic information acquisition and processing, would also be fundamentally helpful to realise real-time in-situ AE instrumentation for the health monitoring of in-service structures.

## 3 ACTIVE SHM

The AE technique described above makes use of PWAS bonded or embedded in a structure and implemented in a passive way. The same attached sensors can also be used in an active way to produce and detect high-frequency vibrations. A transmitter is used to send a diagnostic stress wave along the structure and a receiver to measure the changes in the received signal caused by the presence of a defect or damage in the structure. This wave propagation approach is a natural extension of traditional NDE techniques, and it is very effective in detecting defects and damage in the form of geometrical discontinuities.

### 3.1 Guided wave propagation in composites

The structure under investigation is a CFRP plate consisting of carbon fibre fabric reinforcement in an epoxy resin (Figure 2). The plate dimensions are

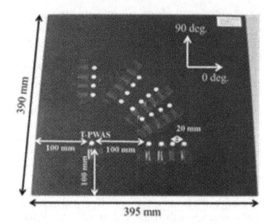

Figure 2. Picture of the network of piezoelectric wafer active sensor bonded on the CFRP.

$390 \times 395 \times 2 \, mm^3$. The CFRP material is HexPly® M18/1/939; this is a woven carbon prepreg manufactured by Hexcel. This material is commonly used in aircraft industry. The plate plies have the orientation $[0, 45, 45, 0]_s$. Twenty one PWAS transducers (Steminc SM412, 8.7 mm-diameter disks and 0.5 mm-thick) were used for Lamb wave propagation experiments. The PWAS network bonded on the CFRP plate is shown on Figure 2. The instrumentation consisted of an HP33120 A arbitrary signal generator, and a Tektronix TDS210 digital oscilloscope. A LabView™ computer program was developed to record the data from the digital oscilloscope, and to generate the raw data files.

The multi-physics finite element (MP-FEM) simulation was carried out to determine the attenuation coefficient. A 150 kHz three-tone burst modulated by a Hanning window with a 20 Volts maximum amplitude peak to peak was applied to the top surface of the T-PWAS transducer, and the other PWAS transducers as receivers. Due to the dispersion curves presented and the tuning effect described on reference (Gresil & Giurgiutiu, 2013a; Gresil & Giurgiutiu, 2014), both S0 and A0 modes are present at this frequency. Figure 3 shows the comparison between the MP-FEM simulation and experimental electric signal measured at R-PWAS placed at 100 mm from the T-PWAS with the stiffness proportional coefficient $\beta = 2.10^{-8}$ which corresponds to the A0 mode attenuation as calculated in the reference (Gresil & Giurgiutiu, 2013c; Gresil & Giurgiutiu, 2014). With this stiffness proportional coefficient $\beta = 2.10^{-8}$, the MP-FEM signal for the S0 and the A0 modes are in very good agreement with the experimental signal.

However, the MP-FEM signal between the S0 mode packet and the A0 mode packet is different from the experimental signal. This different

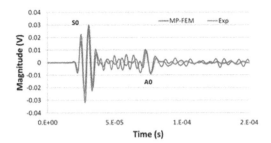

Figure 3. Comparison between the experimental and the MP-FEM received signal at 100 mm from the T-PWAS at 150 kHz with $\alpha = 0$ and $\beta = 2.10^{-8}$ (Gresil & Giurgiutiu 2014).

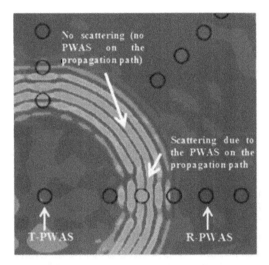

Figure 4. Snapshot of the guided waves propagation showing the scattering due to the bonded PWAS on the guided wave propagation path between the T-PWAS and the R-PWAS (Gresil & Giurgiutiu 2014).

signal may be due to the scattering effect by the other bonded PWAS on the guided wave propagation path between the T-PWAS and the R-PWAS as described on the MP-FEM snapshot on Figure 4 (Gresil & Giurgiutiu, 2014).

### 3.2 Electromechanical impedance spectroscopy

The principles of electromechanical impedance method are illustrated in Figure 5. The drive-point impedance presented by the structure to the active sensor can be expressed as the frequency dependent variable

$$Z_{str}(\omega) = k_{str}(\omega)/j\omega = k_e(\omega) - \omega_m^2(\omega) + j\omega c_e(\omega) \qquad (2)$$

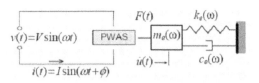

Figure 5. Electromechanical coupling between PWAS and structure for 1-D dynamic model.

Through the mechanical coupling between PWAS and the host structure, on one hand, and through the E/M transduction inside the PWAS, on the other hand, the drive-point structural impedance is reflected directly in the electrical impedance, $Z(\omega)$, at the PWAS terminals

$$Z(\omega) = \left[ j\omega C \left(1 - \kappa_{31}^2 \frac{\chi(\omega)}{1+\chi(\omega)}\right) \right]^{-1} \qquad (3)$$

where $C$ is the zero-load capacitance of the PWAS and $\kappa_{31}$ is the E/M cross coupling coefficient of the PWAS $\left(\kappa_{31} = d_{31}/\sqrt{\bar{s}_{11}\bar{\varepsilon}_{33}}\right)$, and $\chi(\omega) = k_{str}/k_{PWAS}$ with $k_{PWAS}$ being the static stiffness of the PWAS.

The electromechanical impedance SHM method is direct and easy to implement, the only required equipment being an electrical impedance analyser, such as the HP 4192 A impedance analyser. An example of performing PWAS electromechanical impedance spectroscopy is presented in Figure 6. The HP 4194 A impedance analyser (Figure 6a) reads the in-situ electromechanical impedance of the PWAS attached to a specimen. It is applied by scanning a predetermined frequency range in the high kHz band (up to 15 MHz) and recording the complex impedance spectrum.

During a frequency sweep, the real part of the E/M impedance, Re[$Z(\omega)$], follows the up and down variation as the structural impedance goes through the peaks and valleys of the structural resonances and anti-resonances (Figure 6b). By comparing the real part of the impedance spectra taken at various times during the service life of a structure, meaningful information can be extracted pertinent to structural degradation and ongoing damage development. On the other hand, analysis of the impedance spectrum supplies important information about the PWAS integrity. The frequency range used in the E/M impedance method must be high enough for the signal wavelength to be significantly smaller than the defect size. From this point of view, the high frequency EMIS method differs from the low-frequency modal analysis approaches.

In the MP-FEM approach, the mechanical coupling between the structure and the sensor is implemented by specifying boundary conditions of the

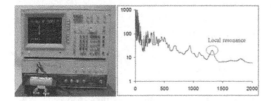

Figure 6. (left) Impedance analyser; (right) example of measured impedance spectrum (Gresil et al. 2012).

sensor while the electromechanical coupling is modelled by multi-physics equations for the piezoelectric material. The first coupling allows the mechanical response sensed by the piezoelectric element to be reflected in its electric signature composite. The Glass Fibre Reinforced Polymer (GFRP) structure considered in this study is modelled as a homogeneous orthotropic material (Gresil, Yu, Giurgiutiu, & Sutton, 2012). The test specimen is numerically modelled with the MP-FEM method using a 3D mesh. The SOLID186 layered structural solid element is used to model the five layers laminated GFRP composite specimen with layer orientation of 0 degree on the x-axis; the adhesive layer is modelled with the SOLID95 element. The PWAS transducer is modelled with the SOLID226 coupled field element. Each element has twenty nodes. At low frequency (below 500 kHz), at medium frequency (500 kHz to 5 MHz) and at high frequency (5 to 15 MHz), the size of the mesh is 1 mm, 0.5 mm and 0.1 mm respectively, to obtain a good convergence of the problem.

The comparison between the simulated and the experimental impedance spectra results are presented in Figure 7. In Figure 7, the results are in the range up to 5 MHz. It is apparent that a relatively good agreement between the experiments and 3D MP-FEM simulation has been achieved. The good matching is achieved by adjusting the damping coefficients used in the structural model. The correlation of the modal frequencies between the experimental and the numerical results is quite good, especially at higher frequencies. However, some discrepancies in the magnitudes of some resonances are observed, especially in the range of 450 to 650 kHz. It is interesting to see that the best match is obtained in the 700 kHz to 2 MHz frequency range. This is very beneficial, because this frequency range has shown the best detection of delamination damage (Gresil et al., 2012).

Figure 8 shows a comparison of the impedance spectra in a very high frequency range (of 5 to 15 MHz).

Only one peak is observed at ~11 MHz; this peak corresponds to the thickness mode resonance of the PWAS transducer. It seems that at high

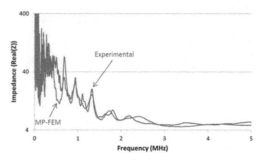

Figure 7. Comparison of experimental and 3D MP-FEM model impedance spectra of laminate GFRP for a frequency range 10 kHz to 5 MHz (Gresil et al. 2012).

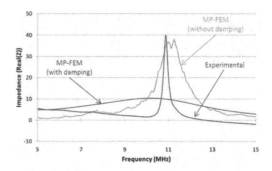

Figure 8. Comparison of experimental and 3D MP-FEM model impedance spectra of laminate GFRP for a frequency range 5 MHz to 15 MHz (Gresil et al. 2012).

frequency (5 to 15 MHz), the vibration is localized near the PWAS so the bonding condition and the PWAS geometry is very important. In the case of the simulation the bonding layer is perfect and also the PWAS geometry. In reality, this is not true so we can explain more difference between the experimental and the simulation results for high frequency. Moreover the magnitude of the vibration pick is very small due to the damping effect, and this effect is very hard to simulate because of the non-linearity of this effect.

The comparison between the 3D simulation and experimental results has revealed two different regions of behaviour: (i) below 5 MHz, the experimental result matches the result from a 3D model with structural damping (Figure 7); (ii) however, above 7 MHz, the experimental result matches better with a 3D model without structural damping (Figure 8). One possible explanation is that at lower frequency the vibration covers a larger area and the overall structural damping is important; whereas at high frequency the vibration is localized in thickness mode resulting that the structural damping has negligible effect. In comparison with other models of the EMIS technique, the model discussed here exhibits remarkable robustness at very high frequency (Gresil et al., 2012).

## 4 SUMMARY AND GUIDELINES FOR FUTURE WORK

PWAS have been shown to be very well-suited for structural health monitoring. The same set of piezoelectric elements can be utilised for ageing monitoring, damage detection, location and identification, and finally to record the AE activity of the structure under test.

The multi-functional character of PWAS is quite relevant in its application to the on-line health monitoring of aeronautical structures. In practice, the use of the same set of sensors to perform at least three kinds of measurement reduces the weight of the monitoring system and hence simplifies the maintenance process.

Furthermore, it will be of great interest to develop wireless frameworks of sensors allowing both weight savings and simplification of the insertion process, especially in the case of the articulated parts of a complex structure. In conclusion, PWAS have great potential for development in the field of SHM. In the near future, more multi-functional SHM systems have to be developed to collect different kinds of signals and physical parameters with the same network of sensors.

## REFERENCES

ACARE. (2001). *Advisory Council for Aeronautics Research in Europe —European Aeronautics: a vision for 2020*.

Diamanti, K., & Soutis, C. (2010). Structural health monitoring techniques for aircraft composite structures. *Progress in Aerospace Sciences, 46*(8), 342–352.

Giurgiutiu, V. (2008). *Structural Health Monitoring With Piezoelectric Wafer Active Sensor*: Elsevier Academic Press.

Gresil, M., & Giurgiutiu, V. (2013a). *Guided wave propagation in carbon composite laminate using piezoelectric wafer active sensors*. Paper presented at the SPIE Smart Structures and Materials+ Nondestructive Evaluation and Health Monitoring, San-Diego, CA, USA.

Gresil, M., & Giurgiutiu, V. (2013c). *Prediction of attenuated guided wave propagation in carbon fiber composites*. Paper presented at the The 19th International Conference on Composite Materials, Montreal, Canada.

Gresil, M., & Giurgiutiu, V. (2014). Prediction of attenuated guided waves propagation in carbon fiber composites using Rayleigh damping model. *Journal of Intelligent Material Systems and Structures*. doi: 10.1177/1045389 × 14549870.

Gresil, M., Soutis, C., & Giurgiutiu, V. (2014). *Ultrasonic Structural Health Monitoring—An Overview*. Paper presented at the 53rd Annual Conference of The British Institute of Non-Destructive Testing, Manchester, UK.

Gresil, M., Yu, L., Giurgiutiu, V., & Sutton, M. (2012). Predictive modeling of electromechanical impedance spectroscopy for composite materials. *Structural Health Monitoring, 11*(6), 671–683.

Gresil, M., Yu, L., Shen, Y., & Giurgiutiu, V. (2013). Predictive model of fatigue crack detection in thick bridge steel structures with piezoelectric wafer active sensors. *Smart Structures and Systems, 12*(2), 001–635.

Grondel, S., Assaad, J., Delebarre, C., & Moulin, E. (2004). Health monitoring of a composite wingbox structure. *Ultrasonics, 42*(1–9), 819–824.

Haase, P., Thomson, R., Bishop, P., & Isambert, E. (2013). EASA.2011.NP.24 "Composite Damage Metrics and Inspection" (CODAMEIN II) *EASA.2011. NP.24 Report*.

Ihn, J.-B., & Chang, F. K. (2008). Pitch-catch active sensing methods in structural health monitoring for aircraft structures. *Structural Health Monitoring, 7*(1), 5–19.

Jong-In, K., Hyung-Joon, B., Chun-Gon, K., & Chang-Sun, H. (2005). Simultaneous measurement of strain and damage signal of composite structures using a fiber Bragg grating sensor. *Smart Materials and Structures, 14*(4), 658.

Koh, Y. L., Chiu, W. K., Rajic, N., & Galea, S. C. (2003). Detection of Disbond Growth in a Cyclically Loaded Bonded Composite Repair Patch Using Surface-mounted Piezoceramic Elements. *Structural Health Monitoring, 2*(4), 327–339.

Martin, T., Hudd, J., Wells, P., Tunnicliffe, D., & DasGupta, D. (2001). The Use of Low Profile Piezoelectric Sensors for Impact and Acoustic Emission (AE) Detection in CFRC Structures. *Journal of Intelligent Material Systems and Structures, 12*(8), 537–544.

Perez, I. M., Cui, H., & Udd, E. (2001). Acoustic emission detection using fiber Bragg gratings.

Pfeiffer, H., & Wevers, M. (2007). *Aircraft Integrated Structural Health Assessment—Structural Health Monitoring and its implementation within the European project AISHA*. Paper presented at the EU Project Meeting on Aircraft Integrated Structural Health Assessment (AISHA), Leuven, Belgium.

Staszewski, W. J., Mahzan, S., & Traynor, R. (2009). Health monitoring of aerospace composite structures—Active and passive approach. *Composites Science and Technology, 69*(11–12), 1678–1685.

Sung, D.-U., Oh, J.-H., Kim, C.-G., & Hong, C.-S. (2000). Impact Monitoring of Smart Composite Laminates Using Neural Network and Wavelet Analysis. *Journal of Intelligent Material Systems and Structures, 11*(3), 180–190.

*Emerging Technologies in Non-Destructive Testing VI – Aggelis et al. (Eds)*
*© 2016 Taylor & Francis Group, London, ISBN 978-1-138-02884-5*

# Guided wave tomography based inspection of CFRP plates using a probabilistic reconstruction algorithm

J. Hettler, M. Tabatabaeipour, S. Delrue & K. Van Den Abeele
*Department of Physics KU Leuven Kulak, Wave Propagation and Signal processing Group,*
*Kortrijk, Belgium*

ABSTRACT: The rapidly increasing usage of Carbon-Fiber Reinforced Plastics (CFRP) calls for novel, cost and time effective NDT inspection methods. The Reconstruction Algorithm for Probabilistic Inspection of Damage (RAPID) has been previously introduced to tackle this challenge. This ultrasonic method utilizes the difference between the signals acquired in an intact state and in a damaged state to identify and localize the damage in the plate-like structures. The present study demonstrates the applicability of the RAPID methodology to structures such as CFRP aircraft wing components. It will be shown that RAPID is able to detect and localize defects such as low-velocity impacts and delaminations in quasi-isotropic CFRP. A baseline-free RAPID method, which combines RAPID with the nonlinear elastic wave techniques, will be proposed as an alternative to the conventional technique. The feasibility of this approach will be demonstrated by the numerical simulations and preliminary experimental results.

## 1 INTRODUCTION

Nowadays, CFRP has already found its way to numerous areas of the industrial design and manufacturing. Thanks to an excellent weight-to-strength ratio, it is gradually becoming one of the key materials in many applications, ranging from automotive components to aerospace structural parts. On the other hand, the excellent strength-to-weight and stiffness-to-weight ratios of the CFRP are balanced with complicated post-damage behaviour and the fact that they are difficult to repair.

The ultrasonic testing of CFRP relies mainly on phased array PA (or single-element) pulse-echo and through transmission inspection. These are usually conducted either in immersion, in air or in a contact manner (Helfen et al. 2012, Habermehl and Lamarre 2008). They offer reasonably good resolution, sensitivity and acceptable inspection times. Piezoelectric Waver Active Sensors (PWAS) also emerge as a promising technique for SHM applications (Giurgiutiu 2003, Giurgiutiu and Soutis 2012, Yu and Giurgiutiu 2005, Santoni et al. 2007, Xu 2009). Despite these efforts, however, there is a constant demand for a faster and equally reliable techniques for the inspection of large plate-like CFRP structures.

Ultrasonic Guided Wave Tomography (GWT) is a suitable candidate that meets the previously mentioned requirements. It is capable to interrogate large areas with a limited number of transducers in a very short time. In the presented work,

the applicability of the GWT to the inspection of CFRP structures is studied.

Classic ultrasonic tomographic methods utilize the Time-Of-Flight (TOF) or attenuation changes to inspect the interior of bulky media (Stotzka et al. 2005). Attempts were made to apply a similar approach to inspect plate-like metal and CFRP structures (Hinders et al. 1998, Malyarenko and Hinders 2000, Malyarenko and Hinders 2001, McKeon and Hinders 1999). However, methods like cross-hole tomography, fan beam tomography or ray-tracing suffer from several problems when it comes to thin plate like structures (Hou et al. 2004). The main issues are: precise TOF estimation, dispersion and sensitivity to the changes in the measuring environment (Hay et al. 2006). Another drawback is the complexity of the experimental equipment.

Unlike classic ultrasonic methods that use TOF or attenuation, the centre point of the RAPID method is the Signal Difference Coefficient (SDC). By using the signals from an ultrasonic sparse array and deriving the mutual SDC values between the transducers in this network, we are able to calculate the damage presence probability in the region of interest.

RAPID was described in detail by Gao et al. (2005), Gao et al. (2005), Hay et al. (2006) and Zhao et al. (2007). Further improvements of the method were made by Velsor et al. (2007), Sheen and Cho (2012). So far, most of the experimental work has been done on rather simple aluminium plate structures. Attempts have been made to overcome the sensitivity of RAPID to the harsh environmental conditions as well as to temperature variations (Zeng and Lin 2013, Hua et al. 2013).

In this paper, we would like to show that the RAPID technique is able to localize delaminations and impact damage in CFRP plate structures. We will also demonstrate, on numerical simulations as well as preliminary experimental measurements, that it can be used for baseline-free tomographic inspection of simple CFRP plate structures. To achieve this goal, nonlinear elastic wave phenomena was considered and appropriate processing and excitation techniques, such as Scaling Subtraction Method (SSM), were explored (Scalerandi et al. 2008, Solodov et al. 2010).

## 2 METHODS

RAPID utilizes the data from an ultrasonic sparse array that is coupled to the interrogated structure. This array usually consists of $N$ permanently attached piezopatches or angle-beam transducers (Sheen and Cho 2012). One of the common array geometries is depicted in figure 1.

### 2.1 Reconstruction Algorithm for Probabilistic Inspection of Damage (RAPID)

The first step in the probabilistic tomographic reconstruction is the calculation of the SDC for each transmitter-receiver pair in the array. The SDC value is basically a measure of the dissimilarity of two specified signals. In the presented work, we calculated the SDC value based on the correlation coefficient and mean square error (Sheen & Cho 2012). The latter was used for the baseline-free RAPID analysis.

Let the signal transmitted from the array element $i$ to element $j$ be denoted $B_{ij}$ and $D_{ij}$ for the intact (baseline) and damage state respectively. Then the correlation coefficient is defined as

$$\rho_{ij} = \frac{Cov(B_{ij}, D_{ij})}{\sigma(B_{ij})\sigma(D_{ij})}$$

$$= \frac{\sum_k (B_{ij}(t_k) - \mu_{ij}^B)(D_{ij}(t_k) - \mu_{ij}^D)}{\sqrt{\sum_k (B_{ij}(t_k) - \mu_{ij}^B)^2} \sqrt{\sum_k (D_{ij}(t_k) - \mu_{ij}^D)^2}}, \quad (1)$$

where $k = 1, 2, ..., n$. $n$ is the number of samples/length of the recorded signal, $t_k$ is a discrete time (sample index), and $\mu_{ij}^B$, $\mu_{ij}^D$ are the mean values of the baseline and damage state signals respectively. The SDC value can then be calculated using $\rho_{ij}$ as

$$SDC_{ij} = 1 - \rho_{ij}, \quad (2)$$

where $i, j = 1, ..., N$ and $N$ is the number of array elements. The total number of employed signals adds up to $\frac{N}{2}(N-1)$ if reciprocity is assumed and $N(N-1)$ otherwise.

The inspected area of the sample is then overlaid with a rectangular mesh as a next step of the algorithm. The *a priory* probability distribution $s_{ij}(x, y)$ is defined for each transmitter-receiver (TR) and point of the mesh with coordinates $[x, y]$ as follows:

$$s_{ij}(x, y) = \begin{cases} \frac{\beta - R_{ij}(x,y)}{1-\beta}, & \text{if } \beta > R_{ij}(x, y) \\ 0, & \text{if } \beta \leq R_{ij}(x, y) \end{cases}, \quad (3)$$

where $\beta$ stands for an free threshold parameter that defines the area influenced by one TR pair (see Figure 2). $R_{ij}(x, y)$ is the geometrical function defined as

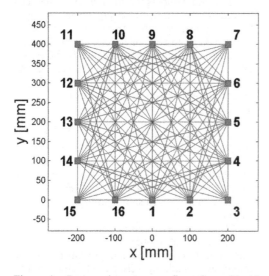

Figure 1. Rectangular array configuration with 16 active transducers. The lines connect different TR pairs and mark the region of interest.

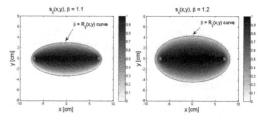

Figure 2. Values of the $s_{ij}(x, y)$ function calculated for a rectangular mesh. Transducers are indicated by the small squares.

$$R_{ij}(x,y) = \frac{\sqrt{(x_i-x)^2+(y_i-y)^2}}{\sqrt{(x_j-x_i)^2+(y_j-y_i)^2}} + \frac{\sqrt{(x_j-x)^2+(y_j-y)^2}}{\sqrt{(x_j-x_i)^2+(y_j-y_i)^2}} \quad (4)$$

To observe the effect of the parameter $\beta$ on the $s_{ij}(x,y)$ see figure 2.

The final probabilistic 2D heatmap $P(x,y)$ that describes the potential damage presence is finally calculated as

$$P(x,y) = \sum_{i=1}^{N} \sum_{j=1, i \neq j}^{N} SDC_{ij} s_{ij}. \quad (5)$$

The full matrix array is considered in this case resulting in $N(N-1)$ TR pairs.

## 2.2 Baseline-free RAPID

The conventional RAPID is a baseline dependent method with all its disadvantages like, e.g. temperature and environmental instability (Hay et al. 2006, Hua et al. 2013). The unavailability of intact state signals for the parts that are already in operation is also a significant drawback. Therefore, a baseline-free version of the RAPID is a much needed improvement.

We used the SSM method in order to replace the baseline signals. In order to eliminate the need for baseline or "defect-free" signals, we replaced these by the low amplitude "defect-free" reference signal acquired on damaged sample, denoted again $B_{ij}$. The other set of signals $D_{ij}$ is then acquired under the same measurement conditions only with the excitation amplitude upscaled by a factor

$$a = \frac{D_{ex}}{B_{ex}}, \quad (6)$$

where $B_{ex}, D_{ex}$ are the amplitudes of the excitation signals. The SDC coefficient is then calculated directly using the mean square difference formula as:

$$SDC_{ij} = \frac{1}{n} \sum_{k=1}^{n} (aB_{ij}(t_k) - D_{ij}(t_k))^2. \quad (7)$$

If the inspected system is purely linear, both response signals scale up perfectly and $SDC_{ij} = 0$. However, if we assume the presence of the nonlinear defect in the interrogated sample, it becomes a non-zero. Using this assumption the SDC in (5) can be replaced by equation (7). $s_{ij}(x,y)$ remains the same, and the method becomes baseline independent.

## 3 NUMERICAL SIMULATIONS

Numerical simulations were carried out using COMSOL® finite element software. The sample was simulated as an orthotropic plate with dimensions of 288 × 296 × 2.7 mm (see Figure 3). The modelled material was a T300/924C CFRPpolymer with density $\rho = 1548 /^3$. The elastic properties of the simulated plate are noted in table 1. Rayleigh damping model was used with the mass damping parameter $\alpha_{dK} = 0$ and stiffness damping parameter

$$\beta = \frac{1}{2\pi fQ}, \quad (8)$$

where $f$ is the excitation frequency and $Q = 20$ the quality factor.

The simulated sparse array consisted of 8 two-dimensional square domains 10 × 10 mm located at the top surface of the plate. The normal displacement $z(t)$ is prescribed at the domain that acts like a transmitting element. More precisely, a Hanning windowedsine burst

$$z(t) = \frac{A}{2}(1 - \cos(\frac{2\pi t}{T}))\sin(2\pi ft), \quad (9)$$

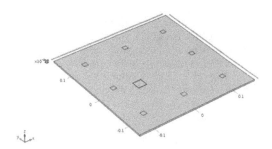

Figure 3. Model of the orthotropic CFRP plate with a linear delamination centered at 1/4 of the plate thickness. 8 elements form the array.

Table 1. Elastic properties of the simulated orthotropic plate.

| Young's mod. [GPa] | Shear mod. [GPa] | Poisson's ratio |
|---|---|---|
| $E_x = 127.11$ | $G_{xy} = 5.0$ | $\nu_{xy} = 0.320$ |
| $E_y = 8.34$ | $G_{yz} = 2.7$ | $\nu_{yz} = 0.461$ |
| $E_z = 8.85$ | $G_{zx} = 4.8$ | $\nu_{zx} = 0.009$ |

where $f = 50$ kHz is the frequency, $A$ is the excitation amplitude, $t$ time and $T$ the duration of the signal, was used to excite the plate. The duration $T$ corresponds to 20 cycles (numerical study) and 3 cycles (experimental study) at the excitation frequency $f$. The excitation amplitude $A$ was varied from 10 nm to 1 μm. The normal displacement was measured at the positions of the receiving elements and 800 samples were acquired at 1 MHz for both linear and nonlinear simulations. The delaminations were simulated in following two ways.

### 3.1 Linear delamination

A linear delamination was modelled as a $20 \times 20 \times 2.7$ mm block subdomain with different material properties centered at 1/4 of the thickness over the plate. Inside the region of the linear delamination, all elastic properties in table 1 were multiplied by a factor of 0.9, so that the delamination appears as a "softer" region in an otherwise homogeneous orthotropic material.

### 3.2 Nonlinear delamination

A nonlinear defect was modelled as a "clapping" delamination under the influence of dynamic boundary conditions. This model, proposed and implemented by (Delrue and Van Den Abeele 2012), is in fact a spring-damper model that accounts for the possibility of opposite delamination's faces touching each other. The local elastic properties of the material are not influenced by the introduction of the "clapping" behaviour of the delamination. For details of the model see (Delrue 2011). This "clapping" delamination is again centered at 1/4 of the thickness of the plate and it has exactly the same dimensions as the linear defect.

## 4 NUMERICAL SIMULATION RESULTS

Several test cases were simulated, e.g. 1 linear delamination, 1 nonlinear delamination and 2 nonlinear delaminations. The baseline signals $B_{ij}$ for the linear RAPID were recorded using the plate model without any delamination inside. The results and output of the tomographic reconstruction for these cases are summarized in the two following sections.

### 4.1 Linear delamination + RAPID

The first simulation is a test case for conventional RAPID, demonstrating that a linear delamination can be easily detected using the algorithm described by equation (5). The location and size of the defect are very well recovered (see Figure 4).

### 4.2 Nonlinear delamination + RAPID

The second simulation result shows the detection of a nonlinear delamination by the conventional RAPID. If the excitation amplitude $D_{ex} = B_{ex} \Rightarrow a = 100$ then there is an indication of the defect in the resulting image (see Figure 5). The nonlinearity ("clapping") was triggered and the delamination can be detected using the conventional RAPID due to the added nonlinearity in the signal. This effect illustrates that $a$ is a critical parameter

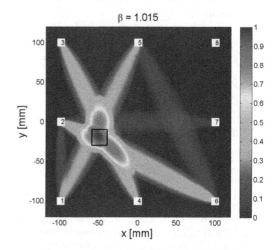

Figure 4. Conventional RAPID with single linear delamination. Excitation parameters: 20 cycles, Hanning window, $f = 50$ kHz, $B_{ex} = 1$ μm, $\beta = 1.015$. Black box indicates the damaged area. The defect is centered at [−50, −20].

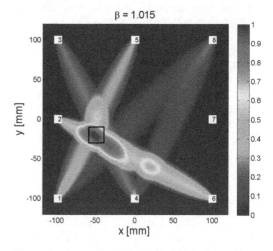

Figure 5. Conventional RAPID with single nonlinear delamination. Excitation parameters: 20 cycles, Hanning window, $f = 50$ kHz, $B_{ex} = 1$ μm, $\beta = 1.015$. Black box indicates the damaged area. The defect is centered at [−50,−20].

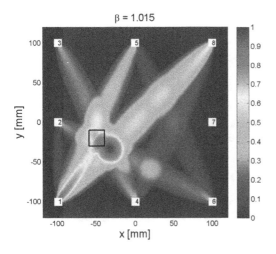

Figure 6. Baseline-free RAPID with single nonlinear delamination. Excitation parameters: 20 cycles, Hanning window, $f = 50$ kHz, scaling factor $a = 100$, $B_{ex} = 10$ nm, $D_{ex} = 1$ μm. Black box indicates the damaged area. The defect is centered at [−50, −20].

for detection of the nonlinear delaminations. It has to be high enough in order to cross the nonlinear threshold and initiate the "clapping" behaviour.

### 4.3 Nonlinear delamination + baseline-free RAPID

The third example shows the detection of a nonlinear defect by the baseline-free RAPID. The excitation amplitude, increased by a scaling factor $a = 100$, activates the "clapping" behaviour and triggers the nonlinear distortion of the propagating signal. This results in a clear indication of the damaged area (see Figure 6). However, it has to be noted that the indication doesn't match with the theoretical position of the defect as precisely as with the baseline-dependent RAPID.

## 5 EXPERIMENTS

### 5.1 Equipment

The experimental setup validating the RAPID tomographic reconstruction algorithm consisted of one arbitrary waveform generator NI-PXI 5411 and 12-channel receiving array connected to 4 NI-PXI 5122 digitizers housed in a single PXI chassis. The excitation signal was amplified using an AR150A100B amplifier (Amplifer Research) and then fed to one of the PI DuraAct® piezopatches via a multiplexing unit. The piezopatches were permanently attached to the structure using Loctite Hysol® glue. The data acquisition process was controlled using LabVIEW®.

### 5.2 Test sample

The tested sample was a large CFRP plate with dimensions of 500 × 500 × 5 mm with a low velocity barely visible impact (BVID) in the upper right hand corner (see Figure 8). 16 elements were arranged in a rectangular array around as depicted in figure 8. The sample was quasi-isotropic. The low frequency $A_0$ mode was selected for the inspection, because $A_0$ is rather nondispersive in the selected region with the phase velocity reaching a constant level (see Figure 7).

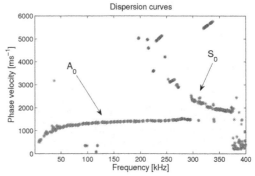

Figure 7. Dispersion curves for the tested sample as measured by laser Doppler vibrometer and processed using the 2D FFT algorithm.

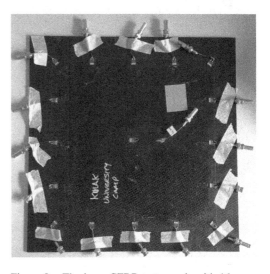

Figure 8. The large CFRP test sample with 16 array elements organized into rectangular array.

### 5.2.1 RAPID

An excitation waveform consisting of a 3-cycle Hanning windowed sine burst at 50 kHz was used. The resulting conventional RAPID result is depicted in figure 9. The position of the impact is well indicated in the upper right hand corner in between array elements 5 and 9. A classic ultrasonic C-scan is shown in Figure 10 for comparison.

### 5.2.2 Baseline-free RAPID

The baseline-free RAPID algorithm was applied to the same test sample with a scaling coefficient $a =$

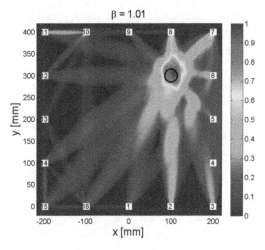

Figure 9. Conventional RAPID image of the large CFRP plate. Excitation parameters: 3 cycles, Hanning window, $f = 50$ kHz, $\beta = 1.01$. The original defect zone is marked with a black circle.

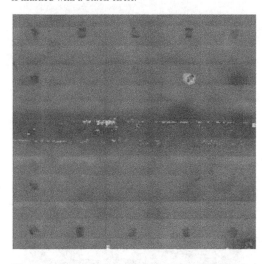

Figure 10. Ultrasonic C-scan of the large CFRP sample. Damage is located in the upper right hand corner. Dark blue strip in the middle of the plate is an embedded thin copper foil in the bottom surface layer.

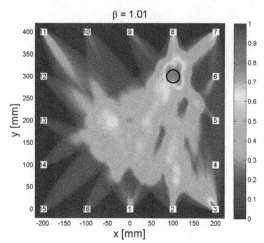

Figure 11. Baseline-free RAPID localization. Excitation parameters: 3 cycles, Hanning window, $f = 50$ kHz, $\beta = 1.01$, $a = 10$. The original defect zone is marked with a black circle.

10. The resulting image is shown in figure 11. The lower excitation voltage was $B_{ex} = 8$ V. The correct defect position was roughly localized. However, the location accuracy is lower than with the conventional RAPID, but still sufficient to reliably discern it from intact areas, without having to rely on the "intact state" reference signals.

## 6 CONCLUSIONS

We have shown that conventional RAPID is capable of damage detection in thin plate-like CFRP components. The technique was able to detect a barely visible low-velocity impact in a 500 × 500 mm CFRP quasi-isotropic sample with a reasonable precision. However, even if there is an ongoing effort to make the calculation of the SDC parameter more robust, it is still very sensitive to the changes in temperature, coupling of the sensors and other environmental parameters. Therefore its applications in harsh industrial environment are very limited. On the contrary, the baseline-free RAPID seems to be more promising for real-world applications, because it overcomes all the problems related to the measuring conditions without sacrificing too much accuracy.

Numerical simulations have confirmed the potential of the baseline-free RAPID technique to detect nonlinear delaminations, which was then verified by preliminary experimental results. However, further research has to be conducted in order to ascertain theperformance of this method on other types of defects and samples with more complicated shape and elastic properties. An especially interesting and challenging question in this contaxt will be the potential of baseline-free RAPID on anisotripic conposites.

## ACKNOWLEDGEMENT

The research leading to these results has gratefully received funding from the European Union Seventh Framework Programme (FP7/2007–2013) for research, technological development and demonstration under the Grant Agreements n 284562 (SARISTU), and n 314768 (ALAMSA).

## REFERENCES

Delrue, S. (2011). *Simulations as a guidance to support and optimize experimental techniques for ultrasonic non-destructive testing*. Ph. D. thesis, Katholic University of Leuven.

Delrue, S. & K. Van Den Abeele (2012). Three-dimensional finite element simulation of closed delaminations in composite materials. *Ultrasonics 52*(2), 315–324.

Gao, H., Y. Shi, & J.L. Rose (2005). Guided wave tomography on an aircraft wing with leave in place sensors. *Review of Quantitative Nondestructive Evaluation 24*, 1788–1795.

Gao, H., Y. Shi, J.L. Rose, X. Zhao, C. Kwan, & V. Agarwala (2005). Ultrasonic Guide Wave Tomography in Structural Health monitoring of an aging aircraft wing. *Proc of the American Society of Nondestructive Testing Fall Conference*, 412–415.

Giurgiutiu, V. (2003). Lamb wave generation with piezolectric wafer active sensors for structural health monitoring. *Smart Structures and Materials* (March), 2–6.

Giurgiutiu, V. & C. Soutis (2012, February). Enhanced Composites Integrity Through Structural Health Monitoring. *Applied Composite Materials 19*(5), 813–829.

Habermehl, J. & A. Lamarre (2008). Ultrasonic phased array tools for composite inspection during maintenance and manufacturing. *17th World conference on nondestructive testing*, 25–28.

Hay, T.R., R.L. Royer, H. Gao, X. Zhao, & J.L. Rose (2006, August). A comparison of embedded sensor Lamb wave ultrasonic tomography approaches for material loss detection. *Smart Materials and Structures 15*(4), 946–951.

Helfen, T.B., R.S. Venkat, U. Rabe, S. Hirsekorn, & C. Boller (2012, February). Characterisation of CFRP Through Enhanced Ultrasonic Testing Methods. *Applied Composite Materials 19*(6), 913–919.

Hinders, M., E. Malyarenko, & J. McKeon (1998). Contact scanning Lamb wave tomography. *The Journal of the Acoustical Society of America 104*(3), 1790.

Hou, J., K.R. Leonard, & M.K. Hinders (2004, December). Automatic multi-mode Lamb wave arrival time extraction for improved tomographic reconstruction. *Inverse Problems 20*(6), 1873–1888.

Hua, J.D., L. Zeng, J. Lin, & W. Shi (2013, July). Ultrasonic Guided Wave Tomography for Damage Detection in Harsh Environment. *Key Engineering Materials 569–570*, 1005–1012.

Malyarenko, E.V. & M.K. Hinders (2000, October). Fan beam and double crosshole Lamb wave tomography for mapping flaws in aging aircraft structures. *The Journal of the Acoustical Society of America 108*(4), 1631–9.

Malyarenko, E.V. & M.K. Hinders (2001). Ultrasonic Lamb wave diffraction tomography. *Ultrasonics 39*(4), 269–281.

McKeon, J.C.P. & M.K. Hinders (1999). Parallel projection and crosshole Lamb wave contact scanning tomography. *The Journal of the Acoustical Society of America 106*(5), 2568–2577.

Santoni, G.B., L. Yu, B. Xu, & V. Giurgiutiu (2007). Lamb Wave-Mode Tuning of Piezoelectric Wafer Active Sensors for Structural Health Monitoring. *Journal of Vibration and Acoustics 129*(6), 752.

Scalerandi, M., A.S. Gliozzi, C.L.E. Bruno, & K. Van Den Abeele (2008, November). Nonlinear acoustic time reversal imaging using the scaling subtraction method. *Journal of Physics D: Applied Physics 41*(21), 215404.

Sheen, B. & Y. Cho (2012, May). A study on quantitative lamb wave tomogram via modified RAPID algorithm with shape factor optimization. *International Journal of Precision Engineering and Manufacturing 13*(5), 671–677.

Solodov, I., N. Krohn, & G. Busse (2010). Nonlinear Ultrasonic NDT for Early Defect Recognition and Imaging. In *ECNDT*.

Stotzka, R., N.V. Ruiter, T.O. Müller, R. Liu, & H. Gemmeke (2005). High resolution image reconstruction in ultrasound computer tomography using deconvolution. In *Ultrasound*, pp. 315–325.

Velsor, J.K., H. Gao, & J.L. Rose (2007). Guided-wave tomographic imaging of defects in pipe using a probabilistic reconstruction algorithm. *Insight - Non-Destructive Testing and Condition Monitoring 49*(9), 532–537.

Xu, B. (2009). *Structural health monitoring instrumentation, signal processing and interpretation with piezoelectric wafer active sensors*. Ph. D. thesis.

Yu, L. & V. Giurgiutiu (2005). Advanced signal processing for enhanced damage detection with piezoelectric wafer active sensors. *Smart Structures and Systems 1*(2), 185–215.

Zeng, L. & J. Lin (2013, October). Chirp-based dispersion precompensation for high resolution Lamb wave inspection. *NDT & E International*.

Zhao, X., H. Gao, G. Zhang, B. Ayhan, F. Yan, C. Kwan, & J.L. Rose (2007, August). Active health monitoring of an aircraft wing with embedded piezoelectric sensor/actuator network: I. Defect detection, localization and growth monitoring. *Smart Materials and Structures 16*(4), 1208–1217.

*Emerging Technologies in Non-Destructive Testing VI – Aggelis et al. (Eds)*
*© 2016 Taylor & Francis Group, London, ISBN 978-1-138-02884-5*

# Modelling the dispersive behavior of fresh and hardened concrete specimens through non-local lattice models

### S.N. Iliopoulos & D.G. Aggelis
*Department of Mechanics of Materials and Constructions, Vrije Universiteit Brussel, Brussels, Belgium*

### D. Polyzos
*Department of Mechanical Engineering and Aeronautics, University of Patras, Patras, Greece*

ABSTRACT: The propagation of longitudinal ultrasonic waves through fresh and hardened concrete materials is strongly affected by dispersion and this is clearly indicated experimentally from the change of phase velocity at low frequencies. Analytically, only few attempts have been made to explain this dispersive behavior through enhanced elastic theories. The most commonly used higher order theory is the dipolar gradient elastic theory which takes into account the microstructural effects in heterogeneous media like concrete. These microstructural effects are described by two internal length scale parameters (g and h) which correspond to the micro-stiffness and micro-inertia respectively. In the current paper, it is shown that Mindlin's dipolar theory can effectively predict the velocity dispersion of fresh cementitious materials and hardened concrete specimens with various water over cement (w/c) ratios and sand contents and additionally inform on the phase change through the monitoring of the relative difference of the g and h values.

## 1 INTRODUCTION

The quality assessment and control of concrete elements and structures is of paramount importance for a prolonged and cost effective civil infrastructure. It is based on the determination of the material's early age properties and the monitoring of their evolution which can be effectively achieved by means of the ultrasonic pulse velocity nondestructive technique (Trtnik & Gams 2014). Although many works have been published on ultrasonic measurements in concrete, only a few of them deal with the dispersive nature of the material. There, it has been observed that pulse velocity is frequency dependent while the phenomenon of wave dispersion is clearly depicted on the change of phase velocity at low frequencies.

Concrete is a strongly heterogeneous material consisting of aggregates, sand, cement and water. It covers a wide range of length scales varying from a few cm to the level of μm rendering the propagation of an ultrasonic wave through it a much complicated process needing further investigation.

The dynamic behavior of concrete has been treated analytically in literature in many different ways. Traditionally, concrete was assumed a macroscopically homogeneous linear elastic material (Reynolds et al. 1978), but although simple and practical, it failed to predict the dispersive nature

of this bulk medium. Concrete was then considered isotropic viscoelastic with frequency dependent Lame constants (Fan et al. 2013), but it could only explain the velocity dispersion of hardened specimens since it provided only increasing phase velocities as a function of frequency. Moreover, viscoelastic considerations did not provide internal length scale parameters that correlate the microstructure with the macro-structural behavior of cement-based materials, even though their application was necessary to account for the significant creep deformation of the latter.

Alternatively, scattering theories were introduced to describe the experimentally observed wave dispersion in fresh and hardened concrete (Aggelis et al. 2005, Chaix et al. 2012). Despite the variety of scatterers (aggregates, light inclusions, air voids) only few of them could closer estimate the dispersive trends at the low frequency regime (Mpalaskas et al. 2014), while, on the other hand, they introduced many material and geometric parameters which rendered any inversion process a very difficult and time consuming task.

Recently, cementitious materials (cement paste, mortar and concrete) were tackled in literature as poroelastic (Ulm et al. 2004) characterized, among others, by the Biot and Skempton coefficients. However, so far, there is no experimental evidence that renders Biot theory adequate to explain wave dispersion in all types of cementitious materials.

Since none of the aforementioned approaches could provide intrinsic parameters connecting the microstructure of the considered heterogeneous media with the dynamic behavior of the macrostructure, researchers focused their attention on generalized continuum theories. The most representative are the theories of couple stresses (Toupin 1964, Koiter 1964), Cosserat elasticity, multipolar elasticity (Green & Rivlin 1964), strain gradient elasticity (Mindlin 1964, 1965), nonlocal elasticity (Eringen 1992) and finally of micro-morphic, micro-stretch and micro-polar elasticity (Eringen 1999).

In the current paper, the dispersion of longitudinal ultrasonic pulses in fresh and hardened concrete is simulated through the strain gradient elastic theory introduced by Mindlin in 1964. The simplest form of Mindlin's theory describes a linear elastic medium where microstructural effects are also considered. It is called dipolar, since those microstructural effects are expressed by two internal length scale parameters, which are the micro-stiffness (g) and micro-inertia (h) respectively.

Although Mindlin's dipolar theory and almost all the aforementioned enhanced theories define the dimensions of the internal parameters, they don't explicitly relate them with the size of the microstructure. To address this issue discrete lattice models have been developed which have the advantage of taking into account microstructural effects more accurately by combining all the geometric and material characteristics of the micro-structure with the dynamic behavior of the matter. Such a discrete lattice model with nonlocal interactions will be presented in the following section.

## 2 THEORETICAL BACKGROUND

### 2.1 1-D Mindlin's dipolar theory

According to Mindlin's celebrated paper in 1964, the classical elastic medium consideration can be enriched with new terms in the expressions of the potential and kinetic energy density. Thus, if the gradients of strains and the gradients of the velocities are introduced in the first and second expression respectively the microstructural effects of materials are taken into account. Applying Hamilton's principle on the variation of the strain and kinetic energy leads to the following one dimensional wave equation describing a continuum with microstructure:

$$\partial_{xx}u - g^2\partial_{xxxx}u = \frac{1}{c^2}\left(\partial_{tt}u - h^2\partial_{xxtt}u\right), c^2 = \frac{E}{\rho} \quad (1)$$

The classical and non-classical boundary conditions are referred to pairs $(u, p)$ and $(\partial_x u, R)$, respectively, with $p$ and $R$ representing traction and double traction, respectively, written in the form:

$$p = E\left(\partial_x u - g^2\partial_{xxx}u\right) + \frac{1}{3}\rho'd^2\partial_{xtt}u \quad (2)$$

$$R = Eg^2\partial_{xx}u \quad (3)$$

where $\partial_x$ and $\partial_t$ denote differentiation over the variable $x$ and $t$, respectively, u is the displacement in the longitudinal direction, c the wave velocity, E the Young's modulus, ρ the mass density, $g^2$ is the new internal length scale parameter (units of $m^2$) corresponding to micro-stiffness, $h^2$ is the second new intrinsic parameter (units of $m^2$) corresponding to micro-inertia, ρ' is the mass density of microstructure and $d$ is half the size of the microstructural cell embedded in a unit material cell.

### 2.2 Discrete lattice model with nonlocal interactions

The physical equivalent of the Mindlin's dipolar theory is the nonlocal lattice model shown in Figure 1. It consists of discrete identical masses M, local and nonlocal springs of stiffness $k_1$ and $k_2$ respectively and dashpots with damping coefficient c. It is interesting to notice that the springs connecting neighboring masses at distances l "carry" also their mass (m) unlike the nonlocal springs. These masses simulate the concrete matrix and are responsible for the micro-inertia term while the masses M are the mechanical similitude of the concrete aggregates. The stiffness of the springs is described by Hooke's law as $k_i = E_i A/l$, i = 1, 2 where l is the unit cell of the heterogeneous material.

Assuming for the sake of simplicity the lattice model of Figure 1 without considering the dashpots concludes to the following one dimensional wave equation, classical and non-classical boundary conditions:

$$\partial_{xx}u - \frac{l^2}{4}b^2\partial_{xxxx}u = \frac{1}{c^2}\left(\partial_{tt}u - \frac{l^2}{3}\frac{\rho'}{\rho}\partial_{xxtt}u\right), c^2 = \frac{E}{\rho} \quad (4)$$

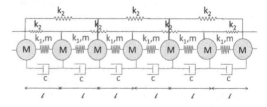

Figure 1. Non local lattice model with two-neighbor interactions.

$$p = E\left(\partial_x u - \frac{l^2}{4}b^2 \partial_{xxx} u\right) + \frac{1}{3}\rho' l^2 \partial_{xtt} u \quad (5)$$

$$R = E\frac{l^2}{4}b^2 \partial_{xx} u \quad (6)$$

where:

$$E = E_1 + 4E_2 = E_1\left(1 + 4\frac{E_2}{E_1}\right) \quad (7)$$

$$b^2 = \frac{E_1 + 16E_2}{E_1 + 4E_2} = \frac{1 + 16\frac{E_2}{E_1}}{1 + 4\frac{E_2}{E_1}} \quad (8)$$

If Equations 1–3 are compared with Equations 4–6 it can be noticed that:

$$g^2 = \frac{l^2}{4}b^2 \quad (9)$$

$$h^2 = \frac{l^2}{3}\frac{\rho'}{\rho} \quad (10)$$

where $\rho = \rho' + \rho_M$.

As it is clearly shown the aforementioned lattice model fully reproduces the Mindlin's differential equation, as well as the classical and non-classical boundary conditions and it additionally provides a physical link between the microstructural coefficients and the geometrical and mechanical properties of the considered medium.

Assuming a plane harmonic wave with angular frequency ω and wavenumber $k_p$ propagating in the gradient elastic medium described by Equation 1 or the nonlocal discrete medium described by Equation 4 leads to the following wave dispersion equation:

$$V_p = \frac{\omega}{\sqrt{\dfrac{-(c^2-\omega^2 h^2)+\sqrt{(c^2-\omega^2 h^2)^2+4c^2 g^2 \omega^2}}{2c^2 g^2}}} \quad (11)$$

where $V_p = \omega/k_p$ is the phase velocity of the dispersive medium characterized by microstructural effects. For more details on the derivation of the presented in this section equations refer to Polyzos & Fotiadis (2012).

## 3 EXPERIMENTAL DETAILS

The velocity measurements on the fresh and hardened material included the use of a waveform generator and two broadband piezoelectric transducers

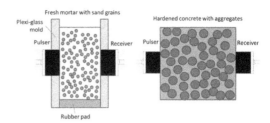

Figure 2. Experimental setup for fresh mortar and hardened concrete velocity measurements in a schematic representation.

in a typical through transmission arrangement. In the first case, a special plexi-glass mold was used where holes of the shape and size of the sensors had been drilled to provide direct contact with the cementitious material (Fig. 2). The propagating distance (thickness of the specimen) was defined by a U-shaped rubber between the two external plates. Concerning the hardened material, the sensors were also directly attached on the tested concrete elements (Fig. 2).

Among the variety of tested fresh mortar specimens, in the current study, those that have a) w/c ratio by mass equal to 0.55 and sand content 0 (cement paste), 25%, 30%, 40% or b) fixed sand content (30%) and w/c = 0.46;0.475;0.525;0.55 will only be considered, while for the hardened concrete specimens those that have aggregate to cement (a/c) ratio by mass equal to 3 and w/c = 0.375;0.425;0.45 will only be taken into account. It needs to be mentioned that the size of sand grains in the fresh mortar tests was ranging from 1 mm to 4 mm. The hardened cubic concrete specimens had a dimension of 150 mm and consisted of cement paste (cement powder and water), sand grains of maximum size 4.75 mm and coarse aggregates of maximum size 37.5 mm.

Further details on the experimental setup, the phase velocity measurements as well as on the composition and the number of tested fresh and hardened concrete materials can be found in Aggelis et al. (2005) and in Philippidis & Aggelis (2005).

## 4 RESULTS

### 4.1 Identification of the g and h coefficients using the Mindlin's wave dispersion equation

The experimentally obtained pairs of phase velocity and frequency of the fresh specimens are presented with red dots in Figures 3–4. The first figure concerns specimens with the same w/c ratio (0.55) but a varying sand content from 0 to 40%, while the second refers to specimens with sand content

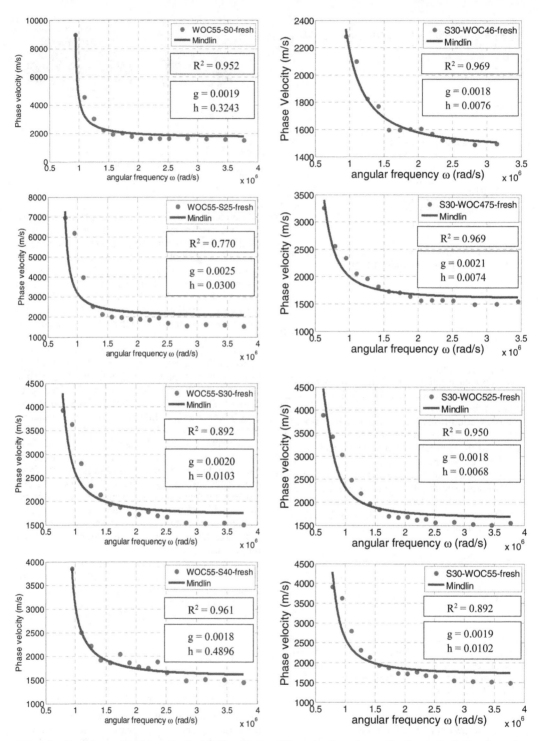

Figure 3. Experimentally observed velocity dispersion of fresh concrete with various sand contents and theoretical predictions using the Mindlin's wave dispersion equation [g and h in m].

Figure 4. Experimentally observed velocity dispersion of fresh concrete with various w/c ratios and theoretical predictions using the Mindlin's wave dispersion equation [g and h in m].

30% and w/c ratio ranging from 0.46 to 0.55 respectively. Both cases show strong dispersion with phase velocities dropping from really high values (even 9000 m/s in the first case) to the value of sound velocity in water (1500 m/s). The drop is seen around 150 kHz ($0.94 \times 10^6$ rad/s), while beyond this frequency the phase velocity remains almost constant. Since none of the fresh mortar ingredients exhibits individually such high velocities it is deemed necessary to search on microstructural effects. In general, it can be said that stronger dispersion is noted for less stiff material (low sand content and high water content), while when the water decreases or sand increases the dispersion drops being always significant

On the contrary, as shown in Figure 5 the dispersion in hardened concrete is the opposite, characterized by a continuous increase of the phase velocity up to 150–200 kHz (angular frequencies around $10^6$ rad/s) beyond which it remains constant (up to the maximum frequency tested—1 MHz). It is therefore made clear that the microstructural effects responsible for the dispersion turn around as the material is transformed from the initial liquid phase to a hardened solid medium.

In order to identify the two "unknown" microstructural coefficients (g and h) of the Mindlin's wave dispersion equation 11 the experimental data of fresh and hardened concrete are separately used. Through a least square minimization procedure between the experimental and theoretical data the values of the micro-stiffness g, the micro-inertia h and the coefficient $R^2$ which shows how good the theoretical curves (continuous lines) fit to the experimental data are found and displayed in Figures 3–5.

It is interesting to notice that when the material is fresh (Figs 3–4), always the microstructural coefficient g is smaller than the microstructural coefficient h. Moreover, the values of the micro-stiffness g are very close to each other. They are of the order of mm which needs to be highlighted since it is the same as the grains used for the fresh mortar tests of this study (maximum size of 4 mm).

Unlike the fresh material, when concrete is hardened (Fig. 5), the inverse relation between the microstructural coefficients is observed; the micro-stiffness parameter g is greater than the micro-inertia parameter h in all w/c cases, suggesting that monitoring of the relative difference of the microstructural coefficients could effectively provide information on the setting process of the above mentioned material in a nondestructive way.

It is equally interesting to notice that an increase of the w/c ratio causes a monotonic decrease of the micro-stiffness g and that the latter is of the order of a few cm, which is in agreement with the aggregate sizes used for the experiments on hardened concrete (maximum size 37.5 mm). Comparing g values of fresh concrete, which vary between 1.8 mm and 2.5 mm, with g values of hardened concrete which vary between 23.8 mm to 26.4 mm it is evident that the first case exhibits an order of magnitude lower values than the second case, similarly to the particle size used in these cases.

Finally, as shown from Figures 3–5, the same Mindlin's wave dispersion equation can closely predict the dispersive behavior of both fresh and hardened concrete material.

### 4.2 Identification of the g and h coefficients of the hardened concrete using the lattice model

It is already stated in section 2.2 that the presented nonlocal lattice model with two-neighbor interactions fully reproduces the theory of Mindlin, while its advantage is that it can attribute properties to the microstructural coefficients. To do so, certain assumptions on the concrete's Young's modulus E and that of the local springs $E_1$ need to be made. The first is defined according to the mixture law, while the second is realistically assumed equal to the Young's modulus of the matrix $E_m$:

$$E = E_a v_a + E_m v_m \qquad (12)$$

where $E_a$, $v_a$ = Young's modulus and volume fraction of the concrete aggregates (Table 1); $E_m$, $v_m$ = Young's modulus and volume fraction of the concrete matrix (Table 1).

As shown in Table 1, the stiffness of the matrix decreases as the w/c increases.

Based on the previous two assumptions the ratio $E_2/E_1$ expressing the intensity of the nonlocal interactions is calculated according to Equation 7 and displayed in Table 2. It is observed that the nonlocal interactions become stronger as the w/c ratio increases.

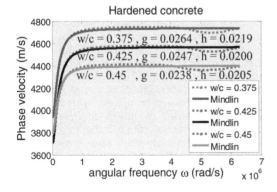

Figure 5. Experimentally observed velocity dispersion of hardened concrete with various w/c ratios (dashed lines) and theoretical predictions (continuous lines) using the Mindlin's wave dispersion equation [g and h in m].

Table 1. Young's modulus of the hardened concrete and physical-mechanical properties of the ingredients.

| w/c | 0.375 | 0.425 | 0.45 |
|---|---|---|---|
| Aggregates | | | |
| $E_a$(GPa) | 70 | 70 | 70 |
| $\rho_a$(kg/m³) | 2650 | 2650 | 2650 |
| $v_a$ | 0.31 | 0.30 | 0.29 |
| Matrix | | | |
| $E_m$(GPa) | 20 | 17 | 14 |
| $\rho_m$(kg/m³) | 2100 | 2080 | 2050 |
| $v_m$ | 0.69 | 0.70 | 0.71 |
| Concrete: $E = E_a v_a + E_m v_m$ | | | |
| E (GPa) | 35.5 | 32.9 | 30.2 |

Table 2. Defined lattice size l, microstructural coefficients g and h and intensity of the nonlocal interactions for the hardened concrete with various w/c ratios using the lattice model.

| w/c | 0.375 | 0.425 | 0.45 |
|---|---|---|---|
| l (m) | 0.0355 | 0.0359 | 0.0363 |
| g (m) | 0.0269 | 0.0281 | 0.0293 |
| h (m) | 0.0164 | 0.0166 | 0.0169 |
| $E_2/E_1$ | 0.194 | 0.234 | 0.290 |

To define the microstructural coefficient g using the lattice model the definition of the lattice size l is also necessary, as described by Equation 9. Assuming a unit cell of size l containing a single mean aggregate of diameter d it can be proved that:

$$l \approx \frac{d}{\sqrt[3]{v_a}} \qquad (13)$$

To answer on the best diameter value of the mean aggregate (d) several values are considered ranging from 20 mm to 30 mm with an interval of 1 mm. It is found that if the diameter is 24 mm then the values of the microstructural coefficients g and h are very close to the ones obtained using the dipolar theory of Mindlin (Fig. 5). Moreover, the diameter of 24 mm is a reasonable dimension which realistically describes the representative microstructure considering that the maximum aggregate size is 37.5 mm.

Knowing the diameter and the aggregate volume fraction (Table 1), the lattice size is calculated according to Equation 13 and displayed on Table 2. It is interesting to notice that the lattice size l is slightly greater (approximately 36 mm) than the mean aggregate size and that an increase of the w/c ratio triggers a slight increase of the size l, since the aggregates are moved slightly further away.

The microstructural coefficient g is calculated according to Equation 9 and the result is also displayed in Table 2. The term $b^2$ which is necessary for the calculation of g is provided by Equation 8 making use of the values $E_2/E_1$ found in Table 2. Finally, the microstructural coefficient h is calculated according to Equation 10 where $\rho_M = \rho_a v_a$, $\rho = \rho_m v_m$ and $\rho = \rho_+ \rho_M$. The result is also displayed in Table 2.

It needs to be emphasized that using the lattice model it is strongly verified that when concrete is hardened the microstructural coefficient g is greater than the micro-inertia coefficient as explained using the dipolar theory of Mindlin.

From the above study, it is made clear that prior knowledge of the mean diameter, mechanical and physical properties of the concrete matrix and aggregates is able to explicitly provide the lattice size l, the intensity of the non-locality through the ratio $E_2/E_1$ as well as the microstructural coefficients g and h that could efficiently be used for quality characterization of concrete.

## 5 CONCLUSIONS

The presented work shows that the wave dispersion equation obtained using the dipolar gradient elastic theory of Mindlin applies to both fresh and hardened concrete and closely describes their dispersive behavior. Moreover, it is found that when concrete is fresh and liquid micro-inertia dominates compared to micro-stiffness, while when concrete is hardened the opposite is true. This implies that monitoring of the setting process is possible through the monitoring of the relative difference of the microstructural coefficients. The result of the least-square minimization procedure between the theoretical and experimental data indicated that micro-stiffness g is indicative of the microstructure, since it was of the order of mm for mortar and cm for concrete similar to the real distribution of the inclusions. Using the lattice model it is found that mechanical and geometrical properties of the considered medium can be attributed to the microstructural coefficients. It is also seen that the microstructural coefficients are similar with the ones obtained with Mindlin's theory and that all the observations made using the latter are verified through the use of the lattice model.

## ACKNOWLEDGEMENTS

The financial contribution of FWO Research Foundation-Flanders is gratefully acknowledged.

## REFERENCES

Aggelis, D.G., Polyzos, D., Philippidis, T.P. 2005. Wave dispersion and attenuation in fresh mortar: theoretical predictions vs. experimental results. J Mech Phys Solids 53:857–83.

Chaix, J-F., Rossat, M., Garnier, V., Corneloup, G. 2012. An experimental evaluation of two effective medium theories for ultrasonic wave propagation in concrete. J. Acoust. Soc. Am. 131 (6): 4481–4490.

Eringen, A.C. 1999. Microcontinuum Field Theories I: Foundations and Solids. New York: Springer.

Eringen, A.C. 1992. Vistas of Nonlocal Continuum Physics. International Journal of Engineering Science 30 (10): 1551–1565.

Fan, L.F., Wong, L.N.Y., Ma, G.W. 2013. Experimental investigation and modeling of viscoelastic behavior of concrete. Construction and Building Materials 48: 814–821.

Green, A.R, Rivlin, R.S., 1964. Multipolar continuum mechanics. Archive for Rational Mechanics and Analysis 17: 113–147.

Koiter WT. 1964. Couple stress in the theory of elasticity I–II. Proceedings Koninklijke Nederlandse Akademie Tan Wetenschappen B67: 17–44.

Mindlin R.D., 1964. Micro-Structure in Linear Elasticity. Arch. Ration. Mech. An. 16: 51–78.

Mindlin RD., 1965. Second gradient of strain and surface-tension in linear elasticity. International Journal of Solids and Structures 1: 417–438.

Mpalaskas, A.C., Thanasia, O.V., Matikas, T.E., Aggelis, D.G. 2014. Mechanical and fracture behavior of cement-based materials characterized by combined elastic wave approaches. Construction and Building Materials 50: 649–656

Philippidis, T.P., Aggelis, D.G. 2005. Experimental study of wave dispersion and attenuation in concrete. Ultrasonics 43: 584–595.

Polyzos, D., Fotiadis, D.I., 2012. Derivation of Mindlin's first and second strain gradient elastic theory via simple lattice and continuum models. International Journal of Solids and Structures 49: 470–480.

Reynolds, W.N., Wilkinson, S.J., Spooner, D.C. 1978. Ultrasonic wave velocity in concrete. Magazine of Concrete Research 30 (104): 139–144.

Toupin RA. 1964. Theories of elasticity with couple-stress. Archive for Rational Mechanics and Analysis 17:85–112.

Trtnik, G., Gams, M. 2014. Recent advances of ultrasonic testing of cement based materials at early ages. Ultrasonics 54: 66–75.

Ulm, F.J., Constantinides, G., Heukamp, F.H. 2004. Is concrete a poromechanics material? A multiscale investigation of poroelastic properties, Materials and Structures/Concrete Science and Engineering 37: 43–58.

*Emerging Technologies in Non-Destructive Testing VI – Aggelis et al. (Eds)*
*© 2016 Taylor & Francis Group, London, ISBN 978-1-138-02884-5*

# The Ultrasonic Polar Scan: Past, present and future

M. Kersemans
*Universiteit Gent, Zwijnaarde, Belgium*

A. Martens, S. Delrue & K. Van Den Abeele
*KULAK-KULeuven, Kortijk, Belgium*

L. Pyl, F. Zastavnik & H. Sol
*Vrije Universiteit Brussel (VUB), Brussels, Belgium*

J. Degrieck & W. Van Paepegem
*Universiteit Gent, Zwijnaarde, Belgium*

ABSTRACT: The Ultrasonic Polar Scan (UPS) technique was developed in the 1980's as a means to nondestructively determine the fiber direction of composites. Although recognized by many scientists as a sophisticated and promising methodology, only limited progress has been made in the past 30 years.

Recently however, the UPS technique experienced a strong revival and various modifications to the original UPS setup have been implemented. As such, several interesting capabilities and applications of the UPS technique have emerged.

Currently, we are investigating several novel research lines in order to bring the UPS technique to the next level of maturity, and to expand its applicability to a broader range of materials.

This paper gives a short historical overview of the results obtained with the UPS technique for investigating (damaged) fiber reinforced plastics, and indicates several currently investigated research lines.

## 1 INTRODUCTION

An Ultrasonic Polar Scan (UPS) is obtained by replacing the translational movement of a classical ultrasonic C-scan setup with a rotational movement. Hence, instead of scanning a surface at normal incidence, the UPS insonifies a predefined material spot from as many oblique incidence angles $\psi(\varphi, \theta)$ as possible. A schematic of the UPS method is presented in Figure 1. Figure 2 shows current state-of-the-art pulsed UPS recordings for aluminum and $[0]_8$ Carbon/Epoxy (C/E) laminate. The vertical incident angle $\theta$ is put on the radial axis, the in-plane polar angle $\varphi$ is represented along the angular axis, while the assigned gray scale is a measure for the transmitted (or reflected) amplitude. Hence, the pulsed UPS image comprises a collection of amplitudes of obliquely transmitted sound. Within the UPS image characteristic contours emerge which relate to the stiffness tensor of the insonified sample. In the example of Figure 2a, the circular symmetry clearly puts on view the isotropic nature of the aluminum sample. Figure 2b on the other hand has a stretched appearance due to the unidirectional character of the $[0]_8$ C/E laminate.

## 2 THE PAST: 1981–2010

In the early 1980's, the UPS technique was introduced by Van Dreumel and Speijer in a pulsed version in order to assess the fiber orientation of composites [1]. The beauty of the intriguing patterns made the pioneering authors state: *"A library of Polar-patterns, stored as "fingerprints", raises the possibility of laminate identification by pattern recognition"*[1]. It is unfortunate that in the years after, only one sequel study has been performed by Van Dreumel and Speyer to further explore the capabilities of the ultrasonic polar scan [2].

It took fifteen years before the technique has been investigated again through the work of Degrieck [3–5]. He used a modernized scanning system to obtain more accurate and detailed UPS experiments. The work of Degrieck has identified several practical applications of the UPS technique for composite materials: (i) estimation of the fiber direction, (ii) determination of fiber volume fraction, (iii) determination of the porosity and (iv) detection of fatigue damage [4–5]. In addition, he and co-worker Van Leeuwen implemented a numerical procedure for simulating UPS images of homogeneous composites [3], in view of

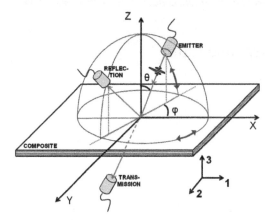

Figure 1. Schematic of UPS method.

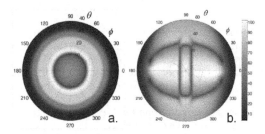

Figure 2. Pulsed UPS experiment for (a) aluminum and (b) [0]$_8$ C/E laminate.

bringing the technique to the next level: full quantitative characterization of the elastic properties of composite materials using a mixed experimental-numerical approach. Although the gap between experiment and simulation has never been bridged, their numerical results did contribute to the physical understanding of the formation of a UPS image. They found that the characteristic patterns are (more or less) a representation of critical bulk wave angles, while the global transmission amplitudes expose attenuation properties. Consequently, a UPS image may be used for characterizing viscoelastic material properties.

The ultrasonic polar scan research has been further extended by Declercq, first as a student of Degrieck [6–8] and afterwards as a professor at Georgia Institute of Technology [9]. He applied the technique for detecting tension-tension induced fatigue damage in glass fiber composites by tracing any changes in the characteristic fingerprint [6]. In addition, Declercq extended the simulation technique towards layered viscoelastic materials having arbitrary anisotropy using a global matrix method [7–8]. He experimentally implemented a Time-Of-Flight (TOF) version of the UPS method, and commented on the superior sensitivity (compared with amplitude recording) to the presence of damage features [9].

Hence, it is clear that the UPS technique is a promising technique for inspecting composites, but it is evenly clear that its capabilities have not yet been fully explored.

## 3 THE PRESENT: 2010–2014

We have identified several barriers impeding the further development of the UPS methodology. Three main barriers may be summarized as:

- Lack of high quality experimental data
- Lack of a computationally efficient simulation model
- Lack of adequate inverse modeling techniques to couple experiment to simulation.

With this background, the PhD research of the principal author (Mathias Kersemans) started in 2010 [10].

### 3.1 The ultrasonic polar scan revisited

The experimental barrier has been tackled by the development of a 5-axis scanner which performs UPS experiments in a fully automated way. The scanner insonifies more than 1,000,000 incidence angles $\psi(\varphi,\theta)$ in less than 15 minutes. In this way, we identified several pitfalls in the experimental procedure which prevented the correct recording of a UPS [11]. The exceptional quality of the obtained UPS experiments can be seen in Figure 2a-b for aluminum and [0]$_8$ C/E laminate.

Secondly, a simulation technique has been implemented to support experimental observations [12]. The simulation model is founded on a cascade matrix technique [13], and allows the simulation of UPS for layered viscoelastic anisotropic media. Compared to previous simulation models [7–8], we were able to significantly reduce the computational time. Figure 3a-b shows the simulated UPS of aluminum and [0]$_8$ C/E laminate. The used parameters

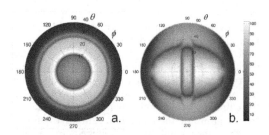

Figure 3. Pulsed UPS simulation for (a) aluminum and (b) [0]$_8$ C/E laminate.

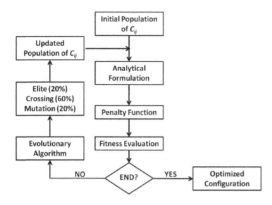

Figure 4. Schematic of inversion procedure to identify composite properties [14].

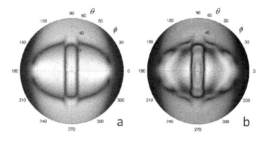

Figure 5. Experimental recordings of a $[0]_8$ C/E laminate: (a) pulsed UPS and (b) harmonic UPS.

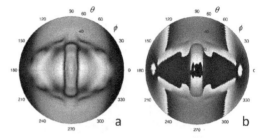

Figure 6. Harmonic UPS recordings for $[0]_8$ carbon/epoxy laminate: (a) amplitude and (b) phase analysis.

are identical to those of the experimental recordings. Hence, the simulations (Figure 3) may be straightforwardly compared with the experiments (Figure 2).

The third barrier is tackled by implementing an inversion procedure to couple experiment to simulation, in view of identifying composite parameters. Basically, the simulated UPS image is fitted to the recorded UPS image while updating the composite material properties by means of a genetic algorithm [14–15]. A schematic of the optimization procedure is shown in Figure 4.

In this way, the (visco)elastic properties for a range of thin (composite) materials have been identified, showing good correspondence with alternative identification techniques [14–15].

### 3.2 Extensions of UPS method

#### 3.2.1 Harmonic ultrasonic polar scan

Our numerical simulations indicate that the global view of a UPS image is not only function of material parameters, but also strongly depends on the (temporal) shape of the employed ultrasound wave. This has led us to the experimental and numerical investigation of a harmonic version of the ultrasonic polar scan [12]. In Figure 5 a comparison is shown between pulsed and harmonic UPS recordings of a $[0]_8$ C/E laminate.

Contrary to the pulsed UPS image, the patterns in the harmonic UPS image depend on the employed ultrasonic frequency. This is easily understood considering that the harmonic UPS image puts on view the stimulation condition of dispersive Lamb waves, while the pulsed UPS image is mainly governed by bulk wave characteristics [16].

In case of harmonic signals, the UPS analysis can be extended to the evaluation of both the amplitude and the phase of the transmitted wave. Figure 6 shows the harmonic UPS images for a $[0]_8$ C/E laminate, considering the analysis of both the amplitude and the phase of the transmission signal.

#### 3.2.2 Ultrasonic Backscatter Polar Scan (UBPS)

During our UPS investigations, we persistently observed a small amount of energy being backscattered to the emitter, even for large incidence angles. Recording of the backscattered signal according to the UPS principle then results in an Ultrasonic Backscatter Polar Scan (UBPS). Depending on the employed wave, we speak of a pulsed UBPS [17] or a harmonic UBPS [18]. A schematic of the UBPS method, together with a harmonic UBPS recording for a $[0]_8$ C/E laminate, is presented in Figure 7.

Typically, such a harmonic UBPS image is characterized by high amplitude spikes, and we found that these spikes expose geometrical characteristics of (sub)surface structures. For the $[0]_8$ C/E laminate, the backscatter spikes originate at the imprint left by the peel-ply cloth during manufacturing of the laminate.

### 3.3 Applications of UPS and UBPS

The UPS and the UBPS, both in pulsed and harmonic version, have been applied for a range of

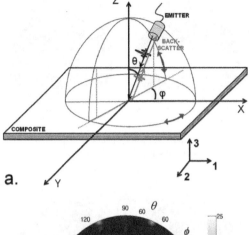

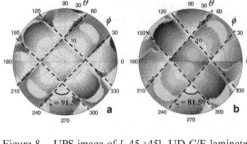

Figure 8. UPS image of [–45,+45]$_s$ UD C/E laminate: (a) virgin state and (b) after applying shear load.

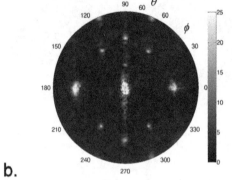

Figure 7. (a) Schematic of UBPS and (b) harmonic UBPS experiment for a [0]$_8$ C/E laminate.

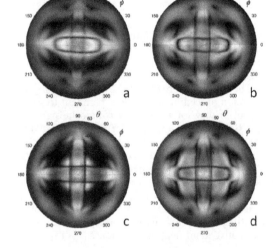

Figure 9. Harmonic UPS recordings of a cross-ply C/E laminate: (a) [0,90,90,0], (b) [0,D,90,90,0], (c) [0,90,D,90,0] and (d) [0,D,90,90,D,0]. 'D' stands for the depth position of the water-filled delamination.

NDT applications involving ceramic, metallic and fiber reinforced materials.

### 3.3.1 Static damage

Various fiber reinforced plastics have been statically loaded to induce material degradation. In Figure 8, UPS image are shown of [–45,+45]$_s$ C/E laminate before as well as after (quasi-)static shear loading. The indicated overlap angle $\zeta$ provides a measure for the actual fiber orientation. Initially $\zeta = 91.5°$ indicating that the stacking of the laminate was not done correctly (see Figure 8a). After loading in shear, the overlap angle reduced to $\zeta = 81.5°$ (see Figure 8b), which thus reveals a fiber distortion of 10°. The evolution of the overlap angle $\zeta$ has been monitored for a range of shear load levels, and we found a clear relation between the applied shear load level and the observed fiber distortion [19].

### 3.3.2 (Multi-) Delamination

As the patterns in a harmonic UPS image are governed by stimulation conditions of Lamb waves, they will shift position in the presence of a delamination. This is easily understood considering that a delamination divides the original laminate in two sub-laminates having different boundary conditions. This change in boundary conditions invokes different Lamb wave stimulation conditions, and as such yields a different harmonic UPS image [20]. This is demonstrated in Figure 9 for a cross-ply C/E laminate (thickness of 1.1 mm) provided with a water-filled delamination (thickness of 50 µm) at different depth positions. Note that such types of delaminations are in general difficult to assess in thin composites using conventional normal incidence techniques. Figure 9 clearly shows that the depth position of the delamination is well represented in the harmonic UPS images. Figure 9d further indicates that also a multi-delamination is easily detected.

Numerical computations even indicate the ability of the harmonic UPS to characterize delamination parameters such as depth position and delamination thickness.

### 3.3.3 *Fatigue damage*

The UPS method was also applied to fatigued composites. Fatigue loading typically leads to the initiation, progression and accumulation of micro defects. At the macroscopic level, these defects manifest themselves in a directional reduction of the stiffness properties. Such a directional stiffness reduction should be visible as a stretching (along the direction of loading) of the UPS contours. This is explicitly demonstrated in Figure 10 for a woven C/E laminate which was fatigued along $\varphi = 0°$. The UPS experiment yields a stiffness reduction of 12.8% along the loading direction, which is in good agreement with extensometer data yielding a stiffness reduction of 11% [21].

### 3.3.4 *Corrugation*

As the characteristics of a periodic (sub)surface structure determine the position of the observed backscatter spikes in a harmonic UBPS experiment (see Figure 7b), we may also reverse this: start from a UBPS experiment and reconstruct the periodic (sub)surface structure. We have demonstrated this for a 2D subsurface corrugation (see Figure 11),

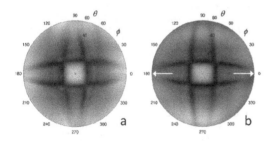

Figure 10. Pulsed UPS: (a) before and (b) after applying tension fatigue.

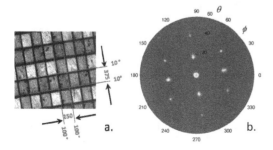

Figure 11. (a) 2D subsurface corrugation and (b) corresponding harmonic UBPS recording.

Table 1. Corrugation parameters: design (column 1), ultrasonic reconstruction (column 2) and optical measurement (column 3).

| Design parameters | Ultrasonic reconstruction | Optical measurement |
|---|---|---|
| 250 µm | 249.8 µm | 250.1 µm |
| 100° | 99.3° | 99.2° |
| 375 µm | 375.3 µm | 374.8 µm |
| 10° | 9° | 9.1° |

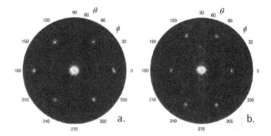

Figure 12. Harmonic UBPS for (a) unstrained and (b) strained DC06 steel coupon.

and obtained excellent agreement with both the design parameters and optical measurements (see Table 1).

The reconstruction procedure was also successfully applied to periodic subsurface structures having a certain degree of geometrical randomness [10].

### 3.3.5 *Strain measurement*

As the UBPS method is capable to reconstruct a (sub)surface structure with excellent accuracy, it should also be capable to detect any change in surface parameters due to strain. Analysis of the transformed surface structure then yields a representation of the applied in-plane strain [22]. Instead of machining a periodic surface structure, which would mechanically weaken the sample, we simply exploit residual surface roughness features. The ultrasonic strain measurement technique has been demonstrated on cold-rolled DC06 steel coupons which were strained at different levels (see Figure 12). One can clearly observe the shifting of the backscatter spikes when strained. This shifting is then used to reconstruct the applied in-plane strain field.

In addition, we also analyzed the response of an ultrasonic pulse at normal incidence in order to detect shifted thickness resonances. As such, we can not only reconstruct the in-plane strain components, but also the out-of-plane strain component. The ultrasonically obtained strain values

Figure 13. Current design of portable UPS scanner.

have been confronted with the results of conventional strain measurements techniques (e.g. stereovision DIC), showing excellent agreement for a wide range of applied strain values (up to 35%) [22]. Interestingly, the developed ultrasonic strain gauge is the only method which is able to determine a 3D local strain field in a single-sided and contactless manner.

## 4 THE (NEAR) FUTURE: 2015 – ...

Despite the recent progress in the U(B)PS technique, there is still plenty left to investigate.

Currently, we are investigating the capabilities of the TOF version of the UPS for characterizing composites. In contrast to common TOF approaches in literature, our TOF-UPS approach is not restricted to bulk wave propagation. In addition, it does not require any prior knowledge about the symmetry planes of the material under investigation. As such, our TOF-UPS approach has particular advantages for characterizing thin composite plates. The current status of our TOF-UPS research will be presented in the Session 'Elastic Waves 3' of the ETNDT6 conference.

As we realize that our current laboratory scanner does not meet in-field requirements, we are in the process of designing a portable scanner. The current status of the portable scanner is presented in Figure 13. However, this mechanical scanner will only serve as an intermediate device as we want to exclude any mechanical movement by using advanced phased matrix technology.

## 5 CONCLUSIONS

This paper has given a short review of the ultrasonic (backscatter) polar scan research for composite characterization and NDT. We started with the pioneering results of the initiators, discussed current state-of-the-art research and finished with several future research lines having high potential.

## ACKNOWLEDGMENTS

MK acknowledges funding of the FWO-Vlaanderen through research grants G012010N and G0B9515N.

## REFERENCES

[1] W.H.M. Van Dreumel and J.L. Speijer, "Non-destructive composite laminate characterization by means of ultrasonic polar-scan," *Materials Evaluation,* vol. 39, pp. 922–925, 1981.

[2] W.H.M. Van Dreumel and J.L. Speijer, "Polar-Scan, a Non-Destructive Test Method for the Inspection of Layer Orientation and Stacking Order in Advanced Fiber Composites," *Materials Evaluation,* vol. 41, pp. 1060–1062, 1983.

[3] J. Degrieck and D. Van Leeuwen, "Simulatie van een Ultrasone Polaire Scan van een Orthotrope Plaat (in Dutch)," presented at the the 3rd Belgian National Congress on Theoretical and Aplied Mechanics, Liege, Belgium, 1994.

[4] J. Degrieck, "Some possibilities of nondestructive characterisation of composite plates by means of ultrasonic polar scans," in *Emerging technologies in nondestructive testing (ETNDT),* Patras, Greece, 1996, pp. 225–235.

[5] J. Degrieck, et al., "Ultrasonic polar scans as a possible means of non-destructive testing and characterisation of composite plates," *Insight,* vol. 45, pp. 196–201, Mar 2003.

[6] N.F. Declercq, et al., "On the influence of fatigue on ultrasonic polar scans of fiber reinforced composites," *Ultrasonics,* vol. 42, pp. 173–177, Apr 2004.

[7] N.F. Declercq, et al., "Ultrasonic polar scans: Numerical simulation on generally anisotropic media," *Ultrasonics,* vol. 45, pp. 32–39, Dec 2006.

[8] N.F. Declercq, et al., "Simulations of harmonic and pulsed ultrasonic polar scans," *Ndt & E International,* vol. 39, pp. 205–216, Apr 2006.

[9] L. Satyanarayan, et al., "Ultrasonic Polar Scan Imaging of Damaged Fiber Reinforced Composites," *Materials Evaluation,* vol. 68, pp. 733–739, Jun 2010.

[10] M. Kersemans, "Combined Experimental-Numerical Study to the Ultrasonic Polar Scan for Inspection and Characterization of (Damaged) Anisotropic Materials," PhD thesis; Ghent University, p. 508, 2014.

[11] M. Kersemans, et al., "Pitfalls in the experimental recording of ultrasonic (backscatter) polar scans for material characterization," *Ultrasonics,* vol. 54, pp. 1509–1521, Aug 2014.

[12] M. Kersemans, et al., "The quasi-harmonic ultrasonic polar scan for material characterization: experiment and numerical modeling," *Ultrasonics,* vol. 58, pp. 111–122, 2015.

[13] S.I. Rokhlin and L. Wang, "Stable recursive algorithm for elastic wave propagation in layered anisotropic media: Stiffness matrix method," *Journal of the Acoustical Society of America,* vol. 112, pp. 822–834, Sep 2002.

[14] M. Kersemans, *et al.*, "Identification of the Elastic Properties of Isotropic and Orthotropic Thin-Plate Materials with the Pulsed Ultrasonic Polar Scan," *Experimental Mechanics*, vol. 54, pp. 1121–1132, 2014.

[15] M. Kersemans, *et al.*, "Quantitative Measurement of the Elastic Properties of Orthotropic Composites by means of the Ultrasonic Polar Scan Method," *JEC Composites*, vol. 75, pp. 48–52, 2012.

[16] M. Kersemans, *et al.*, "Extraction of bulk wave characteristics from a pulsed ultrasonic polar scan," *Wave Motion*, vol. 51, pp. 1071–1081, 2014.

[17] M. Kersemans, *et al.*, "The Pulsed Ultrasonic Backscatter Polar Scan and its Applications for NDT and Material Characterization," *Experimental Mechanics*, vol. 54, pp. 1059–1071, 2014.

[18] M. Kersemans, *et al.*, "Ultrasonic Characterizaion of Subsurface 2D Corrugation," *Journal of Nondestructive Evaluation*, vol. 33, pp. 438–442, Sep 2014.

[19] M. Kersemans, *et al.*, "Nondestructive damage assessment in fiber reinforced composites with the pulsed ultrasonic polar scan," *Polymer Testing*, vol. 34, pp. 85–96, 2014.

[20] M. Kersemans, *et al.*, "Detection and localization of delaminations in thin carbon fiber reinforced composites with the ultrasonic polar scan," *Journal of Nondestructive Evaluation*, vol. 33, pp. 522–534, 2014.

[21] M. Kersemans, *et al.*, "Damage Signature of Fatigued Fabric Reinforced Plastic in the Pulsed Ultrasonic Polar Scan," *Experimental Mechanics*, vol. 54, pp. 1467–1477, 2014.

[22] M. Kersemans, *et al.*, "A Novel Ultrasonic Strain Gauge for Single-Sided Measurement of a Local 3D Strain Field.," *Experimental Mechanics*, vol. 54, pp. 1673–1685, 2014.

*Emerging Technologies in Non-Destructive Testing VI – Aggelis et al. (Eds)*
*© 2016 Taylor & Francis Group, London, ISBN 978-1-138-02884-5*

# Time-Of-Flight recorded Pulsed Ultrasonic Polar Scan for elasticity characterization of composites

**A. Martens**
*Department of Physics, Wave Propagation and Signal Processing, KU Leuven—Kulak, Kortrijk, Belgium*

**M. Kersemans, J. Degrieck & W. Van Paepegem**
*Department of Materials Science and Engineering, Ghent University, Zwijnaarde, Belgium*

**S. Delrue & K. Van Den Abeele**
*Department of Physics, Wave Propagation and Signal Processing, KU Leuven—Kulak, Kortrijk, Belgium*

ABSTRACT: In its orginal configuration, the Pulsed Ultasonic Polarscan (P-UPS) mainly focussed on elastic material characterization through the inversion of amplitude landscape measurements. However, for several materials, special attention is required asminima in the transmission amplitudes do not exactly coincide with critical angles calculated from the Christoffel equations. Consequently, other means to extract the information on elastic moduli from P-UPS measurements are being investigated. In the present paper, we report on the use of Time-Of-Flight Ultrasonic Polarscan (TOF-UPS) simulations as a new means of material characterization. Previous TOF inversions, although successful, were based on bulk wave approximations, which are not longer valid for thin materials. Our first inversion results on numerical cases demonstrate the usefulness of the new developed technique and highlight the added value compared to the bulk wave approximation

## 1 INTRODUCTION

From as early as 1980s, the ultrasonic polarscan (UPS) was considered to be a promising tool in the field of non-destructive testing. However, due to many technological subtleties the technique never reached its full potential. In recent years, most of the existing the pitfalls [1] were resolved and consequently the first applications of the UPS were investigated (e.g. study of fibre misalignment, layering, delaminations, fibre breakage, disbonding, corrosion, strain measurements,etc.). Initially, the amplitude landscape of the Pulsed Ultrasonic Polarscan (P-UPS) was considered as a means of elastic material characterization. Even though the developed inversion scheme resulted in accurate material properties [2], special care had to be taken in some cases as the angular positions of the critical angles do not necessarily match the minima in transmission. Moreover, the amplitude landscape is highly dependent on the damping of the material and as such, amplitude UPS measurements are more suitable for the determination of the viscous part of the material's visco-elasticity.

As an alternative, the Time-Of-Flight (TOF) of the P-UPS can be considered for elastic material characterization. The use of TOF measurements has been succesfully applied for the determination of elastic properties assuming rather thick samples (ranging from 3 to 6 mm at frequencies of 2–5 MHz) in combination with a bulk wave approximation[3][4]. In many cases, however, the actual thickness of plates is smaller and, as such, deviations from the bulk wave approximation are to be expected, as shown in figure 1 where a clear deviation from the bulk wave TOF solution is seen around the longitudinal and shear critical angles for an aluminum plate of 1.5 mm, examined at a frequency of 5 MHz.

In this paper we will illustrate the use of the full Time-Of-Flight Ultrasonic Polarscan (TOF-UPS) simulation to invert the elastic properties of isotropic plate like media. The method automatically accounts for the deviations from the bulk wave approximation and is thus applicable for any thickness. Extension to anisotropic media is straightforward, but requires the simultaneous inversion of several polar orientations at once.

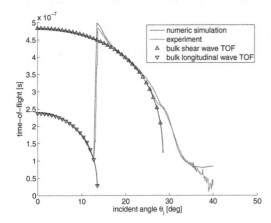

Figure 1. Comparison of the experimentally obtained TOF with the numerically simulated TOF as function of the incidence angle for an aluminum plate of 1.5 mm examined at a frequency of 5 MHz. The bulk wave solutions are represented by the dashed lines.

## 2 THEORETICAL BACKGROUND

### 2.1 TOF determination for plates

Let us suppose the general case of an impinging ultrasonic wave on an anisotropic plate with polar angles $(\phi, \theta)$. Following the theory that leads to the Christoffel equations, the combination of the anisotropic wave equation with Snells law and the subsequent introduction of the formal solution of an ultrasonic wave, leads to an expression, represented in eq.(1) where the vertical component of the wave number, $k_z$, is the only unknown,

$$k_z^6 + A_1 k_z^5 + A_2 k_z^4 + A_3 k_z^3 + A_4 k_z^2 + A_5 k_z + A_6 = 0 \quad (1)$$

where the $A$ coefficients are functions of the plate material properties and the known wave number components $k_x$ and $k_y$ of the impinging wave.

In general, this equation has six solutions. However, in most cases, the material under study will exhibit some kind of symmetry, which simplifies the equation. For example, when laminated composites are studied, an orthotropic symmetry can be assumed as the lowest possible type. This leads to a reduction of equation (1) into a third order polynomial in the variable $k_z^2$, resulting in three independent solutions.

Going back to the most general case, the solution for the ultrasonic displacement vector ($\vec{U}$) of the problem should be represented by a summation over the six solutions of the sixth-order polynomial.

$$\vec{U}(\vec{r},t) = \sum_{n=1}^{6} \vec{u}_n e^{i(k_x x + k_y y + k_z^n z - \omega t)} \quad (2)$$

with $\vec{u}_n$ the polarization vector corresponding to each solution $k_z^n$ [5].

Applying appropriate boundary conditions expressed by continuity of normal stress ($\tau_{xz}, \tau_{yz}, \sigma_{zz}$) and displacement ($U_z$) between the general solution inside the solid (2) and in the surrounding fluid, leads to the transmission and reflection coefficient [5]. Finally, the frequency dependent transmission coefficient is multiplied by the spectrum, $I(\omega)$, of the incident pulse and an inverse FFT is used to obtain the transmitted pulse signal.

In order to determine the time for the wave to arrive at the receiver on the transmitting side, one can calculate the argument of the maximum correlation between transmitted and incident signal. However, to simplify the calculations, we have found and verified [6] that it is equally adequate to define the TOF as the difference in time of the maximum transmitted and incident peak (Eq. (3)).

$$TOF_{plate}(\theta, \phi) = \max_t T(\theta, \phi, t) \quad (3)$$

$$T(\theta, \phi, t) = \int_{-\infty}^{\infty} d\omega T(\theta, \phi, \omega) I(\omega) e^{-i\omega t} \quad (4)$$

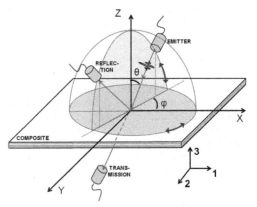

(a) schematic of the UPS technique

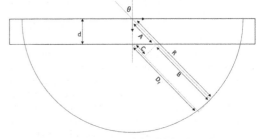

(b) Cross-section of the UPS principle

Figure 2. Top: General scheme of the UPS technique with source and receivers moving on a sphere with radius R; Bottom: Cross-section of the UPS principle in transmission.

## 2.2 TOF-UPS correction

The TOF simulations, as presented in 2.1, are not sufficient to model a TOF-UPS in the correct manner. In a real UPS experiment the transducers move on a sphere with centerpoint on the top surface of the sample (Fig. 2a) and, as such, a correction for the plate-receiver path must be introduced. Given that the transmitted wave in the numerical model is derived with reference to $(x = 0; z = \frac{-d}{2})$, the correction term can be calculated as follows (see also Fig. 2b):

$$t_{cor}(\theta_i) = \frac{D_f}{V_f} = \frac{d}{V_f}(1-\cos(\theta_i)) \qquad (5)$$

$$\text{TOF}_{\text{UPS}}(\theta_i,\phi) = \text{TOF}_{\text{plate}}(\theta_i,\phi) + t_{cor}(\theta_i) \qquad (6)$$

here, $d$ is the sample thickness, and $V_f$ is the sound velocity in the immersion fluid.

As an example, a numerically simulated TOF-UPS experiment is illustrated in Fig. 3a for an

Table 1. Material constants for the hypothetical orthotropic plate.

| Parameter | Value [GPa] |
| --- | --- |
| $C_{11}$ | 122.73 |
| $C_{22} = C_{33}$ | 13.46 |
| $C_{12} = C_{13}$ | 6.57 |
| $C_{23}$ | 6.55 |
| $C_{44}$ | 3.39 |
| $C_{55}$ | 5.86 |
| $C_{66}$ | 6.25 |

orthotropic material with elastic properties given in table 1. Figure 3b displays the corresponding amplitude UPS results. The TOF results are characterized by sharp discontinuities in the polar plot. Interestingly, these discontinuities correspond to minima in the amplitude landscape (Fig. 3b) and indicate angle combinations where the dominant wave inside the layer switches polarization state. The first discontinuity for instance where the shear wave becomes more dominant than the longitudinal wave.

## 3 INVERSION SCHEME

The main objective of the current report is to upgrade the commonly utilized inversion scheme based on bulk wave approximation. Material characterization based on bulk waves has already proven its usefulness [4][3], however, due to the ever decreasing composite thickness, the approximation loses its universality. This loss of validity for thin samples, as illustrated in figure 1 by the large deviations between the actual TOF and the predictions by bulk-wave approximation, primarily occurs at and around the critical angles. The time-of-flight calculation, as presented in this paper, has the advantage of taking these deviations into account.

The presently introduced inversion scheme performs a best fit between the experimental data and the numerical model, based on the full TOF-UPS calculation, and uses the following cost function:

$$F(C_{ij}) = \sum_{\theta_i,\phi}(\text{TOF}_{\exp}(\theta_i,\phi) - \text{TOF}_{\text{UPS}}(\theta_i,\phi))^2 \qquad (7)$$

Minimization of this cost function was performed by way of a Genetic Algorithm (GA). The GA scheme has the advantage of inverting black-box problems with large parameter space, in general. In particular, GA inversion algorithms have already proven their usefulness in inverting amplitude based P-UPS data [2]. In short, a GA uses the

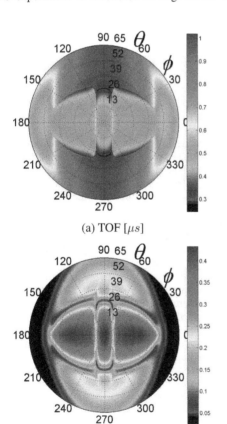

Figure 3. Numeric P-UPS simulation TOF (a) and amplitude (b) of an orthotropic plate ($d = 1.5$ mm) at a center frequency of 5 MHz.

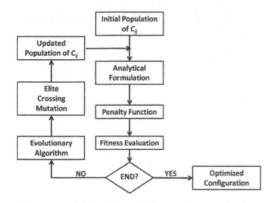

Figure 4. Schematic of the presented inversion technique.

Table 2. Inversion results for a numerical simulation on an aluminum plate of thickness 1.5 mm. The inversions are performed using the TOF approach with different range boundaries. Mean value and standard deviation for twenty independent inversions are considered at each range boundary.

| | | $\lambda + 2\mu$ | $\mu$ |
|---|---|---|---|
| Actual values [GPa] | | 107.74 | 26.00 |
| Inverted values [GPa] | 10% | 108.40 ± 0.34 | 26.07 ± 0.08 |
| | 25% | 108.78 ± 1.05 | 26.09 ± 0.20 |
| | 50% | 108.84 ± 2.08 | 26.05 ± 0.56 |
| | 75% | 107.79 ± 8.30 | 26.27 ± 0.34 |
| | 90% | 108.13 ± 10.28 | 25.61 ± 1.43 |

Table 3. Same as table 2, however, in this case, the inversions are performed using the bulk-wave approximation with different range boundaries.

| | | $\lambda + 2\mu$ | $\mu$ |
|---|---|---|---|
| Actual values [GPa] | | 107.74 | 26.00 |
| Inverted values [GPa] | 10% | 107.67 ± 0.63 | 24.15 ± 0.13 |
| | 25% | 107.04 ± 1.31 | 24.27 ± 0.18 |
| | 50% | 107.01 ± 2.54 | 24.21 ± 0.28 |
| | 75% | 106.98 ± 4.02 | 24.18 ± 0.38 |
| | 90% | 107.94 ± 4.00 | 24.30 ± 0.42 |

principles of evolution to create a parameter set that best minimizes the cost function. This optimization is performed over several generations. Each generation consists of a specific amount of members, and each of the members gets a fitness value based on the cost function (7). Only the members with the best fitness value are chosen to create the members of the next generation. Some pass directly to the new generations (elite) whilst others undergo mutation and crossover. This results in a new generation with most parameter sets iteratively converging to values in the neighbourhood of the best set. However, the presence of mutation can lead to the discovery of an area of better sets which are not in the vicinity of the current best parameter set. The algorithm stops when a preset convergence criterium is fulfilled. A schematic summary of the inversion technique is presented in Fig. 4.

## 4 RESULTS FOR A NUMERICAL TEST-CASE ON ALUMINUM AND DISCUSSION

In this section the above presented inversion algorithm is applied to a numerical test case on aluminum. As input for the algorithm, the targeted TOF data have been determined using the forward transmission model (virtual experiment) for an isotropic medium with $\lambda + 2\mu = 107.74$ GPa and $\mu = 26$ GPa. Several inversions were conducted using different range boundaries on the elasticity parameters. To prove the enhanced accuracy of the new algorithm, the analysis is repeated, for several parameter bounds, with a bulk-wave based inversion scheme. The resulting values of the material's moduli after inversion are presented in tables 2 and 3, for the various range boundaries. The reported values correspond to the mean and standard deviation for twenty independent inversions considered at each range boundary. For the TOF based algorithm, only small deviations ($\approx 1-2\%$) from the actual values are obtained. The bulk-wave based method, on the other hand, has small deviations for one of the two parameters whereas large deviations (6 –7%) are consistently observed for the other one (shear modulus).

The origin of the error on $\mu$, introduced by the bulk-wave method, is obviously attributed to sizeable deviations which are only apparent after the first discontinuity (see also Fig. 1). TOF measurements before the discontinuity are governed by the arrival of the longitudinal wave and as such the parameter $\lambda + 2\mu$ should return a good inversion. As the transmitted signal after the discontinuity is entirely dominated by the shear wave, it is reasonable to expect that the shear modulus parameter $\mu$ leads to rather poor inversion results in the bulk-wave approximation.

## 5 CONCLUSION

In the above presented study, the full TOF-UPS simulation model was introduced in an inversion scheme to extract the elastic properties of a solid plate. This scheme, based on a genetic algorithm, can be considered as an upgrade of existing time-of-flight in version methods in that sense that it takes into account deviations from bulk wave behaviour that arise in thin plates. Moreover, it is a good alternative for the existing amplitude UPS based inversion scheme, where special care has to be taken not to confuse the occurrence of amplitude minima with the positions of critical angles.

The TOF based inversion approach was applied on an aluminum plate and compared to the existing bulk-wave based inversion techniques. It is found that the new inversion scheme is able to successfully invert two material constants. In contrast, the more traditionally used bulk-wave technique is only able to invert one of the parameters as it is not capable to cope with deviations that occur for thin plates. Future prospects are to apply the developed technique to real experimental data and to extend the inversion to other symmetry classes such as orthotropic materials in view of characterization and damage monitoring of composites.

## ACKNOWLEGDEMENTS

The research leading to these results has gratefully received funding from the European Union Seventh Framework Programme (FP7/2007–2013) for research, technological development and demonstration under the Grant Agreement no. 314768 (ALAMSA), and from the fund for scientific research-Flanders (FWO Vlaanderen) through grant G.0B95.15.

## REFERENCES

[1] Mathias Kersemans, Wim Van Paepegem, Koen Van Den Abeele, Lincy Pyl, Filip Zastavnik, Hugo Sol, and Joris Degrieck. Pitfalls in the experimental recording of ultrasonic (backscatter) polar scans for material characterization. *Ultrasonics*, 54(6):1509–1521, 2014.

[2] Mathias Kersemans, Arvid Martens, Nicolas Lammens, Koen Van Den Abeele, Joris Degrieck, Filip Zastavnik, Lincy Pyl, Hugo Sol, and Wim Van Paepegem. Identification of the elastic properties of isotropic and orthotropic thin-plate materials with the pulsed ultrasonic polar scan. *Experimental Mechanics*, 54(6):1121–1132, 2014.

[3] SI Rokhlin and W Wang. Double through-transmission bulk wave method for ultrasonic phase velocity measurement and determination of elastic constants of composite materials. *The Journal of the Acoustical Society of America*, 91(6):3303–3312, 1992.

[4] B Hosten, M Deschamps, and BR Tittmann. Inhomogeneous wave generation and propagation in lossy anisotropic solids. application to the characterization of viscoelastic composite materials. *The Journal of the Acoustical Society of America*, 82(5):1763–1770, 1987.

[5] Adnan H Nayfeh. *Wave propagation in layered anisotropic media: With application to composites*. Elsevier, 1995.

[6] Mathias Kersemans, Ives De Baere, Joris Degrieck, Koen Van Den Abeele, Lincy Pyl, Filip Zastavnik, Hugo Sol, and Wim Van Paepegem. Nondestructive damage assessment in fiber reinforced composites with the pulsed ultrasonic polar scan. *Polymer Testing*, 34:85–96, 2014.

*Emerging Technologies in Non-Destructive Testing VI – Aggelis et al. (Eds)*
*© 2016 Taylor & Francis Group, London, ISBN 978-1-138-02884-5*

# Ultrasonic testing of adhesively bonded joints in glass panels

B. Mojškerc, T. Kek & J. Grum
*Faculty of Mechanical Engineering, University of Ljubljana, Ljubljana, Slovenia*

ABSTRACT: This paper presents a method for evaluating the quality of adhesively bonded joints in glass panels by means of amplitudes of reflected ultrasonic signals. Glass panel specimens consist of three layers: hardened glass, polysulfide or silicone adhesive and a polymer profile. Each layer has its own acoustical impedance, which in turn alters the amplitude of reflected and transmitted ultrasonic signals. Measurements are carried out using the pulse-echo ultrasonic method. Experiments are performed on good and bad adhesively bonded joints in order to detect the presence of defects in interfaces between adhesive-glass and/or adhesive-polymer profile. Defects include air pockets, grease and duct tape. Most of the mentioned defects except smaller areas of grease can be successfully detected. Ultrasonic testing is therefore a reasonable way to evaluate the quality of adhesively bonded joints in glass panels.

## 1 INTRODUCTION

Over the past years, the development of civil engineering has paved the way for frequent use of glass panels in a wide array of buildings. Glass panels usually consist of hardened glass and a polymer profile, joined together by polysulfide or silicone adhesive. Adhesively bonded joints are required to perform with predetermined mechanical properties. Low strength joints present a flaw in the integrity of the building and also pose a danger to bystanders in case of joint failure. The adhesive bond is sometimes difficult to evaluate due to time dependency of its properties, therefore a good method of quality control is needed. In order to prevent low strength joints, their quality has to be evaluated during production and lifecycle.

To best evaluate adhesive bond quality and strength, multiple destructive and non-destructive tests can be used, as presented by Roach et al. (2010). They created various adhesive joint specimens with and without contaminants and confirmed contaminant influence on joint strength by using a tensile failure test. They also performed multiple non-destructive tests and established a relation between bond strength threshold and non-destructive test results. Especially promising were the ultrasonic methods. They present an opportunity to evaluate the quality of a wide palette of adhesively bonded joints without their mechanical failure. Using ultrasonic methods, joint strength cannot be directly measured, but can be indirectly evaluated using specific ultrasonic parameters in combination with destructive tests. These parameters are obtained via laboratory experiments and can be later used as a reference value for future tests.

Pulse-echo ultrasonic method was used by Goglio & Rossetto (1998) to inspect adhesively bonded joints of 0.8 mm thin steel sheets and a 0.5 mm layer of epoxy adhesive. Their quality control method was based on the coefficient of reflection at the metal-adhesive interface. An adhesion index was defined in order to detect and evaluate defective zones of a bonded joint. An automatic procedure based on statistical methods was developed in order to allow discerning acceptable bonds from defective bonds.

Pulse-echo ultrasonic method was also used by Titov et al. (2008) to detect void disbonds at front and rear metal-adhesive interfaces of adhesively bonded joints in automotive assemblies. Specimens consisted of steel and aluminium sheets, ranging from 0.7–2 mm, joined together by epoxy adhesive, ranging from 0.1–1 mm. Ultrasonic signals were measured at points of interest and compared with the previously recorded reference waveforms outside the bond. Absence of disbonding at the front metal-adhesive interface was indicated by large deviations in the ultrasonic signal. At the rear metal-adhesive interface, the sound phase inversion phenomena was used as an indication of good acoustic contact.

Another example of ultrasonic testing of adhesively bonded joints is through thickness first mode ultrasonic resonance method. The method is robust and can be used to test joints with thin adherends and a wide variation of adhesive thickness. It was used by Allin et al. (2003) to detect disbonds of automotive components. Adhesively bonded joints consisted of 2 mm aluminium sheets and a various thickness epoxy adhesive, ranging from 0.1–3 mm. First mode resonance was excited in bonded joint specimens and the received signal was windowed, leaving the ringing of the first mode. If the resonance frequency fell into predicted frequency range, the bond integrity was confirmed. Resonance frequency outside the predicted range indicated a disbond.

Ultrasonic testing can also be used to evaluate the adhesive curing process. As witnessed by Maeva et al. (2006), sound velocity and acoustic impedance of adhesives can change over time, as is apparent during the adhesive curing process. Sound velocity and attenuation measurements can therefore be used as a quality control parameter. Optimal time and curing temperature can be determined.

This paper presents a pulse-echo ultrasonic method in order to evaluate adhesively bonded joints in glass panels and to discern them according to their quality. For this purpose, amplitudes of reflected ultrasonic signals are used. Experiments are carried out on various adhesive bond specimens. In order to determine the applicability of the presented method, flaws and contaminants are introduced.

## 2 EXPERIMENTAL PROCEDURE

Emission and reception of sound by an ultrasonic transducer forms the basis of the pulse-echo ultrasonic method. Sound wave pulses are usually both generated and received by piezoelectric transducers. Emitted sound waves travel through the chosen substance and are reflected at boundaries which express a different acoustic impedance. A higher difference in acoustic impedance relates to a higher percentage of reflection. This indicates that we can ascertain a presence of a certain substance based on its acoustic impedance and the reflected signal amplitude. Quality of adhesively bonded joints can therefore be evaluated by comparing test results with a predetermined reflected signal amplitude. Acoustic impedance is given by equation 1:

$$Z = \rho c \quad (1)$$

where $Z$ = acoustic impedance; $\rho$ = density; and $c$ = sound velocity. The reflected signal amplitude is based on the coefficients of reflection and transmission of sound, which can be calculated using known acoustic impedances of two substances sharing an interface. They are given by equations 2, 3:

$$R = \frac{Z_2 - Z_1}{Z_2 + Z_1} \quad (2)$$

$$T = \frac{2Z_2}{Z_2 + Z_1} \quad (3)$$

where $R$ = coefficient of reflection; $T$ = coefficient of transmission; $Z_1$, $Z_2$ = acoustic impedance of the first and second given substance, respectively.

### 2.1 Equipment and parameters

Experiments were performed using the ultrasonic device Eurosonic UTC-110 with the provided software. A normal incidence Eurosonic T0506 transducer with a 6 mm diameter and a 5 MHz frequency was used. Testing parameters are provided in table 1:

### 2.2 Specimens

A model of the glass panel specimen is presented in figure 1. Specimen dimensions are 300 × 200 × 109 mm. Width of the adhesively bonded joint is 30 mm. The joint is presented schematically in figure 2. It consists of three layers of

Table 1. Testing parameters.

| Parameter | Value and unit |
| --- | --- |
| Pulse frequency | 5 MHz |
| Voltage | 50.15 V |
| Pulse width | 100 ns |
| Frequency of pulses | 1 kHz |
| Sampling rate | 100 MHz |
| Amplification | 13.6–16 dB |

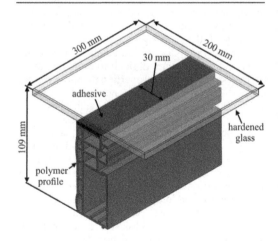

Figure 1. Glass panel specimen.

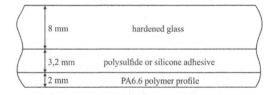

Figure 2. Adhesively bonded joint.

various thicknesses: hardened glass with a thickness of 8 mm, polysulfide or silicone adhesive with a thickness of 3.2 mm and a PA6.6 polymer profile with a thickness of 2 mm.

## 3 RESULTS AND DISCUSSION

Failure of adhesively bonded joints in glass panels can occur on both interfaces glass-adhesive and adhesive-polymer profile. In order to evaluate the joint quality adequately, both interfaces have to be tested. Experiments were performed separately on each interface. Coupling quality effect on the reflected signal amplitudes in the glass-adhesive interface was partially prevented by adjusting signal amplification for each measurement, in order to match the amplitude of the first reflection. However, this was not possible in the case of adhesive-polymer profile interface, as the amplitude of the first reflection was later used as a quality evaluation parameter. Test results are presented in A-scan figures of amplitude over time or length.

### 3.1 Interface glass-adhesive

Polysulfide adhesive was used in testing the glass-adhesive interface. The ultrasonic transducer was positioned on the top side of the glass. Measurements were taken in separate areas with good adhesive presence and with lack of adhesive. The course of the reflected signal in a good glass-adhesive interface is presented in figure 3. The course of the reflected signal in a glass-adhesive interface with an air void is presented in figure 4. Reflected signal amplitude decays faster in case of a good bond, because a larger part of the sound pressure passes through the adhesive than through air, as the acoustic impedance of air is lower than that of adhesive. That also indicates a lower coefficient of

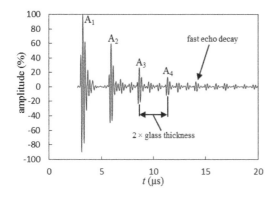

Figure 3. Reflected signal amplitude in glass-adhesive interface.

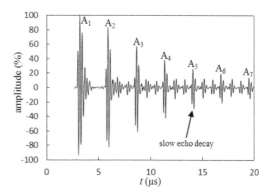

Figure 4. Reflected signal amplitude in glass-adhesive interface, with an air void present.

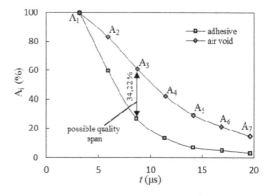

Figure 5. Comparison of echo amplitude decay in glass-adhesive interface.

reflection in case of a good bond. A comparison of the reflected signal decay for both cases is presented in figure 5. Points of decay were determined using the absolute value of the main ultrasonic echoes, marked as '$A_i$', the index '$i$' indicating the number of the main echo.

### 3.2 Interface adhesive-polymer profile

In case of the adhesive-polymer profile interface, silicone adhesive was used. In order to position the ultrasonic transducer on the bottom side of the polymer profile, the profile had to be cut. Same as before, measurements were taken in separate areas with good adhesive presence and with lack of adhesive. The course of the reflected signal in a good adhesive-polymer profile interface is presented in figure 6. The course of the reflected signal in an adhesive-polymer profile interface with an air void is presented in figure 7. The adhesive-polymer profile interface has a more similar acoustic impedance than the glass-adhesive interface,

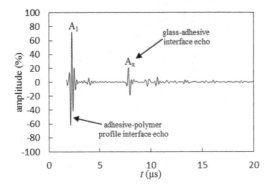

Figure 6. Reflected signal amplitude in adhesive-polymer profile interface.

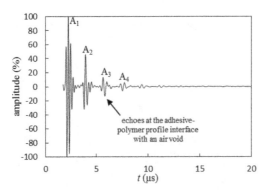

Figure 7. Reflected signal amplitude in adhesive-polymer profile interface, with an air void present.

therefore the amplitudes of the reflected signals are smaller and only one main echo '$A_1$' can be clearly seen in case of a good bond. The other echo '$A_x$' is reflected from the glass-adhesive interface. This echo is not visible in case of a bad joint with an air void, as the ultrasonic waves reflect at the interface adhesive-polymer profile and do not reach the glass-adhesive interface.

### 3.3 Quality evaluation parameter

In order to allow quality evaluation of adhesively bonded joints in glass panels, a parameter from the gathered data had to be chosen. For the evaluation of the glass-adhesive interface, second, third and fourth reflected echo amplitudes were chosen along with attenuation coefficient α, which is given by equation 4:

$$\alpha = \frac{20}{2s} \log \frac{A_1}{A_2} \quad (4)$$

where $s$ = specimen thickness; $A_1$ = first echo amplitude; and $A_2$ = second echo amplitude.

A comparison was made for good bonds and bonds with an air void. Ten measurements were taken. Single factor analysis of variance (ANOVA) was used to identify the largest difference in measured mean values, as presented in tables 2–5. A standard alpha level of 0.05 was used. Critical F value was 4.41 for all cases, based on the degrees of freedom, $df$. The term $SS$ abbreviates the sum of squares and $MS$ abbreviates mean squares. The null hypothesis suggested that the mean values of chosen parameters do not significantly differ in case of a good bond and a bond with an air void. The null hypothesis was rejected for all cases. The third echo amplitude '$A_3$' was chosen as a final quality evaluation parameter for the glass-adhesive interface, as it had the largest mean value difference of reflected signals between both cases of the bond. For the evaluation of the adhesive-polymer profile

Table 2. Analysis of variance for second echo amplitude '$A_2$'.

| Variance source | SS %² | df | MS %² | $F_0$ |
|---|---|---|---|---|
| Between group | 2843.79 | 1 | 2843.79 | 486.67 |
| Within group | 105.18 | 18 | 5.84 | |
| Total | 2948.97 | 19 | $P_{value}$ = 1.76E-14 | |

Table 3. Analysis of variance for third echo amplitude '$A_3$'.

| Variance source | SS %² | df | MS %² | $F_0$ |
|---|---|---|---|---|
| Between group | 5920.37 | 1 | 5920.37 | 2199.26 |
| Within group | 48.46 | 18 | 2.69 | |
| Total | 5968.82 | 19 | $P_{value}$ = 2.85E-20 | |

Table 4. Analysis of variance for fourth echo amplitude '$A_4$'.

| Variance source | SS %² | df | MS %² | $F_0$ |
|---|---|---|---|---|
| Between group | 3889.43 | 1 | 3889.43 | 1046.09 |
| Within group | 66.93 | 18 | 3.72 | |
| Total | 3956.36 | 19 | $P_{value}$ = 2.12E-17 | |

Table 5. Analysis of variance for attenuation coefficient 'α'.

| Variance source | SS %² | df | MS %² | $F_0$ |
|---|---|---|---|---|
| Between group | 0.1830 | 1 | 0.1830 | 151.82 |
| Within group | 0.0217 | 18 | 0.0012 | |
| Total | 0.2047 | 19 | $P_{value}$ = 3.29E-10 | |

interface, the first echo amplitude 'A₁' was chosen, as it was the only clearly present echo in both cases of a good bond and a bond with an air void.

The expected echo amplitude area of good bonds was defined as the area of quality and was determined using the average value of the measured echo signal amplitudes, 'A₁' and 'A₃', each on its own respective interface. Sample standard deviation of the echo amplitudes was calculated for twenty measurements over an area of good adhesive bonding with no present defects. Amplitude limits of the areas of quality were determined with equations 4, 5. Equation 4 was used to assess the area of quality at the glass-adhesive interface, equation 5 was used to assess the area of quality at the adhesive-polymer profile interface.

$$Q_{ga} = A_{3avg} \pm 2s_3 \quad (5)$$

$$Q_{app} = A_{1avg} \pm 2s_1 \quad (6)$$

where $Q_{ga}$ = area of quality for glass-adhesive interface; $Q_{app}$ = area of quality for adhesive-polymer profile interface; $A_{iavg}$ = average echo amplitude; and $s_i$ = sample standard deviation. Area of quality is specific to a given specimen and interface. Same amplitude limits cannot be used for different types of glass panels. They have to be determined for each type separately, if this method of quality evaluation is to be used.

### 3.4 Testing joint quality

In order to test the validity of the chosen quality evaluation parameters, several adhesively bonded joints were tested along their length. The first specimen had a purposely wavy deposit of polysulphide adhesive in order to clearly record the difference of a good adhesive bond and an adhesive bond with an air void at the glass-adhesive interface. The tested area of the specimen is presented in figure 8 together with the test results. Measurements were taken every 5 mm in a straight line. The third echo amplitude 'A₃' was recorded and plotted against joint length. The adhesively bonded joint was unacceptable if the third echo amplitude left the predetermined area of quality. Air voids were easily detected, as they fell out of the area of quality. The difference in amplitude between a good adhesively bonded joint and a joint with an air void was as big as 35%. Small air pockets were also detected and confirmed with visual inspection. Keeping in mind, the transducer diameter was 6 mm, so only a part of the joint width was tested. A matrix array ultrasonic transducer would be recommended to evaluate the quality of the entire joint.

The second specimen had a deposit of silicone adhesive together with some contaminants on the glass-adhesive interface. Contaminants consisted of two areas of grease and an area of duct tape.

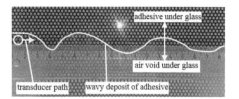

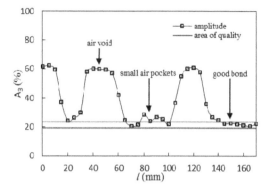

Figure 8. Third echo amplitude along the joint length in glass-adhesive interface.

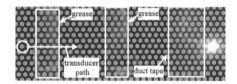

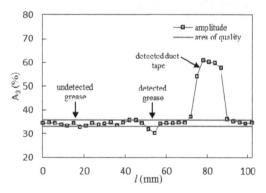

Figure 9. Third echo amplitude along the joint length in glass-adhesive interface for a joint with contaminants.

The tested area of the specimen is presented in figure 9 together with the contaminants and the test results. Measurements were taken every 3 mm in a straight line. As seen by the results, the contaminant detection at the glass-adhesive interface was limited. The second area of grease was detected, while the first area of grease remained undetected. The duct tape area was successfully detected, as witnessed by more than a 27% increase in the third echo amplitude. Flaw detection in the

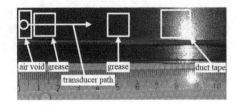
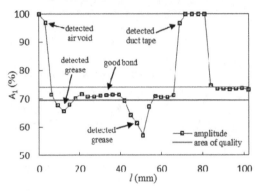

Figure 10. First echo amplitude along the joint length in adhesive-polymer profile interface for a joint with contaminants.

glass-adhesive interface was therefore dependent on contaminant type and its arrangement. Very thin areas of grease were hard to detect.

As mentioned before, both interfaces glass-adhesive and adhesive-polymer profile have to be evaluated. The third specimen had a deposit of silicone adhesive together with some contaminants on the adhesive-polymer profile interface. The tested area of the specimen is presented in figure 10 together with the contaminants and test results. Measurements were taken every 3 mm in a straight line. This time, the first echo amplitude '$A_1$' was recorded and plotted against joint length. Contaminants consisted of two areas of grease and an area of duct tape. There was also an air void present at the start of the joint. Contaminants are not visible with the naked eye, as the polymer profile is opaque. All flaws were successfully detected, even the thin areas of grease. Contaminant detection was therefore more successful in the adhesive-polymer profile interface than in the glass-adhesive interface.

## 4 CONCLUSIONS

This paper presented a non-destructive way of evaluating the quality of adhesively bonded joints in glass panels with the pulse-echo ultrasonic method. Joint quality can be evaluated via the amplitudes of reflected echo signals, which differ in relation with the acoustic impedance of interface materials. Two quality evaluation parameters were chosen. For evaluating the glass-adhesive interface, the third echo amplitude '$A_3$' was chosen on the basis of single factor analysis of variance. For evaluating the adhesive-polymer profile interface, the first echo amplitude '$A_1$' was chosen, because it was the only present echo in both cases of a good bonded joint and a joint with an air void. Acceptable amplitude areas of a quality bond were determined for each specific type of adhesive and interface separately, with the same applied theory. This method of quality evaluation can be used to detect air voids and also contaminants like grease and duct tape on both mentioned interfaces, although very thin areas of grease are hard to detect on the glass-adhesive interface. It is important to note that only flaws at the interfaces can be detected. For the detection of flaws inside the adhesive layer, other methods are needed, e.g. through thickness resonance method. For testing adhesively bonded joints in glass panels, the pulse-echo ultrasonic method should be used in conjunction with matrix array transducers in order to cover larger areas and to produce easily understandable C-scan figures.

## REFERENCES

Allin, J.M. et al. 2003. Adhesive disbond detection of automotive components using first mode ultrasonic resonance. *NDT&E international* 36: 503–514.

Goglio, L. & Rossetto, M. 1998. Ultrasonic testing of adhesive bonds of thin metal sheets. *NDT&E international* 32: 323–331.

Maeva, E.Y. et al. 2006. Monitoring of Adhesive Cure Process and Following Evaluation of Adhesive Joint Structure by Acoustic Techniques. *9th European Conference on Non-Destructive testing, Berlin, 25–29 September 2006*.

Roach, D. et al. 2010. Innovative Use of Adhesive Interface Characteristics to Nondestructively Quantify the Strength of Bonded Joints. *10th European Conference on Non-Destructive Testing, Moscow, 7–11 June 2010*.

Titov, S.A. et al. 2008. Pulse-echo NDT of adhesively bonded joints in automotive assemblies. *Ultrasonics* 48: 537–546.

*Emerging Technologies in Non-Destructive Testing VI – Aggelis et al. (Eds)*
*© 2016 Taylor & Francis Group, London, ISBN 978-1-138-02884-5*

# Guided ultrasonic waves for the NDT of immersed plates

P. Rizzo
*Department of Civil and Environmental Engineering, University of Pittsburgh, Pittsburgh, USA*

E. Pistone
*VCE Vienna Consulting Engineers ZT GmbH, Vienna, Austria*

A. Bagheri
*Department of Civil and Environmental Engineering, University of Pittsburgh, Pittsburgh, USA*

ABSTRACT:   We present a nondestructive testing technique for the inspection of immersed plates. Guided ultrasonic waves were used for the nondestructive evaluation of an aluminum plate immersed in water. Leaky Lamb waves were generated using a pulsed laser to generate broadband stress waves detected by an array of immersion transducers arranged in a semi-circle. The signals were processed to extract some features from the time, frequency, and joint time-frequency domains. These features were then fed to an unsupervised learning algorithm based on the outlier analysis to detect the presence of damage, and to a supervised learning algorithm based on artificial neural networks to classify the types of defect. We found that the hybrid laser-immersion transducers system and both learning algorithms enable the detection of the defects and their classification with good success rate.

## 1   INTRODUCTION

Immersed structures in water are crucial in the modern society. They include vessels, offshore structures, communication cables, water mains, and sea wind farms. The periodic inspection or the permanent monitoring of these structures is critical to guarantee a satisfactory level of serviceability. Usually, the inspection is carried out by applying NonDestructive Evaluation (NDE) approaches that resemble topside NDE methods adequately waterproofed and adapted to the wet environment.

The use of Guided Ultrasonic Waves (GUWs) for the NDE of dry structures such as pipes, plates, and rails is constantly increasing (Kundu 2004, Rizzo & Lanza di Scalea 2007, Bagheri et al. 2012). However, little work has been conducted on the use of these waves in wet systems (Mijarez et al. 2007, Pistone et al. 2013, Bagheri et al. 2014). Sharma and Pathania (2003) investigated the propagation of thermoelastic waves in a plate bordered with inviscid liquids. Finally, Lee et al. (2014) used a continuous laser and a laser Doppler vibrometry to reconstruct the wavefield induced in a plate immersed in water and to compute the velocity-frequency relationship, i.e. the dispersion curves. This non-contact method was also used to demonstrate that the imaging system is capable of damage sizing.

In this paper, we investigated the ability of a moving non-contact system based on a pulsed laser and Immersion Ultrasonic Transducers (IUTs) to scan an immersed plate using leaky Lamb waves. The waves were generated with short-duration laser pulses, and they were simultaneously detected with an array of five IUTs arranged in a semi-circle. The signals were processed with the Gabor Wavelet Transform (GWT) to extract some features from the time, frequency, and joint time-frequency domains. The features were then fed to two learning algorithms for pattern recognition. The first algorithm was based on the Outlier Analysis (OA) to detect the presence of five defects artificially devised on the plate. The second algorithm was an Artificial Neural Network (ANN) used to label four structural conditions.

## 2   EXPERIMENTAL SETUP

A Q-switched Nd:YAG pulsed-laser operating at 0.532 μm was used to test a 750 mm × 1605 mm × 2.54 mm aluminum plate immersed in water. A 1 mm diameter laser beam was delivered to the surface of the plate using a mirror and a plano-convex lens with focal length of 100 mm. A tank contained the plate that sat on four polycarbonate supports and surmounted by 65 mm of water. Figure 1(a) shows the setup. Five immersion transducers resonant at 1 MHz were arranged along a semicircle 15 mm above the plate.

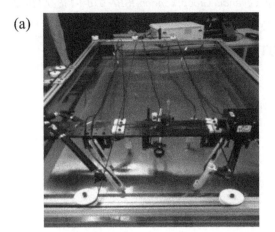

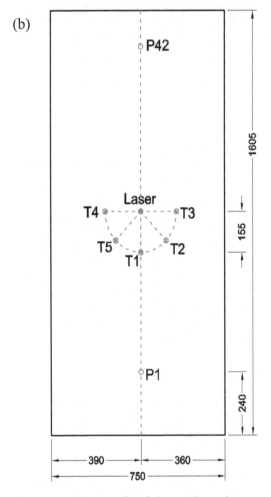

Figure 1. (a) Photographs of the experimental setup. (b) Plan view illustrating the position of the transducers (T1–T5). The drawing is on scale and the quotes are in mm.

As traditionally done with conventional wedge transducers, the alignment angle of the detectors was set to maximize the sensitivity to a particular guided mode, in this case the leaky $S_0$ mode. The location of the transducers as well as their distance from the laser-illuminated point is shown in Figure 1(b). A B-scan system was built in our laboratory to inspect the plate. The system consisted of the laser head and the transducers secured to a moving bench plate. The motion was provided by pulleys and belts driven by a National Instrument PXI unit. A front user interface running under LabVIEW was coded to control the motion of the bench and to select the proper parameters for the generation, detection, and the storage of the ultrasonic waves. The laser irradiated 42 points, hereafter indicated as scanning points. They were linearly spaced by 30 mm and the inspection was conducted five times to evaluate the repeatability of the methodology. To increase the statistical significance of the study, each point was irradiated ten times. This implies that we collected and processed 10500 time waveforms. Figure 1(b) shows the location of the first and the last scanning points, namely P1 and P42.

The following five defects were artificially devised: 1) a 60 mm-long transverse cut; 2) an 80 mm long oblique notch; 3) a 6 mm circular dent; 4) a 10 mm through-thickness hole; 5) a $40 \times 40$ mm² abraded area to simulate corrosion. The defects are shown in Figure 2. They were chosen to provide a wide spectrum of structural anomalies. They are also common in the scientific literature as they are representative of defects artificially machined to validate structural health monitoring and NDE techniques for waveguides. These points were considered damage-free when the area of the array did not overlap any damage. Based on this assumption 20 out of 42 scanning points were considered damage-free and the ultrasonic waves irradiated from those points were expected to be not affected by the flaw induced in the plate.

## 3 SIGNAL PROCESSING

### 3.1 Feature extraction

Figure 3(a) shows the time waveform recorded by transducer T1 when point 27 was irradiated by the laser during the scan 1. This point was considered defect-free as the area of the array did not overlap with any defect. Three main wave packets are visible: the first packet around 40 μsec is the leaky $S_0$ mode; the second packet comprised between 50 μsec and 60 μsec is likely the $A_1$ mode which has a cutoff frequency of 600 kHz; the third wave beyond 80 μsec is the superposition of the quasi-Scholte mode and the bulk wave traveling in

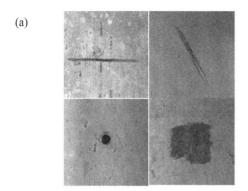

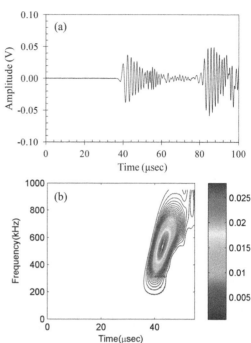

Figure 3. (a) Time waveforms generated at the scanning point 27 and sensed by the immersion transducer T1, and (b) its Gabor wavelet transform (GWT) scalogram.

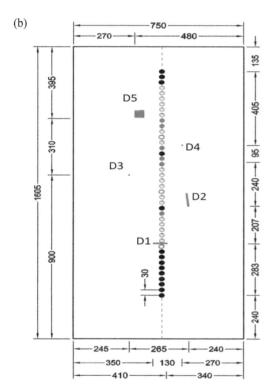

Figure 2. (a) Photographs of four of the artificial defects devised on the plate. (b) Position of the defects with respect to the scanning points. The name of the defect is with respect to the scan direction. For example defect D1 is encountered first.

water. The $A_0$ mode is not observed as most of its energy is converted into the quasi-Scholte wave. In this study, we analyzed the leaky symmetric mode. As such, any signal processing discussed hereafter was conducted by considering the time window 0–55 μsec only.

Figure 3(b) shows the wavelet transforms of the time-series presented in Figure 3(a) in the time interval 0–55 μsec. The transform decomposes the original time-domain signal by computing its correlation with a short-duration wave called the mother wavelet that is flexible in time and frequency. The transform determines the time of arrival and the frequency content of the propagating modes. The scalograms show the broad frequency content of the leaky $S_0$ mode and they unveil the dispersion of the symmetric mode that is faster at lower frequencies. From each scalogram, we computed two subplots, namely the values of the largest wavelet coefficients at each frequency, and the values of the largest wavelet coefficients at each time.

Like previous studies (Pistone et al. 2013), we extracted the eight features $F^{(k)}$. From the time domain, we retained the peak-to-peak amplitude and the Root Mean Square (RMS) which are important information of a signal. From the joint time-frequency analysis, we extracted the scalogram maximum value, the RMS of the curve associated with the largest coefficients in the frequency domain, and the RMS of the curve associated with the largest coefficients in the time domain. In frequency domain, we computed the Fast Fourier Transform (FFT) of the time waveforms to retain the maximum value, the RMS of the bandwidth comprised between 250 and 650 kHz, and the Full Width Half Maximum (FWHM) of the same frequency bandwidth. Then, these features were used to compute a Damage Index (DI) defined as:

$$DI_{(i,j)}^{(k)} = max\left(\frac{F_{(i)}^{(k)}}{F_{(j)}^{(k)}}, \frac{F_{(j)}^{(k)}}{F_{(i)}^{(k)}}\right)$$
$$k = 1, 2, ..., 8, i = j = 1, 2, ..., 5, i \neq j \qquad (1)$$

Our approach hypothesized that the sensing system setup, i.e. the inclination and the lift-off distance of the transducers, remained constant throughout the five scans. This is a common hypothesis in any study related to the laser ultrasound of dry structures.

### 3.2 Outlier analysis

The DIs were collected to form a multi-dimensional vector for the unsupervised learning algorithm. An outlier is a datum that appears inconsistent with a set of data, the baseline, which describes the normal condition of the structure under investigation. Ideally, the baseline should include typical variations in environmental or operative conditions of a structure. When the analysis involves a set of $p$-dimensional (multivariate) data consisting of $n$ observations in $p$ variables, the OA can be performed using the Mahalanobis Squared Distance (MSD) $D_\zeta$. The baseline vector and the covariance matrix in the OA were calculated considering the 140 samples (14 scanning point × 10 measurements) taken from the first scan and associated with the defect-free locations. The baseline dataset was compared to the data relative to all the five scans. The approach mimics the inspection of a structure fivefold longer and periodically subjected to the same type of defects. Once the values of $D_\zeta$ of the baseline distribution were determined, the threshold value was taken as the usual value of $3\sigma$ equal to 99.73% of the Gaussian confidence limit. Knowing the time of flight of the leaky $S_0$ mode and knowing the relative position of the immersion transducers with respect to the laser irradiated points and the location of the artificial damage, we could safely assume that the selected baseline dataset was not affected by the presence of damage.

### 3.3 Artificial neural network

In order to classify unlabeled signals, we coupled the aforementioned eight features to an ANN able to learn from training samples through iterations. A feed-forward back propagation ANN with three layers was used. The input layer received the data vector containing the features, along with the codification of defect types. In this study, we predefined four class labels using a 2-digit ($\alpha$, $\beta$) binary number. The classes (0, 0), (0, 1), and (1, 0) corresponded

respectively to the pristine portion of the plate, the transverse cut, and the oblique notch. Owing to their size, the remaining defects were hypothesized to produce the same effect on the guided waves. Thus, they were grouped into the (1, 1) class. The hidden layer of the network processed the data by multiplying the input vectors by weights and adding biases. The results constituted the argument of a transfer function that squashed the output values into a certain range. For the hidden layer ten neurons were used, because the convergence rate was considered acceptable. The hyperbolic tangent sigmoid transfer function was employed using the *tansig* function in *MATLAB*® software. This function squashes the output values between −1 and +1. For the input and output layers, we used the *purelin* transfer function. To train the network, we used the Levenberg-Marquardt algorithm because of its computational speed.

## 4 RESULTS

We first applied the multivariate analysis using the eight features individually. In this case, the input vector is 10-dimensional since five transducers give rises to ten different pairs. Then, we used all features together. To assess the performance of the OA, we defined the success rate of the algorithm as the sum of proper outliers and proper inliers over the total number of data points. We found that the highest success rate was achieved when the features were used individually. In particular, feature 2 provided the highest success rate equal to 65.95%. Table 1 summarizes the results and it includes the false positives and the false negatives. Figure 4(a) shows the success rate of the multivariate analysis associated with all five scans when all the eight features are considered together. The highest success rate of 85% is associated with scan 1, whereas the MSD of the data relative to scans 2–5 were classified as outliers. However, only 52% of those data were truly subjected to the presence of a defect. We believe that the reason for the underperformance

Table 1. Results (%) of the outlier analysis when a single feature was used.

| Feature | Success | False positive | False negative |
|---------|---------|----------------|----------------|
| 1 | 65.24 | 31.37 | 3.390 |
| 2 | 65.95 | 19.05 | 15.00 |
| 3 | 53.39 | 41.37 | 5.240 |
| 4 | 58.81 | 35.95 | 5.240 |
| 5 | 65.12 | 32.50 | 2.380 |
| 6 | 51.96 | 44.52 | 3.510 |
| 7 | 59.11 | 38.39 | 2.500 |
| 8 | 46.96 | 2.920 | 50.12 |

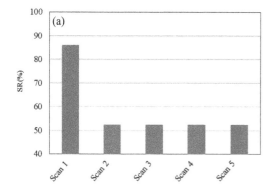

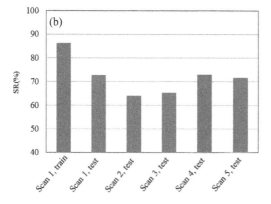

Figure 4. Success rate when eight features were used: (a) OA and (b) ANN.

of the OA is multi-fold. First the OA is an unsupervised algorithm which only provides pristine/damage status; second the dent and the simulated-corroded areas were small to induce a significant variation to the selected features; second the variability of the laser-induced waves may mask any change associated with the presence of the flaws; finally the forward movement of the bench plate and the subsequent return to the start position may have altered the orientation of one or more transducers with respect to the plate. Nonetheless, the study discussed here is unique as it presents a movable system that scan a structure, rather than a static inspection system that gage a small area of the plate comprised between an actuator and a receiver.

We do not exclude that the use of an immersion transducer as transmitter improves the ability to detect even the smallest defect machined in this experiment.

The neural network was trained by using 32 testing points, namely points 1–20 and 31–42, from the first scan. The remaining 178 points were used to test the network. In order to carry a direct comparison between the ANN and the OA, we considered

Table 2. Results (%) of the outlier analysis when a single feature was used.

| Feature | Success | False positive | False negative | Uncertain | Mislabel |
|---|---|---|---|---|---|
| 1 | 81.55 | 0.560 | 17.30 | 0.000 | 0.595 |
| 2 | 78.38 | 0.476 | 16.10 | 0.000 | 5.048 |
| 3 | 79.31 | 0.810 | 16.87 | 0.143 | 2.869 |
| 4 | 79.46 | 0.821 | 17.33 | 0.095 | 2.286 |
| 5 | 77.10 | 1.012 | 16.70 | 0.381 | 4.810 |
| 6 | 78.36 | 1.274 | 16.95 | 0.274 | 3.143 |
| 7 | 77.90 | 1.119 | 16.87 | 0.381 | 3.726 |
| 8 | 78.44 | 0.024 | 16.30 | 0.000 | 5.238 |

the same set of features presented fed to the unsupervised algorithm. Table 2 presents the percentage of data associated to each possible label when the features were used individually: success, false positive, false negative, uncertain, and mislabel. Uncertain data are those outputs that could not be clearly associated to a certain defect class. The column mislabel refers to those classifications where the damage was improperly labeled, i.e. it was assigned to the wrong type of defect. We note that the first damage-sensitive feature provides the best labeling and it has also the lowest mislabels rate. Figure 4(b) shows the success rate of the classifier when the eight features constituted the input vector of the network. The success rate of scan 1 is obviously the highest; this is expected as data from this scan were used to train the network. Interesting the success rates associated with the remaining scan were not identical showing some degree of variability. Overall, the results seem to suggest that the use of many features does not improve significantly the success rate of the classifier.

5 CONCLUSIONS

In this paper, we presented a NDE technique for immersed metallic plates. The technique is based on the noncontact generation and detection of GUWs. The waves are generated by a pulsed laser and are detected by an array of IUTs. The propagation of the leaky first symmetric mode along an aluminum plate subjected to a few representative defects was characterized in the time, frequency, and joint time-frequency domain. A set of eight features were extracted from these domains to create a vector of damage indexes that was then fed to an OA and to an ANN. The first algorithm identifies the existence and location of damage, whereas the neural network identifies also the type of damage.

The OA is the primary class of unsupervised learning algorithms that are applied to data not containing examples from the damaged structure.

The neural network instead allows the classification of the type of damage, since there is the availability of data from the structure with known types and levels of damage. Overall, we found that the ANN provides better success than the OA at detecting the presence of damage and we also found that the number of features selected is not critical to enhance the success rate. Actually, the highest success rate was achieved in the case of single feature.

## ACKNOWLEDGMENTS

This work was supported by the U.S. National Science Foundation, grant number CMMI 1029457.

## REFERENCES

Bagheri, A., Li, K., & Rizzo, P. 2012. Reference-free damage detection by means of wavelet transform and empirical mode decomposition applied to Lamb waves. *Journal of Intelligent Material Systems and Structures* 24(2): 194–208.

Bagheri, A., Pistone, E., & Rizzo, P. 2014. Guided ultrasonic wave imaging for immersed plates based on wavelet transform and probabilistic analysis. *Research in Nondestructive Evaluation*, 25(2): 63–81.

Kundu, T. ed. 2004. *Ultrasonic nondestructive evaluation: engineering and biological material characterization.* CRC press.

Lee, J.R., Jang, J.K., & Kong, C.W. 2014. Fully noncontact wave propagation imaging in an immersed metallic plate with a crack. *Shock and Vibration*.

Mijarez, R., Gaydecki, P., & Burdekin, M. 2007. Flood member detection for real-time structural health monitoring of sub-sea structures of offshore steel oil-rigs. *Smart Materials and structures*, 16(5): 1857.

Pistone, E., Li, K., & Rizzo, P. (2013). Noncontact monitoring of immersed plates by means of laser-induced ultrasounds. *Structural Health Monitoring* 12(5–6): 549–565.

Rizzo, P., & Lanza di Scalea, F. 2007. Wavelet-based unsupervised and supervised learning algorithms for ultrasonic structural monitoring of waveguides. *Progress in Smart Materials and Structures Research*, 227–290.

Sharma, J.N., & Pathania, V. 2003. Generalized thermoelastic Lamb waves in a plate bordered with layers of inviscid liquid. *Journal of Sound and Vibration* 268(5): 897–916.

*Nonlinear elastic waves*

*Emerging Technologies in Non-Destructive Testing VI – Aggelis et al. (Eds)*
*© 2016 Taylor & Francis Group, London, ISBN 978-1-138-02884-5*

# Nonlinear acoustics and acoustic emission methods to monitor damage in mesoscopic elastic materials

M. Bentahar, R. El Guerjouma, C. Mechri, Y. Baccouche, S. Idjimarene,
A. Novak, S. Toumi, V. Tournat & J.-H. Thomas
*LAUM UMR-CNRS 6613, Le Mans, France*

M. Scalerandi
*Institute of Condensed Matter and Complex Systems Physics, Politecnico di Torino, Torino, Italy*

ABSTRACT: Microcracked composites and metals usually exhibit a high level of nonlinearity in their elastic response already at low amplitudes of excitation. In order to quantify these behaviors, different nonlinear indicators can be used depending on the required parameters and the experimental configuration of interest. In this contribution several acoustic nonlinear techniques are presented in order to detect and monitor the presence and evolution of micro-cracks in different kind of materials. In particular, the development of a particular "vibrational/acoustical" arrangement made the use of the robust nonlinear resonance method together with the harmonic generation method possible. Besides, the definition of a proper nonlinear indicator and its dependence on the excitation amplitude, allowed obtaining a power law behavior of cracked materials. Finally, the correlation between acoustic emission and nonlinear relaxation results shows that all mechanisms relax as the logarithm of their corresponding energy but with different dynamics.

## 1 INTRODUCTION

Acoustic nonlinearity corresponding to micro-cracked, or micro-inhomogeneous materials has increased considerably during the last years. It has been established that the classical Landau theory cannot describe the dynamic behavior of these materials since it does not take into account the influence of strain on the elastic constants (Bentahar et al. 2006, McCall et al. 1994). In resonance experiments, many measurements made on micro-cracked materials have shown that real and imaginary components of the elastic modulus are strain dependent. Indeed, resonance frequency and quality factor of the followed resonance modes have proved to be able to monitor damage as a function of the induced dynamic strain. In the nonlinear resonant approach, two hysteretic nonlinear parameters are defined, where the first is related to frequency (elastic modulus) and the second to quality factor (damping). However, the abovementioned NLH parameters are defined for one resonance mode i.e. in a very narrow frequency domain. In order to broaden the frequency domain, we recently developed an experimental method in which we considered different flexural resonances generated in composite plates. In harmonic generation experiments, one can

define different nonlinear parameters and follow their evolution as a function of damage. In this case, many studies focused on the second or third harmonics' amplitudes in order to obtain values of the quadratic or cubic nonlinear parameters (Abeele et al. 2000). Resonance and harmonics generation revealed to be efficient Non-Destructive Testing (NDT) methods. However, the signal processing developed in these methods is performed on a limited range of parameters. One of the possibilities is to gather resonances and harmonics approaches in order to have "new" information such as harmonics of the excited resonances, whose presence and evolution are not analyzed in general. Once the nonlinear indicators obtained, recent studies (Scalerandi et al. 2015) have proved that it is possible to get information about the nature of cracks (closed, open) or its nonlinear mechanism in the framework of a power law behavior.

Despite the sensitivity of acoustical nonlinear techniques to the presence and evolution of the microstructure of materials, they still need to be quantified. The desired quantification aims to link the changes observed on nonlinear parameters to the remaining lifetime of materials. In that sense, one of the proposed methods is the application of acoustic emission, which can separate damage mechanisms and determine their respective energies. Therefore, correlation between nonlinear acoustics parameters of micro-cracked materials and acoustic emission signals is able to bring the necessary information related to the critical damage mechanisms whose presence lead to the failure of the material.

## 2 NONLINEAR FLEXURAL RESONANCES IN COMPOSITE PLATES

### 2.1 Guided flexural waves in thin plates

In the case of a TiC reinforced steel composite composite plate, flexural guided waves were generated. The propagation velocity of a flexural perturbation generated at the frequency $f_i$ is written as a function of the thickness $d$ as well as the mechanical parameters of the thin plate: (Baccouche et al. 2013)

$$V^i = \left( \frac{E\pi^2 d^2 f_i^2}{3\rho} \right)^{1/4} \quad (1)$$

where, $E$ is the young modulus and $\rho$ is the density. Here one should keep in mind that Equation (1), when only the first flexural resonances are considered, gives the same results as the ones developed by Timoshenko and Rayleigh (Graff 1991). At these frequencies, the obtained velocities are the same as the ones corresponding to the fundamental antisymmetric guided mode A0. The frequency below which flexural modes velocities are the same as A0 mode velocity depend on the material properties and dimensions. In the case of the abovementioned metal based composite it has been found to be around ~18 kHz. Therefore, the velocity dispersion can be determined at resonance frequencies $f_i$ as shown in Figure 1.

### 2.2 Nonclassical nonlinear materials

In non-classical nonlinear materials such as micro-cracked composites, the relation linking stress to strain include higher order terms as well as hysteresis and memory effects (Guyer et al. 2009):

$$K(\varepsilon,\dot{\varepsilon}) = K_0 \left(1 + \beta\varepsilon + \delta\varepsilon^2 + \cdots \right) + H(\varepsilon, \text{sign}(\dot{\varepsilon})), \quad (2)$$

where $K_0$ is the linear elastic bulk modulus, $\varepsilon$ is the strain, $\beta$ and $\delta$ represent the classical quadratic and cubic nonlinear parameters, respectively, $H$ is a function describing the hysteretic relation between stress ($\sigma$) and strain ($\varepsilon$). Since the Young modulus can be written as a function of the bulk modulus and the Poisson ratio $\nu$ as $E = 3K(1 - 2\nu)$, Equation (1) can be reconsidered as a function of the strain rate $\varepsilon$:

$$V^i(\varepsilon) = \left( \frac{E(\varepsilon)\pi^2 d^2 f(\varepsilon)_i^2}{3\rho} \right)^{1/4} \quad (3)$$

The presence of Young modulus in equation (3) is explicit on one hand and implicit on the other hand in the frequency. In a previous work (Baccouche et al. 2013), the change of the velocity $V^i$ was followed as a function of $E$ (the frequency $f$ was kept constant) and as a function of $f$ ($E$ was kept constant). Results show that, due to a given strain rate $\varepsilon$, the relative change of frequency $\Delta f/f_0$ could be considered as equivalent to the relative change in Young modulus $\Delta E/E_0$, where $f_0$ and $E_0$ are the linear resonance frequency and the Young modulus.

As a function of an increasing strain (or excitation amplitude) the velocity will decrease with the amplitude as shown in Figure (2).

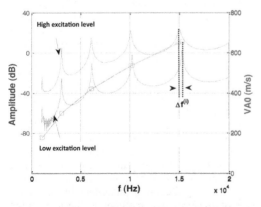

Figure 1. Influence of frequency on amplitude and velocity of A0 mode.

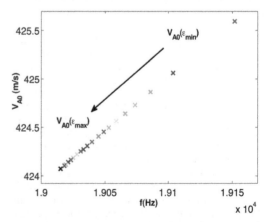

Figure 2. Influence of the excitation on the velocity $V_{A0}$.

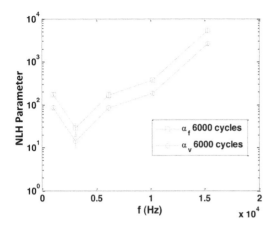

Figure 3. Evolution of the nonlinear hysteretic parameters $\alpha_V$ and $\alpha_f$ (related to the decrease of resonance frequency) as a function of frequency in the case of a fatigued metal composite with 6000 cycles.

It becomes then possible to follow the decrease of velocity using a new nonlinear hysteretic parameter defined as in: (Baccouche et al. 2013)

$$\alpha_V = \frac{1}{\varepsilon_{max}} \frac{\Delta V_{A0}}{V_{A0}(\varepsilon_{min})} \qquad (4)$$

Figure 3 shows clearly that for a diffused damage created within the metal composite using cyclic loads, the NLH parameters related to resonance frequencies or to velocity $V_{A0}$ are frequency dependent. This statement means that the measured dispersion is not geometrical, as it could be imagined when nodes are moving around a localized damage (Baccouche et al., in prep.). In order to have more information about the dispersion of nonlinearity we recommend the following references (Gusev et al. 1998, Zaitsev et al. 2001, Johnson et al. 1996).

## 3 HARMONIC ANALYSIS AT RESONANCE

In nonlinear resonance experiments we usually follow the evolution of a given resonance mode as a function of the excitation level. This result is obtained by analyzing the frequency response of the material under test at the excitation frequency only, without paying attention of the possible existence of higher frequency components. This aim can be achieved by using an adequate signal processing which allows getting harmonics of the excited resonances. The method described in (Farina 2001, Novak et al. 2010) and adapted to NDT applications in (Novak et al. 2012) consists in exciting the material with the help of a swept-sine signal $s(t)$ and before recording synchronously the response $y(t)$. The convolution between $y(t)$ and the inverse filter of $s(t)$ is then determined. This step leads to the estimation of the nonlinear impulse responses of the fundamental frequency (the excitation frequency) and the possible harmonics generated by the material, which can be separated by windowing. Their Fourier transforms are equal to the higher-order frequency response functions, FRFs (See Figure 4).

By analogy to the single nonlinear resonance method in which the hysteretic nonlinear parameter $\alpha_f$ is defined as: $\alpha_f = \frac{1}{\varepsilon_{max}} \frac{\Delta f}{f(\varepsilon_{min})}$, the present method allows to go beyond this definition by introducing higher order nonlinear hysteretic parameters. The latter are defined as:

$$\alpha_{nf} = \frac{1}{\varepsilon_{max}} \frac{f_n(\varepsilon_{min}) - f_n(\varepsilon_{max})}{f_n(\varepsilon_{min})} \qquad (5)$$

$\varepsilon_{min}$ is the strain at which the material behaves linearly and $\varepsilon_{max}$ is the strain at the nonlinear regime, $n$ is the order of the detected harmonic.

Figure 5 shows that the fundamental resonance mode of a glass bar doesn't have a 3rd harmonic at the initial intact state. However, once the material is thermally damaged we notice an important appearance of a 3rd harmonic, which arises from the noise level to reach a dynamic amplitude of approximately ~40 dB.

In the case of a Steel-TiC composite plate, one should keep in mind that $\alpha_f$ was very dispersive as shown in Figure 3. Its value changes from approximately between ~28 and more than ~5270 at the fundamental resonances (the five first flexural resonances). When the nonlinear convolution is applied on this sample, results show that the appearance of harmonics is not straightforward. Indeed, for the resonance modes 1 and 2, we haven't detected any harmonics. At the same time, we noticed the appearance of the 2nd and 3rd harmonics corresponding to the resonance mode 3. This allowed us to evaluate the parameters

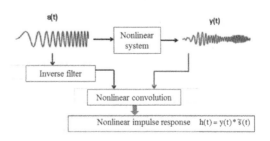

Figure 4. Block diagram of the nonlinear convolution method (Novak et al. 2012).

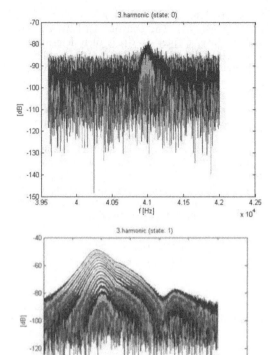

Figure 5. Appearance of a peak corresponding to the 3rd harmonic of the fundamental resonance of a glass bar when thermally damaged.

$\alpha_{2f}(3) = 4810^3$ and $\alpha_{3f}(3) = 2610^3$ whose values are cleary higher than the third order flexural resonance $\alpha_f(3) \cong 168$. The only harmonic detected for the fourth flexural resonance mode is the 3rd harmonic where $\alpha_{3f}(4) \cong 9210^3$ at the moment where $\alpha_f(4) \cong 370$. Finally, the nonlinear hysteretic parameter relative to the 2nd harmonic of the 5th flexural mode was evaluated as $\alpha_{2f}(5) \cong 5710^4$ whose value is greater than $\alpha_f(5) \cong 5270$. Finally, note that the presented results include only harmonics whose frequency changes as a function of excitation. Indeed other harmonics have been detected such as H2 (2), H2 (4), H3 (1), etc., but their characteristics (resonance frequency and quality factor) have not changed as a function of excitation. For what concerns NDT applications the appearance of harmonics is a sign of acoustic nonlinear behavior and could be used as an indicator of micro-cracks presence and evolution as long as the signal-to-noise ratio is good enough to analyze higher order harmonics.

## 4 POWER LAW ANALYSIS

In this section we used a power law analysis in order to characterize the dependence of a nonlinear system response on a forcing variable $y = ax^b$, where the exponent $b$ plays an important role in defining universality classes (Stenull 2008). Here we should notice that the amplitude range limits the use of a power law analysis. Indeed, at low amplitudes noise effects could hide the nonlinear signal, while at large amplitudes saturation effects might appear (Bentahar et al. 2013, Idjimarene et al. 2014). Nonlinear effects, which can appear when microdamage is present breaks the superposition property. When the excitation amplitude $A_0$ is small the response of the material (micro-cracked or intact) is linear. At a higher excitation $A(A > A_0)$ the response of the nonlinear system $u(t)$ is different from the response of the linear system $u_{ref}(t) = \frac{A}{A_0}u_0(t)$ obtained at the same excitation. In order to analyze this effect in the time domain, the Scaling Subtraction Method (SSM) is used. SSM is based on the calculation of the difference between the two responses (Bruno 2009). The SSM signal is defined as $w(t) = u(t) - u_{ref}(t)$. Then, the energy of $w(t)$ is calculated in a short time window as follows:

$$y_{SSM} = \frac{\sqrt{\frac{1}{T}\int_0^T w^2(t_0 + t)\,dt}}{x_u} \qquad (6)$$

where $T$ is the time window length and $t_0$ is the initial window time. $x_u$ is the energy of the time signal recorded at amplitude $A: x_u = \sqrt{\frac{1}{T}\int_0^T u^2(t_0 + t)\,dt}$.

In (Scalerandi et al. 2015) an extensive analysis of data reported in the literature and from our experi-

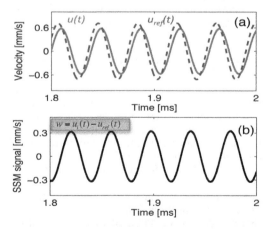

Figure 6. (a) Reference and large amplitude signals, (b) SSM signal.

ments was performed. Analyzed data refers to nonlinear characterization of different materials, where different nonlinear indicators were involved (2nd and 3rd harmonic, sidebands, SSM, etc.). Results showed that in the case of classical nonlinear media (mostly metals and solid media in the intact state) the exponent b is 2. However, in the case of hysteretic nonlinear media b is 1. In this group we find consolidated granular and cellular media (rocks, concrete and bones) or slightly damaged metals and composites. On the other hand the exponent b is found equal to 3 in the case of macroscopically damaged media, which is characterized by the presence of open cracks (ex. consolidated granular). The same procedure was followed in the case of evolving nonlinear media, where damage is induced gradually on the same sample using a mechanical test. Here the exponent b was found to change from 1 to 3 from the state where closed microcracks are present to the open cracks state. The dynamic with which b is changing depends on the performed test, since the evolution of b was found to be very small in the case of a fatigue damage, but the general trend was the same as the previous observations.

## 5 ACOUSTIC EMISSION AND NONLINEAR ACOUSTICS

Nonlinear acoustic parameters revealed to be very sensitive to the evolution of the microstructure of materials. As abovementioned, the consideration of higher order frequency variations allows to increase this sensitivity by several orders of magnitude, which is valuable information for NDT applications. However, the sensitivity of the different parameters is unable to give systematically information about the degree of damage created within the material. This quantification of damage could be made through acoustic emission by linking a given acoustic nonlinear parameter to the elastic energy released by the material during the damage process. Furthermore, the same acoustic emission data can be analyzed in order to link them to the created damage mechanisms. In the case of a polymer-based composite, an appropriate multivariable data analysis based on the time domain characteristics of the AE signals, namely (amplitude, duration, energy, etc.) will allow the monitoring of the different damage mechanisms during a mechanical test for instance (Marec et al. 2008). By doing so, one can calculate the energy related to every damage mechanism at every damage step. Figure 7 shows the main damage mechanisms created in a sheet molding compound SMC composite sample.

At the intact as well as damaged states, the nonlinear relaxation time of the SMC composite plate

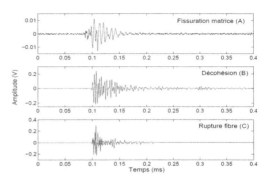

Figure 7. Typical acoustic emission signal created in a glass fiber composite sample (SMC).

is determined (Bentahar et al. 2009). At the beginning, this is performed by exciting the SMC plate around its flexural resonance $f_R = 16.685$ kHz (at the initial intact state) with a very weak level in order to check if it is relaxed. Then, a high amplitude excitation generated with the help of a power amplifier is applied to the SMC plate few minutes. During this conditioning step, the resonance frequency decreases as a consequence of the decrease in the elastic modulus. Right after this conditioning step, the weak amplitude is applied again in order to follow the frequency recovery (or relaxation) of the SMC. Relaxation evolves as a logarithm of time until the SMC reaches its initial frequency. The evolution of the relaxation time as a function of the induced damages is presented in Table 1. In order to make a reliable link between the damage steps (using acoustic emission signals) and the evolution of relaxation time, it is necessary to verify the irreversibility of damage mechanisms, known as Kaiser effect. The latter exists when little or acoustic emission is recorded before any previous maximum stress is achieved. This statement was verified for the gradual damages performed on the SMC plate, which means that all the energy content of the AE signals is a signature of a progressing damage and can therefore be related to the evolution of the relaxation time recorder at every damage step (Bentahar et al. 2009). The correlation between $t_{relax}$ and the cumulative acoustic emission energy recorded at every damage step revealed the existence of a logarithm relationship between these two parameters, where $t_{relax} = a\, Log(E_{abs})$. Here, $a$ is a constant and $E_{abs}$ is the absolute energy of the acoustic emission signals calculated as the sum of every single AE signal along its duration between $[t_1, t_2]$ as: $E_{abs} = \int_{t1}^{t2} A^2 dt$. When the acoustic emission signals are classified as a function of the main damage mechanisms (See Fig. 7) we find also a logarithm relationship but with different coefficients, which means that damage mechanisms relax with

Table 1. Evolution of the relaxation time as a function of damage.

| Damage state | Relaxation time ($t_{relax}$) | |
| | s | % |
| --- | --- | --- |
| Intact (reference) | 3600 | – |
| D1 | 7500 | 108 |
| D2 | 8100 | 125 |
| D3 | 10500 | 192 |
| D4 | 12600 | 250 |

Table 2. Relaxation velocities attributed for the main three damage mechanisms.

| Damage mechanism | Relaxation velocity |
| | s/aJ |
| --- | --- |
| Matrix cracking | 0.5 |
| Fiber/Matrix Debonding | 4.9 |
| Fiber cracking | 53 |

different velocities as presented in Table 2. The latter shows that there is an important change in the relaxation velocities, which means that all mechanisms do not contribute with the dynamic in the relaxation process of the SMC and that the relaxation which is always present seems to be the one attributed to the matrix cracking due to the viscoelastic nature of the epoxy. (Bentahar et al. 2009)

## 6 CONCLUSION

This contribution showed results related to the use of advanced nonlinear acoustic techniques to characterize damage in different composite materials. The advantage in using these nonlinear approaches is to detect the early damage created within these structures on one hand and to understand the physics of the micro-cracks behavior on the other hand. Here, the power law analysis revealed to be very interesting to make such a link. However, the quantification of damage using nonlinear techniques alone is still to be improved. Here we proposed the use of acoustic emission, which can be used under the condition of damage irreversibility. The correlation between acoustic emission and nonlinear acoustics data showed that a quantification of nonlinear parameters is possible. Furthermore, it showed also that the nonlinear behavior of every damage mechanism can be determined in order to better understand the physics that lies behind the macroscopic observations. These encouraging results and techniques are being used and developed by our group and collaborators on different

materials in order to bring more proofs about the necessity to consider in todays world nonlinear acoustics as a regular NDT technique.

## REFERENCES

Baccouche Y., Bentahar M., Mechri C., El Guerjouma R., Ben Ghozlen M-H., 2013. Hysteretic nonlinearity analysis in damaged composite plates using guided waves J. Acoust. Soc. Am. 133(4), EL256–EL261.

Bentahar M., El Guerjouma R., Idjimarene S., Scalerandi M., 2013. Influence of noise on the threshold for detection of elastic nonlinearity.

Bentahar M. EL Guerjouma R. 2009. Monitoring progressive damage in polymer-based composite using nonlinear dynamics and acoustic emission, J. J. Acoust. Soc. Am. 125(1) EL39–EL44.

Bentahar M., Aqra H., El Guerjouma R., Grifa M., Scalerandi M., 2006. Hysteretic elasticity in damaged concrete: Quantitative analysis of slow and fast dynamics. Phys. Rev. B, vol. 73, 2006, 014116.1–014116.10.

Bruno C.L.E., Gliozzi A.S., Scalerandi M., Antonaci P., Analysis of elastic nonlinearity using the scaling subtraction method, Phys. Rev. B 79, 064108.

Farina A. 2001. Nonlinear convolution: a new approach for the auralization of distorting systems. In: AES 108th convention. Amsterdam.

Gusev V., Lauriks W., and Thoan V. 1998. Dispersion of nonlinearity, nonlinear dispersion, and absorption of sound in micro-inhomogeneous materials, J. Acoust. Soc. Am. 103(6), 3216–3226.

Guyer R.A. and Johnson P.A. 2009. Nonlinear Mesoscopic Elasticity: The Complex Behaviour of Rocks, Soil, and Concrete. Wiley-VCH, Germany.

Graff K.F. 1991, Wave Motion in Elastic Solids. Dover Publications, Inc., New York.

Marec A., Thomas J.-H., El Guerjouma R., 2008. Damage characterization of polymer-based composite materials: multivariable analysis and wavelet transform for clustering acoustic emission data. In Mech. Syst. Signal. Process. 22, 1441–1464.

McCall K.R., Guyer R.A., 1994. Equation of state and wave propagation in hysteretic nonlinear elastic materials, Journal of geophysical research, vol. 99(B12), 23887–23897.

Novak A, Simon L, Kadlec F, Lotton P. 2010. Nonlinear system identification using exponential swept-sine signal. IEEE Trans Instrum. Meas. 59(8) 2220–9.

Rasolofosaon P.N.J., 1996. Resonance and elastic nonlinear phenomena in rock, J. Geophys. Res. 101, 11553–11564.

Scalerandi M. Idjimarene S., Bentahar M., El Guerjouma R., 2015. Evidence of microstructure evolution in solid elastic media based on a power law analysis. Commun Nonlinear Sci Numer Simulat 22(1–3)334–347.

Stenull O. 2008. Anomalous elasticity in nematic and smectic elastomer tubule phases, Physical Review E. 78, 031704. J. Appl. Phys. 113, 043516.

Zaitsev V. Yu., Nazarov V.E., and Belyaeva I. Yu. 2001. The equation of state of a microinhomogeneous medium and the frequency dependence of its elastic nonlinearity, Acoust. Phys. 47(2), 178–183. Translated from Akusticheskiœ Zhurnal 47(2), 220–226 (2001).

*Emerging Technologies in Non-Destructive Testing VI – Aggelis et al. (Eds)*
*© 2016 Taylor & Francis Group, London, ISBN 978-1-138-02884-5*

# A simulation study of Local Defect Resonances (LDR)

S. Delrue & K. Van Den Abeele
*Department of Physics KU Leuven Kulak, Wave Propagation and Signal Processing, Kortrijk, Belgium*

ABSTRACT: A highly efficient activation of defects is crucial for nonlinear, ultrasonic, Non-Destructive Testing (NDT) techniques to be able to detect incipient damage (e.g. sub-millimeter size cracks or delaminations). Recently, it has been experimentally demonstrated that the elastic wave-defect interaction can be maximized by employing the concept of Local Defect Resonances (LDR). The LDR concept takes advantage of characteristic frequencies of the defect (defect resonances) and urges the applicationof ultrasonic excitation frequencies which are tuned to the defect. In fact, for nonlinear methodologies, the ultrasonic excitation of the sample should be adjusted at either multiples or integer ratios of the characteristic resonance frequencies. This eventually results in an extreme enhancement of the defect vibration amplitude and substantial rise in efficiency and sensitivity of ultrasonic NDT techniques. To attest the existence of local defect resonances, a simulation study has been performed using a realistic model for the nonlinear dynamic behavior of delaminations. The model allows to define an appropriate excitation method and analysis tool that reveals the local defect resonances.

## 1 INTRODUCTION

Within the different NDT techniques, ultrasonic testing is one of the most used methods (Cheeke 2002, Chen 2007, Krautkrämer and Krautkrämer 1990, Kundu 2004), because ultrasonic testing, although one of the oldest techniques, is relatively simple to use and very effective in the detection of defects. Ultrasonic techniques are based on measuring modifications of ultrasonic waves, with frequencies starting from 20 kHz (i.e. beyond the upper limit of human hearing), upon propagation through the inside of the material. Conventional ultrasonic NDT techniques are commonly based on reflection, diffraction and scattering of acoustic waves by defects (e.g. inclusions) causing amplitude and/or phase variations of the output signal, but the frequency content of the input and output signals remains (Chen 2007, Delrue et al. 2010, Krautkrämer and Krautkrämer 1990, Victorov 1967). The efficiency of the interaction depends on the size of the defects and the degradation of linear material properties caused by the damage. Incipient damage in the form of microcracks or delaminations, which is of our interest, can be detected by traditional linear ultrasonic techniques only if they are open, since the ultrasound is then scattered, reflected or transmitted by the defect. However, these defects tend to be closed at rest, meaning that the adhesion between two regions of the material is affected, even though the two interfaces of the defect are still in contact, causing a lack of acoustic impedance contrast. Nonlinear ultrasonic

techniques (and especially NEWS techniques), on the other hand,have proven to be extremely sensitive to early damage evaluation in materials (Nagy 1998, Guyer and Johnson 2009, Solodov 1998, Solodov et al. 2002, Van Den Abeele et al. 2000, Van Den Abeele et al. 2000). The field of nonlinear acoustics in solids deals with the investigation of the amplitude dependence of material parameters (moduli, velocities, attenuation), which can for instance be evidenced by thorough analysis of modifications in the spectral content. The degree to which these material properties depend on the applied dynamic amplitude can be quantified by various nonlinear parameters. Through the instantaneous detection of the nonlinearity parameters, the internal damage in a material can be measured very efficiently. In case of cracks and delaminations, the closed defect can be consecutively opened and closed by applying a finite excitation amplitude that is able to overcome the defects specific activation threshold. This local contact behavior (commonly referred to as kissing, breathing or clapping) gives rise to a nonlinear stress-strain relation at the defectlocation since the stress-strain response is different in tension than in compression (violation of the simples Hookes law). This is evidenced by the frequency spectrum of the nonlinear structural response which is no longer dominated only by the frequency of the excitation signal. Higher order harmonics and subharmonics, whose frequencies have a prescribed relation with the excitation frequency, appear in the spectrum as a result of the nonlinearity (Delrue and Van Den Abeele 2011,

Delrue and Van Den Abeele 2012b, Delrue and Van Den Abeele 2012a, Krohn et al. 2002, Korshak et al. 2002, Ohara et al. 2007, Ohara et al. 2008, Ohara et al. 2009, Ohara et al. 2010, Sarens et al. 2008, Sarens et al. 2010, Solodov 1998, Solodov et al. 2002).

During the last few years, several studies have been performed to increase the efficiency of the nonlinear ultrasonic detection of incipient damage. By using for instance a resonant excitation of the sample under study, the nonlinear defect response (sub—or higher harmonic generation) can be enhanced (Delrue and Van Den Abeele 2012b, Johnson et al. 1996, Van Damme and Van Den Abeele 2014, Van Den Abeele et al. 2000). A drawback of this method is that defects should be positioned outside the nodal areas of the excited standing wave pattern. To account for this problem, a multi-modal approach was developed, in which multiple resonance frequencies are combined (Van Den Abeele 2007), or sweep signals are used instead of single frequency excitation (Delrue and Van Den Abeele 2011, Delrue and Van Den Abeele 2012b). Another way to enhance the nonlinear acoustic response was recently suggested by the group of Prof. Solodov et al. and is called Local Defect Resonance (LDR)-spectroscopy (Solodov et al. 2011, Solodov et al. 2013, Solodov 2014). LDR-spectroscopy is a novel and efficient NDT technique that takes advantage of characteristic frequencies of the defect (defect resonances) in order to provide maximum acoustic wave-defect interaction. Using the concept of LDR, i.e. applying ultrasonic excitation frequencies tuned to the defect, it has been shown experimentally that the ultrasonic excitation of the defects can be optimized.

In this paper, a numerical study is performed to illustrate the existence of LDR using an existing 3D simulation code for wave propagation in materials containing closed but dynamically active cracks or delaminations, recently developed by the authors (Delrue and Van Den Abeele 2012b, Delrue and Van Den Abeele 2012a). The paper starts with a brief description of the numerical model. Then, an appropriate method to determine local defect resonance frequencies, based on the Scaling Subtraction Method (SSM) (Scalerandi et al. 2008), is discussed. The simulation results discussed in this paper will help us to obtain a better understanding of the concept of LDR and to assist in the further design and testing of LDR-spectroscopy.

## 2 NUMERICAL MODELING OF A DELAMINATED SAMPLE

The material under consideration is a unidirectional (anisotropic) carbon fiber composite with density $\rho = 1800$ kg/m$^3$ containing a circular delamination, parallel to the surface. The composite plate has a length and width of 50 mm, and a thickness of 2 mm, as illustrated in Figure 1. The delamination has a diameter of 10 mm and is centered at $x = 20$ mm and $y = 30$ mm, at a depth of 0.4 mm, with respect to the top surface of the plate. An external acoustic/ultrasonic circular source with a diameter of 10 mm and located at $x = 10$ mm and $y = 10$ mm is introduced in the model by applying an out-of-plane (normal) displacement boundary condition at the top surface of the sample. The 3D numerical model for wave propagation in the considered composite plate uses the commercially available finite element based software package COMSOL Multiphysics, in combination with our own custom-developed clapping model in order to simulate the macroscopic nonlinear clapping behavior of kissing bond flaws (Delrue and Van Den Abeele 2012b, Delrue and Van Den Abeele 2012a). The so-called clapping model considers the introduction of a non-perfect bond between surfaces. This can be described by a set of virtual spring and damper forces at both sides of the defect surface. Above a pre-set separation threshold, the two sides are completely separated (stress-free surfaces). Below the threshold, particular formulations of the Vanderwaals forces are implemented. When the surfaces are close to each other, they will be attracted to each other. However, when they tend to be too close, the force will turn into a repulsive force, trying to separate the surfaces again. Apart from the elastic contact force, damping forces were also implemented as forces that are acting against the velocity of the separation. These forces make sure that the surfaces are not separating too abruptly, avoiding a destruction of the material. At the same time the damping forces assume that the two surfaces are not closing too fast either, so that they cannot overlap. The nonlinear viscoelastic behavior at the defect level, as described above,

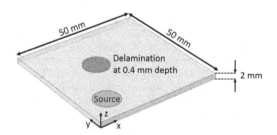

Figure 1. Illustration of the composite plate used in the simulation. The plate has dimensions 50 mm × 50 mm × 2 mm and contains a circular delamination with a diameter of 10 mm, centered at $x = 20$ mm and $y = 30$ mm, at a depth of 0.4 mm, with respect to the top surface of the plate. An ultrasonic circular source with a diameter of 10 mm is located at $x = 10$ mm and $y = 10$ mm.

mimics the clapping behavior of the defect. Depending on the displacement amplitude of an elastic wave passing by the defect, its local nonlinearity will be activated or not, creating a distortion in the wave propagation, which can be measured after appropriate signal processing. Amplitudes must reach a certain threshold for the clapping nonlinearity to be activated. If the wave amplitudes are too low to separate the surfaces in case of a closed defect, there will be no influence in terms of spectral broadening on the wave propagation. The implementation of the nonlinear spring and damper forces is performed by introducing dynamic boundary conditions in COMSOL at those nodes that correspond to the delamination surface. At these positions, the nodes were split in pairs and analytic formulas for the spring and damper forces were defined at each side of the pair. Typical parameters of kissing bond flaws, as for example the stiffness of the defect, can be adapted by changing various parameters in the model (introduced at the level of the defect).

## 3 DETERMINATION OF LDR FREQUENCIES

Resonance frequencies and corresponding vibration patterns in a sample can be determined by exciting the sample with a wide-band ultrasonic signal (e.g. linear sweep) in combination with a scan of the sample surface displacements. However, often, this method will not work to find local defect resonances as the defect faces vibrate at much lower amplitudes when compared to the plate vibrations. A common method used to extract nonlinear features from the elastic response of a sample insonified by an ultrasonic wave is the Scaling Subtraction Method (SSM) (Scalerandi et al. 2008). This method employs a simple difference between signals once a reference signal has been defined for the examined sample. In general, a very low amplitude excitation is used as a comparison signal which, linearly rescaled in amplitude, is subtracted from the signal recorded at larger excitation amplitudes to yield the nonlinear signature.

Applying the concept of SSM to our simulations, we first excite the sample using a linear sweep signal with a low excitation amplitude $A_{low}$:

$$S_{low}(t) = A_{low} \sin\left[ 2\pi\left( f_0 + \frac{Bt}{2T} \right)t \right], \tag{1}$$

where $f_0$ is the starting frequency of the sweep signal, $B$ is the bandwidth determining the total range of frequencies present in the sweep signal and $T$ is the duration of the signal. In the present simulations, we considered a sweep signal that extends over the 10–100 kHz range. The duration $T$ of the signal was set equal to 2 ms. During this low amplitude excitation, the normal displacements at the top surface of the sample are determined. The determined normal displacement amplitudes will not (or only slightly) be influenced by the defect as the excitation amplitude is too low to excite the delamination (i.e. defect will stay closed). The measured signals can thus be used as a 'linear' reference signal. Subsequently, the sample is excited using the same sweep signal, but now with a larger excitation amplitude $A_{high}$ equal to $nA_{low}$. Again, normal displacements are determined at the top surface of the sample. In this case, the excitation amplitude is high enough to excite the delamination and this will cause a nonlinear contribution to the normal displacements. If we then subtract the reference signal (i.e. low amplitude signal), multiplied by a factor $n$, from the normal displacements of the second simulation, the linear contribution is eliminated from the signal, and only the nonlinear contributions at the top surface of the sample remain present in the resulting SSM signal.

In order to analyze and interpret the measured sample surface normal displacements and the scaling subtracted displacement signals in terms of resonance frequencies of the specimen or local defect resonances, global spectral information needs to be extracted from the signals. To do this, we used the following approach. We first temporally Fourier transform the vibration pattern of each point at the surface of the plate. Subsequently, we identify the mean amplitude in the spatial 2D picture for each response frequency. The result of this procedure is plotted in Figure 2 for the high amplitude surface displacement signals (dashed line) and the scaling subtracted signals (dotted line). High amplitude signal peaks and scaling subtracted signal peaks (respectively marked with upward—and downward-pointing triangles) are corresponding to either resonance frequencies of the sample or local defect resonances. By retrieving the mode shapes at the selected resonance frequencies the different modes can be identified.

In order to determine the local defect resonance frequencies, only the peaks in the scaling subtracted response signals should be studied. Indeed, the scaling subtraction method allows to isolate the nonlinear contributions in the time signals calculated at the top surface of the sample and therefore, the peaks observed in the amplitude response of the scaling subtracted signals only occur at frequencies at which the layer above the nonlinear defect is strongly vibrating. Two possibilities still remain:(1) the layer is vibrating at a resonance frequency of the sample because the defect is positioned in or close to an anti-node of the created standing wave pattern (Delrue and Van Den Abeele 2011, Delrue

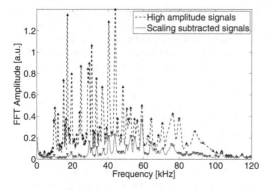

Figure 2. Mean amplitude response at the top surface of the studied composite plate with delamination. The responses are obtained by Fourier transforming both the temporal evolution of the normal surface displacements in case of a high amplitude excitation (dashed line) and the scaling subtracted normal surface displacements (solid line) in each point at the top surface of the sample and calculating the mean amplitude over the entire top surface of the sample at each frequency $f$. High amplitude signalpeaks and scaling subtracted signal peaks, corresponding to either resonance frequencies of the sample or local defect resonances, are respectively marked with upward-pointing and downward-pointing triangles.

and Van Den Abeele 2012a), or (2) the layer above the delamination is vibrating at one of its own resonance frequencies. An example of the first possibility is illustrated in Figure 3, where the simulated mode shapes filtered at 30.8 kHz are plotted for both the high amplitude signals (left figure) and the scaling subtracted signals (right figure). The outer edges of the delamination are each time indicated on the figures to compare them with the displacement amplitude distributions. The maximum values in the scaling subtracted signals occur at positions above the delamination, indicating that the delamination is indeed excited. Even though the observed pattern above the defect resembles that of a higher order LDR mode, the frequency of 30.8 kHz is not a LDR frequency, but a resonance frequency of the sample. The created standing wave pattern in the sample is clearly visible in the vibration pattern of the high amplitude signals. The observed defect vibration pattern is now easily explained by noticing that the parts of the delamination positioned in a nodal area of the standing wave pattern are not vibrating, because the displacements in the vicinity of a node line are very small or even equal to zero. Taking this into account, LDR frequencies are distinguished from sample resonance frequencies by comparing the defect vibration patterns with the sample vibration patterns. If the nodes and antinodes of both patterns do not match, a local defect resonance is found. This is illustrated in Figure 4

for the fundamental LDR mode. Figure 4 shows the simulated vibration patterns filtered at 13.8 kHz for the high amplitude signals (left figure) and the scaling subtracted signals (right figure). From the 'chaotic' vibration pattern observed for the high amplitude signals, it is obvious that the selected frequency is not corresponding to a plate resonance frequency. Moreover, the defect vibration pattern is not linked to possible nodes and anti-nodes found in the sample vibration pattern. Therefore, we conclude that the defect is excited at a LDR frequency, in particular the fundamental LDR frequency as the observed vibration pattern matches the fundamental LDR mode pattern. Figures 5–7 show similar results for some higher

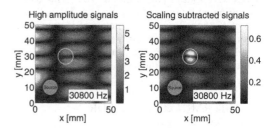

Figure 3. Simulated vibration patterns at 30.8 kHz for the high amplitude signals (left) and the scaling subtracted signals (right). The white circle indicates the position of the delamination.

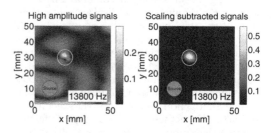

Figure 4. Simulated vibration patterns at 13.8 kHz for the high amplitude signals (left) and the scaling subtracted signals (right). The white circle indicates the position of the delamination.

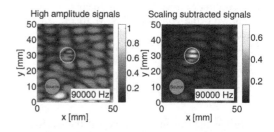

Figure 5. Simulated vibration patterns at 90.0 kHz for the high amplitude signals (left) and the scaling subtracted signals (right). The white circle indicates the position of the delamination.

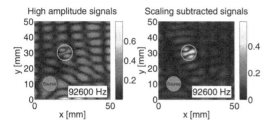

Figure 6. Simulated vibration patterns at 92.6 kHz for the high amplitude signals (left) and the scaling subtracted signals (right). The white circle indicates the position of the delamination.

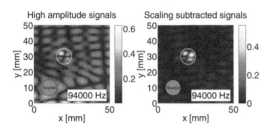

Figure 7. Simulated vibration patterns at 94.0 kHz for the high amplitude signals (left) and the scaling subtracted signals (right). The white circle indicates the position of the delamination.

order LDR modes at respectively 90.0 kHz, 92.6 kHz and 94 kHz. Again simulated vibration patterns at the selected frequencies are plotted for the high amplitude signals (left figure) and the scaling subtracted signals (right figure). In all three cases, the nodal and anti-nodal areas of the defect vibration patterns observed in the scaling subtracted signals cannot be linked to the plate vibrations observed in the high amplitude signals.

## 4 CONCLUSION

A 3D numerical study has been performed in order to better understand the interaction of acoustic waves with a closed but dynamically active delamination in a composite sample. The model uses a self-developed clapping model in order to simulate the macroscopic nonlinear behavior of the delamination. Using the model, it was demonstrated that typical defect vibration patterns can be identified using a sweep excitation in combination with the scaling subtraction method. The scaling subtraction method allows to extract all nonlinear contributions from the signals and therefore, peaks observed in the frequency spectra of the scaling subtracted signals indicate strong defect vibrations.

Using this method, it was shown that efficient excitation of the delamination occurs either at sample resonance frequencies or at local defect resonance frequencies. When the defect is excited at a sample resonance frequency, the observed defect vibration pattern, with nodes and anti-nodes, is strongly linked to the standing wave pattern observed in the sample. When there is no correspondence between the defect and sample vibration patterns, the membrane above the defect is vibrating according to its own eigenmode pattern and the excitation frequency therefore corresponds to a LDR frequency.

In order to further improve the interpretation of defect vibrations, the model needs to be further exploited to be able to better distinguish between plate resonances and defect resonances and to establish the link between several types of defects and their characteristic resonances. Later, the model can be used to study the effect of LDR excitation on the nonlinear ultrasonic defect response.

## ACKNOWLEDGMENTS

The research leading to these results has gratefully received funding from the European Union Seventh Framework Programme (FP7/2007–2013) for research, technological development and demonstration under the Grant Agreements n 315435 (StirScan) and n 314768 (ALAMSA).

## REFERENCES

Cheeke, J. (2002). *Fundamentals and applications of ultrasonic waves*. CRC Press.

Chen, C. (2007). *Ultrasonic and advanced methods for nondestructive testing and material characterization*. World Scientific.

Delrue, S., & K. Van Den Abeele (2011). Finite element simulation of contact acoustic nonlinearities. *Emerging Technologies in Non-Destructive Testing 5*, 295–300.

Delrue, S., & K. Van Den Abeele (2012a). Nonlinear spectroscopy of closed delaminations and surface breaking cracks: Finite element simulation of clapping and nonlinear air-coupled emission. *AIP Conference Proceedings 1474*, 191–198.

Delrue, S., & K. Van Den Abeele (2012b). Three-dimensional finitie element simulation of closed delaminations in composite materials. *Ultrasonics 53*, 315–324.

Delrue, S., K. Van Den Abeele, E. Blomme, J. Deveugele, P. Lust & O. Bou Matar (2010). Two-dimensional simulation of the single-sided air-coupled ultrasonic pitch-catch techniquefor non-destructive testing. *Ultrasonics 50*, 188–196.

Guyer, R., & P. Johnson (2009). *Nonlinear Mesoscopic Elasticity: The complex behaviour of granular media including rocks and soil*. Wiley-VCH.

Johnson, P., B. Zinszner, & N. Rasolofosaon (1996). Resonance and elastic nonlinear phenomena in rock. *Journal of Geophysical Research 101*, 553–564.

Korshak, B., I. Solodov, & E. Ballad (2002). Dc effects, subharmonics, stochasticity and "memory" for contact acoustic non-linearity. *Ultrasonics 40*, 707–713.

Krautkrämer, J. & H. Krautkrämer (1990). *Ultrasonic testing of materials*. Springer-Verlag.

Krohn, N., R. Stoessel, & G. Busse (2002). Acoustic non-linearity for defect selective imaging. *Ultrasonics 40*, 633–637.

Kundu, T. (2004). *Ultrasonic nondestructive evaluation: Engineering and biological material characterization*. CRC Press.

Nagy, P. (1998). Fatigue damage assessment by nonlinear ultrasonic material characterization. *Ultrasonics 36*, 375–381.

Ohara, Y., H. Endo, Y. Hashimoto, Y. Shintaku, & K. Yamanaka (2010). Monitoring growth of closed fatigue crack using subharmonic phased array. *Review of Quantitative Nondestructive Evaluation 29*, 903–909.

Ohara, Y., H. Endo, T. Mihara, & K. Yamanaka (2009). Ultrasonic measurement of closed stress corrosion crack depth using subharmonic phased array. *Jpn. J. Appl. Phys. 48*, 07GD01.

Ohara, Y., T. Mihara, R. Sasaki, R. Ogata, S. Yamamoto, Y. Kishimoto, & K. Yamanaka (2007). Imaging of closed cracks using nonlinear response of elastic waves at subharmonic frequency. *Appl. Phys. Lett. 90*, 011902.

Ohara, Y., S. Yamamoto, T. Mihara, & K. Yamanaka (2008). Ultrasonic evaluation of closed cracks using subharmonic phased array. *Jpn. J. Appl. Phys. 47*, 3908–3915.

Sarens, B., G. Kalogiannakis, & C. Glorieux (2008). Full-field imaging of nonclassical acoustic nonlinearity. *Appl. Phys. Lett. 91*, 264102.

Sarens, B., B. Verstraeten, C. Glorieux, G. Kalogiannakis, & D. Van Hemelrijck (2010). Inonlinearity of contact acoustic nonlinearity in delaminations by shearographic imaging, laser Doppler vibrometric scanning and finite difference modeling. *IEEE T. Ultrason. Ferr. 57*, 1383–1395.

Scalerandi, M., A. Gliozzi, C. Burno, D. Masera, & P. Bocca (2008). A scaling method to enhance detection of a nonlinear elastic response. *Appl. Phys. Lett. 92*, 101912.

Solodov, I. (1998). Ultrasonics of non-linear contacts: propagation, reflection and NDE-applications. *Ultrasonics 36*, 383–390.

Solodov, I. (2014). Resonant acoustic nonlinearity of defects for highly-efficient nonlinear NDE. J. *Nondestruct. Eval. 33*, 252–262.

Solodov, I., J. Bai, S. Bekgulyan, & G. Busse (2011). A local defect resonance to enhance acoustic wave-defect interaction in ultrasonic nondestructive evaluation. *Appl. Phys. Lett. 99*, 211911.

Solodov, I., J. Bai, & G. Busse (2013). Resonant ultrasound spectroscopy of defects: Case study of flat-bottomed holes. *J. Appl. Phys. 113*, 223512.

Solodov, I., D. Döring, & G. Busse (2002). CAN: an example of nonclassical acoustic nonlinearity in solids. *Ultrasonics 40*, 621–625.

Van Damme, B. & K. Van Den Abeele (2014). The application of nonlinear reverberation spectroscopy for the detection of localized fatigue damage. J. *Nondestruct. Eval. 33*, 263–268.

Van Den Abeele, K. (2007). Multi-mode nonlinear resonance ultrasound spectroscopy for defect imaging: An analytical approach for the one-dimensional case. *J. Acoust. Soc. Am. 122*, 73–90.

Van Den Abeele, K., J. Carmeliet, J. Ten Cate, & P. Johnson (2000). Nonlinear elastic wave spectroscopy (NEWS) technique to discern material damage, Part II: Single-mode nonlinear resonance acoustic spectroscopy. *Res. Nondestruct. Eval. 12*, 31–42.

Van Den Abeele, K., P. Johnson, & A. Sutin (2000). Nonlinear elastic wave spectroscopy (NEWS) technique to discern material damage, Part I: Nonlinear wave modulation spectroscopy (NWMS). *Res. Nondestruct. Eval. 12*, 17–30.

Victorov, I. (1967). *Rayleigh and Lamb waves*. Plenum Press New York.

*Emerging Technologies in Non-Destructive Testing VI – Aggelis et al. (Eds)*
*© 2016 Taylor & Francis Group, London, ISBN 978-1-138-02884-5*

# Highly Nonlinear Solitary Waves for the NDT of slender beams

P. Rizzo & A. Bagheri
*Department of Civil and Environmental Engineering, University of Pittsburgh, Pittsburgh, USA*

E. La Malfa Ribolla
*Department of Civil, Environmental, Aerospace and Materials Engineering, University of Palermo, Palermo, Italy*

ABSTRACT: Slender beams subjected to compressive load are common in civil engineering. The rapid in-situ measurement of this stress may help preventing structural anomalies. In this article we describe the coupling mechanism between Highly Nonlinear Solitary Waves (HNSWs) propagating along a granular system in contact with a beam subjected to thermal stress. We evaluate the use of these waves to measure stress in thermally loaded structures and to estimate the neutral temperature, i.e. the temperature at which the stress is null. We investigate numerically and experimentally one and two L-shaped chains of spherical particles in contact with a prismatic beam subjected to heat. We find that certain features of the solitary waves are affected by the beam's stress. In the future these findings may be used to develop a novel sensing system for the nondestructive prediction of neutral temperature and thermal buckling.

## 1 INTRODUCTION

Columns, beam-like structures, and Continuous Welded Rails (CWR) all subjected to axial stress are common in civil engineering. CWR are typically several hundred meters long and they are prone to buckling in hot weather and breakage or pulling apart in cold weather. Reliable nondestructive methods able to determine thermal stress or the Rail Neutral Temperature (RNT), defined as the temperature at which the net longitudinal force in the rail is zero, are needed. The knowledge of the neutral temperature $T_N$ allows for the estimation of the maximum temperature $T_{cr}$ that a structure may withstand before buckling by using the following relationship:

$$T_{cr} = T_N - \frac{\sigma_{cr}}{E\alpha} \tag{1}$$

where $\sigma_{cr}$ is the Euler stress, E is the Young's modulus, and $\alpha$ is the thermal expansion coefficient of the material.

In this paper we propose the Highly Nonlinear Solitary Waves (HNSWs) as a tool to measure thermal stress in prismatic slender beams. HNSWs are compact non-dispersive waves that can form and travel in nonlinear systems such as one-dimensional chains of particles (Nesterenko 1983, 2001, Nesterenko et al. 1995, Lazaridi & Nesterenko 1995,

Coste et al. 1997, Daraio et al. 2005, 2006, Job et al. 2005, 2007, Carretero-González et al. 2009). In the last few years, the propagation of HNSWs in one-dimensional chains of spherical elastic beads was proven successful or the nondestructive testing of structural and biological materials (Yang et al. 2011, 2012a, b, Ni et al. 2012, Ni & Rizzo 2012a, b). The research group by Rizzo investigated the interaction of HNSWs propagating along a chain of particles with beams of different geometric and mechanical properties or subjected to compression due to thermal load (Cai et al. 2013, Bagheri et al. 2015).

In this article, we present some of the experimental work relative to the coupling of one L-shaped chain of particles at the midspan of a prismatic beam held in tension and then heated with a thermal tape. The setup mimics the current practice where CWRs are pretensioned prior to installation in order to counterbalance the thermal load due to warm weather.

To interpret the experimental findings, a discrete particle model was used to predict the propagation of the waves in the chain and to derive the shape and amplitude of the force function (the wave pulse) at the chain-beam interface. The continuous beam theory was then used to predict the vibration of the beam subjected to the solitary pulse and to determine the characteristics of the waves generated by the impact of the vibrating beam bouncing back towards the chain.

## 2 EXPERIMENTAL SETUP

The setup is presented in Figure 1. A 19.05 mm × 9.525 mm stainless steel beam was clamped with a free length of 914 mm to a MTS machine. The properties of the steel were assumed to be density = 7800 kg/m$^3$, Young's Modulus = 200 GPa, yielding stress = 206.8 MPa, and Poisson's ratio = 0.28. From these properties and assuming fixed-fixed boundary conditions, the beam's critical load and the corresponding stress resulted −12.95 kN and −71.395 MPa, respectively. Heat was applied by means of a thermal tape secured to the beam and a thermocouple was used to measure the temperature.

A chain of 24 beads was assembled inside a cast acrylic L-shaped guide with Young's modulus and Poisson's ratio equal to 3.1 GPa and 0.375, respectively. The elbow contained 5 grains and its radius of curvature was equal to 70.4 mm. The particles were stainless steel bearing-quality balls (type 302, McMaster–Carr) with a diameter D = 19.05 mm, mass m = 29 g, elastic modulus E = 200 GPa, Poisson's ratio ν = 0.28, and density ρ = 8000 kg/m$^3$. Seven beads and 12 beads were located in the vertical and in the horizontal segment, respectively, of the guide which was supported by a house-built steel frame.

An electromagnet located above the vertical segment was used to drive the striker from a falling height of about 4.2 mm. The electromagnet was driven by a National Instruments-PXI controlled by LabVIEW. Two sensor particles, hereafter indicated as sensor S1 and sensor S2 were assembled and used to measure the propagation of the HNSWs generated by the mechanical impact of the striker. Each sensor bead contained a 19.05 mm diameter, 0.3 mm thick piezoelectric ceramic disc bonded in between two half-particles according to the inset of Figure 1. S1 and S2 were positioned at the middle of the 14-th and 19-th particle. The sensors were connected to the same NI-PXI and the signals were digitized at 2 MHz sampling rate.

The beam was initially loaded in tension up to 20% of its yield load. Then the MTS machine operated in displacement control held the beam, and heat was applied. We measured the HNSWs during the heating phase at step of 5°C until the beam reached the 60% of its critical load. We took 15 measurements at each step. Two heating processes, hereafter indicated as Test 1 and Test 2, were completed.

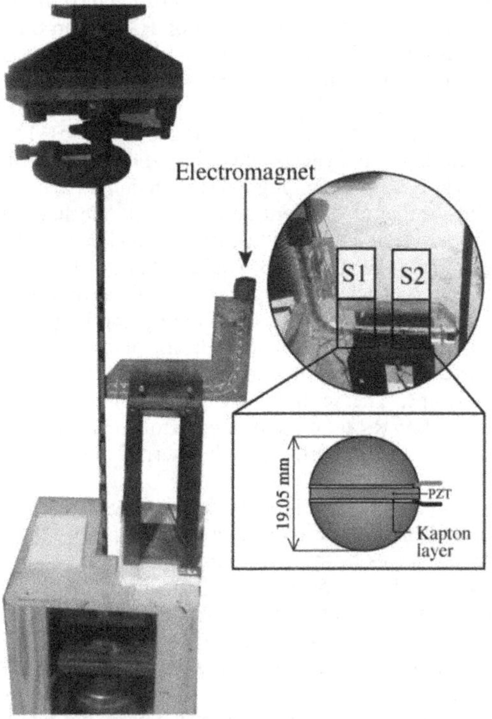

## 3 RESULTS

Figure 2 shows the axial stress as a function of the temperature raise ΔT for both heating ramps. We observe that at zero stress there is a slight change of the slopes. This may be attributed to some settling of the machine's grips. By interpolating the experimental data, a linear relationship between the stress and the temperature is found, in agreement with Eq. (1). By solving the linear fit for the

Figure 1. Photo of the experimental setup. Two sensor particles S1 and S2 were used to measure the waves. Each sensor bead contained a piezoelectric ceramic disc.

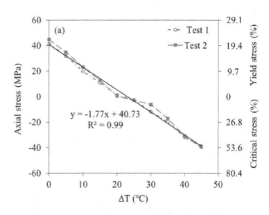

Figure 2. Axial stress as a function of the beam's temperature measured with a thermocouple. The continuous line is the linear interpolation of both heating loads.

zero stress we find that the neutral temperature occurred at $\Delta T = 23.01$ °C.

Figure 3(a) shows the experimental and the numerical solitary waves measured during test 1 in presence of the maximum value of the tensile stress equal to 41.4 MPa by sensor S1. Similarly, Figure 3(b) shows the results relative to the maximum compression stress of −38.7 MPa. The numerical results are overlapped to the first of the corresponding 15 measurements. To ease the comparison between numerical and experimental data, the abscissas were shifted horizontally in order to overlap the arrival of the first pulse. This was necessary since the numerical t = 0 corresponded to the motion onset of the first particle, whereas the experimental trigger was set to the arrival of the first pulse at S1. Moreover, the amplitudes were normalized with respect to the amplitude of the corresponding first peak. This was necessary as the model predicts force whereas the experiments measure voltage.

In Figure 3(a), the first pulse is the Incident Solitary Wave (ISW) generated by the striker. This pulse is tailed by another pulse that is generated at the elbow. We will detail about this second pulse later. The wave visible at about 1.5 ms is the Reflected Solitary Wave (RSW) which is the wave backscattered from the beam-chain interface, hereafter referred simply as the interface. These three waves occur also under large compression as shown in Figure 3(b). Overall, the numerical and the experimental waveforms agree fairly well.

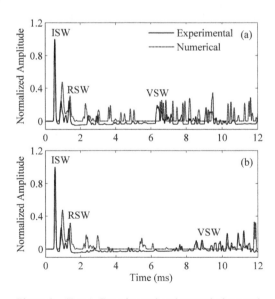

Figure 3. Test 1. Experimental and numerical normalized amplitudes. (a) Time waveform measured by S1 at the largest tensile stress of 41.39 MPa. (b) Time waveform measured by S1 at the largest compressive stress of −38.7 MPa.

The discrepancy may be related to one or more of the following factors: absence of static precompression and dissipation in the numerical model; difference between the real and the selected mechanical properties of the materials. Dissipation occurs at the contact point between the granules and the guides and occurs also due to friction at those contact points. In our study these effects were ignored to simplify the formulation. The agreement between the numerical and the experimental profiles demonstrates that the simplification does not compromise the accuracy of the results.

The origin of the pulse tailing the incident pulse was discussed in (Bagheri et al. 2015).

The variations observed in Figure 3 may not be perceived even by an experienced operator. Thus, in order to quantify the effect of the thermal stress on the wave propagation, four features were extracted from the time series measured at S1 and S2. The first two are related to the amplitude of the pulses normalized with respect to the amplitude of the incident pulse. The normalization was chosen in order to minimize any error relative to the discrepancies mentioned earlier. The remaining features are related to the Time-Of-Flight (TOF) that is the transit time at a given sensor bead between the incident and the reflected or the vibration-induced waves. The features associated with the VSW were computed by picking the maximum amplitude of the cluster and the cluster's front edge.

Figure 4 shows the features measured at S1 as a function of stress. The experimental data are the average value of the 15 measurements and the corresponding $2\sigma$ confidence interval. The amplitude of the reflected wave does not present a monotonic trend and the variations are within the standard deviations. Similar conclusions can be drawn for the amplitude of the vibration-induced wave. The numerical and the experimental results show the same trend but different quantitative results. The mismatch is likely due to the negative offset caused by the compression of the PZT due to the self-weight of the vertical segment of the chain, and to the effect of the dissipative phenomena already discussed. Also, it is known that the precompression in the granules affects the speed of the waves, resulting in a sooner arrival of the experimental reflections. By observing Fig. 4(d), we notice that the arrival of the train of waves associated with the vibration of the beam increases with the temperature in agreement with the fact that the beam's period of vibration is directly proportional to the compression.

The experimental data for both Test 1 and Test 2 and relative to the ToF were interpolated using a second degree polynomial. The results relative to the vibration based feature and sensor S1 are presented in Figure 5.

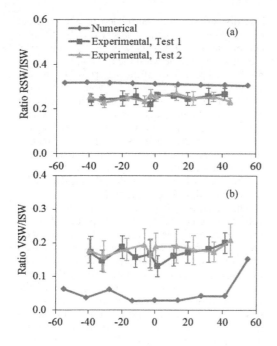

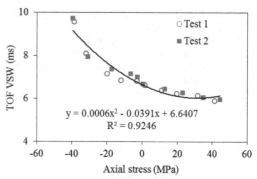

Figure 5. Polynomial interpolation of the experimental data from both Test 1 and test 2. Time of flight of the vibration-induced wave measured at sensor S1.

Ideally, the equation for the TOF of the VSW can be used to calculate the actual stress in any beam identical to the one tested here. By measuring the characteristics of the solitary waves propagating in a chain in contact with the beam, the stress $\sigma_c$ can be inferred while the current temperature $T_C$ can be measured with a thermocouple; these values are then plugged into Eq:

$$T_N = T_c + \frac{\sigma_c}{E\alpha} \quad (2)$$

to determine the neutral temperature $T_N$. For illustrative purpose, let use the polynomial displayed in Figure 5. The average value of the fifteen measurements obtained at $\Delta T = 35°C$ during Test 2 is replaced in the polynomial shown in the figure. The stress of $-15.05$ MPa is retrieved from the polynomial and used in Eq. 1. Using $E = 200$ GPa, $\alpha = 9 \times 10^{-6}$ 1/°C, and the room temperature of 21°C, we determine that the beam's neutral temperature is $T_N = 47.64°C$, which agrees with the experimental value of $T_N = 44.01°C$ ($= 23.01°C + 21.0°C$, the latter being the ambient temperature) found in Figure 2.

## 4 CONCLUSIONS

In this paper, we presented an experimental and a numerical study on the mechanical interaction between a L-shaped granular medium and a slender beam. The chain of spherical particles was used to support the propagation of solitary waves generated by a mechanical striker. The beam was held in tension and subjected to heat to induce thermal stress. We observed the propagation of the solitary waves as the loading induced in the beam crossed from tension to compression until the 60% of its critical load was reached.

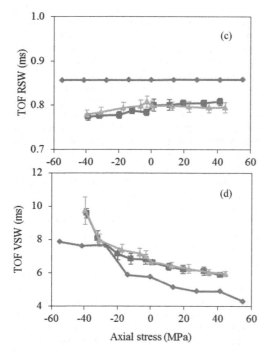

Figure 4. Features extracted from the time waveforms measured at S1 as a function of the axial stress. (a) Ratio of the amplitude of the reflected wave to the amplitude of the incident wave. (b) Ratio of the amplitude of the vibration-induced waves to the amplitude of the incident wave. (c) Time of flight of the reflected wave. (d) Time of flight of the vibration-induced wave.

The results demonstrated that a few features from the solitary waves are consistently able to capture the variation of stress in the beam. From each selected feature we extrapolated an equation that correlates the features to stress. As the relationship is monotonic, we can use this relationship to derive the neutral temperature provided the current temperature of the structure is known.

In the future, the findings of this paper may be used to pave the road towards a novel nondestructive evaluation method for the determination of axial stress in slender beams, columns, and eventually rails.

## ACKNOWLEDGMENTS

This project was supported by the U.S. Federal Railroad Administration under contract DTFR53-12-C-00014. Partial support came from the U.S. National Science Foundation grant CMMI 1200259. We thank Mr. Charles "Scooter" Hager for helping in the design and construction of the house-built steel frame.

## REFERENCES

Bagheri, A., La Malfa Ribolla, E., Rizzo, P., Al-Nazer, L. & Giambanco, G. 2015. On the use of L-shaped granular chains for the assessment of thermal stress in slender structures. *Experimental Mechanics* 55(3): 543–558.

Cai, L., Rizzo, P. & Al-Nazer, L. 2013. On the coupling mechanism between nonlinear solitary waves and slender beams. *International Journal of Solids and Structures* 50: 4173–4183.

Carretero-González, R., Khatri, D., Porter, M.A., Kevrekidis, P.G. & Daraio, C. 2009. Dissipative solitary waves in granular crystals. *Physical Review Letters* 102:024102-1–024102-4.

Coste, C., Falcon, E. & Fauve, S. 1997. Solitary waves in a chain of beads under Hertz contact. *Physical Review E* 56:6104–6117.

Daraio, C., Nesterenko, V.F., Herbold, E.B. & Jin, S. 2005. Strongly nonlinear waves in a chain of Teflon beads. *Physical Review E* 72:016603-1–016603-9.

Daraio, C., Nesterenko, V.F., Herbold, E.B. & Jin, S. 2006. Tunability of solitary wave properties in one-dimensional strongly nonlinear phononic crystals. *Physical Review E* 73:026610-1–026610-10.

Job, S., Melo, F., Sokolow, A. & Sen, S. 2005. How Hertzian solitary waves interact with boundaries in a 1D granular medium. *Physical Review Letters* 94:178002-1–178002-4.

Job, S., Melo, F., Sokolow, A. & Sen, S. 2007. Solitary wave trains in granular chains—experiments, theory and simulations. *Granular Matter* 10:13–20.

Lazaridi, A.N. & Nesterenko, V.F. 1985. Observation of a new type of solitary waves in one-dimensional granular medium. *Journal of Applied Mechanics and Technical Physics* 26:405–408.

Ni, X., Rizzo, P., Yang, J., Katri, D. & Daraio, C. 2012. Monitoring the hydration of cement using highly nonlinear solitary waves. *NDT & E International* 52:76–85.

Ni, X. & Rizzo, P. 2012a. Highly nonlinear solitary waves for the inspection of adhesive joints. *Experimental Mechanics* 52:1493–1501.

Ni, X. & Rizzo, P. 2012b. Use of highly nonlinear solitary waves in NDT. *Materials Evaluation* 70:561–569.

Nesterenko, V.F. 1983. Propagation of nonlinear compression pulses in granular media. *Journal of Applied Mechanics and Technical Physics* 24:733–743.

Nesterenko, V.F., Lazaridi, A.N. & Sibiryakov, E.B. 1995. The decay of soliton at the contact of two "acoustic vacuums". *Journal of Applied Mechanics and Technical Physics* 36:166–168.

Nesterenko, V.F. 2001. Dynamics of Heterogeneous Materials. New York: Springer-Verlag.

Yang, J., Silvestro, C., Khatri, D., De Nardo, L. & Daraio, C. 2011. Interaction of highly nonlinear solitary waves with linear elastic media. *Physical Review E* 83:046606-1–046606-12.

Yang, J., Khatri, D., Anzel, P. & Daraio, C. 2012a. Interaction of highly nonlinear solitary waves with thin plates. *International Journal of Solids and Structures* 49:1463–1471.

Yang, J., Silvestro, C., Sangiorgio, S.N., Borkowski, S.L., Ebramzadeh, E., De Nardo, L. & Daraio, C. 2012b. Nondestructive evaluation of orthopaedic implant stability in THA using highly nonlinear solitary waves. *Smart Materials and Structures* 21:012002-1–012002-10.

# Nonlinear ultrasonic inspection of friction stir welds

M. Tabatabaeipour, J. Hettler, S. Delrue & K. Van Den Abeele
*Department of Physics, Wave Propagation and Signal Processing Group,
KU Leuven KULAK, Kortrijk, Belgium*

ABSTRACT: Kissing bond defects are a worrying kind of root-flaw that may occur in friction stir welding. The tiny size and closed characteristics of such a defect generally result in the inspection incapability of off-the-shelf ultrasonic weld inspection methods. To overcome this problem, we take advantage of a non-linear ultrasonic approach by employing the pulse-inversion method in a contact pitch-catch mode for this purpose. As a result, the obtained pulse-inversion spectral images can be readily interpreted for damage status evaluation.

## 1 INTRODUCTION

### 1.1 Friction stir welding

Friction Stir Welding (FSW) was first introduced in 1991 as a solid-state joint technique by The Welding Institute (TWI-UK), and is nowadays used for a wide range of industrial applications, such as aerospace, automotive, shipbuilding construction (Mishra & Ma 2005; Nandan et al. 2008). The FSW joining procedure is considered to be an environmentally friendly and energy-efficient method. However, FS welded joints are also susceptible to a variety of welding flaws such as kissing bond, wormhole and lazy S-curve. Kissing bonds (potentially caused by the presence of trapped oxide particles from the initial butt contacting surfaces) are specific discontinuities whose adjacent surfaces are in intimate contact with each other, however, with little or no metallic bonding, which makes it hard to detect them using conventional techniques. On the other hand, it is well-known that the fatigue life and strength of the welded joint can be significantly impaired due to presence of kissing bond defects (Oosterkamp et al. 2004; Sato et al. 2005). Therefore, conventional techniques need to be revised and upgraded to enable the detection of these type of defects before they become disastrous.

### 1.2 Pulse inversion technique

Recent experimental and simulation studies have demonstrated the potential of nonlinear ultrasonic techniques for nondestructive testing and evaluation of various materials with diverse types of defects. One of the indicators of nonlinearity is the generation of higher harmonic components in an ultrasonic signal when propagating through a material containing defects. To isolate and intensify the second harmonic in the signal, the so-called Pulse Inversion (PI) technique has been proposed in recent years as a nonlinearity enhancing method. By summing the received signals upon excitation with a negative and a positive polarity, the contribution of the fundamental frequency (linear part of the signal) is suppressed. Indeed, the fundamental signal and all odd harmonics will be counteracted because of their out of phase components and, at the same time, the even harmonic components will be doubled in amplitude (Ma et al. 2005; Simpson et al. 1999; Dos Santos et al. 2010; Kim et al. 2006; Armitage & Wright 2013).

## 2 EXPERIMENTAL SETUP

A contact pitch-catch experiment, as shown in Figure 1, was conducted in order to validate the concept of the pulse-inversion technique for the inspection of kissing bond defects in a friction stir welded butt-joint. Two transducers mounted on the rexolite wedges with angle of 46° degrees, one for

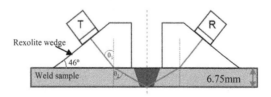

Figure 1. Schematic representation of the contact pitch-catch method for the inspection of friction stir welded butt configuration, T and R stand for Transmitter and Receiver respectively.

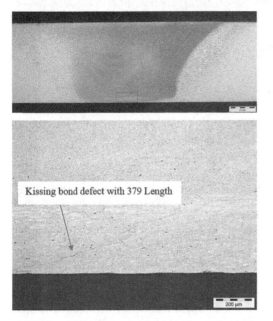

Figure 2. FSW cross-section micrograph (Sample1) showing a kissing bond defect at the weld root, with zoomed-in image (bottom).

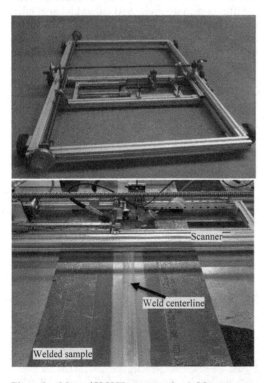

Figure 3. Manual X-Y Flat scanner (top), Measurement with the scanner along the weld centerline (bottom).

the transmitter and the other one for the receiver, were employed in pitch-catch mode, aligned and separated to maximize the reflection from the bottom of the sample. The weld samples in this study are made of aluminum alloy, 7XXX series, with 6.75 mm thickness. Figure 2 illustrates the presence of a kissing bond defect with 379 µm length at the root of a friction stir welded butt-joint.

The measurement was carried out using an AWG NI PXI-5412 and DAQ NI PXI-5122 as generation and data acquisition system respectively. An AR150 A100B amplifier was used for the input signal amplification. In order to examine the samples along the weld centerline, a flat scanner was designed and manufactured as shown in Figure 3. The manual X-Y scanner was used for the A-scan, B-scan and C-scan measurement.

## 3 METHODOLOGY

The excitation signal provided to the transmitter (Panametrics V382 with 3.5 MHz central frequency) consists of two out of phase chirp signals (0 phase and 180 phase) ranging from 2 to 5 MHz with an amplitude of $250V_{pk\text{-}pk}$. Back-wall reflections are received by means of a custom-made Fermat surface transducer with a 7 MHz central frequency. Figure 4 illustrates the typically received out of phase response signals ($R_0$ for phase0 and $R_{180}$ for phase180), while the frequency spectrum of the received positive polarity response ($R_0$) is displayed in Figure 5. As it can be seen, the main frequency band of the response ranges from 2 to 5 MHz.

Subsequently, the two opposite-phase back-wall reflections are summed to obtain the pulse-inversion signal, which is shown in Figure 6. Finally, the normalized frequency spectrum of the pulse-inversion signal is calculated as follows:

$$NPIS = \frac{FFT(R_0 + R_{180})}{FFT(R_0)_{max}^2} \propto \frac{A_2}{A_1^2} \qquad (1)$$

where $R_0$, $R_{180}$ are the phase0 and phase180 response signals respectively, and NPIS stands for the Normalized Pulse-Inversion Spectrum. $A_2$ and $A_1$ are fundamental and second harmonic amplitudes respectively.

The quotient NPIS will represent the ratio of nonlinear to linear scattering. This ratio consequently increases in the presence of nonlinear defects such as kissing bond.

As can be observed in Figure 7, the fundamental frequency has been completely canceled out by the pulse-inversion technique. The main frequency components in the pulse-inversion signal span the region between 4 to 10 MHz.

## 4 RESULTS

The above described frequency spectrum analysis, employing equation (1), was applied to all A-scan signals acquired along the weld centerline. A spectral image was produced by stacking up all *NPIS* data along the weld centerline. As shown in Figure 8, the pulse-inversion spectral image represents a 2D heat map of the presence of nonlinearity along the weld centerline. Horizontal and vertical axes on the 2D heat map represent the frequency and the measurement position along the weld centerline respectively. The colours correspond to the intensity of the pulse-inversion spectrum. In Sample1, the sample with a representative micrograph shown in figure 2, a high degree of nonlinearity can be observed for positions ranging from 150–350 mm, indicating that the severity of the kissing bond defect is dominant in this zone.

As a second illustration, Figure 9 shows the pulse-inversion spectral image for a similar 7XXX aluminum alloy sample, Sample2. Here, two zones can be easily discerned as defect zones, 0–150 mm and around 350 mm, which correspond to severe and mild damage zones.

It is important to note that the colour scale of 2D heat maps should be adjusted to an acceptable quality level of defect. For instance, flaws below 300 μm may be considered as an acceptable quality level of a FSW defect. Therefore, the colour scale should be set to a value representing defect zones greater than 300 μm.

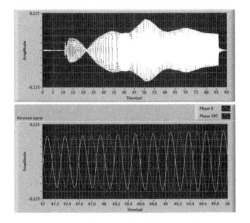

Figure 4. Typical responses of the two out of phase received signals (top); Zoomed-in representation (bottom).

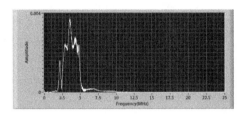

Figure 5. Frequency spectrum of the positive polarity response.

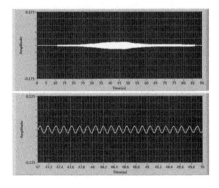

Figure 6. Typical pulse-inversion signal (top); Zoomed-in representation (bottom).

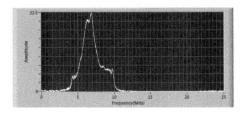

Figure 7. Normalized frequency spectrum of the pulse-inversion signal.

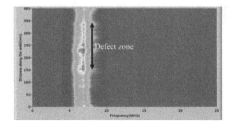

Figure 8. 2D heat map of defect sites in FSW: Pulse-inversion spectral image obtained by stacking up the normalized frequency spectra for each position along the weld centerline, Sample1.

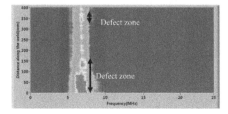

Figure 9. Pulse-inversion spectral 2D heat map of defect sites in FSW, Sample2.

## 5 CONCLUSION

In order to detect kissing bond defects in the friction stir welded joints a contact pitch-catch experiment combined with nonlinear pulse-inversion analysis has been introduced. The resulting pulse-inversion spectral images show that the technique has potential for kissing bond defects examination as the images allow quick and easy inspection interpretation for the damage locations along the weld centerline.

## ACKNOWLEDGEMENTS

The research leading to these results has gratefully received funding from the European Union Seventh Framework Program (FP7/2007–2013) for research, technological development and demonstration under the Grant Agreements n° 315435 (StirScan) and n° 314768 (ALAMSA).

## REFERENCES

Armitage, P.R. & Wright, C.D., 2013. Design, development and testing of multi-functional non-linear ultrasonic instrumentation for the detection of defects and damage in CFRP materials and structures. *Composites Science and Technology*, 87, pp.149–156.

Kim, J.-Y. et al., 2006. Experimental characterization of fatigue damage in a nickel-base superalloy using nonlinear ultrasonic waves. *The Journal of the Acoustical Society of America*, 120, p.1266.

Ma, Q. et al., 2005. Improvement of tissue harmonic imaging using the pulse-inversion technique. *Ultrasound in medicine & biology*, 31(7), pp.889–94.

Mishra, R.S. & Ma, Z.Y., 2005. Friction stir welding and processing. *Materials Science and Engineering: R: Reports*, 50(1–2), pp.1–78.

Nandan, R., Debroy, T. & Bhadeshia, H., 2008. Recent advances in friction-stir welding—Process, weldment structure and properties. *Progress in Materials Science*, 53(6), pp.980–1023.

Oosterkamp, A., Oosterkamp, L.. & Nordeide, A., 2004. Kissing bond phenomena in solid-state welds of aluminum alloys. *Welding journal*, pp.225–231.

Dos Santos, S., Vejvodova, S. & Prevorovsky, Z., 2010. Nonlinear signal processing for ultrasonic imaging of material complexity. *Proceedings of the Estonian Academy of Sciences*, 59(2), p.108.

Sato, Y.S. et al., 2005. Characteristics of the kissing-bond in friction stir welded Al alloy 1050. *Materials Science and Engineering: A*, 405(1–2), pp.333–338.

Simpson, D.H., Chin, C.T. & Burns, P.N., 1999. Pulse inversion Doppler: a new method for detecting nonlinear echoes from microbubble contrast agents. *IEEE transactions on ultrasonics, ferroelectrics, and frequency control*, 46(2), pp.372–82.

*Acoustic emission, damage and lifetime prediction of composites*

*Emerging Technologies in Non-Destructive Testing VI – Aggelis et al. (Eds)*
*© 2016 Taylor & Francis Group, London, ISBN 978-1-138-02884-5*

# Fracture mechanism of CFRP-strengthened RC beam identified by AE-SiGMA

N. Alver
*Department of Civil Engineering, Ege University, Bornova, Izmir, Turkey*

H.M. Tanarslan
*Department of Civil Engineering, Dokuz Eylul University, Buca, Izmir, Turkey*

Ö.Y. Sülün & E. Ercan
*Department of Civil Engineering, Ege University, Bornova, Izmir, Turkey*

ABSTRACT: Externally bonded Carbon Fiber-Reinforced Polymer (CFRP) has been used widely to repair and/or strengthen structures that are deficient in strength. Even though strengthening with CFRP has many advantages, previous studies have shown that sudden failure due to debonding is the major problem for FRP-strengthened RC beams. Thus, it is important to clarify failure mechanisms of a strengthened RC beam with CFRP to hinder this problem. In this study, AE-SiGMA method is applied for determination of cracking development mechanism of RC beams strengthened with CFRP in different widths. In the experimental work, specimens strengthened with CFRP strips were tested under cyclic loading along with AE tests. It was found out that increasing the strip width does not enhance the behavior.

## 1 INTRODUCTION

The lack of shear resistance in Reinforced Concrete (RC) elements causes brittle shear failures. To provide a better behavior for these deficient elements, strengthening techniques based on the use of Fiber Reinforced Polymer (FRP) materials have been proposed and developed in the last two decades (Barros 2005, Cao 2005, Sakar 2008, Tanarslan 2008). Instead of using Carbon Fibre Reinforced Polymer (CFRP) as a sheet, using of CFRP strips as externally bonded reinforcement, is a practically effective method of strengthening and upgrading RC members when economic advantages considered. The ultimate capacity of the strengthened beam is controlled by either flexural shear cracking induced debonding at concrete-CFRP interface, compression crushing of concrete and rupture of CFRP. Rupture indicates that the full capacity is obtained hence the other failure modes show that the full capacity of CFRP is not entirely obtained. Therefore, the reasons causing these kinds of failures have to be investigated in detail.

Debonding is often activated because of the brittle adhesives which cannot resist high stress concentrations at the bonding surface. Thereafter a sudden fracture occurs right after the separation of CFRP from concrete surface. It is crucial for the researchers to clarify failure mechanisms during debonding to hinder this unwanted failure.

The unit strain from the strain gauges situated on the top of CFRP is used to understand the failure mechanism of CFRP strengthened RC structures. It has been interpreted that the increase in the strain gauges results from CFRP being pulled due to cracking. Decrease in unit strain is regarded as damage in the bonding. However, the distance to the strain gauge, the crack orientation and crack type effects the unit strain behavior directly. Therefore, more detailed investigation is needed to clarify the fracture mechanism of CFRP strengthened RC structures. Therefore, based on moment tensor analysis of Acoustic Emission (AE) signals, AE-SiGMA method (Ohtsu 1991, Grosse 2008) is applied to clarify cracking mechanisms of CFRP-strengthened RC beams. By AE-SiGMA, crack kinematics of locations, types and orientations are quantitatively determined. The method is known to be successfully applied for identification of crack kinematics of RC specimens (Ohtsu 1998, Ohno 2010).

In this study, it is aimed to identify the effect of CFRP width on fracture mechanism of CFRP-strengthened RC beams by AE-SiGMA. For this purpose, two specimens with similar spacing but distinct CFRP width were tested in the laboratory. Mechanical observations as well as acoustic emission measurements were taken from the needed locations to identify the types and directions of the cracks.

## 2 AE-SiGMA ANALYSIS

SiGMA analysis is based on the generalized theory of AE and consists of AE source location and moment tensor analysis. It is a simplified form of Green's functions for moment tensor analysis. Following equation is obtained after simplifying the Green's solutions while modeling a crack motion.

$$A(x) = C_s \frac{\operatorname{Re} f(t,r)}{R} r_p m_{pq} r_q DA \qquad (1)$$

where $A(x)$ = amplitude of the first motion; $C_s$ = calibration coefficient of the sensor sensitivity; Ref = reflection coefficient; $R$ = distance from the source to the sensor; and $r$ = its direction vector. $m_{pq}$ = moment tensor; and $DA$ = area of the crack surface.

In the case of isotropic elasticity,

$$m_{pq} = (\lambda l_k n_l \delta_{pq} + \mu l_p n_q + \mu l_q n_p) \Delta V \qquad (2)$$

where $\lambda$ and $\mu$ = Lame constants; $l$ = direction vector of crack motion; and $n$ = unit normal vector of the crack surface.

In an isotropic material, the moment tensor is symmetric and of the second order and thus the number of unknown moment tensor components is six. To solve equation 1, two parameters of the arrival time and amplitude of the first motion are needed from recorded AE waveforms. In order to classify crack type, the eigenvalue analysis of the moment tensor was developed by Ohtsu 1991. The eigenvalues of the moment tensor are represented by combination of shear crack (slip motion) and the tensile crack (crack-opening motion).

## 3 EXPERIMENTAL STUDY

### 3.1 *Design of structurally deficient beams*

Two T-section RC beams were fabricated, strengthened and subjected to cyclic load in the experimental program. Same CFRP spacing with 30 mm and CFRP application as side bonding was ordered for the strengthened RC beams. However, CFRP strip width shows distinction with 50 mm and 100 mm to identify their effect on ultimate behavior and fracture behavior. The details of different ways of strengthening scheme and cross-sectional geometries are shown in Figure 1.

Beams were designed with stated longitudinal reinforcement and concrete strength to hinder flexural problems during the experimental program. Accordingly, three 20 mm diameter deformed steel rebars in the compression zone and three 20 mm diameter deformed steel rebars in the tension zone were used as longitudinal reinforcements. The shear reinforcements of specimens have a spacing of 300 mm which is decided to anticipate a shear failure for the non-strengthened specimen. Stirrups were increased in number to block local cracks that might occur due to applied load.

### 3.2 *Material properties: Concrete, steel bars and CFRP*

To achieve similarity in strength, a single concrete mix was used for all specimens. The concrete was a mixture of water, cement, sand and aggregate with the ratio of 0.68:1:2:3 by mass. Three cylinder specimens were casted and tested at the same time of beam test to determine the compressive strength of concrete. The average compressive strengths of 28-day concrete cylinder for BEAM-1 and BEAM-2 were 28.7 MPa and 29.5 MPa, respectively.

Steel reinforcing bars, which were used in the experimental program, were tested in tension to obtain the mechanical properties. For all specimens, standard deformed reinforcement steel bars with a characteristic strength of 490 MPa and elastic modulus of 208 GPa were used for longitudinal reinforcement. 8 mm diameter deformed steel bars were used as stirrups and their characteristic strength is 510 MPa and elastic modulus is 210 GPa.

SikaWrap 230c (unidirectional) CFRP sheets were used for strengthening. The strengthening procedure includes surface preparation, application of priming adhesive layer and bonding of the CFRP sheets. Sikadur-330, which was prepared in accordance with manufacturer's directions, was applied on the prepared concrete surface. Afterwards, CFRP laminates were attached at the designated places and a roller was used to guarantee impregnation of the sheets by the resin. Then another layer of epoxy was put on top of the fabric and the excess resin was cleaned. All applications were performed at room temperature.

### 3.3 *Experiments*

All specimens were tested under cyclic loading. To perform cyclic load to the specimen, a loading column was designed with hinges by the beam's free end. Load was applied in cycles of loading and unloading. Load cycles were selected as they will help to evaluate the flexure and shear crack propagations and their affect to behavior. Loading was increased up to yield of flexural reinforcements or until the failure of the specimen. Four potentiometers were used to monitor displacements.

Measurement of strain is crucial for designating the contribution of CFRP reinforcements to shear capacity. Eight strain gauges were used for each specimen. Strain gauges were attached at the

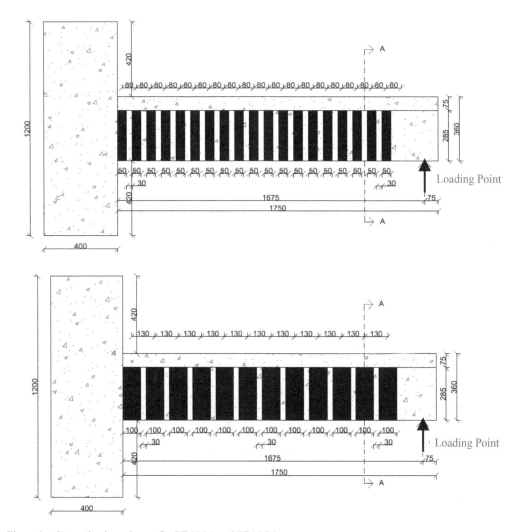

Figure 1. Strengthening schemes for BEAM-1 and BEAM-2.

section mid-height where shear cracks is expected to be developed, between 80 to 1600 mm apart from the beam's support. For the wrapped specimen three more strain gauges were attached at the tension face to evaluate the strain activity at that section.

Eight AE sensors of 150 kHz (R15, PAC) were attached to the surface of the test specimens to capture elastic waves that nucleate due to release of stored strain energy during fracture. An 8-channel DiSP AE system by Mistras Holdings was used to record AE data. AE waves were amplified with 40 dB gain by pre-amplifiers and 40 dB gain by DiSP system. AE sensors were attached to the concrete surface by silicon grease to have a good contact for signal detection.

## 4 EXPERIMENTAL RESULTS AND EVALUATIONS

Shear force-displacement hysteretic curves of the test specimens are presented in Figure 2. First crack always appeared as a flexural crack for the specimens. As the loads increased the firstly developed flexure cracks were progressed through the sides and caused shear cracks between the CFRP strips as it can be visually observed. These cracks then propagated towards the CFRP strips and advanced along the interfacial concrete. It is not possible to see by eye if any other cracks were developed under CFRP strips. Thereafter the interfacial concrete started to weaken, the bond's strength is reduced according to strain gauge out-

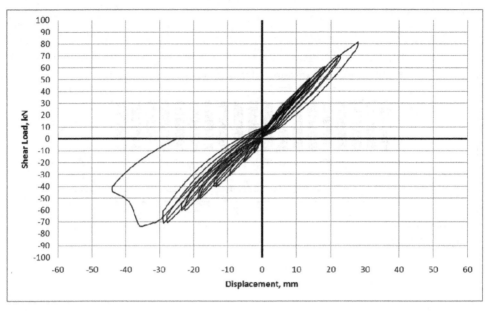

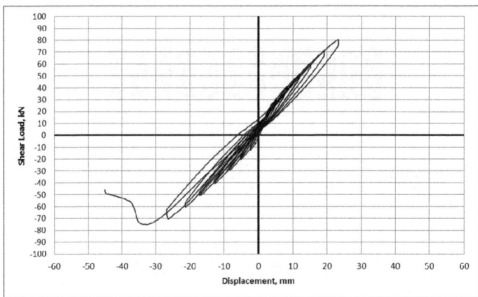

Figure 2. Shear force-displacement hysteretic curves of the test specimens (BEAM-1 and BEAM-2).

comes. Afterwards the CFRP strips were separated from the concrete surface. BEAM-1 and BEAM-2, showed similar behavior and both failed in shear after splitting of one or more CFRP strips (with adherent concrete which denotes the appropriateness of the application). As the load reached to 15 kN, flexural cracks occurred at the bending region of BEAM-1. Thereafter, shear crack developments were observed at the unstrengthened part of the specimen, between CFRP strips. After exceeding 70 kN load level, cracks started developing faster, and widen quickly. 8th CFRP strip was debonded from its upper end while the backward loading. In the next cycle specimen resisted the forward load cycle but debonding experience (since the certain strengthened part lost its resistance) cause the specimen to fail abruptly in shear at 73.73 kN as shown in Figure 2. To indicate the effect of CFRP width on the behavior and fracture mechanism of CFRP-strengthened RC beams, the CFRP width of BEAM-2 was increased to twice of BEAM-1. However, similar behavior was observed according to BEAM-1. This time the 5th CFRP strip was debonded while the backward loading step. Distinctly, shear cracks were propagated at a specific section between the 5th and 7th CFRP strips. Then main crack propagation was occurred suddenly at 75.19 kN.

SiGMA analysis was conducted by using results by 8 AE sensors. Cracks were classified into three type as; tensile, shear and mixed-mode. SiGMA results obtained for both specimens are shown in Figure 3 and 4 for BEAM-1 and 2, respectively. In BEAM-1, first tensile cracks were observed at

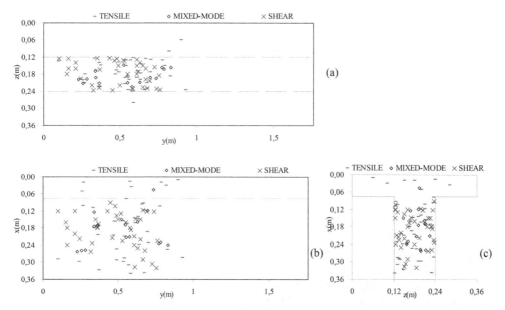

Figure 3. AE-SiGMA results for BEAM-1 after failure ((a) Top view, (b) Side view, (c) Cross-sectional view).

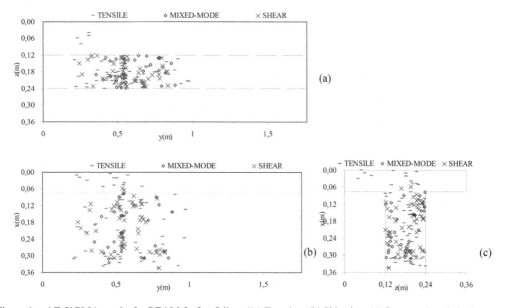

Figure 4. AE-SiGMA results for BEAM-2 after failure ((a) Top view, (b) Side view, (c) Cross-sectional view).

the flange at 0.27 meter from the beam support. First shear crack was observed at the 0.135 meter from the beam support in the middle of the cross-section. Cracks were not seen on the surface and with the increase in load they extended towards the beam surface. Tensile cracks were observed in the beginning and when the load was increased, mixed-mode and shear cracks appeared. After the load exceeded tensile strength of concrete, number of shear cracks increased rapidly. The number of shear cracks is more than mixed-mode and tensile type of cracks. When the fracture pattern of the specimen is considered, it can be said that, there is good correlation between the real crack pattern and the AE activities observed by SiGMA.

In BEAM-2, first tensile cracks were observed at the flange at 0.466 meter from the beam support. First shear crack was observed at the 0.562 meter from the beam support in the middle of the cross-section under the 5th CFRP strip. With increase in load, AE activities increased. In low load levels, tensile cracks were dominant and by increasing the load, mixed-mode and shear cracks were obtained more. A distinctive tensile crack was observed in the specimen at low load levels; however, increasing the load, shear cracks became dominant. Both specimens failed from shear after the critical cracks extended and peeled the CFRP strips off of the concrete surface.

Comparing SiGMA results for both specimens, it is seen that 50 mm strip width (BEAM-1) shows a similar behavior to 100 mm strip width (BEAM-2). Cracks formed in BEAM-1 are distributed; however, cracks formed in BEAM-2 are concentrated on specific locations. Once a crack goes beneath the CFRP strip it continues until it peels off the strip of the concrete surface, however, if the crack is not beneath the strip but encounters the strip, the strip does not let the crack to continue easily. This result is clearly seen from the SiGMA analysis. Tensile cracks observed in the last cycles of loading are good indicatives since they are formed due to CFRP strips trying to pull the concrete surface. Number of these cracks is more in BEAM-2.

## 5 CONCLUSIONS

In order to identify the effect of CFRP width on fracture mechanism of CFRP-strengthened RC beams by AE-SiGMA, two specimens with similar spacing but distinct CFRP width were tested in the laboratory. Mechanical observations as well as acoustic emission measurements were taken from the needed locations to identify the types and directions of the cracks. SiGMA analysis was performed to obtain crack activities along with their locations and crack types. Following conclusions are derived from this study.

Even though the width of the strips was doubled and total area that CFRP strips covered was larger, BEAM-2 showed a similar behavior to BEAM-1. Load bearing capacity did not differ and BEAM-1 showed a slightly more ductile behavior. This result is obtained due to the fact that once a crack goes beneath the CFRP strip it continues until it peels off the strip of the concrete surface, however, if the crack is not beneath the strip but encounters the strip, the strip does not let the crack to continue easily.

CFRP strips enhanced the behavior; however, due to side bonding, final failure for both specimens was shear failure.

SiGMA results were in good agreement with the actual cracks observed by naked eye. Number of shear cracks was dominant for both specimens. SiGMA analysis results were more realistic since nearly all small or large-scaled cracks whether they are crossed by CFRP strip or not, could be monitored.

## ACKNOWLEDGEMENTS

Financial support provided by TUBITAK (The Scientific and Technological Research Council of Turkey) to conduct this research under the grant number 111M559 is greatly acknowledged.

## REFERENCES

Barros, J.A.O., Fortes, A.S. 2005. Flexural strengthening of concrete beams with CFRP laminates bonded into slits, *Cement and Concrete Composites*, 27(4):471–480.

Cao S.Y., Chen J.F., Teng J.G., Hao Z., and Chen J. 2005. Debonding in RC Beams Shear Strengthened with Complete FRP Wraps. *J. Compos. for Constr.*; 9:5:417–428.

Grosse C.U., Ohtsu M. 2008. Acoustic Emission Testing, Springer, Heidelberg.

Ohtsu M. 1991. Simplified moment tensor analysis and unified decomposition of AE source: application in situ hydrofracturing test. *J Geophys Res* 1991;96(B4):6211–21.

Ohtsu M, Okamoto T, Yuyama S. 1998. Simplified moment tensor analysis of acoustic emission for cracking mechanisms in concrete. *ACI Struct J*;95(2):87–95.

Ohno K., Ohtsu M. 2010. Crack classification in concrete based on acoustic emission. *Constr Build Mater*; 24:2339–2346.

Sakar, G. 2008. Shear Strengthening Of RC Beams Subjected To Cyclic Load Using Cfrp Strips, *Advanced Composites Letters*, Vol. 17, Iss. 6.

Tanarslan H.M., Altin S., Ertutar Y. 2008. The Effects of CFRP Strips for Improving Shear Capacity of RC Beams, *Journal of Reinforced Plastics and Composites*, Vol. 27, No. 12, 1287–1308.

*Emerging Technologies in Non-Destructive Testing VI – Aggelis et al. (Eds)*
*© 2016 Taylor & Francis Group, London, ISBN 978-1-138-02884-5*

# Acoustic emission of fiber reinforced concrete under double-punching indirect tensile loading

D. Choumanidis, P. Nomikos & A. Sofianos
*Tunnelling Laboratory, School of Mining Engineering, National Technical University of Athens, Athens, Greece*

E. Komninou, P. Oikonomou & E. Badogiannis
*Laboratory of Reinforced Concrete, School of Civil Engineering, National Technical University of Athens, Athens, Greece*

ABSTRACT: The Acoustic Emission (AE) characteristics of fiber reinforced concrete under double-punching indirect tensile loading, are investigated. Cylindrical concrete specimens are subjected to compression loading along their axis, using two cylindrical steel platens between the loading machine platen and the upper and lower face of the specimens. Thus, indirect tensile stresses are produced within the specimens. The applied load and the platen displacement are continuously recorded during the test. For each concrete mixture preparation, two different synthetic fibers were used, at 0.5% and 1% content by volume of concrete and, two specimens were prepared and tested. The acoustic emission activity was continuously monitored by six acoustic emission transducers attached on the specimen surface. Specific acoustic emission indices, such as the AE average frequency and the RA value, are investigated to characterize the matrix cracking and fiber extrusion or breakage during the test as well as the pre-peak and post-peak loading stages.

## 1 INTRODUCTION

### 1.1 Acoustic emission of fiber-reinforced concrete

As more and more fiber reinforced concrete used in structural engineering, the need for an adequate monitoring and analysis of its behavior increases. In flexural loading tests, tensile cracks generally develop at moderate loading levels, while shear cracks dominate at large loading levels (Yuyama et al. 1999). Nevertheless tensile cracks result in acoustic emissions of higher frequency content and shorter duration (Aggelis et al. 2013a, b, Aggelis et al. 2012). This is demonstrated clearly in many works during four-point bending tests (Aggelis et al. 2013, Aggelis 2011, Ohno & Ohtsu 2010). Acoustic emission indices, such as average frequency of the signals, are correlated with the type of source and the loading stages. It has been found, that shear fractures emit signals that exhibit higher initial Rising Angle (RA)-values with smaller Average Frequency (AF) than tensile ones (Farhidzadeh et al. 2014, Aggelis et al. 2012, Aggelis 2011, Ohno & Ohtsu 2010). The boundaries of these features vary due to several parameters like member geometry, material properties, sensor location and response (Carpinteri et al. 2013). Particularly the average frequency decreases significantly at the moment of main fracture (Aggelis

et al. 2011). Additionally to the mentioned above, RA values and average frequencies plotted shall be calculated from the moving average of more than 50 hits, as it is recommended by RILEM TC 212-ACD (Ohtsu 2010).

Except the abovementioned parameter RA, the "Ib-value" represents the ratio of weak to strong events and is a representative indicator of the upcoming failure (Aggelis et al. 2011). Considerable definitions of acoustic emission parameters:

– Amplitude: is the voltage of the highest peak and is commonly measured either in Volts or dB.
– Peak amplitude: is the largest amplitude value of the waveform.
– Duration: is the time window between the first and the last threshold crossing.
– Rise Time (RT): is the time between the first arrival and the peak amplitude.
– Energy: is the area under the rectified signal envelope.
– (RA): is RT over peak amplitude and is measured in ms/V.
– Counts: is the number of threshold crossings.
– Average Frequency (AF): is the number of counts over duration.
– Peak frequency: is the frequency with the maximum magnitude of the spectrum after FFT of the waveforms.

## 1.2 Scope

This work is an effort to examine flexural strength of fiber reinforced concrete (FRC) in a different test method (double punching test method) than the usual one of bending, by using nondestructive testing technology. Acoustic emission is used not only to investigate the behavior of FRC at discrete loading stages, but also different types of fibers at different percentages by volume of concrete.

## 2 BARCELONA TEST

The Barcelona test is a double Punch Test (DPT) performed on a cylindrical FRC specimen with height equal to diameter equal to 150 mm (AENOR UNE 83515 2010). The specimen is subjected to compression loading along its axis, by inserting two cylindrical steel punches with a height of 24 mm and a diameter of 37.5 mm between the loading machine platen and the center of the upper and lower face of the specimen (Fig. 1). An extensometer chain is placed at half-height of the specimen to measure the Total Circumferential Opening Displacement (TCOD). A constant relative displacement rate of $0.5 \pm 0.05$ mm/min is applied by the press piston. The force and the TCOD are also measured. Studies have shown that it is possible to estimate the TCOD without the need of using the extensometer chain (Carmona et al. 2011) using an experimental correlation between the axial displacement and the TCOD. In addition, Pujadas et al. (2014) presented that Barcelona test may be conducted with cubic specimens with 150 mm of side without the use of an extensometer chain. In this work, the set-up of the generalized double punch test was used, without using an extensometer chain.

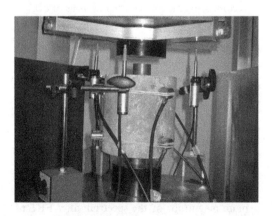

Figure 1. Set-up of Barcelona test.

## 3 EXPERIMENTS

### 3.1 Specimen preparation

In total five different concrete mixtures were prepared. Two different synthetic fibers were used, at 0.5% and 1% content by volume of concrete, producing four different fiber reinforced concrete mixtures, with constant mixture constituents and proportions (see Table 1). All the fiber reinforced concrete mixtures are compared to a plain concrete mixture, without fibers. For each concrete mixture two specimens were prepared and tested. The geometrical and mechanical characteristics of the fibers are given in the Table 2.

All cast specimens were water cured (20°C) after demolding and tested at the age of 28 days. In total 10 cylindrical specimens were prepared.

### 3.2 Experimental setup

A testing machine of 5000 kN capacity with a servo-hydraulic control unit for axial displacement control was used. The displacement rate was set equal to 0.5 mm/min. Each test lasted until axial displacement reached 4 mm after cracking of the specimen at the peak stress. The displacement was measured as the average of three transducers placed

Table 1. Mixture constituents.

| Constituents (kg/m³) | Plain | PPF (Sika) 0.5%* | 1.0% | PPFX (Concrix) 0.5% | 1.0% |
|---|---|---|---|---|---|
| Cement | 400 | 400 | 400 | 400 | 400 |
| Sand (0/4) | 900 | 900 | 900 | 900 | 900 |
| Gravel (8/16) | 700 | 700 | 700 | 700 | 700 |
| Water | 228 | 228 | 228 | 228 | 228 |
| Fibers | 0 | 4.5 | 9.0 | 4.5 | 9.0 |
| Superplasticizer | 0.87 | 1.33 | 0.9 | 1.33 | 3.25 |

*% by volume.

Table 2. Fibers' geometrical and mechanical characteristics.

| | PPF | PPFX |
|---|---|---|
| Type | (Sika fiber PP 940-50) | (Concrix ES) |
| Material | Polyolefine | Polyolefine |
| Length | 50 mm | 50 mm |
| Diameter | 0.75 mm | 0.5 mm |
| Rupturing force | 400 N/mm² | 510 N/mm² |
| Modulus of elasticity | 7000 N/mm² | 10000 N/mm² |
| l/d | 66.6 | 100 |

around the specimen at 120° angle measuring constantly the variation of the distance between the platens of the loading frame.

On the specimen surface six acoustic emission transducers were attached, so that the acoustic emission activity was continuously monitored. The transducers were piezoelectric sensors PAC R15 with dimensions 19 mm × 22 mm (diameter × height), peak sensitivity 69 V/(m/s) and operating frequency range 50–200 kHz. The sensors were attached radially around the specimens per 120°, three of them around the upper part and the other three around the lower part of the specimens. Ultrasonic gel was used as a couplant when attaching the acoustic emission sensors to the specimen surface. The acoustic emission signals were amplified by six PAC 2/4/6 preamplifiers, with amplification capability 20/40/60 dB respectively, adjusted at 40 dB with frequency operating range 10 kHz–2 MHz. A six-channel PAC board, with three PCI-2 cards was used as the monitoring system and AE-Win Software and Noesis 7.0 were used for analysis. The threshold parameter was set to 50 dB.

### 3.3 Results

The failures of the specimens are presented in Figure 2, from left to right: plain concrete, PPF 0.5%, PPFX 0.5%, PPF 1.0%, PPFX 1.0%. The plotted diagrams are separated to three loading stages, the first one until reaching the peak load, the second one after peak load till approaching the residual load and the third one from the end of stage two until the end of testing (Fig. 3), in order to explain more adequately the behavior of each specimen. Some typical results of AF vs. RA for the loading stage until fracture load for a) plain concrete, b) PPF 0.5% and c) PPF 1.0% fibers by volume of concrete are demonstrated in Figures 4a, b and c, respectively. Typical results at the second and third loading stages are demonstrated in Figures 5a and b respectively.

It has been suggested (RILEM TC 212-ACD 3) a classification of cracks into tensile and other-type ones including shear cracks, based on the RA and average frequencies values. The plotted values

Figure 2. Fractures of all specimens.

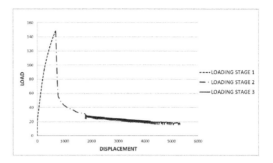

Figure 3. Typical load-displacement curve.

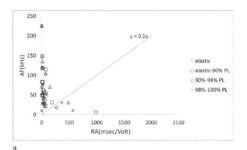

a

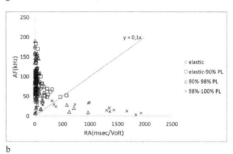

b

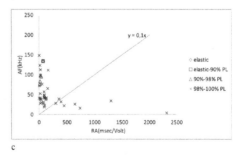

c

Figure 4. AF vs RA of first loading stage: a) plain concrete, b) fibers 0.5% by volume, c) fibers 1.0% by volume.

are calculated from the moving average of 50 hits (in this case it was chosen not to use the moving average, but the whole data set, due to lack of data before the main fracture). For plotting data, the

193

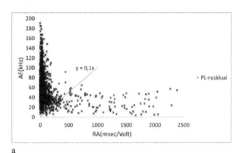

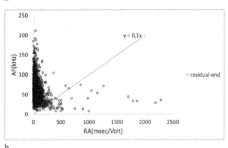

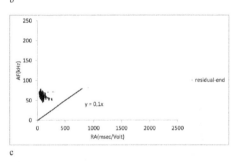

Figure 5. AF vs RA for PPF 0.5%: a) second loading stage, b) third loading stage, c) third loading stage— values calculated from the moving average of 50 hits

ordinate is shown in kHz and the abscissa scale is msec/volt. In Figures 4 and 5, the ratio of the abscissa scale to the ordinate scale is set to 10 as suggested by Ohtsu (2010). Considering the values of RA and AF shown in Figure 4, the initiation of the hits of all specimens has characteristics of tensile cracking (micro cracking loading stage) and as the main fracture approaches (macro cracking loading stage) (Aggelis et al. 2011), the mode of cracking becomes mixed and shear, due to the creation of the conical wedges that are formed during the test and due to the main tensile cracks that are formed along the specimens. After the main fracture, the mode of cracking turns again to mixed, with trend to tensile mode, till the end of testing (Fig. 5). The mixed mode behavior of cracking after the main failure can be explained as the conical wedges are inserting along the longitudinal axis of the specimen, forcing it to dilatation and fracturing the asperities that have been formed both on the surfaces of conical wedges and specimen. At the same time, among the main fracture planes, the tensile rupture of the fibers produce a great amount of hits, while the number of extrusions compared to the number of tensile ruptures is much less (Fig. 6). This phenomenon is clearer at the third loading stage when the main shear fractures have already occurred.

The modes of cracking are not possible to be determined by using the plotted values from the moving averages of 50 hits, because at the stage of the residual load, it is shown that hits have only tensile fracture characteristics (Fig. 5c).

In addition to the correlation AF vs. RA, the behavior of the ratio of cumulative energy over cumulative hits vs. load has been investigated, as an additional method to Ib-value and as a possible prediction method of an upcoming failure. In Figure 7, the energy over cumulative hits vs. load plot is demonstrated, as an additional variation of this ratio. Analysis result for the determination of Ib-value, is not demonstrated in this work, due to lack of adequate data population, that produce diagrams only to the limited loading stage of peak load. The ratio of cumulative energy over cumulative hits is a transient feature, updated with each new hit recorded during the fracture process. Figure 7 shows that at all specimens there is a characteristic point, after which there is a dominant augmentation of the ratio of cumulative energy over the number of hits. This change occurs when the load is between 52% and 89% of the peak load. Further, the effect of the type and percentage of the fibers on the amount of hits and the cumulative energy, which is released during the three loading stages, is demonstrated. It may be observed, that at the first loading stage (Figs. 8a, 9a), from start until the peak load, the amount of recorded hits and cumulative energy decreases as the percentage of fibers increases regardless of the type of

Figure 6. Fiber ruptures and extrusions.

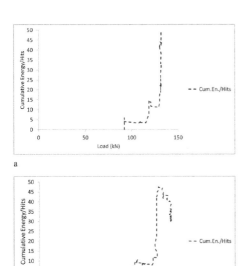

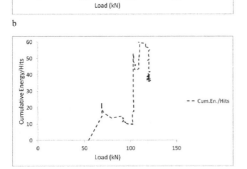

Figure 7. Cumulative energy/Hits vs load: a) Plain concrete, b) PPF 0.5%, c) PPFX 1%.

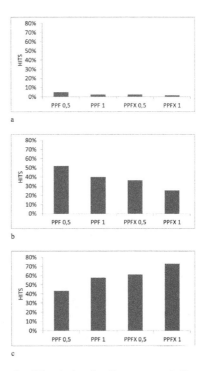

Figure 8. Hits during loading stages: a) Start-peak load, b) Peak load-residual, c) Residual-end.

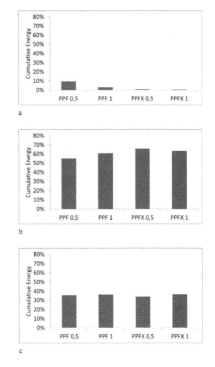

Figure 9. Cumulative energy during loading stages: a) Start-peak load, b) Peak load-residual, c) Residual-end.

fibers. During the second loading stage (Fig. 8b) the amount of hits decreases, as the percentage of fibers by volume increases, for both types of fibers. At the same loading stage, the cumulative energy that is released is independent of the percentage and the type of fibers (Fig. 9b). The third loading stage (Fig. 8c) shows the reversed trend of the behavior of the amount of hits, compared to the second loading stage (Fig. 8b) and the same behavior with respect to the cumulative energy. In summary, for all specimens, the first loading stage, with a range of the ratio of axial displacement over the whole cumulative axial displacement of the test 0.13 to 0.15 produce only a range of 0.1% to 10% of the whole cumulative energy of AE, while the second and the third loading stage, within ranges of same ratios of axial displacement from 0.08 to 0.21 and from 0.64 to 0.78 respectively, produce energy that ranges from 55% to 66% and from 33% to 37% respectively.

It was observed, that, at the first loading stage, as the amount of fibers increases, the integrity of the specimens is maintained more satisfactorily. Furthermore, immediately after peak load, specimens of high fiber content produce signals characterized by higher amplitudes and energies compared to specimens of low fiber content. On the other hand, the large number of signals in the residual loading stage gives a more even distribution of the emitted energy.

# 4 CONCLUSIONS

The AE technology can be applied to the double-punching test, in the same way it is applied to the bending test. With the aim of AE technology, the classification of the mode of fracture during testing is possible, while the definition of initiation time of fracturing of the specimen can be estimated. The number of AE hits and the rhythm of releasing energy at different loading stages could be a classification criterion for fiber reinforced concrete.

# ACKNOWLEDGEMENTS

This work was co-funded by IKY, through the Siemens Programme: "IKY Fellowships of Excellence for Postgraduate Studies in Greece—Siemens Program".

Dr. George Papantonopoulos of the Tunnelling Laboratory of School of Mining Engineering of the National Technical University of Athens is gratefully acknowledged for his help to the execution of the experiments.

Mr. Dimitrios Kotsanis of the Laboratory of Engineering Geology and Hydrogeology of School of Mining Engineering of the National Technical University of Athens is gratefully acknowledged for his help to the acoustic emission experimental setup.

# REFERENCES

AENOR. UNE 83515:2010. Hormigones con fibras. Determinación de la resistencia a fisuración, tenacidad y resistencia residual a tracción. Método Barcelona, Asociación Española de Normalización y Certificación Madrid.

Aggelis, D.G. Soulioti, D.V. Sapouridis, N. Barkoula, N.M. Paipetis, A.S. Matikas, T.E. 2011. Acoustic emission characterization of the fracture process in fibre reinforced concrete. *Construction and Building Materials* 25 (2011): 4126–4131.

Aggelis, Dimitrios G. 2011. Classification of cracking mode in concrete by acoustic emission parameters. *Mechanics Research Communications* 38 (2011): 153–157.

Aggelis, D.G. Mpalaskas, A.C. Ntalakas, D. Matikas, T.E. 2012. Effect of wave distortion on acoustic emission characterization of cementitious materials. *Construction and Building Materials* 35 (2012): 183–190.

Aggelis, D.G. Mpalaskas, A.C. Matikas, T.E. 2012. Acoustic signature of different fracture modes in marble and cementitious materials under flexural load. *Mechanics Research Communications* 47 (2013): 39–43.

Aggelis, D.G. Mpalaskas, A.C. Matikas, T.E. 2013. Investigation of different fracture modes in cement-based materials by acoustic emission. *Cement and Concrete Research* 48 (2013): 1–8.

Aggelis, D.G. Verbruggen, S. Tsangouri, E. Tysmans, T. Van Hemelrijck, D. 2013. Characterization of mechanical performance of concrete beams with external reinforcement by acoustic emission and digital image correlation. *Construction and Building Materials* 47 (2013): 1037–1045.

Carmona, Sergio Malatesta Aguado de Cea, Antonio Molins, Climent Borrell 2012. Generalization of the Barcelona test for the toughness control of FRC. *Materials and Structures* (2012) 45: 1053–1069.

Carpinteri, A. Lacidogna, G. Accornero, F. Mpalaskas, A.C. Matikas, T.E. Aggelis, D.G. 2013. Influence of damage in the acoustic emission parameters. *Cement & Concrete Composites* 44 (2013): 9–16.

Concrix, 2011. Datasheet Concrix ES.

Farhidzadeh, Alireza Mpalaskas, Anastasios C. Matikas, Theodore E. Farhidzadeh, Hamidreza Aggelis, Dimitrios G. 2014. Fracture mode identification in cementitious materials using supervised pattern recognition of acoustic emission features. *Construction and Building Materials* 67 (2014): 129–138.

Ohno, Kentaro & Ohtsu, Masayasu 2010. Crack classification in concrete based on acoustic emission. *Construction and Building Materials* 24 (2010): 2339–2346.

Ohtsu, Masayasu et al. 2010. Recommendation of RILEM TC 212-ACD: acoustic emission and related NDE techniques for crack detection and damage evaluation in concrete: 3. Test method for classification of active cracks in concrete structures by acoustic emission. *Materials and Structures* (2010) 43: 1187–1189.

Pujadas, P. Blanco, A. Cavalaro, S.H.P. de la Fuente, A. Aguado, A. 2014. Multidirectional double punch test to assess the post-cracking behaviour and fibre orientation of FRC. *Construction and Building Materials* 58 (2014): 214–224.

Sika, 2013. Datasheet Sika® Fiber PP 940-50: Construction.

Yuyama, Shigenori Li, Zheng-wang Ito, Yoshihiro Arazoe, Masaki 1999. Quantitative analysis of fracture process in RC column foundation by moment tensor analysis of acoustic emission. *Construction and Building Materials* 13 (1999): 87–97.

*Emerging Technologies in Non-Destructive Testing VI – Aggelis et al. (Eds)*
*© 2016 Taylor & Francis Group, London, ISBN 978-1-138-02884-5*

# Acoustic emission monitoring of high temperature process vessels & reactors during cool down

D. Papasalouros, K. Bollas, D. Kourousis & A. Anastasopoulos
*Mistras Group Hellas ABEE., Athens, Greece*

ABSTRACT: Structural integrity assessment of critical structures in petrochemical and oil industry is of great concern regarding safety, environmental and financial impacts. Especially large-volume industrial structures require a global scale NDT method in order to have a fast and efficient inspection.

Cool-down monitoring with Acoustic Emission is ideal for any high–temperature, thick-walled structure as it is taken out of service and cooled down for turn-around maintenance and inspection. This process always generates high thermal loading inside the structure. As the internal temperature drops, a thermal gradient is established through the vessel wall. The higher the cooling rate, the higher the gradient generated. The thermal gradient gives rise to thermal strains that add to the existing hoop and longitudinal strains. This presents an ideal stressing scenario for the Acoustic Emission Testing for detecting active flaws that may be present. High-frequency piezoelectric transducers, mounted on the wall, usually with the usage of waveguides to avoid direct contact with the high-temperature surface, detect the growth of active discontinuities during cool-down. AE data is processed and stored by advanced equipment. Furthermore, real-time location algorithms estimate the position and graphically depict any present active source on the structure. The intensity of the emissions received from each zone of the vessel is graded from "A" (insignificant) to "E" (most severe) and recommendations for the extent of any follow-up inspection are provided, ("E" sources requiring the most strict follow-up).

The major benefits of AET during cool down are the small interference with production or the turnaround, the 100% volumetric inspection over a short period of time and the fact that the turnaround time and cost are reduced by knowing in advance where to concentrate follow-up inspection and repair.

This paper presents results of AE Cool-Down tests performed by Mistras Group Hellas and comparisons with other NDT methods (e.g. UT Phased Array) within follow-up inspections. A specific monitoring case will be presented with respect to the testing time requirements, overall findings and global inspection capabilities.

## 1 INTRODUCTION

Because of the size and operating conditions of the high temperature vessels, Acoustic Emission Testing (AET) is most commonly used for defects detection, in conjunction with a cool down process. Tests by former Dunegan Testing (currently Physical Acoustics) in the early 1980's on power formers and afterwards on unicrackers and gas reformers, have demonstrated the success and benefits of cool down monitoring. AE monitoring has been applied ever since in a variety of thick-walled, high-temperature structures including reactors, towers, columns and associated hot-circuit piping systems.

## 2 APPLICATION OF THE METHOD

AE monitoring during cool down is a real time, global examination NDT technique, enabling 100% inspection of the vessel. Using temperature decrement as stimulus, AE monitoring detects signals resulting from sources inside the material.

Cool down monitoring is typically performed on a reactor as it is taken out of service and cooled down for turn-around maintenance and inspection (Anastasopoulos *et al.*, Carlos *et al.*, Ternowchek *et al.*, Cole *et al.*). As the internal temperature drops (the internal pressure is held at the operating level), a thermal gradient is established through the vessel wall. The higher the cooling rate, the higher the gradient. The thermal gradient gives rise to thermal strains that add to the existing hoop and longitudinal strains. Any discontinuities that grow as a result of such stressing give rise to AE and are detected by sensitive piezoelectric transducers.

Sufficient detectability (complete coverage of the vessel) and sensitivity are assured by using high-tech AE equipment and performing the test according to ASME-V, Article 12 and MONPAC testing procedures and ASTM, operated under license to Physical Acoustics Corp. The MONPAC procedure, applicable to both new and in-service vessels, provides a method of AE testing and data analysis based on laboratory results and field tests on thousands of metal tanks, pressure vessels and piping systems. The related specific software is used to grade the intensity of any source of acoustic emission recorded during the test. Each grade level is accompanied by a specific follow-up recommendation. Some phenomena that can generate AE during a cool down process are as follows:

- Flaws In both parent metal and weld associated regions.
- Effects of corrosion, cracking of corrosion products.
- Stress corrosion cracking.
- Embrittlement.
- Flaws In weld associated regions.
- Lack of penetration.
- Voids and porosity.
- Crack growth.
- Localized yielding of the material due to wall thickness loss.
- Certain metallurgical changes.
- Pits and gouges.
- Incomplete fusion.
- Undercuts.
- Inclusions

Note that flaws in unstressed areas and passive flaws (structurally insignificant under the applied load) will not generate acoustic emission. The ideal cool down test is one where, for a given internal pressure, the thermal gradient stays within limits that correspond to 110% and 150% of normal operating load.

## 3 TEST CASE

The case study of a reactor vessel of 2.2 m diameter and 9.264 m height, cooled down from 505°C to 80°C is presented in the following pages.

### 3.1 Background & setup

The vessel was normally operating at 520°C. Based on attenuation measurements, a total of 15 AE sensors were mounted on the shell, using waveguides, in order to protect the sensors from high temperatures (Figure 1).

### 3.2 Test procedure

Prior to the test, the background noise level was monitored. The actual Acoustic Emission monitoring was performed from an average temperature of 505 °C to about 80 °C. In general, pressure was held constant during the test to approximately

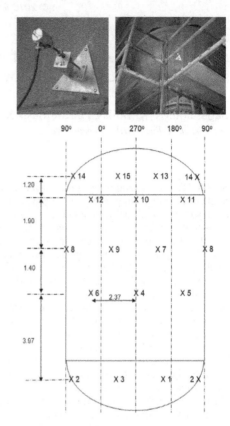

Figure 1. Mounted Sensors with with waveguide on the insulated shell.

7 Kg/cm$^2$. The cool-down test lasted 20 hours and 4 minutes until the temperature of about 80°C. Cool down rate was not constant (Figure 2). Various external noise sources were identified during the test, in the general area of the reactor due to nearby working crews. In addition to that, due to different thermal expansion between the vessel and the insulation, sliding of waveguides relatively to the vessels was detected, by having very distinct signal features, which differentiated from the "real" AE sources due to thermal stressing.

### 3.3 Analysis & results

The above mentioned test sequence (Temperature vs. Time) and the corresponding filtered global AE activity during acquisition is presented in Figure 3. Identification and removal of surrounding noise (nearby working crews, waveguide sliding relatively to the vessel due to contraction etc.) was removed prior to the evaluation. The results from standard analysis with zonal grading typically done in this type of testing, were further investigated using

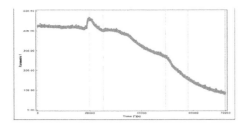

Figure 2. Reactor Temperature Vs. Time.

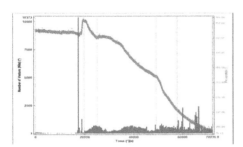

Figure 3. Reactor AE Activity & Temperature Vs. Time (NOESIS software).

| Cluster | ID | Points | Historic | Severity | Grade |
|---|---|---|---|---|---|
| z | 4 | 96 | 1.92 | 10.90 | B |
| y | 11 | 84 | 1.75 | 11.10 | A |
| x | 1 | 71 | 2.73 | 10.50 | B |
| w | 16 | 50 | 1.23 | 17.00 | A |
| v | 13 | 49 | 1.75 | 12.70 | A |
| u | 6 | 36 | 2.50 | 9.30 | B |
| t | 5 | 29 | 3.27 | 4.30 | C |
| s | 27 | 21 | 1.17 | 7.40 | N |
| r | 2 | 20 | 2.58 | 13.50 | B |
| q | 19 | 18 | 1.50 | 13.30 | A |
| p | 32 | 17 | 2.37 | 11.40 | B |
| o | 3 | 14 | 1.88 | 8.20 | B |
| n | 20 | 12 | 2.11 | 5.50 | B |
| m | 23 | 11 | 1.06 | 15.10 | A |
| l | 31 | 10 | 1.47 | 3.00 | A |
| k | 43 | 10 | 2.91 | 9.10 | B |

Figure 4. List of Located/Graded Sources. **N Grade** Source is a Non-Significant Source based on selected criteria.

clustering algorithms that involve the location and the energy grading of signals that were picked up by no less than 3 sensors (or more) in order to have an accurate estimate of the areas that may have damage accumulation (Figure). In addition, a more quantitative view of the activity in these areas is revealed and the cluster criticality can be assessed. Source analysis is performed as a further AE Data Analysis (following MONPAC standard recommendations) as mentioned by Fowler et al. by isolating the activity that was produced by each source, and investigating it.

Source grading is listed from the software (Figure 4) while it focuses on the activity of the source only and not of the general zone (Figure 7). Finally clusters are overlapped on the vessel in order to visually assist the additional NDT that may be performed with more localized methods in the specific areas based on their criticality (Figure 5a).

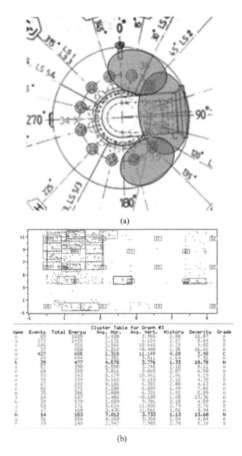

Figure 5. (a) Vessel with source areas of the first testing, (b) Clustering and grading using unfolded sensor positions.

| GRADE | COLOUR | INTERPRETATION |
|---|---|---|
| A | Green | Very minor zone |
| B | Cyan | Minor zone |
| C | Magenta | Active zone |
| D | Yellow | Intensive zone |
| E | Red | Critical zone |

Figure 6. Color description of sources based on selected criteria

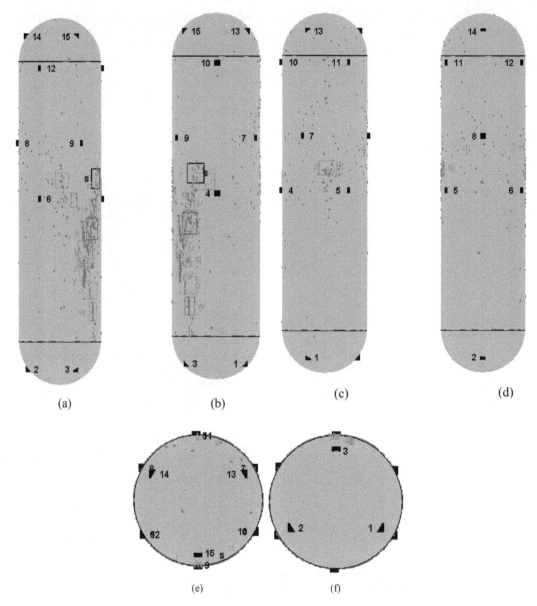

Figure 7. (a to d) West, North, East, South views, (e) Top view, (f) Bottom view of the vessel with overlapped graded sources.

### 3.4 Follow-up

The UT follow-up of the area indicated by the results of the first test of this vessel (Figure 5a & 5b) shown an under surface active crack of 150 mm length 2 mm depth (Top Head). The specific area was closely monitored as the main part of the presented test case in order to verify that the specific source found from the previous test (and repaired) was not active. From the results and located events of the filtered data we indeed conclude that the specific area had no AE activity, therefore no active sources (older or newer one) are present.

For the rest of the vessel, especially for the shell area there are lot of cluster located. More specifically the area on the North side of the vessel (Figure 7b) that appears to have a multiple clusters of located AE events was given to the vessel owner and is pending follow-up inspection, with insulation removal and internal inspection.

In typical cases a follow-up is difficult to be performed would, AE monitoring during start-up combined with continuous and/or permanent on-line monitoring during normal operation, would further verify or reveal the existence of previously identified critical areas. Again by reassessing their criticality with advance clustering algorithms (using a grading scheme for each source), any critical sources (growing indications—accumulated damage) are quickly identified.

## 4 LIMITATIONS

Cool Down method is experience-based and is able to identify and grade AE activity, resulting, for example, from cracking, corrosion damage, overstress, in order to determine if there is a need for detailed local inspection. It does not give information on defect size or orientation but grades the AE activity within the pre-specified zones on the vessel (from "insignificant" to "intense") and gives recommendations for the extent of any follow-up inspection. Location is primarily "zonal" and indicates the "active" zone. Planar location is, sometimes, possible when a source is detected by at least 3 sensors. This defines a smaller and refined search area. The method does not identify manufacturing defects if they have not propagated during subsequent service. However, significant manufacturing defects, including embrittled welds, may be identified during the monitoring of tests to high pressures, for example during a hydro-test to the original proof pressure.

## 5 CONCLUSIONS

In general, the cool down monitoring utilizing AE method offers a unique opportunity for asset owners to inspect globally and with minimum interference the high temperature thick-walled vessels with guaranteed minimum or even no downtime at all. Results of the test can be given almost immediately after test, thus providing a great assistance to the general operations regarding the vessel. Follow-up local inspection in the areas indicated by AE may be, immediately performed after the test, when the vessel will be cool anyway and any repairs, if so required, may be scheduled within the shut-down period.

It is apparent that the swift and accurate identification of the areas that are assessed as critical, result in major monetary savings with respect to maintenance/repairs that might be needed and, perhaps the most important, minimizes the risk of not identifying areas that can result in failure in the long run.

In addition, all the above advantages can be further expanded and enhanced with the option of installing a permanent AE monitoring system with remote access functionality, in order to monitor the vessel during in-service operation and/or prolonged time periods. Typical systems of these capabilities and magnitude are already successfully installed around the globe, adding to a better operational condition of the vessel.

## REFERENCES

ASME Code, Section V, Article 12.

ASTM E 1139–92 *"Standard Practice for Continuous Monitoring of Acoustic Emission from Metal Pressure Boundaries"*.

Anastasopoulos, A. & Tsimogiannis, A. "Evaluation of Acoustic Emission Signals during Monitoring of Thick-Wall Vessels Operating at Elevated Temperatures", JAE, Vol.22, 2004, p.59–70

Anastasopoulos, A. "Pattern Recognition Techniques for Acoustic Emission Based Condition Assessment of Unfired Pressure Vessels", J. of Acoustic Emission, Vol 23, 2005, pp 318–330.

Carlos, M.F. & Wang, D. &Vahaviolos, S. & Anastasopoulos A., *"Advanced acoustic emission for on-stream inspection of petrochemical vessels"*, "Emerging Technologies in NDT", Van Hemelrijck, D. (*ed*) & Anastasopoulos, A. (*ed*) & Melanitis N.E (*ed*), Proceedings Of The 3rd International Conference On Emerging Technologies In Non-Destructive Testing, 26–28 May 2003, Thessaloniki, A.A. Balkema, Netherlands 2004, ISBN 90 5809 645 9 (Volume)-. 90 5809 645 7(CD), pp. 167–172.

Carlos, M.F. *et al.*, "Advanced Acoustic Emission for on-stream inspection of petrochemical vessels", in Emerging Technologies in NDT, A.A. Balkema – ISBN 90 5809 645 9, 2004, pp. 167–172.

Cole, P.T. & Gautrey, S.N. "On Stream monitoring of process plant", in Emerging Technologies in NDT, A.A. Balkema – ISBN 90 5809 645 9, 2004, pp. 163–166.

Fowler, T.J. & Blessing, J.A. & Conlisk, P.J. & Swanson, T.L., "The MONPAC System", Journal of Acoustic Emission, **8** (3), 1989, 1–8.

MONPAC Testing Procedure (of Metal Vessels), Issue 6.

MONPAC-PLUS *Procedure for Acoustic Emission Testing of Metal Tanks/Vessels*, Draft D, August 1992.

NOESIS, "User's Manual", Mistras Group Hellas ABEE. 2015.

Ternowchek, S. & Miller, R. & Cole, P.T. "Web Based Nondestructive Evaluation Using Acoustic Emission", PVP2009–77259 Proc., July 26–30, 2009, Prague, Czech Republic

Yuen, S. & Wang, D. & Bedictus, P. "Acoustic Emission Testing of a Process Reactor", PVP2005–71764, July 17–21, 2005, Denver, Colorado, USA.

*NDT in aerospace*

*Emerging Technologies in Non-Destructive Testing VI – Aggelis et al. (Eds)*
*© 2016 Taylor & Francis Group, London, ISBN 978-1-138-02884-5*

# Thermal strains in heated Fiber Metal Laminates

B. Müller & J. Sinke
*Structural Integrity and Composites, Delft University of Technology, Delft, The Netherlands*

A.G. Anisimov & R.M. Groves
*Aerospace Non-Destructive Testing Laboratory, Delft University of Technology, Delft, The Netherlands*

ABSTRACT: Current trends in aircraft design go towards smart materials and structures including the use of multi-purpose materials. Fiber Metal Laminates (FML) with embedded electrical heater elements in leading edges of aircraft used for anti- or de-icing follow those trends. The laminated structure of FMLs with layers of different materials leads to anisotropic material characteristics. The FML used in this study is GLARE (Glass Laminate Aluminum Reinforced Epoxy). The anisotropic structure raises questions concerning possible effects on the material characteristics when frequently heated by embedded heater elements and cooled by flight conditions. In order to investigate those possible effects on FMLs, knowledge about the thermal strains and stresses is important. Furthermore, non-destructive techniques are likely to be a future requirement to detect defective heater elements and delaminations at heated leading edges. Thus, this research uses a shearography (speckle pattern shearing interferometry) instrument in order to investigate the surface strain components of FMLs during thermal loading with the embedded heater elements. Parallel to the experiments, numerical analyses were conducted in order to investigate the strain-stress state due to thermal loading with embedded heater elements. The results of both, the strain measurement with the shearography instrument and the numerical analyses were analyzed and compared. The numerical results show how the embedded heater element affects the residual stress-strain state and the stresses due to thermal loading.

## 1 INTRODUCTION

The development of Fiber Metal Laminates (FMLs) such as GLARE has been a field of research since decades (Vlot & Gunnink 2001, Sinmazcelik et al. 2011). GLARE is the acronym of glass fiber rein-forced aluminum laminate and is a certified laminate material composed of alternating layers of alumi-num and S-2 fiber glass prepregs (Vlot 2001). The S-2 fiber glass prepregs are unidirectional (UD) layers. GLARE is currently used in the fuselage of the A380 (Hagenbeek 2005). The benefits of using GLARE are a lower fatigue crack growth rate com-pared to monolithic aluminum (Alderliesten & Homan 2006), a better strength bearing capability in combination with favorable impact and lightening resistance (Vlot & Gunnink 2001, Vermeeren 2002) and better UV- and moisture-resistance compared to pure glass fiber laminates (Park et al. 2010). Those features make GLARE to a favourable material to design damage tolerant parts (Alderliesten & Homan 2006).

On the contrary, the laminated structure of FMLs with layers of different materials leads to anisotropic material characteristics. The layered buildup of FMLs in the thickness direction raises questions concerning effects on crack and delamination initiation and growth due to different thermal expansion coefficients. Hence, quantification of the thermal strains and stresses is important in order to investigate the possible effects of thermal cycling, i.e. when frequently cooled and heated by flight conditions and embedded heater elements.

Embedding heater elements in GLARE laminates makes it to a multipurpose material and enables the design of heated leading edges without additional substructures in the wings (FMLC 2015). Thus, heated GLARE shows weight saving potential as no bleed air systems for circulating hot air in the leading edges or structures for pneumatic de-icing have to be added in the wing structure. Electrically powered embedded heater elements used for anti- or de-icing follow current trends towards electric aircraft and multipurpose materials (Mohseni & Amirfazli 2013). Figure 1 shows the cross-section of the GLARE 5-4/3-0.3 layup with an embedded heater element between the UD layers, as used in this study. In addition to residual stresses formed in the curing process (Abou-hamzeh et al. 2015), a (local) temperature increase causes strains and stresses in the laminate material due to different thermal expansion coefficients

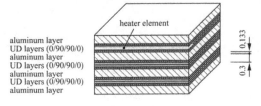

Figure 1. GLARE 5-4/3-0.3 layup with a heater element.

(Hagenbeek 2005). Although, only a few studies has been conducted (Costa et al. 2012, Müller et al. 2014), the temperature related stresses and strains are expected to affect the material characteristics of FMLs.

This research uses a 3D shearography (speckle pattern shearing interferometry) instrument (Hung 1982) to investigate the surface strains due to thermal loading (Corso Krutul & Groves 2011) by embedded heater elements. The surface strains of different shapes and connecting techniques of heater elements are investigated. Furthermore, artificial delaminations (two PTFE foil pieces of size 55 mm) were embedded at both sides of one heater element to be studied with shearography. In addition to the experiments, numerical analyses were conducted in order to investigate in more detail the strain-stress state due to thermal loading with these embedded heater elements. The results of both, the strain measurement with the shearography instrument and the numerical analyses were analyzed and compared.

## 2 EXPERIMENTAL SETUP

### 2.1 Specimens

Two specimens of GLARE 5-4/3-0.3 with embedded heater elements were manufactured (cf. Figure 1) and cured by means of the standard autoclave procedure at 120°C and 6 bar (Hagenbeek 2005). Figure 2 depicts the ultrasonic C-scan results of one specimen including its dimensions. The C-scan results show, apart from the artificial delaminations, no delaminations in the vicinity of the heater elements.

Three heater elements were embedded in each specimen (cf. Figure 2). These are the reference heater element without artificial delaminations or imperfections due to connections techniques, heater elements with different connecting techniques (overlap, soldered, spot welded), one heater element with artificial delaminations and one s-shaped heater element. Further details about the heater elements and the experimental setup are provided in Anisimov et al. (2015).

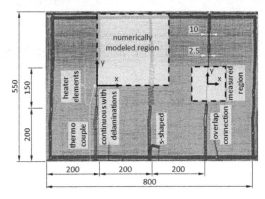

Figure 2. C-scan of specimen with three heater element types: continuous, s-shaped and overlapped connection (units in mm).

### 2.2 Measurement of displacement gradients

3D Shearography is a full-field technique used to measure surface displacement gradients, from which surface strain components are processed (Hung 1982, Steinchen & Yang 2003). The isolation of the surface strain components can be done by using a multicomponent shearography instrument with a minimum of three viewing directions, i.e. three shearing cameras (James & Tatam 1999). Combined processing of phase maps from each camera together with system geometry parameters give an estimation of the surface strain components in the x- and y-derivative directions (Goto & Groves 2010). Six different surface displacement gradients $\partial u/\partial x$, $\partial u/\partial y$, $\partial v/\partial x$, $\partial v/\partial y$, $\partial w/\partial x$ and $\partial w/\partial y$ are obtained. Partially the displacement gradients can be interpreted as strain components frequently used for describing the stress-strain state (Francis et al. 2010).

$$\varepsilon_x = \frac{\partial u}{\partial x}, \varepsilon_y = \frac{\partial v}{\partial y}, \varepsilon_{xy} = \varepsilon_{yx} = \frac{1}{2}\left(\frac{\partial u}{\partial y} + \frac{\partial v}{\partial x}\right) \quad (1)$$

During measurements the speckle pattern is generated by illuminating the object with an expanded laser beam. The object is imaged by multiple cameras each with shearing devices to obtain interferograms at the initial state and after each thermal load step. Each interferogram is a superposition of two divided fields of view which are separated by the shear distance.

Surface strains affect the distribution of the phase and are reconstructed after each load step by capturing three phase-shifted interferograms (Dorrio & Fernandez 1999). The absolute change of phase differences calculated for each load step is an unwrapped phase map which gives the surface deformation.

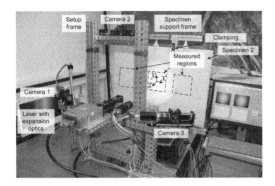

Figure 3. Experimental setup.

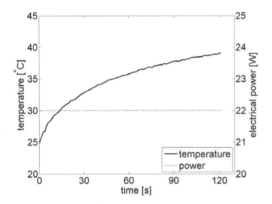

Figure 4. Thermal loading conditions.

## 2.3 Experimental procedure and test conditions

Figure 3 shows the multicomponent shearography experimental setup, i.e. the position of the three cameras and a laser in respect to the vertically positioned specimen which was clamped at the top and bottom. The three cameras were calibrated in pairs as stereovision systems for accurate estimation of setup geometry (Heikkila and Silven 1997).

The thermal loading of the specimens was done using electrical power. Figure 4 shows the electrical power and the temperature increase measured using an embedded thermo couple at a distance of 5 mm left of each heater element. The electrical resistances of all heater elements are similar. The heater elements were heated one by one. The ambient and the specimen temperature were always 24.8 ± 0.5°C.

## 3 EXPERIMENTAL RESULTS

### 3.1 Surface strain distributions

Figure 5 shows the surface strain components $\varepsilon_x$ and $\varepsilon_y$ of the reference, soldered and spot welded heater elements obtained by the shearography measurements. Figure 2 shows exemplarily the measured region of the overlapped heater element. All six subplots show strain concentrations in the vicinity of the heater elements. However, the strain magnitudes differ for the different specimens, although, all specimens were exposed to the same thermal loading conditions (see Figure 4).

Figure 6 compares the measured strains along the x-axis, of all six investigated heater elements due to the local thermal loading of the specimens. Similar to Figure 5, the results show a variation of the strains in the vicinity of the different heater elements due to the thermal loading. Furthermore, the strain peaks between $-0.005 < x < 0.005$ m in Figure 6 (b) show the effect of the heater element connections on $\varepsilon_x$.

The strain components $\varepsilon_x$ along the x-axis of the reference heater element and the s-shaped heater element have similar shapes but different magnitudes (cf. Figure 6 (a)). On the contrary, the strain components $\varepsilon_y$ differ for the reference and the s-shaped specimen, but are similar for the s-shaped heater element and the heater element with artificial delaminations (cf. Figure 6 (c)). For the overlapped, soldered and spot welded heater elements the strain components $\varepsilon_x$ have similar shapes and slightly different magnitudes (cf. Figure 6 (b)). The strain component $\varepsilon_y$ varies slightly more (cf. Figure 6 (d)).

### 3.2 Interpretation of the experimental results

The thermal strain components (cf. Figures 5 and 6) imply that the heated areas of the specimens bend in different directions, i.e. in the negative and positive z-direction. Hence, the measurements show different strain magnitudes during heating for different heater elements. In order to clarify these results, the initial specimen geometry (before thermal loading) was measured using the Digital Image Correlation (DIC) technique (Sutton et al. 2009).

Figure 7 shows the DIC results of the specimen shown in Figure 2. Pre-deformations across the whole specimen can be identified. These pre-deformations result from the manufacturing process due the heater elements which locally introduce a non-symmetric layup (cf. Figure 1). The varying pre-deformations across the specimen, i.e. in the vicinity of each heater element, are likely to cause the different strain distributions when the heater elements are subjected to thermal loading (cf. Figures 5 and 6).

Hence, the predeformations superimpose on the (local) deformations due to thermal loading, affect the bending direction and lead to different overall strain magnitudes in the vicinity of the heater elements.

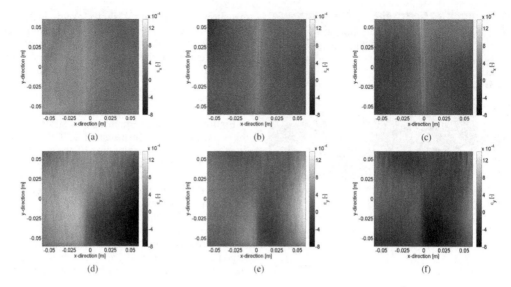

Figure 5. Shearography results: Strain distributions $\varepsilon_x$, $\varepsilon_y$ of the (a,b) reference, (d,e) soldered and (c,f) spot welded heater element.

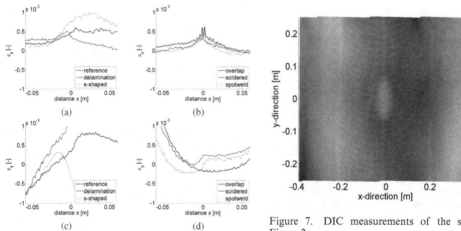

Figure 6. Shearography results: Strain components (a,b) $\varepsilon_x$ and (c,d) $\varepsilon_y$ along the x-axis (perpendicular to the heater element).

Figure 7. DIC measurements of the specimen, cf. Figure 2.

## 4 NUMERICAL ANALYSES

### 4.1 Numerical model of the reference configuration

Transient coupled thermal analyses were conducted in order to provide for additional insights in the structural behavior on top of the experimental measurements. The FE-code Abaqus (2015) was used.

The dimensions, material parameters, thermal and mechanical loading conditions were based on the test setup described in section 2. One quarter of the vicinity of the (reference) heater element was modeled due to the expected symmetry of the (thermal) loading conditions and geometry along and perpendicular to the heater element(s). Hence, the numerical model does not account for possible pre-deformations of the specimen (cf. Figure 7). A reduction of the modeled area was executed in order to reduce the number of elements and thus to reduce the calculation time. The area of 300 × 300 mm with symmetric boundary conditions along the x- and y-axis was modeled. This area is indicated by dashed lines in Figure 2.

The layered material composition of GLARE 5-4/3-0.3 and the heater element depicted in

Figure 1 were accounted for by partitioning of the structure into layers which equal the thicknesses and width of the different materials. Perfect heat conduction was assumed between the layers due to the partitioning. The temperature dependent material data was taken from Kundig and Cowie (2006) and Hagenbeek (2005). The anisotropic material behavior of the UD layers was accounted for by averaging the material parameters of the glass fibers and the epoxy in each direction. Each layer is modeled by means of three C3D8T-elements through the thickness (z-direction). The element size in the x-y-plane was $x_{el.h} = y_{el.h} = 1.25$ mm in the vicinity of the heater element and increased to $x_{el} = y_{el} = 5$ mm at distances of 70 mm.

The initial conditions without taking residual stresses (FEM) into account were applied by introducing a predefined (temperature) field across the numerical model of 24.8°C. The temperature of the predefined field equaled the specimen and room temperature at the start of the experiment. The heating of the heater element was modeled by means of the temperature increase which is shown in Figure 4.

To take the residual stresses into account (FEM RS) the initial temperature was made equal to the curing temperature of 120°C. In the first load step the modeled GLARE laminate was cooled down from 120°C to 24.8°C at a constant pressure of 6 bar (Poon et al. 2008). This linear elastic approach gives a reasonable approximation of the residual stress-strain state (Abouhamzeh et al. 2015). In a second load step the temperature increase of the heating element is modeled as described previously.

The mechanical boundary conditions of the modeled GLARE laminate result from the vertical position and the clamping of the specimen at four positions (cf. Figure 3). Along the x-edges displacements in the y-direction and rotational degrees in the x- and z-directions are locked. Along the y-edges, displacements in the x-directions and rotational degrees in the y- and z-direction are locked.

The thermal boundary conditions, i.e. convection and radiation, are considered by means of interactions. Convection is modelled by employing a surface film condition (sink temperature = 24.8°C) and the radiation by employing surface radiation (emissivity = 0.5). Perfect thermal conductivity between the aluminum and UD layers is assumed.

4.2 *Numerical results*

Figure 8 shows the temperature distribution in the field of $0 \le x \le 0.1$ m and $0 \le y \le 0.1$ m. The surface temperature varies between the ambient temperature and temperatures similar to the maximum heater element temperature (cf. Figures 2 and 4). The temperature gradients above the heater element are largest and constantly decrease with increasing distance from the heater element.

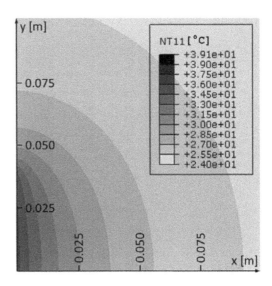

Figure 8. Temperature distribution predicted by the numerical analyses for $0 \le x \le 0.1$ m and $0 \le y \le 0.1$ m.

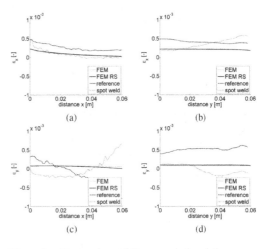

Figure 9. Comparison of the numerical and shearography results. Strain components (a,c) $\varepsilon_x$ and (b,d) $\varepsilon_y$ along the x-axis.

The comparison of the measured and the numerically predicted strain components shows that both, the measured results of the spot welded heater and the numerical results give similar distributions and magnitudes of the strain component $\varepsilon_x$ (see Figure 9). The magnitudes of the numerically predicted strain component $\varepsilon_y$ along the x- and y-axes are smaller than the measured ones.

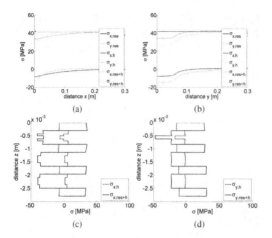

Figure 10. Stress components along (a) x-, (b) y-, (c, d) z-axis.

Furthermore, Figure 9 shows that the numerically predicted strains due to thermal loading when taking the residual stress-strain state into account (FEM RS) are up to $\Delta\varepsilon = 1.4 \cdot 0^{-5}$ larger than in case of neglecting it (FEM). The effect of the residual stresses shows more clearly in Figure 10.

Figure 10 shows the residual stress components $\sigma_{x.res}$ and $\sigma_{y.res}$, caused by the temperature increase of the heater element $\sigma_{x.h}$ and $\sigma_{y.h}$ and their superposition $\sigma_{x.res+h}$ and $\sigma_{y.res+h}$ along the x-, y- and z-axes. The stress distributions along the z-axis emphasis the tensile stresses in the aluminum layers, the compression stresses in the UD layers and the effect of the heater element on the stress distribution. In Figure 10 (c,d) the position of the heater element lies between $-0.506 \geq z \geq -0.626$ mm.

## 5 DISCUSSION

The numerical predictions of the strain components $\varepsilon_x$ and $\varepsilon_y$ lie between the experimental results of the reference and spot welded heater elements (cf. Figure 9). Pre-deformations of the specimens are expected to be the main reasons for the different experimental results and for the larger measured strain component $\varepsilon_y$. Thus, the numerical analyses give reasonable results.

The numerically predicted residual stresses of $\sigma_{x.res} \cong 42$ MPa in Figure 10 are slightly higher than the predictions (30 MPa) of Hofslagare (2003). These slightly higher values result from using GLARE 5, i.e. four UD layers between the aluminum layers in this study. Hofslagare used less UD layers between the aluminum layers. The slight differences of the residual stresses in the x- and y-directions are a consequence of using UD layers and not woven fibers.

The embedded heater elements cause (residual) stress peaks at the x-y-surface and in thickness (z-) direction (cf. Figure 10 (a,c,d)).

The stresses caused by the temperature increase of the heater cause a stress relief (cf. Figure 10). As the temperature increases towards the curing temperature, the (thermal) residual stresses reduce to zero. However, the residual stress magnitudes are more than twice as big as the stress magnitudes caused by a temperature difference of $\approx 15°C$. Hence, accounting for the residual stress-strain state when simulating thermal loading of GLARE is recommended.

## 6 CONCLUSIONS

Using shearography to measure displacement gradients and interpret them as strain components during the thermal loading of GLARE using embedded heater elements leads to reasonable results. The heater element positions during the temperature increase can be detected and the strain components quantified.

All three investigated connection techniques (overlap, soldered, spot welded) cause similar strain peaks. All of them locally double the heater element thickness. Thus, connection techniques which cause minimum thickness variations should be preferred.

The measured strain magnitudes for different embedded heater elements differ. Local predeformations in the vicinity of the heater elements are most likely to be the main reason for the different strain magnitudes during the thermal loading. These pre-deformations cause bending of the specimens in the positive and negative z-directions, respectively. The numerically predicted strains agree to order of magnitude with the measured strains. Numerically predicted magnitudes in the direction of the heater elements $\varepsilon_y$ are smaller than the measured ones.

Numerical analyses which estimate the residual stress-strain state with a linear elastic approach when accounting for temperature dependent material behavior predict similar stresses as provided in the literature. In order to draw conclusions on the stress-strain state or to detect possible stress or strain concentrations it is important to consider the residual stress-strain state in the numerical analyses.

## ACKNOWLEDGEMENTS

This study is partially funded by the Dutch Technology Foundation STW, Fokker Aerostructures and Skoltech.

# REFERENCES

Abaqus (2015). www.simulia.com, accessed 9.2.2015.

Abouhamzeh, M., J. Sinke, & R. Benedictus (2015). Investigation of curing effects on distortion of fibre metal laminates. *Composite Structures* 122, 546–552.

Alderliesten, R.C. & J.J. Homan (2006). Fatigue and damage tolerance issues of Glare in aircraft structures. *International Journal of Fatigue* 28 (10), 1116–1123.

Anisimov, A.G., B. Müller, J. Sinke, & R.M. Groves (2015). Strain characterization of embedded aerospace smart materials using shearography. Proc. SPIE 9435, 943524.

Corso Krutul, E. & R.M. Groves (2011). Thermal loading for shearography: multi-parameter model of the measurement setup. Proc. SPIE 8083, 80831C.

Costa, A.A., D.F.N.R. da Silva, D.N. Travessa, & E.C. Botelho (2012). The effect of thermal cycles on the mechanical properties of fiber metal laminates. Materials and Design 42 (1), 434–440.

Dorrio, B.V. & J.L. Fernandez (1999). Phase-evaluation methods in whole-field optical measurement techniques. Meas. Sci. Technol. 10, R33–R35.

FMLC (2015). Fibre Metal Laminates Centre of Competence, http://www.fmlc.nl, accessed 9.2.2015.

Francis, D., R.P. Tatam, & R.M. Groves (2010). Shearography technology and applications: a review. Meas. Sci. Technol. 21, 102–129.

Goto, D.T. & R.M. Groves (2010). A combined experimental with simulation approach to calibrated 3D strain measurement using shearography. Proc. SPIE 7387, 77871J.

Hagenbeek, M. (2005). Characterisation of Fibre Metal Laminates under Thermo-mechanical Loadings. PhD thesis, TU Delft.

Heikkila, J. & O. Silven (1997). A four-step camera calibration procedure with implicit image correction. Proceedings of IEEE Computer Society Conference on Computer Vision and Pattern Recognition, 1106–1112.

Hofslagare, P. (2003). Residual Stress Measurement on Fibre metal-laminates. J. of Neutron Research 11 (4), 215–220.

Hung, Y.Y. (1982). Shearography: a new optical method for strain measurement and non-destructive testing. Opt. Eng. 21, 213–391.

James, S.W. & R.P. Tatam (1999). Time-Division-Multiplexed 3D Shearography. Proc. SPIE 3744, 394–403.

Kundig, K.J.A. & J.G. Cowie (2006). Copper and copper alloys. In M. Kutz (Ed.), Mechanical Engineers Handbook-Materials and Mechanical Design. Hoboken, New Jersey, USA: John Wiley & Sons, Inc.

Mohseni, M. & A. Amirfazli (2013). A novel electrothermal anti-icing system for fiber-reinforced polymer composite airfoils. Cold Regions Science and Technology 87, 47–58.

Müller, B., S. Teixeira De Freitas, & J. Sinke (2014). Measuring thermal diffusion in fiber metal laminatess. Proc. of the ICAST2014: 25th Int. Conf. on Adaptive Structures and Technologies 76, 1–11.

Park, S.Y., W.J. Choi, & H.S. Choi (2010). The effects of void contents on the long-term hygrothermal behaviors of glass/epoxy and GLARE laminates. Composite Structures 92 (1), 18–24.

Poon, C., T. Morii, J. Lo, A. Nakai, S.V. Hoa, & H. Hamada (2008). Design, Manufacturing and Applications of Composites. Proceedings of the seventh joint Canadian-Japan workshop on composites 7, 124–132.

Sinmazcelik, T., E. Avcu, M.O. Bora, & O. Coban (2011). A review: Fibre metal laminates, background, bonding types and applied test methods. Mat. & Design 32, 3671–3685.

Steinchen, W. & L. Yang (2003). Digital Shearography: Theory and application of digital Speckle pattern shearing interferometry. Bellingham, Wash: SPIE Optical Eng. Press.

Sutton, M.A., J.-J. Orteu, & H. Schreier (2009). Image Correlation for Shape, Motion and Deformation Measurements—Basic Concepts, Theory and Applications. Springer Science+Business Media.

Vermeeren, C. (2002). Around Glare—a new aircraft material in context. Dordrecht, The Netherlands: Kluwer Academic Publishers.

Vlot, A. (2001). Glare—history of the development of a new aircraft material. Dordrecht, The Netherlands: Kluwer Academic Publishers.

Vlot, A. & J.W. Gunnink (2001). In A. Vlot and J.W. Gunnink (Eds.), Fibre metal laminates—an introduction. Dordrecht, The Netherlands: Kluwer Academic Publishers.

*Emerging Technologies in Non-Destructive Testing VI – Aggelis et al. (Eds)*
*© 2016 Taylor & Francis Group, London, ISBN 978-1-138-02884-5*

# Lamb wave dispersion time-domain study using a combined signal processing approach

## P. Ochôa & R.M. Groves
*Aerospace NDT Laboratory, Faculty of Aerospace Engineering, TU Delft, Delft, The Netherlands*

## R. Benedictus
*Faculty of Aerospace Engineering, Structural Integrity and Composites, TU Delft, Delft, The Netherlands*

ABSTRACT: Ultrasonic Lamb wave techniques are described as one of the most encouraging developments for structural health monitoring of aerospace composite structures. The reliability of those techniques is highly dependent on the quality of signal processing algorithms capable of extracting useful information out of complex responses. When damage localization is involved, it is crucial to rigorously determine Time-Of-Flight (TOF) of wave groups. Among the available methods for automated TOF extraction the Akaike Information Criterion (AIC) and the Hilbert Transform (HT) have become very popular. The first one detects the onset-time of a signal based on the minimization of the AIC function. The second one relies on the HT to define the response envelope, allowing maximum amplitude points to be used for time interval measurement. This paper focuses primarily on comparing the aforementioned methods in order to assess their reliability for TOF determination. Additionally, a combined AIC-HT approach is used to further quantify Lamb wave dispersion phenomena.

## 1 INTRODUCTION

The long-term success of composite materials as first choices for aircraft primary structures depends on how strongly they adhere to the airworthiness regulations. A major step towards this is the installation of reliable, automated Structural Health Monitoring (SHM) systems (Wenk & Bockenheimer 2014) in order to cope with the brittle-type behavior which, in the presence of barely-visible impact damage, may lead to unexpected failure under fatigue loading (Schijve 2004).

Some of the solutions already in operation rely on the long-range capabilities of active interrogation of ultrasonic guided waves, (Wenk & Bockenheimer 2014) as they are *"one of the most encouraging tools for quantitative identification of damage in composite structures"* (Su *et al.* 2006). One of the key missions of SHM systems is the localization of damage or detected events (SAE ARP6461 2013), making Time-Of-Flight (TOF) a crucial parameter to be determined for acousto-ultrasonic applications. TOF can be defined as *"the time consumed for a specific wave mode to travel a certain distance"* (Su & Ye 2009). Some authors have explored this definition by computing the difference between onset times of signals. In these cases the focus was placed on identifying the first instant when a specific criterion for the separation of noise and desired signal was met. The simplest

approach is to assign the onset to the point where a pre-defined fixed amplitude threshold is crossed (Gagar *et al.* 2013, Tong 1996). Although effective, the reliability of this technique is easily compromised by the existence of noise or other acoustic phenomena, and it requires trial-and-error tuning. A more robust approach is possible by using a dynamic threshold, which is adapted to the probability of signal occurrences (Youm & Kim 2003). In any case, all the aforementioned techniques cannot avoid relying on some pre-tuned parameters. Other techniques make use of mathematical formulations, such as autoregressive models. The quality of the fit of a fixed-order stationary autoregressive model to a dataset can be measured by the Akaike Information Criterion (AIC) function (Kitagawa & Akaike 1978). If the signal is windowed in a region where the onset certainly exists, then the global minimum of the AIC function indicates the point at which that signal portion can be best divided into the two different stationary segments, i.e. the onset. This principle has been widely used to develop automatic onset pickers for seismic and acoustic emission applications (Carpinteri *et al.* 2012, Kurz *et al.* 2005, Niccolini *et al.* 2012, Sedlak *et al.* 2013, Sedlak *et al.* 2009, Zhang *et al.* 2003).

From a signal processing point of view, the previous definition can, in some circumstances, be extended to a broader one according to which TOF time interval relates to a specific signal feature.

Hilbert Transform (HT) can be applied to the signal and the absolute value of the resulting complex amplitude defines the signal envelope. Then, by picking the point of maximum amplitude (Liu et al. 2013) of the first wave group from different sensors and computing the time difference between them (Bao 2003, Su & Ye 2009) TOF can be measured. However, if dispersion is strongly present it may induce a shift in the relative position of the maximum amplitude point of the first wave group, affecting the reliability of this approach. Moreover, if other modes and reflections overlap, the extraction of the peak itself might even be erroneous.

Therefore the purpose of this article is primarily to compare the AIC and HT methods for automatic TOF determination and to establish the limits of validity of the second one. Additionally, the combination of the HT method with a dynamic threshold algorithm for wave group boundary estimation is described and tested for quantifying Lamb wave dispersion.

## 2 ALGORITHMS

### 2.1 AIC picker

When analyzing a time series it is possible to model the series by using a stationary autoregressive process composed of a backward-deterministic segment, and a non-deterministic segment which is considered to be Gaussian white noise and uncorrelated to the deterministic part (Kitagawa & Akaike 1978, Maeda 1985).

The fit of an autoregressive model to a dataset varies with the process order, and its quality can be evaluated by applying the maximum likelihood estimation to the modelled time series (Kitagawa & Akaike 1978). The solution is an expression that establishes the Akaike Information Criterion (AIC). According to it, the lower the AIC function value, the more adequate is the autoregressive process order and the better is the fit to a dataset. For a fixed order, the AIC function can be written for a signal window (portion) as in Equation 1 (Maeda 1985), and its global minimum indicates the point of optimal separation between the deterministic and non-deterministic segments.

$$AIC(k_w) = k_w \log\{var(x[1,k_w])\}$$
$$+ (n_w - k_w)\log\{var(x[k_w + 1, n_w])\} \quad (1)$$

where subscript $w$ = chosen window; $n_w$ = last sample point of chosen window; $k_w$ = sample point between 1 and $n_w$; and $var$ = variance of the function.

By definition the AIC function is conditioned by the window to which it is applied. If the window is chosen so that the signal onset is contained

therein, then the global function minimum will correspond to the signal onset point, to a certain accuracy level. That accuracy is strongly dependent on the distribution of the deterministic and non-deterministic segments within the window. Therefore the window choice must be guided by a framing operation.

In order to tackle these problems in an optimal way it was decided to combine and adapt the approaches used by Carpinteri et al. (2012), Kurz et al. (2005), Niccolini et al. (2012), Sedlak et al. (2009) and Sedlak et al. (2013), keeping in mind that a trade-off between accuracy and computation time must be achieved. The squared-normalized HT envelope (see section 2.2) is used to guide the framing operation. In the definition of the envelope, the HT part is a version of the signal with a phase shift of $\pi/2$. When computing the absolute value this shift produces a positive numerical anomaly at the first sample points. Thus, the framing is performed by placing the window start some thousands of samples after the beginning of the recorded signal, and the window end right after the first wave-packet maximum (Sedlak et al. 2009) making sure it is positioned after the window start. This approach allows the algorithm to avoid placing the window end before the window start. Furthermore, it aims at minimizing the influence of later wave-packets so that the stationary autoregressive model fits better to the data series and the AIC method performance is maximized. After framing the window the AIC function is calculated for each sample point, and the global minimum is determined and returned as the picked signal onset.

### 2.2 HT envelope picker

The other implemented algorithm makes use of the squared-normalized signal envelope to determine the maximum amplitude point of the first wave group after the onset. First the signal envelope is extracted by calculating the absolute value of the complex amplitude of the Hilbert transformed signal (Equations 2 and 3). Then the square of the envelope is divided by its own absolute maximum value in order to obtain the squared-normalized envelope, which is much more sensitive to changes or singularities in the signal.

$$H\{x(t)\} = \frac{1}{\pi}\lim_{R\to\infty}\int_{-R}^{R}\frac{x(\tau)}{t-\tau}d\tau \quad (2)$$

$$Env(t) = \sqrt{x(t)^2 + H\{x(t)\}^2} \quad (3)$$

The position of the first group peak is extracted by searching for the points which correspond to zero-crossings of the first derivative of the envelope, are above a pre-defined minimum threshold amplitude, and have a relevant level difference (*Ldiff*) relative to the surrounding data defined as a quarter of the difference between the maximum and the minimum amplitudes of the envelope. By doing so for each recorded channel it is possible to compute the time differences between first group peaks in order to arrive at TOF measurements.

## 2.3 Group boundary estimation

The extraction of useful information from Lamb wave signals is strongly dependent on the correct interpretation of modes and groups, a task that can be assisted by determining their boundaries in the time-domain, which is something not fully explored in the literature related to Lamb wave interrogation.

The adopted approach for group time-boundary estimation is conceptually simple, light in implementation and as general as possible. The idea behind the algorithm is to link the end of a group to the global minimum between each two consecutive wave packets in the signal envelope. For that it is first necessary to determine the total amount of relevant groups and their locations. A fixed threshold peak detection method would result in either too few or too many findings. In this case it is crucial to select the relevant peaks after the signal onset. Instead of assigning an expression to model the behavior of the threshold during the peak finding process (Dimou et al. 2005, Youm & Kim 2003), the fixed-threshold sub-routine described in sub-section 2.2 is used recursively to always search for the first relevant peak in an envelope whose starting point is dynamically updated to a point immediately after the previous group peak. Since the envelope is adapted in each iteration, also $Ldiff$ is (indirectly) adapted, creating a (pseudo) dynamic threshold algorithm which is allowed to run until a maximum of 20 peaks is identified or the end of the recorded signal is reached. After this process, each group boundary can be captured by searching for the global minimum between consecutive group peaks using the same fixed-threshold sub-routine (sub-section 2.2).

## 3 EXPERIMENTS

It was decided to conduct the experiments on a panel made from aluminum alloy 6082, in order to isolate the time-domain variables from attenuation and dispersion as much as possible. The dimensions of the panel were $900 \times 1500 \times 1.5$ mm$^3$, so that the wave propagation length would be enough to test the limits of validity of the second method.

A five-cycle ultrasonic Hanning windowed toneburst was produced by an Agilent 33500B Series waveform generator, and transmitted to the plate through a Physical Acoustics WSα Series piezoelectric transducer (actuator) placed 200 mm

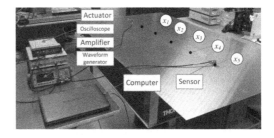

Figure 1. Experimental set-up: equipment and sensor positions.

away from the edge of the panel at the middle of the width. The Lamb wave response was sensed also by a Physical Acoustics WSα Series piezoelectric transducer (sensor 1) whose position was varied from 200 to 1000 mm away from the actuation point, with increments of 200 mm (positions $x_1$ to $x_5$) always measured collinearly with the actuator and parallel to the length of the panel. The coupling of both transducers to the plate was ensured by Sonotech shear gel and adhesive tape. The sensed signals were conditioned in a Thurlby Thandar Instruments WA301 wideband amplifier with a gain of ten times, and acquired via a Pico-Scope 6402 A digital oscilloscope, as depicted in Figure 1.

The excitation amplitude was set to 10 V$_{pp}$, and the center frequency to 250, 500 and 750 kHz, allowing three different dispersion states to be tested for each sensor 1 position.

## 4 RESULTS AND DISCUSSION

### 4.1 TOF analysis

In order to minimize computation time and keep statistical relevance it was decided to analyze the first five waveforms of each set of 32 extracted at each sensor position, for each frequency. The five different sensor 1 configurations enabled the extraction of the TOF for four different propagation distances: TOF$_{12}$ for $d_{12} = |x_1 - x_2|$, TOF$_{13}$ for $d_{13} = |x_1 - x_3|$, TOF$_{14}$ for $d_{14} = |x_1 - x_4|$, and TOF$_{15}$ for $d_{15} = |x_1 - x_5|$.

After computing the signal onset and the time of the first wave-packet peak for each of the analyzed waveforms, the difference between corresponding points was taken to arrive at each of those TOFs. The obtained values were then used to calculate the $S_0$ Lamb mode group velocity at each tested frequency. As shown in Figure 2 the results from the two methods reveal good agreement between them, except for 500 kHz where they differ by 220 m/s.

Overall, the experimental points present an acceptable agreement with the theoretical curve. However, they only allow a very limited evaluation

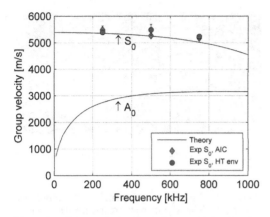

Figure 2. Group velocity dispersion curves for the aluminum panel (computed with DISPERSE™ software).

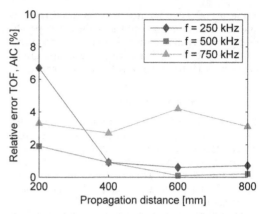

Figure 4. Variation of the relative error of the AIC results with distance, for each excitation frequency.

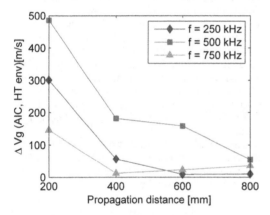

Figure 3. Variation of the $S_0$ group velocity difference between methods with distance, for each excitation frequency.

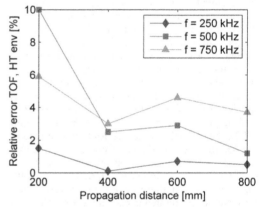

Figure 5. Variation of the relative error of the HT envelope results with distance, for each excitation frequency.

of the performance of the two algorithms. Hence it was decided to plot the difference of group velocity ($\Delta v_g$) obtained with each algorithm as a function of propagation distance, for each excitation frequency (Figure 3). Two main observations can be made from the chart. Firstly, there is a clear steady decrease in $\Delta v_g$ with increasing distance for all frequencies, indicating consistency of both methods when applied to $S_0$ mode analysis. This was expected, since at larger distances the two zero-order modes are more clearly separated and the peak of the (not so dispersive) $S_0$ mode can be more unambiguously detected. Secondly, while the curves for 250 and 750 kHz are relatively close to each other (especially after 200 mm), almost all the points for 500 kHz are at least 150 m/s above the others.

The reason for this discrepancy at 500 kHz can be better understood by looking at Figures 4 and 5, where the relative error of the TOF results from the AIC and HT envelope algorithms (respectively) are plotted as a function of propagation distance, for each frequency.

In both graphs the relative error falls within the same range of values for 250 and 750 kHz at all distances except at 200 mm (see discussion about relative error in the AIC algorithm results on the next page), corroborating the proximity of the $\Delta v_g$ curves for these two frequencies. However, in Figure 4 the relative error for 500 kHz is always below 2%, while in Figure 5 it starts at 10% and only goes below 2% at 800 mm. Thus, the discrepancy of $\Delta v_g$ for this particular frequency is in fact due to poorer TOF results from the HT envelope method at that frequency.

This occurrence can be explained by analyzing the first (or first two) wave-packet(s) from the signals which were captured at $x_1$ and $x_2$ for 500 kHz and used to compute the TOF for 200 mm (Figures 6 and 7). At $x_1$ (200 mm away from the actuator), the $S_0$ and $A_0$ are superimposed as it is

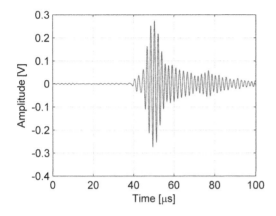

Figure 6. Part of the signal acquired at $x_1$, for 500 kHz.

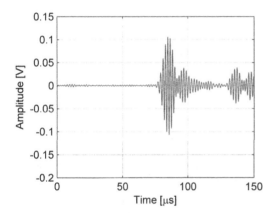

Figure 7. Part of the signal acquired at $x_2$, for 500 kHz.

not possible to distinguish their groups separately. As a result, their amplitude is summed, masking the position of the peaks of each mode. Therefore, the observed first wave-packet peak is not actually the $S_0$ peak. At $x_2$ (400 mm away from the actuator), it is already possible to distinguish two groups, although they are still not completely separate. There is still mode overlapping with part of the $S_0$ mode (algebraically) summed to part of the $A_0$ mode. Therefore, at this position the first wave-packet peak approximately corresponds to the desired mode peak. Nevertheless, the subtraction of the peak times results in $TOF_{12}$ being considerably different from the theoretical value.

This phenomenon is stronger for 200 mm, but it is always present because the (more dispersive) $A_0$ mode always overlaps with part of the $S_0$. However, for 200 mm it has clearly more influence at 500 kHz. The explanation lies in the combination of two tendencies. At 250 kHz the $A_0$ group velocity is lower causing the mode to arrive late enough to be discernible, even though $x_1$ is closer to the actuation

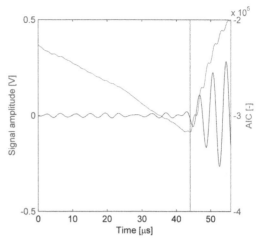

Figure 8. Determination of the signal onset with the AIC function at $x_1$ with a 250 kHz excitation frequency.

point. At 750 kHz the $A_0$ group velocity is higher causing the mode to arrive earlier. But because frequency is higher the mode duration becomes lower, making it easier to discern the presence of two groups, although not as clearly as for 250 kHz.

Previously it was pointed out that for 250 kHz at 200 mm, the results obtained with the AIC method presented a relative error of 6.7% (Figure 4). At that frequency, noise and actual signal have more similar oscillatory characteristics than in the other cases. Moreover, at $x_1$ and $x_2$, the length of the noise segment contained in the frame used for the AIC function computation is smaller than at the other points. These two factors combined make more difficult for the AIC function to reach a sharp global minimum (i.e. to clearly separate noise from actual signal), as seen by its shape in Figure 8.

To fully characterize the performance of the HT envelope method for TOF determination it is envisioned to perform a similar study for the $A_0$ mode. That can be achieved by implementing a time-scale domain mode separation algorithm based on wavelet transform (Se & Ye 2009). The possibility of installing a selective mode generation system is wittingly left aside because the reliability of that strategy can only be assured in laboratory conditions (Su & Ye 2004).

### 4.2 Dispersion analysis

The results from 1st wave-packet duration study proved to be inconclusive, since no clear pattern was observed with increasing propagation distance, for the tested frequencies. This was due to erroneous group end-time determination, which in turn was caused by the mode superposition effect described in the previous sub-section. Therefore, it is valid to say that mode time-boundary estimation cannot solely rely on direct HT signal envelope information. In future developments the algorithm will be improved by using time-scale domain mode separation and the combined analysis of multiple signal features.

## 5 CONCLUSIONS

This article focused primarily on the evaluation of two different methods for TOF measurement, irrespectively of the propagation medium material.

The combined use of the HT envelope for guiding the signal window selection in the AIC algorithm proved to be of value for the extraction of consistent and reliable TOF results. The performance of the AIC-HT method is expected to be unaffected by the stronger dispersive properties of composite materials.

The optimization of the capabilities of the combined AIC-HT method for SHM relies on the use of effective signal filtering, adaptable framing operations, and advanced programming techniques to make the algorithm computationally more efficient. If this is achieved, it is an algorithm with a high level of accuracy and a strong potential for integration.

After this study it is valid to state that the HT envelope method can be reliably used, with computational efficiency advantages, for TOF analysis of weakly dispersive Lamb modes (e.g. $S_0$), as long as mode separation is previously performed by applying some extra signal processing (e.g. time-scale domain analysis with wavelet transform).

The use of the HT envelope method for the TOF analysis of strongly dispersive Lamb modes (e.g. $A_0$) is dependent on the accuracy of predictions of the dispersion effect in order to compensate for the time spreading of the mode. This case is still to be tested and is expected to provide additional information on the performance of the HT envelope method when applied to TOF measurement in composite materials (i.e. highly dispersive media).

## REFERENCES

Bao, J. 2003. Lamb wave generation and detection with piezoelectric wafer active sensors. *PhD Thesis*, University of South Carolina.

Carpinteri, A., Xu, J., Lacidogna, G. & Manuello, A. 2012. Reliable onset time determination and source location of acoustic emission in concrete structures. *Cement & Concrete Composites* 34: 529–537.

Dimou, A. Nemethova, O. & Rupp, M. 2005. Scene change detection for H.264 using dynamic threshold techniques. In *Proceedings of 5th EURASIP Conference on Speech and Image Processing, Multimedia Communication and Service, June 29–July 2 2005, Slovak Republic*.

Gagar, D., Martinez, M., Yanishevsky, M., Rocha, B., McFeat, J., Foote, P. & Irving, P. 2013. Detecting and

Locating Fatigue Cracks in a Complex Wing-Box Structure Using the Acoustic Emission Technique: A verification Study. In Fu-Kuo Chang (ed.), *Structural Health Monitoring 2013—A Roadmap to Intelligent Structures; Proc. 9th Intern. Workshop on SHM, Stanford, 10-12 September 2013*. Lancaster: DEStech Publication, Inc.

Kitagawa, G. & Akaike, H. 1978. A procedure for the modeling of non-stationary time series. *Ann. Inst. Statist. Math.* 30 (Part B): 351–363.

Kurz, J.H., Grosse, C.U. & Reinhardt, H.W. 2005. Strategies for reliable automatic onset time picking of acoustic emissions and of ultrasound signals in concrete. *Ultrasonics* 43: 538–546.

Liu, P., Groves, R.M. & Benedictus, R. 2013. Signal processing in optical coherence tomography for aerospace material characterization. *Optical Engineering* 52(3), 033201, 1–7.

Maeda, N. 1985. A method for reading and checking phase times in auto-processing system of seismic wave data. *Zisin* 38: 365–379.

Niccolini, G., Xu, J., Manuello, A., Lacidogna, G. & Carpinteri, A. 2012. Onset time determination of acoustic and electromagnetic emission during rock fracture. *Progress In Electromagnetics Research Letters* 35: 51–62.

SAE ARP6461. 2013. Guidelines for Implementation of Struc-tural Health Monitoring on Fixed Wing Aircraft.

Schijve, J. 2004. *Fatigue of Structures and Materials*. Dordrecht: Kluwer Academic Publishers.

Sedlak, P. Hirose, Y. & Enoki, M. 2013. Acoustic emission localization in thin multi-layer plates using first-arrival determination. *Mechanical Systems and Signal Processing* 36: 636–649.

Sedlak, P., Hirose, Y., Khan, S.A., Enoki, M. & Sikula, J. 2009. New automatic localization technique of acoustic emission signals in thin metal plates. *Ultrasonics* 49: 254–262.

Su, Z. & Ye, L. 2004. Selective generation of Lamb wave modes and their propagation characteristics in defective composite laminates. In *Proceedings of the Institution of Mechanical Engineers, Part L: Journal of Materials: Design and Application* 218 (2): 95–110

Su, Z. & Ye, L. 2009. *Identification of Damage Using Lamb Waves: From Fundamentals to Applications*. In Pfeiffer, F. & Wriggers, P. (eds), *Lecture Notes in Applied and Computational Mechanics* 48. Springer.

Su, Z., Ye, L. & Lu, Y. 2006. Guided Lamb waves for identification of damage in composite structures: A review. *Journal of Sound and Vibration* 295: 753–780.

Tong, C. & Kennet, B.L.N. 1996. Automatic Seismic Event Recognition and Later Phase Identification for Broadband Seismograms. *Bulletin of the Seismological Society of America* 86(6):1896–1909.

Wenk, L. & Bockenheimer, C. 2014. Structural Health Monitoring: A real-time on-board 'stethoscope' for Condition-Based Maintenance. In *Airbus technical magazine, Flight Airworthiness Support Technology* 54: 22–29.

Youm, S. & Kim, W. 2003. Dynamic threshold method for scene change detection. In *Proceeding of Intern. Conf. Multimedia and Expo* 2: 337–340, 6–9 July 2003.

Zhang, H, Thurber, C. & Rowe, C. 2003. Automatic P-Wave Arrival Detection and Picking with Multiscale Wavelet Analysis for Single-Component Recordings. *Bulletin of the Seismological Society of America* 93(5): 1904–1912.

*Emerging Technologies in Non-Destructive Testing VI – Aggelis et al. (Eds)*
*© 2016 Taylor & Francis Group, London, ISBN 978-1-138-02884-5*

# Fuse-like devices replacing linear sensors—working examples of percolation sensors in operational airliners and chemical installations

H. Pfeiffer, I. Pitropakis, M. Wevers
*Department of Materials Engineering, KU Leuven, Leuven, Belgium*

H. Sekler
*Lufthansa Technik AG, Frankfurt, Germany*

M. Schoonacker
BASF, *Antwerp, Belgium*

ABSTRACT: In order to detect diverse failure modes and damage sizes, sensors in traditional Non-Destructive Testing (NDT) for aircraft and other engineering structures usually work in a quasi-linear mode. Therefore, also when used in a Structural Health Monitoring (SHM) system, their performance is limited by operation-dependent baseline variations as well as by signal interference in complicated structures. Further technologies are thus required such as elaborated data processing and high-end hardware components. These sophisticated countermeasures however hinder in turn the fast and efficient implementation of SHM. Our answer to this challenge is the use of sensor materials that operate like fuses, i.e. they start to be highly sensitive just when a pre-set degree of damage has been reached. In this way, baseline variations when no relevant damage is present are naturally suppressed. The physical principles for the detection applied are electrical conductivity, electrochemical impedance and optical transmission.

## 1 INTRODUCTION

In order to cover the detection of diverse failure modes and damage sizes, sensors in traditional Non-Destructive Testing (NDT) for aircraft usually work in a quasi-linear mode. Therefore, when sensors are permanently embedded in aircraft structures to be part of a Structural Health Monitoring (SHM) system, their performance is usually limited by operation-dependent baseline variations as well as by signal interference with the intrinsically complicated airframe structures. In those cases, further technologies are required such as elaborated data processing as well as high-end hardware components operating during flight requiring on-board power supply and data storage/processing. These sophisticated countermeasures however hinder in turn the fast and efficient implementation of SHM due to the high cost, the limited coverage, the complex certification strategy and finally the uncertain return-on-investment.

A possible answer to this challenge is the use of sensor materials that operate like fuses, i.e. they start to be highly sensitive just when a pre-set degree of damage has been reached. In this way, baseline variations when no relevant damage is present are naturally suppressed. Detection systems following this principle already exist in isolated cases, such as the well-known electrical fuses, crack gauges or percolation sensors (Pfeiffer et al., 2012), but a systematic approach for SHM was not yet presented. The most promising physical principles proposed in the literature for the detection are electrical conductivity, electrochemical impedance and optical transmission.

Also in our case, since four years, three operational airliners are equipped with percolation sensor networks (Boeing 737–500, two Boeing 747–400) reporting wetness problems in the floor structure below galley and service doors areas (Pfeiffer, 2012). Moreover, a couple of outside pipelines in chemical installations were installed with an analogous sensor network (Pfeiffer et al., 2013). The results obtained are convincing, and the concept of the percolation sensor was further extended to other liquids that are frequently used in an aircraft and other engineering structures (Pfeiffer et al., 2014).

The primary impact can result in early gains for aircraft and plant maintenance operations while avoiding complicated implementation issues resulting in clear and measurable reduction of maintenance costs and increase safety by radically increasing the frequency of interrogations.

## 2 STATE-OF-THE-ART

Sensing systems in diverse structures following the fuse-principle were already reported for isolated implementation areas, and their performance can be tailored to specific requirements, such as thresholds and environmental constraints, operation and reversibility. To obtain fuse-like behavior it is required to search for mechanical or physico-chemical processes where a physical quantity changes its value over many orders of magnitude within a very narrow range that needs to be physically related to the occurrence of a certain degree of damage. Those three principal processes are very promising candidates, 1) the well-known fracture of materials related to e.g. the complete loss of electrical conductivity, 2) phase transitions, such as melting processes or modifications of solid-state conformations as well as 3) solution processes. The last two options are of special interest when fuse-like behavior needs to be reversible, e.g. if the irreversible fracture of a material is not desired.

### 2.1 Fracture-based fuse systems

A prominent example of fracture-based systems are small-sized crack propagation gauges made from tiny metal stripes embedded in polymers to be glued on the material under investigation. Those systems are already commercially available (Pfeiffer et al., 2008), however, they are not appropriate for SHM because of their limited size and adaptability to complex surfaces. Furthermore, at some occasions, closed cracks cannot be detected reliably because the electrical contact remains intact, and partially after rupture, the reversibility of those systems is not granted. A solution to this problem is partially offered by crack gauges that are directly painted or cold-sprayed on the respective part using specific, conductive ink, adaptable to different complex surfaces and shapes. This principle has already been proposed in the past for building structures and can be combined with Radio-Frequency Identification (RFID) technology (Morita and Noguchi, 2008). The feasibility of this concept to aircraft structures was also proven by appropriate lab tests performed on Al-2024 fuselage material and A320 slat tracks (Pitropakis et al., 2012). The typical fuse-like response is clearly visible (Fig. 1).

### 2.2 Phase transitions for monitoring

An interesting example for the application of phase transitions regards an application in the frame of the well-established Positive Temperature Coefficient (PTC) technology which is used to protect electric circuits against overcurrent. Although according to our best knowledge not yet applied in structural health monitoring, it is

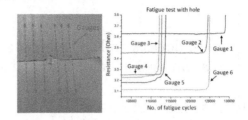

Figure 1. Painted crack gauges under standard coating (left) and resulting electrical resistance obtained indicating progressing crack growth (right).

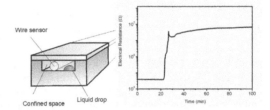

Figure 2. Detection of water leakage by a fuse-type sensing material using a lyotropic phase transition.

an interesting example for the desired operation mode. After a certain temperature increase arising from overcurrent, the conducting component (carbon black) changes its phase state resulting in a strong enhancement of the electrical resistance, i.e. also here a phase transition enables a change of the percolation conductivity. This process is essentially reversible and the performance can be tuned according to the operator needs. Here, the damage signal is approximately 6 orders of magnitude above baseline variations (Fundamentals of PolySwitch Overcurrent and Overtemperature Devices, 2008). (Device, 2008).

Another system using phase transitions are the leak detection wires detecting harmful hot leaking bleeding air, such as installed in different Airbus aircraft, called Fenwal sensing elements. Here, eutectic salt mixtures are used that melt and become electrically conductive at the pre-set detection temperature. The system operates in a reversible mode because when the temperature decreases, the salt solidifies again. In this way, the difference between baseline variations and damage signal is also practically infinite. (Aviation Maintenance Technician Handbook, 2012). (Handbook, 2012).

A solvent-dependent, thus lyotropic glass transition in polyvinyl-alcohol serves as the tool to detect the presence of water in engineering structures (Pfeiffer et al., 2012). Here, the electrical conductivity is strongly reduced after the uptake of water (Fig. 2).

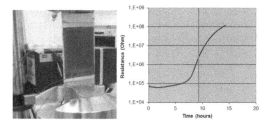

Figure 3. Printed "fuse" for detecting hydraulic liquids (left) in aircraft and response curve for engine oil or kerosene (right).

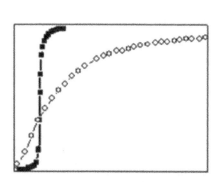

Figure 5. Highly simplified POD (a.u.) for linear sensor (circles) and fuse-like sensors (squares).

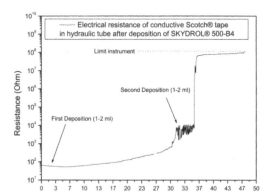

Figure 4. Deposition of aircraft-type hydraulic liquids on conductive acrylic material.

## 2.3 Solution—based sensing systems

When the presence of certain harmful liquids needs to be detected, also the partial or complete solution of a sensing material can be utilized, here an electrical conducting material composite is used where the isolating matrix is susceptible to the liquid to detect (Pfeiffer et al., 2014).

For instance, for the detection of hydraulic liquids in aircraft, usually based on phosphate ester compounds, an acrylic matrix can be used (Fig. 3, left, Fig. 4), when the presence of jet engine oil of fuel needs to be determined, electrically conductive rubber is applied (Fig. 3, right).

## 3 PROBABILITY OF DETECTION

### 3.1 General considerations

The general idea can also be explained by the Probability of Detection (POD). The POD concept gives an estimate of the performance of a detection technology in terms of a probability (Fig. 5). Basically, the POD is defined by the ratio of detected damage divided by the number of all tests performed. It is expected that the shape of the resulting POD curve for the proposed structural fuses will be a convolution of the POD curve of the linear reference case combined with the response curve of a highly nonlinear sensor, i.e. at low damage sizes, POD will be approximately zero, and at the "working point (vertical line)" it will approach POD = 1 within a very narrow range. Finally, very small defects will be missed (always acceptable in the case of the usual "damage tolerance" approach) but relevant damage will always be detected in a very reliable way. Last but not least, the possible loss of wide-range sensing will be highly compensated by the gain of simplicity and robustness when using this concept (see also Fig. 5).

## 4 EXAMPLES IN ENGINEERING STRUCTURES

### 4.1 Wetness in floor structures of aircraft

The possibility to detect the presence of wetness in floor structures was practically tested in a couple of aircraft. Until today, three operational airliners are equipped with percolation sensors networks (Boeing 737–530, Boeing 747–430) protecting the floor structure below galley and service doors areas (Pfeiffer, 2012).

A general problem to tackle are the long intervals between scheduled inspections, i.e. it is possible that corrosive liquids in those structures remain undetected for several years. The result is that heavy corrosion can occur that leads not only to partial safety issues but especially to very expensive repair costs. The sensor is thus situated in a confined space within the aircraft (Fig. 6) and a very important feature of this device is that it also detects the presence of liquid that is not touching the device but if it is only present in its neighborhood (Fig. 2, left). This is possible because evaporated liquid is

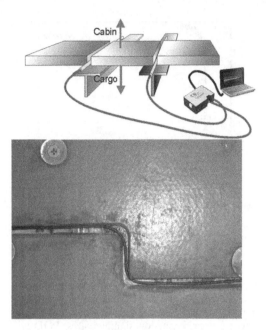

Figure 6. Position of fuse-type sensing device in the floor structures of an aircraft and real implementation in a Boeing 737-530.

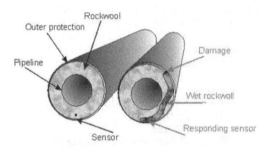

Figure 7. Positioning of the wired sensor in pipelines in chemical installations (top) and real implementation area.

able to enter the sensing material by absorption. The obtained curves look essential like the response function given by (Fig. 2, right).

4.2 *Presence of corrosive liquids in chemical plants*

Similar to the presence of liquids in floor structures of aircraft, water-like compounds in pipelines in chemical installations can lead to heavy corrosion causing leakage of harmful compounds and expensive repair costs. Also here, the sensing material, performed as a sensing wire, was able to reliably detect the presence of corrosive liquids. In detail, the sensing wire is situated within the mineral wool that is used to thermally isolate the pipeline (Fig. 7). Comparable to the floor structures, also here, liquid in the neighborhood of the sensing material is detected because water vapor is diffusing through the whole isolation. A certain attention need to be given to the temperature ranges, i.e. the sensitivity is a function of the temperature because absorption and diffusion processes are depending on thermal boundary conditions. However, this is not a principal problem because a responding sensor always indicated a defect isolation, and it is always a question of time when those liquids will cause heavy corrosion. Also here, the response curves are similar to the examples given by Figure 2, right.

ACKNOWLEDGEMENTS

The research leading to these results has partially received funding from the European Community's Seventh Framework Programme [FP7/2007–2013] under grant agreement n°212912.

REFERENCES

Aviation Maintenance Technician Handbook (2012) Federal Aviation Administration, 2, 17–4.
Fundamentals of PolySwitch Overcurrent and Overtemperature Devices (2008) Menlo Park, Raychem Circuit Protection Products.
Morita, K. & Noguchi, K. (2008) Crack detection methods using radio frequency identification and electrically conductive materials. *The 14th World Conference on Earthquake Engineering, October 12–17.* Beijing, China.
Pfeiffer, H. (2012) Structural Health Monitoring makes sense. *LHT Connection–The Lufthansa Technik Group Magazine.*
Pfeiffer, H., De Baere, D., Fransens, F., Van Der Linden, G. & Wevers, M. (2008) Structural Health Monitoring of Slat Tracks using transient ultrasonic waves. *EU Project Meeting on Aircraft Integrated Structural*

*Health Assessment (AISHA).* Leuven, Belgium, www.ndt.net.

Pfeiffer, H., Heer, P., Pitropakis, I., Pyka, G., Kerckhofs, G., Patitsa, M. & Wevers, M. (2012) Liquid detection in confined aircraft structures based on lyotropic percolation thresholds. *Sensors and Actuators B-Chemical,* 161, 791–798.

Pfeiffer, H., Heer, P., Sekler, H., Winkelmans, M. & Wevers, M. (2013) Structural Health Monitoring Using Percolation Sensors-New User Cases from Operational Airliners and Chemical Plants. *Structural Health Monitoring 2013, Vols 1 and 2,* 2130–2137.

Pfeiffer, H., Heer, P., Winkelmans, M., Taza, W., Pitropakis, I. & Wevers, M. (2014) Leakage monitoring using percolation sensors for revealing structural damage in engineering structures. *Structural Control & Health Monitoring,* 21, 1030–1042.

Pitropakis, I., Pfeiffer, H., Gesang, T., Jannsens, S. & Wevers, M. (2012) Crack Detection on an Airbus A320 Slat-Track and on Aluminium 2024-T3 Plates using Crack Propagation Gauges. *8th World Conference on Non-Destructive Testing (18th WCNDT),* Durban, South Africa, ndt.net.

*Applications*

*Emerging Technologies in Non-Destructive Testing VI – Aggelis et al. (Eds)*
*© 2016 Taylor & Francis Group, London, ISBN 978-1-138-02884-5*

# Non-Destructive Testing techniques to evaluate the healing efficiency of self-healing concrete at lab-scale

E. Gruyaert, J. Feiteira & N. De Belie
*Magnel Laboratory for Concrete Research, Ghent University, Ghent, Belgium*

F. Malm, M. Nahm & C.U. Grosse
*Non-Destructive Testing Laboratory, Technische Universität München, München, Germany*

E. Tziviloglou & E. Schlangen
*Technische Universiteit Delft, Delft, The Netherlands*

E. Tsangouri
*Vrije Universiteit Brussel, Brussels, Belgium*

ABSTRACT: Within the European FP7 project HEALCON, Non-Destructive Testing (NDT) and monitoring techniques are developed and combined to characterize the effects of self-healing mechanisms in small and full-size specimens.

In the first stage, healing mechanisms were evaluated at lab-scale. Specimens containing encapsulated polymer precursors were cracked and reloaded after the healing period. During loading, healing and reloading, NDT techniques (acoustic emission analysis, vibration analysis and ultrasonic measurement) were applied to help understanding the cracking behavior, capsule breakage and healing efficiency. Moreover, the effect of the flexibility of the polymeric healing agent on the crack re-opening during reloading was investigated on cracked and healed mortar specimens, using acoustic emission and digital image correlation techniques.

The results show the applicability of NDT methods to evaluate the self-healing efficiency for small specimens. Comparing the NDT techniques, some of them (e.g. ultrasound) seem to be good candidates for in situ monitoring of the healing efficiency.

## 1 INTRODUCTION

Within the European FP7 project HEALCON, different self-healing techniques are developed and optimized. For the evaluation of the performance, it is essential to study the efficiency of the different self-healing methods. The efficiency is mostly investigated through performance of permeability tests or by reloading and investigating the regain in mechanical properties. However, these techniques have several disadvantages i.e. that the specimen are not available for further investigations after destruction and that the techniques are of limited use in practice. Therefore, project partners are developing non-destructive testing and monitoring techniques and are combining existing techniques in order to characterize the effects of different self-healing mechanisms in small and full size specimens. In the first stage of the project, healing mechanisms were observed at lab-scale. Techniques as acoustic emission analysis (Van Tittelboom et al. 2012, Malm & Grosse 2014), vibration analysis (Neild et al. 2003)

and ultrasonic measurements were applied. The most interesting results obtained for specimens containing encapsulated polymer precursors are reported in this paper.

## 2 MATERIALS AND METHODS

### 2.1 *Healing agents*

The precursors used in this study are commercially available products used for crack injection of concrete. These are polyurethane-based products that cure upon contact with moisture. Products with low viscosity were chosen, as this is a critical property to guarantee a large crack coverage and good healing performance. Good adhesion to concrete was also considered essential, otherwise detachment of the polymer from the crack walls would occur before considerable crack widening. The precursor here designated by SLV (super low viscosity) has a viscosity of 200 mPa·s and a maximum elongation of 100% according to its technical sheet. The precursor designated by CUT has a viscosity of 350 mPa·s and was chosen for comparison purposes due to its considerably lower maximum elongation (not available), according to the manufacturer. For the tests on mortar specimens, the precursors were mixed with 2 wt% of an accelerator, to reduce the time for full curing. Additionally, the accelerator also induces a foaming effect upon contact with moisture.

## 2.2 Encapsulation

The polymer precursors were encapsulated in glass tubes with external and internal diameters of 3.35 mm and 3.00 mm respectively, cut to a length of 50 mm (mortar specimens) or 100 mm (concrete specimens) and sealed with Poly(methyl methacrylate) glue.

## 2.3 Tests on concrete beams

Reference concrete beams (containing no healing agent), control concrete beams (containing empty glass capsules) and self-healing concrete beams (containing SLV encapsulated in glass capsules) with dimensions of 150 mm × 150 mm × 550 mm were cast (Figure 1). The capsules in the self-healing concrete beams were fixed to nylon strings which were connected with the mould. Two reinforcement bars (Ø 6 mm) were positioned in the compressive zone in order to prevent brittle failure of the specimen during the three-point bending test.

The concrete composition used and the compressive strength of the concrete, measured on cubes 150 mm × 150 mm × 150 mm at 28 days, is given in Table 1.

Table 1. Concrete composition.

| Constituent | Composition kg/m³ |
| --- | --- |
| Cement CEM I 42.5 N | 300 |
| Water | 150 |
| Sand 0/4 | 686 |
| Gravel 4/8 | 502 |
| Gravel 8/16 | 809 |
| Mean compressive strength (28 days) | 48.3 N/mm² |

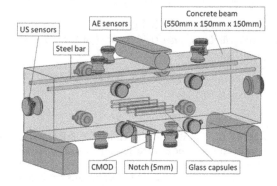

Figure 2. Set-up for the three-point bending test – positioning of the acoustic emission sensors and US transmitter and receiver.

After a storage period of 28 days under water, the specimens were moved to a climate room at 20°C and 60% RH and prepared for the first loading test. A notch with a depth of 5 mm was sawn in the middle of the specimens, Vibration Analysis (VA) measurements were performed and the Acoustic Emission (AE) sensors (15 in total) and Ultrasonic (US) transducers were fixed with hot glue to the concrete surface (Figure 2). After the crack width controlled three-point bending test (speed of 0.6 mm/min and crack width of ~ 350 μm after unloading), vibration analysis tests were performed and the polymer was allowed to cure for 3 days. During that time, the evolution of the S-waves was continuously monitored. Before reloading, vibration analysis tests were performed and acoustic emission sensors were glued to the concrete surface intentionally at the same position as during the first loading test. During reloading, acoustic emission and ultrasonic measuring techniques were applied. The settings for the acoustic emission and ultrasound transmission measurements and analysis are given in Table 2 and Table 3 respectively.

## 2.4 Tests on mortar specimens

Mortar specimens with dimensions 40 mm × 40 mm × 160 mm were moulded with 2 glass capsules containing the mix of the polymer precursor

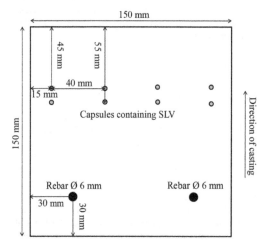

Figure 1. Cross section at the middle of a self-healing concrete beam.

Table 2. Settings for acoustic emission analysis.

| | Amplifier |
|---|---|
| Number of channels | 16 |
| Gain | 60 dB |
| Filter | |
| High-pass | 10 kHz |
| Low-pass | 1 MHz |
| | Elsys TranAX |
| Operation Mode | ECR Multichannel |
| Sample rate | 5 MHz |
| Pretrigger | 500 S |
| Posttrigger | 5 kS |
| HoldOff | 5.5 kS |
| Range | 20 V |
| Slew Rate Trigger | 0.2V/10S |

Table 3. Settings for ultrasound transmission.

| Pulser voltage | 800 V |
|---|---|
| Pulser width | 5 µs |
| Signals per test | 1 |
| Sample rate | 10 kHz |
| Recorded time | 1024 µs |

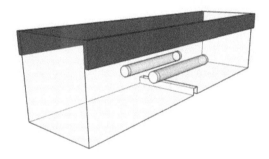

Figure 3. Cracked mortar beam reinforced with a FRP strip.

and 2 wt% of accelerator. The capsules were positioned with nylon strings crossing the mould at the mid section of the specimen, where a crack is expected during 3-point bending. The specimens were also reinforced with a 10 mm wide FRP strip along its length. The FRP was not used in a typical reinforcement configuration but as a way to stop the crack from developing across the whole height of the specimen, and thus was positioned in the compressive zone during bending. Figure 3 shows the final specimen configuration.

The mortar specimens were painted white and a black speckle pattern was sprayed with an air brush for Digital Image Correlation (DIC) measurements. After cracking the specimens in a crack width controlled three-point bending test and after curing of the polymeric healing agent, one AE sensor was attached with tape at the side of the specimen, close to the bending crack. The specimens were reloaded at a speed of 0.04 mm/min for the piston's displacement of the bending machine.

## 3 DETECTION OF CRACKING AND HEALING WITH NON-DESTRUCTIVE TEST METHODS ON CONCRETE BEAMS

### 3.1 Strength regain

From the loading and reloading curves (Figure 4), no strength regain is detected for the control specimens (ECAPS on Figure 4), while the strength regain obtained for the specimens which were self-healed with a polymeric healing agent SLV, amounts to 60% (CAPS on Figure 4).

### 3.2 Vibration analysis

An important dynamic property of any elastic system in material science and engineering technology is the dynamic modulus of elasticity. For a concrete beam of given dimensions, the natural frequency of vibration is mainly related to the dynamic modulus of elasticity and density. Therefore, the modulus of elasticity and also the shear modulus of a material can be determined from the measurement of the longitudinal, flexural (transverse) and torsional frequency of vibration of the object geometry and the corresponding mathematical relationship. Originally, these relationships were derived for solid media considered to be homogeneous, isotropic, and perfectly elastic like steel bars, but it is shown that they can be applied to heterogeneous systems, such as concrete.

Figure 5 shows the moduli of elasticity in the shear, longitudinal and flexural mode before crack formation, after cracking and after healing ( = before reloading).

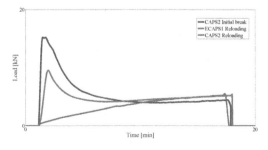

Figure 4. Loading and reloading curves of a self-healing specimen (CAPS2) and reloading curve of a control specimen (ECAPS1).

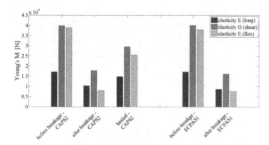

Figure 5. Moduli of elasticity in longitudinal, shear and flexural mode. Control specimen (ECAPS1) - self-healing specimen (CAPS2).

As can be seen, the moduli of elasticity drastically decrease after cracking and the values obtained for the control (ECAPS1) and self-healing concrete (CAPS2) specimens are similar. A considerable increase is noticed for the self-healed specimens after curing of the SLV. Relative to the initial value, the modulus of elasticity in the flexural mode increases during the healing process from ~ 20% to ~ 65%, in the shear mode from ~ 45% to ~ 75%, and in the longitudinal mode from ~ 60% to 80%. Compared to the control specimen (ECAPS1), the self-healing specimen (CAPS2) reveals a partial regain in mechanical properties.

### 3.3 Acoustic emission

Acoustic emission techniques have proven to be useful to observe the performance of cracks in concrete (Grosse & Ohtsu 2008). The technique is used to observe the activation of the healing agents through breakages of the capillaries and to record the cracking behavior in concrete prior and after healing (Grosse et al. 2013, Van Tittelboom et al. 2012). The cumulated number of events and the average signal energy of the acoustic events that have been detected in the reference and self-healing specimens during loading and reloading are presented in Figure 6.

During the first loading of the reference specimens, a lot of events, which are related to concrete cracking, have been detected, and some of them have a high energy level. However, during reloading, a limited number of events was recorded, indicating that besides a re-opening of the crack almost nothing else happens. The sudden increase in events that can be seen also for the self-healing specimens at the moment of unloading is due to shrinkage effects induced by the presence of the reinforcement in the specimens.

When we consider the self-healing specimen, much higher energy events are present during the first loading cycle. Localization of these events

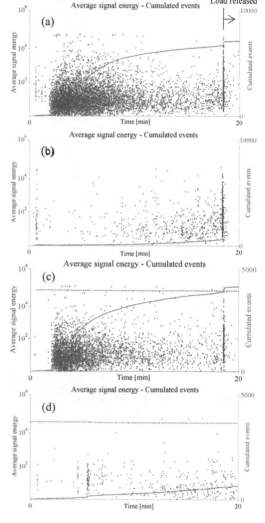

Figure 6. Average signal energy of the events during three-point bending: (a: loading – b: reloading) reference specimens, (c: loading—d: reloading) self-healing specimen.

clearly indicates that they are correlated to capsule breakage (Figure 7). However, during reloading, again a limited number of events is recorded, but a sudden increase in the number of events appears after ~ 6 minutes. All of this is due to the fact that a flexible polymer is used. This polymer can follow the crack widening during reloading up to a certain extent and then partly fails (by detachment from the crack wall or internal failure).

The events which can be seen in Figure 6 with an almost constant energy that are detected for the self-healing specimens are caused by the US pulses generated by the FreshCon system and registered

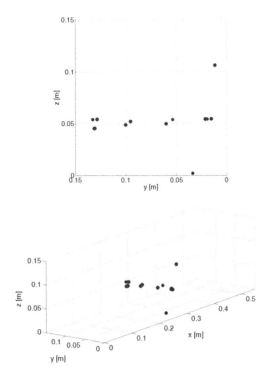

Figure 7. Localized events with an average signal energy higher than $1.6 \times 10^5$. The position of these events corresponds with the position of the capsules.

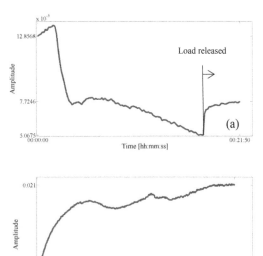

Figure 8. Evolution of the S-wave amplitude during loading (a) and healing (b).

by the acoustic emission sensors. The fact that the energy level of that signal is different during loading and reloading can be explained since the AE sensors were decoupled after the first loading and recoupled just before reloading the beam.

### 3.4 Ultrasonic measurements

Real-time monitoring of the ultrasonic pulses by the FreshCon system allows to extract information about cracking of the concrete and healing of the crack. For the transmission measurement, shear wave sensors (Olympus Panametrics V150-RB) were used. With these sensors, the shear wave is dominant in comparison to the longitudinal wave. Moreover, it could be confirmed, based on a theoretical calculation of the shear wave velocity through a concrete sample and the predicted arrival time, that the amplitude of the selected, significant wave peak, as presented below, corresponds with the shear wave.

In Figure 8, the S-wave amplitude is shown in function of time. The amplitude is much more sensitive than the velocity that is usually measured. The S-wave amplitude clearly decreased with increasing crack depth and width. When the load was released, there was a small increase in amplitude. The reaction of the PU over the first 18 h after cracking was also clearly seen in the increasing S-wave amplitudes.

## 4 DETECTION OF FAILURE MODES FOR BRITTLE AND FLEXIBLE HEALING AGENTS WITH NON-DESTRUCTIVE TEST METHODS

### 4.1 Loading curves

During loading of the healed specimens, it was clear that both precursors lead to similar strength regain (~50% compared to sound specimens) and bending stiffness (as the load was displacement-controlled and kept constant across all specimens).

The main difference in the reloading curves shown in Figure 9 is the failure mode, which is brittle in case of the CUT precursor (sudden significant load drop) and progressive in the case of SLV, implying flexible behaviour. The load drop in the case of SLV is believed to be due to both ductile deformation of the polymer and its progressive detachment from the crack walls.

### 4.2 Digital image correlation

With DIC measurements, it was possible to accurately monitor the crack opening during testing. Figure 10 shows the crack opening across the full

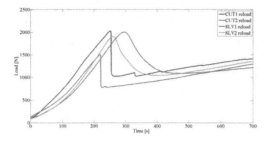

Figure 9. Load drop differences between cracked specimens healed with rigid (CUT) and flexible (SLV) polymers.

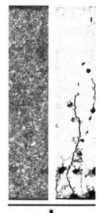 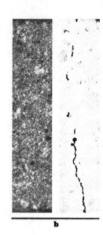

Figure 11. Crack mouth profiles of specimens CUT1 (a) and SLV2 (b).

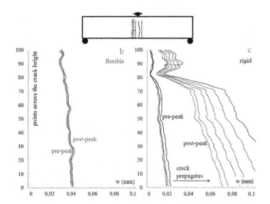

Figure 10. The brittle failure results from rupture of the rigid polymer, as shown by the sudden opening of the crack immediately after the post-peak of loading.

height of the specimen, based on the displacement of 100 points on both sides of the crack, positioned as close to it as possible. It can be seen that during failure, i.e. after peak load, the original crack reopens instantly in the case of the specimen healed with the CUT agent. During the same time span, the crack in the specimen healed with the SLV agent opened just slightly and progressively due to its flexibility.

It was also noticed after testing that the specimens healed with the brittle polymer showed new crack branches beside the original, reopened crack. The specimens were impregnated with epoxy containing fluorescent dye and a thin layer of mortar was removed from the bottom part, revealing any voids and cracks filled with epoxy. The dye in the epoxy then allowed easy highlighting of voids, new crack branches and sections of the original crack that were not healed. It can be seen in Figure 11 that at the crack mouth there were sections not healed, but also the clear opening of new crack branches in the case of the specimens healed with the CUT precursor, which were absent from the specimens healed with the flexible SLV precursor.

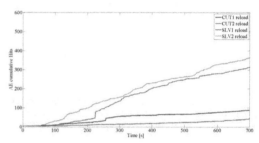

Figure 12. The brittle fracture generates a high number of hits at the same time (vertical discontinuity in the cumulative hits curve).

### 4.3 Acoustic emission

The data acquired by AE during loading of healed specimens also allowed identifying features that distinguish the acoustic events occurring during failure of both polymers. As shown in Figure 12, there was no correlation between the total number of acoustic events reaching the acoustic sensors and the type of precursor or failure. There are however vertical discontinuities in the cumulative hits curves of the specimens healed with the CUT precursor occurring during brittle failure (Figures 9 and 12 are time-aligned). These discontinuities represent a large number of events happening in a very short time and are an unambiguous feature of the catastrophic failure occurring when the brittle polymer was used to heal a crack. Additional features of this type of failure, which are absent when the flexible polymer was used, are a much higher number of acoustic events with energy above 100 (Figure 13) and, more clearly, the occurrence of hits with a Rise Angle (RA) above 3 (Figure 14).

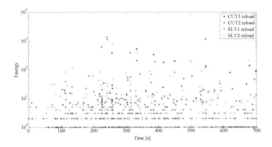

Figure 13. The specimens healed with the rigid polymer generate considerably more high energy hits (100–1000), associated with large crack propagation.

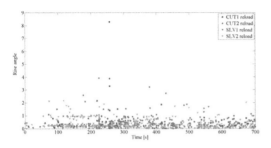

Figure 14. Only the specimens healed with the rigid polymer generate hits with high rise angle (ratio between rise time and amplitude of the first part of the waveform).

From this data, it was not clear whether the high RA and energy hits result from failure of the brittle polymer or the additional crack branches occurring during catastrophic failure, as hits with similar characteristics were also registered during cracking of sound specimens (where only cracking of the mortar matrix occurs).

## 5 CONCLUSIONS

Different NDT techniques were applied on small size specimens and it is shown that cracking as well as healing can be clearly monitored. With acoustic emission, concrete and capsule rupture and failure of the healing agent can be detected and localized. Real-time monitoring of the evolution of US waves allows to follow continuously the cracking and healing process. With a view to monitoring of the healing efficiency in situ, US measurements seem to be promising.

At lab scale, the influence of the flexibility of the polymeric healing agents on the crack re-opening during reloading was investigated using acoustic emission and digital image correlation techniques. The failure mode, which is brittle in case of the rigid polymers and progressive in case of flexible polymers, could be visualized.

## ACKNOWLEDGEMENTS

The research leading to these results has received funding from the European Union Seventh Framework Programme (FP7/2007–2013) under grant agreement n° 309451 (HEALCON).

## REFERENCES

Grosse, C.U. & Ohtsu, M. 2008. *Acoustic Emission Testing in Engineering—Basics and Applications.* Heidelberg: Springer publ.

Grosse, C.U., Van Tittelboom, K. & De Belie, N. 2013. Non-destructive testing techniques for the observation of healing effects in cementitious materials—an introduction. In N. De Belie, S. van der Zwaag, E. Gruyaert, K. Van Tittelboom & B. Debbaut (eds.), *ICSHM 2013, Ghent, 16–20 June 2011.*

Malm, F. & Grosse, C.U. 2014. Examination of Reinforced Concrete Beams with Self-Healing Properties by Acoustic Emission Analysis. *EWGAE, Dresden, 3–5 September 2014,* http://www.ewgae2014.com/Proceedings.

Neild, S., Williams, M. & McFadden, P. 2003. Nonlinear Vibration Characteristics of Damaged Concrete Beams. *J. Struct. Eng.* 129(2): 260–268.

Van Tittelboom, K., De Belie, N., Lehmann, F. & Grosse, C.U. 2012. Acoustic emission analysis for the quantification of autonomous crack healing in concrete. *CBM* 28(1): 333–341.

Emerging Technologies in Non-Destructive Testing VI – Aggelis et al. (Eds)
© 2016 Taylor & Francis Group, London, ISBN 978-1-138-02884-5

# Assessment of grouted connection in monopile wind turbine foundations using combined non-destructive techniques

A.N. Iliopoulos, D. Van Hemelrijck & D.G. Aggelis
*Department of Mechanics of Materials and Constructions, Vrije Universiteit Brussel, Brussels, Belgium*

C. Devriendt
*Department of Mechanical Engineering, Acoustic and Vibration Research Group, Vrije Universiteit Brussel, Brussels, Belgium*

ABSTRACT: Many large-scale offshore wind farm projects use monopile foundation to obtain a cost effective design. These foundations contain pile-sleeve connections that basically consist of two concentric steel pipes cast together by means of high strength grout. The harsh offshore conditions due to combined action of wind and wave excitations strongly influence the structural integrity as they can lead to potential detachment of the grout. Moreover, at several windfarms, during the construction phase it is noticed that the transition piece (sleeve) is slipping downwards relative to the monopile. This slippage is attributed to a weakening of the adhesion between the grout and the steel causing possible further damage to the grout during extreme operational conditions requiring mitigation to avoid further degradation. In order to eliminate the ambiguities and answer the question as to what extent the damage in the grout is significant for the structural integrity of the monopile, a thorough investigation of grout samples taken from the site is performed. For this purpose, four offshore wind turbines were selected for core sampling. All the samples were subjected to Ultrasonic Pulse Velocity (UPV) Non-Destructive Testing (NDT) and compression testing while a set of samples was also subjected to X-ray analysis. This paper presents the results of the combined use of the X-ray and UPV techniques as well as correlations with compressive strength outputs focusing on the parameters that seem promising for in-situ application.

## 1 INTRODUCTION

The abundance of wind resource potential in swallow and deep-water seas is of worldwide growing interest and continuously forms a high motivation for increasing installations of new offshore wind farms. Until the end of 2014, a total of 2488 wind turbines had been installed and connected to the electricity grid in 74 offshore wind farms in 11 countries across Europe reaching a cumulative total installed capacity of 8 GW (EWEA 2015). Most of these large-scale offshore wind farm projects use monopile foundation to obtain a cost effective design. More precise, 78.8% of the total number of installed substructures are monopiles (EWEA 2015). These foundations contain pile-sleeve connections that basically consist of two concentric steel pipes cast together by means of high strength grout (Fig. 1). The primary loads related to this type of offshore wind turbines are the wind and wave induced vibrations and consequently overturning moments. Thus, the loads can fundamentally be transferred as a force couple in the top and the bottom of the grouted region.

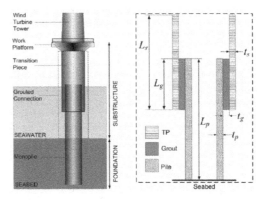

Figure 1. Lower part of the tower-foundation system of a wind turbine on a monopile foundation (left) and detailed view of the MP/TP interface.

Due to this alternating moment loading, there is a potential wear of the sliding surfaces and hence a reduction in the axial load capacity. Moreover, at several windfarms, during the construction phase it is noticed that the transition piece (sleeve) is slipping downwards relative to the monopile. The ovalisation of the transition piece (TP)/monopile (MP) is considered as major cause according to various Root Cause Analyses undertaken amongst others by DNV et al (DNV 2010). This ovalisation is either caused by the upscaling which reduces the stiffness of the members increasing their susceptibility to load-induced deformation, or by the unintentional tapering of the grout annulus due to the misalignment of TP and MP.

As the adhesion between the grout and the steel is weakened, this can cause possible further damage to the grout during extreme operational conditions requiring mitigation to avoid further degradation. In order to eliminate the ambiguities and assess the strength of the grout and thus its capability of withstanding the loads imposed on it, a thorough investigation of grout samples taken from the site is performed. For this purpose, four offshore wind turbines were selected for core sampling. All the samples were subjected to Ultrasonic Pulse Velocity (UPV) Non-Destructive Testing (NDT) and compression testing while a set of samples was also subjected to X-ray analysis. These techniques are well known for assessment of materials, but to the authors' knowledge it is the first time that they are applied for assessment of the grouted connection in offshore wind turbines on monopile foundations.

## 2 CORE SAMPLING AND TESTING PROCEDURE

A core sampling campaign was initiated as this is the only practical way of assessing in-situ grout. On each of the four selected turbines A, B, C and D which correspond to different locations at the wind farm, six cores were drilled along the circumference. A water cooled diamond drill was utilized to drill cylindrical cores starting from the top of the grout annulus with a depth ranging from 350 to 750 mm. The cores were subsequently cut into parts of size 40 × 80 mm (diameter × length) and were analyzed by means of tests which are herein discussed.

### 2.1 Ultrasonic Pulse Velocity test

The UPV technique has been used for several years for the inspection of elements and structures (Anderson & Seals 1981, Qasrawi 2000, Popovics 2001, Kaplan 1960). It allows the user to gain insight on the mechanical properties and damage condition in a non-destructive essence. The test instrument consists of a means of producing and introducing a wave pulse into the mortar cylinder (pulse generator and transmitter) and a means of sensing the arrival of the pulse (receiver). The transmitter is a resonant piezoelectric acoustic emission transducer which transforms the electric pulse given by the pulse generator into a mechanical wave of a frequency of approximately 150 kHz. On the other side of the measured specimen there is a similar transducer which acts as a receiver of the mechanical propagating wave and inversely transforms it into electrical signal. This signal is pre-amplified by 40 dB and is stored in the acquisition board which has a sampling rate of 10 MHz. When measuring the time taken by the pulse to travel through the mortar cylinder, a time delay due to the electromechanical conversion in the transducers is introduced. In order to avoid this time delay and obtain a reference signal, a direct line between the pulse generator, the transducers and a different channel of the acquisition board is used. Vaseline is applied between the sensors and the specimen for acoustic coupling. The cylinders were inspected in three positions along the axis and the average value of these measurements was used for the post-processing of the results. In Figure 2 the working principle of the UPV technique is summarized and accompanied with the realistic representation of the experimental setup.

The basic idea on which pulse velocity method is established, is that the velocity of a pulse of compressional waves through a rod depends on the elastic properties and density of the medium, as shown in the following equation:

$$V_{longitudinal} = \sqrt{\frac{E}{\rho}} \quad (1)$$

Pulse velocity $V$ is measured by the time delay $\Delta t$ between the reference signal and the received through mortar, see Figure 3.

$$V_{longitudinal} = \frac{X}{\Delta t} \quad (2)$$

where X is the exact length of each specimen (in this case approximately 80 mm).

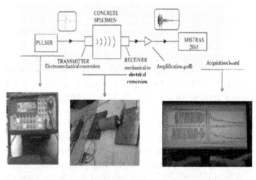

Figure 2. UPV working principle and realistic representation of the experimental setup.

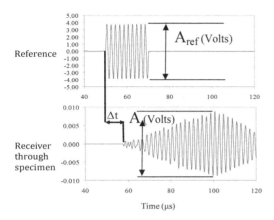

Figure 3a. Time delay between reference and received signals.

Figure 4. Process of compression until fracture of a grouted cylinder.

Figure 3b. X-ray analysis of a sample. From left to right: the sample; only the largest cavities; all cavities color codes with respect to the size.

The onset of the signals is typically considered as the first visible disturbance and is measured by a "threshold" crossing, where the threshold is equal to the noise level of the system. Due to the sampling rate of 10 MHz (0.1 µs) and the propagation time of approximately 17 µs, the typical error is approximately 0.6%.

Apart from the delay of the received signal that is used for pulse velocity calculations, the amplitude A of the wave is also indicative of the material's quality. If there are a lot of cavities and pores, the wave is scattered to other directions than the original and does not reach the receiver. Thus, the amplitude of the received waveform is lower.

Amplitude can be used directly or the attenuation coefficient α, can be calculated, based on the distance travelled and the amplitudes of the received signal $A$ and the reference signal $A_{ref}$.

$$a = -\frac{1}{X} 20 \log\left(\frac{A}{A_{ref}}\right) \left[\frac{dB}{mm}\right] \quad (3)$$

It has been shown that attenuation is a parameter even more sensitive to damage than pulse velocity (Punurai et al. 2006, Selleck et al. 1998).

## 2.2 X-Ray analysis

X-rays have been used in conjunction with Acoustic Emission (Suzuki et al. 2010) and with UPV technique for the characterization of the concrete porosity (Kocur et al. 2010). The X-ray analysis allows the user to make any arbitrary cross-section, which can be used to evaluate cavity size, shape, patterns, interconnectivity and distribution. The images below (fig. 3) show an indicative result of the X-ray investigation. The total content of cavities, hereafter called "porosity" can be calculated.

## 2.3 Compression test

The most common performance measure of concrete is its compressive strength. The compressive strength is measured by breaking cylindrical concrete specimens in a compression-testing machine. The compressive strength is calculated from the failure load divided by the cross sectional area resisting the load and is reported in MPa. It is used to check if the concrete mixture as delivered meets the requirements of the specified strength in the job specification. The crushing strength of the concrete is strongly dependent on the size, shape and method of fabrication. The samples examined within the scope of this paper had a diameter of approximately 40 mm and a length of 80 mm. Figure 4 shows the two snapshots of the process of compression until fracture of the grouted cylinders.

## 3 RESULTS

The set of samples was subjected to UPV, X-ray analysis and compressive test. The results are shown and discussed hereafter.

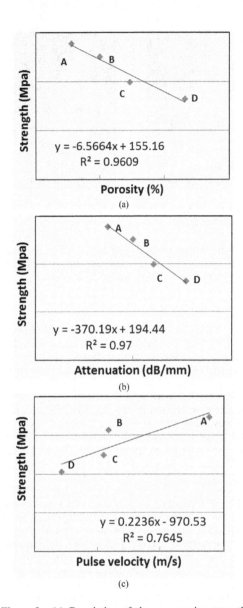

Figure 5. (a) Correlation of the compressive strength with the void content identified from X-rays, (b) correlation of the compressive strength with the attenuation calculated from the UPV technique and (c) correlation of the compressive strength with the pulse velocity calculated from UPV technique.

In Figure 5 the absolute values of all measured parameters cannot be disclosed due to industrial requirements, so the discussion focuses on the trends and correlations. Fig. 5a shows the correlation between the compressive strength and the porosity as calculated by the X-ray method. Each dot is the average of all specimens from the same wind turbine. According to Figure 5, porosity as calculated by X-rays is very well correlated to the strength. The lowest porosity is accompanied by the highest strength (A specimens) and the highest porosity (D specimens) is observed for the less strong specimens. Furthermore, the attenuation of ultrasound matches very well the strength. Specimens with the highest strength (A specimens) exhibit the lowest attenuation whereas specimens with the lowest strength (D specimens) exhibit the highest attenuation. Finally, the pulse velocity is also in good correlation with the strength in the sense that an increase in strength between the successive groups, is escorted by an increase of UPV as well. The observed correlations between the different NDT measurements and the strength are based on the fact that voids decrease the effective load bearing cross-section resulting in lower strength; however, at the same time they increase the scattering attenuation, indirectly allowing an estimation of strength based on ultrasonic parameters. This remark opens the path for indirect strength measurements of grouted connections of offshore wind turbines from external in-situ measurements of the pulse velocity and attenuation via non-destructive UPV testing.

4 CONCLUSION

This paper presents the results of the combined use of the X-ray and UPV techniques as well as correlations with compressive strength outputs in grout samples from the TP/MP interface of offshore wind turbines. The cementitious connection between the components of the monopiles is a very sensitive part which accumulates damage due to dynamic bending loading and harsh environment. It is shown that there is a clear correlation between the void content (porosity) as measured by X-ray tomography and the compressive strength of the grout cylinders. More precisely, the higher the void content the lower the strength and vice versa. In addition, it is noticed that there is a clear correlation between the ultrasonic attenuation and the compressive strength governed by the void content. These findings are promising for in-situ application for the assessment of grouted connections in monopile wind turbine foundations and could also be applied in regular intervals for the evaluation of the deterioration of structural condition.

ACKNOWLEDGEMENTS

This research has been performed in the framework of the Offshore Wind Infrastructure Project (http://www.owi-lab.be) and the IWT SBO Project, OptiWind. The authors therefore acknowledge the financial support from the Agency for innovation by Science and Technology (IWT).

## REFERENCES

Anderson, D.A. & Seals, R.K. 1981. Pulse Velocity as a Predictor of 28- and 90-Day Strength. *ACI JOURNAL* 78(2): 116–122.

Det Norske Veritas (DNV) 2010. Summary report from the JIP on the capacity of grouted connections in offshore wind turbine structures. Report n° 210–1053.

European Wind energy Association (EWEA) 2015. The European offshore wind industry-key trends and statistics 2014. *Report by the European Wind Energy Association.*

Kaplan, M.F. 1960. The Relation between Ultrasonic Pulse Velocity and the Compressive Strength of Concretes Having the Same Workability but Different Mix Proportions. *Magazine of Concrete Research* 12(34): 3–8.

Kocur, G.K., Saenger, E.H. & Vogel, T. 2010. Elastic wave propagation in a segmented X-ray computed tomography model of a concrete specimen. *Construction and Building Materials* 24: 2393–2400.

Popovics, S. 2001. Analysis of the Concrete Strength versus Ultrasonic Pulse Velocity Relationship. *Materials Evaluation* 59(2): 123–130.

Punurai, W., Jarzynski, J., Qu, J., Kurtis K.E. & Jacobs, L.J. 2006. *NDT&E International* 39: 514–524.

Qasrawi, H.Y. 2000. Concrete Strength by Combined Nondestructive Methods Simply and Reliably Predicted. *Cement and Concrete Research* 30:739–746.

Selleck, S.F., Landis, E.N., Peterson, M.L., Shah, S.P. & Achenbach, J.D. 1998. Ultrasonic Investigation of Concrete with Distributed Damage. *ACI Materials Journal* 95(1): 27–36.

Suzuki, T., Ogata, H., Takada, R., Aoki, M. & Ohtsu, M. 2010. Use of acoustic emission and X-ray computed tomography for damage evaluation of freeze-thawed concrete. *Construction and Building Materials* 24: 2347–2352.

*Emerging Technologies in Non-Destructive Testing VI – Aggelis et al. (Eds)*
© *2016 Taylor & Francis Group, London, ISBN 978-1-138-02884-5*

# Detection of damaged tool in injection molding process with acoustic emission

### T. Kek
*Faculty of Mechanical Engineering, University of Ljubljana, Ljubljana, Slovenia*

### D. Kusić
*Slovenian Tool and Die Development Centre, TECOS., Celje, Slovenia*

### J. Grum
*Faculty of Mechanical Engineering, University of Ljubljana, Ljubljana, Slovenia*

ABSTRACT: This paper presents the experimental results regarding acoustic emission signals measured during the injection moulding of those standard test specimens commonly used for examining the shrinkage behaviour of various thermoplastic materials. In daily industrial production of different plastic products we often have to deal with various errors that practically occur on the mold primarily as a result of tool wear and tear, improper storage and improper settings on the injection molding machine. In the testing phase of plastic materials we use many times different inserts that are made from standard tool steels, such as OCR12VM. In case of tool steel inserts after some years of usage a few micro-cracks can occur in the early stage, which can later quickly spread according to the applied loading. With the help of different non-destructive testing methods we know that we can most certainly detect possible formation of cracks on the tool steel inserts. The acoustic emission was measured on an injection mould with the visible sign of a crack on the cavity's surface, using two contact PZT sensors under normal and increased injection pressure loads. In this paper, we focused exclusively on the acoustic emission signal acquisition by using two resonant 150 kHz piezoelectric AE sensors on such tool steel inserts that are already affected by macro-cracks. On such tool steel insert the obtained acoustic emission results were compared with those obtained from a brand new tool steel insert. The final obtained acoustic emission results on the crack defected tool steel insert revealed as expected that the energy and intensity of the captured AE signals is higher compared with the ones that were captured on the brand new engraving insert under same processing conditions.

## 1 INTRODUCTION

Injection molding is a well-known plastic manufacturing process where heated molten plastic material is forced into a mold cavity under high pressure. The plastic material solidifies into a shape that has conformed to the contour of the mold. Nowadays it is still regarded as the most important and very popular manufacturing process because of simple operation steps. A typical production cycle begins when the mold closes, followed by the injection of the plastic into the mold cavity. Once the cavity is filled, additional pressure compensates the material shrinkage. In the next step, the screw turns, feeding the next shot to the front screw tip. This causes the screw to retract as the next cycle is almost prepared. Once the molded part is sufficiently cooled, the mold opens and the part is finally ejected. An example of a typical

injection molding machine is shown in Figure 1. Like in all today's industrial manufacturing processes, injection molding process can produce plastic parts which have poor quality and therefore are characterized as bad ones. In the field of injection molding during long term production first signs of micro-cracks can occur on the mold that usually causes production of bad parts. In such cases if a proper inspection inside the company is not provided then a whole series of freshly produced parts can be rejected. For a production company this can lead to a heavy economical loss. Location and advancement of a possible crack on the tool steel insert can be detected by the use of acoustic emission technique as already reported by many researchers (Cao et al. 2012, Biancolini et al. 2006; Mukhopadhyay et al. 2012, Berkovits & Fang 1995. The major AE signal sources detected by NDT are crack growth and plastic deformation of the steel

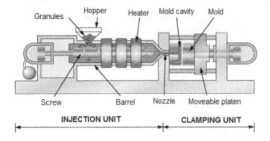

Figure 1. Example of a typical injection molding machine.

(Miller et al. 2005). Acoustic emission (AE) as transient elastic waves has drawn a great attention because of its applicability to on-line surveillance and capability to acquire data with high sensitivity (Vallen 2006). The primary objective of monitoring the crack advancement with acoustic emission method is to obtain useful information about the quality of tool steel inserts which guarantees good quality of the produced test specimens.

## 2 EXPERIMENTAL PROCEDURE

Acoustic emission signals were captured during the production cycle of standard test specimens that are intended for shrinkage evaluation of various plastic materials. The main aim was to analyze the influence of a possible crack located in tool steel insert on the captured acoustic emission signals. After finishing the first experimental part on a tool steel insert with a macro-crack we repeated the experiment under the same processing conditions also on the brand new tool steel insert. In this way the captured acoustic emission signals could be compared. The captured acoustic emission signals were then correlated with the quality of the produced test specimens. Acoustic emission measurement system AMSY-5 from Vallen-Systeme GmbH was used for capturing and analyzing the AE signals. Two piezoelectric AE sensors VS150-M (resonant at 150 kHz) were mounted with silicone grease with two sensor holders on the tool steel insert from both sides as shown on Figure 2. Both PZT sensors were connected via two preamplifiers AEP4 with a fixed gain of 40 dB on the first and second channel of AMSY-5 measurement system. In the evaluation stage of acquired AE signals we focused closely on their maximal amplitudes and energy values during filling and packing stage.

If we want to produce standard test specimens with good quality then the process parameters must be correctly set. Before the start of the experiment, it was necessary to select and fix the follow-

Figure 2. Tool steel insert with AE sensors.

ing process parameters: injection pressure was set to different levels between 1100 and 1200 bar in the course of the experiment. Holding pressure was between 300 and 500 bar. The melt temperature in cylinder has been set to 230 °C and 240 °C and the injection speed was set between 40 and 50 mm/s. The main criterion for the quality of produced test specimens was chosen to be the size of the shrinkage in longitudinal and transverse direction of the melt flow.

## 3 EXPERIMENTAL RESULTS

In the first part, we carried out the first experiments on a brand new tool steel insert. Later in the second part of experiments a tool steel insert with visible sign of macro-crack was used. In both cases to provide similar processing conditions, we used the same polypropylene material from Sirmax manufacturer (H40 C2 FNAT), which is mainly used in the automotive industry. After the test specimens were produced, they were scanned with an optical 3D digitizer ATOS II SO after 24 hours. This industrial 3D optical digitizer has two high-resolution CCD cameras (1280 × 1024 pixels) as shown on Figure 3.

Once the test specimens are digitized the measured data can be saved and used later on. Usually we are interested in individual measuring values and sections across the test specimen. Larger deviations and/or dimensional changes compared to the nominal ones are easy to verify and to control. Optical 3D digitizer is based on the principle of capturing

Figure 3. Optical 3D digitizer ATOS II SO with two high-resolution CCD cameras.

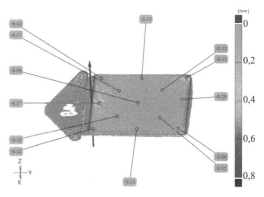

Figure 5. Measuring surface deviations on a test specimen produced on tool steel insert with visible sign of macro-crack.

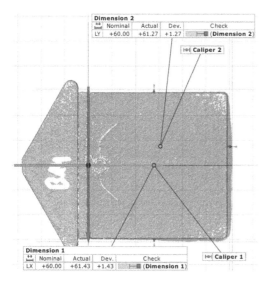

Figure 4. Scanned test specimen produced on a tool steel insert with visible sign of macro-crack.

images through the camera, which then through an appropriate program prepares a computer model. The accuracy that can be achieved with these digitizers depends largely on the quality of the camera, which records the desired object. Of course, in a very small precision scale the accuracy itself also depends on the wavelength of the light. The biggest advantage of 3D digitizer from Figure 3 lies in extremely rapid procedure of digitizing and excellent precision (precision declared for very small objects are up to 2 microns). Example of produced and later scanned test specimens with insert with a visible macro-crack is shown in Figure 4.

As can be seen from Figure 4 the test specimen has a clearly visible macro-crack. Also both dimensions are outside of the nominal 60 mm in the longitudinal (1.27 mm) and transverse (1.43 mm) direction. Test specimens produced with an undamaged tool are within the prescribed tolerances.

A good advantage of scanning the test specimens is the fact that quality of surface and its deviation can be compared to the flat (ideal) surface as shown on Figure 5. We can notice significant surface deviations with regard to the ideal surface, which are within the range of 0.17 mm exactly on the place of macro-cracks.

Figure 6 shows the amplitudes and AE signal energy during the production of test specimens with an intact tool and damaged tool for injection molding. Presence of cracks in the insert can be detected by increase of amplitudes and energies of measured AE signals. On the tool steel insert with macro-crack a significant increase in the number of detected AE signals at the end of filling phase and beginning of holding phase can also be seen. Increase of AE burst rate is caused by increase of mechanical stresses in the injection molding machine. Figure 7 shows a distinctive increase of AE burst rate during filling phase with a use of damaged mold.

The analysis of the AE signal based on the spectrogram has been utilized to evaluate the behavior of injection molding process. A spectrogram is a visual representation of the spectrum of frequencies in an AE signal as they vary with time. Time–frequency analysis is based on the decomposition of one-dimensional signals into two dimensions, time and frequency, represented through the respective axes. The time–frequency diagram allows observation of the distribution of energy of the acoustic signal. Colors of the time frequency

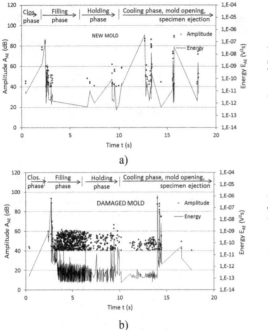

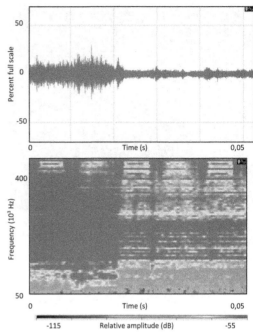

Figure 6. AE signals during production of (a) test specimen 8 (new insert), (b) test specimen 14 (insert with macro-crack).

Figure 8. Time frequency diagram (spectrogram) of AE signal during filling phase with a new mold.

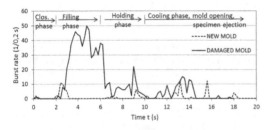

Figure 7. Burst rate during production of a test specimen with a new mold and a test specimen with a damaged mold.

diagram represent the most important acoustic peaks for a given time frame, with red representing the highest energies, then in decreasing order of importance, orange, yellow, green, cyan, blue, and black areas below a threshold decibel level. The result of Fourier analysis of AE signal waveform measured during filling phase at 3 s after the beginning of injection molding of specimen with a new mold is a spectrum or time frequency diagram shown on Fig 8. Fig. 9 shows AE signal waveform measured during filling phase at 6,5 s after the beginning of injection molding of specimen with a damaged mold and spectrogram of this signal.

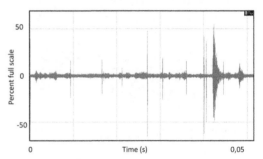

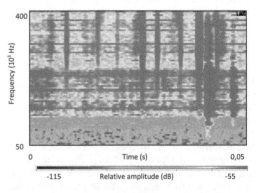

Figure 9. Time frequency diagram (spectrogram) of AE signal during filling phase with a damaged mold.

During the injection molding of specimen with a new mold we measure predominantly signals with longer duration and not pronounced exponential decay of amplitude. These signals can be characterized as process orientated signals. During the injection molding in a tool with a crack we measured signals with an instant increase of amplitude and pronounced exponential decay of the amplitude. The acoustic emission burst rate during injection molding of specimen with a damaged mold is considerably higher and is a consequence of crack growth in steel tool insert. Pronounced frequency in the measured AE signals is 150 kHz that is resonant frequency of the sensor.

## 4 CONCLUSION

The aim of this research work was to determine to which extent is it possible with AE method to detect the presence of cracks on engraving tool steel inserts by conducting a closer comparison of captured AE signals obtained from new tool steel insert under the same processing conditions. From our experimental results we were able to obtain useful information about the presence of macro-cracks during production of standard test specimens. We have found in the filing stage during production of test specimens on both inserts that there was more apparent difference in the number of detected AE signals during the injection phase of commercial polypropylene material in favor of the tool steel insert with visible sign of macro-cracks. The results clearly show also the difference in the maximum amplitudes during filling and holding phase, as well as in energies of captured AE signals. As could be seen from the

practically obtained results from captured AE signals we can successfully use this method for detecting fractures on tool steel inserts, which is one of the important advantages of this nondestructive testing method.

## ACKNOWLEDGEMENT

The research work was partly financed by the European Union, European Social Fund, and by the Slovenian Ministry for Higher Education, Science and Technology.

## REFERENCES

Berkovits, A., Fang, D. 1995. Study of fatigue crack characteristics by acoustic emission, Engineering Fracture Mechanics, Vol. 51, pp. 401–416.

Biancolini, M.E., Brutti, C., Paparo, G., Zanini, A. 2006. Fatigue cracks nucleation on steel, acoustic emission and fractal analysis, International Journal of Fatigue, Vol. 28, pp. 1820–1825.

Cao, J., Luo, H., Han, Z. 2012. Acoustic emission source mechanism analysis and crack length prediction during fatigue crack propagation in 16Mn steel and welds, Procedia Engineering, Vol. 27, pp. 1524–1537.

Miller, R.K., Hill, E.V.K., Moore, P.O. 2005. Nondestructive Testing Handbook, Third edition, Vol. 6, Acoustic Emission Testing, American Society for Nondestructive Testing, Inc..

Mukhopadhyay, C.K., Sasikala, G., Jayakumar, T., Raj, B. 2012. Acoustic emission during fracture toughness tests of SA333 Gr.6 steel, Engineering Fracture Mechanics, Vol. 96, pp. 294–306.

Vallen, H. 2006. Acoustic Emission Testing: Fundamentals, Equipment, Applications, Castell Publication Inc., Wuppertal, Vol. 6.

*Emerging Technologies in Non-Destructive Testing VI – Aggelis et al. (Eds)*
*© 2016 Taylor & Francis Group, London, ISBN 978-1-138-02884-5*

# An experimental investigation of Electromagnetic Acoustic Transducers applied to high temperature plates for potential use in solar thermal industry

M. Kogia, L. Cheng, A. Mohimi, V. Kappatos, T.-H. Gan, W. Balachandran & C. Selcuk
*Brunel Innovation Centre, Brunel University, Cambridge, UK*

ABSTRACT: Solar thermal industry is a promising renewable energy source and an environmentally friendly mean of power generation, which converts sunlight into electric power. Parabolic Trough Concentrated Solar Plants (CSPs) is the most widely used type of CSP and its main components are the parabolic trough shaped reflectors and the absorber tubes. The absorber tubes have a complex structure and operate at high temperatures (400°C-500°C); as a result, they are likely to suffer from creep, thermo-mechanical fatigue, hot corrosion etc. Hence, both their inspection and monitoring is of crucial importance and very challenging tasks as well. Electromagnetic Acoustic Transducers (EMATs) are a promising, non-contact technology that can be used for the inspection of large structures at high temperatures by exciting Guided Waves. In this paper, a study regarding the basic principles of EMATs, their potential use in this application and their performance at high temperature is presented. In addition, a Periodic Permanent Magnet (PPM) EMAT with racetrack coil, exciting Shear Horizontal waves, has been experimentally evaluated both at room and high temperatures. During the experiments these EMATs were tested on a stainless steel plate so that an experimental characterization of their ultrasonic potentials would be accomplished. Additionally, the EMATs were tested regarding the impact of shielding on their performance, their power supply requirements, their lift-off limitations and their performance at high temperatures as well.

## 1 INTRODUCTION

Solar thermal industry is an environmentally friendly mean of power generation, converting solar energy into electricity. Concentrated Solar Plants (CSPs) are employed in solar thermal industry for collecting the sunlight and converting it firstly into heat and later into electricity. There are several types of CSPs such as the solar tower, the dish concentrator and the parabolic trough CSPs; the latter is the most widely used (Poullikkas, 2009) and is shown in Figure 1a.

Parabolic trough CSPs are mainly composed of long, parabolic trough shaped reflectors and absorber tubes. The reflector focuses the sunlight to the absorber tube. The absorber tubes are long, thin, stainless steel tubes covered by a glass envelope under vacuum, as shown in Figure 1b. They are located at the focal point of the reflectors and run the whole length of the reflector. Their entire surface should be exposed to the sunlight and therefore there are few access points for any extra hardware to be attached to them. Inside the tubes, either water/molten salt or synthetic oil is flowing, absorbing the heat of the sunlight and transmitting it into a heat engine for the final stages of the power generation procedure. The high temperatures (400–

500°C), the contraction/expansion endured due to the fast cooling of the tube at high temperatures and the vibrations can result in the appearance of defects either on the surface or inside the absorber tube. Problems with absorber have been reported by Guillot *et al.* (2012) and Herrmann *et al.* (2002), such as creep, thermo-mechanical fatigue and hot corrosion and Non Destructive Testing (NDT) techniques are suggested to be applied for the structural health monitoring or inspection of the absorber tubes.

Several NDT techniques can be used for either the inspection or the monitoring of high temperature structures; Acoustic Emission (AE), Eddy Current (EC), Holographic Interferometry, Laser Ultrasonic, Long Range Ultrasonic Testing (LRUT) and Thermography are some of those. These techniques have been reported for operating at temperatures up to 300°C, with shortcomings though. Some of the drawbacks are qualitative results, sensitivity to noise, cost and need of coupling mediums (Hutchins *et al.,* 1990, Kirk *et al.,* 2005, Malmo *et al.,* 1988, Momona *et al.,* 2010, Shen *et al.,* 2007).

The inspection of absorber tubes requires a non-contact NDT technique that can be applied at high temperatures without the need for couplant

Figure 1. (a) Parabolic Trough Concentrated Plants (Staff, 2013) (b) Absorber Tube (Stewart Taggart, 2008).

and can inspect the whole length of the tube from a single point. Therefore, LRUT can be applied with the use of non-contact transducers that can withstand high temperatures and excite/receive guided waves and more particularly T(0,1) wave mode (or Shear Horizontal) (Cawley et al., 1996, Hirao et al., 1999). The Electromagnetic Acoustic Transducer (EMAT) can be used for this application, since it is a non-contact technology and can be applied for the inspection of structures under hostile conditions such as moving specimen and elevated temperatures.

In the following section the main operating principles of EMATs are described and emphasis is also given on the potentials and the limitations of EMATs at high temperatures. In Sections 3 and 4 the experimental setup and the results of the experimental evaluation of a pair of EMATs are presented. During the experiments, emphasis was given to the EMAT design; if this design is suitable for exciting/receiving Shear Horizontal waves for axial propagation. The parameters that affect EMAT performance at both room and high temperatures were also investigated. Hence, the EMATs were tested at both room and high temperatures and for different lift-offs and power supply levels. A high temperature EMAT exciting shear horizontal waves and operating at 550–600°C for either monitoring or long term inspection is to be accomplished in the near future.

## 2 EMAT TECHNOLOGY

EMAT is a non-contact technology of transducers that can be used for the inspection of moving structures or at high temperatures. Their response can travel through vacuum as well, making them even more suitable for this application compared to the piezoelectric transducers. EMATs can excite/receive Guided Waves and thus they can be employed in GWT for either the inspection or the monitoring of large structures. Hence, high temperature EMATs may be used for the inspection or the monitoring of the absorber tubes.

### 2.1 Brief introduction to EMATs

A typical EMAT transducer is composed of either a permanent magnet or an electromagnet for the generation of a static magnetic field, and a coil. The coil is driven by alternating current which generates a dynamic magnetic field. This dynamic magnetic field induces eddy current in the specimen placed below the coil. When the specimen being tested is an electrical conductor and non-ferromagnetic, only Lorentz force is exerted upon the material particles; Equation 1 shows that Lorentz force is equal to the product of eddy current density and the overall magnetic field.

$$F_L = J_e \times (B_{st} + B_{dyn}) \qquad (1)$$

where $F_L$ is Lorentz force, $J_e$ is the eddy current density, $B_{st}$ stands for the static magnetic field and $B_{dyn}$ refers to dynamic magnetic field.

Both the coil design and the magnet configuration affect the distribution of the Lorentz force in space, its direction and amplitude. As a result, the wave mode the EMAT is to generate and/or receive and its performance at high temperatures get affected by the configuration and the material properties of its two main components. A wave mode with in-plane displacement, propagating all along the absorber tube is suitable for this application; $T(0,1)$ (or alternatively called Shear Horizontal in plates) has this characteristics and it is also non dispersive and therefore simplifies the signal processing interpretation. An EMAT configuration that can be used in this case is the Periodic Permanent Magnet (PPM) EMAT with racetrack coil (Dixon et al., 2001, Jian et al., 2006, Hernandez-Valle, 2011, Idris et al., 1994).

Another important parameter that influences the performance of EMATs is its shielding mechanism. EMATs are sensitive to electromagnetic noise and therefore, an electrostatic shielding and/or a magnetic shielding is required. Shielding concentrates the electromagnetic power generated by the EMAT to a specific area and prevents any interference; therefore, it enhances the performance of EMATs.

### 2.2 EMAT at high temperatures

Temperature affects the performance of EMATs and the material properties of the structure being tested. Both the permanent magnets and the coil have a maximum operating temperature. Material properties such as Young's Modulus, poisson ratio and density of a specimen change with temperature rise, resulting in variations in the velocity of the propagating wave mode. Hence, a careful study regarding all the above should be carried out before the design of a high temperature EMAT.

Young's Modulus and density of the specimen decrease with increase in temperature, while poisson ration increases. These changes result in a decrease in the velocity of the T(0,1). Hence, the temperature should be kept stable during the inspection or temperature compensation should take place.

Moreover, the material the coil is made of also influences the EMAT performance. The resistance of the coil affects the energy transmitted to the specimen and the SNR/quality of the signal received; Equation 2 shows how the resistance increases with temperature rise and Equation 3 demonstrates the relationship between the temperature and the noise level of the signal received.

$$R(T) = R_0[1 + \alpha(T - T_0)] \qquad (2)$$

where $T_0$ is the room temperature and $R_0$ is the resistance of a single turn coil.

$$V_{Noise} = (4K\beta TR_{EMAT})^2 \qquad (3)$$

where $K$ is the Boltzmann constant, $T$ is the temperature measured in Kelvin, $\beta$ is the bandwidth and $R_{EMAT}$ is the resistance of the EMAT coil per turn. Consequently, the coil should preferably be made of a low resistant material, such as copper. However, copper is known to oxidize at high temperatures and, therefore, other materials or alloys could be used for this application such as silver, platinum, nickel or nickel alloys (alumel, chromel, kulgrid); especially the nickel alloys like kulgrid do not get corrode at high temperatures (500°C) and their resistance is almost equal to copper. The thermal insulation of the wire is an another issue; in fact, apart from the heat transferred from the specimen to the EMAT, the coil itself also produces an additional amount of heat that may lead to the coil break down. Hence, ceramic insulated wires, such as Ceramawire, or ceramic encapsulated coils should be employed for temperatures higher than 250°C. Hernandez-Valle (2011) has already made an electromagnet EMAT that can operate up to 600°C without any cooling system for thickness measurements and apart from designing a high temperature electromagnet he has also tested several coil designs at high temperatures.

The permanent magnets also have limitations regarding their maximum operating temperature. The magnetization (M) of a magnet is proportional to the density of the ordered magnetic dipoles inside a magnet; unfortunately, these dipoles change and become disordered at temperatures higher than the Curie Temperature of the magnet and so the strength of the magnet starts decreasing rapidly with respect to temperature. Hence, the Maximum Operating Temperature (MOT) of a magnet is equal to half of its Curie Point. Consequently, high Curie Point magnets are preferable for high temperature applications. Nevertheless, the magnetic strength of the main two types of high temperature magnets, Alnico (with maximum operating temperature of 500°C) and SmCo (with maximum operating temperature of 300°C), is smaller than the magnetic strength of Neo (NdFeB) magnets. However, NdFeB magnets cannot be used at temperatures higher than 200°C and they are also expensive.

A cooling system may be required so that both the magnets and the coil will operate efficiently. Idris et al. (1994) have designed and tested a water cooled EMAT that can obtain signals up to 1000°C for thickness measurements. The EMAT was exposed to the heat source for as much time as it needed for the signal to be recorded and then it was removed. Oil and air cooled EMATs have also been designed. Generally, the reported high temperature EMATs seem to operate at high temperatures for short periods of time, which are suitable for inspection, but they cannot be used in long term condition monitoring. Consequently, an EMAT operating at high temperatures (500°C) for long periods of time is required and could be used for the structural health monitoring of high temperature structures such as the absorber tubes.

## 3 EXPERIMENTAL PROCEDURE

A pair of EMATs manufactured by Sonemat Limited was evaluated at both room and high temperature. Their design is a PPM EMAT with racetrack coil as it was mentioned previously. The pitch of the magnets is 12 mm and equal to the wavelength of the shear horizontal. The wavelength is half the minimum defect size, which is 24 mm (Kogia et al., 2014). Shielding, Lift-off and power supply requirements were some of the parameters that were investigated.

The experimental setup used is shown in Figure 2. The specimen used is a 316Ti stainless steel square plate of 1,25 m and 3 mm thickness. The EMATs tested use a PPM configuration and a racetrack coil, made of lacquered copper wires of 0.315 mm diameter and with maximum operating temperature of 250°C. Ritec RAM 5000 SNAP pulser/receiver was used for driving the EMATs with a 6 cycle, Hanning windowed pulse of 256 kHz frequency. Ritec was also used for amplifying the signal received with a gain of 80dB and filtering it within the bandwidth of 10 kHz and 20 MHz.

During the room temperature experiments both free of defect and defective areas were tested. The

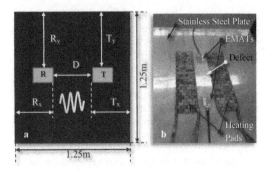

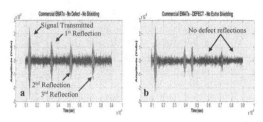

Figure 3. Signal received from (a) defect-free area with no extra shielding (b) defective area with no extra shielding.

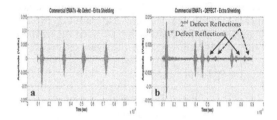

Figure 4. Signal received from (a) defect-free area with extra shielding (b) defective area with extra shielding.

Figure 2. (a) Schematic of the experimental setup in a free of defects area (b) Experimental setup—EMATs—Heating pads-Defect.

distance between the transmitter and the receiver was equal to 30 cm. Five defects were created with different length and mass loss each and located 10 cm away one from the other. The defect tested was 20 mm long and with 66.6% mass loss; the transmitter was 15 cm away from one edge of the defect and the receiver were 15 cm away from the other edge of the defect. The second defect was 10 cm away from the bottom edge of first defect. The effect of extra shielding on the quality of the signal received has been also investigated; in fact, an additional thin, carbon steel cover was also placed all around the transducers for providing extra shielding. The influence of lift-off on EMAT response was investigated from zero to 1 mm lift-off with a step of 0.1 mm. A study regarding the power supply requirements of these EMATs was also accomplished by gradually decreasing the power output of Ritec with a step of 5% starting from its maximum power level (5000 Watts) and stopping at 20% of its maximum power where no useful information could be retrieved anymore from the signal received. For the high temperature experiments the distance between the transmitter and the receiver was equal to 30 cm with the first defect located 15 cm away from the transmitter and 15 cm away from the receiver as well; similarly with the room temperature setup. The temperature was increased from ambient to 180°C with a step of 10°C. During the high temperature experiments the EMATs were continuously in contact with the specimen.

## 4 EXPERIMENTAL RESULTS & DISCUSSION

### 4.1 Room temperature experiments

From the first set of experiments, there are four case studies that should be investigated. Figure 3a shows the signal received from a defect-free area with no extra shielding having been used. In this figure the first reflection is the signal transmitted from the transmitter to the receiver, the second reflection comes from the edge of the plate closer to the receiver (first plate edge reflection), the third reflection bounces at the edge closer to the transmitter and goes back to the receiver (second plate edge reflection) and the fourth reflection is the same wave with the third reflection propagating further and bouncing at the edge close to the receiver as well (third plate edge reflection). In this case the noise level is high and the amplitude of all the reflections is low, leading to a low SNR. Figure 3b illustrates the signal received when the defective area was tested without the shielding. Similarly with above, the signal transmitted and the three reflections from the edges of the plate are clearly obvious in the signal received. However, no reflections from the defect are obvious. Figure 4a demonstrates the signal received when the EMATs test the defect-free area with the use of shielding. In this case, the amplitude of the reflections increased, the noise level decreased and as a result the SNR increased five times more. Figure 4b shows the signal received from the defective area when shielding was also used. Similarly with Figure 4a, the SNR of the signal increased four times more compared to the signal shown in Figure 3b. Actually, in Figure 4b both reflections from the crack are clearly obvious as well as the reflections from the adjacent defect as well. Consequently, shielding proved to be significantly beneficial as far as quality of the signal received is concerned and improves the probability of defect detection. Nevertheless, both attenuation of the signal and the Time of Flight of the main four reflections from the edges of the plate remain features that enables us to distinguish the defective from the defect-free areas when no extra shielding is used. Although the velocity of shear horizontal was not to change during the experiments, for no temperature rise was occurred, the Time of Flight of both the first and the second reflection from the edges of the plate change from one case study to the other. It is likely that the defect causes this time delay by trapping portion of the energy/spectrum of the wave propagating inside the plate.

Figure 5 illustrates how the magnitude of the first reflection changes (%) with respect to the lift-off. According to literature EMATs are sensitive to lift-off (Jian et al., 2006) and therefore their efficiency decreases with the increase of lift-off. The magnitude of the first reflection decreases almost linearly with the lift-off increase; when the lift-off is equal to 1 mm only the first reflection can be clearly observed. This confirms the high sensitivity of EMATs to lift-off. A parameter that influences the performance of EMAT regarding lift-off is the impedance of the coil. The impedance changes with lift-off as well as with the material properties of the specimen.

An experimental evaluation of these EMATs regarding their power requirements was also conducted. The power level decreased gradually from 100% to 20% with a step of 5%; as no useful information could be retrieved from the signal received when the power level was smaller than 20%. Figure 6 shows how the magnitude of the first reflection increases with power supply increase; it can be observed that the magnitude increases almost linearly with power increase. Similarly with lift-off, the impedance of the coil affects the power requirements of EMATs. If the pulser/receiver unit drives the EMAT transmitter through an output resistor, the magnitude of the impedance of the coil should be equal to this output resistor of the pulser unit, so that the voltage drop in the coil will be minimized. Therefore, impedance matching is

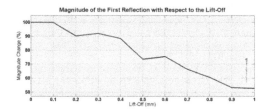

Figure 5. Magnitude change of the first reflection with lift off.

Figure 6. Magnitude change of the first reflection with power supply increase.

always required, which means that either a unit with zero output impedance should be chosen or an impedance matching network should be added between the pulser and the EMAT transmitter. Impedance matching should be used for the coil to be driven with the maximum power possible so that strong signals will be obtained. Nevertheless, the power supply level of EMATs remains high, leading to the conclusion that EMATs are considered to be more efficient as receivers rather than transmitters.

4.2 *High temperature experiments*

The EMATs were tested from ambient to 180°C with a step of 10°C; the maximum operating temperature of these EMATs is 250°C and therefore they were tested up to 180°C only, so that any serious and irreversible damage will be avoided. The EMATs were continuously exposed to the heat source with zero lift-off during the rise in temperature. Figure 7a-d show the signal received at room temperature, 60°C, 100°C and 180°C respectively. Firstly, in the signal obtained at room temperature both the reflections from the plate edges and the reflectiosn from the first defect are clear. However, it is obvious that the amplitude of the second reflection from the plate diminishes greatly after 60°C, while at 100°C and 180°C it can hardly be noticed. Similarly, the amplitude of the first reflection from the defect and the forth reflection from the plate decreases with the increase in temperature. Figure 8 shows how the amplitude of the signal transmitted decreases with temperature rise; it is clear that the amplitude dwindles almost linearly with rise in temperature. Time shifting is also observed, as it was expected due to the change in the wave velocity at high temperatures. The third reflection from the plate shifts in time and starts coming closer to the second reflection from the crack leading to the increase of the magnitude of the latter. Consequently, temperature compensation should take place and both the Time of Flight of the reflections as well as the drop of their amplitude may be two features that can be further

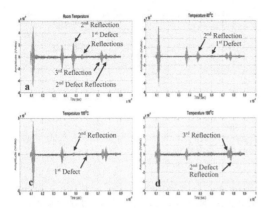

Figure 7. Signal received (a) at room temperature (b) at 60°C c) at 100°C d) at 180°C.

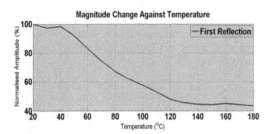

Figure 8. Amplitude of the signal transmitted with temperature.

processed and used for the indentification of the temperature of the structure being tested.

## 5 CONCLUSION & FUTURE WORK

Absorber tube is an essential part of Parabolic Trough CSPs and is very likely to get damaged due to its hostile operating conditions. Hence, NDT techniques are required for their monitoring and/or inspection. A promising technique for this application is the use of high temperature EMAT transducers for the excitation of guided waves and more particularly of T(0,1) wave mode. A detailed study concerning EMAT operation and more especially their performance at high temperatures as well as an experimental evaluation of a pair of EMATs were accomplished. It was proved that a PPM racetrack EMAT design has the potentials to be successfully employed in this application.

The design and manufacturing of a high temperature PPM EMAT with racetrack coil operating efficiently at temperatures higher than 300°C for a long term inspection or monitoring is our next step. Several thermal, electromagnetic and mechanical simulations are already carried out and have given encouraging results. Moreover, a further experimental investigation regarding the impedance of the EMAT and its relationship with the lift-off the material properties of the specimen being tested, is also another part of our future research.

## ACKNOWLEDGMENT

The authors are indebted to the European Commission for the provision of funding through the INTERSOLAR FP7 project. The INTERSOLAR project is coordinated and managed by Computerised Information Technology Limited and is funded by the European Commission through the FP7 Research for the benefit of SMEs programme under Grant Agreement Number: GA-SME-2013–1–605028. The INTERSOLAR project is a collaborative research project between the following organisations: Computerised Information Technology Limited, PSP S.A., Technology Assistance BCNA 2010 S.L., Applied Inspection Limited, INGETEAM Service S.A., Brunel University, Universidad De Castilla—La Mancha (UCLM) and ENGITEC Limited.

## REFERENCES

Cawley, P. & Alleyne, D. 1996. The use of Lamb waves for the long range inspection of large structures. Ultrasonics 34 (1996): 287–290.

Dixon, S. Edwards, C. & Palmer, S.B. 2001. High-accuracy non-contact ultrasonic thickness gauging of aluminium sheet using electromagnetic acoustic transducer. Ultrasonics 39: 445–453.

Guillot, S. Faika, A. Rakhmatullina, A. Lambert, V. Verona, E. Echegut, P. Bessadaa, C. Calvet, N. & Py, X. 2012. Corrosion effects between molten salts and thermal storage material for concentrated solar power plants. Applied Energy 94(2012): 174–181.

Herrmann, U. & Kearney, D.W. 2002. Survey of Thermal Energy Storage for Parabolic Trough Power Plants. Journal of Solar Energy Engineering 124(2): 145–152.

Hernandez, V.F. 2011. Thesis: Pulsed-electromagnet EMAT for high temperature applications. Warwick: University of Warwick. Department of Physics.

Hirao, M. & Ogi, H. 1999. A SH-wave EMAT technique for gas pipeline inspection. NDT&E International 32(1999): 127–132.

Hutchins, D.A. Saleh, C. Moles, M. & Farahbahkhsh, B. 1990. Ultrasonic NDE Using a Concentric Laser/EMAT System. Journal of Nondestructive Evaluation 9(4).

Idris, A., Edwards, C. & Palmer, S.B. 1994. Acoustic wave measurement at elevated temperature using a pulsed laser generator and an Electromagnetic Acoustic

Transducer detector. Nondestructive Testing and Evaluation 11: 195–213.

Jian, X. Dixon, S. Edwards, R.S. & Morrison, J. 2006. Coupling mechanism of an EMAT. Ultrasonics 44(2006): 653–656.

Kirk, K.J. Lee, C.K. & Cochran, S. 2005. Ultrasonic thin film transducers for high-temperature NDT. Insight 47(2).

Kogia, M. Mohimi, A. Liang, C. Kappatos, V. Selcuk, C. Gan, T.H. 2014. High temperature Electromagnetic Acoustic Transducer for the inspections of jointed solar thermal tubes, The First Young Professionals International Conference. Budapest Hungary.

Malmo, J.T. Jøkberg, O.J. & Slettemoen, G.A. 1998. Interferometric testing at very high temperatures by TV holography (ESPI). Experimental Mechanics 28(3): 315–321.

Momona, S. Moevus, M. Godina, N. R'Mili, M. Reynauda, P. Fantozzi & Fayolle, G. 2010. Acoustic emission and lifetime prediction during static fatigue tests onceramic-matrix-composite at high temperature under air. Composites: Part A 41 (2010): 913–918.

Poullikkas, A. 2009. Economic analysis of power generation from parabolic trough solar thermal plants for the Mediterranean region—A case study for the island of Cyprus. Renewable and Sustainable Energy Reviews 13 (2009): 2474–2484.

Shen, G. & Li, T. 2007. Infrared thermography for high temperature pipe. Insight 49(3).

Staff C.W. 2013. Industry experts trumpet untold solar potential. http://www.constructionweekonline.com/article-23701-industry-experts-trumpet-untold-solar-potential/. [Accessed 12 March 2015]

Taggart, S. 2008. Parabolic troughs: Concentrating Solar Power (CSP)'s quiet achiever. Available at: http://www.renewableenergyfocus.com/view/3390/parabolic-troughs-concentrating-solar-power-csp-s-quiet-achiever, [Accessed 12 March 2015]

*Emerging Technologies in Non-Destructive Testing VI – Aggelis et al. (Eds)*
*© 2016 Taylor & Francis Group, London, ISBN 978-1-138-02884-5*

# Monitoring of self-healing activation by means of Acoustic Emission and Digital Image Correlation

P. Minnebo, D.G. Aggelis & D. Van Hemelrijck
*Mechanics of Materials and Construction (MEMC), Vrije Universiteit Brussel, Brussels, Belgium*

ABSTRACT: In self-healing of concrete it is necessary to know if the self-healing is activated. With Acoustic Emission (AE), a Non-Destructive Technique (NDT), this can be observed. AE measurements need to be described and compared to realize the possibility of using AE as a useful technique. A comparative study between mortar beams with glass capsules, glass-IPC-coated capsules and gelatin-IPC-coated capsules was conducted. By means of three-point bending, the beams were loaded until failure by one predominant crack going through the self-healing capsule, to confirm this Digital Image Correlation (DIC) was used to measure the strain-field. This document summarizes results of this comparative study, including the experimental setup, the used sensors and the AE-signal output analysis. It was observed that AE in combination with DIC serves as good methods to detect the activation of self-healing and that AE therefore is a promising and useful NDT in the field of self-healing concrete.

## 1 INTRODUCTION

The success of self-healing concrete is respectively measured as the activation of the self-healing and the self-healing efficiency by means of the recovery of mechanical properties (Van Tittelboom et al., 2011), (Granger, Loukili, Pijaudier-Cabot, & Chanvillard, 2007). This paper focuses on the former. Self-healing mechanisms are usually embedded in concrete and the activation of the mechanism, i.e. breaking of embedded capsules or embedded tubes, is not always detectable by visual inspection. Using non-destructive techniques (NDT's) the activation of self-healing mechanisms can be detected. In this paper several new approaches in self-healing concrete will be investigated on their breaking behavior, by means of AE and (Digital Image Correlation) DIC, as was also used by other researchers (Granger et al., 2007; Tsangouri, 2015), (Tang, Kardani, & Cui, 2015). Inorganic Phosphate Cement (IPC) will be used in combination with glass and gelatin.

### 1.1 Self-healing in concrete

Research on self-healing concrete is increasing in the last decades. Different approaches are investigated, such as self-healing using bacteria, capsules, vascular system and Super Absorbent Polymers (SAP's) (Jonkers, Thijssen, Muyzer, Copuroglu, & Schlangen, 2010), (Van Tittelboom & De Belie, 2013; Wu, Johannesson, & Geiker, 2012). This paper focusses on capsules and a vascular system.

A distinction between autogenic and autonomous self-healing needs to be made. Autogenous self-healing is the ability of the material to heal itself without embedding or adding extra materials. It mainly consists out of promoting further hydration of unhydrated cement particles or the formation of $CaCO_3$ or sealing a crack by fine particles whom get stuck inside the crack. Autonomous self-healing is obtained by embedding an extra system (bacteria, capsules, vascular) or extra materials in the concrete. In the case of capsules and vascular self-healing a healing agent is released inside the cracks.

### 1.2 Non-destructive techniques

Two NDT techniques will be used. The first is Acoustic Emission (AE) which is based on elastic waves emitted by plastic deformations within a material. With this method it is possible to locate the crack and to differentiate the material(s) that are cracking by means of signal post-processing as was done by several researchers in engineering fields (Granger et al., 2007; Tsangouri, 2015). The post—processing was done using dedicated software from AE Mistras Holding. The second NDT is Digital Image Correlation (DIC), this is an optical technique where a sequence of photos is taken on regular time interval. Post-processing by dedicated software of correlated solutions measures the deformations on the surface of the material, and enables the location of cracks. If the position of the self-healing system is known, DIC can confirm if the crack path will go around, away or through the self-healing system.

## 2 MATERIALS AND METHODS

Three groups of mortar specimen were prepared. Mortar was chosen over concrete for the ease of preparation and the small dimensions of the experimental specimen (40 mm×40 mm×160 mm). The differences between the three groups of specimen concern the capsule material. Specifically three types were used glass, glass-IPC, and gelatin-IPC.

### 2.1 Preparation of the self-healing mechanism

The self-healing mechanisms investigated in this paper are capsules (i.e. length-radius ratio <10) and tubes (i.e. length-radius ratio >10).

For the first case, gelatin capsules are coated with IPC. As can be seen on Figure 1. both a cylindrical shape and granulate shape are made. First the gelatin capsules were filled with a liquid, then they are put in a mold which is then filled with liquid IPC. The filled mold is then put in the oven for 2 hours at 60°C. The average dimensions of the capsule-shaped gelatin-IPC capsules are 28,9 mm length with a radius of 0,98 cm. Note that 100% of these capsules and granulates survived the concrete mixing process.

The tubes are made out of glass or a combination of glass with IPC coating. Borosilicate glass is used.

### 2.2 Preparation of the mortar specimen

For each group of specimen a minimum of three specimens were made together with three reference beams. The mortar's composition is water/cement/sand respectively 0,4/1/2 using CEM 52,5 N. The size of the specimen is 40 mm×40 mm×160 mm. It is important that the self-healing mechanisms are placed in the tensile zone. In order to do so, the mortar was poured into the mold in two phases,

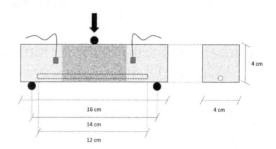

Figure 2. Three-point bending setup with two AE sensors and a speckle pattern on the front of the mortar beam for the DIC, inside the beam the self-healing system is embedded, in this case a capsule.

first the mold was filled up till one third of the height of the mold, then the self-healing system is manually placed, after which the rest of the mold was carefully filled up.

The mortar was unmolded after 24 h and then put into a water batch for the remaining 27 days in order to ensure optimal hydration conditions. 28 days after casting, the beams were tested by means of three-point-bending.

## 3 EXPERIMENTAL METHODS

Three-point bending test were performed on mortar samples using an Instron 5885 H. The displacement rate was set at 0,2 mm/min. The supports were at a distance of 14 cm as can be seen on Figure 2.

During the experiment the strains on the face of the specimen were measured by DIC using a unique speckle pattern applied on the front surface (Lecompte, Bossuyt, Cooreman, Sol, & Vantomme, 2007). This allowed for detailed monitoring of crack propagation. The release of elastic energy during the experiment was measured by means of AE.

Positioning of the measuring equipment can be seen on Figure 2. Two AE pico PZT transducers were positioned on the back side of the beam, each 4 cm from the middle. The DIC system was applied on the side-face of the beam. A two camera system was used together with the acquisition and post processing software of Correlated Solutions.

## 4 RESULTS AND DISCUSSION

### 4.1 Group I: self-healing capsules

The activation of the self-healing in this paper is defined as breaking of the embedded self-healing system, i.e. capsules or tubes.

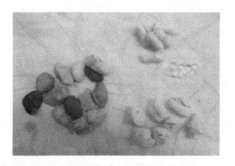

Figure 1. Gelatin capsules coated with IPC, both in a cylindrical shape and granulate shape. One can see the gelatin capsules (middle on the right side) and the original granulates (left).

A total of 21 experiments were conducted. On 100% of the specimens, debonding between the capsules and concrete was observed (Figure 5). Comparing the signal of a reference mortar beam and a mortar beam with a capsule, the latter delivers a stronger signal (Figure 3).

With DIC it was observed that the crack path went towards the position of the capsules. This is visually confirmed on the samples, Figure 4 and Figure 5.

By means of AE it was confirmed that the crack appeared towards the center of the mortar beam. This can be seen on Figure 6 and is confirmed by the DIC images. On the DIC figures it can be seen that a first big strainfield arises at a different loca-

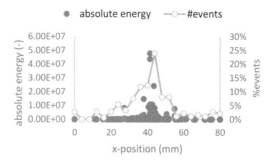

Figure 6. Light: absolute energy versus the x-position of the signal, dark: the relative amount of signals originated from one position.

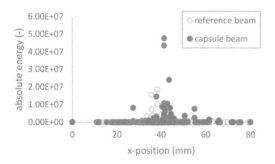

Figure 3. The energy from the reference beam, with only mortar, is smaller than the energy from a beam containing a self-healing capsule.

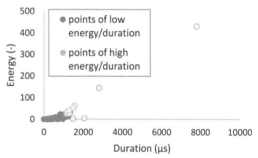

Figure 7. Energy versus duration graph showing AE-signals with greater energy due to cracking.

Figure 4. Two DIC images just before cracking (a), the biggest strain are at the right side, moments after the crack appears at the center beneath the capsule (b).

Figure 5. (a) Cross section of the two parts after three-point bending. The capsule did not break. (b) Plan view on the same specimen as figure 3 where the debonding of the capsule and mortar can be seen.

tion of where the crack will actually appear. On Figure 7 points with high energy and duration are selected.

From Figure 7, a duration versus energy plot made by the software NOESIS, it can be observed that mortar beams containing a capsule tend to have more AE-hits with much longer duration and energy in comparison to the reference (no capsule) beams. A distinction is made between datapoints with a higher energy and longer duration than was found in the reference beams.

Comparing the DIC images from Figure 4 and Figure 8, proof of the shift of the crack location can be found. Due to the accuracy of the sensors this is only a small indication. The events with high energy are concentrated near the center, the events with low energy or more scattered.

On Figure 9 it can be seen that the signals with a high duration and energy only appear after crack initiation.

To conclude this, some indications of cracking of the beam can be found, compared to the visual inspection during the experiment the signals mainly originate from cracking of the mortar. The recorded activity by AE includes two mechanisms:

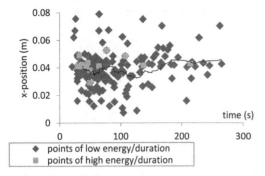

Figure 8. Displaying the event source location in time. The trend line takes the mean of 50 successive data points.

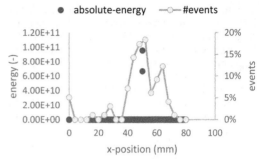

Figure 10. Absolute energy versus the x-position of the signal, the relative amount of signals originated from one position. The crack position evaluating both parameters is around 50 mm.

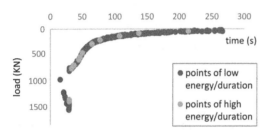

Figure 9. Load—time graph showing that the points with high energy occur just before and after the load-jump.

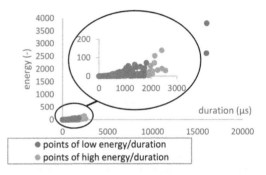

Figure 11. The energy versus duration graph shows 2 points with a very high energy and duration comparing to the other signals. A zoom on the remaining signals identify a group of signals with a duration above 2000µs.

mortar cracking and debonding between the capsule and the matrix. Notice the difference in energy with the reference beam, which suggests another mechanism than the mortar cracking. The location of the crack as indicated with AE and the visual location of the crack corresponds very well. AE is therefore suited to locate cracks in situations were no visual inspection is possible.

4.2 *Group II: self-healing tubes*

A total of 8 experiments were conducted and breaking of the tubes happened in 100% of the cases. The experimental setup is the same as in Figure 2 but with a longer tube (approximately 100 mm). In comparison to the specimen containing a self-healing capsules, the higher rate of successful breaking of the embedded self-healing mechanism is due to the bigger stresses that can be achieved on a long tube than on a short capsule. Breaking of these tubes caused strong and characteristic wave-signals picked up by AE. By means of DIC it was also confirmed that the crack path goes through the position of the tubes.

On Figure 10 it can be seen that the crack is around the center of the specimen based on the AE signal location, as was expected with a three-point bending experiment.

On Figure 11 points with a higher duration and/or energy are highlighted. In order to compare this graph with Figure 7 the circled area was zoomed into. Most of these points are due to fracture of the mortar as can be seen on the load versus time graph (Graph 8). This location shift was also observed in other specimen but was not witnessed in the reference beams without an embedded tube.

On the time versus x-position of the AE-signal (Figure 12) a shift can be seen after 120 s. This shift is confirmed by the DIC strain-fields in Figure 14.

Figure 13 the load versus time graph shows that the points of high energy occur after 50 seconds. Most of the points are registered after the peak load.

To conclude, giving the above graphs and pictures it shows very clear that AE in combination with DIC can be used as a NDT to check for tube breakage in the material. For breaking of the capsule both the energy and absolute energy are signif-

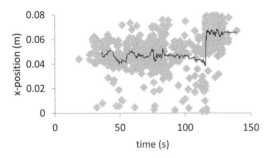

Figure 12. Displaying the AE hits in time and its correlated position. This graph shows a move of position of the signal's origin visualized by the trend line which takes the mean of 50 data points.

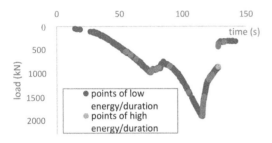

Figure 13. Load versus time graph, comparing to the reference and the capsule beams the load jump takes longer due to the tubes acting as tensile reinforcement.

Figure 14. A developing crack seen by the DIC technique. The picture on the left shows the beginning of the crack and the picture on the right shows the further crack development going through the known position of the embedded self-healing tube.

icantly bigger than in the first case were debonding was observed. This makes it possible to make a distinction based on the signal of what is happening. Note that the same results are found for glass-IPC tubes.

## 5 CONCLUSION

By means of DIC it can be visualized if the crack path goes through the position of the capsules or tubes but it is difficult to confirm if the capsules are opened. Together with AE it can be confirmed if the capsules or tubes are broken, which in this paper is considered to be the activation of the self-healing process. Specifically, a clear distinction in the AE-signal can be made between breaking of the capsules or tubes which emits large amounts of acoustic energy and debonding which carries low amount of energy similar to mortar cracking. Source localization by AE is compatible to the strain map produced by DIC and thus it is concluded that AE provides a non-destructive technique of inspecting real-life self-healing structures in the future.

## REFERENCES

Granger, S., Loukili, a., Pijaudier-Cabot, G., & Chanvillard, G. (2007). Experimental characterization of the self-healing of cracks in an ultra high performance cementitious material: Mechanical tests and acoustic emission analysis. *Cement and Concrete Research*, *37*(4), 519–527. http://doi.org/10.1016/j.cemconres.2006.12.005.

Jonkers, H.M., Thijssen, A., Muyzer, G., Copuroglu, O., & Schlangen, E. (2010). Application of bacteria as self-healing agent for the development of sustainable concrete. *Ecological Engineering*, *36*(2), 230–235. http://doi.org/10.1016/j.ecoleng.2008.12.036.

Lecompte, D., Bossuyt, S., Cooreman, S., Sol, H., & Vantomme, J. (2007). Study and generation of optimal speckle patterns for DIC. *Proceedings of the Annual Conference and Exposition on Experimental and Applied Mechanics.*, 1643–9.

Tang, W., Kardani, O., & Cui, H. (2015). Robust evaluation of self-healing efficiency in cementitious materials—A review. *Construction and Building Materials*, *81*, 233–247. http://doi.org/10.1016/j.conbuildmat.2015.02.054.

Tsangouri, E. (2015). *Experimental assessment of fracture and autonomous healing of concrete and polymer systems.* Ph.D. thesis. Brussels: Vrije Universiteit Brussel.

Van Tittelboom, K., & De Belie, N. (2013). Self-healing in cementitious materials-a review. *Materials* (Vol. 6). http://doi.org/10.3390/ma6062182.

Van Tittelboom, K., De Belie, N., Van Loo, D., & Jacobs, P. (2011). Self-healing efficiency of cementitious materials containing tubular capsules filled with healing agent. *Cement and Concrete Composites.* http://doi.org/10.1016/j.cemconcomp.2011.01.004.

Wu, M., Johannesson, B., & Geiker, M. (2012). A review: Self-healing in cementitious materials and engineered cementitious composite as a self-healing material. *Construction and Building Materials*, *28*(1), 571–583. http://doi.org/10.1016/j.conbuildmat.2011.08.086.

*Emerging Technologies in Non-Destructive Testing VI – Aggelis et al. (Eds)*
*© 2016 Taylor & Francis Group, London, ISBN 978-1-138-02884-5*

# Finite element model updating using thermographic measurements: Comparison between materials with high and low thermal conductivity

### J. Peeters
*Op3Mech Research Group, University of Antwerp, Antwerp, Belgium*

### B. Ribbens
*Op3Mech Research Group, University of Antwerp, Antwerp, Belgium*
*Acoustics and Vibration Research Group, Vrije Universiteit Brussel, Brussels, Belgium*
*Laboratory of Biomedical Physics, University of Antwerp, Antwerp, Belgium*

### J.J.J. Dirckx
*Laboratory of Biomedical Physics, University of Antwerp, Antwerp, Belgium*

### G. Steenackers
*Op3Mech Research Group, University of Antwerp, Antwerp, Belgium*
*Acoustics and Vibration Research Group, Vrije Universiteit Brussel, Brussels, Belgium*

ABSTRACT: In non-destructive evaluation the use of finite element models to evaluate structural behavior and experimental setup optimization can complement experience of the inspector. A new adaptive response surface methodology, especially adapted for thermal problems, is used to update the experimental setup parameters in a finite element model to the state of the test samples measured by pulsed thermography. As optimized parameters ambient temperature, the local plate thickness and the thermal pulse time are used. Poly Vinyl Chloride and Aluminum test samples are used to examine the results for a thermal insulator and a material with high conductivity. A comparison of the results is made by changing the target values by updating the raw data of pulsed thermography. The results of the described technique are compared between both materials. A time efficiency increase with a factor of more than 20 and an accuracy of over 99.5% is achieved by the choice of the correct parameter sets and optimizer for thelow conductive test sample. For materials with high thermal conductivity, the lateral heat flow has a higher influence on the stability of the optimization accuracy. The dependency of accuracy on the thermal conductivity of the material is discussed.

## 1 INTRODUCTION

The study of Pulsed Infrared Thermography (PT) has become an important aspect of Non-Destructive Evaluation techniques (NDE) for damage detection and material updating in metallic structural elements Chaudhuri & Santra (2006), Vollmer & Möllmann (2010), as well as CFRP (Carbon Fiber Reinforced Polymer) and GFRP (Glass Fiber Reinforced Polymer) composites Susa & Ibarra-Castanedo (2007). High-end post-processing algorithms are essential for failure detection, but the interpretation of the influence of a specific defect on the global behavior of the structure remains difficult and depends on the skills of the inspector.

Finite Element (FE) analysis has been applied as a verification tool in some applications involving infrared thermography Chaudhuri & Santra (2006), Louaayou et al. (2008), Peeters, Arroud & et al. (2014). Besides thermography, FE modeling has proven its success in multiple application areas for worst case predictions and quantitative failure detection. FE models are considered as approximation models as the parameter values are estimated theoretical values of data sheets. Most of the FE models used in these situations need to be updated to the real conditions of the structure to deliver accurate results. There are several methods published in the field of structural dynamics by Friswell & Mottershead (1995), Marwala (2010) but the analytic modeling of thermal physics differs as it numerically solvesa partial differential equation with complex non-linear boundary conditions. This results in a different approach to implement FE model updating in thermal problems with a reduced amount of data.

In this paper the developed method of Peeters, Arroud & et al. (2014) for thermal insulator materials is adapted and validated to develop an accurate FE model by optimization with reduced PT thermal data for high thermal conductive materials. The presented technique has a far less time expensive training time as the method described in Darabi & Maldague (2002) by Darabi but has also its drawbacks on robustness and stability for conductive materials.

## 2 TEST-CASE DESCRIPTION

We used experimental measurements on different test cases of different materials. In order to verify the used parameter set we developed a FE model for updating with the experimental setup to the actual measurement setup. The experiments are based on active Pulsed Thermography (PT) measurements on different test samples. These measurements are performed in an anechoic chamber without any lights or reflective hot spots except the heat sources in front of a thermal absorptive background. The configuration is shown in Fig. 1.

We used a flat bottom hole plate to validate the results found in literature Peeters, Arroud & et al. (2014), Ibarra-Castanedo et al. (2007), Maldague et al. (2002). The test sample is made of two different materials, first a low conductive material, PVC (PolyVinyl Chloride) to approximate a 1 dimensional heat flow as a similar model for carbon fibre reinforced plastics and second a high conductive Aluminum alloy (3003) which is often used for lightweight constructions. With this high conductive material it is possible to investigate the influence of the lateral heat flow on the measurements.

The used thermographic measurement technique is based on Pulsed Thermography (PT). An automatic external triggering for correct timekeeping is used and the frame grabbing is synchronized on the same external clock as the excitation trigger.

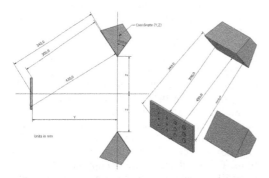

Figure 1. Technical drawing of the measurement setup and measured distances in mm.

To define the start time of the pulse of the excitation source there is made use of the methodology described in Peeters, Arroud & et al. (2014) which is based on the derivative of the temperature propagation in one point. The pulse length of the heat sources are adapted for each different material, for the Aluminum test samples there is made use of a 2 s excitation time and a total measuring time of 12 s. There is chosen to excite with this high excitation time to investigate the lateral heat dissipation and validate the influence on the FE model updating. After 10 s of cooling the thermal ambient conditions are achieved. The PVC test plates are excited in the same setup but with an excitation time of 5 s and a total measurement time of 3 m and 5 s. The used frame rate is reduced to 10 Hz.

The governing differential equation is a combined heat transfer equation with conduction, convective cooling and radiation of an external heat source, formulated in Eq.(1) and Eq.(2) where $\rho$ is the density, $C_p$ is the heat capacity at constant pressure, $T$ is the absolute surface temperature, $\kappa$ is the thermal conductivity, $t$ is the time, $Q(t)$ is the time dependent radiating heat source and $h$ is the heat transfer coefficient Chapman (1984), Stotter et al. (2014). The air between the lamps and the surface is neglected. The simulation set-up is shown in Figure 1. The numerical solution can be considered sufficiently accurate for FE model updating of thermographic NDE applications as described in Darabi & Maldague (2002).

$$\rho C_p \frac{\partial T}{\partial t} + \nabla.(-\kappa \nabla T) = Q$$
$$\text{with } T(x, y, z, 0) = T_\infty = 292.88\,[K] \qquad (1)$$

$$\vec{n}(k\nabla T) = h(T_\infty - T) \qquad (2)$$

There are made a few simplifications:

- The air velocity is assumed to be 0 allowed by the laboratory conditions.
- The thermal conductivity $\kappa$ of PVC and Aluminum is isotropic Dashora et al. (2005).
- The test sample is opaque and behaves like an ideal grey body as a result of a black paint coating.

The physics are designed with 2 external uniform radiating heat sources defined as described in Peeters, Steenackers & et al. (2014), Peeters, Arroud & et al. (2014).

## 3 METHODOLOGY

Both the quality of the experimental data and the FE model data will contribute to the evolution of the optimization algorithm. The implemented optimization algorithm is based on the adaptive response surface method described in Peeters, Arroud & et al. (2014). The method is expanded for the use of high conductive materials.

## 3.1 Adaptive response surface method

For the optimization algorithm there is made use of noise filtered experimental data as target for the FE model. The optimization uses the relative temperature differences $\theta = T - T_0$ to reduce the measurement uncertainties. The algorithm has no fixed amount of iterations in contrast to the method described in Steenackers et al. (2009). The parameter stability control is increased in contrast to the method described in Peeters, Steenackers & et al. (2014) by the choice of independent parameter sets and the use of multiple initial value for each parameter set.

In a first stage there is created a meta model of a response surface built out of $2*n-2$ data points for $n > 2$, retrieved from random distributed FE simulations across the design space. The design space is built out of the physical boundary values for each parameter of the objective function. This meta model is built of n-dimensional polynomials, each dimension represents an unknown parameter and in total they form one single n-dimensional response surface. The optimization algorithm searches for a global minimum of the objective function in the meta model which results in estimated values for each parameter. The estimated parameter values are the input of a new FE model, called the updated FE model. The results of this updated FE simulation have a higher correlation with the experimental target values and deliver new input for the meta model which could replace the least accurate data point inside the meta model with the higher accurate data point of the last simulation. This iterative process reforms the response surface locally in the region of the global minimum. The local conversion of the response surface is a local solution of the optimization problem. The use of different initial values at the same time delivers a conversion to a global optimum. This process continues until convergence is achieved for the error value between the experimental value and the minimum of the meta model.

## 3.2 Optimized parameters choice

The choice of a suitable parameter set is one of the most critical steps in the optimization routine as mentioned in Peeters, Arroud & et al. (2014). This is also shown in Figure 2 which is a result of the first parameter set of Table 1. As we can see, there are several values which result in a minimal error value. This means that the solution is not unique

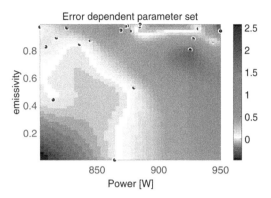

Figure 2. Error plot of the first parameter set of Table 1 with in white all solutions with equal error value.

Table 1. Dependent and independent optimization parameters with their boundary conditions for PVC and Aluminum Granta (2013).

| Dependent parameter | Lower | Upper |
|---|---|---|
| Heat source power [$W$] | 900 | 1000 |
| Conductivity PVC [$W/(m \cdot K)$] | 0.12 | 0.209 |
| Conductivity Al $k[W/(m \cdot K)]$ | 155 | 250 |
| Specific heat capacity PVC $C_p[J/(Kg \cdot K)]$ | 1000 | 1800 |
| Specific heat capacity Al $C_p[J/(Kg \cdot K)]$ | 840 | 1080 |
| Density PVC $\rho[kg/m^3]$ | 1290 | 1560 |
| Density Aluminum $\rho[kg/m^3]$ | 2500 | 2950 |
| Heat transfer coefficient $W/(m^2 \cdot K)$ | 5 | 30 |
| Emissivity | 0.80 | 0.95 |

| Independent parameter | Low | Up |
|---|---|---|
| Ambient Temperature $T_\infty[K]$ | 273.15 | 300.15 |
| Hole thickness $d[mm]$ | 0.10 | 4.80 |
| Pulse Time Al $t_{pulse}[s]$ | 0.5 | 3.5 |
| Pulse Time PVC $t_{pulse}[s]$ | 3.5 | 6.5 |

ant there is a dependency between the 2 parameters which are presented.

$$T_{exp}(t) = T_{FE}(t, k, C_p, \varepsilon, \rho, P) \qquad (3)$$

The optimization strategy is divided in three stages:

1. The pulse start time is a result of the second derivative calculation of the experimental data as shown in Fig. 3, which forms the input of the global optimization routine. The start pulse is located 0.11 s after the beginning of the measurement, which corresponds at a framerate of 180 frames per second (fps) with a delay of 20 frames. The lamp coordinates are calculated out of triangulation measurements.

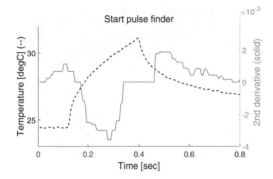

Figure 3. Mathematical detection of the start of the pulse time from experimental Aluminum test case.

2. The second parameter set of Table 1 is optimized to a solution following Eq.(3) which delivers a correct curve fitting between experimental data and the FE model for each value of the parameters of Table 1.
3. The exact values of the independent parameter set of Table 1 are calculated using the optimization algorithm.

This results in a 1D heat transfer with a 3D global optimization problem whereby independent parameters as the source coordinates and the pulse start time are calculated separately out of the experimental measurement data.

## 3.3 Objective function

The used objective function is based on locating the minimum of the response surface. The minimization is defined in Eq. (4) where $\ell(\vartheta)$ is defined as the objective function which should be minimized, $T$ is the relative temperature for respective the experimental target data set and the approximated model, whereby the relative temperature results of the model are dependent of the optimized parameters $\vartheta$. Further on there is tested in the objective function if the boundary conditions aren't violated.

$$\ell(\vartheta) = T_{model}(\vartheta) - T_{Ex} \quad (4)$$

### 3.3.1 Optimizers

Three gradient based optimizers are compared and investigated to find which one is the fastest and the most accurate. The optimization strategies are all designed for non-linear, continuous problems with a continuous first derivative and are chosen for their stability and robust objective function adaptation. The detailed information for the choice of this optimizers could be found in Peeters, Arroud & et al. (2014) and is based on The MathWorks (2013). The algorithms of Peeters, Arroud & et al. (2014) are adapted for the use with high thermal conductive materials.

The best expected optimizer (LCFTRR) is especially designed for curve fitting problems and defined in Eq. (5) whereby the objective function delivers predictions of the temperature values instead of the error function as described in Eq. (4). The absences of the use of weighting factors makes it a stable and robust optimizer as mentioned in bullet 4 in section 1 referred to V. Vavilov. The MathWorks (2013)

$$\min_{\vartheta} \sum_{i=0}^{t} (T_{FE,i}(\vartheta,t) - T_{Ex,i})^2 \text{ with } lb \leq \vartheta \leq ub \quad (5)$$

## 4 RESULTS AND DISCUSSION

First a low conductive material is investigated as described in section 2. The results are presented in Table 2. A representation of the response surface for the 3 optimized parameters with the LCF TRR optimizer is seen in Figs. 4 and 5. Figure 4 shows the response surface for the ambient temperature and the thickness parameter and in Fig. 5 the

Table 2. Results experimental optimization method for PVC sample.

| Optimization method | Target | FE-LCF TRR | FMC | Lsq TRR | Lsq LM | LCF TRR | LCF LM |
|---|---|---|---|---|---|---|---|
| Temperature $T_\infty[K]$ | 297.66 ± 0.02 | 293.17 | 297.68 | 297.67 | 297.67 | 297.67 | 297.67 |
| Hole thickness $d[mm]$ | 0.9875 ± 0.01 | 0.9980 | 0.95425 | 0.98592 | 0.98567 | 0.98797 | 0.98577 |
| Pulse Time $t_{pulse}[s]$ | 5.5 ± 0.5 | 2.5802 | 5.1761 | 5.2761 | 5.2760 | 5.2683 | 5.2761 |
| Iterations | / | 54 | 1276 | 1045 | 1068 | 857 | 1034 |
| Residuals | / | 4.9906e-3 | 1.9458e1 | 4.9996e-2 | 5.0000e-2 | 4.9999e-2 | 4.9993e-2 |
| Time [s] | / | 10 432.8 | 11 670 | 9 441.4 | 9 819.6 | 7 671.1 | 8 860.5 |

response surface is plotted for the ambient temperature and the pulse time parameter.

Finally the results of the Aluminum sample in 1D are presented in Table 3. These results are calculated with 2 dimensional lateral heat loss implemented for a high conductive material. The method adapts the amount of used frames according to the used frame rate whereby the excitation keeps good approximation. With the implementation of the lateral diffusion of the heat in the test sample, the expected accuracy of the method which should be 0.12 for 2160 frames is not achieved. As could be seen in Table 3, non of the optimizers delivers this accuracy except the FE updating, even with the use of a 10 times larger amount of iterations. A possible explanation could be that the influence of the lateral diffusion influences the FE model more than expected. This results in a model with too much noise to update and should be further explored in the future. We can see that the use of the response surface methodology flattens the accuracy on 0.27.

The choice of optimizer for the descending cooling curve shows a discrepancy between the least square methods and the minimal constraint. The suggested conclusion is that the minimal constraint optimizer is less appropriate than the least square optimizers to use for FE updating of thermal data. A possible reason therefore could be the non-linear partial differential characteristic of the modeled physics. The summation made by the minimal constraining linearizes between 2 time steps which delivers a larger error value and instability. Further, we can conclude that the largest difference between the optimizers is situated in the region between the uphill and downhill temperature difference. This can conclude that the choice between Lsq or LCF is unimportant when only the cooling is monitored but with the implementation of the heating characteristic as done in the PVC testcase, the choice of a trust region algorithm and the curve fitting optimizer would improve the results. An other conclusion we can make is that the lateral diffusivity in the aluminum test piece makes it very difficult to estimate the heating characteristic. A possible explanation therefore could be that the combination of the neglected heat loss in the heat sources, the reduced emissivity and the lateral heat dissipation

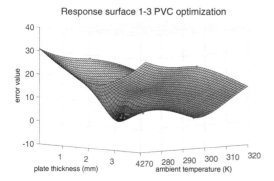

Figure 4. The response surface of the ambient temperature versus the material thickness.

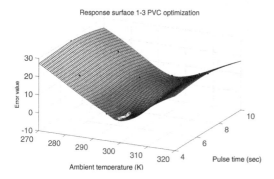

Figure 5. The response surface of the ambient temperature versus the pulse time.

Table 3. Results experimental optimization method for Aluminum sample.

| Optimization method | Target | FE-LCF TRR | FMC | Lsq TRR | Lsq LM | LCF TRR | LCF LM |
|---|---|---|---|---|---|---|---|
| Temperature $T_\infty[K]$ | 299.54 ± 0.05 | 299.21 | 298.93 | 299.19 | 299.19 | 299.19 | 299.19 |
| Hole thickness $d[mm]$ | 1.025 ± 0.01 | 0.9437 | 1.107 | 0.9661 | 0.9657 | 0.9660 | 0.9658 |
| Pulse Time $t_{pulse}[s]$ | 1.9277 ± 0.5 | 1.9092 | 2.8632 | 1.7967 | 1.7765 | 1.7760 | 1.7804 |
| Iterations | / | 146 | 99 | 79 | 88 | 76 | 82 |
| Residuals | / | 0.11462 | 0.4595 | 0.2748 | 0.2750 | 0.2750 | 0.2749 |
| Time [s] | / | 1 079.9e3 | 4 735.4 | 3 505.8 | 3 986.6 | 3 494.7 | 3 780.3 |

in the test object makes the difference between the FE model and the experimental results too large.

## 5 CONCLUSIONS

This paper presents the use of a new adaptive response surface method for thermal optimization of geometrical parameters to the experimental values, measured by pulsed thermography on materials with low and high thermal conductivity. From the results we can conclude that the use of the optimization algorithm speeds up the convergence and results in a time reduction of factor 20 with respect to an optimization without the use of a response surface for the full time window (heating and cooling) of a low thermal conductive material. Further we can conclude that the optimization algorithm decreases the calculation time over 300 times for high conductive materials, but the accuracy decreases from 0.11 to 0.27 residual value. The dependency of the used optimization algorithm and optimizer were investigated. The use of a least square curve fitting optimizer with a trust region algorithm was found to be the most appropriate for low thermal conductive test cases. With the implementation of lateral diffusivity the accuracy decreases but the parameter estimation keeps sufficient for FE updating. The difference between the least square optimization algorithms diffuses but the methods remain more stable than the minimal constraining method. The optimized model could be used in failure detection or other IR based image analysis to assist the IR measurements and combine the benefits of surrogate modeling (fast and economical efficient) with IR NDE for quantitative detection of the influence of changed material conditions and FE model updating. It is also made possible to adapt the setup conditions of the FE model to the realistic state of the test sample or to define the most efficient measurement setup for a specific structure. This concludes that the described method could be used to improve the experimental test setup with the use of an accurate FE model to improve the non-destructive testing of complex shaped structures.

## ACKNOWLEDGEMENTS

This research has been funded by the University of Antwerp and the Institute for the Promotion of Innovation by Science and Technology in Flanders (IWT) by the support to the TETRA project 'Smart data clouds' with project number 140336. Furthermore, the research leading to these results has received funding from Industrial Research Fund FWO Krediet aan navorsers 1.5.240.13 N. The authors also acknowledge the Flemish government (GOA-Optimech) and the research councils of the Vrije Universiteit Brussel (OZR) and University of Antwerp (fti-OZC) for their funding.

## REFERENCES

Chapman, A.J. (1984), *Heat transfer*, 4 edn, Macmillan, New York.

Chaudhuri, P. & Santra, P. e. a. (2006), 'Nondestructive evaluation of brazed joints between cooling tube and heat sink by IR thermography and its verification using FE analysis', *NDT E Int.* 39(2), 88–95.

Darabi, A. & Maldague, X. (2002), 'Neural network based defect detection and depth estimation in TNDE', *NDT E Int.*.

Dashora, P., Gupta, G. & Dashora, J. (2005), 'Thermal conductivity, diffusivity and heat capacity of plasticized polyvinyl chloride', *Indian J. Pure Appl. Phys.* 43(February), 132–136.

Friswell, M. & Mottershead, J. (1995), *Finite Element Model Updating in Structural Dynamics*, Springer. Granta (2013), 'PVC ( chlorinated, molding and extrusion)'.

Ibarra-Castanedo, C., Genest, M. & et al. (2007), Active Infrared Thermography Techniques for the Nondestructive Testing of Materials, *in* 'Ultrason. Adv. Methods Nondestruct. Test. Mater. Charact.', Chen, CH, pp. 325–348.

Louaayou, M., Naït-Saïd, N. & Louai, F.Z. (2008), '2D finite element method study of the stimulation induction heating in synchronic thermography NDT', *NDT E Int.* 41(8), 577–581.

Maldague, X., Galmiche, F. & Ziadi, A. (2002), 'Advances in pulsed phase thermography', *Infrared Phys. Technol.* 1(418), 1–11.

Marwala, T. (2010), *Finite Element Model Updating Using Computational Intelligence Techniques: Applications to Structural Dynamics*, Springer.

Peeters, J., Arroud, G. & et al. (2014), 'Updating a finite element model to the real experimental setup by thermographic measurements and adaptive regression optimization', *MSSP* p. 15.

Peeters, J., Steenackers, G. & et al. (2014), Finite element optimization by pulsed thermography with adaptive response surfaces., *in* 'Quant. Infrared Thermogr.', University of Antwerp, Bordeaux, p. 10.

Steenackers, G., Presezniak, F. & Guillaume, P. (2009), 'Development of an adaptive response surface method for optimization of computation-intensive models', *Comput. Ind. Eng.* 57(3), 847–855.

Stotter, B., Gresslehner, K. & et al. (2014), Quantitative application of pulse phase thermography to determine material parameters, *in* 'Quant. Infrared Thermogr.', number 1, Bordeaux, p. 10.

Susa, M. & Ibarra-Castanedo, C. (2007), Pulse thermography applied on a complex structure sample: comparison and analysis of numerical and experimental results, *in* 'IV Conf. Panam. END Buenos Aires', p. 12.

The MathWorks (2013), 'The Matlab user manual'.

Vollmer, M. & M̈ollmann, K. (2010), *Infrared thermal imaging: fundamentals, research and applications*.

*Emerging Technologies in Non-Destructive Testing VI – Aggelis et al. (Eds)*
*© 2016 Taylor & Francis Group, London, ISBN 978-1-138-02884-5*

# Portable automated Radio-Frequency scanner for Non-Destructive Testing of Carbon-Fibre-Reinforced Polymer composites

B. Salski, W. Gwarek & P. Kopyt
*Institute of Radioelectronics, Warsaw University of Technology, Warsaw, Poland*

P. Theodorakeas, I. Hatziioannidis & M. Koui
*Materials Science and Engineering Department, National Technical University of Athens, Athens, Greece*

A.Y.B. Chong, S.M. Tan, V. Kappatos, C. Selcuk & T.-H. Gan
*Brunel Innovation Centre, Brunel University London, London, UK*

ABSTRACT: A portable automated scanner for Non-Destructive Testing (NDT) of Carbon-Fibre-Reinforced Polymer (CFRP) composites has been developed. Measurement head has been equipped with an array of newly developed Radio-Frequency (RF) inductive sensors mounted on a flexible arm, which allows the measurement of curved CFRP samples. The scanner is also equipped with vacuum sucks providing mechanical stability. RF sensors operate in a frequency range spanning from 10 up to 300 MHz, where the largest sensitivity to defects buried below the front CFRP surface is expected. Unlike to Ultrasonic Testing (UT), which will be used for reference, the proposed technique does not require additional couplants. Moreover, negligible cost and high repeatability of inductive sensors allows developing large scanning arrays, thus, substantially speeding up the measurements of large surfaces. The objective will be to present the results of an extensive measurement campaign undertaken for both planar and curved large CFRP samples, pointing out major achievements and potential challenges that still have to be addressed.

## 1 INTRODUCTION

Future generation of engineering structures will consist mostly of carbon composite materials, notably in the civil, aerospace, automotive and marine industries due to their performance and structural efficiency. The modes of failure in composite intensive structures such as the increasingly popular Carbon-Fibre-Reinforced Polymer (CFRP) composite reinforced concrete beams [1–3] or the recent operational aircraft, Boeing 787 Dreamliner (with 50% composite) [4], are not fully known as they are still near the beginning of their design life. However, it is clear that these CFRP materials are susceptible to internal impact damage, not visible with an un-aided eye at the surface. In spite of this, inspections at the point of manufacture and in service is largely manual with consequent low area coverage. Operation downtime is usually inevitable during scheduled or unscheduled inspection. Common Non-Destructive Testing (NDT) techniques utilized for CFRP include Ultrasonic Testing (UT) [5–6], Eddy Current Testing (ECT) [7–8], shearography [9], microwave and millimeter wave [10–11]. However, results obtained are difficult to interpret for most NDT techniques due to the intrinsic anisotropy of the structure. Moreover, there are

requirements for specific techniques which may be easily applicable in-situ. As a result of the recent successful development of the planar coupled spiral inductors tailored for the NDT of CFRP composites [12–13], the sensor and auxiliary modules were subject to further development in which a line array of sensors is realized. The sensor array is integrated with a portable automated scanner. The fundamental principle of operation is based on Radio-Frequency (RF) inductive approach where the coupling between the two inductors tangential to the surface of the Material Under Test (MUT) is measured. For brevity, this technique is referred as RF Inductive Testing (RFIT) in this paper. Experiments were carried out to validate the system prototype on both planar and curved CFRP composites with typical buried and surface defects. The results clearly demonstrate the applicability of such automated RFIT system for reliable and efficient CFRP inspection.

## 2 SCANNER

The whole NDT system, as shown in Fig. 1, consists of the scanner with a sensor arm, data acquisition board (DAQ), and a PC station.

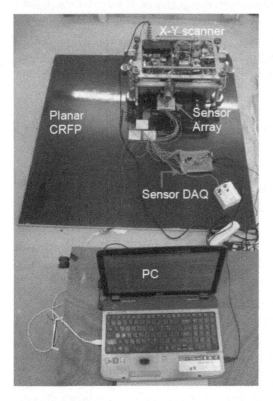

Figure 1. NDT scanner setup.

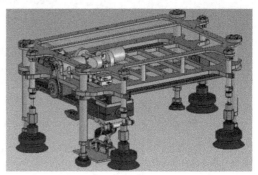

Figure 2. Scanner platform with XY precision translators, sensor arm and vacuum sucks.

## 2.1 Chassis

One of major objectives was to minimize a scanner's weight, while providing high scanning rates, good resolution, positioning repeatability in both absolute and relative coordination systems, involving the minimum degrees of freedom. For that reason, an important functional requirement was the development of an ergonomic portable chassis with the weight less than 10 kg. Furthermore, it should provide the necessary space for seamless integration of all modalities that will implement the requested functional requirements of the scanner. Based on the aforesaid, the scanner frame was designed and manufactured through the use of two aluminum plates as presented on a schematic design in Fig. 2. The chassis was secured via the application of four tubes with collars that interconnect the two layers of aluminum plates. The aluminum plates were pre-cut with all necessary cut-outs and bores so as to be able to attach all necessary parts, gantry, motor supports, electronics, and pneumatic systems. The chassis structure has the advantage that all parts are accessible and can be removed readily for repair or replacement. It can be easily disassembled, machined if modifications are necessary during service period and

reproduced on request. The overall size of the chassis frame is $600 \times 400 \times 250$ mm$^3$.

The main function of the x-y gantry system is to implement a scanning motion of a sensor arm. For that purpose, an Igus Drylin stage drive product was selected. It provides lubricant-free linear axles that are driven either by trapezoidal thread, steep thread or toothed belt. The user can choose a suitable individual solution from lightweight solid plastic units up to massive stainless steel solutions. For our application trapezoidal threads with 2 mm pitch were chosen for both X and Y directions. Along with robust design of these components, their main features include ruggedness and insensitivity to dirt, water, chemicals, heat or impacts.

Magnetic encoders have a resolution of 1024 ticks per shaft revolution. One revolution is converted to a linear motion of 2 mm with a resolution of 10 μm per revolution of the DC motors. The power source is mounted within the chassis boundaries and provides the scanner with voltage supply options of 5–12–24V DC. The scanner is controlled via RS-485 protocol converted to USB before it is connected to the PC.

Another function of the scanner is the attachment to the composite's surface. Due to operational principles of the inspection method and the composite material characteristics, it was not possible to implement magnets and clamping systems for the robust mounting of the scanner. For that reason, plastic suction cups were developed which satisfy three crucial preconditions: a) there is no magnetic field interference with RF sensors, and b) they cause absolutely no surface damage to the MUT. The mounting of the scanner is performed by the following procedure. Initially, the operator places the scanner on the MUT's surface, ensuring that the supporting legs and the plastic cups are conformed on the surface. Subsequently, pressure can be produced to the cups by pressing an air compressor activation button. As a result of a network pressure of 8 bar, the scanner becomes firmly attached to the surface.

## 2.2 Sensor arm

Sensor arm, as shown in Fig. 3, is a mounting platform for the array of RF sensors. It is made of aluminum housed with bearings, pivot brackets, and springs to produce a lab jack mechanism with better conformance of the sensors to the MUT's surface. RF sensors are placed in PA6 plastic protection pads, which are mounted to a holder arm with the aid of a plastic holder plate to prevent any large metallic parts in the vicinity of the sensors. Although mounting pads indicated in Fig. 4 fit precisely to the plastic protection pads, the attachment can be further enhanced with thermoplastic adhesive. The dimensions of the holder plate are $100 \times 120$ mm$^2$ with a thickness of 8 mm.

Vertical force pushing the sensors to the CFRP surface is provided by the lab jack mechanism that drives a passive spring on the outside of a telescopic pair shown in Fig. 4. An exerted force over the sensor plate is adjusted by the use of springs in

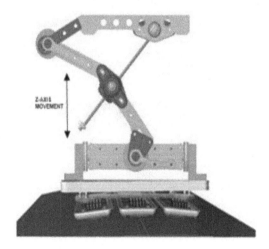

Figure 3. Sensor flexible arm.

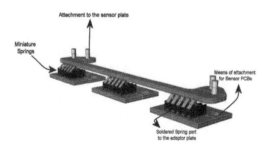

Figure 4. Mounting pad of RF sensors.

Figure 5. Photograph of a sensor arm.

order to maintain a firm contact with the MUT's surface during a sliding movement of the sensors. The total travel distance on the z-axis is 40 mm, enabling the application of proper pressure to the sensor plate, in order to ensure its smooth conformance on curved surfaces. The vertical displacement can be monitored with an integrated absolute encoder.

## 2.3 RF sensors

Sensor array is made in the form of a line of RF inductors manufactured on a printed circuit board as shown in Fig. 6. Single measurement utilizes two adjacent inductors treated as primary and secondary windings of a transformer. As a result, there are $N$-1 measurement points for $N$ aligned inductors.

If RF sensors are attached to the MUT's surface with the aid of the sensor arm, a magnetic field penetrates the MUT provided that the penetration depth $d_p$ is large enough at a given frequency. For instance, conductivity of CFRP composites is usually in the range of $\sigma = 10^4$ S/m, which means $d_p$ is over 1 mm for frequencies below ca. 30 MHz. It implicitly determines the frequency range for RF inductors if the thickness of CFRP panels is given.

## 2.4 Data acquisition board

In the view of the above considerations of the frequency range of operation of the RF sensors, electronic circuitry, dedicated to the measurement of the number of individual sensor channels, has been developed (see Fig. 7). The main role in that system is played by a Direct Digital Synthesizer (DDS), controlled with a 32-bit microcontroller, which also serves as an interface with a PC host station. DDS generates signals spanning from 20 MHz up to 300 MHz, as this is the spectrum where the largest sensitivity to defects buried in ca. 1 mm thick CFRP panels is expected. The signal is multiplexed to all measurement channels in a sequential form.

Figure 6.  Exemplary RF sensors.

Figure 7.  Data acquisition board.

Each channel is also equipped with its individual logarithmic wideband power detector. As a result, each measurement shot provides the whole spectrum of power transmission through each coupled sensor pair, as indicated in Fig. 6.

## 3  MEASUREMENTS

Characterization of flat and curved CFRP panels will be presented in this Section. Both panels, manufactured by ATARD [14], consist of four layers of CFRP twill immersed in epoxy resin. All the samples have been measured with a single line of RF inductive sensors shown at the top of Fig. 6 mounted in the scanner depicted in Fig. 1 with 12 measurement pairs. As the scanning step has been set to 2 mm, an image of a $300 \times 200$ mm$^2$ surface consists of over 15 000 steps, which means that the scanner has to undertake 1250 incremental shifts across the surface. That number can be further decreased if additional sensor lines are exploited in the scanner (see Fig. 3), which will be the subject for future enhancements of the system.

### 3.1  Flat CFRP panel

The flat CFRP panel is shown in Fig. 8. As it can be noticed, a few types of defects have been intentionally introduced, such as holes, bubbles, cracks, and delamination. Figure 9 shows RF images stored at two distinct frequencies. The image at 50 MHz clearly indicates a large hole, which is not visible at

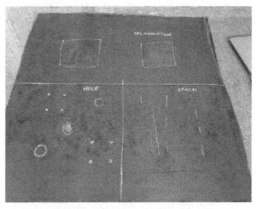

Figure 8.  Flat CFRP panel.

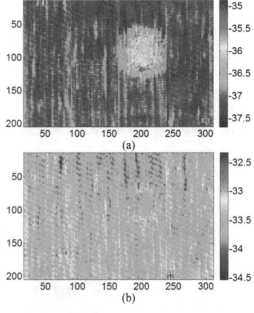

Figure 9.  Hole buried in the CFRP sample shown in Fig. 8a measured at (a) 50 MHz and (b) 200 MHz.

the front surface of the panel. It can be partially confirmed by the lack of the hole at the RF image at 200 MHz, which due to a substantially decrease penetration depth is sensitive mostly to the CFRP features, which are shallowly buried under the top surface of the panel. In particular, regular fracture appearing in Fig. 9b is correlated with the orientation of the front CFRP twill.

Figure 10 indicates an RF image measured at 20 MHz at the areas of the flat CFRP panel shown in Fig. 8, where buried bubbles are introduced by the manufacturer. Indeed, several tiny fluctuations can be observed in the central part and the right side of the image, while the changes at the left side of the image are rather smooth and results mostly due to slight misalignments of the CFRP twill.

Subsequently, Fig. 11 shows vertical cracks buried at selected regions of the CFRP sample. As the cracks are visible even at 300 MHz, it can be concluded that the defects are spanning almost across the total thickness of the sample, although those are not visible with a naked eye.

Eventually, Fig. 12 shows delamination buried in the CFRP sample shown in Fig. 8a measured at two distinct frequencies. As it can be noticed in Fig. 12a, there is a large deep minimum at a left side of the image, where the defect is expected and which cannot be noticed at 300 MHz indicating its deeper location.

## 3.2 Curved CFRP panel

Figure 12 shows the sample of a curved CFRP panel with several types of defects introduced in a similar way as in the flat sample. Curvature radius of the panels is ca. 1.5 m. As it will be shown, measurement of curved panels brings additional issues related to uneven attachment of the sensor line to the surface, which has to be de-embedded from raw measurement data.

Figure 13 depicts specific bubbles occurring at a front surface of the curved CFRP panel. The corresponding RF image consists of four individual

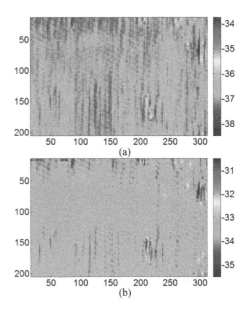

Figure 11. Cracks buried in the CFRP sample shown in Fig. 8a measured at (a) 20 MHz and (b) 300 MHz.

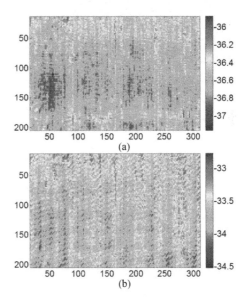

Figure 12. Delamination buried in the CFRP sample shown in Fig. 8a measured at (a) 20 MHz and (b) 300 MHz.

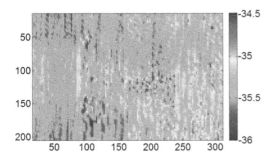

Figure 10. Specific bubbles buried in the CFRP sample shown in Fig. 8a measured at 20 MHz.

stripes due to the chosen direction of the sensor indicated in Fig. 13. Figure 14 shows the obtained RF image, where those four stripes are well visible due to the aforementioned uneven attachment of the sensor line to the curved surface. Fortunately, it results in a linear trend occurring in power transmission measured across the sensor line, so it can

Figure 12. Curved CFRP panel.

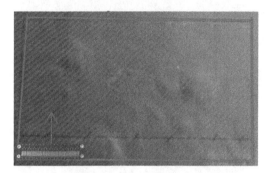

Figure 13. Specific bubbles on the curved CFRP sample shown in Fig. 12.

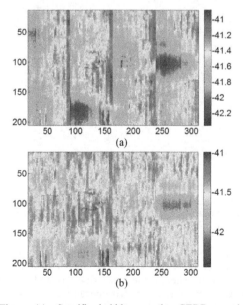

Figure 14. Specific bubbles on the CFRP sample shown in Fig. 13 and measured at 20 MHz. (a) Raw data. (b) Linear trend removed.

be estimated and removed at a post-processing stage. Consequently, Fig. 14b is smoother enabling better recognition of the bubbles, the location of which corresponds very well with those shown in Fig. 13.

Figures 15–17 show another types of defects measured with the RF scanner at 20 MHz. In the case of delamination, it can be seen in the middle of the image in Fig. 15. In addition to that, there is also some vertical defect visible at the left side of the RF image shown in Fig. 15, which may be some buried crack unintentionally introduced by the manufacturer.

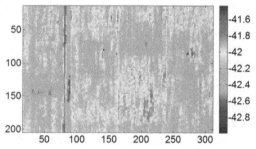

Figure 15. Delamination buried in the CFRP sample shown in Fig. 13 measured at 20 MHz. Linear trend is removed.

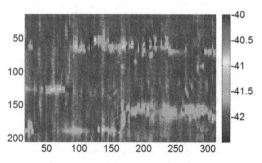

Figure 16. Cracks buried in the CFRP sample shown in Fig. 13 measured at 20 MHz. Linear trend is removed.

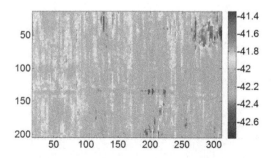

Figure 17. Holes buried in the CFRP sample shown in Fig. 13 measured at 20 MHz. Linear trend is removed.

Subsequently, Fig. 16 indicates several cracks measured on the curved CFRP sample, while a hole can be clearly observed in the top right corner of the image in Fig. 17. Similarly to the unintentional crack depicted in Fig. 15, there is a horizontal fault visible in Fig. 17, which has not been explicitly highlighted by the manufacturer.

## ACKNOWLEDGEMENT

The research leading to these results has received funding from EU 7th Framework Programme managed by REA under grant agreement no. 314935. Part of this work was also funded by the Polish National Ministry of Science and Higher Education under grant agreement no. 2826/7. PR/2013/2.

The research has been undertaken as a part of the project entitled "Radio Frequency Sensing for Non-Destructive Testing of Carbon Fibre Reinforced Composite Materials for Structural Health Monitoring" – Comp-Health, in the collaboration between the following organizations: E.T.S. Sistemi Industriali srl, Kingston Computer Consultancy Limited, Nemetschek ood, Atard a.s., UAB Elmika, Warsaw University of Technology, National Technical University of Athens and Brunel Innovation Centre of Brunel University London.

## REFERENCES

[1] Y. Zhou, M. Guo, F. Zhang, S. Zhang, and D. Wang, "Reinforced concrete beams strengthened with carbon fibre reinforced polymer by friction hybrid bond technique: Experimental investigation," *Material and Design*, vol. 50, pp. 130–139, 2013.

[2] R. Garcia, Y. Helal, and K. Pilakoutas, "Bond behavior of substandard splices in RC beam externally confined with CFRP," *Construction and Building Materials*, vol. 50, pp. 340–351, 2014.

[3] X.L. Zhao and L. Zhang, "State-of-the-art review on FRP strengthened steel structures," *Engineering Structures*, vol. 26, no. 8, pp 1808–1823,2007.

[4] Available: www.compositesworld.com/articles/boeing-787-update, last accessed on 26/03/2015.

[5] C. Scarponi and G. Briotti, "Ultrasonic technique for the evaluation of delaminations on CFRP, GFRP, KFRP composite materials," *Composites Part B: Engineering*, vol. 31, no. 3, pp. 237–243, 2000.

[6] M.D. Farinas, T.E.G. Alvarez-Arenas, E.C. Aguado, and M.G. Merino, "Non-Contact Ultrasonic Inspection of CFRP Prepregs for Aeronautical Applications during Lay-Up Fabrication," IEEE IUS Book Series, pp. 1586–1589, 2013.

[7] J. Chen, H.L. Ji, J.H. Qiu, T. Takagi, T. Uchimoto, and N. Hu, "Novel electromagnetic modeling approach of carbon fibre-reinforced polymer laminate for calculation of eddy currents and eddy current testing signals," *Journal of Composite Materials*, vol. 49, no. 5, pp. 617–631, 2015.

[8] G. Mook, R. Lange, and O. Koeser, "Non-destructive characterization of carbon-fibre-reinforced plastics by means of eddy-currents," *Composites Science and Technology*, vol. 61, no. 6, pp. 865–873, 2001.

[9] F. Taillade, M. Quiertant, and K. Benzarti, "Shearography and pulsed stimulated infrared thermography applied to a nondestructive evaluation of FRP strengthening systems bonded on concrete structures," *Construction and Building Materials*, vol. 25, pp. 568574, 2011.

[10] M. Ravuri, M. Abou-Khousa, S. Kharkovsky, R. Zoughi and R. Austin, "Microwave and Millimeter Wave Near-Field Methods for Evaluation of Radome Composites," *Review of Progress in Quantitative Nondestructive Evaluation Quantitative Nondestructive Evaluation*, vol. 27B, pp. 976–981, 2008.

[11] S. Kharkovsky and R. Zoughi, "Microwave and Millimeter Wave Nondestructive Testing and Evaluation—Overview and Recent Advances," *IEEE Instrumentation and Measurement Magazine*, vol. 10, no. 2, pp. 26–38, 2007.

[12] B. Salski, W. Gwarek, and P. Korpas, "Electromagnetic inspection of carbon-fiber-reinforced polymer composites with coupled spiral inductors," *IEEE Trans. Microw. Theory Techn.*, vol. 63, no. 7, pp. 1535–1544, 2014.

[13] B. Salski, W. Gwarek, P. Korpas, S. Reszewicz, A.Y.B. Chong, P. Theodorakeas, I. Hatziioannidis, V. Kappatos, C. Selcuk, T.H. Gan, M. Koui, M. Iwanowski, and B. Zielinski, "Non-destructive testing of carbon-fibre-reinforced polymer materials with a radio-frequency inductive sensor," *Compos. Struct.*, vol. 122, no. 104, Nov. 2014.

[14] ATARD, Defence and Aerospace Industry Advanced Technology Applications Research and Development Inc., private communication, February 2013.

*Emerging Technologies in Non-Destructive Testing VI – Aggelis et al. (Eds)*
*© 2016 Taylor & Francis Group, London, ISBN 978-1-138-02884-5*

# Detection of the interface between two metals by DC current stimulated thermography

N.J. Siakavellas & J. Sarris

*Department of Mechanical Engineering and Aeronautics, University of Patras, Patras, Greece*

ABSTRACT: The capability of DC current thermography in detecting the interface between two (different) metals is investigated numerically and experimentally. The investigations concern the detection of the boundary between two halves of a plate, each half made of a different material. The numerical and the experimental results show that the sharpness of the interface depends on how dissimilar the two materials are, on the number of gradations in the temperature field (which is a function of the sensitivity of the infrared camera as well as the maximum increase in temperature within the work-piece) and finally on the position of the two electrodes that apply the current.

## 1 INTRODUCTION

In most practical cases metallic structures are covered with paint or other dielectric coating. Corrosion, cracks and other defects hidden under the coating is a major problem as it is difficult to detect them by visual inspection. Conventional inspection techniques like eddy current testing (Smith & Hugo 2001, Smith 1995a, Hagemaier & Nguyen 1994) and ultrasonic (Smith 1995a, Silva et al. 2003) are widely used for such non-destructive evaluations. Other techniques employed are X-ray radiography (Smith 1995b, Wang et al. 2011), neutron radiography, thermal wave interferometry, etc. In many cases however the above techniques are either time-consuming or expensive, or both. Recent developments in infrared cameras and computer technology have made infrared thermography popular for various engineering applications, including non-destructive evaluation (Grinzato & Vavilov 1998, Grinzato et al. 2007, Genest et al. 2009).

In recent years, several researchers have focused on the detection and identification of cracks by active thermography in a wide range of materials (Sakagami & Kubo 2001, Oswald-Tranta 2007, Tsopelas & Siakavellas 2010, Noethen et al. 2010, Tsopelas & Siakavellas 2011, Walle & Netzelmann 2006). In most of the cases, these investigations were limited to the detection of rather idealized cracks that have the shape of an insulating line. In realistic situations however, a defect can be thought of as a region of lower electrical and thermal conductivity which slows down the rate of flow of current and heat respectively, across the test piece. The formation of such regions may be due either to the ageing of the material (e.g. fatigue, corrosion etc),

or to a repaired metallic structure. Such a situation may arise for example in case where the repair of a defective metallic component is necessary: Instead of replacing the whole component, one may replace solely the defective region by using a metallic piece (patch) made of a dissimilar metal. It is therefore necessary to detect the different regions below the coating, without having to remove it. In all the above cases however, the changes of the electrical and thermal properties of the defective material in respect to the surrounding (sound) one are less abrupt than in the case of a crack.

In a previous work of ours the detection of the plate boundary separating two different materials by active thermography was explored (Siakavellas & Sarris 2014). For the excitation two heat sources were used, either a hot air jet, or a quartz halogen emitter. The detection of the plate boundary was based on the heat transfer rate within the workpiece: depending on the thermal diffusivity of the material, heat propagates into the medium faster or slower. A key parameter is the orientation of the heat flow within the plate in respect to the boundary separating the two materials. Eddy current or DC current excitation have the advantage that the detection of the boundary is based on the fact that it disturbs not only the heat flow but the current flow as well. So, it is of particular interest to investigate the capability of inductive (eddy current) and conductive (dc current) thermography in detecting regions of different or similar composition to the surrounding material. Excitation by eddy current thermography has been investigated numerically (Siakavellas & Tsopelas, in press), while an experimental validation is still pending. In that case however, if the induction coil and the infrared camera are placed on the same side of the tested sample, then the coil occludes the sample and the camera will not have a view. Therefore, the coil and the camera must be located on opposite sides of the work-piece, i.e. the measurements are carried out only in transmission mode. This however may create a problem in case where the rear side of the work-piece is not accessible. On the contrary, in case of DC current excitation, the measurements may be carried out either in transmission or in reflection mode.

The present work evaluates experimentally and numerically the capability of conductive thermography in detecting the boundary separating two sections of different composition.

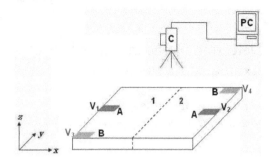

Figure 1. Schematic diagram of the experimental facility: **1**: Section 1 of inspected work-piece, **2**: Section 2 of the work-piece, $V_1$ and $V_2$: position **A** of the electrodes, $V_3$ and $V_4$: position **B** of the electrodes, **C**: infrared camera, **PC**: computer.

## 2 EXPERIMENTS AND NUMERICAL SIMULATIONS

In a first approach of the problem, we performed numerical simulations and experiments concerning the detection of the boundary between two halves of a plate, each half made of a different material. A schematic diagram of the experimental facility is illustrated in Figure 1. The work-pieces under inspection are rectangular plates. Namely two work-pieces were used in the experiments and the numerical simulations:

1. A rectangular (30 cm x 15 cm) plate with 1 mm thickness. One half of the plate is made of copper, the other half of brass.
2. A rectangular (30 cm x 15 cm) plate with 1 mm thickness. One half of this plate is made of iron, the other half of stainless steel.

The conductive excitation of the work-piece under inspection is achieved by direct galvanic contacts, placed at position A ($V_1$, $V_2$) or B ($V_3$, $V_4$), as it is shown in Figure 1.

The face of the plate facing the IR camera was coated by a flat black paint to maintain a high emissivity (except the small area occupied by the electrodes). Thus, the surface of the plate has the same emissivity (about 0.95). In this way the thermal field developed inside the plate was not affected significantly by the environmental thermal radiation.

For the capture of IR images, *Flir SC660* camera was employed connected to a PC through the software package *FLIR ResearchIR*. The detector analysis was 640 × 480 pixels and the thermal sensitivity 0.05 °C. The image update rate (frames per second) was 15–30 Hz.

For the work-pieces considered, numerical simulations were performed as well. For the numerical investigations, the galvanic contacts are placed not only at position A, but also at opposite corners (position B in Fig.1). In the former case, both the current flow and the heat flow are perpendicular to the separating boundary, while in the latter the current and heat flow are oblique to the separating boundary.

The model adopted for the numerical analysis is described in the next section.

## 3 MODEL FOR THE NUMERICAL ANALYSIS

For the numerical simulations, it is assumed that the material under inspection has the shape of a plate and lies in the x-y plane (Fig.1). Since the thickness of the work-pieces is $w = 1$ mm, it is further assumed that the plate is electrically and thermally thin in the z direction. This implies that the current density established within the plate by the voltage applied has two components, which may be expressed in terms of the electric potential ($V$) and the electrical conductivity ($\sigma_i$) of the material as: $J_x = -\sigma_i \, \partial V/\partial x$ and $J_y = -\sigma_i \, \partial V/\partial y$. The electric potential is a solution of Laplace equation:

$$\frac{\partial^2 V(x,y)}{\partial x^2} + \frac{\partial^2 V(x,y)}{\partial y^2} = 0 \qquad (1)$$

Equation (1) is subjected to the following boundary conditions:

i. At the galvanic contacts of the electrodes on the plate (Fig.1):

$$V = V_1 \text{ and } V = V_2 \qquad (2a)$$

ii. At the plate boundary:

$$\partial V/\partial x = 0 \text{ and } \partial V/\partial y = 0 \qquad (2b)$$

iii. At the interface separating the two materials:

$$\sigma_1 \frac{\partial V}{\partial n}\bigg|_1 = \sigma_2 \frac{\partial V}{\partial n}\bigg|_2 \qquad (2c)$$

where $n$ is the normal to the boundary.

The power per unit volume generated by the DC currents and used as the thermal source of the process is expressed as:

$$p_i(x,y,t) = \frac{J_x^2 + J_y^2}{\sigma_i} = \frac{1}{\sigma_i}\left[\left(\frac{\partial V}{\partial x}\right)^2 + \left(\frac{\partial V}{\partial y}\right)^2\right] \qquad (3)$$

The assumption that the plate is thin in the z direction implies that the temperature is uniformly distributed over the thickness $w$ of the plate and the heat losses through the circumferential surface, $2Pw$ ($P$ is the perimeter of the plate), are negligible. Then, the energy balance yields the following equation for the plate temperature $T(x, y, t)$:

$$\rho_i c_i \frac{\partial T(x,y,t)}{\partial t} = k_i \left(\frac{\partial^2 T}{\partial x^2} + \frac{\partial^2 T}{\partial y^2}\right) - \frac{h_{i,up} + h_{i,d}}{w}$$
$$\times (T - T_a) - \frac{\varepsilon_{i,up} + \varepsilon_{i,d}}{w}\sigma(T^4 - T_a^4) + p_i(x,y,t) \quad (4)$$

Here $T_a$ is the ambient temperature, $h_{i,up}$ and $h_{i,d}$ are mean heat transfer coefficients for the two faces of the plate (up and down), $\varepsilon_{i,up}$ and $\varepsilon_{i,d}$ are their emissivities, $\sigma_{SB}$ is the Stefan-Boltzmann constant while $k_i$ is the thermal conductivity, $\rho_i$ the material density, and $c_i$ the specific heat. The subscript $i$ takes the values 1 or 2, depending on the material. The term $(h_{i,up} + h_{i,d})$ is approximated by (Holman 1992): $h_{i,up} + h_{i,d} = 1.91\,(\Delta T/L)^{1/4}$, where $\Delta T = T - T_a$ and $L = A/P$ ($A$ is the area and $P$ the perimeter of the heat transfer surface), as suggested by Goldstein et al. (1973) and Lloyd & Moran (1974). Equation (4) is subjected to homogeneous boundary conditions of the 2nd kind on the circumferential surface. The boundary condition at the interface separating the two materials is:

$$k_1 \frac{\partial T}{\partial n}\bigg|_1 = k_2 \frac{\partial T}{\partial n}\bigg|_2 \qquad (5)$$

where $n$ is the normal to the boundary. Finally, the initial condition is taken as:

$$T(x,y,t) = T_0 = T_\alpha \quad \text{at } t = 0 \qquad (6)$$

i.e. it is assumed that initially the temperature of the plate is equal to the ambient one.

For the numerical analysis a finite difference mesh has been considered. The discretization of equations on the mesh has been performed by employing the Crank—Nicolson scheme.

In order to simulate numerically the sensitivity of an infrared camera, the temperature values that result from the computations were rounded off by retaining only one or two decimal digits. For example, if a sensitivity of 0.05°C is assumed, the value 21.4385°C is round off to 21.45°C, while for sensitivity 0.02°C the rounded value is 21.44°C.

## 4 NUMERICAL AND EXPERIMENTAL RESULTS

We present now numerical and experimental results for the detection of the interface separating the two metals. In order to improve the sharpness of the cracks, the numerical experiment results are processed by various techniques. So, besides the isotherms, the depiction of the temperature derivative norm distributions is given in terms of iso-$D_1T$ lines and iso-$D_2T$ lines at various instants. The norm of the 1st and 2nd spatial derivatives of temperature is given respectively as:

$$D_1 T(x,y,t) = \sqrt{\left(\frac{\partial T}{\partial x}\right)^2 + \left(\frac{\partial T}{\partial y}\right)^2} \qquad (7)$$

$$D_2 T(x,y,t) = \sqrt{\left(\frac{\partial^2 T}{\partial x^2}\right)^2 + \left(\frac{\partial^2 T}{\partial y^2}\right)^2} \qquad (8)$$

### 4.1 Numerical results

The observation of a sequence of numerical snapshots (isotherms) indicates that heat flows faster within the left part of the plate and slower within the right part. This is due to the fact that the thermal diffusivity of copper or iron is higher than the corresponding one of brass or stainless steel respectively.

Isotherms in the case of the copper—brass plate at $t = 2$ s, when the electrodes are placed at position A (see Fig.1), are illustrated in Figure 2.

The depiction of the temperature derivative norm distributions removes (totally or partially) the effect of non-uniform heating of the plate. In that case the interface between the two metals is clearer. Iso-$D_1T$ lines (norm of the 1st spatial derivative of temperature, given by eq.7), and iso-$D_2T$ lines (norm of the 2nd derivative of temperature, given by eq.8), at $t = 2$ s, are illustrated in Figures 3 and 4, respectively.

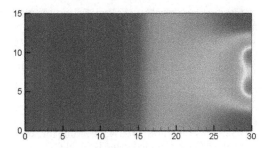

Figure 2. Isotherms in the case of the copper—brass plate at $t = 2$ s (electrodes at position A).

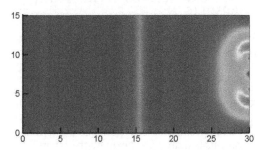

Figure 3. Iso-$D_1T$ lines in the case of the copper—brass plate at t = 2 s (electrodes at position A).

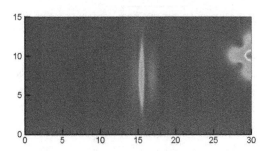

Figure 4. Iso-$D_2T$ lines in the case of the copper—brass plate at t = 2 s (electrodes at position A).

When the electrodes are placed at position A, the heat flow within the plate is parallel to the X–axis (Fig. 1), i.e. it is mainly perpendicular to the interface between the two metals. On the contrary, when the electrodes are placed at position B (Fig. 1), the heat flow is mainly oblique to the interface. In that case, the isotherms give only a vague indication about the interface. The interface is sharper if the temperature derivative norm distributions are depicted. Iso-$D_2T$ lines at $t = 2$ s, are illustrated in Figure 5 and iso-$D_1T$ lines at $t = 50$ s, are illustrated in Figure 6. Notice that at $t = 2$ s (Fig.5) the indication of two existing regions is limited within the central region of the plate, while at $t = 50$ s (Fig.6) the indication extends over the whole length of the interface.

Isotherms, iso-$D_1T$ lines and iso-$D_2T$ lines at $t = 10$ s in the case of the iron—stainless steel plate, when the electrodes are at position A (Fig. 1) are illustrated in Figures 7, 8 and 9 respectively.

The isotherms, which are illustrated in Figure 7, indicate two different regions in the plate. The numerical experiment results are improved when the temperature derivative norm distributions are depicted. The iso-$D_1T$ lines (Fig. 8) and iso-$D_2T$ lines (Fig. 9) show clearly the interface between the two metals.

If now the electrodes are placed at position B (Fig. 1), the heat flow is mainly oblique to the interface. In that case, similarly to the corresponding case for the copper—brass plate, the isotherms give only a vague indication about the interface. The indication is clear when the temperature derivative norm distributions are depicted. Iso-$D_2T$ lines at $t = 50$ s, in the case of the iron—stainless steel plate are illustrated in Figure 10.

### 4.2 Experimental results

In the case of the copper—brass plate, the experimental results give only a vague indication that two different regions exist in the plate. On the contrary, the indication is stronger in the case of the iron—stainless steel plate. This is due mainly to the fact that, in the former case, the heating of the plate is not adequate to create detectable temperature gradients. Indeed, the conductivity of copper and brass

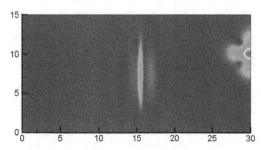

Figure 5. Iso-$D_2T$ lines in the case of the copper—brass plate at $t = 2$s (electrodes at position B).

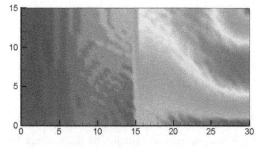

Figure 6. Iso-$D_1T$ lines in the case of the copper—brass plate at $t = 50$ s (electrodes at position B).

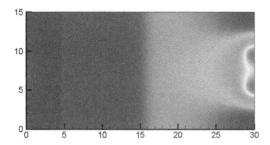

Figure 7. Isotherms in the case of the iron—stainless steel plate at $t = 10$ s (electrodes at position A).

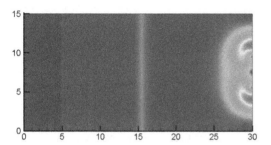

Figure 8. Iso-$D_1T$ lines in the case of the iron—stainless steel plate at $t = 10$ s (electrodes at position A).

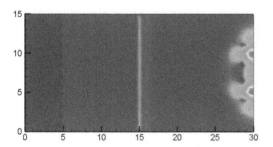

Figure 9. Iso-$D_2T$ lines in the case of the iron—stainless steel plate at $t = 10$ s (electrodes at position A).

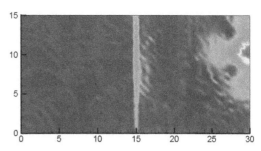

Figure 10. Iso-$D_2T$ lines in the case of the iron—stainless steel plate at $t = 50$ s (electrodes at position B).

Figure 11. Thermogram in the case of the iron—stainless steel plate, at t = 3 s (electrodes at position A).

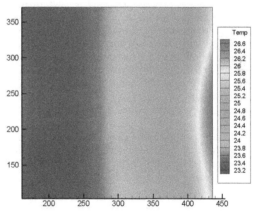

Figure 12. Mean value of temperature over the time period 3–4 s in the case of the iron—stainless steel plate (the central part of the plate is depicted).

are elevated (in respect to the corresponding ones of iron and stainless steel); therefore, the power per unit volume generated by the DC currents (eq. 3) is relatively low. Furthermore, the ratio of the properties of brass (i.e. electrical conductivity, thermal conductivity, thermal diffusivity) to the corresponding ones of copper is about 0.3, i.e. the difference is 70%. On the contrary, the ratio of the properties of stainless steel to the corresponding ones of iron is about 0.2, i.e. the difference between the properties of the two metals is 80%. In other words, the plate that is made of iron and stainless steel is made of two metals that are more dissimilar than the metals of the other plate (copper—brass).

Isotherms (thermogram) at $t = 3$ s in the case of the iron—stainless steel plate, when the electrodes are placed at position A, are illustrated in Figure 11.

In order to compact the most useful information of a sequence of thermal images in a single image, the mean value of temperature over a time period is depicted. If $t_w$ is the time window and $m$ is the frame rate, then the number of images to be analyzed is $N = m\, t_w$. Let $T(x, y, n)$ be the temperature at a particular location $(x, y)$ of the $n$th thermogram in the image sequence ($1 \leq n \leq N$). The mean value of temperature $T_{mean}(x, y; t_w)$ at $(x, y)$ over the time period $t_w$ is computed as:

$$T_{mean}(x,y;t_w) = \frac{1}{N}\sum_{n=1}^{N} T(x,y,n) \qquad (9)$$

The mean value of temperature over the time period: $3\,\text{s} \le t \le 4\,\text{s}$, in the case of the iron—stainless steel plate, is illustrated in Figure 12.

## 5 CONCLUSIONS

The numerical and the experimental results show that the sharpness of the interface depends on how dissimilar the two materials are, on the number of gradations in the temperature field (which is a function of the sensitivity of the infrared camera as well as the maximum increase in temperature within the work-piece) and finally on the position of the two electrodes that apply the current.

The discrepancy between experimental and numerical results may be attributed mainly to the following factors: (i) As the plate materials are not pure (e.g. 100% copper or 100% iron), but they have a low content of impurities, their physical properties (thermal conductivity, thermal capacity, mass density etc) are not given in high precision by the manufacturer (i.e. the materials used in the experiments were not standard). (ii) The contact between the electrodes and the work-piece was not perfect. In fact, there is a thermal contact resistance, which, in the numerical simulations, was not taken into consideration.

The present numerical an experimental investigations show that the detection of the interface between two (different) metals is possible by dc current thermography. Alternative approaches of active thermography may also be used, i.e.: ac current thermography, eddy current thermography, pulsed phase thermography etc. Furthermore, besides the data processing techniques used in the present work, the effectiveness of techniques like Fourier transform, wavelet transform, etc. must be explored in combination to the mode of excitation.

## REFERENCES

Genest, M., Martinez, M., Mrad, N., Renaud, G. & Fahr A. 2009. Pulsed thermography for non-destructive evaluation and damage growth monitoring of bonded repairs. *Composite Structures* 88: 112–120.

Goldstein, R.J., Sparrow, E.M. & Jones, D.C. 1973. Natural convection mass transfer adjacent to horizontal plates. *Int. J. Heat Mass Transfer* 16:1025.

Grinzato, E. & Vavilov. V. 1998. Corrosion evaluation by thermal image processing and 3D modelling. *Revue Generale de Thermique* 37: 669–679.

Grinzato, E., Vavilov, V., Bison, P.G. & Marinetti, S. 2007. Hidden corrosion detection in thick metallic components by transient IR thermography. *Infrared Physics & Technology* 49: 234–238.

Hagemaier, D. & Nguyen, K. 1994. Automated eddy current scanning of aircraft for corrosion detection. *Materials Evaluation* 52 (1): 91–95.

Holman, J.P. 1992. *Heat Transfer*. London: McGraw Hill.

Lloyd, J.R. & Moran, W.R. 1974. Natural convection adjacent to horizontal surface of various planforms. ASME Pap. 74-WA/HT-66.

Noethen, M., Wolter, K.J. & Meyendorf, N. 2010. Surface crack detection in ferritic and austenitic steel components using inductive heat thermography. 33rd Int. Spring Seminar on Electronics Technology, IEEE, p. 249–254.

Oswald-Tranta, B. 2007. Thermo-inductive crack detection. *Nondestruct. Test. Eval.* 22: 137–153.

Sakagami, T. & Kubo, S. 2001. Development of New Crack Identification Technique Based on Near-Tip Singular Electrothermal Field Measured by Lock-in Infrared Thermography. *JSME International Journal, Series A* 44: 528–534

Siakavellas, N.J. & Sarris, J. 2014. Detection of the Plate Boundary Separating Two (a) Different Materials, and (b) Different Thickness Sections by Active Thermography. *Sensor Letters* 12 (1): 102–112.

Siakavellas, N.J. & Tsopelas, N. Detection of the interface between two metals by eddy current thermography. *Nondestruct. Test. Eval.* DOI: 10.1080/10589759.2015.1034716.

Silva, M.Z., Gouyon, R. & Lepoutre, F. 2003. Hidden corrosion detection in aircraft aluminum structures using laser ultrasonics and wavelet transform signal analysis. *Ultrasonics* 41: 301–305.

Smith, R.A. & Hugo, G.R. 2001. Transient eddy current NDE for ageing aircraft—capabilities and limitations. *Insight* 43(1): 14–25.

Smith, RA. 1995. Non-destructive evaluation for corrosion in ageing aircraft. Part 1. Introduction, ultrasonic and eddy current methods. *Insight* 37 (10): 798–807.

Smith, RA. 1995. Non-destructive evaluation for corrosion in ageing aircraft. Part 2. Radiography, mechanical impedance and alternative methods; comparison of techniques. *Insight* 37 (11): 884–891.

Tsopelas, N. & Siakavellas, NJ. 2010. Eddy current thermography in circular aluminium plates for the experimental verification of an electromagnetic-thermal method for NDT. *Nondestruct. Test. Eval.* 25: 317–332.

Tsopelas, N. & Siakavellas, N.J. 2011. Experimental evaluation of electromagnetic-thermal non-destructive inspection by eddy current thermography in square aluminium plates. *NDT&E Int* 44: 609–620.

Walle, G. & Netzelmann, U. 2006. Thermographic Crack Detection in Ferritic Steel Components Using Inductive Heating. *Proceedings of the Ninth European Conference on NDT*, Berlin, Germany, Sept. 25–29, Paper No Tu.4.8.5.

Wang, X., Wong, B.S., Tan, C.S. & Tui, C.G. 2011. Automated crack detection for digital radiography aircraft wing inspection. *Research in Nondestructive Evaluation* 22: 105–127.

*Emerging Technologies in Non-Destructive Testing VI – Aggelis et al. (Eds)*
*© 2016 Taylor & Francis Group, London, ISBN 978-1-138-02884-5*

# Water-Absorption-Measurement instrument for masonry façades

## M. Stelzmann & U. Möller
*University of Applied Sciences, Leipzig, Germany*

## R. Plagge
*University of Technology, Dresden, Germany*

ABSTRACT: This paper introduces an in-situ device for measuring the water absorption coefficient of masonry façades. The "Water-Absorption-Measurement" instrument (WAM) allows non-destructive determination of the time-based function of water penetration. In a dense stack box a masonry façade area of $0.40 \times 0.51$ m is watered by nozzles. The gravimetric measuring system reaches a repeatable accuracy up to $\pm 8.3 \times 10^{-4}$ kg / (m² × √s). Especially on fair-faced masonry, it is possible to measure the water penetration on an average area with bricks and mortar. This report contains detailed information about functionality of the developed measuring concept. Measuring results with laboratory—and a masonry façade are demonstrated. Customary instruments like the RILEM tube are compared to the developed measuring instrument.

## 1 INTRODUCTION

Protection against rain water penetration is an important factor in building physics for the evaluation of historic masonry façades. If rain water hits the building materials of the masonry façade, it can infiltrate into the building structure by capillary power. Damages caused by moisture such as damage by frost can be the result of too much water penetration. The resistance to capillary water penetration is described with characteristic values of the basic material water absorption coefficient ($A_w$). The lower the water absorption coefficient, the higher the protection against capillary water penetration and associated moisture based damage. Limits for the $A_w$—value for masonry façade materials are defined in various standards (e.g. plaster standard EN 998–1, DIN 4108–3 for climate-related moisture protection or WTA Guideline 6–4 for planning inside insulation). Depending on rain load, position, orientation and design of the masonry façade, these values are ranged between $3.3 \times 10^{-2}$ and $1.7 \times 10^{-3}$ kg/(m² × √s). In a region of high rainfalls for example, a maximum value $A_w$ of $8.3 \times 10^{-3}$ kg/(m² × √s) is prescribed for exterior plaster. Furthermore, the $A_w$—value is required for hygrothermal simulations on façades. If the $A_w$—value of a masonry façade is known, the moisture behavior can be simulated by using numerical programs.

The $A_w$—value is determined in the laboratory by using a standardized test. The EN ISO 15148 describes the necessary procedure. It provides the basis for all further considerations. In the experiment, an air-dried sample is immersed a few millimeters in a water bath. At defined time intervals, the sample has to be removed from the water and weighted. The weight gain of the test specimen corresponds to the amount of water absorbed. The medium rise of weight gain referring to the surface of the immersed test specimen (1) and the root of the duration of the test (2), the water absorption coefficient of the sample construction material in kg/(m² × √s) is calculated therefrom, with:

$$\Delta m_t = \frac{(m_t - m_i)}{A} \tag{1}$$

where $m$ = weight of test specimens; $A$ = surface of the immersed test specimen and

$$A_w = \frac{\Delta m'_{tf} - \Delta m'_0}{\sqrt{t_f}} \tag{2}$$

where $\Delta m'_{tf}$ = the value of $\Delta m$ at medium rise by weight of test specimens to surface of the immersed test specimen of test specimens; $t_f$ = testing time.

For a precise examination of historic masonry façades destructive taking of samples is necessary. In order to make a representative statement about an entire façade, several points should be investigated accordingly. This is not always possible in terms of historic masonry façades. Here it is necessary to use other non-destructive measurement techniques that can be used directly on masonry façades.

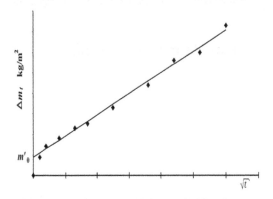

Figure 1. Graph of water absorption plotted to root of time (EN ISO 15148).

## 2 PREVIOUS NON-DESTRUCTIVE MEASURING DEVICE

There are already a number of methods available for determination the water absorption coefficient of masonry façades. The main methods for non-destructive testing are described in this section. In (Möller & Stelzmann 2013) the pros and cons of the following examination methods are presented.

### 2.1 RILEM tube

The RILEM tube (RILEM 1980) is a classical tool for consultants for determination of water absorption of façades. The RILEM tube is nearly 0.13 m tall and has an absorption surface area about $4.91 \times 10^{-4}\,m^2$. Using permanent elastic sealing putty the device made from glass or plastic is attached to a sample area. The unit is filled with water by a small tube with a printed measurement scale.

Thru reading the water level in the tube, the water consumption of material is concluded. Usually a measurement period is takes 15 minutes. The readings in ml/min are to be compared with limit values in the end. The RILEM tube is an easy manageable, inexpensive and fast device. Due to a smaller absorption surface area it is necessary to do many measurements. With a large scattering the water activity of the façade can be valued. Besides, the sealing putty is affecting the examination surface. Another disadvantage of this procedure is to achieve a permanent hydraulic water pressure, which pressurizes the test surface. Thereby micro cracks are rated too critically in practice.

### 2.2 Testing panel for water permeability as per Prof. Franke

A further development of the RILEM tube is the testing panel for water permeability as per Prof. Franke (Franke & Bentrup 1991). This device has been invented for testing a masonry hydrophobic treatment. The absorption surface area of the device is 0.25 m wide and 0.081 m high. Therefore the test panel for water permeability has a larger examination area as the RILEM tube. The measurement duration is typically 15 minutes. Attaching and measuring process have nearly the same functionality as the RILEM tube. In case of the water absorption testing panel after Franke the readings will be compared with prescribed limit values. Hydraulic water pressure also appears on the examination surface underneath the device.

### 2.3 ASTM C1601

The international standard ASTM C1601 describes a field study to determination of water absorption of masonry façades. The sample is implemented in a testing box with an absorption surface area of about 1 $m^2$. The enclosed masonry façade surface

Figure 2. Vertical application of Rilem test method no. 11.4 RILEM tube is fixed on wall.

Figure 3. Testing panel for water permeability as per Prof. Franke is used on a historical fair-faced masonry façade.

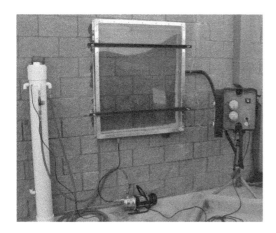

Figure 4. Measuring instrument determination of water absorption of masonry facades after ASTM C1601 (BDG-USA 2015).

gets artificially sprinkled with water. The loss of water corresponds to the amount resorbed water by the masonry façade. It is necessary to anchor the testing box on the masonry façade. The representative absorption surface area of 1 m² is being examined. The measurement duration is about 4 hours. Inside the device the air pressure is being artificially increased by a blower. Towards the reading of the amount of water absorbed by the façade, the water level is monitored in a container. The accuracy during a reading is ± 0.1 l. In contrast to the two methods, described before, the field study is working with significant lower pressure. The ASTM C1601 is a complex and time-consuming procedure. Furthermore it is a partially destructive testing method.

3 DEVELOPED MEASURING METHOD

Since 2012 a new in-situ device called "Water-Absorption-Measurement" instrument, has been developed at the University of Applied Sciences in Leipzig. This new measuring method is able to determine the water absorption coefficient of masonry façades.

A schematic sketch of the principle is shown in figure 5. This new developed measurement method is based on the principle of ASTM C1601. Contrary to ASTM C1601 it has a smaller reference surface of 0.2 m². Furthermore a gravimetric principle of measurement is used in the "Water-Absorption-Measurement" instrument. For measuring process the "Water-Absorption-Measurement" instrument is fixed directly on the surface of the masonry façade with a special sealing compound. Additionally the instrument can be attached to the façade directly on the scaffold or on a hook.

The prototype of the "Water-Absorption-Measurement" instrument ist shown in figure 6. The result is a waterproofed area with the dimensions of 0.40 × 0.51 m. With the help of a pump the area of the façade can be artificially sprinkled. A closed water film is formed in the surrounded façade area. Depending on the quality of the masonry façade a part of the water gets absorbed. The rest flows back through a vent into the water container. A water circulation with just one way out over the surface of the façade is consisting. The weight of

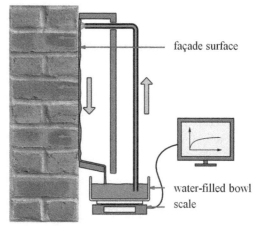

Figure 5. Schematic sketch of the principle of developed "Water-Absorption-Measurement" instrument.

Figure 6. Prototype of developed "Water-Absorption-Measurement" instrument at a historical fair-faced masonry façade.

283

water container is measured permanently with a scale underneath. The loss of mass corresponds to the water absorbed by the masonry façade. The gravimetric measurement principle achieves a reproducible accuracy of $\pm 8.3 \times 10^{-4}$ kg/(m² × √s). With the innovative procedure a representative area of $0.40 \times 0.51$ m is measured. Especially for fair-faced masonry façades this possibility of an integral measurement of the water absorption coefficient yields over more than one stone and joint layer. In the end of a measurement the "Water-Absorption-Measurement" instrument is removed without leaving a trace on the façade. One measurement duration is taking usually 3.600 sec, however this is not depending on the measurement principle. For instance a strongly sucking underground can already lead to a significant result after 1.200 sec.

During a subsequent evaluation the mass of the container before and after measurement as well as the continuously difference in weight of the water container were calculated up to a function of the water absorption. In consideration of the additional loss of water throughout the system, such as moistening within the measuring chamber, evaporation rate and water edge effects. Out of the function of the capillary water absorption is the water absorption coefficient analogue to an evaluation according to EN ISO 15148 determined. During different laboratory tests the new developed measuring method achieves a reproducible precision up to $\Delta A_w = 8.3 \times 10^{-4}$ kg/(m² × √s). In (Stelzmann et al. 2013, Möller & Stelzmann 2013) more studies describe the edge effects, the instrument calibration and measuring results.

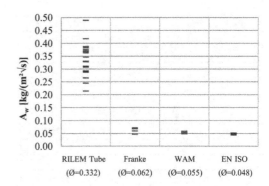

Figure 7. Measuring results of calcareous sandstone by laboratory experiment.

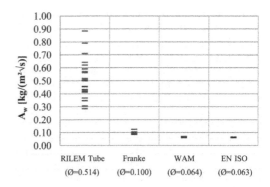

Figure 8. Measuring results of autoclaved aerated concrete by laboratory experiment.

## 4 LABORATORY EXPERIMENTS

The following section contains results of tests with in-situ-devices that are explained before. In a laboratory experiment the RILEM tube, the testing panel for water permeability as per Franke and the "Water-Absorption-Measurement" instrument were compared to the standard test EN ISO 15148. The Surrounding temperature was about $22 \pm 3°C$, relative air humidity ranges between 35% —r.h. and 50% —r.h. Calcareous sandstone and autoclaved aerated concrete material were used. For this test bricks without separations were utilized. The size of the test specimens is about $0.51 \times 0.57 \times 0.115$ m. Both materials are artificial and homogeny, therefore ideal conditions for experiments in a laboratory are created. For fixing the devices an elastic sealing was used. Before the devices were added, the test specimens were dried in a furnace. The testing time of each individual measurement of the various devices was all in all 60 minutes. The $A_w$—values were determined in accordance to EN ISO 15148. It means in effect that the different size of the measuring area of the specific examination method is used for calculation.

Results of the laboratory experiment are shown in figure 7 and 8. The number of single measurements of each material was by RILEM tube more than 20, by other methods more than 5. Results by RILEM tube present a high variation of single values. Also the average values do not associate with standard test EN ISO 15148. The testing panel for water permeability as per Franke and the "Water-Absorption-Measurement" instrument have little scattering and nearly correspond to EN ISO 15148.

## 5 FIELD EXPERIMENTS

In the next step the RILEM tube, the testing panel for water permeability as per Franke and the "Water-Absorption-Measurement" instrument were tested on a building façade. At these in-situ-tests it was not possible to take samples for testing in labratory. The results of all three in-situ-devices can be compared to each other. The tested façade is a historical fair-faced masonry in Leipzig, Germany (Figure 9). The house was renovated before the tests. The clinkers were cleaned and mortar joints were changed. For renovating the façade, first the old joints were removed completely about 2 cm deep inwards the façade-surface. Then a low hydrophobic effect was applied to the old clinker.

Now the façade was flatly grouted with a mortar-sludge. Finally the clinkers were cleaned by rests of mortar. The main aim of these tests is to quantify the quality of wind-driven-rain-resistance. Therefore the main criterion is the averaged water absorption coefficient of the façade surface materials. The research object got a renovated historical façade as fair-faced masonry. To define the water absorption coefficient of the façade, it is necessary to measure clinker and mortar. Especially at the RILEM tube the measuring results are weighting by part of clinker and mortar. During the measurement no difficulties occurred. Figure 10 contains the results of the in-situ-tests. A typical graph of measuring data by the "Water-Absorption-Measurement" instrument is shown in Figure 11.

The measured values for the RILEM tube are excessive scattering. In contrast to results of the laboratory experiment the average value for the RILEM tube is substantial smaller. The scatter of the testing panel for water permeability as per Franke and the "Water-Absorption-Measurement" instrument are significantly smaller. The WAM instrument has, compared to the method as per Franke, a lower absolute value. Nevertheless the absolute $A_w$—value of the method as per Franke is twice as big as the absolute $A_w$—value of the WAM instrument. As a cause for these results the different sized measuring surfaces within the methods can be named. A relatively larger measuring size

Figure 9. Tested historical fair-faced masonry façade in Leipzig, Germany.

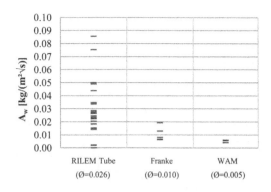

Figure 10. Measuring results of field experiments for determinate water absorption coefficient at historical masonry façade.

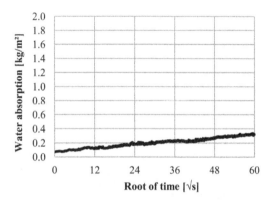

Figure 11. Raw data of "Water-Absorption-Measurement" instrument at testing described masonry façade, resulting $A_w$ value is 0.004 kg / (m² × √s).

is the result of less influence of edge effects und a representative sample. A lower testing-pressure is the result of less influence of fine cracks and tear off by clinker and mortar. In addition, more examination results of masonry façades are shown in Stelzmann 2013, Möller & Stelzmann 2013).

As a Result, the RILEM tube is to be utilized for estimation of water absorption of masonry facades. Consequently, it is not possible to measure the water absorption coefficient with a needed accuracy thereby. The testing panel for water permeability as per Franke has a bigger absorption surface area. For specific application it is possible to assess the water absorption coefficient of masonry facades by the method as per Franke. In contrast to the RILEM tube and the testing panel for water permeability as per Franke, the "Water-Absorption-Measurement" instrument is using another measuring principle. A larger measuring size and a waiver of a measuring pressure results in a lower scatter and a higher accordance to standard test.

Figure 12. Further development of „Water-Absorption-Measurement" instrument, ready for series production version, WAM 100 B.

## 6 CONCLUSION

In this paper, the relationships and the problems of wind-driven-rain-resistance and water absorption coefficient of masonry façades are briefly explained. Following is a presentation of known in-situ test methods for determining the water absorption coefficient. Subsequently, the developed "Water-Absorption-Measurement" instrument is presented. The composition, the measurement principle, the measurement process and the measurement evaluation of the developed instrument are discussed. Subsequent are results from laboratory and field experiments.

The developed "Water-Absorption-Measurement" instrument allows non-destructive determination of the water absorption coefficient of masonry façades. The gravimetric measuring system reaches a repeatable accuracy up to $\pm 8.3 \times 10^{-4}$ kg/(m² × √s). Especially on fair-faced masonry a representative sample about $0.40 \times 0.51$ m is tested.

In cooperation with a local partner, the University of Applied Sciences in Leipzig also developed a version of the "Water-Absorption-Measurement" instrument which is ready for series production. The "WAM 100 B" is a bit smaller than the prototype. In Figure 12 the "WAM 100 B" is shown during a measurement.

For the WAM instrument a mainly usage is to check the existing wind-driven-rain resistance in light of planning an inside insulation on historical fair-faced masonry façades. In practice the WAM instrument can be used for preliminary investigation or quality control for renovating masonry facades. For example after finishing a hydrophobic impregnation the created effect can be checked. Furthermore the optimal solution for renovating a masonry façade can be selected by using different test surfaces. The measured values also can be used for research and development as input for simulation programs.

## REFERENCES

ASTM C1601:2014, Standard Test Method for Field Determination of Water Penetration of Masonry Wall Surfaces.

BDG-USA 2015: Building Diagnostics Group, Inc (BDG)—Corporate Office Online: 2015-03. http://www.bdg-usa.com/astm-1601.html.

DIN 4108–3:2014–11, Thermal protection and energy economy in buildings—Part 3: Protection against moisture subject to climate conditions—Requirements and directions for design and construction.

EN 998–1:2010, Specification for mortar for masonry—Part 1: Rendering and plastering mortar.

EN ISO 15148:2002, Hygrothermal performance of building materials and products Determination of water absorption coefficient by partial immersion (ISO 15148:2002).

Franke, L. & Bentrup, H. 1991. Einfluß von Rissen auf die Schlagregensicherheit von hydrophobiertem Mauerwerk und Prüfung der Hydrophobierbarkeit. *Bautenschutz + Bausanierung. nr 14*: 98–101 and 117–121.

Möller, U. & Stelzmann, M. 2013a. Neue Messmethode zur Bewertung der kapillaren Wasseraufnahme von Fassaden. *wksb nr. 69:* 62–65.

Möller, U. & Stelzmann, M. 2013b. In-Situ-Messgerät für die zerstörungsfreie Messung der Wasseraufnahme. *2th International Interior insulation concress 12th and 13th April 2013.* Dresden, Germany: 188–197.

RILEM 1980, Water absorption under low pressure (pipe method—Test No. 11.4. RILEM Commission 25-PEM. Tentative Recommendations.

Stelzmann, M. 2013. In-situ-Messgerät für die zerstörungsfreie Messung der kapillaren Wasseraufnahme von Fassaden. *Messen, Planen, Ausführen: 24. Hanseatische Sanierungstage 7th to 9th November 2013.* Ostseebad Heringsdorf Usedom, Germany: 175–184.

Stelzmann, M. Möller, U. & Plagge, R. 2013. Messverfahren zur zerstörungsfreien Bewertung der kapillaren Wasseraufnahme von Fassaden. *Bauphysiktage Kaiserslautern 27th and 28th November 201.* Kaiserslautern, Germany: 89–91.

WTA Guideline 6–4:2009–05, Inside insulation according to WTA I: planary guide.

WTA Guideline 6–5:2014–04, Interior insulation according to WTA II: Evaluation of internal insulation systems with numerical design methods.

*Emerging Technologies in Non-Destructive Testing VI – Aggelis et al. (Eds)*
*© 2016 Taylor & Francis Group, London, ISBN 978-1-138-02884-5*

# Assessment of eSHM system combining different NDT methods

M. Strantza & D.G. Aggelis
*Department of Mechanics of Materials and Construction, Vrije Universiteit Brussel, Brussels, Belgium*

D. De Baere & P. Guillaume
*Department of Mechanical Engineering, Vrije Universiteit Brussel, Brussels, Belgium*

D. Van Hemelrijck
*Department of Mechanics of Materials and Construction, Vrije Universiteit Brussel, Brussels, Belgium*

ABSTRACT: A new concept of structural health monitoring, the effective Structural Health Monitoring (eSHM) system, has the potential to improve life safety and decrease the direct operational cost. The basic physical principle of the eSHM system is checking the absolute fluid pressure changes in a 3D-curved network of capillaries that are integrated with 3D printing. A pressure change in the capillary indicates the presence of a crack. The main objective of this study is to evaluate the eSHM system and the crack location by means of various Non-Destructive Testing techniques (NDT). In this approach, liquid penetrant inspection, eddy current, radiography and acoustic emission were used in order to confirm the damage and indicate the damage location on unnotched four-point bending metallic beams with an integrated eSHM system. Prior to this study, those beams have been subjected to four-point-bending fatigue testing and they are damaged as detected by the eSHM system. It will be shown that the eSHM system in a combination with NDT can provide accurate information on the damage condition of 3D printed metals.

*Keywords*: structural health monitoring; liquid penetrant inspection; radiography; acoustic emission; eddy current; additive manufacturing

## 1 INTRODUCTION

In the last three decades, numerous SHM techniques have been investigated, with the focus on damage identification in components (Farrar & Worden 2007) and some techniques were inspired by the biological nervous system (Sharp et. al 2014). The main scientific challenge for SHM is to detect damage, which is a very local phenomenon, by measuring the global response of a structure. Currently, visual inspection and a large number of non-destructive evaluation techniques with a local inspection capability are available to e.g. the aerospace industry.

Failures of various materials that are used in aerospace are revealed and inspected by the use of Non-Destructive Testing techniques (NDT). There are numerous NDT methods that can examine a material like visual inspection, eddy current, ultrasonic inspection, acoustic emission,

eddy current, radiography etc. (Kamsu-Foguem 2012).

On the other hand, 3D printing or Additive Manufacturing (AM) is increasingly gaining momentum the last decade. AM is material efficient manufacturing of (near) net shape products and has the capability to create complex geometrical structures which cannot be produced by conventional production techniques (Vayrea et al. 2012).

In that aspect, a new concept of effective structural health monitoring system takes the advantage of the greatest strength of AM—complex structures. The basic physical principle of the effective Structural Health Monitoring (eSHM) system is checking the absolute fluid pressure variations in a 3D network of capillaries or cavities that are integrated by AM techniques in the interior of a part. The pressure inside the capillary is initially significantly lower (or higher) than the ambient pressure.

A sudden pressure change in the capillary indicates the presence of a crack between the outer surface and the capillary. More details on the eSHM system development can be found in (De Baere et al. 2014).

One of the main advantages of the eSHM system is that a whole structure can be monitored by a single pressure sensor. The pressure sensor is monitoring the pressure variations inside the capillary. The capillary or cavities that are integrated in the metallic structure by means of AM, could mimic the biological nervous system. The size of the capillaries currently can vary between 1 and 3 mm in diameter.

This work is a continuity of previous research on the response of the eSHM system during four-point bending fatigue testing (Strantza et al. 2014). The main objective of the current work is the NDT evaluation of the crack location that is detected by the eSHM system and the evaluation of the damage detection of the eSHM system. The cracks that are going to be evaluated in the current study are cracks on unnotched specimens which have been initiated and propagated during four-point fatigue testing. Before the eSHM system can be implemented in real structures, its results concerning the sensitivity towards damage initiation and propagation should be validated. In this respect, we applied different NDT techniques (liquid penetrant, eddy current, radiography and acoustic emission) to monitor the crack location and development and their results are discussed in relation to the eSHM.

## 2 EXPERIMENTAL STUDY

### 2.1 Test specimen

Four metallic specimens were used for this study. The specimens were produced by additive manufacturing, two Ti-6Al-4V produced by Selective Laser Melting (SLM) and two AISI 316L produced by Laser Metal Deposition (LMD). SLM and LMD are additive manufacturing processes and they differ in the way the layers are deposited, in order to build a material (Levy et al. 2003). All the specimens had an integrated eSHM system by means of a 3D sinusoidal shape of capillary which was implemented during manufacturing. The integrated capillary had a diameter of 3 mm for the LMD samples (AISI 316L) and 1 mm for the SLM samples (Ti-6Al-4V). The sinusoidal shape had amplitude of 2 mm and a period of 25.7 mm. The dimensions of the samples are shown in Figure 1.

The specimens were equipped with a pressure transducer (Kulite XTL-123C-190M- 1,7-BAR-A) at one side and with a check valve (Clippard MCV-1-M5) at the other side. Before the installation of the check valve, a thread sealant (Loctite 577) was applied on the first few threads of the valve as an adhesive between the specimen and the check valve. Thread sealants are filling the spaces between threaded metal parts in order to prevent and stop leakages. After sealing, the capillary in the specimen was partially evacuated of air in a vacuum chamber, keeping it at an under-pressure of 0.5 bar. As the final step, a stop (Clippard 11755-M5-PKG) was installed on the check valve as an additional occlusion. Prior to this study, the specimens were subjected to four-point-bending fatigue testing until the damage was detected by the sensor of the eSHM system.

### 2.2 Static loading test procedure

In the current study, the specimens which had been subjected to crack initiation by fatigue were subjected to four-point bending static loading. This test setup produces a uniform moment without shear force between the two inner loading rollers in the specimen, which gives rise to a uniform maximum tensile stress at the lower side of the specimen's outer surface. During the static loading procedure the specimen were monitored by the pressure transducer and two AE broadband transducers (Fig. 2). The acoustic emission sensors were also placed between the supporting pins in a distance of 80 mm. The sensors were of "pico" type of Mistras Holdings having a broadband response and peak frequency at 450 kHz. The signals were pre-amplified by 40 dB and

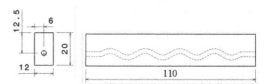

Figure 1. Four-point bending specimen, 2D side and front views.

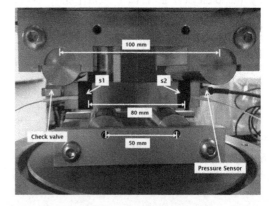

Figure 2. Four-point bending setup with the AE and the pressure sensors placed on the specimen.

recorded by the acquisition board with sampling rate of 10 MHz (micro-II, 8 channels). The threshold was set at 40 dB and the acoustic coupling was improved by vaseline grease between the sensors and the metal specimen. The sensors were secured by means of tape during the loading.

Two AISI 316L specimens (LMD) and two Ti-6Al-4V specimens (SLM) were subjected to static loading. The loading for each specimen was increased in steps of 5 kN starting from zero up to the maximum load of the crack detection (according to the eSHM system). The crack detection in both LMD samples happened at the load level of 34 kN (530 MPa). On the other hand, for the first sample of SLM at 20 kN (322 MPa) and for the second SLM sample at the load level of 36 kN (570 MPa).

## 3 RESULTS AND DISCUSSION

### 3.1 NDT by means of Liquid Penetrant Inspection (LPI), radiography and eddy current

After the crack detection of the samples by the eSHM system, all the samples were inspected by means of LPI level 2 and level 3 in unloaded conditions (ARDROX fluorescent penetrants 9704 and 9705). Initially, the specimens were coated by liquid penetrant by means of a brush. After 15 minutes the penetrant was removed and then the developer was applied. When the developer was dry on the surface, the crack was visible by means of UV light. It can be concluded that for the current material and the current crack size the cracks cannot be clearly located by means of LPI level 2. This may be associated with the fact that the crack in unloading conditions is sufficient closed. The penetration of the level 2 liquid penetrants is not possible compared to the penetration of level 3 liquid penetrants, where the location is easier to be determined but not completely clear. It is possible that the same sensitivity of LP would yield better results when the material is still under load.

Eddy current inspection is used in order to locate surface and near-surface (few mm below the surface) defects (Garcia-Martin et al. 2011). In general, the eddy current technique is used to examine rather small surface regions. The probe design and test parameters must be determined with a good understanding of the defect. In order to locate the cracks by eddy current, a NORTEC 500 Series eddy current flaw detector from Olympus was chosen with a probe of 2 MHz (Figure 3). The locations of the cracks were nicely detected and this will be also confirmed later by LPI level 2 under loading conditions.

Additional to eddy current and LPI, radiography was also performed to the specimens. Radiography is a non-destructive technique which can provide information on the internal state of a sample (Rometsch et al. 2014). It requires the projection and penetration of radiation energy through an inspected material. The radiation energy is absorbed homogenously by the material, except in the regions where thickness or density variations arise. The energy that cannot be absorbed passes through to a sensing medium (film) that captures an image of the specimen. For the current study the equipment that was used is of Baltospot 200 kV type. In Figure 4 the radiographic images of the first sample of AISI SS316 and the

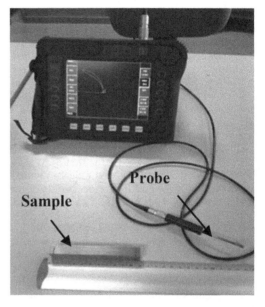

Figure 3. Eddy current stet-up during detection of the location of a crack on one of the samples that are inspected.

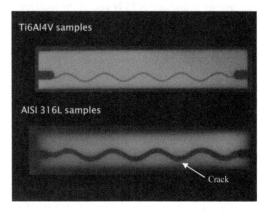

Figure 4. Radiographic images of the four-point bending specimens after the crack detection of the eSHM system.

first sample of Ti-6Al-4V is shown. The crack is clearly shown in AISI 316L and it is located on the lower part of the sinusoidal capillary. However, the crack in Ti-6Al-4V could not be detected by radiographic images. The difficulties on the detection may associate with the angle between the crack and the radiation. If the radiation is not parallel with the defect, the defect may appear distorted, out of position and less visible in the image.

### 3.2 NDE by means of Acoustic Emission (AE) and Liquid Penetrant Inspection (LPI) level 2 during static loading of the samples

Prior to the static loading, a pressure of approximately 0.5 bars was applied in the capillaries. When the specimens are not under stress, the crack is closed so the internal pressure was maintained. Additionally, the specimens were monitored by AE during test. For the first two samples of LMD, the loading was increased in steps of 5 kN from 0 kN until the final loading of crack detection of 34 kN and then unloading in steps of 5 kN. For the other two SLM samples the same procedure is followed until the crack detection load during fatigue is reached. Which was 20 kN for the first SLM sample and 36 for the second SLM sample. In Figure 5 the behaviour of the presure during loading of the first specimen of LMD can be clearly observed. It is depicted that at the load level of 19,5 kN a rise of the pressure is observed, where at 34 kN the pressure is reaching the atmospheric pressure which is approximately 1.02 bar according to the pressure sensor.

The same behaviour was observed also in the other 3 samples, which are not depicted graphically in this paper. For the second LMD sample the increase of the pressure was detected at 20 kN and for the second SLM sample at 14.5 kN. On the contrary, for the first SLM sample a pressure increase was noticable at 7 kN and at the load level of 18.4 kN a brittle failure has occured. This sudden failure was caused by porosity or unmelted regions and leads the SLM processed material faster to the failure. In the literature it is demonstrated that the porosity within the SLM samples have a radical effect on the failure and specificaly on the fatigue behaviour (Leuders et al. 2013).

Concurrently with the static loading and the pressure behavior, AE was also monitoring the sample behavior. The AE technique has been conducted in several applications for damage characterization of metallic specimens (Roberts et al. 2003). AE is defined by the literature as the transient elastic wave which propagates in a continuous media due to a sudden release of energy after nucleation or propagation of a crack. In the current study the samples were already cracked so the stage of crack initiation cannot be monitored; however other mechanisms like possible crack propagation and friction of the crack faces can be recorded.

In Figure 6 the cumulative AE hits and the load vs. time are depicted for the first LMD sample. The hit rate is moderate throughout the initial stages of loading and suddenly reaches a maximum value at the load of 34 kN which is the load level of the crack detection during fatigue loading according to the eSHM system. It is also characteristic that the cumulative number of AE hits is 28 hits until 32 kN and up to 34 kN the population has already increased to 120.

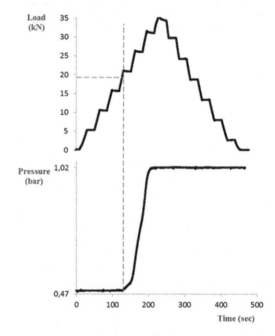

Figure 5. Load history and pressure vs. time for one of the two LMD specimens.

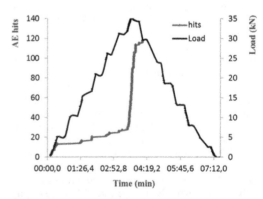

Figure 6. Load history and AE history vs. time for the first LMD sample.

Similar results are also observed in Figure 7, where the amplitude and the load history vs. time are presented. Until the load of 32 kN hits exceeding the amplitude of 50–60 dB were scarce. However, after that point there is also a large population of signals above that level. It is clearly observed that the AE hits with the highest amplitudes are appearing on the moment where the load is reaching the maximum value of the load of the crack detection (between 30 and 34 kN). This behavior indicates possible crack propagation during static loading. It is mentioned that the absence of strong AE for the low levels of load is in relation to the "Kaiser effect" (Kaiser 1950), that states that while reloading an intact specimen, no AE is recorded until the previous maximum load has been overpassed. In this case the activity during the static reloading started at 32 kN, 2 kN lower with respect to the previously applied load. This is a logic behaviour considering the fact that the initially introduced damage originated from a fatigue test.

After the static loading, all specimens (except the one that failed during static loading) were subjected again to another round of static loading on the load level of the crack openning. According to the eSHM system during the previous static loading and not during fatigue, the crack opening happened in both LMD samples at the load level of 20 kN and for the SLM sample at the load level of 14.5 kN. This new round of loading is important in order to confirm the presence of the crack by LP that was not possible in stress-free conditions. Initially, the specimens were loaded staticly. At a certain load level, the specimens were coated by liquid penetrant. All the cracks were acurately located as it is demonstrated for the second LMD specimen in Figure 8.

## 4 CONCLUSION

This study presents the assessment of the crack detection of a new concept of structural health monitoring by means of various NDT methods. Additionally, NDT evaluations are used in the current study in order to confirm the damage existence and location. Four metallic samples produced by 3D printing with an integrated eSHM system were subjected to liquid penetrant inspection (level 2 and 3), eddy current, radiography and acoustic emission. It is presented that radiography and liquid penetrant inspection (level 2 and level 3) cannot give an accurate location of the crack in all the cases. However, the liquid penetrant inspections on samples in loaded condition were able to locate the defect precisely. Eddy current can be used as an NDT method for the location of the current defects since it is always indicating the right defect location. Acoustic emission parameters like cumulative hits and amplitude are also sensitive to the damage identification in order to lead to a warning against the final fracture. It seems that AE parameters and well known phenomena developed for metals, like the Kaiser effect, are valid to the additive manufacturing components as well. It is promising that additional NDT techniques can support the findings of the eSHM system and can assist in its development. It is concluded that the eSHM system can successfully detect damage on a metallic structure. Since crack location is also of high importance, further investigations are conducted on the eSHM system to enable the extra ability of crack location.

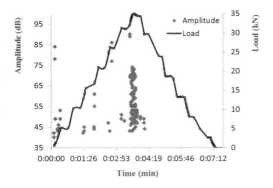

Figure 7. Load history and amplitude vs. time for the first LMD sample.

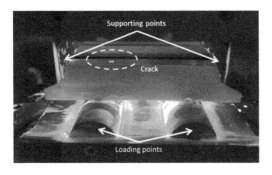

Figure 8. Crack location on additive manufactured samples by means of LPI level 2 under static loading.

## ACKNOWLEDGEMENTS

Research funded by an SBO Project grant 110070: eSHM with AM of the Agency for Innovation by Science and Technology (IWT). Part of this work (eddy current, radiography and LPI) was performed in the CC V&C, Department Aeronautical Material—NDI of the Belgian military facilities. The authors would like to acknowledge

Capt Jurgen Vereecken and his colleagues for the assistance during the inspections and their valuable feedback.

## REFERENCES

De Baere, D. Strantza, M. Hinderdael, M. Devesse, W. & Guillaume, P. 2014. Effective Structural Health Monitoring with Additive Manufacturing. *EWS-HM-7th European Workshop on Structural Health Monitoring.*

Farrar, C.R. & Worden, K. 2007. An introduction to structural health monitoring. Philosophical Transactions of the Royal Society A: Mathematical, Physical and Engineering Sciences, 365(1851), 303–315.

García-Martín, J. Gómez-Gil, J. & Vázquez-Sánchez, E. 2011. Non-destructive techniques based on eddy current testing. *Sensors, 11*(3), 2525–2565.

Kaiser J. 1950 Results and conclusions from measurements of sound in metallic materials under tensile stress: technische hochshule Munich.

Kamsu-Foguem, B. 2012. Knowledge-based support in Non-Destructive Testing for health monitoring of aircraft structures. *Advanced Engineering Informatics, 26*(4), 859–869.

Leuders, S. Thöne, M. Riemer, A. Niendorf, T. Tröster, T. Richard, H.A. & Maier, H.J. 2013. On the mechanical behaviour of titanium alloy TiAl6V4 manufactured by selective laser melting: Fatigue resistance and crack growth performance. *International Journal of Fatigue, 48*, 300–307.

Levy, G.N. Schindel, R. & Kruth, J.P. 2003. Rapid manufacturing and rapid tooling with Layer Manufacturing (LM) technologies, state of the art and future perspectives. *CIRP Annals-Manufacturing Technology, 52*(2), 589–609.

Roberts, T. & Talebzadeh, M. 2003. Acoustic emission monitoring of fatigue crack propagation. *Journal of Constructional Steel Research, 59*(6), 695–712.

Rometsch, P.A., Pelliccia, D., Tomus, D. & Wu, X. (2014). Evaluation of polychromatic X-ray radiography defect detection limits in a sample fabricated from Hastelloy X by selective laser melting. NDT & E International, 62, 184–192.

Sharp, N., Kuntz, A., Brubaker, C., Amos, S., Gao, W., Gupta, G. & Mascareñas, D. 2014. Crack detection sensor layout and bus configuration analysis. *Smart Materials and Structures, 23*(5), 055021.

Strantza, M., De Baere, D., Rombouts, M., Clijsters, S., Vandendael, I., Terryn, H. & Van Hemelrijck, D. 2014. 3D Printing for Intelligent Metallic Structures. *EWS-HM-7th European Workshop on Structural Health Monitoring.*

Vayre, B., Vignat, F. & Villeneuve, F. 2012. Metallic additive manufacturing: state-of-the-art review and prospects. *Mechanics & Industry, 13*(02), 89–96.

*Emerging Technologies in Non-Destructive Testing VI – Aggelis et al. (Eds)*
*© 2016 Taylor & Francis Group, London, ISBN 978-1-138-02884-5*

# Non-destructive evaluation of an infusion process using capacitive sensing technique

Yang Yang, Thomas Vervust, Frederick Bossuyt & Jan Vanfleteren
*Center for Microsystems Technology, IMEC and Ghent University, Ghent, Belgium*

Gabriele Chiesura, Geert Luyckx & Joris Degrieck
*Department of Materials Science and Engineering, Ghent University, Ghent, Belgium*

Markus Kaufmann
*Sirris Leuven-Gent Composites Application Lab, Leuven, Belgium*

ABSTRACT:  In this study, a capacitive sensing based non-destructive evaluation technique is applied to a vacuum assisted resin infusion process for the fabrication of glass fibre reinforced composites, as such different steps of the fabrication process (the injection of resin, the curing and the post curing) can be better understood to increase the quality of the fabricated part and reduce the fabrication costs. An interdigital coplanar capacitive sensor was designed, fabricated, and embedded in the glassfibre reinforced composites. Experimental data clearly shows different stages of the resin infusion process: wetting of the glass fibres marked by rapid increase of capacitance; domination of ionic conduction at the early stage of the cure when the resin is still in a liquid state; the vitrification point, indicating a transition of the resin from a gelly state to a glassy state, marked by the relatively big decrease in capacitance; further polymerization during post-curing, marked by a peak in capacitance at the beginning of post-curing cycle, and finally the completion of the cure marked by the saturation of capacitance to a final value. The different phenomena observed during the experiment can be used as a tool for *in situ* on-line monitoring of composites cure.

## 1  INTRODUCTION

Fibre Reinforced Polymer (FRP) is becoming a valid alternative to many traditional heavy metal industries because of its high specific stiffness over the more classical construction metals. Among the various manufacturing techniques, Liquid Composite Moulding (LCM), and especially the Vacuum Assisted Resin Infusion process (VARI) is becoming a common tool for large composite part production. For instance, the traditional hand-layup process is getting replaced by VARI with the key motivation being health improvement of working environment of manufacturing plants due to lower volatile component emissions and less hazard contact between the operator and the materials. Besides, better control of the final component thickness is another advantage of VARI in comparison to hand layup process. Other benefit of VARI includes more economically advantageous than other process such as autoclave, since no expensive tooling is needed and cheaper raw materials in comparison to expensive pre-pregs (Poodts, Minak, Dolcini, & Donati 2013).

With the increasing difficulty in composite manufacturing process arising from a demand for more complex geometry and better quality of composite structures, a method to optimize the manufacturing process is imposed such that different manufacturing steps (e.g. the injection of resin, the curing and post curing stage) can be understood to increase the quality of the produced part while reducing manufacturing costs. An important stage during manufacturing is certainly the curing phase: during which the polymer constitute of the composites cross-links, results in the consolidation of whole composite structure. By controlling the curing phase, one can obtain a superior quality component and, as a consequence, will increase the reliability and optimize the design of the part, leading to a decrease of the life cycle cost. Several methods have been proposed to characterize the cure kinetics and monitor the cure process, for instance Differential Scanning Calorimetry (DSC), Dynamic Mechanical Analysis (DMA), Ultrasonic Techniques, Raman Spectroscopy, and Dielectric Analysis (DEA)(Menczel & Prime 2009) (Monni, Niemel, Alvila, & Pakkanen 2008)

(Lionetto, Montagna, & Maffezzoli 2011). DEA is identified as an effective tool for on-line monitoring of composites cure, as the technique can be applied to different processing environment such as ovens, presses and autoclaves in a non-destructive manner (Maistros & Partridge 1998). Most of the other techniques are better suited for a laboratory environment under ideal conditions. For instance, DSC is useful to characterize the material only with a small quantity (50 mg), prohibiting it from being used in a real structure. Ultrasonic technique applies two transducers (one as excitation and one as detection) and is only capable of measuring the bulky property of the material. The distance between the opposing sensors is crucial for getting accurate measurement, as a result an additional distance control unit is required. For a more complex composite structure, this will limit the further use of ultrasonic technique.

In this work, a flexible interdigital capacitive sensor was designed, fabricated, and embedded in the glass fibre reinforced composites fabricated by a vacuum assisted resin infusion process. Experimental data clearly shows different stages of the resin infusion process: wetting of glass fibres marked by rapid increase of capacitance; domination of ionic conduction at the early stage of the cure when the resin is still in a liquid state; the vitrification point, indicating a transition of the resin from a gelly state to a glassy state, marked by the relatively big decrease in capacitance; further polymerization during post-curing, marked by a peak in capacitance at the beginning of post-curing cycle; and finally the completion of the cure marked by the saturation of capacitance to a final value. The different phenomena observed during the experiment can be used as a tool for *in situ* on-line monitoring of composites cure.

## 2 EXPERIMENTAL

### 2.1 *Working principle of capacitive sensing technique*

Capacitive sensing technique is often utilized to study the dielectric properties of a material, which is commonly referred to as dielectrometry or dielectric analysis (Menczel & Prime 2009). This principle can be extended to cure degree monitoring of thermoset matrix composites: during the curing of the composites, the mobility of dipoles (since most reactive molecules are dipolar) and free-moving ions (introduced during the synthesis of the resin) gets restricted, leading to decreased mobility of rotational dipoles and translational ions, and they are represented in the change of real relative permittivity and loss tangent. Monitoring of these changes allows us to identify different stages of the curing process. Materials dielectric property $\varepsilon^*$, is represented in a complex form:

$$\varepsilon^* = \varepsilon' - j\varepsilon'' \qquad (1)$$

The real part $\varepsilon'$ is referred to as the real relative permittivity or dielectric constant: it expresses the capacity of a material to store the electrical energy. The imaginary part $\varepsilon''$ is referred to as the dielectric loss factor: it expresses the energy loss caused by the dielectric medium. They are dependent on polarization of the permanent dipole moment and translational ionic groups.

Most common capacitive sensors are based on the use of parallel plate electrodes or interdigital electrodes. For a parallel plate system, the Material Under Test (MUT) is sandwiched between two electrodes to form a capacitor. Care should be taken when using parallel plate system: the distance between the electrodes should be kept constant throughout the experiment; A/d (A:area of the plate(s), d:distance between the plates) ratio should be reasonably large for the impedance measurement system to give enough sensitivity; the electrodes should be present on both sides of the mould (or other production environment). Interdigital electrode refers to a digit-like or finger-like periodic pattern of parallel in-plane electrodes used to build up the capacitance or impedance associated with the sensitive coating or embedding material, into which the electric field penetrates. In contrast to parallel plate electrodes, the geometry of the system is more stable as both anode and cathode are on the same plane supported by the substrate layer. Moreover, due to the coplanarity of electrodes, one side access to the material is possible, this makes the embedding of the sensor less troublesome in many cases. Figure 1 shows the transition of parallel plate capacitor to a coplanary interdigital sensor and illustrates interdigital sensor's principle of operation.

The simplified geometry of an IDC is shown in Figure 2(left). When an electric potential is applied between the anode and cathode of the sensor, an

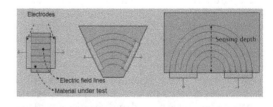

Figure 1. Transition from parallel plate capacitor (left) by opening up the face-to-face parallel electrodes(middle) until coplanar fringing field capacitor (right).

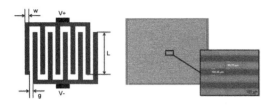

Figure 2. Simplified schematic of an interdigital capacitive sensor (left): where w stands for the width of each finger, g stands for the gap (spacing) between fingers, and L stands for the finger length, and a home-made sample (right) realized by flexiblecircuit board technology.

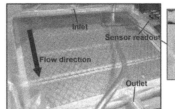 

Figure 3. Experimental setup for the vacuum assisted resin in fusion process, with the capacitive sensor embedded in between glass fibre reinforcement of the composites.

electric field is created between the fingers, and extended to the space on both sides of the finger structure to a certain distance. The distance that the majority of the fringing electric field is able to reach is usually called the penetration depth of the sensor, determined by the spatial wavelength (w+g) of the structure. Further details on interdigital sensor can be found in the paper of Mamishev et al (Mamishev, Sundara-Rajan, Fumin, Yanqing, & Zahn 2004). To establish a relationship between capacitance value and permittivity, these parameters were studied by mathematical models, Finite Element Analysis and experimental measurements. FEA and the model proposed by Igreja et al (Igreja & Dias 2004) agree well with the experimental work; model proposed by H. Engan (Engan 1969) has a large deviation from experimental result, as the model includes only one dielectric in the formula, whereas for the capacitive sensor at least two dielectric were present (substrate of the capacitive sensor and material under test).

The interdigital sensor, used in this study, is fabricated in-house by patterning copper structure on flexible substrate (PI/PET/PEN) via lithography and wet etching process, Organic Solderability Preservative (typical thickness 0.15–0.3 micron) coating is then applied on top of electrodes to protect the sensor surface from oxidation. Figure 2(right) shows the fabricated capacitive sensor. The resulted sensor is 68 $\mu m$ in thickness, and 180 $mm^2$ in size. The area and thickness of the sensor is quite critical for initiating delaminations and local damage within the laminate. In order to evaluate the impact of our sensor on this and find the optimal substrate for the application, sensor on different flexible substrate (PI/PET/PEN) has been made. As a future work, we will evaluate their adhesion to the matrix materials.

## 2.2 Experimental setup

A vacuum assisted resin infusion process to fabricate glass fibre reinforced composites is shown in Figure 3. Six layers of UDO ES 500 UD Glass Fibre textile were layered up on top of a flat glass mould, the plate was sealed on the 4 edges with vacuum bag. An interdigital sensor and a temperature sensor were embedded between the middle layers. The epoxy matrix system used in this study (from Momentive), is a two part system consisting of the EPIKOTE MGS RIMR 135 resin and the EPIKURE MGS RIMH 137 hardener. The mixing ratio is 100:30 by weight. 300 g of RIMR135 (epoxy) and 90 g of RIMH137 (hardener) were thoroughly mixed by a mechanical blender, degassed, and infused into the vacuum bagging through the resin inlet, the outlet was connected with a tube to the vacuum line, and once the resin reaches the outlet, both tubes were clamped in order to stop the resin flow and keep the bag under vacuum. The composite plate is cured at room temperature for 24 h, followed by a post cure in an oven at 80 °C for 16h as recommended by the datasheet. A capacitive sensor, and two thermocouple were installed for capacitive and temperature measurement respectively throughout the curing. An HP 4284 A Precision LCR meter performed the capacitive measurement with frequency sweep ranging from 100 Hz to 1 MHz with 10 points per decade. The time required per sweep was approximately 16 seconds. One thermocouple was installed next to the capacitive sensor for local temperature measurement, and another one outside the vacuum bag for ambient temperature measurement.

## 3 RESULTS AND DISCUSSIONS

Capacitive data for the resin cured at 20 °C is shown in Figure 4. At the beginning of cure, we observe a large increase in capacitance value at around minute 35 indicating that the resin has reached the capacitive sensor, and flown completely over the capacitive sensor (marked by the maximum value). The large increase in capacitance at the beginning is caused by electrode polarization (Lvovich 2012). As the resin starts off as liquid with low viscosity, the ionic impurities (introduced

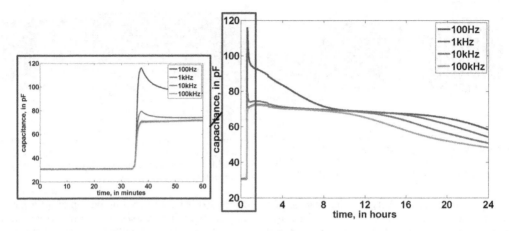

Figure 4. Capacitive measurement of beginning of curing (left) and whole curing (right).

into the system during the synthesis of the resin) are able to move freely. Ion impurities accumulate at the electrode surfaces, causing interfacial polarization at the electrodes, leads to the large increase in capacitance dominating at lower frequencies. As the cure progresses, cross-linking between monomers or oligomers is continued, the polymer network is formed and expanded. As the resin system becomes more viscous, the translational diffusion of ionic component and the rotational diffusion of dipole moment get restricted. They need longer time for the re-orientation and the relaxation, and are not able to follow the excitation signal of higher frequencies. Electrode polarization phenomenon disappears first at higher frequencies, and gradually shifts to lower frequencies (Fig. 4). At around 12 hours into the cure, a relatively steep decrease in capacitance is observed and is referred to as the gelation point or the vitrification point (Maistros & Bucknall 1994). This point indicates the transition of the resin from a gelly state into a glassy state. Worth mentioning during the 24 h curing stage, no obvious change of resin temperature is observed: resin temperature is following ambient temperature and no indication of degree of cure can be observed by temperature measurement.

Capacitive and temperature data for the post curing are shown in Figure 5. As suggested by the datasheet, a heat treatment is required to achieve a fully cured sample. When the oven is raised from room temperature to 80 °C, further cure or polymerization happens within the laminates. A large increase in capacitance, which possibly indicates further polymerization within the resin, is observed in the beginning of post curing. After the peak, the capacitance decreases and saturates to nearly stable values, suggesting the slowdown of post-curing. After 16 h, the oven is turned off. The decrease in temperature causes the drop in capacitance, as the

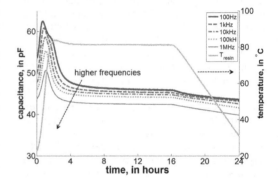

Figure 5. Capacitive and temperature measurement from post curing phase.

dielectric property is temperature dependent. However, the change in absolute value is much smaller compared to the peak observed at the beginning of post-curing or the change during the curing stage. On the other hand, temperature measurement follows the programmed heating profile of the oven during the whole post curing process. There is a lack of correlation between temperature changes and degree of cure. As can be seen from the plot, the capacitance measurement gives information about resin that is mostly related to degree of cure rather than the influence of the ambient environment, whereas the temperature measurement reveals little information about cure itself.

## 4 CONCLUSIONS

In this work, a flexible interdigital capacitive sensor was designed, fabricated, and studied. The developed capacitive sensor is applicable not only to a laboratory scale for research purpose, but also to an industrial scale for real manufacturing monitoring purpose. We successfully embedded the capacitive sensor in a vaccum assisted resin infusion process for the fabrication of glass fibre reinforced composites. Experimental data clearly shows different stages of the resin infusion process: wetting of the glass fibres marked by rapid increase of capacitance; domination of ionic conduction at the early stage of the cure when the resin is still in a liquid state; the vitrification point, indicating a transition of the resin from a gelly state to a glassy state, marked by the relatively big decrease in capacitance; further polymerization during post-curing, marked by a peak in capacitance at the beginning of post-curing cycle; and finally the completion of the cure marked by the saturation of capacitance to a final value. The different phenomena observed during the experiment can be used as a tool for *in situ* on-line monitoring of composites cure.

## ACKNOWLEDGEMENTS

The authors thank S. Dunphy, S. Van Put, and K. Dhaenens for the fabrication of capacitive sensors, J. Windels for his advice and assistance with hardware related issues. The research leading to these results has received funding from the Flemish Agency for Innovation by Science and Technology (IWT) through the program for Strategic Basic Research (SBO) under grant agreement n 120024 (Self Sensing Composites).

## REFERENCES

Engan, H. (1969). Excitation of elastic surface waves by spatial harmonics of interdigital transducers. *IEEE Transactions on Electron Devices 16*(12), 1014–1017.

Igreja, R. & C.J. Dias (2004). Analytical evaluation of the interdigital electrodes capacitance for a multi-layered structure. *Sensors and Actuators A: Physical 112*(2–3), 291–301.

Lionetto, F., F. Montagna, & A. Maffezzoli (2011). Ultrasonic transducers for cure monitoring: design, modelling and validation. *Measurement Science and Technology 22*(12), 124002.

Lvovich, V.F. (2012). *Impedance spectroscopy: applications to electrochemical and dielectric phenomena*. Hoboken, N.J.: Wiley.

Maistros, G.M. & C.B. Bucknall (1994). Modeling the dielectric behavior of epoxy-resin blends during curing. *Polymer Engineering and Science 34*(20), 1517–1528.

Maistros, G.M. & I.K. Partridge (1998). Monitoring autoclave cure in commercial carbon fibre/epoxy composites. *Composites Part B-Engineering 29*(3), 245–250.

Mamishev, A.V., K. Sundara-Rajan, Y. Fumin, D. Yanqing, & M. Zahn (2004). Interdigital sensors and transducers. *Proceedings of the IEEE 92*(5), 808–845.

Menczel, J.D. & R.B. Prime (2009). *Thermal analysis of polymers: fundamentals and applications*. Hoboken, N.J.: John Wiley.

Monni, J., P. Niemel, L. Alvila, & T.T. Pakkanen (2008). Online monitoring of synthesis and curing of phenolformaldehyde resol resins by raman spectroscopy. *Polymer 49*(18), 3865–3874.

Poodts, E., G. Minak, E. Dolcini, & L. Donati (2013). {FE} analysis and production experience of a sandwich structure component manufactured by means of vacuum assisted resin infusion process. *Composites Part B: Engineering 53*(0), 179–186.

*Emerging Technologies in Non-Destructive Testing VI – Aggelis et al. (Eds)*
*© 2016 Taylor & Francis Group, London, ISBN 978-1-138-02884-5*

# In-situ testing using combined NDT methods for the technical evaluation of existing bridge

**N.V. Zoidis**
*Geotest S.A Management Director, Ioannina, Greece*

**E.N. Tatsis**
*Geotest S.A & Materials Science and Engineering Department, University of Ioannina, Greece*

**E.D. Manikas**
*Civil Engineer (Msc), Ioannina, Greece*

**T.E. Matikas**
*Materials Science and Engineering Department, University of Ioannina, Greece*

ABSTRACT: A survey was conducted in order to investigate the causes that led to extensive cracking on the surface of the concrete of the bridge. The objective of the study was the evaluation of the condition of the concrete of the bridge and the examination of the static efficiency of the bridge at its current state, through systematic tests using NDT methods: Visual inspection and recording of the cracks, evaluation of deck waterproofing integrity (Impulse Response method), Petrographic Analysis of concrete, concrete quality estimation with (cores cutting, Pull-out Testing), estimation of the crack depth (ultrasound pulse velocity measurements) and corrosion evaluation with (Half Cell Potential, Total Chloride Ion Concentration and Carbonation depth). The results of the survey showed that Alkali—Silica Reaction was the predominant reason for this very extended and severe cracking on the elements of the bridge.

## 1 INTRODUCTION

The tests were conducted by GEOTEST S.A. in order to investigate the causes that led to extensive cracking on the surface of the concrete of the bridge. The bridge has three openings and consists of two abutments (A0 & A3), two piers (M1 & M2) and the 72 m deck. The Abutments and piers of the bridge are numbered from west to east. According to the design drawings, the concrete of the deck of the bridge is B35 type, the concrete of the piers is B25 type, the reinforcement steel is s420/s220 type and the pre-stressed steel is 1500/1770 type.

During the visual inspection, the visible cracks were recorded. The pattern and direction of the cracks indicated Alkali-Silica Reactions. Even though the phenomenon is extremely rare in Greece, a petrographic analysis was ordered. Furthermore, an Impulse-Response test was performed to evaluate the waterproofing of the bridge deck, to clarify if water was a factor. Ultrasonic Pulse velocity measurements) took place. Ultrasonic pulse velocity measurements were also used for the crack depth estimation. visual inspection (recording of all the appearing the cracks by position, width and length) and to determine corrosion of reinforcement concrete with Carbonation depth, Total Chloride Ion Concentration and Half Cell Potential. Reinforcement location and cover depth estimation was performed to examine the accordance with the construction drawings.

According to the on-site testing and petrography analysis results, Alkali-Silica Reaction was the predominant reason for this very extended and severe cracking.

## 2 TESTING METHODOLOGY

### 2.1 *Evaluation of deck waterproofing integrity using the impulse-response method (ASTM C1740)*

The road surface of the bridge was examined using the impulse-response method on a grid of 4 test positions on each lane, every 2,5 m, in order to check the waterproofing of the deck. This test investigated the seating uniformity of the asphalt layer on the concrete structure and the existence of flaws in the membrane between them that might have caused local anomalies.

The operating principle is based on a low strain impact produced by a hammer with a rubber tip. The impact causes vibrations in the element and stimulates primarily flexural form. A velocity receiver set adjacent to the point of impact, records the response. The load cell and the velocity receiver are connected to a laptop computer and by using appropriate software the results are analyzed.

Specifically, the function of the force in time, produced by the hammer and the measured velocity response is transformed in the frequency domain using the Fast Fourier Transformation (FFT). The range of velocity response divided by the range of force is called "mobility" and is a function of frequency. The mobility is given in units of speed per power (m/s)/N.

The parameters of the mobility diagram (Figure 1) used for assessing integrity are:

- The average mobility (lined bar)
- The dynamic stiffness, the inverse of the initial slope of the mobility spectrum from 0 to 40 Hz, expressed in units of N/m
- The mobility slope between 100 and 800 Hz
- The voids index (the ratio between the width of the initial maximum mobility to the value of the average mobility).

The s'Mash device, from Germann Instruments, was used for testing.

### 2.2 Petrographic analysis of concrete

A full petrographic analysis comprising a macroscopic and a microscopic examination was carried out by Ramboll (Denmark), on three cores (C2C, C1Bn-U1, C3Bs-U3). Special focus is given to an evaluation of the causes for the severe cracking of the concrete, including an eventually presence of alkali silica reactions.

The petrographic analysis was performed on thin sections taken from concrete cores. The concrete cores used for analyzing concrete are impregnated with a fluorescent dye. This procedure renders pores and microscopic cracks easily visible. The thin sections are examined in an optical microscope in transmitted light. The microscope is operated either in the polarizing or in the fluorescent mode.

Microanalysis provides data regarding the aggregates, particularly the type of sand and coarse aggregates and the extend of Alkali—Silika reactions.

### 2.3 In place concrete strength determination

- ores strength examination

The concrete compressive strength was determined by testing concrete cores that were cut off from suitable sampling points on the deck and piers of the bridge. Before this procedure, the reinforcement bars were located in order to avoid their injury.

The cores were then transferred to the laboratory where they were tested in compression. Finally, their compressive strength was related to the compressive strength of cubic ($15 \times 15 \times 15$ cm) specimens using the corrections of the E7 Circular (Greek Specification).

- Pull-Out testing (ASTM C900)

In the process of the test, a recess (slot) is routed in the hole to a diameter of 25 mm and at a depth of 25 mm, through an 18.4 mm hole that is made perpendicular to the surface. A split ring is expanded in the recess and pulled out using a pull machine reacting against a 55 mm diameter counter pressure ring. The concrete in the strut between the expanded ring and the counter pressure ring is

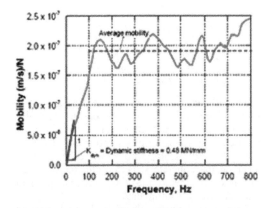

Figure 1. Mobility Diagram.

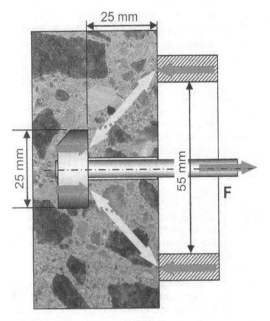

Figure 2. Pull out test method and damage.

in compression. Hence, the ultimate pullout force F is related to the compressive strength of cubic or cylindrical specimens, via a correlation curve.

## 2.4 Crack depth estimation

The estimation of the depth of cracks was carried out using ultrasonic pulse velocity measurements and complementary cores were cut off for the calibration of the results. The measurement of the crack depth with ultrasonic relies on the existence of cracks or internal discontinuities in concrete that cause diffraction to the propagation of ultrasound. This increases the propagation time and the velocity appears to be reduced.

The process of calculating the depth of an ultrasonic surface crack perpendicular to the surface is as follows:

- Measuring the ultrasonic velocity at a healthy part of the concrete.
- Measuring the ultrasonic velocity by placing the transmitter and the receiver on both sides of the crack, equidistant
- Calculation of the crack depth by figure 3.

$$d = \frac{L}{2}\sqrt{\left(\frac{t_c}{t_p}\right)^2 - 1}$$

where:

$d$ = crack depth,
$L$ = distance between transmitter and receiver,

Figure 3. Equation for the calculation of the crack depth.

$t_c$ = propagation time to healthy part of concrete, and
$t_p$ = propagation time across the crack.

For the measurements, the portable Surfer's device from ACS Company was used.

## 2.5 Corrosion evaluation

- Carbonation depth measurements

The carbonation depth of concrete was measured by spaying the cores, right after they were drilled, with a solution of phenolphthalein.

If a change of color is noticed after spraying the core, it means that the specimen retains the natural alkalinity of cement paste in concrete that results in a protective oxide coating on steel reinforcement that prevents the steel from rusting. If the color of the specimen does not change, this is an indication that its PH is lower than 9 (PH < 9, carbonated concrete) and the reinforcement is not protected adequately.

- Total chloride ion measurements

The percentage of the acid soluble chlorides free chlorides and the physically bounded chlorides was determined using the GERMANN INSTRUMENTS' RCT method that includes two stages. At first the concrete powder is obtained by drilling and then, the chloride ion percentage is calculated using the RCT device. The sample is mixed into a distinct amount of extraction liquid and shaken. The extraction liquid removes disturbing ions, and extracts the chloride ions in the sample. A calibrated electrode is submerged into the solution to determine the amount of chloride ion, which is expressed as percentage of concrete mass. The calibration is done by submerging the electrode into solutions of known chloride ion percentage.

The measurements took place using powder samples of hardened concrete that were obtained by drilling in various depths from the surface of the concrete (0–1,5 cm, 1,5–3 cm and 3–4,5 cm).

The percentage of the total chlorides was calculated by using the GERMANN INSTRUMENTS' RCT method.

According to the international literature, a common threshold chloride ion content is 0,10% by mass of concrete. Sometimes, the stricter threshold of 0,05% by mass of concrete is used.

- Half cell potential measurements (ASTM C 876)

For the probability of corrosion of reinforcement, Half Cell Potential measurements of the reinforcement will be performed in accordance with ASTM C876.

The operating principle of the method is based on voltage measurements through an electric circuit between the reinforcement bars (local

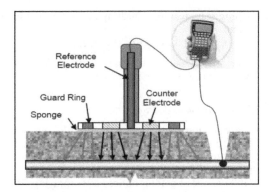

Figure 4. Galva Pulse device setup.

exposure of the reinforcement) and the reference electrode, which contacts the outer surface of the concrete. The measurements are taken by a voltometer which is also connected to the circuit. The reference electrode used during the test can be Cu/CuSO$_4$, Ag/AgCl, Hg/Hg$_2$Cl$_2$ etc. The contact with the concrete is achieved by using a sufficiently wet sponge. The setup of the device is shown in figure 2.

The test will be conducted with the Galva Pulsedevice of GERMANN INSTRUMENTS, which uses an Ag/AgCl reference electrode. During the analysis of the results the prices will be converted to the equivalent of reference electrode Cu/CuSO$_4$, referred to by ASTM C 876.

According to the international bibliography and the ASTM C 876 the following possibilities are distinguished:

- When E > −200 mV (CSE) at 90% probability, no corrosion occurs.
- When −200 mV > E > −350 mV (CSE) is not sure if corrosion occurs.
- When E < −350 mV (CSE) at 90% probability corrosion occurs.

### 2.6 Reinforcement location and cover depth measurements

For the reinforcement cover depth estimation in the concrete, the device Ferroscan PS200 of the company HILTI was used. The method is based on the application of a magnetic field on the outer surface of the concrete through the testing device in accordance with the BS 1881:204 standard.

Reinforcement scans and cover depth measurements will be executed in order to confirm the reinforcement layout based on the construction drawings and to investigate any relation between the reinforcement layout and the appearing cracks. The scans will be conducted at random, in sufficient test area sat all three openings of the bridge.

## 3 RESULTS

### 3.1 Evaluation of deck waterproofing integrity using the impulse-response method (ASTM C1740)

The top of the bridge deck was tested using the Impulse—Response method to evaluate the waterproofing integrity and the condition of the concrete of the top of the deck, on both lanes. The testing grid was designed four positions (hammer hits) wide, for each lane, and 29 positions (hammer hits) long (one position for every adjacent railing column, around 2,5 m apart). The westbound lane grid was 30 positions long because axis 19 was inserted in the area of a surface crack on the pavement, transverse to the bridge deck.

The Impulse—Response method is a comparative, Non Destructive Testing method that has the ability to map big areas in a short time.

The results of the method were calibrated with three rectangular cuts (around 80cm × 80cm) on the pavement, on test positions that showed different response to the method according to the average mobility results. The results of the mapping using the Impulse—Response method, as well as the specific results and the findings after the cuts at the three calibrating positions are presented below.

Out of the 236 total test positions, 9,3% had average mobility over 20.

### 3.2 Petrographic analysis of concrete

The petrographic analysis, performed by RAMBOLL (Denmark) showed many signs of ongoing alkali—silica reactions in all three thin sections. Many of the potentially reactive grains showed

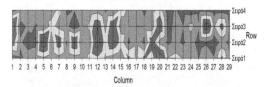

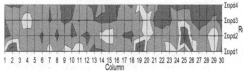

Figure 5. Impulse—Response method testing direction.

Figure 6. Petrographic analysis indicative photos.

extensive reactions causing severe cracking and gel formation in the concrete.

The main conclusions of the petrographic analysis are summarized below:

The concrete is composed of fine grained cement, crushed limestone/dolomite and a fine aggregatewhich contains a relatively high amount of potentially reactive grains (ASR).

The capillary porosity of the concrete is in general, relatively medium to high corresponding to a water to cement ratio of 0.55, however, variations of 0.40–0.70 are seen in every thin sections. The variation is probably caused by an inadequate mixing and/or compaction/vibration of the concrete.

The concrete is not air-entrained, and the air content is general very low.

The examined concrete is generally of a poor quality and apparently with a reduced durability if exposed to moisture.

Based on the examinations, the concrete can be characterized as a highly Alkali Silica Reactive (ASR) concrete with a relatively porous cement paste.

The concrete in the cores contains many external and internal cracks caused by severe ASR in the concrete. The alkali silica reactions are related to a potential reactive mix of porous/semi porousflint-like grains (probably containing opal) and metamorphic rock types with microcrystalline quartz (similar to rhyolite) in the fine fraction.

With unchanged level of exposure it is very likely that the ASR will continue and causing more cracking in the concrete.

It is also likely that local corrosion of reinforcement due to crack-carbonation would be able to occur in the vicinity of the many deep surface cracks over a period of a couple of years.

### 3.3 In place concrete strength determination

The equivalent strength of $15 \times 15 \times 15$ cm cube specimens from the cores is according to the E7 Circular (Greek Specification). According to the E7 Circular, taking into account the 6 worst results of core strength, the concrete of the bridge deck is C30/37 type, which complies with the design (B35 type). The cores that were drilled at the piers were Ø75 mm diameter, because of the dense reinforcement, and the corrections used by the E7 Circular do not comply.

The equivalent strength of $15 \times 15 \times 15$ cm cube specimens from the pull out tests and the Schmidt Hammer strength estimation are according to the manufacturer's nomogram.

### 3.4 Corrosion evaluation

- Carbonation depth measurements

The carbonation depth was measured on site, right after the cores were drilled. The cores drilling positions were 23 and the carbonation depth range were 5–25 mm.

- Half cell potential measurements (ASTM C 876)

Half-cell potential measurements were executed at five areas of the soffit of the bridge. The test positions were chosen at areas of the bridge that had visible signs of humidity and are the most suspicious areas concerning reinforcement corrosion.

The test areas and the results, according to the ASTM C876, are presented below.

### 3.5 Reinforcement location and cover depth measurements

From the examination of the results of the numerous scans to locate the reinforcement bars and

Figure 7. Carbonation depth measurement on concrete core.

Table 1. Half Cell Potential measurements results.

| Test positions | Total number of potentials measured | P < −350 mV (%) | −350 < P < −200 mV (%) | P > −200 mV (%) |
|---|---|---|---|---|
| Hcp_1 | 24 | 0 | 92 | 8 |
| Hcp_2 | 20 | 0 | 100 | 0 |
| Hcp_3 | 20 | 0 | 5 | 95 |
| Hcp_4 | 20 | 0 | 75 | 25 |
| Hcp_5 | 20 | 30 | 70 | 0 |

*P is potential.

estimate their cover depth, the following conclusions were extracted:

- Deck

The location of the horizontal and vertical bars that were measured was according to the construction drawings in the sides of the deck, outside of piers area.

The location of the horizontal bars that were measured was according to the construction drawings in the sides of the deck, piers area. The vertical bars were found 192 mm to 196 mm apart, while the design puts those 100 mm apart.

The location of the transverse bars that were measured was according to the construction drawings in the soffit of the deck. The longitudinal bars were found to be 2 to 4 under the tendons and 4 under the extruded polystyrene "barrel" voids, while the design describes 4 under every tendon and 5 under every extruded polystyrene "barrel" void.

- Piers

The location of the vertical bars that were measured was according to the construction drawings (100 mm apart). The horizontal bars were found 149 mm to 200 mm apart, while the design puts those 100 mm apart.

### 3.6 Deck draining holes drilling

Holes (Ø16 mm) were drilled at the bottom side of the deck, under every extruded polystyrene "barrel" void, over both the Service Roads and at the middle and both ends of the central opening of the bridge. A total of 42 holes were drilled.

Small amounts of water (a few drops up to a few liters) were found at the holes H15, H19, H26, H29, H36, that are marked with magenta color.

Big amounts of water (holes draining at maximum supply for up to 20 minutes) were found at the holes H1, H7, H8, H14, H28, H35, H42, that are marked with green color.

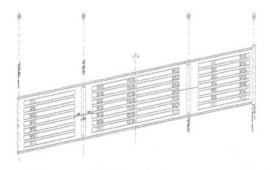

Figure 8. Draining hole positions.

Figure 9. Large amounts of water draining from drilled holes.

## 4 CONCLUSIONS

### 4.1 General

According to the results of the on-site testing and the petrographic analysis that was performed by Ramboll (Denmark), Alkali—Silica Reaction is the predominant reason for this very extended and severe cracking on the elements of the bridge. The alkali silica reactions are related to a potential reactive mix of porous/semi porous flint-like grains (probably containing opal) and metamorphic rock types with microcrystalline quartz (similar to rhyolite) in the fine fraction.

The big amounts of water that were found inside the extruded polystyrene "barrel" voids (especially the ones on the edges of the deck) seem to have been a deteriorating factor.

Alkali Silica Reactions (ASR) has occurred in the concrete in all three cores/thin sections to a very high extent. However there are still many potentially reactive grains with moderate sign of reactions which still has a significant residual expansion. The risk of future alkali-silica reactions is considered to be very high due to a significant residual expansion of the potential reactive grains in the aggregates—in case of unchanged exposure conditions.

Taking into account the results from the various testing methods that were used for the evaluation of the structural integrity of the existing structure the conclusions are the following.

### 4.2 Deck waterproofing integrity

The waterproofing of the deck has failed to protect the concrete of the deck. The waterproofing has serious problems at the areas of rain inlets and the membrane appears to be aged.

The extended cracking shown in your previous reports was confirmed, as well as the existence of cracks on the upper side of the bridge deck.

### 4.3 Concrete strength

The concrete's compressive strength which was determined by cores and NDT testing is C30/37 quality, which is satisfactory, concerning the bridge deck design (quality B35). The strength of the cores that were drilled at the piers and from the NDT methods that were used seems to comply with the required quality (B25).

## 4.4 *Crack depth estimation*

The crack depth estimation that was carried out using ultrasonic pulse velocity measurements and was calibrated using cores showed that the maximum crack depth is around 15cm and that the crack depth is related to the crack's surface width.

## 4.5 *Corrosion evaluation*

The carbonation depth of the concrete is low (0–20 mm) compared to the average cover depth of the reinforcement.

The total chloride ion concentration is around 0,005% (or lower) per concrete mass, which is the lower traceable percentage, at all test depths of every sampling position. The stricter corrosion risk threshold used is 0,05% by mass of concrete.

The HCP measurements are mostly between 200 mV and −350 mV, where, according to ASTM C876, it is uncertain if corrosion occurs. However some specific measurements were < −350 mV which means that (according to ASTM C876) it is 90% certain if corrosion occurs.

Taking the above into consideration, as well as the fact that carbonation of the cement paste is observed along the surface of the cracks to a depth behind the reinforcement it is possible that reinforcement corrosion could occur in the future.

The reinforcement bars that were revealed showed no visible signs of corrosion.

## REFERENCES

Sansalone MJ & Street WB. Impact-echo nondestructive evaluation of concrete and masonry. Ithaca (NY): Bulbrier Press: 1997.

ASTM C1740-10. Standard practice or evaluation the condition of concrete plates using the impulse-response method.

ASTM C597-09. Standard test method for pulse velocity through concrete.

Davis, A.G. 2002. *Industrial floors on ground DTI.* Copenhagen: Testing of structures NDT-Methods.

Zoidis, N., Tatsis, E., Vlachopoulos, C., Clausen, GS., Matikas, TE. & Aggelis, DG. 2001. *Visual inspection and evaluation using NDT testing methods of extensively cracked concrete floor Emerging Technologies in Non-Destructive Testing V.* Ioannina: Proceeding of the 5th conference on emerging technologies in NDT.

Grelk, B. & Pade, C. 1997. *Petrographic Analysis of 3 Cores from Concrete Masts.* Danish: RAMBOLL No. 971450.

*Biological applications*

*Emerging Technologies in Non-Destructive Testing VI – Aggelis et al. (Eds)*
*© 2016 Taylor & Francis Group, London, ISBN 978-1-138-02884-5*

# Kinect sensor for 3D modelling of logs in small sawmills

J. Antikainen & D. Xiaolei
*Natural Resources Institute Finland, Joensuu, Finland*

ABSTRACT: In small circular sawmills the sawing pattern and the positioning of the log is based on the evaluation of the sawyer. The evaluation is based on the expertise of the sawyer which may not produce the optimal sawing setting for the log and yield value. This decision making can be improved using 3D model of the log which includes necessary information about the dimensions and the shape. Produced 3D model can be used to optimize the sawing patterns and positioning of the log. However, commercial systems for 3D log scanning are very expensive and they cannot be used in the small sawmills because of the low production rates. This study proposes a new low cost 3D scanning method based on Kinect sensor. The system produces a 3D model of the log which can be used to guide the sawyer in a real-time.

The one sensor measurement system was tested with Scots pine and Norway spruce logs in real sawmill environment and the results were compared to caliper measurements. With one sensor system, the log had to be measured in several pieces because of the scanning limitations of the sensor. The average difference between the calipers and the proposed method was 1.07% with 0.97% deviation. These results show that the system could be used in real-time and it produces good quality models of the logs. This study expands the one sensor system to contain multiple Kinect sensors which can be used at once. This will provides higher resolution for the model and the logs can be measured more efficiently.

## 1 INTRODUCTION

### 1.1 *3D scanning for logs*

Three dimensional (3D) information of the log is one of the most important factors in the sawing industry. The 3D model contains information of the shape, dimensions and other quality factors such as knots and how crooked the evaluated log is. The 3D model can be used to determine the optimal sawing patterns, positioning and predicting the quality of the sawn products. Using the 3D model and the wanted sawing pattern, the log can be rotated into the correct position according the sawing blades. (Wenshu & Jingxin 2011, Skog & Oja 2007)

Most of the current 3D measurement systems are based on laser triangulation method where the log is moved under the projected laser line and the reflected light is detected by camera (Dashner 1993 & 1999). In many practical cases the 3D shape is formed using several laser lines and multiple cameras to obtain the full model of the log with higher accuracy. Laser triangulation method is very fast and it works well on the production lines where log travel along its lengthwise direction. However, this method cannot be used in small circular sawmills and the costs of the commercial measurement systems will rise too high compared to total costs of the sawing mill.

### 1.2 *3D scanning at small circular sawmills*

In small circular sawmills the log is handled and sawed different way than in large scale sawing mills. The sawing parameters and positions are selected individually to each log which affects to the quality and the value of the sawed product. Position of the log will affect to the sawed product dimensions and several quality factors such as strength properties. Therefore, the optimal log positioning will increase the value (better timber grades) and decrease the material losses for sawn timber. The rotation and positioning of the log is mainly based on the evaluation of the sawyer which may affect the result. In this case the traditional laser triangulation methods do not work as wanted without the lengthwise movement. Only possible measurement place and time is during the log rotation on the sawing table.

Previous study with one Kinect sensor has shown that the sensor can be utilized to work in a real sawmill environment. (Antikainen & Verkasalo 2013). Inspired by the previous study, this work will expand the single sensor system to multiple sensor system which will speed up the measurement process and enables longer specimens to be measured at once (Xiaolei 2014).

## 1.3 The Kinect sensor

The first version of the Kinect sensor was announced by the Microsoft at 2009 (Fig. 1a) and the main target for the sensor was gaming industry. The sensor was originally developed and patented by PrimeSense Corporation (Shpunt et al. 2008). The low cost of the sensor is the one significant advantage compared to other commercial three dimensional scanners. Usually, the 3D scanning tools costs thousands of euro but the Kinect sensor price is only a couple of hundreds euro. Therefore, the low price enables more interesting applications were normal commercial applications would be too expensive.

The Microsoft published first developer packages for the sensor at 2011 for non-commercial use. This increased the usage of the sensor expanded rapidly from gaming industry to other fields as well. It has been used for example for human body (Jing et al. 2012) and face scanning (Hernandez et al. 2012) applications and even for controlling altitude of quadrotor helicopters (Stowers 2011).

The Microsoft has also worked on the Kinect Fusion project (Newcombe et Al. 2011, Izadi et al. 2011) which helps the application development even more. It does not only provide library functions to control the sensor but it provide full functionality to create 3D meshes easily.

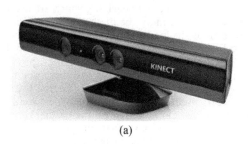

(a)

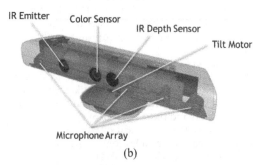

(b)

Figure 1. a) The Microsoft Kinect sensor and b) the structure of the sensor.

## 2 MATERIALS AND METHODS

### 2.1 Log materials

The single sensor study was made for green Scots pine (*Pinus sylvestris*) and Norway spruce (*Picea abies*) logs with a park. Length of the logs was between 3.7 and 5.5 meters and they were mainly gathered from the butt end of the tree but there were a couple of middle logs also. Multiple sensor measurement system was tested with Silver birch (*Betula pendula*) logs.

### 2.2 The measurement principle of the Kinect sensor

The Kinect sensor includes several different components which can be used at the measurement process. (Fig. 1b) The most important parts are the IR emitter and IR depth sensor which are used in the depth map measurements. The depth map is calculated from known speckle pattern which is formed using a diffractive element and an infrared laser (IR Emitter). The laser beam is scattered to dense point cloud which is projected on the surface of the target. The used laser is on an infrared region so it cannot be detected by an eye; therefore the projected image is captured using an infrared camera (IR Depth Sensor). The resolution of the IR depth sensor is 640 × 480 pixels with 11 bits dynamics which defines the scanning accuracy of the sensor. The final 3D object can be formed by registration of different depth maps using the Iterative Closest Point (ICP) method (Besl & McKay 1992).

The depth map calculations are done inside the sensor hardware and it can produce 30 depth map frames per second which is enough for real-time analysis. The Kinect sensor also provides a RGB video sensor (Color Sensor), which can be utilized for texture and color analysis. The texture and color information can be used to detect external defects of the log such as knots and scars.

### 2.3 3D model analysis

The model analysis can be done using open source software called MeshLab. The software can be used to analyze dimensions of the log and it can also be used to combine separate models to one single model. It supports various 3D model formats which Kinect Fusion packages provide. However, the developed multi sensor measurement software can analyze the data automatically after the scan. It can automatically calculate the diameters at different locations and also the volume of the log.

### 2.4 Single sensor measurement system

In the single sensor measurement system, the measurements were done with the software provided by the Kinect Fusion developer package. In the single sensor version the sensor is connected above the sawing table where the log can be rotated along it vertical axes. (Fig. 2) The measurement distance between the sensor and the log is 1.5 m and the measurement region varies from 1 m to 1.6 m depending of the log diameter. The resolution of the used model was 1.5–2.5 mm per pixel.

## 2.5 Multiple sensor measurement system

Developed measurement system with multiple Kinect sensors was programmed using KinectFusion 1.7 library. The model reconstruction is done using a *Graphics Processing Unit* (GPU). The GPU provides extremely fast parallel computation for simple arithmetic operations which enables real-time modelling of the log. If the measurement time increases too high it will affect to accuracy of the measured model. In this study, NVIDIA GTX560 graphics card was used which provides enough computation capacity for several sensors at the same time.

### 2.5.1 Measurement setup

The measurement locations are divided to several blocks. (Fig. 3) Each Kinect sensor measures one single block at the time. The size of the block

Figure 2. The single sensor setup at the sawmill (Antikainen & Verkasalo 2013).

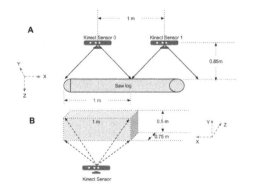

Figure 3. Measurement setup for the multiple sensor system.

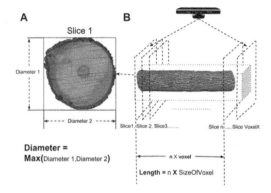

Figure 4. Dimensional measurement methodology (Xiaolei 2014).

changes if the distance of the sensor changes. After the measurement has completed the software combines acquired blocks to one single model. This model can be analyzed in the software or the model can be stored and analyzed afterwards with external software such as MeshLab.

### 2.5.2 Automatic dimension calculation

The developed measurement system will automatically calculate the dimensions and the volume of the measured model. For the dimensional calculation the measured model is divided into multiple slices as in Figure 4 and each slice is analyzed individually. Both diameters are saved and the average or the maximum diameters are calculated. The volume can be estimated using the calculated diameters or using the surface areas of the slices along the full model. The distances between slices are depended of the used voxel size which can be selected from 128 (~7.8 mm/voxel) to 512 (~2 mm/voxel) per meter. The surface area method will give more accurate result than the average diameter based method because it does not reduce the dimensional information.

## 3 EXPERIMENTS AND RESULTS

### 3.1 Experiments with single sensor

The single sensor version was tested in real sawing mill environment. The used logs were measured using the calipers (Fig. 5) before the actual 3D measurements. The 3D measurements were done during the log rotation on the sawing table. Because of the limited scanning region of one Kinect sensor the logs were measured in four separate parts to achieve reasonable resolution. The full model of the log was reconstructed from the separate scans using the Meshlab software. The same software

Figure 5. Logs were measured using calipers before the 3D scanning.

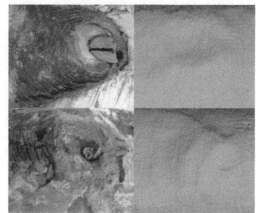

Figure 7. Left side shows the RGB images of small knots and the right side shows the result of the measurement system.

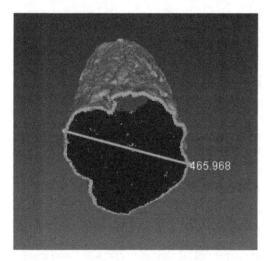

Figure 6. Measured model was analyzed with MeshLab software.

was used to analyze the dimensions of the measured model (Fig. 6).

The diameter difference between the model analysis and the calipers varied between 1–10 mm depending of the measurement place and orientation. The average error with all logs was 1.07% with the standard deviation 0.97% which shows that the measurement method is accurate enough for real environment purposes.

3.2 *Experiments with multiple sensors*

Multiple sensor system was tested in the laboratory environment with birch logs. The main objective for the multiple sensor system was the develop multi threated software which can handle several sensors at once. Results of the implementation were evaluated against the physically measured dimensions and visually determined defects. The error for the diameter between the physical measurements and model analysis was 2.3% which was higher than single sensor setup. Reason for this might be the smooth surface of the log (birch) or the reconstruction was delayed due the higher computation time compared to single sensor system. The multiple sensor measurement system was also tested for external defect detection. Figure 7 shows one result of the accuracy evaluation for small knot detection.

As seen from the Figure 7, the measurement method cannot efficiently detect small knots directly from the 3D model. However, the measurement method can be improved utilizing the RGB camera in the sensor. It will give us more information of the surface and it can be combined directly to the 3D model.

4 DISCUSSION AND CONCLUSION

The main idea of this study was to develop a low cost 3D modelling system to log scanning. The scanning speed is fast enough for a real-time analysis which is needed for the real environment. The graphics card shows a crucial part of the method because low computation efficiency will decrease the accuracy of the model or even make it to fail. In the multiple sensor system the measurement speed decreased compared to single sensor method because the computation capacity of the used graphics card.

It was noticed that the automatic registration of the separate frames worked better surfaces with rough surfaces such as in Scots pine logs.

The registration algorithm works better if the surface contains high variety of different shapes. In a case of smooth surfaces the tracking of the different frames is relatively hard because the smooth surface may be mistakenly interpreted as lack of (horizontal) movement by the system. This was a result of the relatively small resolution of the scanning system which cannot detect small details. Registration problem occurred with some spruce logs because the log surfaces were too smooth. With pine logs, the measurement was more robust.

The efficiency of the used GPU card was on its limit with multiple sensors which also affected to the measurement accuracy. This problem can be minimized using a faster GPU and the registration algorithm.

The Kinect sensor includes normal RGB camera and it can be used simultaneously with the depth sensor. The RGB images can be used to detect defects and park using the texture analyses methods as well. There is a huge research potential with the low cost Kinect sensor in a field of 3D object scanning and quality feature extraction. In addition, the next generation Kinect V2 sensor has been reported to be more accurate than the first version. This will enable more robust and efficient scanning applications in the future. The proposed measurement method works

In the future work, the proposed method will be tested and evaluated for biomass volume estimation. The sensor is going to be used to measure wood chip loads inside the truck containers.

## REFERENCES

Antikainen J. & Verkasalo E. 2013. A real-time 3D log modelling for sawing optimization in a small circular sawmill, In *Proceedings of the 21st* International Wood Machining Seminar (*IWMS*-21), Tsukuba, Japan.

Besl, P.J., McKay, N.D. 1992. A method for registration of 3-D shapes" *in IEEE Transactions on Pattern Analysis and Machine Intelligence*, vol. 14, 239–256.

Dashner B. 1993. 3D Log scanning—the next generation. Proceedings of the 5th Int. Conference on Scanning Technology and Process Control for the Wood Products Industry, Atlanta, 25–27.

Dashner, B. 1999. Three-dimensional scanning of logs on carriages. In Szymani, R. (Ed.): *Scanning Technology & Process Optimization: Advances in the Wood Industry* (pp. 124–132). Miller Freeman Books, San Francisco.

Hernandez, M., Choi, J. & Medioni, G. 2012. Laser scan quality 3-D face modeling using a low-cost depth camera, In *Proceedings of 20th European Signal Processing Conference*.

Izadi, S., Kim, D., Hilliges, O., Molyneaux, D., Newcombe, R., Kohli, R., Shotton, J., Hodges, S., Freeman, D., Davison, A. & Fitzgibbon, A. 2011. KinectFusion: Real-time 3D Reconstruction and Interaction Using a Moving Depth Camera, In *Proceedings of ACM Symposium on User Interface Software and Technology*, October.

Jing, T., Jin, Z., Zhigeng, P. & Hao Y. 2012. Scanning 3D Full Human Bodies using Kinects, In *Transactions on Visualization and Computer Graphics*, vol. 18(4).

Newcombe, R., Izadi, S., Hilliges, O., Molyneaux, D., Kim, D., Davison, A., Kohli, P., Shotton, J., Hodges, S., & Fitzgibbon, A. 2011. KinectFusion: Real-Time Dense Surface Mapping and Tracking" In *Proceedings of the 10th IEEE International Symposium on Mixed and Augmented Reality*, pp. 172–136.

Shpunt, A., et. al. 2008. Depth-varying light fields for three dimensional sensing", US 2008/0106746 A1 patent.

Skog, J. & Oja, J. 2007. Improved log sorting combining X-ray and 3D scanning—a preliminary study, *COST E53 Conference—Quality Control for Wood and Wood Products*.

Stowers, J., Hayes, M., Bainbridge-Smith, A. 2011. Altitude control of a quadrotor helicopter using depth map from Microsoft Kinect sensor. In *Proceedings of the IEEE International Conference on Mechatronics*, ICM 2011, Istanbul, Turkey, 13–15, pp. 358–362.

Wenshu, L. & Jingxin, W. 2011. Development of a 3D log sawing optimization system for small sawmills in central Appalachian, US, *Wood and Fiber Science*, 379–393.

Xiaolei Du. 2014. Kinect-based system for log surface visualization and 3D reconstruction, *Master's thesis work*, University of Eastern Finland.

*Emerging Technologies in Non-Destructive Testing VI – Aggelis et al. (Eds)*
*© 2016 Taylor & Francis Group, London, ISBN 978-1-138-02884-5*

# A noninvasive approach for the assessment of dental implants stability

P. Rizzo
*Department of Civil and Environmental Engineering, University of Pittsburgh, Pittsburgh, USA*

E. La Malfa Ribolla
*Department of Civil, Environmental, Aerospace and Materials Engineering. University of Palermo, Palermo, Italy*

A. Di Cara
*Department of Chemical, Management, Computer and Mechanical Engineering, University of Palermo, Palermo, Italy*

ABSTRACT: We propose the Electromechanical Impedance (EMI) technique to monitor the stability of dental implants. The technique consists of bonding one wafer-type piezoelectric transducers to the implant system. When subjected to an electric field, the transducer induces structural excitations which, in turn, affect the transducer's electrical admittance. The hypothesis is that the health of the bone surrounding the implant affects the sensor's admittance. In this paper we present the results of an experiment where a sensor is bonded to an abutment screwed to implants secured into bovine bone samples. The results show that the EMI technique can be used to monitor the stability of dental implants although more research is warranted to examine the repeatability of the results and the advantage with respect to existing commercial systems.

## 1 INTRODUCTION

A dental implant is an artificial titanium alloy device that is placed in contact with the jawbone and oral connective tissues to replace functionally and aesthetically the natural tooth. Brånemark and co-authors (Brånemark et al. 1969) defined osseointegration as the direct, structural and functional contact between the living bone and implant. The stability of an implant occurs in two stages. The primary stability is achieved immediately after the surgical procedure as the outcome of the mechanical interlock with the bone. This stability is a critical factor in the success of the implant integration (Ahmad & Kelly 2013).

To achieve safe and reliable therapy, it is necessary to evaluate the stiffness of the osseous tissue-implant interface quantitatively. Typically, dental surgeons use empirical methods based on palpation and patient sensation to determine when to load the implant with the prosthesis (Sullivan et al. 1996). Alternatively, noninvasive imaging and biomechanical techniques can be used. Biomedical techniques are preferred because they are portable, more economical, and do not produce harmful radiations. There are two commercial biomechanical systems. The Periotest® (Schulte et al. 1983) consists of a handheld tappet accelerated by

an electromagnet; the tappet impacts the implant several times and an accelerometer, embedded in the tappet's head, measures the contact time between the probe and the implant. A loose implant induces a longer contact time and the corresponding Periotest Value (PTV) is higher than an osseointegrated implant. PTV ranges from −5 to +5, when used to assess implant mobility, while it ranges from −8 to +50, when used to assess natural teeth. The other commercial system is based on the Resonance Frequency Analysis (RFA) and commercialized by Osstell (Meredith 1997). A small L-shaped transducer screwed to the implant or the abutment is excited by a sinusoidal signal typically in the range of 5–15 kHz. The resonance frequency of the implant is used as an indicator of the stability; the higher the resonance frequency, the more stable is the implant. The measure of the first resonance peak is converted into a value called Implant Stability Quotient (ISQ), ranging from 0 to 100 and practitioners consider values below 45–50 risky for sufficient stability.

In this paper we present a different technology to assess the primary stability of dental implants and to assess the surrounding tissue during healing. We hypothesize that the Electromechanical Impedance (EMI) method is able to determine the progress of the osseointegration. In general the method consists of bonding a Piezoelectric Transducer (PZT) to the structure to be monitored, the host structure. When subjected to an electric field, the transducer induces low to high frequency structural excitations that affect the transducer's electrical conductance. As the structural vibrations depend on the mechanical impedance of the structure, the measurement of the conductance can be exploited to assess the health of the host element.

In the study presented in this paper we evaluated the repeatability of the EMI method at assessing the stability of dental implants. The study expands earlier studies (Boemio et al. 2011, Tabrizi et al. 2012) in which the EMI method was experimentally applied to implants entrenched in Sawbones® and bovine bone samples. In those studies, the samples were degraded in nitric acid to simulate at large the inverse of the healing process that occurs naturally. Recently (La Malfa Ribolla et al. 2015) used the EMI method to monitor numerically and experimentally simulated osseointegration by using a root canal sealer to secure commercial dental implants in in vitro bovine bone samples.

Figure 1. Photo of one test sample.

## 2 MATERIALS AND METHODS

Solid standard polyurethane from Sawbones® (Pacific Research Laboratories) is widely used to mimic human bone. Four 180 × 130 × 40 mm$^3$ blocks were used. The density of the blocks were equal to 240.3, 320.4, 480.6 and 640.7 kg/m$^3$ (15, 20, 30, and 40 lb/ft$^3$), and they are hereafter denoted with the letter A (the lightest) to D (the heaviest). From each block eight 40 × 35 × 30 mm$^3$ specimens were obtained and they are indicated herein with a progressive number 1 to 8.

Each specimen was drilled using the Nobelbiocare OssoSet™ 100 device, and following the Nobelbiocare standard clinical procedure. The implant was manually placed and screwed using the Nobelbiocare Torque Wrench, applying a torque of 35 N*cm measured using a wrench handle. Once the implant was inserted, the permanent prosthetic abutment carrying the PZT was screwed to the implant, and the transducer was wired to the EMI measurement system.

A square piezoelectric transducer (PSI-5A4E 2 × 2 × 0.267 mm$^3$) was glued to the implant using the JB Weld epoxy resin. The photo of one sample ready for testing is shown in Fig. 1. To approximate the process of osseointegration that naturally occurs at the interface between the implant and the bone after the surgical therapy, the implant itself was removed and screwed in blocks of different densities. To build a pool of data statistically meaningful the eight blocks per density were considered.

The mechanical impedance of a point on a structure is the ratio of the force applied on the point to the resulting velocity at that point. The general principles of the EMI applied to the dental implant problem were presented by Rizzo and co-authors (Boemio et al. 2011, Tabrizi et al. 2012, La Malfa Ribolla et al. 2015). The transducer was driven by an up-chirp that is a signal in which the frequency increases with time. A National Instruments® PXIe-1062Q unit running under LabView® software was used and the frequency increased from 0 to 1 MHz with 50 Hz frequency resolution.

The following experiments were conducted. The test specimens sat on a bench. All thirty-two specimens were considered, namely A1 trough D8.

Build-in Matlab® functions were used to compute the average and the standard deviation of the conductance signatures and of the features selected in this study.

## 3 RESULTS

To ease the comprehension of the methodology, the conductance as a function of the actuation frequency of the free PZT and of the free superstructure is presented in Figure 2. Here, the superstructure represents the PZT bonded to the abutment screwed to the implant. For the free PZT one main peak and two side peaks are visible at around 920 kHz. They are associated with the resonance characteristics of the piezoelectric

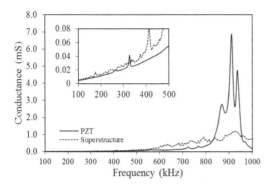

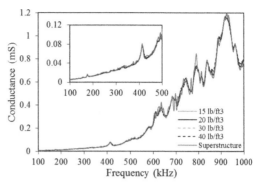

Figure 2. Conductance as a function of the excitation frequency under free boundary conditions for the PZT patch and the superstructure.

Figure 3. Conductance as a function of frequency. Each plot is the average of the measurements associated with the eight samples for each density. The conductance of the free-superstructure is overlapped. The inset presents the close-up view of the same graph in the sub-range 100–500 kHz.

material. The inset presented in Fig. 2 represents a close-up view of the subrange 100–500 kHz. In the inset, the vertical scale is two orders of magnitude smaller. Two peaks at 260 kHz and 335 kHz are visible. The electrical conductance of the superstructure drastically decreases because the force provided by the PZT is the same but the resulting velocity is smaller. More peaks also appear as can be seen observing the close-up view at the 100–500 kHz; these new peaks are associated with the vibration of the superstructure and are not related to the resonance of the piezoelectric material.

Figure 3 shows the conductance as a function of the excitation frequency when the specimens sat on the work bench. The values relative to the four densities are the average of the conductance associated with the eight samples. The conductance of the free superstructure is superimposed. Overall the interlock between the implant and the Sawbones block reduces the number of structural peaks although the frequency at which these peaks occur is similar. A close-up view of the 100–500 kHz range is presented in the inset of Fig. 3.

As the differences between the curves cannot be ascertained, a few statistical features were extracted from Fig. 3. They were the root mean square (RMS) across the frequency range of interest, and the amplitude and the corresponding frequency of the largest peak. Figure 4 shows these features as a function of the sample density. The average value of the eight samples per density and the corresponding standard deviation are presented. The results do not reveal any monotonic trend that univocally associates the EMI-based feature to the density of the sample, i.e. the features cannot consistently indicate that the stability of the implant increased with the foam density. This occurs despite the fact that the compressive and shear moduli increase with greater foam density. We believe that the effect of the local stiffness on the structural vibration of

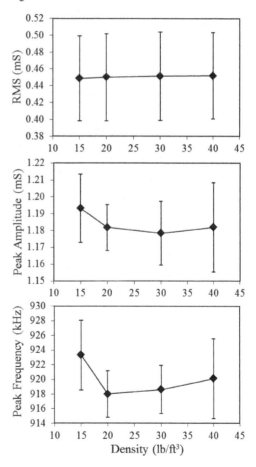

Figure 4. Selected features as a function of the density calculated in the excitation range 100–1000 kHz. (Top) RMS, (Middle) Peak amplitude, (Bottom) Peak frequency.

the superstructure is compensated by the increase of the sample's mass.

## 4 CONCLUSIONS

Within the circumstances of this study, the results seem to suggest that the method is able to discriminate monolithic bones of different densities. We hypothesize that the sensitivity of this EMI-based approach could be enhanced by including other features in the analysis or by considering structural peaks other than those selected in this study. Conclusive opinion about the effectiveness of this methodology can be made only after conducting a comparative analysis with respect to the two most popular commercial mechanical devices, namely Osstell and Periometer, which have not yet received consensus opinion regarding usefulness or clinical validity as remarked by Ahmad and Kelly (Ahmad & Kelly 2013).

## ACKNOWLEDGMENTS

The second and the third authors conducted this research as visiting scholars at the University of Pittsburgh's Laboratory for Nondestructive Evaluation and Structural Health Monitoring studies. The second author acknowledges the scholarship "Borse di studio finalizzate alla ricerca e Assegni finanziati da Programmi Comunitari" provided by the University of Palermo, Italy. The third author acknowledges the sponsorship of the Italian Ministero dell'Istruzione dell'Universita' e della Ricerca under the program Messaggeri della Conoscenza project ID 322.

Most of the equipment used in this study was purchased through a U.S. National Science Foundation grant (CMMI 1029457).

The authors declare no conflicts of interest related to this study.

## REFERENCES

Ahmad, O.K. & Kelly, J.R. 2013. Assessment of the primary stability of dental implants in artificial bone using resonance frequency and percussion analyses. *International Journal of Oral & Maxillofacial Implants* 28(1): 89–95.

Boemio, G., Rizzo, P. & De Nardo, L. 2011. Assessment of dental implant stability by means of the electro-mechanical impedance method. *Smart Materials and Structures* 20(4): 045008.

Brånemark, P.I., Breine, U., Adell, R., Hansson, B.O., Lindström, J. & Ohlsson, Å. 1969. Intra-osseous anchorage of dental prostheses: I. experimental studies. *Scandinavian Journal of Plastic and Reconstructive Surgery and Hand Surgery* 3(2): 81–100.

La Malfa Ribolla, E., Rizzo, P. & Gulizzi, V. 2015 On the use of the electromechanical impedance technique for the assessment of dental implant stability: Modeling and Experimentation. *Journal of Intelligent Materials and Structures* (available online, *1045389X14554129*).

Meredith, N. 1997. Assessment of implant stability as a prognostic determinant. *The International journal of prosthodontics* 11(5): 491–501.

Schulte, W., d'Hoedt, B., Lukas, D., Muhlbradt, L., Scholz, F., Bretschi, J., Frey, D., Gudat, H., Konig, M., and Markl, M. 1983. Periotest—a new measurement process for periodontal function. *Zahnarztliche Mitteilungen* 73(11): 1229–40.

Sullivan, D.Y., Sherwood, R.L., Collins, T.A. & Krogh, P.H. 1996. The reverse-torque test: a clinical report. *The International journal of oral & maxillofacial implants* 11(2): 179.

Tabrizi, A., Rizzo, P. & Ochs, M.W. 2012. Electromechanical impedance method to assess dental implant stability. *Smart Materials and Structures* 21(11): 115022.

# Monitoring techniques for nanocrystalline stabilized zirconia from some medical prosthesis

A. Savin & M.L. Craus
*National Institute of R&D for Technical Physics, Iasi, Romania*

V. Turchenko
*Joint Institute for Nuclear Research, Dubna, Russia*

A. Bruma & S. Malo
*CRISMAT, National Graduate School of Engineering, University of Caen on Normandy, Caen, France*

T.E. Konstantinova
*Donetsk Institute of Physics and Technology named after O.O. Galkin, Ukraine*

ABSTRACT: The paper proposes the investigation of influence of doping with Ce ions to the phase stability and mechanical properties of zirconium dioxide. For this investigation, we used X-ray and neutron diffraction, electron microscopy and microstructure characterization, including Resonant Ultrasound Spectroscopy (RUS) for nondestructive evaluation of physical-mechanical properties of ceramic made from zirconia. The RUS method has the advantage of being rapid, has a good accuracy and can be used in quality control of the object.

## 1 INTRODUCTION

Mechanical properties and bioinert behavior of ceramic materials based on zirconium oxides were extensively investigated in connection with their possible applications hip implants. Ceramics based on the tetragonal modification T-$ZrO_2$ of zirconium dioxide are now one of the strongest and inert ceramics available.

Most of the implants are modular (Fig.1a), assuring the possibility to adapting the geometry of the prosthesis to the joint morphology of the patient, providing more flexibility during primary surgery and simplified revision procedures (Dzaja et al. 2014).

At the moment, there are no long-term studies evaluating the newest ceramic hip replacements. Furthermore, the metal and plastic implants are work very well and have an excellent track record. While the ceramic hip replacements may be better, there is a little proof to support that idea. Generally, in total hip implants, the ceramic femoral heads are prepared with a tapered bore in which the projected male counter face of the metallic stem is filled to fix the head with stem (Fig. 1b).

Oxide based ceramics are formed by close packed crystals of oxides of zirconium metals. The pres-

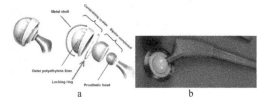

Figure 1. Hip implants: a) modular principle; b) photo.

ence of impurities determine the modifications of mechanical properties. The medical grade ceramics are most chemically and biologically inert of all materials, the main disadvantage being their fragility. The ceramic materials cannot deform under the stress, as plastic and metals can do. When the stress acting on medical grade ceramics exceeds a certain limit, ceramic material burst and formally explode in many splinters.

Mechanical properties and bio inert behavior of ceramic materials based on zirconia requirements imposed to medical grade zirconia are standardized in (BS EN ISO 13356/2013).

The paper proposes the investigation of the phase stability and mechanical properties of zirconium dioxide following a process of doping with

Ce ions. For this investigation, we used X-ray and neutron diffraction, scanning electron microscopy and microstructure characterization, including Resonant Ultrasound Spectroscopy (RUS) for nondestructive evaluation of physical-mechanical properties of ceramic made from zirconia. The RUS methods have the advantage of being fast, have a good accuracy and can be used in quality control of the object.

## 2 THEORETICAL PRINCIPLES OF RESONANT ULTRASOUND SPECTROSCOPY

RUS uses normal modes of elastic bodies to infer material properties such as elastic moduli. The elasticity of a material is the predisposition of a material to return to a minimum energy configuration. For an isolated body, $v$, that is, one bounded by a closed, stress free surface, $C_{ijkl}$ its elastic stiffness tensor and $\rho$ its density. $C_{ijkl}$ and $\rho$ may vary with position in $v$.

Let $\omega$ be a non-negative real number, and $\bar{u}(\bar{r})$ be a real valued function at position $\bar{r}$ in $v$. Then, the combination $\{\omega,\bar{u}\}$ is a free oscillation or resonance if the real-valued displacement field

$$S(\bar{r},t) = \Re(\bar{u}(\bar{r})\exp(j\omega t))$$ 
(1)

where $j = \sqrt{-1}$, satisfies the elastic equations of motion in $v$ and the stress-free boundary condition on its surface.

The potential energy $E_p$ associated with the displacement field $\bar{u}$ is given by the strain energy (Zadler et al. 2004)

$$E_p = \frac{1}{2}\int_\Omega C_{ijkl}\partial_j u_i \partial_l u_k dV$$ 
(2)

where $u_i$, $i=1,2,3$ are the Cartesian components of $\bar{u}$ and we are using the summation convention for repeated indices and $C_{ijkl}$ constants from the elastic modulus tensor.

The properties of an elastic medium are determined by the stiffness tensor ($C_{ijkl}$) which relates the stress ($\tau$) applied to a sample with the strain (e) experienced $\tau_{ij} = C_{ijkl}e_{kl}$. (Landau & Lifschitz 1982).

The corresponding kinetic energy $E_k$ is given by

$$E_k = \omega^2 k; \quad k = \frac{1}{2}\int_V \rho u_i u_i dV$$ 
(3)

The quantity

$$I = \omega^2 k - E_p$$ 
(4)

is stationery if and only if $\omega$ and $\bar{u}$ are a resonance of $v$ (Dahlen & Tromp 1998), i.e. the value of $I$ does not change if we replace $\bar{u}$ by $\bar{u}+\delta\bar{u}$, where $\delta\bar{u}$ is any small displacement. This is the Rayleigh-Ritz method. Let $\{\Phi_\lambda(\bar{r}), \lambda=1,2,...,N\}$ be a set of $N$ specified functions over $v$. We represent $\bar{u}$ by a prescribed basis

$$u_i = a_{i,\lambda}\Phi_\lambda$$ 
(5)

The choice of basis functions $\Phi_\lambda$ is a significant issue when using Rayleigh-Ritz methods are chosen. Various trigonometric functions and orthogonal polynomials were used (Demarest 1969) substantial effort was required to formulate and program the integrals for each new sample or geometry (Zadler et al. 2004). The essence of Rayleigh-Ritz method consists into a convenient way of writing the set of basis function $\Phi_\lambda$. The $xyz$ algorithm

$$\Phi_\lambda = x^{n(\lambda)}y^{\theta(\lambda)}z^{\xi(\lambda)}$$ 
(6)

where $\eta,\ \theta,\ \xi$ are positive integers, is numerically stable and very flexible. It allows a general anisotropic tensor which any position dependence and any shape with arbitrary density variation (Visscher et al. 1991). Inserting equation (5), (3), (2) into (4) we obtain

$$I = \omega^2\alpha\cdot\bar{k}\cdot\alpha - \alpha\cdot\bar{E}\cdot\alpha$$ 
(7)

where $\alpha$ is a vector comprising the juxtaposed components of $a_{i,\lambda}$. $I$ is stationery if and only if

$$\omega^2\bar{k}\alpha = \bar{E}\alpha$$ 
(8)

This is the standard form for the generalized symmetric eigenvalue problem. Here, $\alpha$ is a vector containing all of the coefficients for all the components of $\bar{u}$. If $f(x,y,z)$ is either density or one the Cartesian components of the elastic stiffness tension, in any region $S \subseteq v$ in which a function $f(x,y,z)$ can be written as a sum of monomial basis functions, a condition that includes the nearly universal case of being constant S is that integrals we need are of the form

$$I_S(l,m,n) = \int_S x^l y^m z^n dV$$ 
(9)

For simple and regulated geometrical shapes of region S (parallelepiped, cylinder, ellipsoid, etc.), the integral (9) can be analytically calculated and for complicated shapes, its evaluation can be made only numerically.

# 3 MATERIALS AND METHODS

In recent years, significant progress has been achieved in the synthesis of size, shape, crystal structure and composition controlled nanoparticles (Schmid 2004, Xie et al. 2011). Along with Y-TZP ceramics, ceria—stabilized zirconia Ce-TZP (ceria tetragonal zirconia polycrystals) are often used in the manufacture of plain bearing. Zirconium dioxide has attracted attention due to its unique physical properties: high fracture toughness and bulk modulus, low thermal conductivity, extremely refractory, chemically inert, corrosion resistant, high dielectric constant.

Zirconia based compounds are formed by closed packed crystallites. The chemical composition and the cations distribution in the unit cell influence the characteristics of the resulting material. The medical grade ceramics are the most chemically and biologically inert of all materials. In order to validate the RUS method and examine the way in which the cerium oxides stabilize the zirconia, sinterized cylindrical samples were taken into study.

The samples were obtained through the standard ceramic technology, a mixture of ceria and zirconium oxides being used, in proportions established *a priori*. After grinding and pressing, the samples have been obtained in the shape of cylinders. The samples were sintered for 6 hours at 1500°C temperature in air. Some crystallographic and mechanical features of the samples are presented in Tables 1 and 2.

Table 1. Dependence of crystallographic structure (space group, SG), unit cell volume (V), observed density ($\rho_{obs}$) and X ray density ($\rho_{XRD}$) with Ce concentration (x) for $Zr_{1-x}Ce_xO_2$ samples.

| Sample ID | x | SG | V ($Å^3$) | $\rho_{obs}$ (kg/m³) | $\rho_{XRD}$ (kg/m³) |
|---|---|---|---|---|---|
| П 2 | 0.09 | P 2₁/c | 144.14 | 5378.3 | 5880.0 |
| | | P 4₂/nmc | 68.476 | 5378.3 | 6188.7 |
| П 3 | 0.13 | P 4₂/nmc | 69.235 | 5885.2 | 6214.8 |
| П 4 | 0.17 | P 4₂/nmc | 69.720 | 6263.0 | 6264.5 |

Table 2. Dependence of mechanical properties with Ce concentration (x) for $Zr_{1-x}Ce_xO_2$ samples.

| Sample ID | Elasticity modulus (GPa) | Shear modulus (GPa) | Poisson ratio | Molecular Mass (g/mole) |
|---|---|---|---|---|
| П 2 | 145.85 | 56.49 | 0.291 | 127.62 |
| | 145.85 | 56.49 | 0.291 | 127.62 |
| П 3 | 168.53 | 64.47 | 0.307 | 129.58 |
| П 4 | 193.43 | 73.16 | 0.322 | 131.53 |

The structural analysis of ceramic samples of zirconium dioxide $Zr_{1-x}Ce_xO_2$ (x = 0 ÷ 0.17) was performed at room temperature using Time-of –Flight (TOF) High Resolution Fourier Diffractometer HRFD at the IBR-2 pulsed reactor in Joint Institute for Nuclear Research (JINR), Dubna, Russia (Balagurov 2005).

At HRFD, the correlation techniques of data acquisition is used, which provides a very high resolution ($\Delta d/d \approx 10^{-3}$). HFRD resolution is practically constant in a wide interval of $d_{hkl}$ interplanar distances. The high resolution diffraction patterns were collected by detector placed at back scattering angles ($2\theta = \pm152°$, $d_{hkl} = 0.6 ÷ 3.6Å$).

It should be specified that via X-ray diffraction, a layer of typical thickness of ≈10 μm is investigated, whereas via neutron diffraction, a layer of nominal thickness of the order of centimeters is typically investigated.

A discussion concerning the crystallographic data concerning the structure type, lattice constants, cations and anions positions in unit cell, average size of crystalline blocks and microstrains will be published elsewhere.

# 4 EXPERIMENTAL SET-UP

## 4.1 *Experimental set-up for RUS*

Sample geometry affects data acquisition. For cylinders with high ratio of length-diameter we have few excitation modes and the spectra are simple. In the case of the studied samples the ratio is around the unit and the spectrum requires more analysis than a long bar. The increasing of spectral complexity supposes adjacent analyses of the samples before concluding about these samples.

In order to determinate a parameter, more resonance frequencies must be searched. This requires large time of computational calculation and repeated tests.

The same number of normal modes for a short cylinder comparing with a long cylinder requires a narrow frequency range.

The measurements are carried out with transducers with contact. The coupling of the transducer with the specimen influences which modes are measured. When the sample is pinned on its edge, more modes are excited and the modes are better defined than when the transducers are placed on the ends of the cylinder (Fig. 2).

The probe is fixed between the emission and reception transducers in order to accomplish the condition of stress free surface. The equipment allows the setting so that for the established position of the cylindrical sample, the contact on the edge assures the excitation of a maximum number of possible resonances for the fixed geometry.

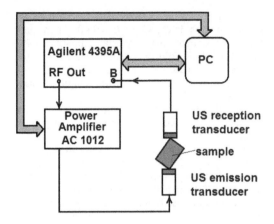

Figure 2. Experimental set-up for RUS.

A Network/Spectrum/Analyzer 4395 A Agilent USA generates a sweep frequency between 80 kHz and 250 kHz in 1 kHz step. The signal is amplified into a Power amplifier AC 1012 AG&TC Power Inc. USA and applied to an US emission transducer P111.O.06P3.1 type, selected for the large bandwidth. The signal delivered by the reception transducer, identical to the emission one, is applied to the B port of the 4395 A Agilent, the spectrum being acquired by a PC that is used also to program the functioning of the equipment with a numerical code developed in Matlab, via PCIB interface. The command of the power amplifier is made with the same PC via RS232 interface.

### 4.2 Ultrasound characterization of samples

The Young modulus, shear modulus and Poisson ratio were determined based on the propagation speed of longitudinal and transversal ultrasonic waves using impulse—echo method. In order to determinate the elastic and shear moduli for the samples Π2-Π4, the propagation speed of longitudinal and transversal ultrasound waves were measured using transmission procedure.

For the measurement of the propagation speed of longitudinal wave, the transducer G5 KB GE with central frequency of 5 MHz was used, the coupling being assured by ZG-F Krautkramer gel and for the transversal waves, the transducer employed was MB4Y GE with central frequency of 4 MHz. The PR 5073 Pulser Receiver Panametrics NDT USA is used for the emission impulses and the reception of the signals. The digitizing of the signals and the measurements of the time of flight was made with the digital oscilloscope Wave Runner 64Xi Le Croy USA.

## 5 RESULTS AND DISCUSSIONS

According to X-ray and neutron data, the $Zr_{1-x}Ce_xO_2$ samples are not homogeneous and the structure of investigated samples is dependent on the Zr/Ce concentration (s. Fig.3 and Tabs.1 and 2).

We should take into account that only a thick layer at the surface of the sample which can be studied by means of XRD. For $x = 0.09$ we observed a mixture of tetragonal phase (P4$_2$/nmc, GS 131) and monoclinic phase (P2$_1$/nmc, GS 14), Figure 4a.

At larger Ce concentration, a tetragonal phase, together a strange phase were identified in X ray diffractograms (Figures 4b and c). The increase of concentration of Ce ions leads to changing of crystal structure from monoclinic (P 2$_1$/c) to tetragonal (P 4$_2$/nmc) and an increase of the volume of tetragonal unit cell from 68.5 Å$^3$ to 69.2 Å$^3$ ($x = 0.13$) and, respectively, 69.7 Å$^3$ ($x = 0.17$) (s. Tab.1).

The microstructure of the material is displayed in Figures 5 and 6.

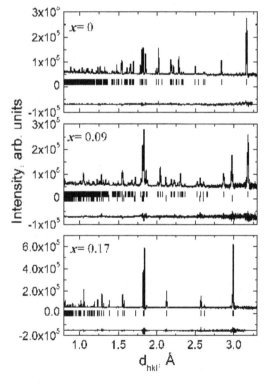

Figure 3. Diffraction pattern of $Zr_{1-x}Ce_xO_2$ samples measured at the HRFD at room temperature and treated with Fullprof programe.

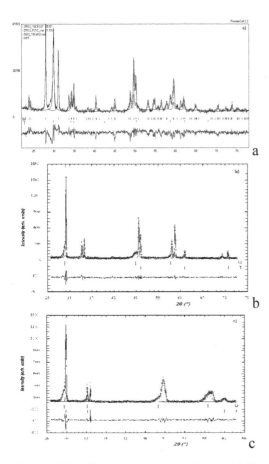

Figure 6. Electron microscopy (SEM) for the sample corresponding to x = 0.17.

Figure 4. The difractograms correspond to x = 0.9 (a), x = 0.13 (b) and x = 0.17 (c). The vertical bars correspond to monoclinic phase (M) and tetragonal phase (T) (4a, PowderCell), to tetragonal phase (T), (4b and 4c) and unknown phase (U), (4b and 4c, Fullprof).

Figure 5. Electron microscopy (SEM) for the sample corresponding to x = 0.13.

In the SEM micrographs, it can be clearly observed that from the cutting procedure has resulted into a thin layer small amount of residue depositing between/on the crystallites (Figs 4–6). SEM micrographs emphasize, in all cases, the three dimensional aspects of the crystallites. The crystallites sizes vary considerably, between 2 and 4 µm and are characterized by a high degree of faceting, a consequence of the heating procedure thermal treatment at 1500°C. Moreover, in agreement with Carter & Norton (2013), the 3D characteristics of the crystallites belonging to the samples which were treated at high temperature, are depending on the chemical composition of the sample and the treatment temperature.

The average size of the crystallites is between 0.5 and 3 µm. The properties of the sintered cylindrical samples presented in Table 1, noted Π2–Π4 are validated by RUS method. The resonance spectra were traced for zirconia cylindrical samples and because Π2–Π4 are axisymmetric, isotropic and homogeneous (Zadler et al. 2004), therefore we can conclude that every mode must fall into one of the three classes, presented below. The resonance spectrum has been traced for the ceramic zirconia cylindrical samples, noted Π 2–Π 4, presented in Tables 1 and 2.

These samples are axisymmetric, considered isotropic and homogeneous and we can conclude that every mode of such sample must fall into one of three classes (Zadler et al. 2004):

– Torsional axisymmetric pure shear motion consisting of rigid rotation of rings or material around the simple axis. The frequencies of these modes depends entirely upon the samples shear velocity
– Extensional axisymmetric mixtures of compression and shear motions

– Flexural modes through along pass that are tilted with respect to the cylinder axis. The flexural modes occur in pairs named doublets, both members of which have the same resonance frequency.

Excess loading can cause the two members of the doublet to experience different frequencies shifts so that a single peak can be distorted into two distinct peaks in a loaded sample. This can also occurs due to the poor sample preparation or inhomogeneities such as small cracks or inclusions. Figure 7 presents the resonance spectra for the studied samples.

In Figure 8 are presented the simulation of deformation for two extensional modes (Fig.8 a and b) and three flexural modes (Fig.8 c, d and e) for the sample Π2, using finite element method in SolidWorks 2011. The resonant frequencies depend non-linearly on the elastic parameters. The resonance frequencies obtained by simulations correspond very well to those experimentally obtained. For determination in the basis of experimentally measured resonance spectra, of the main elastic properties of the sample, the inversion of data was used, implying conjugate gradient method (Zadler et al. 2004) minimizing the objective function (s. Eq. 10).

$$F = \sum_i w_i \left( f_i^{(p)} - f_i^{(m)} \right)^2 \qquad (10)$$

In Eq.(10) $f^{(p)}$ are the computed frequencies, $f^{(m)}$ are the measured frequencies, $w_i$ are the weights, which characterize the confidence we have in the measurements. The optimization problem was numerically solved using Matlab 2013a. Due to the fact that the number of peaks and corresponding, the resonance frequencies, is relatively small, the inversion was applied only for determination of $E$ and $G$, and not for geometrical dimensions and respectively for densities of the cylindrical samples made from zirconia.

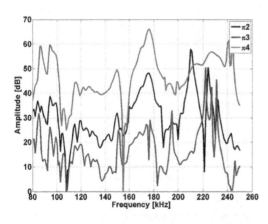

Figure 7. Resonance spectrum for the samples Π2-Π4.

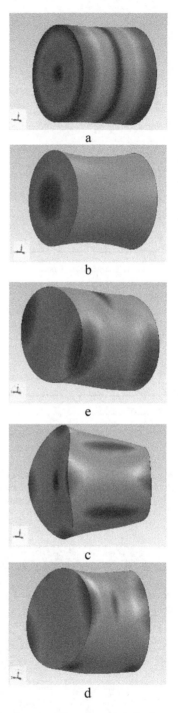

Figure 8. Aspects of few resonant modes for sample Π2:
a) extensional mode f = 163 kHz;
b) extensional mode f = 236 kHz;
c) flexural mode f = 185 kHz;
d) flexural mode f = 208 kHz;
e) flexural mode f = 214 kHz

## 6 CONCLUSIONS

The substitution of Zr with Ce in $Zr_{1-x}Ce_xO_2$ leads to a change of the phase composition. The sample Π 2 corresponding to x = 0.09 contains two phases, a monoclinic to tetragonal phase, while the samples with larger Ce concentration contain only tetragonal phase. The unit cell volume of the tetragonal phase increases with the increase of Ce concentration, for the inner part of the samples. An important result is an increase of the observed density with the increase of the cerium concentration.

The difference of unit cell parameters corresponding to the surface layer and bulk as defined by X-ray and, respectively, neutron diffractions, could be attributed to the difference between its chemical compositions. A small amount of an unknown phase appears at the surface of the samples in connection with a change in chemical composition of the tetragonal phase. On other hand a small amount of residue appears at the surface of the samples due the cutting process. It was observed at electronic microscope (s. Figs 5 and 6). RUS can be used for real-time evaluation of change in elastic properties of ceramics (such as Young's modulus, Poisson ratio, etc.) elements of hip prosthesis as femoral heads. The differences produced by molding (dry pressing, injection molding, extruding, and tape casting) and clean processing of the semi fabricates, followed by the thermal treatment and of surface (grinding, lapping and polishing) require knowledge about the entire spectrum of ceramics. The initiation of fracture of ceramic elements of hip prosthesis can be the presence of low density zone containing dispersed high density agglomerates in the inner part of the ball head and, any deformation will be immediately apparent through changes in the resonance modes, with deviations from the normal spectrum.

## 7 AUTHOR CONTRIBUTIONS

The manuscript was written through contributions of all authors. All authors have given approval to the final version of the manuscript and contributed equally.

## ACKNOWLEDGEMENTS

This paper is partially supported by Romanian Ministry of Education and Scientific Research under projects PN-II-ID-PCE-2012-4-0437 and Nucleus Program—Contract PN 09 43–01–04 and JINR Dubna, Russia Bilateral Cooperation Protocol no.4414–4-1121–2015/2017.

## REFERENCES

Balagurov A.M. 2005 Scientific Reviews: High-Resolution Fourier Diffraction at the IBR-2 Reactor. *Neutron News.* 16 (3):8–12.

BS EN ISO 13356/2013 - Implants for surgery – Ceramic materials based on yttria – stabilized tetragonal zirconia.

Carter, C.B., & Norton, M.G., 2013. *Ceramic Materials,* 2nd edition, Springer, New York.

Dahlen, F.A., & Tromp, J., 1998. *Theoretical Global Seismology,* Princeton University Press.

Demarest H.J., 1969. Cube-resonance method to determine the elastic constants of solids. *Journal of the Acoustical Society of America,* 49:768–775.

Dzaja I., Lyons M.C., McCalden R.W., Naudie D.D., & Howard J.L. 2014. Revision hip arthroplasty using a modular revision hip system in cases of severe bone loss.*J Arthroplasty.* 29(8):1594–1597.

Landau L.D., & Lifshitz E.M. 1982. *Theory of Elasticity,* 2nd ed., Pergamon, N.Y.

Schmid G. 2004. Nanoparticles: *From Theory to Application,* Wiley-VCH, Weinheim.

Visscher W.M., Migliori A., Bell T.M., & Reinert R.A. 1991. On the normal modes of free vibration of inhomogeneous and anisotropic elastic objects. *Acoustical Society of America,* 90:2154_2162.

Xie, Z.G., Meng, X.F., Xu, L.N., Yoshida, K., Luo, X.P., & Gu, N. 2011. Effect of air abrasion and dye on the surface element ratio and resin bond of zirconia ceramic. *Biomedical Materials,* 6(6), 065004.

Zadler B.J., Le Rousseau J.H.L., Scales J.A., & Smith M.L.2004. Resonant ultrasound spectroscopy: theory and application. *Geophysics Journal International,* 156:154–169.

*Emerging Technologies in Non-Destructive Testing VI – Aggelis et al. (Eds)*
*© 2016 Taylor & Francis Group, London, ISBN 978-1-138-02884-5*

# Raman spectroscopy: An emerging technique for minimally-invasive clinical testing

M.Z. Vardaki, P.S. Papaspyridakou, G.L. Givalos, C.G. Kontoyannis & M.G. Orkoula
*Department of Pharmacy, University of Patras, Patras, Greece*
*Foundation for Research and Technology Hellas, Institute of Chemical Engineering Sciences, Patras, Greece*

ABSTRACT: Human skeleton is degraded through the years. Bones and joints are affected by diseases which reduce their strength and functionality, causing pain and mobility problems. Osteoporosis, osteopetrosis and osteoarthritis are some of them. Osteopetrosis is a bone disorder leading to increased risk of bone fracture. Osteoarthritis is a joint disease caused by degeneration of articular cartilage. Current diagnostic methods involve clinical examination, radiographic measurements (X-rays) and histopathologic evaluation of excised specimen. Most of them are subjective, include health risk due to harmful radiation, or are invasive.

Raman spectroscopy is a modern analytical technique for chemical characterization. It is not constrained by water which makes it applicable to biological samples. It is non-destructive, quick and easy to use. In the present work, Raman spectroscopy was employed for the detection of the chemical changes induced to tissues (bone and cartilage) by relevant diseases (osteopetrosis and osteoarthritis) compared to healthy controls. Raman spectra acquired from apoE-deficient mice lumbar vertebrae revealed disturbed mineral to collagen matrix proportion as well as significant alterations in collagen network non-detectable by clinical techniques. Human osteoarthritic menisci from knee joints were overwhelmed by lipids absent in control samples, while degeneration of articular cartilage in femoral heads from human hips uncovered the subchondral bone as well as plethora of lipids.

The fact that Raman spectroscopy is non-destructive as well as the use of harmless radiation in visible or near-infrared region, makes the in-vivo application plausible. A bundle arrangement of fiber optics can be employed in order to acquire spectrum of bone in non-invasive way i.e. without removing or piercing skin.

## 1 INTRODUCTION

### 1.1 Skeleton and skeleton diseases

Human skeleton provides the framework which supports the body and maintains its shape. The skeleton serves several functions: support, movement and protection of vital internal organs. It consists of bones which is tissue tough enough to provide endurance to stress and strain. The joints between bones allow movement, smooth sliding of bones and reduce friction between them.

The skeleton degrades through the years. Bones and joints are affected by diseases which reduce their strength and functionality, causing pain and mobility problems. Osteoporosis, osteoarthritis, osteopetrosis, osteomalacia are some of them.

The term «bone quality» is often used to describe the well-being of bone and its ability to operate properly. It is an ensemble of properties such as composition, architecture and mechanical performance (Morris et al., 2011). Osteoporosis and osteopetrosis are bone abnormalities where

the relative proportion of mineral (bioapatite) and collagen content is disturbed. Osteoporotic bone is less dense than normal bone, while, osteopetrotic is a much denser bone, but, both are very fragile. They are bones of degraded bone quality.

Osteoarthritis is a disorder of the joints. When affected by osteoarthritis, the cartilage of the joint will degrade, soften and wear away. The space between bones decreases. The joint may be painful to move, may move in unusual directions or may be immobile completely (Creamer and Hochberg, 1997).

### 1.2 Current clinical diagnostic tests

Current diagnostic methods involve bone mineral density measurements (DEXA), physical examination, X-rays, arthroscopy, MRI or histopathologic evaluation of excised specimen. All of these methods are either invasive, include health risk due to harmful radiation, are subjective or provide inadequate data for the evaluation of the disease and the detection of early stages. For these reasons there seems to be a need for a technique which will be able to provide sufficient information concerning the bone and cartilage quality in a non-invasive, reliable and safe way.

### 1.3 Raman spectroscopy

Raman spectroscopy is a modern analytical technique for chemical characterization. It is not constrained by water which makes it applicable to biological samples. It is non-destructive, quick and easy to use.

There have been several publications for osteoporotic bone samples from animal models (Orkoula et al., 2012) or human excisions (Mandair et al., 2010) but it is the first time, to the authors knowledge, that osteopetrosis is studied with Raman spectroscopy.

Results from Raman studies on subchondral bone from human osteoarthritic hips (Kozielski et al., 2011) or knee joint fluids (Yavorskyy et al., 2008) have been quoted in the literature. In the present work, the chemical changes of osteoarthritic cartilage (femoral head and knee meniscus) is analyzed.

## 2 EXPERIMENTAL PART

### 2.1 Lumbar vertebrae (gene deficient mice)

Lumbar vertebrae from apoE-deficient C57BL/6 mice were used for the analysis as well as age-matched controls (Karavia et al., 2011). The animals were sacrificed at 24 weeks. Soft tissue was removed using a scalpel. Lumbar vertebrae were wrapped in a bandage soaked in PBS (Phosphate Buffered Saline) and stored at -20°C until its use.

Weighted vertebrae samples were immersed in $HNO_3$ 0.5 N for dissolution of mineral. Calcium concentration was measured employing Atomic Absorption spectrometry analysis (AAS). Another set of vertebrae samples were used for collagen content measurements by Thermal Gravimetry analysis (TGA). Bone density was evaluated immersing vertebrae in water and carefully weighing the displaced amount.

### 2.2 Osteoarthritic femoral heads

Human osteoarthritic femoral heads, removed during replacement surgeries, were used. Thin sections (5 mm thickness) were sliced parallel to the longitudinal axis of the femur with the use of an adamantine wheel. The sections were cleaned from blood, wrapped in a bandage soaked in PBS and stored at −20°C until its use.

### 2.3 Osteoarthritic menisci

Osteoarthritic human menisci were excised during surgery. The samples were rinsed with ultrapure water and maintained at −20°C until use.

Before spectra recording, all samples were thawed to room temperature and kept moist throughout the measurements by a drip of PBS (Raghavan et al., 2010).

### 2.4 Spectral recording and processing

Spectra recording was accomplished employing a micro-Raman spectrometer (Renishaw, inVia Raman microscope, UK). A laser line at 785 nm was focused through the objective lens (20x) of an optic microscope onto the sample's surface which was placed in a vertical direction. The incident radiation power was 35 mW on the objective lens edge. Each spectrum was the result of 2 accumulations from 750 to 1750 $cm^{-1}$. The time required for each recording was 1.5 min and the resolution of the instrument was 2.0 $cm^{-1}$.

Raman spectra were processed using specialized software (Peakfit© v4.0, Jandel Scientific, San Rafael, CA) for the determination and separation of peaks under each band envelope. The intensity (height and area) of each subband was measured and recorded (Orkoula et al., 2012).

### 2.5 Raman metrics

Bone compositional data were obtained using spectral data from the characteristics peaks of mineral and matrix. For the mineral to matrix ratio (MMR) the phosphate at 959 $cm^{-1}$ and the proline and hydroxyproline peaks at 855, 877 and 922 $cm^{-1}$ peaks (959 $cm^{-1}$/(855 $cm^{-1}$ + 877 $cm^{-1}$ + 922 $cm^{-1}$)), the carbonate peak at 1072 $cm^{-1}$ for the carbonate to phosphate ratio (CPR), and the inverse of phosphate 959 $cm^{-1}$ bandwidth at half-maximum for mineral crystallinity. Furthermore, the peaks at 1660 and 1685 $cm^{-1}$ were used (1660 $cm^{-1}$/1685 $cm^{-1}$) to evaluate the cross-linking and subsequent quality of collagen (Orkoula et al., 2012).

## 3 RESULTS AND DISCUSSION

### 3.1 Evaluation of bone quality (lumbar vertebrae)

Lumbar vertebrae from apoE-deficient mice were studied with Raman spectroscopy in order to evaluate the quality of bone in comparison with age-matched controls. Mineral to matrix ratio calculated from Raman spectra (figure 1a) exhibited a small but statistically significant increase. Calcium (per mass unit) measurements with Atomic

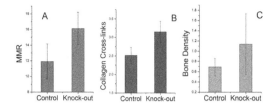

Figure 1. Raman metrics for the lumbar vertebrae: (a) Mineral to matrix ratio (MMR), (b) Collagen cross-links, (c) Bone density.

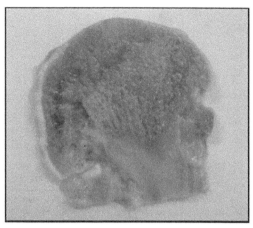

Figure 2. Section of a human femoral head.

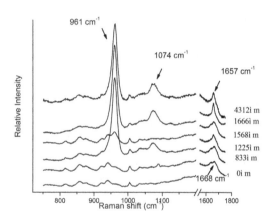

Figure 3. Raman spectra from a healthy area on the femoral head section.

Absorption spectrometry (AAS) showed no considerable difference between knock-out and control samples. On the other hand, Thermal Gravimetry analysis (TGA) revealed reduced collagen content (per mass unit) for the gene deficient mice. AAS and TGA results are in accordance with the Raman results. Mineral content (calcium) remains stable but matrix (collagen) reduces so the MMR ratio finally appears increased.

A change in the shape of the amide I envelope in Raman spectra which corresponds to collagen secondary structure was also observed. A substantial increase of the non-reducible/reducible collagen cross-links was detected for the second animal group (figure 1b). This suggests that collagen network gets stiffer and less elastic due to denser connective bonds.

Bone density measurements, done by means of immersion of vertebrae in water, disclosed a remarkable increase in the case of knock-out mice (figure 1c).

All the previous results consort to the features of an osteopetrotic bone (Wojtowicz et al., 1997) i.e. a denser bone which is actually more fragile than normal.

### 3.2 Osteoarthritis (femoral heads from hip joint)

Healthy and osteoarthritic areas on the femoral head section were tracked down by visual inspection (figure 2). Away from the center and near the neck, a layer of intact cartilage can be seen. Moving towards the center of the ball, the layer appears worn out or disappears completely, leaving the subchondral bone exposed. This is the result of the friction due to movement in the hip joint and the evidence of osteoarthritis.

Raman spectra recorded from successive spots in healthy area of the section, revealed the existence of pure unmineralized collagen (0–833 μm). When the laser moved from the unaffected cartilage towards the adjacent interior, an abrupt transition from cartilage to subchondral bone was recorded. This change involves a sudden increase of bioapatite phosphates (961 cm$^{-1}$) and carbonates (1074 cm$^{-1}$) peak intensity (1225–4312 μm). The transition from cartilage to subchondral bone is also revealed by the change in the amide I band (1550–1750 cm$^{-1}$) morphology. This change refers to the replacement of the wide peak at 1668 cm$^{-1}$ in articular cartilage by a narrow peak at 1657 cm$^{-1}$ which is attributed to lipids (figure 3).

In the osteoarthritic area, around the center of the femoral head, the spectral characteristics of bone mineral (vibrations of phosphate and carbonate) are recorded in the superficial layers (figure 4). This is in agreement with the observation of the total disappearance of the protective cartilage layer and the exposure of subchondral bone in osteoarthritis. An additional spectral feature is the characteristic

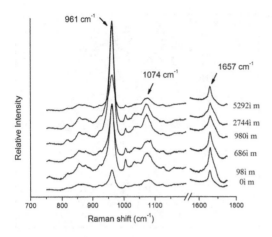

Figure 4. Raman spectra from an osteoarthritic area on the femoral head section.

 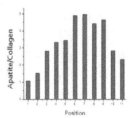

Figure 5. Raman metrics for distinguishing different stages of osteorthritis in femoral heads.

peak of lipids which overwhelms the amide I spectral area. Overall, the spectral characteristics of the osteoarthritic area are in accordance with those of the subchondral bone underlying articular cartilage in healthy area of the femoral head.

### 3.3 Osteoarthritis (meniscus of knee joint)

In knee joint meniscus, osteoarthritis is usually localized on the outer areas, expanding towards the interior. Raman spectra from healthy spots confirmed the existence of collagen, whereas the intense appearance of lipids mark osteoarthritic areas.

### 3.4 Raman spectroscopy: possibility for in vivo application?

Raman spectroscopy has a potential for in-vivo application to study hip or knee joints using fiber optics which are compatible to arthroscopes (Morris et al., 2011). For the case of osteoporosis, Morris and coworkers have reported a home-made device with a bundle of fiber optics 3 for the collection of Raman spectrum of bone without piercing skin (Okagbare et al., 2010).

### 3.5 Simulating the in vivo process

Placing the laser line parallel to the sample section, to simulate an in vivo process where a Raman fiber optic approaches the femoral head to collect spectra from the outer layer, in areas where intact cartilage layer exists (healthy area) the spectral features of collagen were recorded. In osteoarthritic areas, where cartilage is worn out, mineral, i.e. phosphate and carbonate peaks as well as lipids were traced. It is concluded, thus, that the two situations can be discerned. The question is whether various degrees of the disease, i.e. osteoarthritis in a premature stage, can also be diagnosed. The intensity ratio of the phosphate to amide I peak was calculated for various spots on the perimeter of the femoral head and found to receive lower values for the healthy spots and increased in osteoarthritic spots. Furthermore, the higher values correspond to more damaged spots (figure 5). It can be concluded that different degrees of the disease progress can be discerned in case of in vivo application.

## 4 CONCLUSIONS

It was proved that Raman spectroscopy can probe chemical changes induced to bone due to disease (osteopetrosis). These changes refer to the relative proportion of bone constituents i.e. mineral and collagen, as well as alterations in collagen network consisting of increased connective bonds between collagen fibers which result in stiffer and less elastic organic matrix more susceptible to cracks.

Chemical differences in osteoarthritic compared to healthy zones for femoral heads and knee joint menisci were also quoted. In the case of osteoarthritic hip joints the disease state can be discernible with Raman spectroscopy and a metrics was proposed for the evaluation of the disease process. An intense appearance of lipids in all cases of osteoarthritic samples, femoral heads and knee joint menisci was also noted.

Possibility of in-vivo application in a non or minimally invasive way was also discussed.

## ACKNOWLEDGMENTS

The authors wish to thank Prof. Panagiotis Megas and Prof. Kyriakos Kypraios for providing the samples for the study.

# REFERENCES

Creamer, P. & Hochberg, M.C. 1997. Osteoarthritis. *Lancet* 350: 503–509.

Orkoula, M.G., Vardaki, M.Z. & Kontoyannis, C.G. 2012. Study of Bone Matrix Changes Induced by Osteoporosis in Rat Tibia Using Raman Spectroscopy, *Vibrational Spectroscopy*, 63: 404–408.

Raghavan, M., Sahar, N.D., Wilson, R.H., Mycek, M.A., Pleshko, N., Kohn D.H. & Morris, M.D. 2010. Quantitative polarized Raman spectroscopy in highly turbid bone tissue. *Journal of Biomedical Optics*, 15(3): 037001-1–7.

Mandair, G.S., Esmonde-White, F.W.L., Akhter, M.P., Swift, A.M., Goldstein, S.A., Recker, R.R. & Morris, M.D. 2010. Potential of Raman Spectroscopy for Evaluation of Bone Quality in Osteoprosis Patients. Results of a Prospective Study. *Proc. of SPIE*, 7548: 754846-1–9.

Okagbare, P.I., Esmonde-White, F.W.L., Goldstein, S.A. & Morris, M.D. 2010. Development of non-invasive Raman spectroscopy for in-vivo evaluation of bone graft osseointegration in a rat model. *Analyst*, 135: 3142–3146.

Morris, M.D. & Mandair, G.S. 2011. Raman Assessment of Bone Quality. *Clinical Orthopaedics and Related Research*, 469, 2160–2169.

Karavia, E.A., Papachristou, D.J., Kotsikogianni, I., Giopanou, I. & Kypraios, K.E., 2011. Deficiency in apolipoprotein E has a protective effect on diet-induced nonalcoholic fatty liver disease in mice. *FEBS Journal*, 278(17):3119–29.

Kozielski, M., Buchwald, T., Szybowicz, M., Blaszzak, Z., Piotrowski, A. & Ciesielczyk. B., 2011, Determination of composition and structure of spongy bone tissue in human head of femur by Raman spectral mapping, *J Mater Sci: Mater Med*, 22:1653–1661.

Wojtowicz, A., Dziedzic-Goclawska, A., Kaminski, A., Stachowicz, W., Wojtowicz, K., Marks, S.C.Jr. & Yamauchi, M., 1997. Alteration of Mineral Crystallinity and Collagen Cross-Linking of Bones in Osteopetrotic Toothless (tl/tl) Rats and Their Improvement After Treatment with Colony Stimulating Factor-1, *Bone*, 20(2):127–132.

Yavorskyy, A., Hernandez-Santana, A., McCarthy, G. & McMahon G., 2008. Detection of calcium phosphate crystals in the joint fluid of patients with osteoarthritis—analytical approaches and challenges, *Analyst*, 133: 302–318.

*Combination of NDT techniques*

*Emerging Technologies in Non-Destructive Testing VI – Aggelis et al. (Eds)*
*© 2016 Taylor & Francis Group, London, ISBN 978-1-138-02884-5*

# Concrete compressive strength estimation by means of combined NDT

G. Concu, B. De Nicolo, N. Trulli & M. Valdés
*Department of Civil Engineering, Environmental and Architecture, Cagliari, Italy*

ABSTRACT: Non Destructive Testing (NDT) is an effective tool for assessing the condition of existing buildings and for estimating materials physical and mechanical parameters, such as concrete compressive strength, on-site. Several regulations recommend on-site estimating concrete compressive strength by using a combination of two or more NDT methods in order to improve the accuracy and to reduce errors dependent on materials, aggregate dimension and size, environmental parameters. An experimental campaign has been carried out on concrete cubic specimens casted using different design concrete strength classes. Rebound Hammer Test, Ultrasonic Testing and Pull Out Test have been carried out and compression tests have been performed on cores extracted from cubic specimens. Correlations between NDT parameters, including ultrasonic shear wave velocity, and concrete compressive strength are proposed and discussed.

## 1 INTRODUCTION

Non Destructive Testing (NDT) is an effective tool for assessing the condition of existing buildings and for estimating materials physical and mechanical parameters such as concrete compressive strength on-site.

A combination of two or more NDT methods is generally recommended for on-site estimating concrete compressive strength in order to improve the accuracy and to reduce errors dependent on materials, aggregate dimension and size, environmental parameters (Breysse 2012, Concu 2014). To this end, combined NDT methods most frequently used are Rebound Hammer Test (RHT), Ultrasonic Testing (UT) and Pull Out Testing (PT).

Despite some correlations between NDT parameters and concrete compressive strength are available in literature, these correlations should be performed by statistical regression of experimental data, and they should be calibrated through compression tests carried out on cored samples (Concu, 2011, RILEM, 1993 Puccinotti, 2015).

In order to better exploit the correlation between compressive strength, rebound index, ultrasonic velocity and pull out force, an experimental campaign has been carried out on concrete cubic specimens, 300 mm sized, having different design concrete strength classes. RHT, UT and PT have been carried out on the specimens and then compression tests have been carried out on cores extracted from the same specimens.

UT usually considers longitudinal waves, because their acquisition is easier than that of shear waves,

which are much slower and difficult to detrmine, so that fewer researchers have sice now focused on the application of shear waves in the structural diagnosis of concrete elements. In this study both longitudinal and shear waves have been used.

Some correlations between NDT parameters, including ultrasonic shear wave velocity, and mechanical parameters, such as compressive strength and static elastic modulus of elasticity, are proposed, and their reliability in predicting concrete compressive strength is discussed.

## 2 MATERIAL AND METHODS

The experimental campaign has been carried out on 9 concrete cubic specimens with a side length of 300 mm and casted by using design concrete strength classes ranging from 10 N/mm$^2$ to 45 N/mm$^2$. Specimens have been sorted into three groups, A, B and C, according to the design strength class. Each group is made of three specimens.

UT has been carried out on the specimens and ultrasonic pulse velocity of both longitudinal (V) and shear $(V_s)$ waves has been measured. Next, RHT and PT have been carried out on the specimens in order to evaluate the rebound hammer index Ir and the pull out force F. After that, 15 cores, 1 or 2 cores from each cubic specimens, having diameter of 100 mm and length of 260 mm, have been extracted and put through compression test, in order to determine the concrete compressive strength and the static elastic modulus of elasticity.

## 2.1 Ultrasonic testing

Non destructive testing techniques based on ultrasonic waves propagation are generally used in building structural diagnosis.

UT is often used on-site for investigating a wide range of structures and infrastructures and in laboratory for materials characterization, for example concrete and rock elements (EN 12504-4, 2004 and EN 14579, 2004).

UT traditional application is based on measurements of the velocity of waves propagating through the material. The velocity is obtained from the ratio $l/t$, where $t$ is the time the wave needs to travel along the path of length $l$ (EN 12504-4, 2004, EN 14579, 2004).

Longitudinal waves are usually utilized, because their acquisition is easier than that of shear waves, which are much slower and difficult to detrmine, so that fewer researchers have since now focused on the application of shear waves in the structural diagnosis of concrete elements. In this study both longitudinal and shear waves have been used.

UT has been performed by using the ultrasonic test equipment Pundit Lab+, developed by Proceq®. The testing equipment includes:

- a pair of standard transducers with natural frequency of 54 kHz for emitting and receiving longitudinal waves;
- a pair of shear transducers with natural frequency of 250 kHz for emitting and receiving shear waves;
- a unit for signals generation, acquisition and preliminary analysis;
- a PC for data storage and further signal processing;
- a dedicated software, Pundit Link, which unlocks the full capabilities of the ultrasonic test system.

The Direct Transmission Technique, in which the wave is transmitted by a transducer through the specimen and received by a second transducer on the opposite side, has been applied. For each specimen, ultrasonic signals travel time has been measured along five selected paths for each couple of opposite sides (Fig. 1), and ultrasonic measurements have been repeated five times.

From the basic theory of propagation of ultrasonic waves on concrete (Jones, 1962) the dynamic modulus of elasticity has been calculated as follows:

$$E_d = 2 \cdot \rho \cdot V_S^2 \cdot (1 + v_d) \quad (1)$$

$$v_d = \frac{V^2 - 2 \cdot V_S^2}{2 \cdot (V^2 - V_S^2)} \quad (2)$$

where $v_d$ is the dynamic Poisson's ratio, V and $V_s$ are the velocities of longitudinal and shear waves respectively.

Figure 1. Grid points for UT measurements.

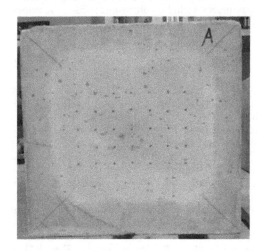

Figure 2. Grid points for RHT measurements.

## 2.2 Rebound hammer test

RHT is based on the principle that the rebound of an elastic mass strongly depends on the hardness of the concrete surface against which the mass strikes. When the rebound hammer plunger is pressed against the concrete surface, the spring controlled mass in the hammer rebounds. The amount of the mass rebound depends on the concrete surface hardness. (EN 12504-2, 2012).

The rebound index has been measured on grids of 30 points for each side of the specimens (Fig. 2).

## 2.3 Pull out testing

Pull out testing is a standardized partially-destructive testing method for estimating a mean value of concrete compressive strength. PT measures

Figure 3. Pull out testing on concrete cubic specimen.

the force required to extract from a hardened concrete element a steel insert with an enlarged head (Fig. 3).

European Standard EN 12504-3 provides a procedure for the determination of the pull out force. The force required for pulling out the embedded metal insert is related to the compressive strength of the concrete, so that an estimation of on-site concrete compressive strength can be provided by using suitable correlations.

## 3 RESULTS AND DISCUSSION

Table 1 summarizes the average values of V, $V_s$, Ir, F—determined on cubic specimens—and of the compressive strength $f_{ck}$ determined on the extracted cores.

A regression analysis has been performed aiming at predicting on-site compressive strength by using the NDT parameters. All the correlations have been represented by means of a linear regression.

Figures 4, 5, 6 and 7 show the correlations between $f_{ck}$ and V, $V_s$, Ir and F respectively. The NDT parameters well fit with the concrete compressive strength as shown by the values of $R^2$.

In order to obtain a more suitable indicator of the compressive strength a multiple correlation including together V, $V_s$, Ir and F has been studied.

A linear model (Eq. 3) has been analyzed

$$f_c = a + b \cdot V + c \cdot V_s + d \cdot Ir + e \cdot F \quad (3)$$

Table 1. Average values of experimental data.

| Parameter | A | B | C |
|---|---|---|---|
| V [m/s] | 4005 | 4310 | 4440 |
| SD* [m/s] | 257.05 | 53.30 | 26.35 |
| CoV** [%] | 6.40 | 1.25 | 0.60 |
| $V_s$ [m/s] | 2299 | 2515 | 2605 |
| SD* [m/s] | 159.88 | 18.70 | 13.85 |
| CoV** [%] | 6.95 | 0.74 | 0.53 |
| Ir | 29 | 41 | 44 |
| SD* | 1.80 | 2.45 | 0.50 |
| CoV** [%] | 6.10 | 5.95 | 1.10 |
| F [kN] | 23 | 33 | 38 |
| SD* [kN] | 3.45 | 2.55 | 3.95 |
| CoV** [%] | 15.35 | 7.75 | 10.40 |
| $f_{ck}$ [N/mm²] | 20.10 | 38.60 | 58.60 |
| SD* [N/mm²] | 5.15 | 5.25 | 4.65 |
| CoV** [%] | 25.70 | 13.65 | 7.85 |

*SD = Standard Deviation; **CoV = Coefficient of Variation.

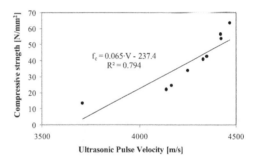

Figure 4. Correlation between concrete compressive strength and ultrasonic longitudinal pulse velocity.

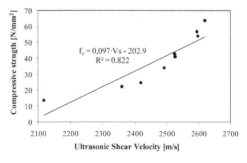

Figure 5. Correlation between concrete compressive strength and ultrasonic shear velocity.

where $f_c$ is the expected compressive strength, V and $V_s$ are the ultrasonic velocities of longitudinal and shear waves respectively, Ir is the rebound hammer index, F is the pull out force, while a, b,

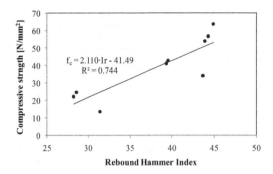

Figure 6. Correlation between concrete compressive strength and rebound hammer index.

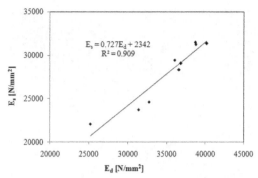

Figure 8. Correlation between dynamic and static modulus of elasticity.

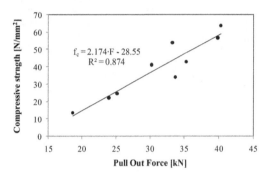

Figure 7. Correlation between concrete compressive strength and pull out force.

c, d and e are coefficients determined in order to minimize the sum of the squares difference between expected and experimental concrete compressive strength. The coefficients values are the following: a = −146.89954; b = 0.05507; c = −0.04126; d = 0.82542; e = 0.72259.

The multiple regression model shows a coefficient of determination $R^2 = 0.9022$.

It can be noted that the combined NDT approach leads to a correlation higher than that achieved by using each NDT parameter solely.

The static modulus of elasticity $E_s$ has been evaluated on the extracted cores by applying the Method B described in EN 12390-13, 2013:

$$E_S = \frac{\Delta\sigma}{\Delta\varepsilon} \qquad (4)$$

where $\Delta\sigma$ is the difference between the measured stresses and $\Delta\varepsilon$ is the strain difference during the test.

Figure 8 shows the correlation between the static modulus of elasticity $E_s$ (cores) and the dynamic modulus of elasticity $E_d$ (cubic specimens). A very high correlation has been achieved.

It is worth noting that UT leads to a prediction of $E_s$ better than that of $f_c$.

## 4 CONCLUSION

An experimental campaign has been carried out aiming at evaluating the reliability of combined NDT in predicting concrete mechanical parameters such as compressive strength and static elastic modulus of elasticity.

Some cubic specimens having different design concrete strength classes have been tested by means of UT, RHT and PT. In addition, compressive tests have been carried out on 15 cores extracted from the cubic specimens.

UT usually considers longitudinal waves, because their acquisition is easier than that of shear waves, which are much slower and difficult to detrmine, so that fewer researchers have since now focused on the application of shear waves in the structural diagnosis of concrete elements.

In this study both longitudinal and shear waves have been used.

Throughout the achieved results several conclusions can be drawn as follows:

– the NDT parameters V, $V_s$, Ir and F increase as concrete compressive strength increases;
– F leads to an estimation of the on-site compressive strength better than the other NDT parameters;
– the combination of the NDT parameters including ultrasonic shear waves velocity allows a higher correlation coefficient to be reached;
– a very good correlation between static and dynamic modulus of elasticity has been achieved.

## ACKNOWLEDGEMENTS

The authors would like to thank Unical S.p.A. for the support.

## REFERENCES

Breysse, D. 2012. Nondestructive evaluation of concrete strength: an historical review and a new perspective by combining NDT methods. *Construction and Building Materials* 33: 139–163.

Concu, G., De Nicolo, B., Pani, L. 2011. Non-Destructive Testing as a tool in reinforced concrete buildings refurbishments. *Structural Survey* 29 (2): 147–161.

Concu, G., De Nicolo, B., Pani, L., Trulli, N., Valdés, M. 2014. Prediction of Concrete Compressive Strength by Means of Combined non-destructive Testing. *Advanced Materials Research* 894: 77–81. Trans Tech Publications, Switzerland doi:10.4028/www.scientific.net/AMR.894.77.

EN 12390-13. 2013. Testing hardened concrete. Part 13: Determination of secant modulus of elasticity in compression.

EN 12504-2. 2012. Testing concrete in structures—Part 2. Non-destructive testing. Determination of rebound number.

EN 12504-3. 2005. Testing concrete in structures—Part 3. Determination of pull-out force.

EN 12504-4. 2004. Testing concrete—Part 4. Determination of ultrasonic pulse velocity.

EN 14579. 2004. Natural stone test methods. Determination of sound speed propagation.

Jones, R. 1962. Non Destructive Testing of Concrete. Cambridge University Press, London.

Pucinotti, R. 2015. Reinforced concrete structure: Non destructive in situ strength assessment of concrete. *Construction and Building Materials* 75: 331–341.

RILEM NDT 4. 1993. Recommendation for in situ concrete strength determination by non destructive combined methods, Compendium of RILEM Technical Recommendations, E&FN Spon, London.

*Emerging Technologies in Non-Destructive Testing VI – Aggelis et al. (Eds)*
*© 2016 Taylor & Francis Group, London, ISBN 978-1-138-02884-5*

# Development of a Condition Monitoring system for tidal stream generators rotating components combining Acoustic Emission and vibration analysis

Juan Luis Ferrando Chacon, Vassilios Kappatos & Cem Selcuk
*Brunel Innovation Centre, Brunel University, Uxbridge, Middlesex, UK*

Antonio Romero, Jesus Jimenez, Slim Soua & Tat-Hean Gan
*TWI Ltd., Great Abington, Cambridge, UK*

ABSTRACT: Tidal stream power is a very environmentally attractive renewable energy providing many advantages over other types of renewable energies, mainly in predictability and efficiency. However, the exploitation is mainly being retarded by operation and maintenance problems. To reduce the maintenance costs of tidal stream generators it is essential to employ cost-effective maintenance based on prediction rather than correction, namely Condition Monitoring (CM). Among the different components subjected to failure of tidal turbines, rotating machinery is one of the most critical. Currently, different CM techniques are being used to assess the integrity of rotating machinery. Among them, vibration analysis is the most stablished and robust. However, it presents some drawbacks such as (i) low sensitivity to fault detection in low speed rotating machines and (ii) limited capability of defect detection in early stage. On the other hand, AE is gaining ground as a CM technique as overcomes the two drawbacks previously presented for vibration analysis. This paper presents the development of a CM system combining the advantages of vibration analysis and Acoustic Emission (AE) capable of assessing the integrity of rotating parts of tidal turbines. A test rig designed to simulate the conditions experienced by a real tidal turbine was submerged and operated underwater to evaluate the performance of the system. The results obtained in these experiments are presented in this paper.

## 1 INTRODUCTION

Tidal stream power is a very environmentally attractive renewable energy source that has great potential to be a major contributing source of renewable energy worldwide (S. Galloway, V. M. Catterson,C. Love and A. Robb,, 2014). As stated in the European Union Committee 27th Report of Session 2007–08,20% of energy generation is requisite to come from renewable resources by 2020. The accessible tidal stream power resource is huge. For example, taken as a cyclic time average, it is 90GW i.e. 800TWh/y globally if it is assumed to be only 3% of the estimated total power in the world's tidal streams. This is sufficient to satisfy 5% of the present world electricity consumption, which is 1875GW i.e. 16424 TWh/y. The energy density stored in a tidal current is 800 times that in a normal wind stream for a given speed of current/wind. In addition, tidal currents are especially attractive because of their predictability, being generated by the gravitational interaction between moon and earth (O. Rourke Fergal, B. Fergal and R. Anthony, 2009).

However, tidal stream power exploitation is being retarded by operation and maintenance problems which lead to availability times as low as 25%. In particular, underwater rotating components in the generator drive train are subject to random and periodic wave loading involving stress cycles of much larger amplitude than arises with wind turbines. They can also suffer from loading from marine fouling. The replacement of failed components is relatively expensive because of the underwater environment, especially if catastrophic failure is involved through the breakdown of a component in the generator drive train causing collateral damage to other components.

In order to overcome these problems a remote continuous Condition Monitoring (CM) system has been developed to continuously assess the integrity of the rotating components of tidal stream generators. The goals of this system are as follows:

- Growth of electrical power produced by tidal stream generators to commercial levels.
- Availability level of tidal stream generators exceeding 96%.

- All replacements and maintenance made during scheduled downtime, thus eliminating all forced unplanned shut downs.

To achieve these goals Acoustic Emission (AE) monitoring and vibration analysis have been combined. AEs are defined as transient elastic waves generated by a rapid release of strain energy, caused by a deformation or damage within, or on the surface of a material (M. Leahy, D. Mba, P. Cooper, A. Montgomery and D. Owen, 2006). This mechanical process can be produced by different sources such as cracks, plastic deformation, rubbing, cavitation, and leakage (Y-H Pao, R.R Gajewski and A.N. Ceranoglu, 1979). Particularly, AE in rotating machinery is produced by asperity contact between two surfaces in relative motion. An extensive review of AE applied to rotating machinery (bearings, gearboxes and pumps) was carried out by Mba and Rao (D. Mba and R. B.K.N. Rao, 2006).

The use of vibration analysis is one of the fundamental tools for CM. It has been developed extensively over a period of approximately 35 years (al, 2003). This technology is currently the most widespread method for CM of rotating machines. It can be utilized in all rotating equipment on site. It is used to isolate the location of damage to rotating components, and also to inform quantitatively about the amount of damage produced. Unlike AE, The main limitation of this technique is the low sensitivity to faults at low rotating speeds (Lu, 2009). For this reason, the combination with AE allows for fault detection with high reliability in both the low and high speed shafts found in tidal turbines.

The developed system permanently monitors tidal wave generators, determining the vibrational and acoustic signature of a healthy turbine and informing well in advance if any of the components is subject to failure. An experimental investigation has been performed to examine the capability of the system to detect gearbox defects using a mock up representative of a tidal stream generator. The results obtained in this investigations show positive results using both AE and vibration analysis.

## 2 EXPERIMENTAL SETUP AND MEASUREMENTS

### 2.1 Tidal Turbine mock-up

A mock-up has been designed to generate AE and vibration representative of a full scale tidal power system (Figure 1). The rotating equipment of a main power take off and a pitch control system were represented by two motor/gearbox assemblies, driving rotors which were designed to replicate the operational loading on these assemblies when working under water. All these components were supported by a frame and the whole structure formed a test bench for defect detection on rotating machinery using AE and vibration techniques. All the parts of the mock-up are described below.

In order to evaluate the performance of the CM system under conditions similar to the operation of a real tidal stream turbine, the mock up was submerged in a 2.8 m diameter and 3 m length cylindrical water tank. Figure 2 shows the water tank in which the experiments were carried out.

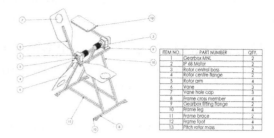

Figure 1. Mock up designed to validate the CM system.

Figure 2. Water tank used in the experiments.

Figure 3. 3-phase motor/gearbox.

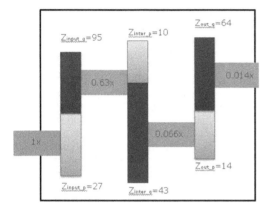

Figure 4. Shaft and gearbox forcing frequencies.

Table 1. Gearbox stage description.

|  | Orders | Frequency (Hz) |
| --- | --- | --- |
| Shaft 1 | 1x | 24.47 |
| Shaft 2 | 0.63x | 15.36 |
| Shaft 3 | 0.066x | 1.62 |
| Shaft 4 | 0.014x | 0.35 |
| GM Stage 1 | 27x | 660.79 |
| GM Stage 2 | 6.3x | 153.65 |
| GM Stage 3 | 0.92x | 22.64 |

2.1.1 *3-phase motor/gearbox*

Figure 3 shows a three phase motor connected to a 3 stage gearbox. This set was waterproof and hence suitable for trials under water.

The speed of the motors was controlled by two different Variable Frequency Drives (VFD). The motors were wired to the parametric output acquisition card which sent a voltage signal to the VFD. The signal which controlled the motors was proportional to the frequency of the high speed shaft (10V = 50 Hz). When the motor was working at nominal speed, the high speed shaft spun at 1500 rpm which was reached when the VFD operated at 50 Hz. The gearbox coupled to the motor reduced the speed through three stages, so the speed of the blades was lower than 1500 rpm.

Figure 4 shows a graphical description of the gearbox. Each gear set generated a unique profile of frequency components that need to be monitored in order to have control the integrity of the gearbox. Accordingly, the forcing frequencies of the gearbox were calculated using the number of teeth provided by the supplier and the frequency of the high speed shaft. Another important feature to take into account was the gear-mesh frequency which was equal to the number of teeth in the pinion or drive gear multiplied by the rotational shaft speed. Table 1 shows these values in orders (frequency explained in terms of running speed or its multiples) and the values for the case when the motor was working at nominal speed (50 Hz).

## 3 EXPERIMENTAL SETUP AND MEASUREMENTS

As shown in Figure 5, two AE and vibration sensors were attached to each gearbox in order to monitor its condition when working underwater. Accelerometers were located on the output shaft bearing, one vertical and one horizontal. Selection of these locations was due to the fact that there had to be a good mechanical path to the bearing so that vibration only travelled through solid metal with no gaps or joints. For AE measurements Vallen VS900RIC sensors were employed with a bandwidth from 100 kHz up to 1 MHz. These included an internal preamplifier of 34dB and were operative in a range from −40 to 85 °C. The signal was band-pass filtered from 100 kHz to 1000 kHz. The preamplifier was powered through a decoupling box connected to a linear power supply. The decoupling box was connected to a commercial data acquisition card through a coaxial cable.

One of the gearboxes of the mock-up was kept in healthy condition while multiple defects were created in the gear teeth of the other gearbox using an engraving machine (Figure 6).

The aim was to detect the changes in the signatures when both gearboxes were working with load (blades attached). The tests accomplished are presented in Table 2. Firstly, to investigate the impact of speed on the AE and vibration signals a test was conducted without blades installed in the mock up. After that the blades were installed on the healthy gearbox and 50 measurements were captured from AE and vibration signals. The length of the waveforms acquired was 0.5 s and 25 s for AE and vibration monitoring respectively. Then, the blades were removed from the healthy gearbox and installed on the faulty gearbox repeating the same acquisitions for AE and vibration signals.

a) Faulty Motor

b) Healthy Motor

Figure 5. Location of the sensors.

Figure 6. Defect creation and example of defect created on the gear.

Table 2. Field trials test plan.

| | Load | Position of the blades |
|---|---|---|
| Test 1 | Shaft running free | No blades |
| Test 2 | Blades | Healthy Gearbox |
| Test 3 | Blades | Faulty Gearbox |

## 4 FEATURE EXTRACTION

Several features were extracted from the vibration and AE signals. The Root Mean Square (RMS), peak value Value, Counts and Crest factor (CF) were extracted from the AE and vibration waveforms using the equations given below.

$$RMS = \sqrt{\frac{1}{N}\sum_{i=1}^{N}X(i)^2} \qquad (1)$$

$$Peak\,value = \max(|X|) \qquad (2)$$

$$CF = \frac{MaxValue}{RMS} \qquad (3)$$

where $x(i)$ the $ith$ data point of the AE signal and $p(i)$ is the probabilities computed from the distribution X and N the number of samples.

The 'counts' parameter is defined as the number of times the signal crosses a given threshold value. Information Entropy (IE) was also extracted from the AE signals. IE is defined as a measure of uncertainty of a process. As the working condition of a machine system deteriorates due to the initiation and/or progression of structural defects, the number of frequency components contained in the AE signal increases, resulting in a decrease in its regularity and an increase in its corresponding information entropy value. The Information Entropy (EI) of a discrete random variable x is defined as:

$$IE = -\sum_{i=1}^{N}P_i \cdot \log(P_i) \qquad (4)$$

## 5 EXPERIMENTAL OBSERVATIONS AND RESULTS

### 5.1 AE module results

Figure 7 (a) and (b) show the AE signals in the time domain captured from the healthy and faulty gearboxes. Continuous AE dominated the signals without significant discrete AE events present. Visually, the signals presented similar patterns and there were no evident differences between them.

A typical frequency domain AE signal acquired from the sensors installed in the healthy and faulty gearbox is shown in Figure 8 (a) and (b) respectively. These graphs show that most of AE activity was concentrated in the 30–200 kHz range. Therefore, to increase the SNR of the AE signals analysed a band-pass filter was applied in that frequency band.

Table 3 shows the evolution of the different features with increasing rotating speed (Test1). It clearly shows an increasing trend in all features with increasing speed but CF feature, which shows a decreasing trend. In particular, the counts feature showed the highest sensitivity to the increase of rotating speed.

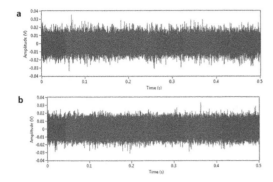

Figure 7. AE time domain signals captured from a) Healthy gearbox b) Faulty gearbox.

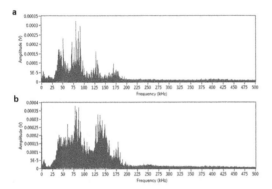

Figure 8. AE Frequency domain signal captured from a) Healthy gearbox b) Faulty gearbox.

Table 3. Progression of AE features as a function of speed.

| Speed | RMS | Peak | Counts | IE | CF |
|---|---|---|---|---|---|
| 40 Hz | 0.0046 | 0.0395 | 73.53 | 105.00 | 8.50 |
| 60 Hz | 0.0073 | 0.0464 | 212.83 | 244.79 | 6.31 |
| 80 Hz | 0.0106 | 0.0573 | 419.50 | 476.94 | 5.36 |

Table 4. AE features captured from the healthy and faulty gearbox.

|  | RMS | Peak | Counts | IE | CF |
|---|---|---|---|---|---|
| Healthy | 0.0044 | 0.021 | 73.53 | 105.00 | 8.50 |
| Faulty | 0.0059 | 0.027 | 212.83 | 244.79 | 6.31 |

Table 4 shows the result of the experiments with the healthy and faulty gearbox (test 2 and test 3). Clearly all the indicators increased from healthy condition to faulty condition apart from CF as occurred with speed variation. The most sensitive feature was also counts but IE also displayed positive results.

## 5.2 Vibration module results

The first test run underwater was without loading in both of the gearboxes. The output shafts were run free at three different speeds (40 Hz, 60 Hz, 80 Hz and 100 Hz). Figure 9–11 show how the RMS value, the peak value and the counts parameter increased with motor speed. These three graphs are divided into three areas representing the features at different speeds.

Table 5 shows the values calculated for the different features in both healthy and faulty gearboxes. All the indicators clearly showed an increase from healthy to faulty condition. It is evident that the most sensitive indicator was counts.

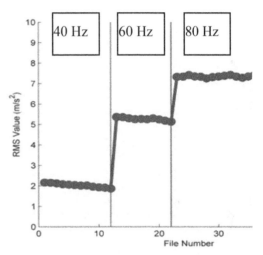

Figure 9. RMS' variation as the speed is increased.

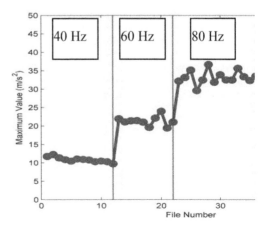

Figure 10. peak value's variation as the speed is increased.

345

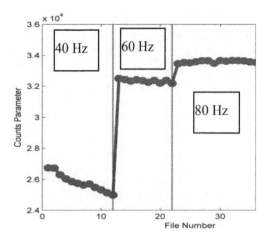

Figure 11. Counts parameter's variation as the speed is increased.

Table 5. Vibration features captured from the healthy and faulty gearbox.

|  | RMS | Peak | Counts | CF |
|---|---|---|---|---|
| Healthy | 0.254 | 1.322 | 196 | 5.204 |
| Faulty | 0.292 | 1.643 | 3256 | 5.626 |

## 6 CONCLUSIONS

The CM system developed has been tested under laboratory conditions using an experimental mock-up specifically designed for this task. In these experiments, the system has proved capable of working underwater. The system was able to detect changes in the rotation speed and condition of the gearbox in both AE and vibration signals. Although the CF value extracted from the AE signals decreased in faulty condition, RMS, counts and information entropy values showed high sensitivity for both speed variation and gearbox fault progression using AE signals. The results show that using vibration signals the increase in counts is clearly the most sensitive indicator for detecting gear defects. Thus, both techniques have shown value in detecting defects in gearbox tidal stream generators.

## ACKNOWLEDGEMENTS

The investigations and results presented have been obtained from REMO Project, which is a European Research & Development Project partly funded un-der the FP7 Framework Research for SME Associations under Grant Agreement Number 314839. REMO project is collaboration between the following organizations TWI Limited, Brunel University, Engitec limited, WLB limited, Desarrollo y Gestion Industrial del Medio Ambiente S.A., Corporate Ser-vices International and Stirling Dinamics Limited.

## REFERENCES

Galloway, S., V.M. Catterson, C. Love and A. Robb,. (2014). Anomaly Detection Techniques for the Condition Monitoring of Tidal Turbines. *Annual conference of the prognostics and health management society.* Texas.

Leahy, M., D. Mba, P. Cooper, A. Montgomery and D. Owen. (2006). Acoustic Emission for the detection of shaft-to-seal rubbing in large power generation turbines. *Advanced Materials Research*, pp. 433–438.

Lu, B. (2009). A review of recent advances in wind turbine condition monitoring and fault diagnosis. *Power Electronics and Machines in Wind Applications*, pp. 1–7.

Mba, D. and R.B.K.N. Rao. (2006). Development of acoustic emission technology for condition monitoring and diagnosis of rotating machines: bearings, pumps, gearboxes, engines, and rotating structures. *The Shock and Vibration Digest*, pp. 3–16.

Pao, Y-H., R.R Gajewski and A.N. Ceranoglu. (1979). Acoustic emission and transient waves in an elastic plate. *Acoust. Soc. Am.*, p. 19.

Reimche, W. et al. (2003). Basic of vibration monitoring for fault detection and process control. *III Pan-American Conference for Nondestructive Testing.* Rio de Janeiro.

Rourke Fergal, O., B. Fergal and R. Anthony. (2009). Tidal energy update. *Appl Energy*.

*Emerging Technologies in Non-Destructive Testing VI – Aggelis et al. (Eds)*
*© 2016 Taylor & Francis Group, London, ISBN 978-1-138-02884-5*

# Detection of incipient SCC damage in primary loop piping using various novel NDE technologies

B.K. Jackson, J.L.W. Warwick & W. Li
*Intertek AIM, Sunnyvale, California, USA*

J.J. Wall
*Electric Power Research Institute, Charlotte, North Carolina, USA*

ABSTRACT: In this paper, different Nondestructive Examination (NDE) technologies to detect the incipient stages of Stress Corrosion Cracking (SCC) damage are evaluated. *In-situ* NDE detection of SCC is finding increasing interest owing to its economic benefits of preventing forced shut down failures in the nuclear industry. This work describes the continued development of an *in-situ* monitoring technique using fibre optic temperature compensated strain gauging mounted on the outer diameter of the pipe, described in a previous papers by the present authors [1–4]. Further investigation includes the suitability and sensitivity of acoustic emission and the use of thermography for SCC detection. This paper details the setup and results of refined stable experimental setup to perform full scale accelerated SCC cracking in boiling Magnesium Chloride ($MgCl_2$). The results show that both the strain gauge and acoustic emission technology are successful in detecting SCC crack initiation and propagation at the pipe Inside Diameter (ID).

## 1 BACKGROUND & MOTIVATION

Stress Corrosion Cracking (SCC) has been observed for decades in austenitic alloys and weldments such as types 304 and 316 stainless steel. SCC is also prevalent in Nickel based alloys. One of the biggest challenges for the nuclear power industry today is SCC in aging materials as plants continue to run beyond their original design lifetimes. This occurs in structural materials of Pressurized Water Reactors (PWRs) and Boiling Water Reactors (BWRs) primarily in the sensitized Heat Affected Zones (HAZ) of weldments, but is also prevalent in materials affected by sensitization, cold work, high residual stresses and processes which reduces material toughness. SCC is a sudden and difficult to detect severe degradation mode. Current NDE technology is limited in its ability to detect SCC cracks in their incipient stages. Previous laboratory-based studies undertaken by EPRI suggest that component lifetime in the field is more often governed by the initiation and growth of short stress corrosion cracks than by the growth of long (deep) cracks [5]. SCC is especially dangerous as it can propagate rapidly once initiated with little or no change in external environment. The inspection interval for each individual weld may be as long as 10 to 15 years of operation. This is likely sufficient time

for a through wall crack to form from an initiation point that currently is too small to detect.

The goal of the work presented in this paper is to further investigate the hypothesis that the initiation and growth of ID stress corrosion cracks can be detected at the sample surface as the local residual stress from fabrication reconfigures during crack growth. This phenomenon is monitored during accelerated SCC tests using OD strain fibre optic gauges, OD mounted acoustic emission sources and thermography.

## 2 HISTORICAL SUMMARY OF STRAIN GAUGE MONITORING

The proof of concept stages of the OD mounted strain gauging technology is documented in detail in references [1–3], and used Electrical Discharge Machine (EDM) to simulate cracks.

Subsequent work, presented in [3–4], two 12-inch OD 304 L pipe sections with 1-inch wall thicknesses were welded together using a standard v-groove circumferential weld, and was subjected to a boiling $MgCl_2$ environment for 12 days, based upon a scaled-up version of ASTM G36-94 (2013) [6] using an immersion heater.

The current work documents a refined experimental technique and evaluates the results of two further more stable full-scale experiments.

## 3 EXPERIMENTAL METHOD

The material was supplied similar to the previous experiment, as two 12-inch OD 304 L pipe sections with 1-inch wall thicknesses that were welded together using a standard v-groove circumferential weld, typical of service-welded pipe.

The experimental setup was modified to remedy heating symmetry and stability difficulties in the previous experiment [4]. The current investigation uses a slightly different experimental setup schematized in Figure 1. Heating to 155°C is achieved using 3 individual heat sources controlled from a single temperature controller. The heat is applied with the use of two BriskHeat® manufactured heat tapes wrapped circumferentially around the top and bottom of the pipe sample and one heat blanket which is attached to the bottom plate using high temperature adhesive. These locations were selected as to not overheat the areas where the strain gauges and acoustic emission are located. The heating locations were then insulated using standard fibreglass insulation. Four monitoring thermocouples were mounted on the OD surface of the pipe on the weld crown to report OD weld temperatures. One centralized fluid thermocouple is utilized to monitor fluid temperature. All temperature data is recorded using control logics data acquisition unit.

Improvements in the strain gauge technology over the first iteration of this experimental technique incorporated the temperature compensation into each strain gauge, which eliminated the need for separate gauges in different locations. This reduces the placement error and increases accuracy of the temperature compensation capability of the gauges. As in the previous experiment, eighteen gauges were equally distributed around the pipe circumference.

The acoustic emission and thermography experiments utilized the same accelerated SCC experimental setup to achieve stable boiling of $MgCl_2$. Though, replacing the strain gauges are four acoustic emission sensors positioned at 90° intervals, as shown in Figure 2. During this experiment thermography was performed on regular basis to evaluate its effectiveness at detecting tight propagation crack such as those formed during SCC.

## 4 RESULTS

### 4.1 Fibre optic strain gauge SCC detection

The results of the second iteration of the boiling $MgCl_2$ testing using fibre optic Strain Gauges (SG) to monitor the change in residual stress on the pipe OD may be seen in Figures 3–5.

Figure 3 shows the ID surface of the pipe after teardown of the experiment and removal of the $MgCl_2$. The ID of the pipe was inspected using Dye Penetrant (DP) to reveal indications of cracking, and the indications found overlaid with paint marker. It can clearly be seen that cracking is indicated over the majority of the ID circumference, centered around the Heat Affected Zone (HAZ) of the welds, as would be expected for SCC cracking. Note that cracking occurred both in the circumferential and axial directions.

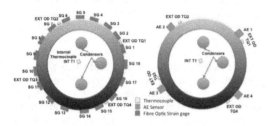

Figure 1. Layout of fibre optic strain gauges, acoustic emission sensors, & thermocouples (plan view).

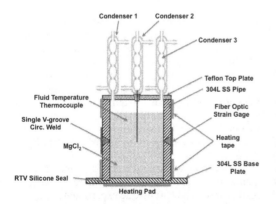

Figure 2. SCC $MgCl_2$ Cell set-up (X-section).

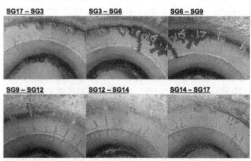

Figure 3. Close-up of DP results from second full scale SG run—Indications overlaid with paint markings.

Several OD locations also exhibited cracking, and it was evident that through wall cracking had been achieved in these experiments. However, as SCC cracks are so tight, the small quantity of MgCl$_2$ that reached the OD solidified when exposed to the ambient environment, thus plugging the leak.

Figures 4–5 show a summary of the results from the experiment. Figure 4 highlights the early portion of the experiment from 0 h–~154 h. This is the most significant section of the results. The Figure shows the affect of the MgCl$_2$ solution being warmed to temperature, with an associated strain response. Note that portions of the data are missing from the start up due to limitations in the data recording software.

The first drop in temperature was due to the opening of the observation window in the top plate to check MgCl$_2$ fluid levels. Again, there is a strain response to the drop in temperature. The most significant portion of the Figure is the third pane, from ~57.5 h–81.5 h. This is where SCC initiation occurred in this experimental run. Note the temperature independent increases in strain from all of the gauges during this time period. There are no corresponding thermal events occurring during this time period. DP inspection indicated cracking had occurred around the majority of the ID surface, hence a response of roughly equal magnitude from all the gauges around the OD would be expected upon initiation. This response is similar to the "steady state growth" response observed in the first experimental run [4]. The difference in this experiment may come from the more symmetrical, and uniform temperature control afforded by the heating pad and tapes, and fewer temperature excursions caused by adding additional MgCl$_2$. As with the first experimental run using the strain gauges to detect SCC, the key to detecting a strain event is the lack of thermal events and a relatively large increase in strain with respect to time (c.f. a water hammer (mechanical) or a long term general temperature rise/fall in the internal fluid). The response indicated for initiation in this experiment is a measurable and flag-able response that may easily be automated in an *in-situ* on-line detection system.

After this time period, the strain values level off. This is shown throughout the remainder of the experiment. Figure 5 shows snippets of the data from the middle and at the termination of the experiment. At all times, the strain response is steady with little/no changes observed. This suggests that the strain gauge technique employed may be insensitive to continuous SCC growth after initiation. This distinction was not possible in the previous experimental run [4], as the temperature control was inferior to this experiment, and the fact that it only ran for a period of 12 days, whilst this experiment was run for approximately 2 months.

### 4.2 *Acoustic emission & thermography SCC detection*

Figures 6–8 summarize the results of the third experimental run with MgCl$_2$. Again, this experiment was conducted using the heating pad and

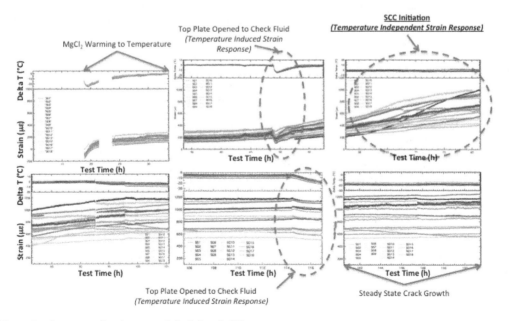

Figure 4. Summary of early stages of the full scale SG run.

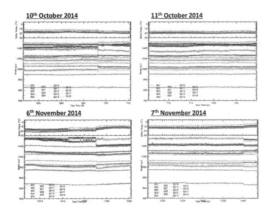

Figure 5. Full scale SG experiment 2: Mid—end experiment.

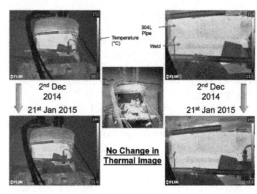

Figure 6. Summary of thermography data (25th Nov. 2014–21st Jan. 2015).

tapes to provide a symmetrical, stable, and uniform heating to the fluid.

The experiment was broken down and the $MgCl_2$ removed. Again, DP was employed to highlight any indications of cracks. For each AE sensor, an external thermocouple was also placed on the OD to monitor OD temperatures in each quadrant. A fifth thermocouple was placed in the fluid to monitor the internal fluid temperature. The DP results clearly showed linked ID cracking around the entire circumference of the pipe, all centered on the HAZ, as would be expected for SCC cracking. In contrast to the previous experimental run, the cracking appeared to be almost entirely circumferentially orientated, with only isolated axially orientated cracks observed. There was insufficient evidence to determine the reason for this, but it is likely simply down to slightly differing weld procedures between the two pipe sections.

Figure 6 summarizes the results of the thermography data. An infra-red image of the same region of the weld at the pipe OD on a roughly daily basis throughout the course of the experimental run (~2 months). Both an overall and a zoomed image were taken of this region each time. The theory was that as an ID crack initiated and progressed through the wall of the pipe, then the effective wall thickness from the tip of the crack to the OD of the pipe would become smaller and smaller. The aim of the thermography experiment was to determine if such a localized "wall thinning" could produce a "hot spot" visible on the OD of the pipe owing to the higher internal fluid temperature than general OD pipe temperature. Unfortunately, as Figure 6 shows, this was not observed. It can be concluded that the infra-red camera had insufficient resolution to capture such a small and highly localized temperature change, and would not be an effective method with which to detect SCC.

Figure 7 summarize the results of the AE monitoring of the pipe section during this 3rd experimental run with $MgCl_2$. The AE system output is a running count of acoustic events detected by the AE sensors (all sensor events are summed together in these figures) during the experimental run. Note that the event count is cumulative throughout the run time.

Figure 7 shows a summary of the entire experimental run (~2 months time). Both the acoustic emission events (shown in number of cumulative counts), and the external quadrant, and fluid thermocouple readings are shown. The initial stages of the experiment involved the addition of a second load of $MgCl_2$ (initially solid) to add to the melt pool to achieve a $MgCl_2$ fluid level sufficiently above the weld & HAZ. These operations, plus the initial few hours of heating the fluid caused a number of mechanically and thermally induced acoustic events.

These events are shown at the initial stages of the experiment, and is evidenced as almost vertical lines on the acoustic event count vs. time plot due to the large number of events in a very short space of time caused by these operations. Other instances in which the top plate was removed are similarly characterized by "vertical steps" in the acoustic emission plot, as highlighted in the Figure. The most significant point in Figure 7 is the detection of the initiation of SCC cracking, as highlighted on the Figure at ~260 h into the run. This point marks a significant increase in acoustic events with respect to time, or mechanical events (compare the outputs of the thermocouples during this time).

Also of note from Figure 7 that an increase in slope is seen that we attribute to initiation and growth. However, a gradual decrease is also seen, and eventual smoothing out of the AE activity. The authors believe that this can be attributed to the cracks entering compressive internal stress fields and arresting. Figure 8 highlights the first 600 h of the experiment. This Figure shows greater detail of the initial warming of the MgCl$_2$, followed by several mechanical interventions. As stated before, these are characterized by a sharp step change in acoustic events, and/or an associated temperature event. The point of SCC initiation is clearly marked on the Figure, and is a temperature and mechanical event independent acoustic response. As with the SG response to SCC initiation, this is an easily flag-able and measureable response for an *in-situ* on-line SCC monitoring system to detect the

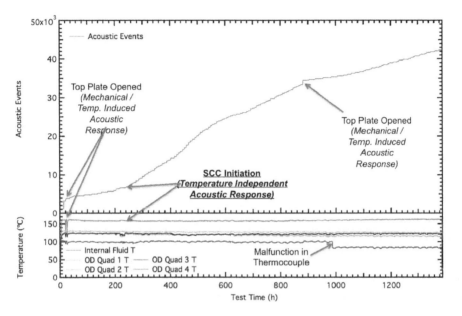

Figure 7. Summary of AE data (25th Nov. 2014–21st Jan. 2015).

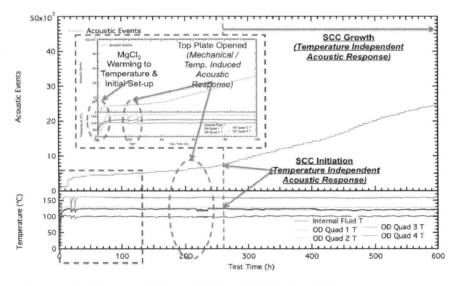

Figure 8. Summary of initial stages of AE data (25th Nov. 2014–20th Dec. 2014).

onset of SCC. An important contrast between the AE method of SCC detection and the SG method, is that steady state growth of the SCC cracks is recorded, owing to the AE events being counted as a cumulative quantity, rather than measuring a relative change. However, there is insufficient evidence from this experiment that other external events (such as a water hammer, thermal, or mechanical event in a cooling loop) would not be also counted by this method, necessitating further experimentation to filter such signals from the AE results.

## 5 CONCLUSIONS

The following conclusions may be drawn from the work conducted in this paper:

- The revised experimental method employing heating pads and tape is more stable, with more uniform and symmetrical heating, allowing experimental runs of 2 or more months.
- In both experiments SCC was initiated and progressed through wall during the course of the 2-month experiments. Cracking occurred around the majority of the ID circumference, initiating in the HAZ of the weld. One experiment yielded a mixture of axial and circumferentially oriented cracks, while the other yielded almost entirely circumferential cracks. There was insufficient evidence to determine the reason for this, but it is likely simply down to slightly differing weld procedures between the two pipe sections.
- The experimental SG run detected ID SCC initiation successfully at ~57.5 h–81.5 h into the run. Temperature independent increases in strain from all of the gauges were observed during this time period. DP inspection indicated cracking occurred around the majority of the ID surface, hence a response of roughly equal magnitude from all gauges around the OD would be expected upon initiation.
- The SG response in this experiment run was similar to the "steady state growth" response observed in the first experimental run [4]. The difference in this experiment may come from the more symmetrical, and uniform temperature control afforded by the heating pad and tapes.
- As with the first experimental SG run [4], the key to detecting a strain event is the lack of thermal events and a relatively large increase in strain with respect to time. The response indicated for initiation in this experiment is a measurable and flag-able response that may easily be automated in an *in-situ* on-line detection system.
- Thermography using an infra-red camera had insufficient resolution to capture such a small and highly localized temperature change, and is not an effective method with which to detect SCC.

- The AE measurements successfully detected initiation of SCC cracking ~260 h into the run. with a significant increase in acoustic events with respect to time, which are completely independent of either thermal or mechanical events.
- Post-initiation, a gradual decrease in slope of AE activity with respect to time is observed. This can be attributed to the cracks entering compressive internal stress fields and arresting.
- The AE response to SCC initiation is, again, an easily flag-able and measureable response for an *in-situ* on-line SCC monitoring system to detect the onset of SCC.
- Steady state growth of the SCC cracks is recorded only by the AE method, owing to the AE events being counted as a cumulative quantity, rather than measuring a relative change (SG).
- Further experimentation is required to characterize potential external events to allow filtration of these from the AE signals.
- Overall this work has demonstrated that both SG and AE methods can successfully detect the initiation stages of SCC growth using OD mounted instrumentation. The measurements may both be automated, and be incorporated into an on-line *in-situ* early warning detection system to improve the reliability and safety of nuclear power plant cooling loops and high energy piping applications.

## REFERENCES

[1] Jackson, B.K. 2010. Interim Status Update: EDM Crack Simulations, *EPRI Project Status Report AES 10067465-2-1.*

[2] Jackson, B.K. et al. 2011. Detection of Incipient SCC Damage in Primary Loop Piping using Fiber Optic Strain Gages, *EPRI Conference: Welding & Fabrication Technology for New Power Plants & Components, Orlando, Fl, USA.*

[3] Jackson, B.K. et al. 2011, Detection of Incipient SCC Damage in Primary Loop Piping using Fiber Optic Strain Gages. *MS & T 2011, Columbus, OH, USA.*

[4] Jackson, B.K. et al. 2014. Detection of Incipient SCC Damage in Primary Loop Piping using Fiber Optic Strain Gages. *Paper PVP 2014-28979, ASME Pressure Vessels and Piping Conference, July 2014, Anaheim, CA, USA.*

[5] J. Hickling et al. 2005. Status Review of Initiation of Environmentally Assisted Cracking and Short Crack Growth. *Electric Power Research Institute Report 1011788, December 2005.*

[6] ASTM G36–94(2013): "Standard Practice for Evaluating Stress-Corrosion-Cracking Resistance of Metals and Alloys in a Boiling Magnesium Chloride Solution".

[7] Materials Safety Data Sheet (MSDS) for Magnesium Chloride Hexahydrate ($MgCl_2 \cdot 6H2O$), CAS# 7791-18-6, ACC# 13365, Revised 8/02/2000.

*Emerging Technologies in Non-Destructive Testing VI – Aggelis et al. (Eds)*
*© 2016 Taylor & Francis Group, London, ISBN 978-1-138-02884-5*

# Comparative study of NDT inspection methods in carbon fiber composite laminates

### G. Steenackers
*Op3Mech Research Group, University of Antwerp, Antwerp, Belgium*
*AVRG, Research Group, Vrije Universiteit Brussel, Brussels, Belgium*

### J. Peeters
*Op3Mech Research Group, University of Antwerp, Antwerp, Belgium*

### G. Arroud
*AVRG, Research Group, Vrije Universiteit Brussel, Brussels, Belgium*

### S. Wille
*SABCA Limburg N.V., Lummen, Belgium*

ABSTRACT: In 2012 SABCA Limburg N.V., a Belgian aerospace company, started with the production of the Flap Support Fairings for the A350–800 commercial aircraft. These parts are manufactured using a combination of honeycomb and carbon fiber pre-preg laminates. After part curing all fairings are inspected for defects. The most straightforward and common inspection method is the 'tap test'. As this inspection method is labor intensive and also subject to interpretation, different alternative methods for inspection will be compared. Key requirements for these alternative methods remain validity, reliability and objectivity. This paper will focus mainly on two inspection methods that are used in the aerospace industry: ultrasonic inspection and thermography. Ultrasonic inspection on composite parts is largely used in the composite manufacturing industry to detect porosities, delamination, cracks etc. Thermography has only been introduced more recently to inspect composite aerospace components. Research shows that it is possible to detect subsurface defects with the aid of thermography. To validate this research, a comparison is made between these two inspection methods using the A350 Flap Support Fairings as test pieces. Both methods are also compared using IRP panels.

## 1 INTRODUCTION

Non-Destructive Testing (NDT) techniques have been used for over 30 years, as analyzing tools in scientific and industrial applications to evaluate the properties of a material, component or system without compromising the integrity of the inspected object (Maldague 2001). However, all these methods have some kind of limitation with respect to the type of objects that can be inspected or the kind of defects that can be detected [1]. In recent years, the detection of defects with NDT has gained prominence (Maldague 1994). When compared to other classical NDT techniques such as C-Scan ultrasonic or X-Rays, Infrared (IR) thermography is safe, nonintrusive and noncontact, allowing the detection of relatively shallow subsurface defects under large surfaces and in a faster manner (Giorleo & Meola 2002). There are many advantages of thermography, these include the fact

that no coupling is required, it can be performed in-situ, it can cover large areas and it can be made a quantitative method by comparing with a Finite Element (FE) model (Busse 2001). Thermography techniques use IR cameras to monitor the surface temperature of a structure whilst in use. Heat is injected into the structure using heat lamps and the surface temperature distribution is recorded and analyzed. In general all geometrical anomalies or material discontinuities influencing the propagation of the heat flow are detected by a suitable IR camera (Jaluria & Atluri 1994)(Wrobel 1992). IR thermography can be applied through active or passive approaches, depending on whether the inspected part is in thermal equilibrium or not. In passive thermography no artificial heating or cooling to the object is applied, thus allowing the evaluation of its temperature distribution due to both natural phenomena or heating developing during its ordinary life. Active thermography involves a deliberate change in temperature. A stimulus is directly applied to the object to cause heating or cooling (Moore & Maldague 2001).

This paper includes a comparison of inspection methods used to track defects in fiber-reinforced plastics. In this paper the inspection methods are divided in two research fields: ultrasonic inspection and thermographic inspection. The purpose of both inspection methods is to detect, in an efficient and effective way, defects on monolithic as well as sandwich structures. Most of the defects seen in composite materials are:

- Hairline fractures (microscopic fractures commonly caused by fatigue);
- Delamination (places where different layers of the monolithic structure are separated);
- Porosities (cavities formed during the manufacturing process);

## 2 ULTRASONIC DETECTION

The principle of ultrasonic detection uses sound waves to search for internal and external defects in different materials. These sound waves are produced by high frequencies (1 MHz–20 MHz). It is possible to determine the thickness of objects using this technology. Two types of ultrasonic detection are commonly used to examine composite materials: the C-Scan and the OmniScan method.

### 2.1 *OmniScan*

Both C-scan and OmniScan methods use an ultrasonic transducer and a piezo-electric crystal to produce the ultrasonic sound needed for the detection. The crystal deforms according to the voltage. This creates an ultrasonic sound wave signal with a certain pattern. In the OmniScan method, the signal is reflected by the back wall of the material (back-wall echo) or a possible defect, and returns to the transducer. The signal now deforms the piezo-electric crystal and thus creates a voltage that can be interpreted. By measuring the time between sending of the pulse and receiving of the reflection it is possible to detect the location of the defect in the material.

The OmniScan method uses 64 transducers instead of the strictly necessary amount of two. This makes the method more efficient as it enlarges the area of measurement. To be detected, a defect must have a minimal area of 36 mm$^2$ or $6 \times 6$ mm$^2$. For materials with a high ability to absorb sound waves a low frequency and high energy potential are used to cut down the effect, e.g. detection of defects in omega profiles, which are filled with foam that has an absorbing effect on the sound waves. The surface roughness of the tested material makes it impossible to get a good acoustic bounce. To improve this a couplant is used. With the choice of couplant, one has to take into account that:

- It has to be easy to produce;
- There should not be any air bubbles, this has an influence on the measurement;
- It should not cause any damage to either the material nor the measurement probe;
- It should improve the bounce of the surface.

A delay line is used with the transducer to create a zero point shift. This has some advantages:

- It has a good resolution in near field;
- The high transducer frequency gives a better resolution;
- It has a positive effect on detecting defects in small boards;

### 2.2 *C-scan*

In the C-scan method, the transducers are placed on opposite sides of the material. The emitter sends a signal through the test piece, guided by a water jet. Any anomalies in the test piece will affect the amount of energy that reaches the receiver.

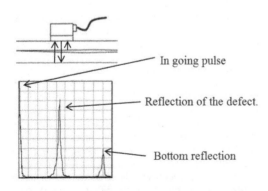

Figure 1. Working principle of the echo pulse technology (Arroyo et al. 2014).

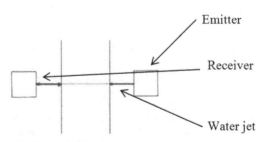

Figure 2. Working principle of the C-Scan method (Pieczonka et al. 2014).

## 3 THERMOGRAPHIC DETECTION

Pulsed thermography is a common measurement technique. In pulsed thermographic research the surface of the material is being heated by a heat pulse, generated by one or more lamps or a heat blower which are placed on a certain distance of the test object. The heat source and the distance at which they are placed have an effect on the rate at which heat will be absorbed by and travel through the test object. The time period of the heat pulse depends on the thickness of the material. The heat placement in the material is monitored with an infrared camera, and can be visualized on a computer afterwards.

When using thermographic research, the measurements of the temperature differences are important. After a heat pulse the initial surface temperature changes. This is due to the diffusion effect of the heat in the material. The heat on the surface is spread by the conduction effect to the depth of the material. The thicker the material, the longer it takes for the heat to spread over the whole surface. At first the temperature change in the damaged area will go the same way it would in an undamaged area. This is due to the fact that the heat didn't reach the defect yet. When the heat reaches the damaged area, the diffusion coefficient will change. This results in a variation of temperature between the damaged and the undamaged areas.

## 4 RESULTS

### 4.1 *OmniScan inspection method*

The IRP-panel consists of 18 defects that are manually applied. When looking at the image above the difference is shown as illustration. On the C-scan images there is a clear distinction to the areas with and without defects in the structure. The parts on the image, which are colored blue, don't consist of any defects, while the oval, red colored parts are the defects. Looking at the B-scan image, the red/yellow collared parts display the back-wall echo, this gives a step shape. When the step shape is interrupted and is filled with a step next to the next step we can conclude that that's the place where there is a defect. Because of the defect, the longitudinal sound wave is not able to reach the bottom of the IRP-panel, this causes the image to show an open space where there should be a step. The diameter of the first defect is 9 mm.

### 4.2 *C-Scan inspection method*

The entire preparation process takes approximately an hour before the inspection can begin. When the preparation is finished there has to be a check from a level 2 NDT inspector. Carrying out the C-scan takes approximately six hours, the process was stopped regularly due to splash in the measurements. As seen on the figure showed here there are several red dots (those dots are visualized as red lines on the figure). These dots show where

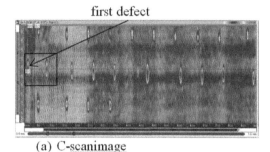

(a) C-scanimage

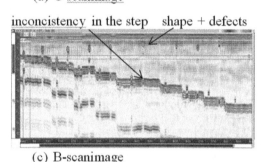

(c) B-scanimage

Figure 3. Showing the differences between the different scans.

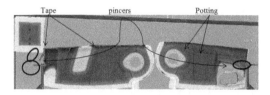

Figure 4. C-scan image of the measurement.

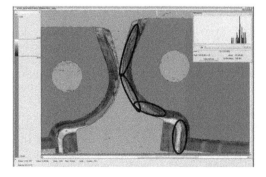

Figure 5. Defects on the monolithic parts.

the longitudinal sound waves couldn't get through the surface of the material.

First we concentrate on IRP-1SL-A350–003–1. Using the C-scan there are four known defects that are clearly visible on the measurements. To scan the monolithic parts, firstly we will scan the defect free parts of the material (29dB). We can use the 6dB general rule for this. When changing the dB-value to 36dB the following image in Figure 5 is visualized.

There isn't any use of the red parts where we can find the honeycomb structure. We only keep the parts of the monolithic area. Here we see that almost the whole monolithic area is red, this meaning that the whole area is covered with defects.

## 5 THERMOGRAPHIC RESEARCH

### 5.1 Results using a tripod and a lamp

Examining the IRP panel with 18 holes, we have to divide the panel in 5 segments. This is in contrast with the OmniScan method where it is possible to directly scan the whole plate. The segments that are created are now scanned separately. To get a good comparison between the thermographic and ultrasonic research we need to form a grid with the results from the thermographic research of the IRP panel. Not every defect can be found with the thermographic research, although the defects right underneath the surface where all but a few found.

### 5.2 Results when holding the camera

On the IRP panel with four defects, the defects where found by the method with the lamp as well as by the method with the hot air blower. Since the method with the tripod is seen as more accurate, and this method couldn't establish all the defects in the IRP panel with 18 defects, it is neglected to do the same test while holding the camera, since this wouldn't give a better result.

## 4 CONCLUSIONS

This paper includes a comparison of inspection methods used to track defects in fiber-reinforced plastics. In this paper the inspection methods are di-vided in two research fields: ultrasonic inspection and thermographic inspection. The purpose of both inspection methods is to detect, in an efficient and effective way, defects on monolithic as well as sandwich structures. Performing a C-scan of an IRP panel revealed all existing defects. In a more realistic application, a degree of uncertainty exists with respect to identification of all existing defects remains until elaborate destructive testing is

performed. The C-scan method has been found to be reliable as an inspection method, but one needs to take into account the following disadvantages:

– Very long scan times;
– Very long computer processing time;
– Not precise enough to detect foil in a monolithic structure;

During an OmniScan on an IRP panel, all defects were detected as well. With this method, as with the C-scan method, a degree of uncertainty remains, though the OmniScan method was able to detect a existing foil between monolithic layers.

Thermographic research of IRP panels with monolithic structure revealed 9 out of 18 defects, but on a panel consisting of a honeycomb structure it revealed all defects. The same remark about uncertainty if all defects have been identified also remains in this case. Thermographic research did not reveal a foil in the monolithic structure. The thermographic research method can be substantially faster if the camera is held in hand.

Statistical validation using random samples shows that the ultrasonic and thermographic research methods do not match up. Thermographic research only succeeded in detecting all defects in honeycomb structures, whereas ultrasonic research detected all defects in both monolithic and honeycomb structures.

## ACKNOWLEDGEMENTS

This research has been funded by the University of Antwerp and the Institute for the Promotion of Innovation by Science and Technology in Flanders (IWT) by the support to the TETRA project 'Smart data clouds' with project number 140336. Furthermore, the research leading to these results has received funding from Industrial Research Fund FWO Krediet aan navorsers 1.5.240.13 N. The authors also acknowledge the Flemish government (GOA-Optimech) and the research councils of the Vrije Universiteit Brussel (OZR) and University of Antwerp (fti-OZC) for their funding. Last but not least the authors want to thank Niels Segers and Sabca Limburg N.V. for providing the necessary measurement data and analysis results.

## REFERENCES

Arroyo, P. et al., 2014. Porosity characterization of CFRP parts by ultrasonic and X-ray tomographic non-destructive methods. In *6th International Symposium on NDT in Aerospace*. Madrid, Spain, p. 7.

Busse, G., 2001. Techniques of Infrared Thermography: Part 4. Lock-in thermography. In X.P. Maldague &

P.O. Moore, eds. *Nondestructive Handbook, Infrared and Thermal Testing, Volume 3*. Columbus, Ohio: ASNT Press, p. 718.

Giorleo, G. & Meola, C., 2002. Comparison between pulsed and modulated thermography in glass-epoxy laminates. *NDT and E International*, 35(5), pp. 287–292.

Jaluria, Y. & Atluri, S.N., 1994. *Computational heat transfer*, Taylor and Francis.

Maldague, X.P., 1994. *Infrared Methodology and Technology*, Taylor & Francis Ltd. Available at: http://books.google.com/books?id = xPOpMYKCTBgC&pgis = 1.

Maldague, X.P., 2001. *Theory and Practice of Infrared Technology for Nondestructive Testing*, Springer; Softcover reprint of the original 1st ed. 1993 edition (20 Nov. 2011).

Moore, P.O. & Maldague, X.P. eds., 2001. *NDT Handbook on Infrared technology* ASNT Handb., ASNT Press (American Society for NonDestructive Testing) Press.

Pieczonka, L. et al., 2014. Nondestructive testing of composite patch repairs. In *11th European Conference on Non-Destructive Testing (ECNDT 2014)*. Prague, p. 6. Available at: http://ww.ndt.net/search/docs.php3?MainSource = –1&showForm = OFF&KeywordID = 9 & MainSource = –1&SearchDocs = Ultrasonic C-scan&restrict = keywords&searchmode = Phrase.

Wrobel, L.C., 1992. Handbook of numerical heat transfer.

*Embedded sensors*

*Emerging Technologies in Non-Destructive Testing VI – Aggelis et al. (Eds)*
*© 2016 Taylor & Francis Group, London, ISBN 978-1-138-02884-5*

# Investigations on the structural integrity and functional capability of embedded piezoelectric modules

### S. Geller, A. Winkler & M. Gude
*Institute of Lightweight Engineering and Polymer Technology, TU Dresden, Dresden, Germany*

ABSTRACT: Fibre-reinforced composites with integrated sensor and actuator elements based on piezoceramic transducers are predestined to realize active lightweight components. Due to moderate process conditions, polyurethane processing technologies are particularly suited for embedding piezoceramic transducers. The development of piezoceramic modules capable for integration and the manufacture of active fibre-reinforced polyurethane composite structures are described. Embedding studies of modules using different processing parameters and composite specifications followed by non-destructive analysis of their structural integrity using X-ray technique are presented. In addition, the vibration behaviour of generic structures is characterized by laser vibrometry. The investigations show that process-related stresses in terms of temperature and pressure do not exceed 100 °C and 3 bar respectively. Further, it can be shown, that robust embedding of the piezoceramic modules is provided and vibration transmission by integrated actuators with thermoplastic carrier film onto fibre-reinforced polyurethane composites is possible.

## 1 INTRODUCTION

Cross-industry efforts for an efficient consumption of resources lead to extensive lightweight constructions and often to an increasing use of fibre-reinforced plastics. In this purpose, associated questions regarding condition or structural health monitoring, damage detection and additional requirements for the realization of actuator functions are present (Edery-Azulay 2006, Belloli 2007). To fulfil these requirements, the development of active composite structures is gaining importance. In recent years, numerous activities focused on the integration of piezoelectric transducers into composite components, mainly using prototypical procedures and structures, such as laminates based on epoxy resins (Melnykowycz 2005, Konka 2012). Transferring existing methods for the integration of piezoelectric modules into structural components from laboratory to large-scale production is subject of the research activities in the Collaborative Research Centre/Transregio 39. Here, the damage-free embedding of brittle ceramic transducers into composite structures is a main challenge for the integration of such modules. Since fibre-reinforced polyurethane composites are characterised by moderate processing conditions, they are favoured for the integration of functional elements (Weder 2013, Hufenbach 2014). In addition, the well-known superior adhesion properties of polyurethanes are motivation for embedding of specially designed thermoplastic-compatible piezomodules (TPM) with thermoplastic carrier films to realize active composite structures.

## 2 MATERIALS

### 2.1 *Fibre-reinforced polyurethane composites*

In the studies presented, TPM with thermoplastic carrier films are embedded into glass fibre-reinforced polyurethane (PUR) composites. For this purpose, a two-component polyurethane system from BASF Polyurethane GmbH is used. Due to its high degree of expansion, different resultant composite densities can be defined within a wide range. The foaming process is carried out chemically, enabled by the addition of water into the A-component (polyol) of the polyurethane. This reacts with the B-component (isocyanate) with separation of carbon dioxide, which acts as a blowing agent. The main processing data of the used matrix system are summarized in Table 1. As reinforcing component, assembled glass fibre rovings (E-glass with silan-based sizing) of supplier Mühlmeier (type EC15 4800 254 (90)) with a yarn count of 4800 tex are used.

### 2.2 *Piezomodules*

Novel thermoplastic-compatible piezoceramic modules (TPM) consist in principle of thermoplastic carrier films, metallized with an electrode structure, and a piezoceramic functional

Table 1. Processing data of the used polyurethane matrix system.

| Property | Polyol | Isocyanate | Mixture |
|---|---|---|---|
| Density [g/cm³] | 1.07 | 1.23 | |
| Viscosity [mPas] | 1600 | 220 | |
| Mixing ration per weight | 100 | 215 | |
| Starting time [s] | | | 85 |
| Free rise density [g/cm³] | | | 125 |

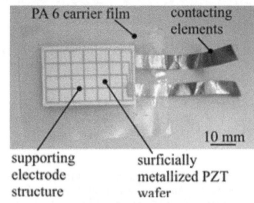

Figure 2. Contacted TPM for embedding into composite structures.

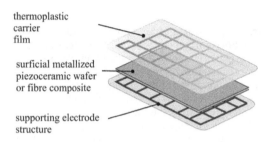

Figure 1. Schematic build-up of used TPM (d31-type).

layer (wafer or piezofibre composites) as illustrated in Figure 1.

Primarily these modules were developed for a material homogeneous integration into fibre-reinforced thermoplastic composite structures. Therefore, the carrier film material is equal to the composite matrix system, so that these two components will melt together by the manufacturing of the active part (Heber 2014). TPMs are developed for different actuation modes (f. e. d33, d31). In this context the differences in the build-up of such modules differ in principle in the design of the electrode structure (interdigitated or surficial electrode) and in the functional PZT layer (PZT fibre composite or wafer).

These modules are manufactured in an efficient roll-to-roll process, which bases on an automated printing of the electrodes to the carrier films, the feeding of the piezoceramic layers, and an adapted hot pressing process for the consolidation of the modules by melting the covering carrier films. In a subsequent step these modules are electrically contacted and polarized. For the presented studies the modules are poled by a voltage of 360 V for 5 min. The TPMs used for the presented investigations include a monolithic PZT wafer with a width of 22 mm, a length of 35 mm and a thickness of 0.18 mm. This wafer is surficial metallized by a screen printed silver electrode. Furthermore, this metallization is supported by a printed electrode structure at the thermoplastic carrier films. Due to the surficial electrodes these TPMs are working in d31 mode, which means a principal deflection normal to the polarisation direction. In Figure 2 a contacted TPM as used in the investigations is illustrated.

## 3 SAMPLE MANUFACTURE

The composite structures are manufactured using the automated long fibre injection (LFI) process. This spray coat method is characterized by processing of reactive two-component PUR systems and long fibre reinforcement made of glass (Frehsdorf 2001). Due to superior adhesive properties of the polyurethane matrix systems, composite structures for exterior applications can be manufactured by direct back-foaming of thermoplastic high-glossy films using the LFI process (Stratton 2006). This suitability is used here for embedding the TPM with thermoplastic carrier films.

Both basic components of the polyurethane (polyol, isocyanate) are mixed within a high pressure mixing head using the counterflow injection procedure. At the same time, glass fibre rovings are chopped in a cutting unit directly above the mixing head to defined lengths (between 12.5 mm and 100 mm) and fed by means of compressed air to the mixing heads outlet to be discharged simultaneously with the matrix system. Hereby, the fibre mass content can be adjusted continuously by the rotational speed of the cutting unit.

For the investigations presented, the TPM to be integrated is placed centric onto the lower mould just before the manufacturing process and back-foamed directly with the glass fibre polyurethane mixture. Composite samples with varying densities at constant fibre mass content have been prepared. Since the dimensions of the cavity have been kept constant through all the investigations, the density is defined by the amount of the discharged PUR and glass fibres, resulting in different degrees of expansion. The fibre length has been set 25 mm for all samples. The composite specifications and used process parameters are summarized in Table 2.

Table 2. Processing parameters and composite specifications.

Processing parameters

| Sample dimensions [mm] | $650 \times 650 \times 5$ |
| Mould temperature [°C] | 80 |
| Clamping force [kN] | 500 |
| Reaction time [s] | 300 |

Composite specification

| | Mass of PUR g | Fibre mass content % | Density g/cm³ |
|---|---|---|---|
| TPM 1 | 1480 | 30 | 1.0 |
| TPM 2 | 1180 | 30 | 0.8 |
| TPM 3 | 890 | 30 | 0.6 |

## 4 EXPERIMENTAL

### 4.1 *Process-related stresses*

In Order to determine the cavity pressure as a result of the varying degrees of expansion, defined areas of the cavity are lined with pressure-sensitive films (Fig. 3). These films include micro hollow spheres which are filled with a coloured liquid. Depending on the applied pressure, these spheres collapse and the discoloration of the film is used as an indicator of the amount of pressure (Fig. 4). Additionally, a sensor for detecting the reaction temperature is placed near the TPM. This sensor detects and saves the resulting temperature within the composite during curing and enables wireless transmission of saved data after demoulding the samples.

### 4.2 *Structural integrity and functional capability*

In regard to the estimation of the integrity and coupling of the TPM to the surrounding composite, the characteristic resonance frequencies of the composite structures are determined. Therefore, a laser scanning vibrometer (LSV) (type PSV-400 produced by Polytec) was used which enables a contactless measurement without distorting the mechanical properties of the investigated object by the additional mass of typically used vibration sensors. The excitation of the plates was realized by the TPM. A harmonic chirp voltage signal from 0 V to 360 V was applied to the TPM in a frequency range of 0.01 kHz to 0.2 kHz with a resolution of 0.25 Hz. The LSV has been positioned normal to the analysed plate because of the direct proportionality of the laser scanning vibrometer output signal to the velocity of the targeted surface along the laser beam direction. For the scanning of the surface a pattern of $11 \times 11$ discrete measuring

Figure 3. TPM placed onto mould surface, surrounded by pressure sensitive films and sensor for temperature measurement.

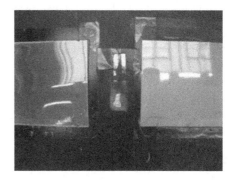

Figure 4. Demoulded composite structure with back-foamed TPM and discoloured films.

Figure 5. Test setup for analyzing the vibration behavior of the samples.

points was used. The free mounting of the investigated plate was realized by vertically hanging using thin and light ropes, see test setup illustrated in Figure 5.

## 5 RESULTS AND DISCUSSION

### 5.1 Process-related stresses

First subject of investigation after manufacturing the composite samples is the evaluation of the measured reaction temperatures. The values determined by the embedded sensor show comparable temperature profiles for all specifications (Fig. 6). However, depending on the composite specification, different maximum temperatures result inside the composite.

To evaluate the cavity pressure, the pressure-sensitive films are scanned after the manufacturing process and by means of a special evaluation software the average pressures are calculated. Hereby, pressures from 2.4 to 2.9 bars are determined. In Table 3 the process-related stresses for the three composite specifications are summarized.

The absolute maximum temperatures between 80°C and 100°C as well as the average pressures are at a relatively low level, so that impairment of the thermoplastic carrier films and the piezoceramic functional layer can be excluded and gentle embedding of the modules is provided.

### 5.2 Structural integrity and functional capability

In order to detect potential damage, the TPM integrated in the respective composites are scanned using X-ray analysis. As expected, no damage to the piezo modules has been found here. Thus it can be stated, that for the chosen conditions damage free embedding of TPM with thermoplastic carrier film into glass fibre-reinforced polyurethane composite structures is possible. In Figure 7 the comparison of an X-ray scan of TPM 1 before and after embedding is illustrated.

The coupling of the embedded modules to the composites structure has been tested by laser vibrometry. Figure 8 shows the spectrum of the surface velocities in dependency to the excitation frequency. The peaks for the resonance frequencies are clearly recognizable.

Figure 7. X-ray scan of TPM integrated in composite specification 1 before (above) and after embedding (bottom).

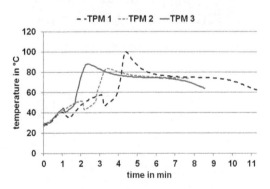

Figure 6. Temperature profile during curing of the composites.

Table 3. Determined process-related stresses.

| Specification | Average pressure bar | Maximum temperature °C |
|---|---|---|
| TPM 1 | 2.9 | 100 |
| TPM 2 | 2.8 | 84 |
| TPM 3 | 2.4 | 88 |

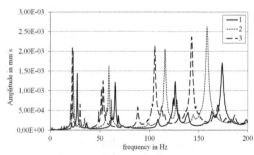

Figure 8. Determined spectrum of surface velocities for different composite specifications.

Due to the clear curve progression a coupling of the TPM to the structure is given. Furthermore, the shift of the resonance frequencies to higher values corresponds to the density of the investigated samples. By this experimental test the functionality and excitation of the TPM, as well as coupling to the structure is distinguishable.

# 6 CONCLUSIONS

Piezoceramic modules with thermoplastic carrier film have been integrated into fibre-reinforced polyurethane composites by means of the LFI process and using varying processing parameters. The determined process conditions during manufacturing of the composites are moderate and facilitate damage-free embedding of the modules. This could be proved by X-ray analysis of the modules before and after embedding. Using the embedded modules, vibrations can be transmitted to the composite structures. The vibration behaviour of samples with different specification has been characterized by laser vibrometry. Here, varying resonance frequencies are determined in dependency of the composite specifications.

# ACKNOWLEDGEMENTS

This research is supported by the Deutsche Forschungsgemeinschaft (DFG) in context of the Collaborative Research Centre/Transregio 39 PT-PIESA, subproject B6.

# REFERENCES

Belloli, A. et al., 2007. Structural vibration control via R-L shunted active fiber composites, Journal of Intelligent Material Systems and Structures 18: 275–287.

Edery-Azulay, L. & Abramovich, H. 2006. Active damping of piezo-composite beams. Composite Structures 74: 458–466.

Frehsdorf, W. & Söchtig, W. 2001. High Requirements— Low Investment Costs. Kunststoffe plast europe, Vol. 91, 3: 23–25.

Heber, T. et al., 2014. Production process adapted design of thermoplastic-compatible piezoceramic modules, Composites Part A: Applied Science and Manufacturing 59: 70–77.

Hufenbach, W. et al., 2014. Cellular fibre-reinforced polyurethane composites with sensory properties, Advanced Engineering Materials 16, 3, 272–275.

Konka, H.P. et al., 2012, The effects of embedded piezoelectric fiber composite sensors on the structural integrity of glass-fiber-epoxy composite laminat. Smart Materials and Structures 21, 015016, 1–9.

Melnykowycz, M.M. et al., 2005, Integration of active fiber composite (AFC) sensors/actuators into glass/epoxy laminates. In: SPIE5761, Smart Structures and Materials: Active Materials: Behavior and Mechanics, doi: 10.1117/12.599109.

Stratton, D. et al., 2006. A Modular Automotive Roof System Design Concept Based on Polyurethane Composite Technology. In Proc. of the 6th Annual SPE Automotive Composites Conference, Troy, USA, 12–14 September 2006: 475–491.

Weder, A. et al., 2013. A novel technology for the high-volume production of intelligent composite structures with integrated piezoceramic sensors and electronic components, Sensors & Actuators A: Physical 202: 106–110.

*Emerging Technologies in Non-Destructive Testing VI – Aggelis et al. (Eds)*
*© 2016 Taylor & Francis Group, London, ISBN 978-1-138-02884-5*

# Healing performance monitoring using embedded piezoelectric transducers in concrete structures

G. Karaiskos, E. Tsangouri, D.G. Aggelis & D. Van Hemelrijck
*Department Mechanics of Materials and Constructions (MeMC), Vrije Universiteit Brussel (VUB), Brussels, Belgium*

A. Deraemaeker
*Building, Architecture and Town Planning (BATir), Université Libre de Bruxelles (ULB), Brussels, Belgium*

ABSTRACT: Concrete structures are prone to degradation due to operational and ambient loadings, as well as to a series of environmental effects which can seriously decrease their anticipated operational service life. In the present study, the use of ultrasonic pulse velocity method based on embedded small-size and low-cost piezoelectric transducers for the online monitoring of small-scale notched unreinforced concrete beams with autonomous self-healing embedded encapsulation system, is experimentally verified. Initially, the concrete beams are subjected to a three-point bending test and the cracking formation and extension trigger the release of the encapsulated healing agent into the void of the cracks. After a 24-hour healing agent curing period, the beams are reloaded following the same loading protocol. The results demonstrate the excellent performance of the proposed monitoring technique during the early-age, the cracking formation and recovery periods.

## 1 INTRODUCTION

### 1.1 *The concrete and its monitoring challenges*

Concrete is still the leading structural material due to a series of advantages such as the low production cost, the great structural design flexibility, the high strength and durability. It is a highly-complex quasi-brittle heterogeneous and anisotropic composite material that is made by binding aggregates together using cement paste and it is characterized by a low capacity for deformation under tensile stress. Except the ageing, a concrete structure is vulnerable to a series of operational and ambient degradation factors which can seriously decrease its anticipated operational service life.

The infrastructure agencies are shifting efforts for achieving a reliable prolongation of the concrete structures service life and this could be efficiently achieved by using reliable and low-cost monitoring techniques and Autonomous Self-Healing Systems (ASH). Traditionally, epoxy and cement based healing agents are manually injected into the void of the generated cracks of concrete, providing local material restoration (Smoak 1997). In that way, only visible and mainly surface macroscopic damages at accessible locations of the structure can be inspected and rehabilitated. In the recent years, the so-called self-healing materials, as a class of smart materials that have the structurally incorporated ability to autonomously rehabilitate the generated damages, such as cracking, have appeared. In concrete, embedded capsules filled with healing agent is one of the most promising material recovery system for filling the void of the generated cracks with appropriate agent, providing healing and sealing (Dry 1994).

Current manual inspection techniques of concrete structures are costly and time demanding and their effectiveness is limited due to the lack of accessibility of several locations of the structure. Reliable, low-cost and automated tools are required for error-free concrete strength and integrity performance evaluation, so that the most cost effective strategy of rehabilitation can be adopted (Bungey et al. 2006). In the recent years, there is an increasing interest in the development of Non-Destructive Testing (NDT) methods able to estimate the quality and integrity of both early-age and hardened concrete not only in the laboratory, but also in in-situ applications (ACI 228.2R-98 1998, Malhorta & Carino 2004). One of the most commercially developed NDT inspection techniques is the Ultrasonic Pulse Velocity (UPV) method, which is able to monitor concrete during

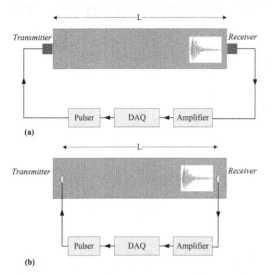

Figure 1. UPV system using (a) bulky external piezoelectric transducers and (b) small-size and low-cost piezoelectric transducers.

early age, as well as to estimate its strength and detect the developed damage in its hardened state. Monitoring the time-of-flight T of the longitudinal ultrasonic stress wave (i.e P-wave), it is feasible to estimate the mechanical properties of the concrete such as the dynamic modulus of elasticity. The commercially available monitoring system is based on bulky external piezoelectric transducers which need to be placed on two opposite faces of the tested concrete specimen (Fig. 1a). The main limitations of the technique are the need for a coupling agent and flat surfaces as well as the lack of flexibility in the transducers arrangement. This usually limits the application to through thickness transducers arrangement and the technique is not adapted for on-line applications. Instead of using the bulky external piezoelectric transducers, embedded small-size and low-cost piezoelectric transducers can be successfully used for monitoring the concrete hydration as well as tracking the cracking formation and its healing (Fig. 1b).

## 2 MATERIALS AND METHODS

### 2.1 Ultrasonic pulse velocity monitoring method

In the present study, the monitoring system used consists of the embedded small-size and low-cost piezoelectric (lead-zirconate-titanate (PZT)) ceramic transducers, a high frequency data acquisition (DAQ) system, a high voltage pulser and a voltage amplifier. A short-duration (2.5 μs) low-amplitude rectangular voltage pulse is generated in the DAQ system and it is amplified up to 800 V using the high voltage pulser and then it is introduced into one of the transducers (i.e. transmitter). The generated P-wave propagates through the concrete and it is detected by the other transducer (i.e. receiver) which is placed at a distance L. The transmitted wave is subjected to loss and scattering because of the reflection, refraction and mode conversion caused by the aggregates, the possible voids and steel reinforcement found along the travel path. The time-of-flight T of the P-wave is measured by a synchronized electronic timer installed in the DAQ system and by taking also into account the calculated time delay of the transducers, computed through a calibration test before embedding them in the concrete

The stable coupling between the transducers and the material matrix (Fig. 1b) can ensure accurate and reliable measurements with high signal to noise ratio. This technique offers the great possibility of overcoming the limitations of traditional method which prevent the application of specific boundary conditions during the measurement. It provides a great flexibility in the choice of the position placing the transducers and thus it makes an ideal in-situ monitoring technique. It can be equally used for the hydration monitoring as well as for monitoring the structural integrity of the concrete during its entire service life. (Gu et al. 2006, Song et al. 2007, Dumoulin et al. 2012, Dumoulin et al. 2014).

### 2.2 Embedded piezoelectric transducers

Following the concept of 'smart aggregates' (Gu et al. 2006) (SMAG), a few small-size and low-cost piezoelectric transducers were designed and fabricated in BATir at ULB in order to be embedded in the tested unreinforced concrete beams with ASH systems. Each transducer used in the present study consists of a flat piezoelectric PZT (lead zirconate titanate) ceramic patch (Fig. 2a) which electrodes are conductively glued with the cable (Fig. 2b). In order to mechanically protect the fragile ceramic patch and avoid capacitive coupling interference between the transducers due to the presence of water in the fresh concrete, the patch should be properly insulated by a waterproof epoxy-based coating (Fig. 2c). Extra electromagnetic shielding is provided by applying a thin layer of conductive paint (Fig. 2d).

Figure 2. Manufacturing steps of the embedded piezoelectric ceramic transducers.

## 2.3 Design of the experimental set-up

A series of three three-point bending tests were designed and executed in MeMC at VUB. The set-up used in the present study follows the Rilem Technical Report FMS-50 specifications (RILEM TC-50-FCM 1985). The test specimens consist of unreinforced and simple concrete made beams with embedded ASH system and SMAGs. The concrete mix proportions used are given in Table 1.

Three notched unreinforced concrete beams of 840 mm × 100 mm × 100 mm with embedded ASH system and SMAGs were prepared (Fig. 3). The ASH system used is based on embedded fragile glass capsules (50 mm long with 3 mm inner diameter) filled with a two-component polyurethane based healing agent. Four, eight and eight couples of glass capsules filled with the healing agent were embedded in the first, second and third concrete beam respectively. The capsules were equally distributed in the monitored area between the SMAGs through a grid made of very low stiffness plastic wire which was attached on the walls of the mold. In the first and second beams, one couple of SMAGs was embedded in each of them (SMAGs 1, 2 and SMAGs 3, 4 respectively) with a distance between them equal to 100 mm and at a height of 35 mm from the bottom of the beams. In the third beam, the monitoring system consists of two couples of SMAGs which were embedded with the same distance of 100 mm. The first couple (SMAGs 5 and 6) was placed at a height of 20 mm and the second one (SMAGs 7 and 8) at a height of 50 mm from the bottom of the beam.

After the concrete casting in the wooden molds and until the demolding of all the beams (almost two days later), all the specimens were subjected to membrane curing in order to prevent moisture loss.

Table 1. Concrete mix proportion.

| Material | Mix proportions kg/m³ |
|---|---|
| Sand 0/4 | 670 |
| Gravel 2/8 | 490 |
| Gravel 8/16 | 790 |
| Cement CEM I 52.5 N | 300 |
| Water | 150 |

## 3 RESULTS

### 3.1 Early-age monitoring

Before loading and during the curing period, pulse velocity of the first concrete beam was monitored in the mold for the first forty hours after casting. A high-voltage pulse was sent to SMAG-1 (transmitter) and the transmitted P-wave is detected by the SMAG-2 (receiver). Taking also into account the time delay (i.e. 4 µs) and the distance between them (i.e. 100 mm), the evolution of the UPV as a function of time was easily calculated (Fig. 4a). A couple of samples of the recorded signals at different times are also shown in Figures 4b, c.

The signal is difficult to be measured at very early ages as it tends to be dominated by noise. It becomes clearer starting around 5 h after the dormant period. The signal to noise ratio and the maximum amplitude of the recorded signals gradually increase and the time-of-flight T of the P-waves is gradually decreased during the transition from fluid to solid.

### 3.2 Cracking and healing monitoring

After demolding the beams and until their loading tests (twelve days later), all the beams were subjected to water curing in order to improve the setting and hardening procedure. Two weeks after the concrete casting, all the three beams were subjected to three-point bending tests, under slow speed (0.04 mm/min) of the crosshead of the

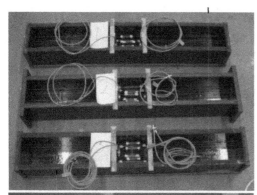

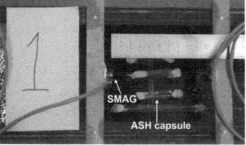

Figure 3. Preparations of the beams with the ASH system (glass capsules filled with healing agent) and the SMAGs.

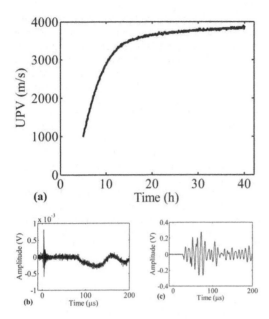

Figure 4. (a) Evolution of the P-wave velocity as a function of time and the recorded time signals (b) at the 5th hour and (c) at the 40th hour after concrete casting in the first beam.

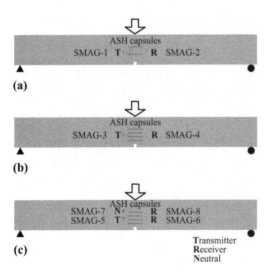

Figure 5. Testing set-up of (a) the first beam, (b) the second beam and (c) the third beam.

testing machine, until wide cracking opening, rupture of the glass capsules and filling of the cracking voids with the healing agent were achieved (Fig. 5). Following that loading rate, ten ultrasonic wave measurement tests every fifteen seconds were performed and the mean value of each received group of results is used in the following presented results. Then, the beams were unloaded and left at rest for a 24-hour curing period of the released healing agent. Finally, the beams were subjected to three-point bending following the same loading protocol, in order to evaluate the NDT monitoring and ASH system performance used in the present study.

The transmitted compressional ultrasonic stress waves are subject to a high level of scattering due to the aggregates and voids found in the concrete. By the time the mechanical wave reaches the SMAG receiver, it is transformed into a complex waveform. The early part of that waveform mainly contains the contribution of a direct wave between the SMAG transmitter and SMAG receiver and therefore carries information about the state of its microstructure. The damage index (d.i.) shown in Equation 1 is therefore based on the early part of the measured waves, as in (Karaiskos et al. 2013):

$$D.I. = \sqrt{\frac{\int_{t_n}^{t_p}(x_j(t)-x_0(t))^2\,dt}{\int_{t_n}^{t_p}x_0^2(t)\,dt}} \qquad (1)$$

where $x_j(t)$ corresponds to the amplitude of the damaged signal (loaded structure) and $x_0(t)$ is the amplitude of the healthy one (sound material), $t_n$ is the arrival time of $x_0(t)$, and $t_p$-$t_n$ corresponds to the duration of the first half-period of the healthy signal (Fig. 6).

The d.i. used in the present study, considered both the shift of the wave arrival time and its amplitude. Thus any increase of the calculated d.i. is indicative of crack formation and extension. This d.i. is a good quality indicator for detecting the cracking generation and its propagation but it does not give any information about the size of the damage.

In Figure 7, the applied load (kN) vs strain (mm/mm) curves corresponding to the loading and reloading tests of the first beam are shown. Strain is the normalized data (mm/mm) received by the crack mouth opening displacement (CMOD) gauge which was placed on the notch limits of the beams. Before testing, the initial opening of the gauge is fixed at 10 mm. Additionally, the gradual evolution of the early part of the recorded signals with increasing load corresponding to the reloading test of the first beam is shown in Figure 8.

The shifts in the wave arrival time and amplitude of the received signals, during the loading and reloading tests in all the three tested beams, are well

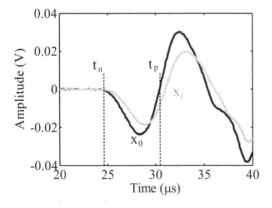

Figure 6. Definition of the damage index: arrival time of the healthy and damaged signals and the first half-period of the healthy signal.

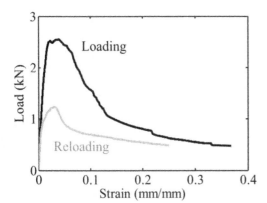

Figure 7. Load vs strain of the loading and reloading tests of the first beam.

captured by the proposed d.i., which is plotted as a function of the applied load (Figs 9–11).

In all the loading tests, the d.i. evolution graphs clearly consist of three similar phases. In 'Phase I' (early damage initiation) the d.i. is very low but non-zero and this is partially attributed to a certain level of noise in the received signals. The d.i. starts to increase noticeably at around 2 kN. The initiation and evolution of the cracking correspond to 'Phase II'. The severe failure of the concrete corresponds to 'Phase III', where the d.i. is close to one. Additionally, in the third beam, the d.i. calculated by the lower placed SMAG-6 (Fig. 5c) grows faster and earlier than the one calculated by the SMAG-8. Taking into account that the cracking initiated closer to the SMAG-6 than the SMAG-8, that is a solid proof that the present NDT technique could be also successfully used to qualitatively monitor the cracking extension.

In the reloading test of the first beam (Fig. 9), the initial d.i. value is close to one and that is a sign of no healing of the cracked concrete and that could be attributed to the deficient quantity of the healing agent. In the second beam (Fig. 10), initially and until almost 1 kN of loading, the d.i. value ($\approx 0.6$) reveals that the beam is partially healed. Then, a sudden increase of the d.i. value leads the beam to the complete failure. Concerning the third beam (Fig. 11), the results seem quite reasonable. The higher placed SMAG-8 measured lower damage than the lower placed SMAG-6, because the crack was thinner at the top. This is a good indicator that the present method could be also used to track the healing of the developed cracking.

It is worth mentioning that in the second beam, the obtained strength and stiffness regain are 36% and 88% respectively. Additionally, in the third beam, it is measured that up to 21% of the initial

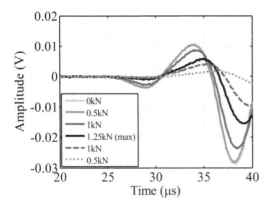

Figure 8. Evolution of the early part of the measured signals of the first beam.

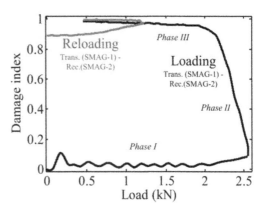

Figure 9. Damage index vs load of the loading and reloading tests of the first beam.

371

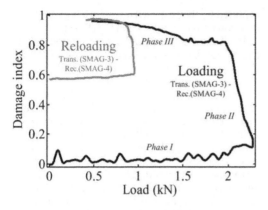

Figure 10. Damage index vs load of the loading and reloading tests of the second beam.

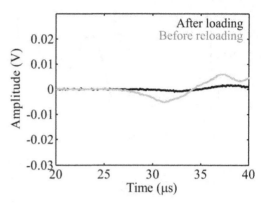

Figure 13. Early part of the received signals by SMAG-8 of the third beam.

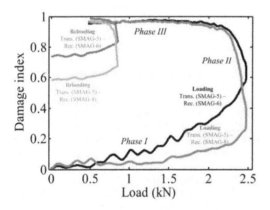

Figure 11. Damage index vs load of the loading and reloading tests of the third beam.

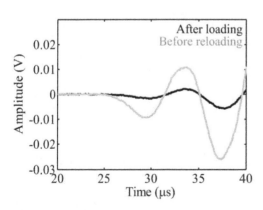

Figure 12. Early part of the received signals by SMAG-6 of the third beam.

strength and 45% of the initial stiffness of healthy concrete are achieved after healing (Tsangouri et al. 2015).

Through the recorded signals of the third beam shown in Figures 12, 13, in the end of the loading test and just before the reloading one, the presence of the healing agent could be reliably detected. In both 'before reloading' graphs shown in Figures 12, 13, the wave arrival time has decreased and its absolute amplitude has increased compared with the respective 'after loading' graphs, due to the presence of the dry healing agent which has filled the voids of the monitored cracking area.

## 4 CONCLUSIONS

The current study is focused on the implementation of the ultrasonic pulse velocity method based on embedded small-size and low-cost piezoceramic transducers in three small-scale notched unreinforced concrete beams with embedded autonomous self-healing system. After the concrete casting in the molds, the P-wave velocity was monitored for almost two days in one of the tested concrete beams. After a two-week curing period, three three-point bending tests were carried out. The proposed non-destructive testing method is able not only to monitor the concrete during its hydration, but also to track the cracking formation. Additionally, the proposed technique is able to evaluate the performance of the very promising autonomous self-healing system which was embedded in the tested concrete beams. A remarkable strength and stiffness regain were achieved in the cracked beams using adequate healing agent through the proposed autonomous self-healing technique.

## ACKNOWLEDGEMENTS

The research described in this paper has been performed in the frame of the SIM program on Engineered Self-Healing Materials (SHE). The authors are grateful to SIM (Strategic Initiative Materials Flanders) and FNRS (Fonds de la Research Scientifique) for providing financial support.

## REFERENCES

ACI 228.2R-98. 1998. *Nondestructive test methods for evaluation of concrete in structures*. Farmington Hills Michigan: American Concrete Institute.

Bungey, J.H., Millard, S.G. & Grantham, M.G. 2006. *Testing of concrete in structures*. Taylor & Francis.

Dry, C.M. 1994. Matrix cracking repair and filling using active and passive modes for smart timed release of chemicals from fibers into cement matrices. *Smart Materials and Structures* 3(2): 118–123.

Dumoulin, C., Karaiskos, G., Carette, J., Staquet, S. & Deraemaeker, A. 2012. Monitoring of the ultrasonic P-wave velocity in early-age concrete with embedded piezoelectric transducers. *Smart Materials and Structures* 21(4): 047001.

Dumoulin, C., Karaiskos, G., Sener. J.Y. & Deraemaeker, A. 2014. Online monitoring of cracking in concrete structure using embedded piezoelectric transducers. *Smart Materials and Structures* 23(11): 115016.

Gu, H., Song, G., Dhonde, H., Mo, Y.L. & Yan, S. 2006. Concrete early-age strength monitoring using embedded piezoelectric transducers. *Smart Materials and Structures* 15(6): 1837–1845.

Karaiskos, G., Flawinne, S., Sener, J.Y. & Deraemaeker, A. 2013. Design and validation of embedded piezoelectric transducers for damage detection applications in concrete structures. *10th International Conference on Damage Assessment of Structures, 8–10 July 2013*. Trinity College Dublin, Ireland.

Malhorta, V.M. & Carino, N.J. 2004. *Handbook on nondestructive testing of concrete*. CRC Press.

RILEM TC-50-FCM 1985. Determination of the fracture energy of mortar and concrete by means of three-point bend tests on notched beams. *Materials and Structures* 18(4): 285–290.

Smoak, G. 1997. *Guide to concrete repair*. USA: Books for Business, Department of Interior.

Song, G., Gu, H., Mo, Y.L., Hsu, T.T.C. & Dhonde, H. 2007. Concrete structural health monitoring using embedded piezoceramic transducers. *Smart Materials and Structures* 16(4): 959–968.

Tsangouri, E., Karaiskos, G., Aggelis, D.G., Deraemaeker, A. & Van Hemelrijk, D. 2015. Crack sealing and damage recovery monitoring of a concrete healing system using embedded piezoelectric transducers. *Structural Health Monitoring* (accepted for publication).

*Emerging Technologies in Non-Destructive Testing VI – Aggelis et al. (Eds)*
*© 2016 Taylor & Francis Group, London, ISBN 978-1-138-02884-5*

# Fibre-reinforced composites with embedded piezoelectric sensor-actuator-arrays

A. Winkler, M. Dannemann, E. Starke, K. Holeczek & N. Modler
*Institute of Lightweight Engineering and Polymer Technology, Technische Universität Dresden, Dresden, Germany*

ABSTRACT: The paper presents unique manufacturing processes and preliminary functionality testing of active Textile-Reinforced Thermoplastic Composites (TRTC) with embedded sensor-actuator arrays. The main part of the manuscript describes the technological aspects of automated assembly of thermoplastic films containing adapted Thermoplastic-compatible Piezoceramic Modules (TPMs) and conducting paths. Herein, the configuration of the specially developed TPMs, the assembly of transducer arrays as well as the TPM integration into textile-reinforced thermoplastics are presented. The initial functionality investigations were carried out with the objective to obtain a directional radiation of ultrasonic waves generated using the integrated sensor-actuator arrays. Finally, the findings were used for the assessment and discussion of appropriate applications and limitations of active TRTC.

## 1 INTRODUCTION

Textile-reinforced thermoplastic composites show a high potential for serial manufacturing of innovative function-integrating lightweight constructions. Such materials enable due to the textile structure, the layered construction, and the associated specific production processes, the possibility for the material-integration of functional elements, like sensors, actuators or even electronic circuit boards. Signals from such integrated sensor-actuator systems are required to broaden the application scope of TRTCs, e.g. through active vibration damping, condition or structural health monitoring.

State of the art technologies utilise conventional piezoelectric transducers, e.g. macro fibre composites or active fibre composites, which are mainly adhesively bonded to the structural components (Hufenbach et al. 2011, Moharana & Bhalla 2014). The associated assembly and bonding processes are characterised by several labour-extensive work steps.

In order to form a sensor-actuator-array numerous transducers have to be positioned in a defined pattern, which realisation can be influenced by the manufacturing process. In dependence of the actuator-sensor-array size, the assembly process could become very time-intensive. Therefore, an automated assembly is necessary for an economic manufacturing. In the following studies an assembly method is used, which bases on an automated thermal or ultrasonic fixing of functional elements on thermoplastic carrier films. The assembly process

has been named *ePreforming* and the respective outcome—a functionalised film—is called *ePreform*. A vital prerequisite for this process is the use of Thermo-plastic-compatible Piezoceramic Modules (TPMs). These modules are based on a piezoceramic functional layer (wafer or fibre composite) enclosed between two thermoplastic carrier films that are metallised with electrode structures (Hufenbach et al. 2009).

To successfully transform the current manual production process of active textile-reinforced thermoplastic composites into a fully automated mass production process, novel piezoelectric modules and adapted manufacturing processes are necessary.

This paper presents a contribution to the manufacturing process in terms of providing an adapted preformed actuator-sensor-layer (*ePreform*). The possibility to *ePreform* whole patterns/arrays of actuators and sensors enables new application fields of these active structures with a high potential for ultrasonic measuring tasks like the measurement of flow rate or distance (see Fig. 1).

## 2 PHENOMENOLOCICAL DESCRIPTION

Embedded piezoelectric transducer arrays are suitable for radiation and reception of plate waves. This property can be used for ultrasound based measuring tasks, condition monitoring or structural health monitoring (SHM) applications. For condition monitoring or SHM the influence of

Figure 1. Schematic illustration for ultrasonic distance measurements.

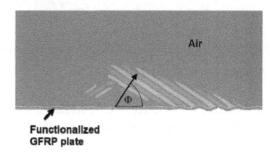

Figure 2. Radiation of the ultrasound wave from the functionalised TRTC plate at an angle Φ.

flaws, defects or damage on the plate wave propagation characteristics can be used (see Kostka et al. 2015). In contrary, for ultrasound measurement tasks the plate waves are used for an efficient radiation of sound waves into the adjacent media, e.g. air. This effect can also be reversed in order receive sound waves.

### 2.1 Ultrasound radiation and reception

The sound radiation and reception is based on the interaction of fluid and solid waves. Especially the bending wave, also known as the first asymmetric plate wave mode ($A_0$-mode), can be efficiently used for sound radiation due to its short wavelength and high out-of-plane displacement. The interaction is shown in Figure 2 for an example of numerically obtained results.

The radiation of the plate waves into the adjacent fluid takes place at an angle $\Phi$ which results from the ratio of the wavelength in the fluid $\lambda_F$ to the wavelength of the plate wave $\lambda_P$. The angle can be calculated from the wavelengths or from the velocities of the waves by

$$\phi = \arccos\frac{\lambda_F}{\lambda_P} = \arccos\frac{c_F}{c_P}$$

where in $c_F$ is the speed of sound in the fluid and $c_P$ is the phase velocity of the bending wave in the plate. For the investigated glass fibre-reinforced polyamide 6 (GF/PA6) plate (thickness 2 mm) the dispersion curves—describing the relation between the wave speed and the excitation frequency—can be calculated analytically (see e.g. Rose 1999) by assuming isotropic material properties (Young's modulus E = 20 GPa, mass density ρ = 1800 kg/m³, Poisson's ratio ν = 0.3). Figure 3 shows the dispersion curves with the phase velocities and the group velocities of the $A_0$- and the $S_0$-modes.

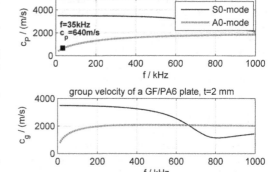

Figure 3. Dispersion diagram for a GF/PA6 plate.

The wavelength calculated from the phase velocity of the $A_0$-mode is shown in the upper chart in Figure 4 as a function of the excitation frequency. For the radiation in the air ($c_F$ = 340 m/s) the radiation angle is shown in the lower chart in Figure 4.

For example, at a frequency of 35 kHz the plate wave phase velocity is 640 m/s and therefore the wavelength of the bending wave is 18.2 mm. The wavelength in air is 9.7 mm which leads to a radiation angle of 58°. As Figure 4 shows, this angle increases with increasing frequency. The angle can be adjusted to specific applications by adapting the material and the thickness of the plate.

### 2.1 Transducer arrays for the generation of directional plate waves

The first step to obtain highly directional ultrasound radiation is the generation of directional bending waves in the plate. This can be achieved by using transducer arrays instead of a single transducer. By adaption of the actuator array configuration, the mainly excited or received plate wave mode can be precisely controlled. Furthermore, the directivity of the plate waves can also be

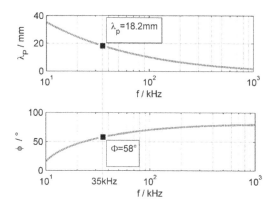

Figure 4. Wavelength of the $A_0$-mode in a 2 mm thick GF/PA6 plate and resulting angle for the radiation of sound waves in air.

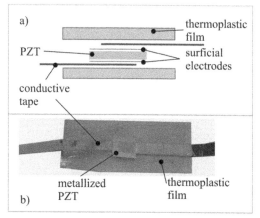

Figure 5. Prototypic TPM-configuration: a) built up, b) consolidated TPM (d31 mode).

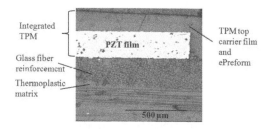

Figure 6. Micrograph of a material homogeneously integrated TPM.

adjusted. The simplest setup is a one-dimensional array on the surface of the plate. Depending on the size and the distance between the individual transducers and the time-delays in the electrical driving signal, the directional characteristics can be adjusted for example to amplify the wave generation in one direction. It is also possible to use an electronic beam steering (see e.g. Moulin et al. 2003) in order to vary the direction of the dominant wave generation.

For an one dimensional array the optimal delay time for an additive superposition in the array longitudinal direction can be analytically calculated using the transducer distances $\Delta l$ and the phase velocity $c_P$. This leads for the GF/PA6 plate (details see above) and a distance of 15 mm between the actuators to

$$\Delta t = \frac{\Delta l}{c_P} = \frac{15 \text{ mm}}{640 \text{ m/s}} = 23.4 \text{ } \mu s$$

The results using this time delay will be discussed in section 5.

## 3 MANUFACTURING PROCESS

For investigations regarding the manufacturing of the active TRTC, the application of TPM on the surface of a fibre reinforced semi finished plate (organic sheet) was necessary. The organic sheet consists of a glass fibre-reinforced polyamide 6 (type: 102-RG600(x)/47% Roving Glass—PA 6 Consolidated Composite Laminate produced by Bond-Laminates GmbH). The *ePreform* consists of a polyamide 6 (PA 6) film with a thickness of 100 µm. In regard to the built up of the TPM, commercially available monolithic wafers (type: PZT 5 A1) with a square area of 100 mm² and a thickness of 0.2 mm were used (see Figure 5b). The outer surfaces of the wafers were metallised by silver printed electrodes and contacted by conductive copper tapes. This lay-up was embedded into two PA 6 carrier films with a thickness of 0.1 mm (see Fig. 5a). The TPMs are polarised in thickness direction and work in $d_{31}$ mode, which means a principal deflection normal to the polarisation direction.

A major precondition for the embedding of the TPMs into the composite structures is the use of identical thermoplastic materials for the TPM carrier films and the matrix of the fibre-reinforced structure. During the consolidation, the module will be material homogeneously integrated into the composite structure. Compared to adhesively bonded modules the integration of TPM enables efficient coupling of the piezoceramic layer to the reinforcement as shown in Figure 6.

The conceptual manufacturing process for active TRTC, which bases on a press technology, starts with the first process step, the so called *ePreforming process*. It comprises the rollup and cutting of

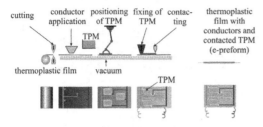

Figure 7. Schematic *ePreforming* process.

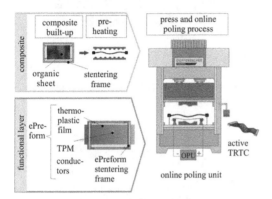

Figure 8. Schematic manufacturing process of active TRTC.

a thermoplastic film, the precise positioning and fixing of TPMs, the application of conductive paths and the accompanied electrical contacting (see Fig. 7).

In the developed manufacturing process (see Fig. 8) the *ePreform* is positioned in the pressing die and covered by preheated and melted organic sheet and subsequently the press closes. During the consolidation the TPMs are simultaneously poled.

In regard to the manufacturing of the *ePreform* an especially developed *ePreforming* unit was used. in order to assure the automated assembly of thermoplastic films with TPM and the necessary conductors. The transfer of TPM from storage to the predefined position is realised by a vacuum gripper whereas the fixation to the thermoplastic film can be realised by thermal stapling or ultrasonic welding. In this investigations the TPM were fixed by thermal stapling. The conductors are automatically rolled up from the wire coil, fixed by thermal welding points and cut at the end of the conductive paths. In these studies the conductors consists of tin coated copper wires with a diameter of 0.21 mm. Furthermore the TPMs, shown in Figure 5, are arranged to a linear pattern of six elements which have an offset of the piezoceramic wafers of 15 mm (see Fig. 9).

For the initial prototypic tests the *ePreform* was positioned and fixed at the organic sheet plate

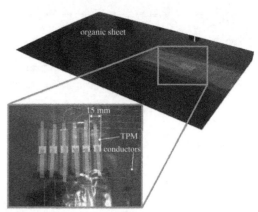

Figure 9. Organic sheet, *ePreform* with TPM pattern and conductors (thermally stapled).

manually using thermal resistant polyimide tape. The plate had dimensions of 1000 mm length, 600 mm width and 2 mm thickness and the TPM pattern was positioned 250 mm from the short side in the middle of the plate. In contrast to the introduced manufacturing process of such active parts, the consolidation of the investigated plate was performed by an autoclave process, because of its prototypic configuration. The main process parameters are a maximum temperature of 230 °C, a dwell time at maximum temperature of 2 min, a consolidation pressure of 5 bar, and a vacuum pressure of 20 mbar.

## 4 EXPERIMENTAL INVESTIGATIONS

The aim of the experimental studies was to characterise the generation of a directed acoustic wave using the integrated actuator array. Herein, two experimental techniques were utilised, i.e. laser Doppler vibrometry (LDV) and microphone measurements in order to identify the actuator-induced wave propagation of the investigated plate and the corresponding acoustic wave.

### 4.1 Laser Doppler vibrometry

To assess the mechanical wave induced by the actuators, a series of experiments using the scanning laser Doppler vibrometer (type PSV-400 produced by Polytec GmbH) were conducted. The application of a contactless measurement system guarantees that the mechanical properties of the investigated object are not distorted by the additional mass of typical vibration sensors.

Since the output signal of the laser scanning vibrometer is directly proportional to the velocity of the targeted surface along the laser beam direc-

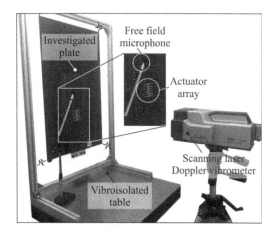

Figure 10. Experimental setup for the determination of propagation of induced mechanical and acoustical waves.

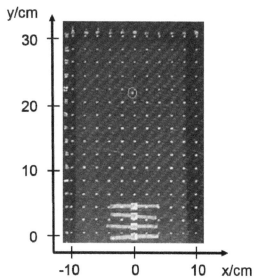

Figure 11. Coordinate system for the microphone measurement (origin at the first transducer of the array).

tion, the LDV has been positioned normal to the analysed plate. In order to assure to high reflection of the laser light and hence a high signal-to-noise ratio, large section on the plate surface was covered with a reflective spray paint. A regular spatial distribution of 55 x 97 discrete measuring points in approximately 2 mm distance has been selected to guarantee reliable capturing the wave spatial and time development. The investigated plate was hung vertically using a thin light rope (Fig. 10)

The actuators were excited with a typical burst signal as it can be used for different measurement tasks. Based on the plate dimensions and the array setup a centre frequency of 35 kHz was chosen. The signal consists of a four-periods-long sinusoidal signal which was windowed with a Hann window to limit the bandwidth. The triggered 2.5 ms long time series of out-of-plane velocities were recorded 20 times in every point and subsequently averaged to reduce the noise content in the acquired signals. The time resolution of the LDV was set to 4.883 μs to assure at least 5 samples per measured signal period.

### 4.2 Microphone measurements

Due to the excitation of the integrated actuator-array mechanical bending plate waves were generated. In regard to the achievement of a directed radiation of plate waves, the excitations of the transducers were delayed 24 μs relatively to the previous transducer. In regard to the directional characteristic of the acoustic wave and the defined radiation angle, the spatial distribution of sound pressure was recorded. For this purpose a ¼" free-field microphone (MK301, Microtech Gefell) was used and the sound pressure was determined at different heights (1 cm, 3 cm, 7 cm) over the plate. The raster of the discrete measuring points in each height was set to 2 cm. Figure 11 shows the planar coordinate system in the plate plane, whereupon the origin was set to the end of the first transducer element.

## 5 RESULTS

The recorded time series of out-of-plane velocities and sound pressures obtained using laser Doppler velocimetry technique and microphone found a basis for the subsequent analysis regarding the mechanical and acoustic wave propagation. The results of this analysis are presented in this section.

### 5.1 Propagation of structural mechanical waves

Mainly, the impact of the time delay between the actuator initialisation has been studied. The wave propagation front for cases with zero time delay and optimal time delay are presented in Figure 12. In this context, an optimal time delay of 24 μs was experimentally determined. The difference of 0.6 μs between theoretical and experimental delay times (see section 2) are effected by uncertainties and tolerances of the geometry and the material properties of the experimental setup.

It is clearly observable that while for the zero time delay the wave propagates equally in positive and negative y-direction, for the case with optimised

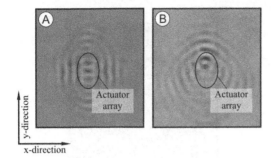
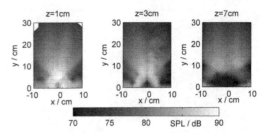

Figure 12. Out-of-plane velocities for different time delays. (A) for zero seconds. (B) for 24 milliseconds.

Figure 13. Measured maximum sound pressure level of the radiated ultrasound burst in different heights (z = [1, 3, 7] cm) over the plate.

time delay the wave propagates mainly in positive y-direction. Additionally the wave amplitude of the latter case was twice as much as in the first case. In order to generate an evidence, that such wave fronts cause a directed sound wave, the signals recorded using the microphone were analysed.

### 5.2 Propagation of acoustic waves

Figure 13 shows the maximum sound pressure of the received bursts for the scanned heights. Herein, the results for the plane 1 cm above the structure show an obvious maximum of about 90 dB at the end of the array. In a height of 7 cm the sound pressure level is about 82 dB to 84 dB and extended over a larger area. The measured data fits quite well to the expected radiation angle of approximately 60°.

On closer examination of the results the lateral emission of ultrasound (side lobes) is recognisable. This is caused by the side lobes of the plate waves discussed in the previous section. These side lobes can be suppressed or prevented by a modified array design.

## 6 CONCLUSIONS AND OUTLOOK

The integration of sensors and actuators like piezoceramic modules exhibits a high potential to functionalise composite materials. Especially a manufacturing technology ready for serial production can reduce the costs of active structures and thus lead to a wider range of possible applications.

The developed *ePreforming* technology gives the possibility to integrate sensors and actuators into composites using established manufacturing technologies like press processes. Furthermore, the process enables the integration of sensor-actuator-arrays with a reproducible positioning. Both possibilities could be validated in the performed manufacturing studies.

Experimental studies were done, showing the high potential of piezoelectric transducer arrays for an angular radiation of ultrasound waves. In future, this effect can be used to realise new material-integrated concepts for measuring distances or flow rates.

## ACKNOWLEDGEMENTS

The presented work is part of the research within the context of the Collaborative Research Centre/Transregio (SFB/TR) 39 PT-PIESA, subproject B04, and the Collaborative Research Centre (SFB) 639, subproject D3. The transducers and transducer arrays were kindly provided by the subprojects A05 and T03 of the SFB/TR39.

The authors are grateful to the Deutsche Forschungsgemeinschaft (DFG; eng. German Research Foundation) for the financial support of the SFB/TR39 and the SFB639.

## REFERENCES

Hufenbach W., Gude M. & Heber T. 2011. Embedding versus adhesive bonding of adapted piezoceramic modules for function-integrative thermoplastic composite structures, *Composites Science and Technology* 71, 1132–1137

Hufenbach W., Gude M. & Heber T. 2009. Development of novel piezoceramic modules for adaptive thermoplastic composite structures capable for series production, *Sensors and Actuators A*, 156, 22–27

Moharana S. & Bhalla S. 2014. Influence of adhesive bond layer on power and energy transduction efficiency of piezo-impedance transducer, *Journal of Intelligent Material Systems and Structures* 26: 247–259

Moulin E., Bourasseau N., Assaad J. & Delebarre C. 2003. Lamb-wave beam-steering for integrated health monitoring applications, *Proc. SPIE 5046, Nondestructive Evaluation and Health Monitoring of Aerospace Materials and Composites II*, 4 August; doi:10.1117/12.484101.

Kostka P, Holeczek K & Hufenbach W. 2015. A new methodology for the determination of material damping distribution based on tuning the interference of solid waves *Engineering Structures* 83, 1–6.

Rose, J.L. 1999. "Ultrasonic Waves in Solid Media," Cambridge University Press.

*Application of NDT/SHM techniques to cultural heritage*

# Comparative surface damage determination at a Jewish grave, found in front of the central building of Aristotle University of Thessaloniki, using two different mobile ultrasonic velocity equipments

B. Christaras
*School of Geology, Aristotle University of Thessaloniki, Thessaloniki, Greece*

A. Moropoulou
*School of Chemical Engineering, National Technical University of Athens, Athens, Greece*

M. Chatziangelou, L. Dimitraki & K. Devlioti
*School of Geology, Aristotle University of Thessaloniki, Thessaloniki, Greece*

ABSTRACT: The non-destructive methods are necessary in the investigation of the physical and mechanical properties of the materials in monuments. In this framework the ultrasonic velocities were used in situ for the elaboration and evaluation of the weathering on the surfaces of monuments. Additionally, the P-wave velocities were used for the estimation of the depth of weathered zone, as well as the depth of cracks at the surface of the monument. This estimation was performed on a tomb placed in the AUTH university campus between the building of Law and Economic Sciences and the Administration building, of the Aristotle University of Thessaloniki.

## 1 THE MONUMENT INVESTIGATED

The ultrasonic velocity method is an in situ, Non-Destructive technique which is used for the evaluation of surface conditions on historical monuments, without receiving testing materials. The results give the opportunity of recording and evaluating data, consisting major factors of quality and weathering of the surfaces. The implementation of this method took place in one of the most impressive monuments in the Aristotle University of Thessaloniki. This tomb dates in the first quarter of 4th century and belongs to a member of Israelite com-

Figure 2. The left side of the entrance of the tomb.

Figure 1. A general option of the monument.

Figure 3. The right side of the entrance of the tomb.

munity in Thessaloniki and was founded in university campus between the central administration building and the building of Law and Economic Sciences. It is about a well-preserved historical monument with favorable position for taking the ultrasonic measurements, (Marki E., 2001).

## 2 METHODOLOGY

For evaluating the weathering degree of the monument surfaces were used two ultrasonic devices, the SURFER ultrasonic detector (Figure 4) and the PUNDIT ultrasonic non-destructive digital tester (Figure 5), (Bruneau, C., Forrer, A. & Cuche A., 1995). The SURFER detector is applicable for measurement of ultrasound velocity and time of longitudinal ultrasound waves propagation in solid materials at surface sounding testing method for estimating the quality of the materials. This estimation is based on the correlation of ultrasound waves velocity in material to its physic and mechanical characteristics and physical statement. The SURFER gives the chance of estimating the depth of the crack (mm) on the surface by calculating the time of ultrasonic waves propagation. This device maintains a stable distance between the two ultrasonic probes (15 cm) which gives the ability of mapping and allocating the weathering of specific surfaces. On the other hand, the PUNDIT ultrasonic non-destructive digital tester is a good index characteristic for determining the physic-mechanical behavior and evaluating the weathering degree of the rocks. For the measurements a pair of small edge transducers was used for estimating the P-wave velocity for the determination of the depth. The tests held out by using the indirect method referring to arrangement of the transducers (transmitter and receiver) on the same surface of the stone. The transmitter was stable along a calibrated line and the receiver was reinstalling at every 5 cm. The final result was the travel time in correlation with the distance in each place. Each pair (travel time-distance) is displayed on a diagram. Given the fact that the most weathered surface demands longer travel time propagation, the change of slope in the plot indicates the quality of the material. The lower velocity in the surface zone is the result of the stone surface weathering. According to this plot, the thickness of the weathered surface layer is estimated as follows: [Vs: pulse velocity in the sound rock (Km/s), Vd: velocity in the damaged rock (Km/s), Xo: horizontal distance at which the change of slope occurs (mm), D: depth of weathering (mm)] (Bruneau, C., Forrer, A. & Cuche A., 1995):

$$D = \frac{Xo}{2} \sqrt{(V_S - Vd)/(V_S + Vd)} \qquad (1)$$

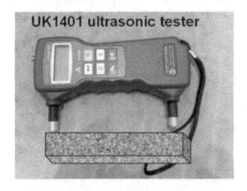

Figure 4. The SURFER ultrasonic detector.

Figure 5. The PUNDIT ultrasonic device.

## 3 APPLICATION OF THE METHOD

First of all, the SURFER device was used for mapping 4 specific surfaces of the monument (the back surface, the convex one upon the back surface, the dome and the entrance of the tomb), by taking a big amount of measurements, approximately 459, (Figure 6).

The application of this method was implemented for the allocation of the weathering along these surfaces, by the construction of 4 maps. Each map depicts the ultrasonic velocities Vp (m/sec) and the way they are distributed depending on the weathering conditions of each surface. The most weathered parts take lower values (longer travel time propagation), so in Figure 7 (back surface), it is obvious that the biggest part (in the right part and left half part of the tomb) is displayed with the lightest colors, assigning the most weathered areas, with Vp = 1200–2000 m/sec in the first most weathered part and Vp = 800–2000 m/sec in the second most weathered part of the back surface.

Figure 6. The use of SURFER device on the monument.

At the same way, along the convex part, upon the back surface appears clearly the weathered part, the values of which range in lower levels than those in the back surface, with Vp = 500–2000 m/sec (Figure 8).

In figure 9, the dome of the monument appears to have a wide weathered area which ranges between 1000–2000 m/sec.

Finally, at the entrance of the tomb, there are distinguished 2 weathered areas that appear to have the same values both in the right and in the left semicircular part. The values of the velocity range at a width of 1800–2000 m/sec, (Figure 10,11).

Except for the SURFER device, PUNDIT Ultrasonic non-destructive digital tester also was

Figure 7. The back surface of the monument (the maps were made with ARC GIS 2010 and the depiction of the maps was made upon the model, by the use of the AUTOCAD 2012. The legend is the same for all the maps, showing the ultrasound velocities Vp (m/sec)).

Figure 8. The convex part upon the back surface, (the maps were made with ARC GIS 2010 and the depiction of the maps was made upon the model, by the use of the AUTOCAD 2012). The legend is the same for all the maps showing, the ultrasound velocities Vp (m/sec)).

Figure 9. The dome of the monument, (the maps were made with ARC GIS 2010 and the depiction of the maps was made upon the model, by the use of the AUTOCAD 2012). The legend is the same for all the maps showing, the ultrasound velocities Vp (m/sec)).

Figure 10. The right part of the entrance, (the maps were made with ARC GIS 2010 and the depiction of the maps was made upon the model, by the use of the AUTOCAD 2012). The legend is the same for all the maps showing, the ultrasound velocities Vp (m/sec)).

used for determining the physic and mechanical characteristics and evaluating the weathering depth. For the purposes of the measurements, the indirect method was used. The final result was the travel time in correlation with the distance in each place. As it is known, the most weathered surface needs longer travel time propagation, so the change of slope in the plot indicates the quality of the material.

According to the equation (1.1), was estimated the depth of each weathered zone of the surfaces of the monument. In detail, the back surface was devised in two areas: in the first area the depth was calculated in D = 66 mm, while in the second one was D = 92.4 mm. At the same way, the depth of the weathering at the convex part upon the back surface is about D = 83 mm and at the dome as it is shown in the following plot (Figure 12), the travel time of P–waves is constant in depth higher than D = 35 mm.

Furthermore, the entrance of the tomb consists of the right side with weathering depth D = 79 mm and the left one with the same weathering depth.

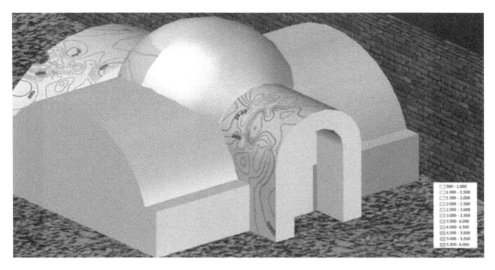

Figure 11. The left part of the entrance, (the maps were made with ARC GIS 2010 and the depiction of the maps was made upon the model, by the use of the AUTOCAD 2012). The legend is the same for all the maps showing, the ultrasound velocities Vp (m/sec)).

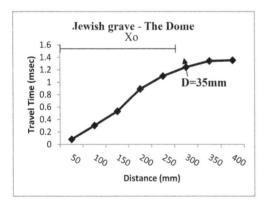

Figure 12. Estimation of the weathering depth on the surface of the dome in the Jewish grave.

Figure 13. The two cracks on the back surface.

As it was mentioned before, the SURFER device gives the opportunity of determining the depth of the cracks of the surfaces

On the back surface of the monument are distinguished two cracks (Figure 13), the depths of which were measured as follows: for the first crack the depth is D = 88 mm, while for the second one is D = 45 mm.

## 4 CONCLUSIONS

First of all, held out the mapping of specific surfaces on the monument by allocating the ultrasonic velocities (Vp m/sec), using the SURFER device detector. It is observed a wide range of values of the velocities between 500 and 6000 m/sec depending on the case.

On the other hand, the depth of the weathered zones of the tomb were estimated by using the PUNDIT ultrasonic non-destructive digital tester. The weathering depth has o minimum value, D = 35 mm and a maximum one, D = 92,4 mm with corresponding minimum velocities 800–1000 m/sec.

Finally, two cracks were located and measured on the back surface of the monument, with the SURFER device and the depth of which was determined in D = 88 mm and D = 45 mm.

## ACKNOWLEDGEMENT

The authors would like to thank the Rector of AUTH for his support to perform the present investigation, and the civil engineer Arampelos

Nikolas for his valuable help for the construction of the 3D model of monument in Autocad.

## REFERENCES

[1] Bruneau, C., Forrer, A. & Cuche A., 1995. Une méthode d'investigation non destructive des materieaux pierreux: les mésures à l' ultrason. Proceedings of the Congr. LCP '95: Preserv. and Restor. of Cultur. Heritage, Montreux, 187–195.

[2] Christaras B., 2009. P-wave velocity and quality of building materials. Proceedings of the 3rd IASME/WSEAS International Conference on GEOLOGY and SEISMOLOGY (GES'09).

[3] Marki E., 2001. Palaiochristian and Byzantine ancient monuments in the AUTH campus area. Thessalonikewn Polis 6.

*Emerging Technologies in Non-Destructive Testing VI – Aggelis et al. (Eds)*
*© 2016 Taylor & Francis Group, London, ISBN 978-1-138-02884-5*

# A new visual-based diagnostic protocol for cultural heritage exploiting the MPEG-7 standard

Anastasios Doulamis, Anastasia Kioussi & Antonia Moropoulou
*National Technical University of Athens, Zografou-Athens, Greece*

ABSTRACT: To reach a new sustainable cultural dimension, we should develop a strategy that harmonizes tools and methods of different criteria and sensory data able to track environmental changes of tangible assets. We exploit the recently developed integrated protocol. In this protocol, we accompany links to sensory data, like optical, thermal and hyper-spectral information so as to assist conservator's analysis and documentation. We need, however, to incorporate description of visual metadata to enable a content-based search. This is achieved using the interoperable schemas of MPEG-7 standard. We incorporate color descriptors, like global color histograms (scalable color), regional dominant color and color layout schemes, texture information (including homogeneous and edges information), and shape/contour descriptions. Selection of the MPEG-7 schema is to guarantee interoperability, universal accessible, portability and exchangeability of the visual content.

## 1 INTRODUCTION

Europe's cultural legacy is one of the world's most diverse. It is a beacon that draws millions every year to our churches and monuments, to our museums and libraries, as well as to concert halls and festivals. It is both the accumulation of past artistic achievements and the expression of continuing tradition and creativity. It is also a very dynamic trigger of economic activities and jobs, reinforcing social and territorial cohesion of European Union (Pickard, 2002).

Cultural Heritage (CH) is steering Europe's social as well as economic development (Vos and Meekes, 1999). The continent is the world's top tourist destination and the direct contribution of Travel & Tourism to GDP was 2.8% in 2011. In 2011, Europe accounted for over half of all international tourist arrivals worldwide and was the fastest-growing region, both in relative and absolute terms. European tourism represents the third largest economic sector in the EU. 1.8 million companies, including Small to Medium-sized Enterprises (SMEs), account for five percent of the EU Gross Domestic Product (GDP) and employment. This means that between 12 and 14 million Europeans find jobs tied in some way to tourism.

Although there are continuing concerns over the effects of the Eurozone crisis, the sector is forecast to increase its influence and contribution to the overall employment rate in the EU, as well as to economic development and social cohesion. Over the next ten years the industry is expected to grow by an average of 4% annually. European Cultural Heritage inevitably plays a central role therein, as it is estimated that cultural tourism accounts for around 40% of all European tourism.

However, European tangible content is under different influences of environmental impacts caused by natural and man-initiated actions. The southern half is located on the earthquake prone area, while the northern part is more endangered by floods and winds (Binda et al., 2000). To reach a new sustainable cultural dimension, we should develop a strategy that harmonizes tools and methods of different criteria and sensory data able to track environmental changes of tangible assets, including "natural" deterioration processes and human interventions (Moropoulou et al., 2012). Towards this direction, we have recently developed an integrated protocol for every cultural heritage building incorporating and supplying with information during its entire life-time such as, specialized building documentation, building materials and structural documentation, diagnosis techniques, materials and structure degradation mechanisms (decay & damage) documentation, environmental factors, etc (Moropoulou et al., 2012), (Koussi et al, 2011).

In these protocols, we accompany links to sensory data, like optical, thermal and hyper-spectral information so as to assist conservators analysis and documentation. Visual data are very important in assessing content and/or cultural heritage procedures (Farley, 2008). The aim of these protocols is to provide with new documentation procedures, advancing the data level in comparison to the current documentation methodologies, providing with criteria and indicators for risk assessment responding to advanced diagnostics and data management, standard documentation procedures applying same methods and tools, standardized outputs and clearly defined database entry without any further need for definition and application of a unified documentation terminology.

The proposed integrated documentation protocols serve the need to feed decision making support systems and can become a useful tool for conservation, management, strategic planning and promotion of cultural heritage. The system has a potential to support sustainable maintenance, preventive conservation and the rehabilitation of historic sites and monuments. It could assist in the application of newly developed strategies that would be designed to evaluate efficiency and be user friendly in their approach. It enabled screening, and monitoring over time, of progressive changes to the physical heritage as a result of recurring human interventions and environmental impacts.

We need, however, to incorporate description of visual metadata to enable a content-based search. This way, we can build mechanisms that retrieve not only relevant textual information but also information regarding visual description of the sensory data. Our description exploits interoperable schemas as the ones proposed in MPEG-7 standard. We incorporate color descriptors, like global color histograms (scalable color), regional dominant color and color layout schemes, texture information (including homogeneous and edges information), and shape/contour descriptions. 3D shape description schemes are also supported. Selection of the MPEG-7 schema is to guarantee interoperability, universal accessible, portability and exchangeability of the visual content.

This is the purpose of this paper. To describe a new standardized framework through which visual descriptors can afford the recognition stage of a CH assessment process. This framework is relied on visual description using *Non-Destructive Techniques (NDT)*.

This paper is organized s follows: Section 2 discuses the integrated Cultural Heritage protocol for tangible objects. The new protocol includes all relevant parameters like historic assessment, previous restorations, environmental factors, etc. In section 3, we present an interoperable model for visually representing the related information. The model is based on MPEG-7 standard to guarantee interoperability, universal accessibility, portability and exchangeability of the content. Rights description is also supported. Conclusions are drawn in Section 4.

## 2 AN INTERGRATED CULTURAL HERITAGE PROTOCOL FOR TANGIBLE OBJECTS

It was made clear during the state of the art survey that a currently applied documentation focuses mainly on the macroscale aspects of the building, such as the vulnerability of its structure, environmental dangers, human impact and natural hazards. However, information regarding the materials' and structural aspects, architectural, historic and aesthetic value, current state of preservation, conservation interventions is seldom considered as a whole entity. An approach that incorporates all these aspects can document more efficiently the complete history of the monument and allow safer conclusions regarding the best way of addressing the problems the monument faces.

Nevertheless, since the materials' and structure's state of conservation depends on their physicochemical and physicomechanical parameters, the documentation level should expand in the direction of revealing the specific active decay mechanism with an integrated decay study both in mesoscale [type of decay (morphology)] and microscale [kinetics of the phenomenon (decay rate) and thermodynamics of the phenomenon (susceptibility to decay)] level, through a diagnostic survey (following a standardized methodology for optimum results) (Kioussi, et al., 2011). This is illustrated in Fig. 1.

Such protocols should be built upon current experiences and existing knowledge and encompass all parameters and criteria, potential hazards and risk indicators regarding protection—management—decision making (Kioussi, et al., 2011).

Integrated documentation protocols for every building are knowledge based with a dynamic

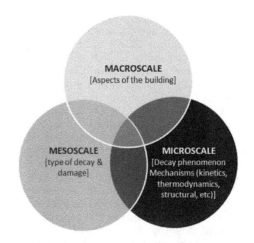

Figure 1. A schematic diagram of the three scales used for applying the new integrated model.

open structure, incorporating and supplying with information on the building, during its entire lifetime regarding documentation of: a) history of the monument, b) geographic localization, c) architecture, d) building materials e) structural analysis, f) state of conservation (diagnosis techniques, methods and results, materials and structure degradation mechanisms), g) risk indicators (environmental hazards and risk factors, human impact & socioeconomic parameters), h) intervention works & their assessment (Kioussi, 2005), (Chandakas, 2004), (Binda, et. al, 2000), and (Farley, 2008).

## 2.1 Protocols proposed structure

Due to the great variety of information included, a three—level classification of data is required to create integrated documentation protocols to organize and manage data (Fig. 2). These three levels reflect the overall data collection on the cultural heritage asset provided by the integrated protocols: (a) Basic data, (b) Knowledge data, (c) Decision Making data.

Each data category begins at the center of the concentric circular structure and ends at the periphery of the circles with increasing complexion of information.

The first level of the protocols delivers basic data on the asset reflecting the primary information about the building status. The second level comprises more extended data on a building's ensemble offering a more detailed knowledge base. At the third level, specialization of data is further increased. At this level, data are correlated within the aforementioned categories and within all levels, for the successful implementation of specific "tools" that allow prioritization of monuments' needs, supporting decision making procedures and allowing for overall management of the monument.

## 2.2 Protocols analysis

In the framework of increasing complexity with increasing level, main categories of data are further analyzed into subcategories which are further elaborated and enriched, to serve the needs of integrated documentation protocols. All categories and subcategories described in Table 1 are present at all three levels of the integrated protocols structure. The analysis of information, however, is dictated by the complexity of each level. The main categories and subcategories (Table 1) cover the whole spectrum of information needed to allow effective management and decision making.

Historical Documentation describes the historic framework of the building including historic description, related dates and historic characteristics sources and external references, relevant data, historical use or function of the heritage item as well as other subsequent uses, contextualization in relation to the structure of the item and in relation to its environment and time of construction, etc.

Geographic documentation delivers information of the spatial localization of the monument. This provides the exact spatial coordinates using Geographic Information Systems (GIS) that illustrate these coordinates in a map or by using geographical descriptors to identify spatial features of the item or its context, maps and relevant surveys, etc.

Architectural documentation refers to the typology of the building such as the building type, its morphology, the architectural style, the physical attributes of the building, internal or external decoration, etc. Also it registers information regarding characteristic local typologies.

Building Materials Characterization includes all the necessary information that identifies the structural and non-structural materials used in the building. It investigates the type of materials, the provenance of these materials, their physical and mechanical properties, their composition as well as processing parameters.

Structural analysis information focuses on the investigation of structure, on its structural state and its problems, and also detailed dimensions and references to its performance, that can be used in computer aided tools.

State of conservation relates to all data regarding the diagnosis of the monuments decay using visual observations, non-destructive techniques, analytical testing, GIS decay mapping, etc., both on structural and non structural elements of the monument. Information on mechanism of the decay as well as on the vulnerability diagnosis are also collected and stored to complete assessment of the overall state of conservation.

Another category refers to the correlation of the various risks and hazards threatening the monument as well as the impact of outer effects. Risk and hazards include accidental actions, human impact, social parameters, economic parameters and others.

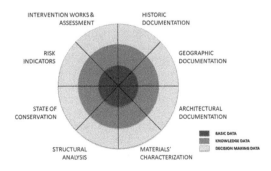

Figure 2. The integrated protocol units.

Table 1. Categories and Subcategories of integrated documentation protocols.

| Categories | Subcategories |
| --- | --- |
| Historic documentation | General information [e.g. Name, unique reference number, etc.] |
| | Stakeholders information |
| | Testimonies |
| | Bibliography |
| | Uses |
| | Legal status |
| Geographic documentation | General data [e.g. Location, GIS data, geographic coordinates, etc.] |
| | Topographic survey |
| | Registry |
| Architectural documentation | General typology |
| | Local historic typology |
| | Architectural plans and drawings |
| | Photogrammetric documentation |
| Building materials characterization | Type of materials [e.g. stone, mortar, ceramic, metal, etc.] |
| | Physical properties |
| | Composition [chemical, petrographic, mineralogical] |
| | Mechanical properties |
| | Processing [e.g. cost and availability, production, etc.] |
| Structural analysis | Investigation of structure [e.g. description of bearing structure, construction techniques, distribution of building materials, etc.] |
| | Structural state—problems [e.g. mapping of structural cracks] |
| State of conservation | Diagnosis of decay [e.g. using visual observations, non-destructive techniques, analytical testing, GIS decay mapping, etc.] |
| | Mechanism of decay |
| | Vulnerability diagnosis [e.g static analysis, etc.] |
| Outer effects impact/hazards/ risk indicators | Environmental factors |
| | Accidental actions |
| | Human impact |
| | Social parameters |
| | Economic parameters |
| Intervention documentation | General intervention data |
| | Intervention materials |
| | Special interventions data |
| | Interventions assessment |

The final category documents all previous interventions performed on the monument, such as: past intervention works, including detailed information about the intervention materials and the techniques used, documents related to these works and basic protection works, and data regarding interventions assessment.

The integrated protocols serve the need to feed suitable decision making support systems for cultural heritage protection. They provide with standard procedures for data collection and storage enabling the establishment of comparable databases on national and international level. The integrated protocols support the classification of buildings according to the priorities for inspection, diagnosis and intervention, leading to knowledge based decision making procedures. (Fig. 3) (Kioussi et al., 2013), (Kioussi, 2016).

## 3 INTEROPERABLE VISUAL REPRESENTATION

As we have stated above, a salient part of the integrated protocol is linking to visual data coming either from optical cameras or thermal devices or even from hyper-spectral sensors. These allow for a NDT analysis of the status and the conditions of a monument. The current protocols just link these data without adopting a common, interposable, accessible, portable and exchangeable framework to describe the visual features of this content. This is proposed in this paper relaying on MPEG-7 standard.

### 3.1 *The MPEG-7 standard*

MPEG-7 is an ISO initiative called "MPEG-7 Multimedia Description Language". The target of this activity was to issue an international MPEG-7 Standard, defining standardized descriptions and description systems that allow users to search, identify, filter, and browse audiovisual content (Manjunath et al., 2011), (Chang et al., 2001).

The ultimate goal and objective of MPEG-7 Visual Standard is to provide standardized descriptions of streamed or stored images or video—standardized header bits (visual low-level Descriptors) that help users or applications to identify, categorize or filter images or video. These low-level descriptors can be used to compare, filter, or browse image or video purely based on non-text visual descriptions of the content, or if required, in combination with common text-based queries. Because of their descriptive features, the challenge for developing such MPEG-7 Visual non-text descriptors is that they must be meaningful in the context of various applications. They will be used

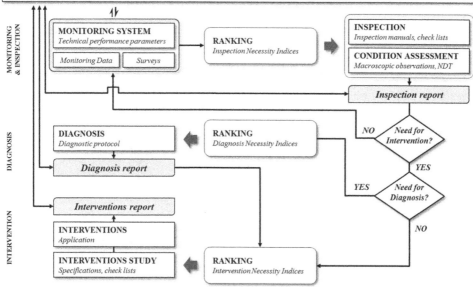

Figure 3. Decision making methodology.

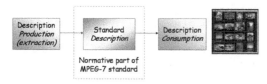

Figure 4. MPEG-7 scope.

differently for different user domains and different application environments (Sikora, 2001).

The Scope of MPEG-7 standard is depicted in Figure 4. A typical application scenario involves that MPEG-7 descriptors are extracted (produced) from the content. It is important to understand that for most visual descriptors, the MPEG-7 Standard only partly describes how to extract these features, to leave a maximum of flexibility to the various applications.

### 3.2 Description Definition Language

The Description Definition Language (DDL) forms a core part of the MPEG-7 standard. It provides the solid descriptive foundation by which users can create their own Description Schemes (DSs) and Descriptors (Ds). The DDL defines the syntactic rules to express, combine, extend and refine DSs and Ds (Hunter, 2001). The DDL is not a modeling language, such as Unified Modeling Language (UML), but a schema language to represent the results of

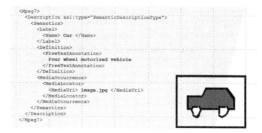

Figure 5. A simple example of the DDL language by depicting a red car.

modeling audiovisual data, (i.e., DSs and Ds) as a set of syntactic, structural, and value constraints to which valid MPEG-7 descriptors, description schemes, and descriptions must conform.

Figure 5 shows a simple example of the DDL in case of red car as depicting at the right side of this figure. We can see that DDL is a description language that allows an interoperable representation of different types of visual content.

### 3.3 Visual content representation

The colour, texture, shape and motion information are adopted as appropriate visual features by the MPEG-7, each of which is characterized by appropriate descriptors.

As far as colour information is concerned, *Scalable Colour Descriptor* (SCD) is defined as the colour distribution over the entire image. Another feature is the *Dominant Colour Descriptor* (DCD) aims at describing local and global spatial distribution of the colour. Finally, the *Colour Layout Descriptor* (CLD) is used to describe spatial distribution of colours in an arbitrary-shape image region.

For texture information, two types of descriptors are extracted. The first refers to the *homogenous texture descriptors*, while the second to the *non-homogeneous* ones (*edge histogram*). The homogeneous descriptor aims at representing directionality, coarseness, and regularity of patterns of visual content, while the second captures the histogram of the edges. The edge information is classified into five categories, vertical, horizontal, 45°, 135° and non-directional edge.

Shape descriptors are categorized into two main classes; *region-based shape descriptors* and *contour-based descriptors*. The former uses the entire shape region to extract meaningful information, which is useful when the objects have similar spatial properties. The latter exploits only the boundary (contour) information of the objects to describe entities of similar contour characteristics. Shape examples of MPEG-7 area shown in Figure 6.

Finally, motion descriptors are classified to the descriptor of the *motion activity* and to the one corresponding to the *camera motion*. The motion activity descriptor, for a video segment, represents the overall motion of the respective segment. This descriptor describes whether a scene is likely to be perceived by a viewer as being slow, fast paced, or action paced. High motion activities scenes are for example the athletic movies or the music video clips. On the contrary, scene of TV news are characterized by low-level actions. The motion intensity is measured by the descriptor, along with the standard deviation of the motion magnitude. Finally, the camera motion descriptor describes the movement of the camera or of a virtual point of the scene. An example of MPEG-7 colour representation is shown in Figure 7.

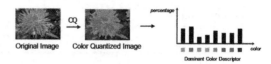

Figure 7. An example of colour representation.

### 3.3.1 Scalable descriptor organization

The extracted audio visual descriptors are organized in *a scalable way* to provide self adaptation of the media content in terms of context variations and users' preferences. This means that the descriptors are organized into *different levels of significance* (*content scales*). Initially, the significant descriptors are used, like for example the dominant colour of a region if we are talking about visual content. Then, the descriptors are decomposed into more detailed level of content resolution (e.g. colours associated with the dominant colour). In this way, *a hierarchical representation* is accomplished. More specifically, the audio visual representation is organized in a *pyramidal form* to allow the multimedia content to be *self-managed*. At the top the pyramid, the dominant content descriptors are used, while as the representation reaches the pyramid base, a more detailed content description is accomplished.

An example of the proposed tree-based hierarchical scalable organization of the extracted descriptors is depicted in Figure 8. In this figure, we have assumed three levels of hierarchy, the *dominant, major and detailed level*. In this example, we assume that the descriptors is analysed into three (3) other descriptors in the following semantic level.

### 3.4 Content rights description

This section aims at describing user's rights associated with the content. To model user's rights the MPEG-21standard will be followed. The standard provides two types of XML-like files able to define the user's rights and manages these rights

- *Rights Expression Language (REL)*. REL specifies a machine-readable language that can declare rights and permissions using the terms as defined in the Rights Data Dictionary (RDD).
- *Rights Data Dictionary (RDD)*: RDD specifies a dictionary of key terms required to describe users' rights.

#### 3.4.1 Rights expression language

REL can declare the rights and permissions using legal terms defined in the RDD. It provides mechanisms to support the use of digital resources in publishing, distributing, and consuming of the media content and honours the rights, conditions and fees specified for digital contents. It also specifies

Figure 6. Example of various shapes which can be indexed using the MPEG–7.

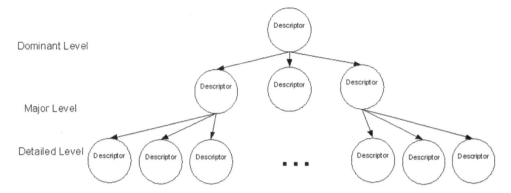

Figure 8. An example of scalable descriptor organization.

access and use controls for digital content in cases where financial exchange isn't part of the terms of use and supports exchange of sensitive or private digital content. REL also provides a flexible, interoperable mechanism to ensure that personal data is processed in accordance with individual rights. REL supports guaranteed end-to-end interoperability, consistency, and reliability between different systems and services. As such, it offers richness and extensibility in declaring rights, conditions, and obligations; ease and persistence in identifying and associating these with digital contents; and flexibility in supporting multiple usage/business models.

The core of MPEG-21 REL is the following four elements: principal, resource, right and condition (shown in Figure 9):

- Principal: identifies an entity such as the person, organisation, or device to whom rights are granted. Each principal identifies exactly one party. Typically, this information has an associated authentication mechanism by which the principal can prove its status.
- Right: specifies the activity or action that a principal can be granted to exercise against some resource. Example rights include play, print, issue, obtain, etc.
- Resource: identifies an object which the principal can be granted a right. It can be a digital work, a service or a piece of information that can be owned by a principal. A Uniform Resource Identifier (URI) can be used to identify a resource. For example, a video file that a principal may play.
- Condition: specifies one or more conditions that must be met before the right can be exercised. For example, a principal may need to pay a fee to exercise a right, a limit to the number of times, a time interval within which a right can be exercised, etc.

An example of the REL language is presented in Figure 10. In this example, Alice has got *a license to print an e-book 3 times. Intuitively, the subject*

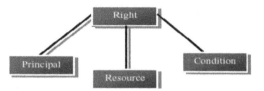

Figure 9. Core elements of the MPEG-21 REL.

Figure 10. An example of REL.

*of this example is "Alice", the object is "book", the right is "print" and the condition is "3 times".*

### 3.4.2 Rights data dictionary

The RDD comprises a set of clear, consistent, structured, integrated, and uniquely identified terms to support REL. The RDD specifies the structure and core terms and specifies how further terms may be defined under the governance of a registration authority. Use of the RDD will facilitate the accurate exchange and processing of information between interested parties involved in the administration of rights in, and use of the content. The RDD also supports the mapping and transformation of metadata *from the terminology of* one namespace (or authority) into that of another namespace (or authority) in an automated or partially automated way, with the *minimum ambiguity or loss of semantic integrity*.

The RDD is a *prescriptive dictionary*, in the sense that it defines a *single meaning for a term* represented by a particular RDD name (or headword), but *it's also inclusive* in that it recognizes the *prescription of other headwords and definitions by other authorities* and incorporates them through mappings. The RDD also supports the circumstance that *the same name may have different meanings under different authorities*. The RDD has audit provisions so that we can track additions, amendments, and deletions to terms and their attributes. The RDD recognizes legal definitions as, and only as, terms from other authorities that we can map into the RDD. Therefore, terms that directly authorized by the RDD neither define nor prescribe intellectual property rights or other legal entities. The RDD standard and the RDD are distinct things.

### 3.5 *MPEG-& towards a NDT analysis*

Colour/texture and edges representations of the visual, thermal and multi-spectral data contain a rich framework for a NDT analysis of the status of cultural heritage monuments. The descriptors can automatically extract features from the visual data that can provide to the conservators some clues about the effectiveness of the method.

In addition, such analysis allows for a quick and efficient content-based retrieval so that we can lay out to the conservators similar images and content so as to examine the status and the effect of NDT techniques.

### 4 CONCLUSIONS

In this paper, we extend the integrated protocol used for cultural heritage buildings using interoperable vision-based descriptors. This way, we allow for the users for a quick representation of visual content with respect to decay properties.

The adoption of the MPEG-7 standard allows for an interoperable description of the content making it accessible, portable and exchangeable among different providers and platforms. Thus, incorporating an ISO standard into the integrated protocol allows us to take advantage regarding visual description and assists the conservator and/or chemical engineer to take correct decisions about material decay and their properties.

### ACKNOWLLDGEMENT

This work was supported by the 3D ORO Project "Pervasive 3D Computer Vision for increasing the efficiency of 3D Digitalization" funded under PAVET Programme of the Greek Secretary of Research and Technology.

### REFERENCES

Binda, L, Saisi, A., Tiraboschi, C., 2000. Investigation Procedures for the diagnosis of the historic masonries. *Construction and Building Materials,* Vol. 14, pp. 99–233.

Chandakas, V., 2004 Criteria and Methodology for the quality control of restoration works – protection of monuments and historic buildings, PhD Dissertation, National Technical University of Athens, Greece.

Chang, S.-F., Sikora, T., Puri, A., 2001. Overview of MPEG-7 Standard. *IEEE Trans. Circuits Syst. Video Technol.*, Vol. 11, pp. 688–695.

Farley, J.M., 2008. "EFNDT Guidelines on the overall NDT quality system in Europe." *17th World Conference on Nondestructive Testing,* Shanghai, China.

Hunter, J, 2011. An Overview of the MPEG-7 Description Definition Language (DDL). *IEEE Trans. Circuits Syst. Video Technol.*, Vol. 11, pp. 765–772.

Kioussi, A. 2016. "Standard methodology of materials and conservation interventions documentation, emphasizing on European cultural heritage identity", PhD thesis in progress, scientific responsible Prof. A. Moropoulou, National Technical University of Athens, Greece.

Kioussi, A., Karoglou, M., Bakolas, A., Moropoulou, A., 2013 "Integrated Documentation Protocols enabling Decision Making in Cultural Heritage Protection", Journal of Cultural Heritage, 14 141–146.

Kioussi, A., Labropoulos, K., Karoglou, M., Moropoulou, A., Zarnic, R. 2011. "Recommendations and strategies for the establishment of a guideline for monument documentation harmonized with the existing European standards and codes", Journal Geoinformatics FCE CTU 6, 178–184.

Kioussi, A. 2005. Development of a data base for the documentation of monuments and historic buildings. Integration into a system of total quality control, diagnostic and conservation intervention study. MSc Dissertation, scientific responsible Prof. A. Moropoulou, National Technical University of Athens, Greece.

Manjunath, B.S., Salembier, P., Sikora, T, 2001. *Introduction to MPEG-7: Multimedia Content Description Standard.* New York: Wiley.

Moropoulou, A, Kioussi, A., Karoglou, M., Bakolas, A., Georgousopoulos, G., Chronopoulos, M. 2012. Innovative protocols for integrated diagnostics on historic materials and structures, in: V. Radonjanin, K. Crews (Eds), *Proceedings of Structural Faults + Repair-2012, Edinburgh,* 3rd–5th July 2012.

Rickard, R, 2002. *European Cultural Heritage.* Google books.

Sikora, T., 2001.The MPEG-7 Visual Standard for Content Description—An Overview. *IEEE Trans. on Circuits and Systems for Video Technology,* Vol. 11, No. 6, pp. 696–702.

Vos, W., Meekes, H., 1999. Trends in European cultural landscape development: perspectives for a sustainable future. *Landscape and urban planning,* Elsevier press, Vol. 46, Nos. 1–3, pp. 3–14.

*Emerging Technologies in Non-Destructive Testing VI – Aggelis et al. (Eds)*
*© 2016 Taylor & Francis Group, London, ISBN 978-1-138-02884-5*

# Comprehensive energy diagnosis methodology integrating non destructive testing

M.A. García-Fuentes & J.L. Hernández
*CARTIF., Technology Centre, Energy Division, Valladolid, Spain*

A. Meiss
*G.I.E., Arquitectura y Energía., University of Valladolid, Valladolid, Spain*

C. Colla
*DICAM Department, School of Engineering and Architecture, University of Bologna, Bologna, Italy*

ABSTRACT: Historic buildings are responsible for a large amount of energy consumption and $CO_2$ emissions, besides presenting comfort condition problems, due to, among others, pathologies derived of their constructive conditions. There is a strong need to bridge the gap between conservation of historic buildings and climate protection, through a comprehensive method for diagnosis. This paper deals with this diagnostic method applied to one of the case studies of the FP7 EU funded project "Efficient Energy for EU Cultural Heritage" (3ENCULT) project where the specific techniques integrating the diagnosis procedure are depicted, concluding how the development of this scientific based diagnosis and evaluation tool can benefit the whole process.

The methodology proposed covers three main phases to drive these interventions: comprehensive diagnosis, evaluation of retrofitting strategies and assessment of the interventions after their implementation, based on a set of targets in the fields of historical value conservation, energy balance and comfort conditions improvement. The diagnostic phase, through a scientific procedure, is based on the utilization of building energy performance simulation tools combined with non-destructive tests, such as the IR thermography, blower door test, and wired and wireless monitoring networks used to set up the baseline of the current conditions.

## 1 INTRODUCTION

European initiatives establish the objective of reducing the energy demand and $CO_2$ emissions due to buildings. However, in the set of the EU-27, about 40% of the housing stock was built before the 60 s and another 40% between 1961 and 1990 (Buildings Performance Institute Europe 2011). Therefore, a large portion of the building stock is old constructions belonging to the historical heritage and its related constraints. Some of them present a high level of energy consumption and low comfort conditions. Moreover, the aesthetic and the heritage must be conserved.

Historic building and cultural heritage are particular cases where the conservation and preservation issues restrict the possibility of actuation. Even more, as mentioned before, these buildings are not designed under energy efficiency patterns. Hence, the analysis of the energy performance is pivotal through available diagnosis tools. Nevertheless, the complexity should be taken into consideration.

Within this context, the current paper is based on the results from the research European Project 3ENCULT, which is funded by the 7th Framework Programme from the European Commission, and aims to establish a methodology for improving the energy efficiency and comfort conditions in European historical buildings. Thus, through the discussion of energy efficiency, comfort and conservation of historical value issues, this project has worked on a procedure for diagnosis and evaluation of the interventions based on these three issues. However, several times, the interventions are strictly needed for the building conservation from the structural point of view, such as the moisture problems in the historical wood structures or the maintenance of the environmental conditions in the spaces with unique paintings or architectural elements.

Having the aforementioned issues in mind, the present paper presents a novel methodology based on Non-Destructive Tests in order to determine the building pathologies to support the intervention decision. This methodology is demonstrated in one of the eight case studies of the project, which is described analysing its main problems, while the techniques integrating the diagnosis procedure (simulation, non-destructive testing and monitoring) are explained.

## 2 DIAGNOSIS METHODOLOGY

Aiming at evaluating the improvement potential and the benefits related to carry out interventions in these buildings, the proposed methodology establishes an evaluation and validation framework based on defining a set of strategies to reduce the energy consumption and improve the comfort conditions, while ensuring the protection of the historical features related to their boundary conditions.

This multi-objective evaluation is only possible through a consistent evaluation method based on assessing the current conditions of the building and allowing the quantification of the benefits related to the specific Energy Conservation Measures (henceforth ECMs) that can be applied given the constraints related to the historical value conservation.

### 2.1 Evaluation and assessment framework

Through the combination of a set of evaluation techniques, it is possible to generate an accurate baseline that allows virtually implementing specific ECMs and calculate their related benefits in terms of energy savings, $CO_2$ emissions reduction, Life Cycle Analysis, economic evaluation, comfort conditions improvement and historical value protection.

In the specific field of energy assessment, this methodology proposes the combination of Energy Performance Simulation (EPS) engines, monitoring and measurement techniques and other non-destructive testing methods. This combination allows performing an Energy model representing the current conditions of the building, and easily evaluating the combination of ECMs that can be implemented under a retrofitting scenario. EPS models usually introduce many uncertainties due to the considerations needed to perform the model, and in the case of historical buildings this is even more remarkable, given the vagueness of the information available to perform them.

Therefore, an important phase of information collection is needed before performing the model, in order to insert more accurate data that allows having a more precise model. This information collection campaign has to be based on different techniques, as depicted in the following sections, where Non-destructive testing methods and monitoring and measurement strategies are essential.

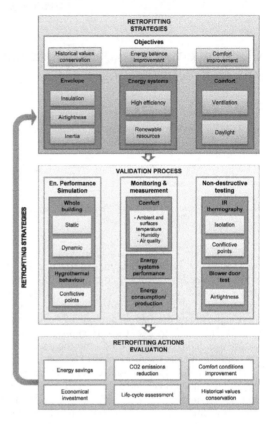

Figure 1. Diagram of the proposed methodology.

The validation process of these proposals combines the use of three diagnosis resources (Fig. 1): the energy performance simulation, monitoring and measurement processes and non-destructive testing in the building. Finally, it is proposed a benchmarking system based on energy savings, $CO_2$ emissions reduction, comfort conditions improvement, economical investment, life-cycle assessment and, one of the most important due to the character of the project: the conservation of the historical value of the building.

The main objective of this methodology is to establish a quantifiable ground, based on scientific foundations in order to assess the impact that some energy efficient interventions may have in this type of building.

### 2.2 Investigation path and knowledge level

Assessment of an existing building and especially of a historical construction–that is the evaluation of its matter, its organism and its performance in qualitative and quantitative terms—is recognized to be a complex task, requiring a multi-step approach and multi-disciplinary competences in a continuous and progressively deepening dialogue between expert skills from various scientific sectors and from different viewpoints. The objectives of the studies and analyses are to ascertain the parameters characterizing the building components and the site where the construction is placed, including its environment both in terms of possible natural hazard risk and microclimate. Tremendously important is also the use destination of the object, be that at present, in the past and in the foreseeable future.

Traditionally, a building would undergo structural assessment at the time when modification to its load-bearing organism would be foreseen but it is only in the last decades that the technical legislation has imposed assessment of the whole structure also for preventive knowledge gathering and for mitigation of inherent deficiencies highlighted by the assessment process.

More recently, due to climate change and energy saving issues, energy performance evaluation of the building together with comfort and life quality for its users have become popular topics in the perspective of a sustainable live-style.

One of the major contributions of the 3ENCULT research project has been to highlight the necessity to unite the structural with the energetic + comfort viewpoints in a combined and comprehensive diagnostic methodology that would share between the two aspects some of the information collected about the building, some of the expertise required and even some of the investigation techniques.

Bearing clear in mind what the destination of the building is or will be, that is its use function and the desires and necessities of its inhabitants, should drive the correct planning and design of the intervention choices, being those either a light maintenance action of upgrading or a major structural/energetic rehabilitation.

Each historic building is a unique case in which its century-long history of construction, use, major events undergone, and specificities of the site make up what could appear as an entangled collection of data that the assessment has to unravel and transform in numerical values, often without the aid of archive documentation. That explains why the structural analysis of historical buildings can be such a demanding task, requiring complex investigation phases to obtain an adequate pre-intervention level of knowledge. Recent Italian construction legislative technical documents (Italian Cultural Ministry, 2011) have set down what can be considered as the most advanced international guidelines for structural assessment, risk reduction and preservation purposes, specifically developed for cultural heritage buildings but equally valid for any existing construction. Among the main steps of this systematic investigative methodology to be carried out—already successfully tested on many occasions of emergency measures, such as after seismic disasters—are: documentary research including tracing the construction's history, developments and their extension; multilevel visual inspections; identification of soil and foundations type, identification of the structural skeleton and/or the various structures forming a complex aggregate building; geometrical surveys; materials and damage surveys (including crack pattern, out of plane deformations of walls and ceilings, tilting, etc.); testing on site and in the laboratory on material samples for direct determination of chemical, physical, mechanical parameters. Necessarily, such testing will be partially destructive and semi-destructive. This is still unavoidable for obtaining reliable and certain quantitative values, therefore has to be kept to a minimum. Nonetheless, a number of Non Destructive Testing[1] (NDT) high-resolution, imaging techniques, exploiting different types of signals and testing procedures, are nowadays available for indirect measurements of characteristics of interest over large portions of structural elements (Colla, 2011). Among the applicable NDT techniques are Infrared thermography (Giuliani e Colla, 2011), Ground-Penetrating Radar (GPR), sonic and ultrasonic tests, sonic tomography and GPR tomography. NDT provides a valuable increase in building knowledge and, without causing damage, is a great aid in reducing the number and in defining the most appropriate positions for later sampling or destructive testing.

In addition to considering the construction principles and practices current at the time and at the location of building's erection, to evaluating the materials' quality and preservation state and to locating points of weakness, discontinuities in the structure and types of connections, it is part of the assessment to verify the integrity and structural homogeneity of the masonry building as a whole, ascertaining also damage mechanisms present and responsible for the present configuration of the building and its likely future developments.

Upon completion of the analysis phases, based on quality and quantity of information collected, one of three knowledge levels (limited, adequate, accurate) can be established. Safety factors (confidence factors) are so distinguished and used in the numerical calculations, the next step of the assessment.

---

1. Investigations not altering the material or object of study. Provided information may include geometry and geometric variations of structural elements, construction details, constituent materials and their condition, presence of defects or heterogeneities and their extent, interventions and modifications as well as indirect measure of a number of parameters.

Drawing from experience with the above-summarised procedure, the project team from University of Bologna has developed and set a similar investigation path for knowledge extraction from the energy and indoor comfort microclimate viewpoints. The researches and testing aimed at these scopes would need to take place at the same time and together with those for the structural assessment, partially sharing the procedure and methodology, thus partially overlapping but also widening the assessment scopes (Colla et al., 2015). The sources of information for the energetic investigations may include different disciplines, investigating and taking into account the past living styles and societal-, work-, supplying-organisations, the historical lighting-, ventilation—and heating-systems and their technology, the site climate. From the findings of this combined assessment method applied on different case studies, it was observed a reciprocal benefit of the two perspectives, the structural and energetic one. It was also noticed that historic buildings may often present hidden or forgotten resources, valuable for a modern and sustainable retrofitting intervention. Further, it was noted that, as partially expected, weak points of the structural system can be used as resources for the energetic retrofitting, and vice-versa. For example, the location in a wall thickness of an abandoned chimney flue can be considered a weak point in the wall stiffness but also an already available space for the planned passage of a new ventilation, heating or electrical pipe/cables without need for damaging the original matter of the historic building hence with advantages from the sustainable preservation of historic constructions.

### 2.3 NDT tools for energy efficiency diagnosis

In this section, some NDT investigation tools useful for energy-related diagnose and for better understanding of the building energy performance are presented. Although non-invasive and flexible, these testing techniques still require a not-neglectable experience in equipment set-up and data collection procedures therefore specialized personnel is required both in these phases and at data analysis and interpretation stages. Some of the techniques listed in the following find in energy-related issues new application fields, others are new to the assessment of historic buildings (Paci et al., 2012). None of these testing techniques is sufficient in itself thus they should possibly be applied in a combined way after a thorough visual inspection of the building.

- Infrared Thermography (IRT) allows a contactless measurement of radiant heat flow from the surface of elements producing 2-dimensional temperature distribution maps over an area, called thermograms, recorded via a thermocamera. IR cameras are nowadays light and portable tools with resonably high resolution. IR investigation depth is very limited and radiant heat flow is influenced by material characteristics (such as smoothness/roughness of surface, presence of humidity, surface and substrate materials, discontinuities...) and by the surrounding environment. Heat will accumulate in case of a local increase in material thermal resistance, thus permitting to record in the thermogram a local temperature increase allowing to detect relevant characteristics (masonry pattern under rendering, dimension of units and mortar joints, timber or metal inclusions in masonry and concrete...) and defects (Colla et al., 2011). Thermal bridges in the structure or in non-structural parts are also detectable via IR (Franzen et al., 2012).

- The measurement of <u>heat flow</u> through building elements (ceilings, walls, doors...) permits the evaluation of their thermal transmittance, the so-called U-value, strictly related with the overall energy performance of a building. Software available for U-value calculations were developed for modern construction materials and building types, rather than for traditional buildings with thick walls and building materials of high thermal resistance (resulting in a low U-value). Therefore, in-situ measures are to be preferred. Instrument setup includes surface temperature sensors and air temperature sensors, both for the inside and outside of the element, in addition to one heat-flux sensor. A preliminary investigation of the measurement positions by other testing techniques (i.e. GPR, IR) would identify the real inner construction and stratigraphy and should exclude defective areas, heterogeneities, ventilated cavities which, affecting the heat flow, would make the measurement not representative.

- <u>Sonics</u> are acoustic signal propagation tests of low frequency (< 20 kHz) particularly suitable for thick elements and lossy materials typical of historic constructions. Equipment consists of an instrumented hammer to generate the signals and an accelerometer for receiving it after propagation in the medium; a central unit for signal conditioning, visualization and storing. Signal travel time is function of material characteristics and strength. Sonic tests can be used for material evaluation, thickness measurements, flaw and crack detection. Plaster removal over an area is necessary if present on wall.

- <u>Ground Penetrating Radar</u> (GPR) exploits electromagnetic signals in the frequency from 600 MHz to 2.6 GHz for high-resolution and in-depth profiling. The movement of an antenna along survey lines permits investigating the stratigraphy of ceilings and walls and their thickness, the arrangement of beams, the presence of voids, stone and metal inclusions, as well as detecting other defects and details. Through GPR, moisture and salt distribution in walls and foundation is also identifiable (Gabrielli and Colla, 2014).

- Blower door test: by means of a dedicated large fan mounted generally on a window or door frame, it is possible to measure the air-tightness of a building after having imposed a determined pressure difference between inside and outside: evaluating the rate of air exchange permits discovering air loss of the building envelope. Although low values of this parameter are preferable because it has an impact on indoor temperature and humidity and thus on energy consumption (air movement is also highly relevant for comfort), it is important not to stop completely the air infiltrations because a certain rate of ventilation maintains building components in healthier conditions and improves indoor air quality by diluting exhausted air and pollutants in it. Hence, the level of air tightness is also an indication of air quality inside the building (IEQ).

## 3 CASE STUDY: SCHOOL OF INDUSTRIAL ENGINEERING OF BEJAR

One of the case studies of the Project 3ENCULT, where the assessment methodology and some of the aforementioned NDT are applied, is the Industrial Engineering School located in Béjar (Spain).

It is not a catalogued building as cultural heritage, but its historical and architectural values lie on its formal character and the social and economic impact that its construction had in the region since it is the first building of the University of Salamanca in the city, adding value to an institution previously established (19th century). This fact reinforced the textile industry developed in the area of Béjar.

The main detected problems, however, are due to low comfort conditions in the indoor spaces (both thermal and lighting) and high energy consumption because of a design little concerned about passive conditioning strategies, as it is usual on all buildings designed before than the energy crisis of 1973.

### 3.1 *Energy performance simulation*

Energy storage in the building envelope (thermal inertia) makes the existence of some critical thresholds that recommend the use of dynamic

tools. In heating mode, these are when outdoor conditions during some part of the day are above the indoor set point. However, this is not the case with the climate in Béjar, where the winter is very cold, therefore the static and dynamic simulations should give reasonable similar results. However, in the disaggregation of heat losses and gains, dynamic tools give more precise results. Therefore, a TRNSYS model was carried out and validated with real monitoring data after including the information collected through NDT for thermal bridges and airtightness level evaluation.

Comparing real consumption data ($\approx 70$ kWh/m$^2$a) and simulated energy demand ($\approx 105$ kWh/m$^2$a), it is verified that the covered range is not sufficient to keep the indoor parameters in comfort conditions during the heating period. By simulating the existing heating systems, the energy consumption is similar to the simulated demand, and under this premise, temperature conditions are out of comfort thresholds approximately over 30% of the heating period.

### 3.2 *Monitoring and measurement*

Monitoring and measurement are well-established techniques to value the energy performance of buildings. Both are focused on twofold objectives: creating a baseline for the assessment through Key Performance Indicators (KPI) as energy use, comfort conditions, etc., and deploying control strategies with the aim of improving the energy usage or avoiding malfunctioning of the facilities.

However, usually, monitoring and measurement are intrusive techniques inside buildings because the most common protocols are wired. This implies the need of the integration of cables into the building structure, which could be damaged, or, at least, it could affect to the aesthetic features of historical buildings and surfaces. Hence, a novel and non-intrusive wireless sensor network has been deployed in the demonstrator. These devices are ZigBee (ZigBee, 2006) based, which *"is a standards-based wireless technology designed to address the needs of low-cost, low-power wireless sensor and control, cost-effective and energy-efficient mesh network"* (ZigBee, 2006). Taking into account the features of the technology, it perfectly covers the requirements in these peculiar buildings where the constraints are too restrictive in terms of conservation and preservation of the cultural heritage.

With the purpose of covering the requirements in the 3ENCULT project, a novel ZigBee sensor, hereafter 3ENCULT WSN, was developed by DEI, Bologna University. This system is able to sense the building climate parameters (temperature, lighting, humidity and $CO_2$, among others), energy consumption through extension cards and collect the

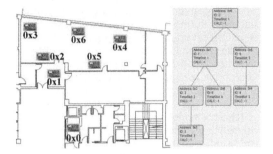

Figure 2. Deployed sensor network in the case study.

measurements in a central server unit (Balsamo, 2013). 3ENCULT WSN is designed under the plug-and-play pattern and its small size reduces the aesthetical impact in cultural heritage. Moreover, the system is embedded, working in a long operative time (up to two years), with reduced maintenance and remote configuration, management and update possibilities (Balsamo, 2013).

The sensor network has been deployed in the Engineering School of Béjar, in the library, whose scheme is illustrated in Fig. 2. A seven nodes network is established where the 0 × 0 device is the central unit and the remaining sensors follow a tree structure for the collection of data.

### 3.3 Blower door test

Under certain environmental conditions, air will penetrate uncontrollably through points where construction elements converge. The randomness of the environmental variables (the pressure gradient) involves a nonlinear phenomenon in which an exact calculation is not possible but can only be simplified and approximated. The only term whose value can be reliably obtained is the air tightness of the enclosure via the fan pressurisation technique.

The pressurization test, commonly called the *Blower Door Test*, assesses the air tightness of the building, which is expressed via a series of objective parameters obtained from the relationship between the flow rate that traverses the enclosure and different pressure gradient intervals.

Before starting the test, it is useful to detect the infiltration foci to document and facilitate subsequent sealing. The process consists of creating an increased depressurisation during a period of time to allow for flow stabilisation and the thermal action of the infiltrated air over the entry points. The simplest method for locating air leakage is the use of infrared thermography.

The tests results show how the envelope presents a very low level of airtightness ($q_{50} \approx 10.0$ m³/m²·h), due to three main points of entrance:

- in the external walls, the different rigidity of the structural elements made of reinforced concrete (pillars and slabs) and the brick walls without anchoring elements, caused longitudinal cracking in the joints;
- the joints of the windows and blinds boxes with the walls are not sealed: in this type of historic buildings this is due to the degradation of the sealing material but, as windows were recently replaced in this building, it makes sense to think that it is due to a construction deficiency;
- there is circulating air coming from adjacent locals through the camera above the ceiling. This problem becomes important when there are neighbouring heated and non-heated rooms.

Finally, in order to calculate the annual rate of infiltration it is necessary to develop a model of the driving forces (wind, indoor-outdoor temperature differences, and mechanical ventilation systems) and their interaction with the building. Thus, the pressures can be calculated separately and then combined by quadratic superposition.

## 4 CONCLUSIONS

Some conclusions can be learned from the use of the proposed methodology. Regarding the simulation tools, its use is essential in order to evaluate the potential ECMs that can be implemented in a consistent manner. In this case, the use of TRNSYS gives a more precise description of the disaggregated energy demands, and the model, validated with real data and adjusted through inserting more accurate data resulting from the information collection phase through methods as NDT, can be used as baseline for evaluation of the proposed interventions.

The approach of a scientific based evaluation tool is very useful in order to quantify the improvements, where the establishment of the historical value is the most subjective and complex one. Through this tool, different strategies can be compared, and can be shown the impact of specific measures to find the best balanced scenario in terms of historic building protection and energy consumption reduction, while improving the indoor comfort conditions.

### ACKNOWLEDGEMENTS

This research work was partially financed by the European Commission through the 7th Framework Programme, under the research project 3ENCULT, Efficient Energy for EU Cultural Heritage, Grant Agreement n°260162.

## REFERENCES

Balsamo D., Paci G., Benini L., Brunelli D. (2013), "Long term, low cost, passive environmental monitoring of heritage buildings for energy efficiency retrofitting", in: Workshop on Environmental Energy and Structural Monitoring Systems (EESMS), IEEE, Trento, 11–12th Sept 2013. ISBN 978-1-4799-0628-4, pp. 1–6.

Colla, C., (2011), "Comparative testing for improved diagnosis of historic structures", In Cultural Heritage Preservation, Proc. of European Workshop on Cultural Heritage Preservation EWCHP-2011, Berlin, September 26–28, Fraunhofer IRB Verlag, ISBN 978-3-8167-8560, pp. 140–147.

Colla, C., Gabrielli, E., Largo, A., Angiuli, R., (2011), "Experimental studies by combined NDT of capillary rise monitoring in masonry specimens", In Cultural Heritage Preservation, Proc. of European Workshop on Cultural Heritage Preservation EWCHP-2011, Berlin, Germany, September 26–28, ed. M. Krüger, Fraunhofer IRB Verlag, ISBN 978-3-8167-8560, pp. 131–139.

Colla, C., Gabrielli, E., Giuliani, M., Paci, G. (2015), "The 3ENCULT project—Case study 3, Palazzina della Viola, Bologna, Italy", In: Energy Efficiency Solutions for Historic Buildings—A handbook, Edited by A. Troi, Z. Bastian, Birkhäuser, Basel, Switzerland, pp. 250–263, ISBN 978-3-03821-646-9 (e-book: ISBN PDF 978-3-03821-650-6).

Franzen C., Baldracchi P., Colla C., Esposito E., Gaigg G., Pfluger R., Troi A. (2011), "Assessment of historic structures by IRT", in: Cultural Heritage Preservation, EWCHP 2011, Fraunhofer IRB Verlag, pp. 101–109.

Gabrielli E., Colla, C., (2014), "Investigation of damp and salt distribution in outdoors full-scale masonry wall via wireless monitoring and radar testing", Key Engineering Materials 624, 155–162, DOI: 10.4028/www.scientific.net/KEM.624.155.

García-Fuentes, M., Hernández, J., de Torre, C., García-Gil, D., Meiss, A. (2013). Energy efficiency and comfort conditions improvement on historic buildings: a methodology approach for diagnosis and interventions. Conference paper. Proceedings of Central European Symposium on Building Physics (CESBP) 2013. Vienna, Austria.

Giuliani, M., Colla, C., (2011), "Thermographic analysis: examples of application developments of the IR technique for the structural diagnose of historical buildings", Proc. of AIMETA 2011—XX Congr. Associazione Italiana di Meccanica Teorica e Applicata", Bologna, Italy, 12–15 Sept, Idn 364, 10 pp., ISBN 978-88-906340-1-7.

Italian Cultural Ministry (2011), "Linee Guida per la valutazione e riduzione del rischio sismico del patrimonio culturale—allineamento alle nuove Norme tecniche per le costruzioni".

Paci, G., Gabrielli, E., Colla, C., (2012), "On-site dynamic wireless sensor monitoring in the historic building of Palazzina della Viola, Bologna, Italy", In Cultural Heritage Preservation, Proc. 2nd European Workshop on Cultural Heritage Preservation, EWCHP 2012, Kjeller, Norway, September 24th -26th, ed. E. Dahlin, NILU—Norwegian Institute for Air Research, pp. 74–81, ISBN 978-82-425-2525-3.

ZigBee Alliance, Zigbee specification, http://www.zigbee.org, ZigBee Alliance official Web site (2006).

*Emerging Technologies in Non-Destructive Testing VI – Aggelis et al. (Eds)*
*© 2016 Taylor & Francis Group, London, ISBN 978-1-138-02884-5*

# Non-contact contemporary techniques for the geometric recording of Cultural Heritage

A. Georgopoulos
*Laboratory of Photogrammetry, School of Rural and Surveying Engineering,*
*National Technical University of Athens, Athens, Greece*

ABSTRACT: Documenting Cultural Heritage is a necessity nowadays and an obligation of civilized countries. Recently, with the development of ICT techniques, this task has tremendously progressed. Cultural Heritage documentation involves the cooperation of Users, e.g. archaeologists, architects and Providers, e.g. ICT specialists, surveyors, photogrammetrists. This progress is met with scepticism because of lack of information and "fear" of the unknown from the part of the Users and lack of the necessary sensitivity from the part of the Providers. Contemporary techniques involve mostly non-contact methods. It is of utmost importance to familiarize all experts with these contemporary techniques, by identifying and understanding the Users' needs combined to available tools for the Providers for serving Restoration, Preservation and Management actions. In this paper reference is made to the latest developments in non-contact data acquisition and processing methods and contemporary documentation products. Examples of good practices of Users/Providers cooperation are given and a future outlook is attempted while the role of CIPA-Heritage Documentation in this effort is also identified.

## 1 INTRODUCTION

Monuments, including immovable structures of any kind and movable artifacts, are undeniable documents of world history. Their thorough study is an obligation of our era to mankind's past and future. Respect towards cultural heritage has its roots already in the era of the Renaissance. Over the recent decades, international bodies and agencies have passed resolutions concerning the obligation for protection, conservation and restoration of monuments. The Athens Convention (1931), the Hague Agreement (1954), the Venice Charter (1964) and the Granada Agreement (1985) are some of these important resolutions in which the need for the protection, study and conservation of the monuments is stressed. Nowadays, all countries of the civilized world are using all their scientific and technological efforts towards protecting and conserving the monuments within or even outside their borders assisting other countries to that end. These general tasks include geometric recording, risk assessment, monitoring, restoring, reconstructing and managing Cultural Heritage. It was in the Venice Charter (1964) that the necessity of the Geometric Documentation of Cultural Heritage was firstly set as a prerequisite. In Article 16 it is stated "… *In all works of preservation, restoration or excavation, there should always be precise documentation in the form of analytical and critical reports, illustrated with drawings and photographs…*"

## 2 GEOMETRIC DOCUMENTATION OF CULTURAL HERITAGE

The integrated documentation of monuments includes the acquisition of all possible data concerning the monument and which may contribute to its safeguarding in the future. Such data may include historic, archaeological, architectural information, but also administrative data, past drawings, sketches, photos etc. Moreover these data also include metric information which defines the size, the form and the location of the monument in 3D space and which document the monument geometrically. The geometric documentation of a monument, which should be considered as an integral part of the greater action, the integrated documentation of Cultural Heritage may be defined as (UNESCO, 1972) "*The action of acquiring, processing, presenting and recording the necessary data for the determination of the position and the actual existing form, shape and size of a monument in the three dimensional space at a particular given moment in time*".

The geometric documentation of a monument consists of a series of necessary measurements, from which visual products such as vector drawings, raster images, 3D visualizations etc. may be produced at small or large scales. These products have usually metric properties, especially those being in suitable orthographic projections. Hence one could expect from the geometric documentation a series of drawings, which actually present the orthoprojections of the monument on suitably selected horizontal or vertical planes. Scale and accuracy should be carefully defined at the outset, before any action on the monument.

Another important issue is the level of detail, which should be present in the final product. For a justified decision on that matter the contribution of the expert who is going to be the user is indispensable. A survey product, a line drawing or an image, implies generalization to a certain degree, depending on the scale. Hence, the requirements or the limits of this generalization should be set very carefully and always in co-operation with the architect or the relevant conservationist, who already has deep knowledge of the monument (Georgopoulos & Ioannidis 2007).

For the geometric recording several surveying methods may be applied, ranging from the conventional simple topometric methods, to the elaborated contemporary surveying and photogrammetric ones, for fully controlled surveys. Surveying and photogrammetric methods are based on direct measurements of lengths and angles, either on the monument or on images thereof. They determine three-dimensional point coordinates in a common reference system and ensure uniform and specified accuracy. Moreover they provide adaptability, flexibility, speed, security and efficiency. To this measurement group belong also the Terrestrial Laser Scanners (TLS) as they use light to collect a huge number of points in 3D space, in a very limited time frame. Most importantly, these methods are, by default, non-contact methods; hence they are highly suitable for cases where contact with the object is either dangerous, or not allowed, as is very often the case with cultural heritage objects.

## 3 IMPACT OF ICT ADVANCES ON CULTURAL HERITAGE DOCUMENTATION

Nowadays, the rapid advances of Digital Technology also referred to as Information Communication Technologies (ICT), have provided scientists with new powerful tools. We are now able to acquire, store, process, manage and present any kind of information in digital form. The available contemporary digital tools include mainly instrumentation for non-contact digital data acquisition, software for processing and managing the collected data and computer hardware, for running the demanding software, storing the data and presenting them in various forms.

The introduction of digital technologies has transformed the traditional work of archaeologists and museums as they are so far known. In other words digital technologies redefine the relationship to Cultural Heritage, as they enable universal access to it and they also connect cultural institutions to new "audiences". Finally they appeal to new generations, as the latter are computer literate. In this way we experience a "democratization" of cultural information across geographic, religious, cultural and scientific borders.

The introduction of Digital Technologies may contribute to all traditional steps of Archaeological practice. It goes without saying that the degree of contribution of ICT is different in the various stages and in the various cases. Modern technologies of remote sensing and archaeological prospection assist the touch less and rapid detection of objects of interest. Spectroradiometers or ground penetrating radars or even the simple processing of multispectral satellite images may easily lead to the rapid location of underground or submerged objects of interest. Contemporary non-contact survey technologies, such as photogrammetry, terrestrial laser scanning and digital imaging, may be used to produce accurate base maps for further study, or 3D virtual renderings and visualizations. The collected data may be stored in interactive databases, georeferenced or not, and be managed according to the needs of the experts. Finally ICT may assist in the presentation stage, by producing virtual models, which may be displayed in museums or be included in an educational gamification, or serve purposes of enabling handicapped persons to admire the treasures of the World's cultural heritage.

With the Charter on the Preservation of the Digital Cultural Heritage (UNESCO 2003) the basic principles of Digital Cultural Heritage for all civilized countries of the world are proclaimed. At the same time numerous international efforts are underway with the scope to digitize all aspects of Cultural heritage. The impact of digital technologies to the domain of Cultural Heritage has increased speed and automation of the procedures which involve processing of the digital data and presentation of the results. At the same time accuracy and reliability has been substantially enhanced. However, most important is the ability to provide to the users new and alternative products, which include two dimensional and three dimensional products, such as orthophotos and 3D models. All in all, the digitization of the world's Cultural Heritage either tangible or intangible is now possible.

## 4 ICT TOOLS—DIGITAL INSTRUMENTATION

ICT tools may be of different kinds. Contemporary digital instrumentation for data acquisition, powerful computers with high resolution monitors capable of 3D presentations and specially developed software to process the acquired data in the most suitable and effective way. The instrumentation necessary to support heritage conservation activities should always be at the technological edge nowadays. Modern instrumentation includes data acquisition instruments, processing software and powerful computers. Data acquisition instruments should include devices which are capable of digitally collecting (a) images or image sequences, (b) points in 3D space and (c) other pieces of information related to Cultural Heritage objects. Rapid technological progress has provided scientists with sophisticated instrumentation, which includes calibrated high resolution digital cameras, digital high resolution video recorders, accurate angle and distance measuring devices, GPS receivers, terrestrial laser scanners, 3D non laser scanners for small artifacts, film scanners and printed document scanners. Moreover instrumentation such as thermal and range cameras, material sampling devices and ultrasonic non-destructive inspecting instruments are also contributing to data acquisition.

Processing of all acquired multi-source data includes actions of position calculations, processing of the digital images or image sequences and working with point clouds. For these actions related software has been developed to cover all possible needs. The processing stage is supported by powerful computing units, which are available today. Processing usually aims to store, archive, manage, visualize, present and publish the collected data and the information derived. In the recent years many research efforts are directed to multi image matching techniques, thus complementing terrestrial laser scanning technology. This interaction of heritage objects with their geographic location is a well-known fact nowadays. It has bridged Geoinformatics and Monument Preservation. Geographic Information Systems is the scientific tool with which monuments and related information has been connected to place. In this way the Monument Information Systems (MIS) have evolved. In addition, the relation of intangible information with tangible Cultural Heritage is highly important and definitely required. Hence intangible Cultural Heritage may be linked to location, while at the same time important attributes of both forms of Cultural Heritage are preserved and interrelated.

## 5 3D DIGITAL DATA

Digital data acquisition is nowadays performed with (a) geodetic digital total stations, which produce 3D coordinates of single points in space, (b) digital image processing, which produces 2D or 3D products and (c) with digital scanning devices, which produce 3D point clouds of the objects. The common attribute of the above is the three dimensional data acquisition, which enables the development of 3D virtual models. It is these 3D models that have made the 3D virtual reconstructions possible. 3D models can be simple linear vector models or they can consist of complex textured surfaces depending on the object and their final use. As the specific technology advances, 3D models are used for multiple purposes. Initially they simply served as means for visualization. Gradually, however, they contributed to other uses, such as study, description purposes and restoration interventions and lately for virtual reconstruction and engineering applications (Valanis et al. 2009).

Technological advances have provided 3D modelling software with numerous capabilities, which enable them to go beyond the simple representation of an architectural structure. They can provide information regarding the materials used and the realistic texture of the surfaces and also be interconnected with a data base for storing, managing and exploiting diverse information. The organization of all these pieces of information, quantitative and qualitative, into one Information System is very much enabled by the Building Information Modelling (BIM) technology. Implemented to Cultural Heritage this technology is usually referred to as HBIM (Agapiou et al. 2010, Brumana et al. 2013, Bregianni et al. 2013). Typical implementations of 3D modelling can already be found in modern museums and educational foundations helping their visitors and students to communicate in a special way with the monument or site of interest as they can 'walk' through it or fly over it and thus examine it better, having always in mind its level of accuracy and likelihood.

## 6 INTERDISCIPLINARY APPROACH

The geometric documentation has been the responsibility of experts concerned with the care of the Cultural Heritage. Traditionally they mainly belonged to the field of archaeology and architecture. However, over the past thirty or forty years more and different specialists developed an interest for the monuments, as they were definitely able to contribute to their study, maintenance and care. Among them are surveyors, photogrammetrists and geomatics engineers in general, as the tech-

Figure 1. The interdisciplinary contribution to Cultural Heritage.

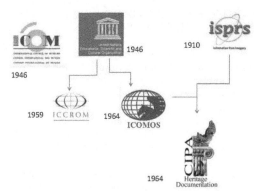

Figure 2. International Organizations involved in Cultural Heritage.

nological advances described above have enabled them to produce interesting, alternative and accurate geometric documentation products. Today the mentality is gradually changing and traditionally involved experts, like Archaeologists and architects, tend to accept and recognize the contribution of other disciplines to the agenda of Cultural Heritage. Hence it is rapidly becoming an interdisciplinary and intercultural issue (Fig. 1).

## 7 INTERNATIONAL COOPERATION

UNESCO (1946) and the Council of Europe have formed specialized organizations for taking care of mankind's cultural heritage. ICOMOS is the most important one, but also CIPA-Heritage Documentation (initially Comité International de Photogrammétrie Architecturale), ISPRS (International Society for Photogrammetry & Remote Sensing), ICOM, ICCROM and UIA are all involved in this task (Fig. 2). The Venice Charter was born from the need to create an association of specialists of conservation and restoration independent of the already existing associations of museologists, ICOM. In 1957, in Paris, the First Congress of Architects and Specialists of Historic Buildings recommended that the countries which still lack a central organization for the protection of historic buildings provide for the establishment of such an authority and, in the name of UNESCO, that all member states of UNESCO join the International Centre for the Study of the Preservation and Restoration of Cultural Property (ICCROM) based in Rome.

The Second Congress of Architects and Specialists of Historic Buildings, in Venice in 1964, adopted 13 resolutions, the first one being the International Restoration Charter, better known as Venice Charter, and the second one, put forward by UNESCO, provided for the creation of the International Council on Monuments and Sites (ICOMOS). CIPA Heritage Documentation was founded in 1964 as an International Scientific Committee (ISC) of ICOMOS and ISPRS (International Society for Photogrammetry and Remote Sensing) and hence is a dynamic international organization that has twin responsibilities: keeping up with technology and ensuring its usefulness for cultural heritage conservation, education and dissemination. These two sometimes conflicting goals are accomplished in a variety of ways, through (cipa.icomos.org):

- Encouraging and promoting the development of principles and good practices for recording, documentation and information management of cultural heritage;
- Leading and participating in international training programs for conservation and informatics professionals, students and site personnel;
- Advising government bodies, regional authorities, non-profit groups and institutions on tools, technology and methods for using technology;
- Sponsoring an international network of professionals in both the fields of technology and cultural heritage for scientific research but also applied practical experience;
- Providing a platform with the bi-annual International Conference for the exchange of ideas, best practices as well as scientific research papers.

In the recent past CIPA undertook the RecorDIM initiative, recognizing that there are critical gaps in the fields of heritage recording, documentation and information management between those who provide information for conservation and those who use it, i.e. providers and users of contemporary documentation information. In response, the International Council on Monuments and Sites (ICOMOS), the Getty Conservation Institute (GCI) and CIPA together created the RecorDIM (for Heritage Recording, Documentation and Information Management) Initiative partnership. The purpose of the initiative was to bring information users and providers together to identify the nature of the gaps between them, to develop strategies to close the gaps and to recommend a framework for action (http://cipa. icomos.org/index. php?id=43).

This current effort concerned with the 3D virtual reconstruction of monuments is motivated exactly by this endeavour to bridge this gap. This will only be done through deep understanding of each other's needs and through proper exploitation of ICT with the benefit of Cultural Heritage always in mind. In addition, the notion of virtual reconstruction is introduced and its use for bringing the reconstructed monuments into a museum environment is investigated. This interdisciplinary approach to the issue of Cultural Heritage has opened vast new possibilities and led to new alternative products for the benefit of monuments. These new possibilities include, among others, the production of 3D models, virtual reconstructions, virtual restorations, monitoring of constructions and the applications of serious games for educational and dissemination purposes.

## 8 SELECTED PRACTICAL EXAMPLES

### 8.1 *3D Models for Virtual Restoration*

The monument of Zalongon is a huge complex of sculptures, situated on top of an 80 m high and steep cliff in Northwestern Greece. The sculpture is about 18m long and 15m high and is a composition of several female figures and commemorates the sacrifice of the Souli village women, who in 1803 preferred death from humiliation by the Ottoman conqueror. Although fairly recently constructed (1957–61), the sculpture has suffered severely from frost and strong winds. The restoration work that was carried out involved the replacement of large pieces that have suffered damages from harsh weather and were deteriorating quickly, and the restoration of parts that have been destroyed by frost. A detailed and accurate geometric documentation and a three dimensional model of the construction were required. Given the size and complexity of the monument, contemporary digital non-contact techniques were employed for this purpose.

The digital data required to build high resolution 3D textured models were 3D scans, geodetic measurements and a significant number of digital images. The detailed geometric documentation of the current situation of the monument included the production of 2D drawings, orthophotos and an accurate 3D model (Valanis et al. 2009). The most interesting product, possible only with the use of contemporary digital methods, was the 3D model. For the creation of the surface of the 3D model all of the original scans were registered into a common reference system by applying a method that was specially developed. The creation of high resolution textured 3D models is undoubtedly a non-trivial task as it requires the application of advanced data acquisition and processing techniques, such as geodetic, photogrammetric, scanning, programming, surfacing, modeling, texturing and mosaicking (Valanis et al. 2009).

For achieving restoration, the basic steps are identifying the destroyed parts, interact with the 3D model and extract the geometry of the parts to be restored, insert them virtually into the 3D model and finally assess the result, before final decision. In the present case the main points of interest were of course the destroyed figure heads, but there were also many other damages to be restored (Figure 6). Engineers who are involved in restoration are greatly facilitated if they can interact with a 3D model and immediately obtain various kinds of information by measuring various distances, areas, volumes, by creating cross-sections, outlines or even by formulating and adding missing parts. However, in cases where the formulation and addition of 3D data is desired different methods and algorithms are required. This was also the case for the monument of Zalongon, where the upper parts of the two tallest figures were almost destroyed. Two main categories of data were extracted, namely the part of the surface that was healthy and would be retained and the broken part that was recorded only in order to help reconstruct what was missing. Efforts were also made to completely restore the original surface virtually. However, in order to obtain a better result, another approach was preferred. The partially completed surfaces were used for the creation of analogue models of a scale 1:5 and an artist, a sculptor, was assigned with the task of completing the forms based on the existing model and old photographs and sketches made by the original sculptor. The new plaster models were scanned with an XYZRGB SL2™ structured light scanner and the data acquired were registered with the 3D model (Ioannidis et al. 2010). The final mesh was exported, appropriately scaled and in such a form to enable masonry experts to reproduce exactly the missing parts and to actually restore the monument (Figure 3).

Figure 3. Virtual restoration: Before (left) and after (right). Notice the repaired heads of the tall figures and the restoration on the missing parts.

### 8.2 Virtual reconstruction of the Middle Stoa in the Athenian Agora

The Ancient Athenian Agora is one of the most important archaeological sites in Athens and is situated at the northern foot of the Acropolis hill. The Middle Stoa was an elongated building 147m by 17.5m, which ran east–west across the old square, dividing it into two unequal halves. This large building was constructed with Doric colonnades at both the north and south sides as well as an Ionic colonnade along the middle. The original steps and three columns remain *in situ* at its eastern end; to the west, only the heavy foundations of reddish conglomerate survive. The Middle Stoa was built between ca. 180 and 140 B.C. and it was continuously used even during of the Roman era. Foreign architects were responsible for its construction; hence it presents particular design and construction elements not usual for that time in the area. Today only the foundations of this majestic building and some individual parts of it are visible in the site (Figure 4).

For the virtual reconstruction several different data were available. Artifacts from the initial construction in the museum, drawings from scholars who had studied thoroughly the monument, survey measurements of the foundations which are visible today and artists' reproductions of pertinent descriptions from travellers of the past. All data were for their reliability and accuracy before usage for the final virtual reconstruction (Figure 5).

This final product reconstructs a building that does not exist today. The visitor may only see the foundations of the building, which at the time of its peak (2nd c. BCE) were buried in the ground. Consequently the virtual reconstruction is a combination of existing detailed architectural drawings, of sketches, descriptions, digitization of real artefacts and other minor sources of information. Of utmost importance were the discussions and suggestions of scientists who have studied the monument from an historical and archaeological point of view, proving once again that a reconstruction is a multi-disciplinary process.

Figure 4. The foundations of the Middle Stoa.

Figure 5. The reconstructed Middle Stoa.

## 9 CONCLUSIONS

Advances in of digital technology together with the Information and Communication Technologies have enabled providers to produce more accurate, more reliable and wealthier in terms of information content documentation products. Moreover the contemporary technologies of data acquisition are practically non-contact and hence non-destructive, which makes them ideal for Cultural Heritage applications. Specialized international organizations, like CIPA-Heritage Documentation are striving to disseminate this information and make these interesting results even more known and accessible to the users, thus "democratizing" this special knowledge and skills up to now reserved only for the experts.

### REFERENCES

Agapiou A., Georgopoulos A., Ioannidis C., Ioannides M., 2010. A Digital Atlas for the Byzantine and Post Byzantine Churches of Troodos Region (central Cyprus). Proceedings of the 38th Conference on

Computer Applications and Quantitative Methods in Archaeology, Granada, Spain, April 2010.

Bregianni, A., Raimondi, A., Georgopoulos, A., Brumana, R., Oreni, D., (2013). HBIM for Documentation, Dissemination and Management of Built Heritage. The Case Study of St. Maria in Scaria D'Intelvi. International Journal of Heritage in the Digital Era, Vol. 2, No. 3, 2013, pp. 433-451, ISSN 2047-4970.

Brumana, R., Oreni, D., Cuca, B., Georgopoulos, A., (2013). HBIM for conservation and management of built heritage: towards a library of vaults and wooden beam floors. ISPRS Annals – Volume II-5/W1, 2013 TC V XXIV International CIPA Symposium 2–6 September 2013, Strasbourg, France, pp. 215–221. http://cipa.icomos.org/fileadmin /template/doc/ STRASBOURG/ANNALS/isprsannals-II-5-W1-215-2013.pdf.

CIPA Recordim http://cipa.icomos.org/index. php?id=43 (last accessed 08.03.2015).

Georgopoulos, A., Ioannidis, Ch. (2004) Photogrammetric and Surveying Methods for the Geometric Recording of Archaeological Monuments FIG International Week, Athens 2004, http://www.fig.net/pub/athens/ papers/wsa1/ /WSA1_1_Georgopoulos_Ioannidis. pdf.

Ioannidis, Ch., Valanis, A., Tapinaki, S., Georgopoulos A. Archaeological Documentation and Restoration using Contemporary Methods. Proceedings of the 38th Conference on Computer Applications and Quantitative Methods in Archaeology. Granada, Spain, April 2010.

Kontogianni, G., Georgopoulos, A., Saraga, N., Alexandraki, E. and Tsogka, K., 2013. 3D Virtual Reconstruction of the Middle Stoa in the Athens Ancient Agora, Int. Archives of Photogrammetry, Remote Sensing and Spatial Information Science, XL-5/W1, 125–131, doi:10.5194/isprsarchives-XL-5-W1-125-2013, 2013.

UNESCO, (2003). Guidelines for the Preservation of Digital Heritage, CI-2003/WS/3.

Valanis, A., Georgopoulos, A., Tapinaki, S., Ioannidis, Ch., 2009. High Resolution Textured Models for Engineering Applications. Proceedings XXII CIPA Symposium, October 11–15, 2009, Kyoto, Japan, http:// cipa.icomos.org/ /text%20files/KYOTO/164.pdf.

*Emerging Technologies in Non-Destructive Testing VI – Aggelis et al. (Eds)*
*© 2016 Taylor & Francis Group, London, ISBN 978-1-138-02884-5*

# Numerical methods for the interpretation and exploitation of AE monitoring results

S. Invernizzi, G. Lacidogna & A. Carpinteri
*Department of Structural, Geotechnical and Building Engineering, Politecnico di Torino, Torino, Italy*

ABSTRACT: The paper reviews some recent numerical applications for the interpretation and exploitation of Acoustic Emission (AE) monitoring results. Among possible numerical techniques, the Finite Element Method and the Distinct Method are considered. The analyzed numerical models cover the entire scale range, from microstructure and meso-structure, up to full size real structures. The micro modeling includes heterogeneous concrete-like material as well as the meso-structure of the masonry texture, where each brick and mortar joint is modeled singularly. The full size models consider different typology of historical structure such as masonry towers, cathedrals and chapels. The main difficulties and advantages of the different numerical approaches, depending on the problem typology and scale, are critically analyzed. The main insight we can achieve from micro and meso numerical modeling concern the scaling of AE as a function of volume and time, being also able to mimic the *b*-value temporal evolution as the damage spread into the structure. The finite element modeling of the whole structure provides useful hints for the optimal placement of the AE sensors, while the combination of AE monitoring results is crucial for a reliable assessment of the structural safety.

## 1 INTRODUCTION

Non-destructive and instrumental investigation methods are currently employed to measure and check the evolution of adverse structural phenomena, such as damage and cracking, and to predict their subsequent developments. The choice of a technique for controlling and monitoring reinforced concrete or masonry structures is strictly correlated with the kind of structure to be analyzed and the data to be extracted (Anzani et al., 2010). For historical buildings, Non-Destructive Evaluation (NDE) techniques are used for several purposes: (1) detecting hidden structural elements, such as floor structures, arches and piers; (2) determining masonry characteristics, mapping the heterogeneity of the materials used in the walls (e.g. use of different bricks during the life of a building); (3) evaluating the extent of the mechanical damage in cracked structures; (4) detecting voids and flaws; (5) determining moisture content and rising by capillary action; (6) detecting surface decay phenomena; and (7) evaluating the mechanical and physical properties of mortar and brick, or stone.

In the assessment of structural integrity, the Acoustic Emission (AE) technique has proved particularly effective, in that it makes it possible to estimate the amount of energy released during the fracture process and to obtain information on the criticality of the process underway. Strictly connected to the energy detected by AE is the energy dissipated by the monitored structure. The energy dissipated during crack formation in structures made of quasi-brittle materials plays a fundamental role in the behavior throughout their entire life. Recently, according to fractal concepts, an ad hoc method has been employed to monitor structures by means of the AE technique. The fractal theory takes into account the multiscale character of energy dissipation and the strong size effects associated with it. With this energetic approach, it becomes possible to introduce a useful damage parameter for structural assessment based on a correlation between AE activity in the structure and the corresponding activity recorded on specimens of different sizes.

The effectiveness of NDT monitoring can be greatly improved when combined with appropriate numerical modeling. Numerical modeling is useful both for better structural assessment and for deeper comprehension of phenomena at the meso and microstructural material levels. In the following, a number of applications of combined AE and numerical analysis performed by the authors are reviewed, starting from real size structure, down to model structures, and meso or microstructural material modeling.

## 2 HISTORICAL STRUCTURES

In the last fifteen years, the authors have investigated a number of relevant existing buildings belonging to the Italian cultural heritage. In this section these structures are briefly described, in order to illustrate the combination of numerical modeling and AE monitoring.

### 2.1 *The medieval towers of Alba*

These masonry buildings from the XIIIth century are the tallest and mightiest medieval towers preserved in Alba (Figure 1) (Carpinteri et al., 2006).

Complete three-dimensional FEM models of the three towers have been built using the CAD drawings meshed with 20-node isoparametric solid brick elements, in order to perform the analysis with the commercial code DIANA. At least five nodes are present in the thickness of the towers wall.

The models take into account the presence of openings and the variation of the wall thickness at different levels. On the other hand, the presence of wood floors has been disregarded. Each structure is mainly subjected to its dead load. As far as Torre Sineo is concerned, also the effect of an increasing tilt of the foundation has been considered, combined to the load provided by the wind action exerted to the upper region of the tower (Figure 2). The cracking processes taking place in some portions of the masonry towers were monitored using the Acoustic Emission (AE) technique. Crack advancement, in fact, is accompanied by the emission of elastic waves, which propagate within the bulk of the material. These waves can be captured and recorded by transducers applied to the surface of the structural elements.

For the Sineo Tower (Carpinteri et al., 2005), through AE monitoring, two cracks were detected in the inner masonry layer at seventh floor level.

Figure 1. Overall view of the medieval towers of Alba.

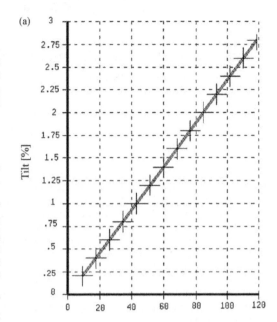

Figure 2. Torre Sineo crack pattern relative to the 3% tilt of the foundation and detail of the crack pattern in the foundation zone.

The monitoring process revealed an ongoing damaging process, characterized by slow crack propagation inside the brick walls. In the most damaged zone, crack spreading had come to a halt, the cracks having achieved a new condition of stability, leading towards compressed zones of the masonry. In this particular case, it can be seen that, in the monitored zone, each appreciable crack advancement is often correlated to a seismic event. In the diagram shown in Figure 3a, the cumulative AE function relating to the area monitored is overlaid with the seismic events recorded in the Alba region during the same time period; the relative intensity of the events is also shown.

A similar behavior was observed for Torre Astesiano. This structure was monitored by means of two transducers applied to the inner masonry layer of the tower, at the fourth floor level near the tip of the large vertical crack. The results obtained during the monitoring period are summarized in the diagram in Fig. 16b. It can be seen how the damage to the masonry and the propagation of the crack, as reflected by the cumulative number of AE counts, evolved progressively over time. A seismic event of magnitude 4.7 on the Richter scale occurred during the monitoring period: from the diagram we can see how the cumulative function of AE counts grew rapidly immediately after the earthquake. The monitoring of Torre Bonino

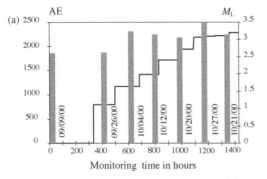

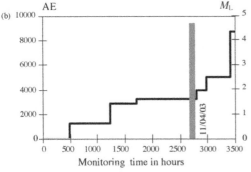

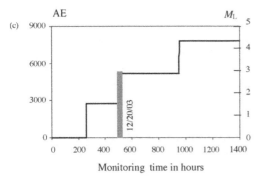

Figure 3. AE counting number and seismic events in local Richter scale magnitude ($M_L$): (a) Torre Sineo, (b) Torre Astesiano, (c) Torre Bonino.

was performed the at first floor level, where the effects of restructuring works have affected the masonry most adversely. Under constant loading, a progressive release of energy is observed, due to a pseudocreep phenomenon in the material. A seismic event of magnitude 3 on the Richter scale occurred during the monitoring period (Figure 3c). The time dependence of the structural damage observed during the monitoring period can also be correlated to the rate of propagation of the micro-cracks. If we express the ratio of the cumulative number of AE counts recorded during the monitoring process, $N$, to the number obtained at the end of the observation period, $N_d$, as a function of time, $t$, we get:

$$\eta = \frac{E}{E_d} = \frac{N}{N_d} = \left(\frac{t}{t_d}\right)^{\beta_t}. \quad (1)$$

In Eq. (1), the values of $E_d$ and $N_d$ do not necessarily correspond to critical conditions ($E_d \leq E_{max}$; $N_d \leq N_{max}$) and the $t_d$ parameter must be construed as the time during which the structure has been monitored. By working out the $\beta_t$ exponent from the data obtained during the observation period, we can make a prediction as to the structure's stability conditions. If $\beta_t < 1$, the damaging process slows down and the structure evolves towards stability conditions, in as much as energy dissipation tends to decrease; if $\beta_t > 1$ the process becomes unstable, and if $\beta_t \cong 1$ the process is metastable, i.e., though it evolves linearly over time, it can reach indifferently either stability or instability conditions. In order to obtain indications on the rate of growth of the dam—age process in the towers, as given in Eq. (1), the data obtained with the AE technique were subjected to best-fitting in the bilogrithmic plane. This yielded a slope $\beta_t \cong 0.648$ for the Sineo Tower, $\beta_t \cong 1.041$ for the Astesiano Tower and $\beta_t \cong 0.491$ for the Bonino Tower (Figure 4). These results confirm how the damage process stabilized in the Sineo and Bonino Towers during the monitoring period, whereas for the Astesiano Tower, it evolved towards a metastable condition.

## 2.2 Cathedrals and vaulted structures

A monitoring campaign and numerical analysis has been performed on the ancient Cathedral of Syracuse in Sicily (Carpinteri et al., 2009) (Figure 5a). The Acoustic Emission (AE) technique is adopted to assess the damage pattern evolution. The localization of the propagating cracks is obtained using six synchronized AE sensors. A clear correlation between the regional seismic activity and the AE acquisition data has been obtained. In fact the AE count rate presents peaks corresponding to the main seismic events. During the observation period (Figure 5b), the number of AE counts was $\cong 4300$.

In order to obtain indications on the rate of the damage process, as given in Eq. 1, the data obtained with the AE technique were subjected to best-fitting in the bilogarithmic plane. This yielded a slant $\beta_t \cong 0.98$. The result confirms that the damage process in the pillar is in metastable conditions according to a quasi-linear progression over time. A detailed 3D model of the most damaged pillar has been obtained according to the geometrical survey. The geometry of each block, as well as the presence of masonry inserts, have been considered.

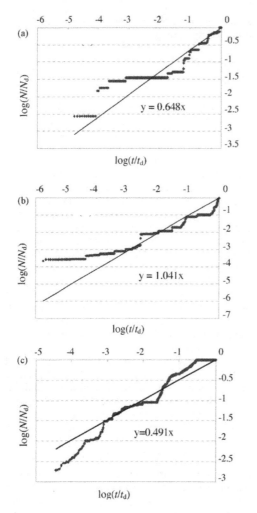

Figure 4. Evolution of damage: (a) Torre Sineo, (b) Torre Astesiano, (c) Torre Bonino.

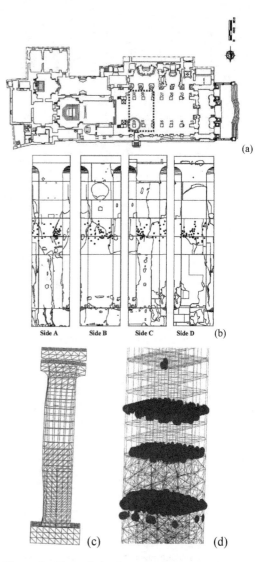

Figure 5. Plan of the Syracuse Cathedral and most critical pillars (a), Cracking pattern and localization of AE sources (b), deformed mesh under the seismic acceleration (c), cracking in the pillar when subjected to the seismic load (d).

Two main loads were considered: the dead load (of the pillar and of the surrounding structure), and a horizontal seismic load provided as horizontal ground acceleration. Due to the horizontal acceleration, cracking can take place in the pillar. A detail of crack nucleation is shown in Figure 5d. The crack occurrence provided by the analysis agrees quite well with the crack localization provided by the AE recording. Cracking corresponds both to diffuse cracking in the continuum elements of the sandstone blocks and to opening or sliding of the discrete interfaces between blocks.

Another case was the Ospedale San Giovanni in Turin (Italy) (Carpinteri et al., 2007), a masonry building complex initiated in 1680, under the design of the Italian architect Amedeo di Castellamonte (1610–1683) (Figure 6). The first floor of the building was recently chosen to host an important fossil collection of the Regional Museum of Natural Science.

Due to this change of use, an assessment of the structural load capacity of the masonry vault beneath the first floor was necessary because the fossil collection involves a sensible increase in the vault load. During the in situ load test, we recorded the acoustic emissions from the vault, as well as

Figure 6. Ospedale San Giovanni: photograph inside the vault.

displacements of the vault and strains in the steel rods. We compared the experimental data with the numerical results obtained from finite element modeling of cracking and crushing. After validation, the model allowed us to assess the ultimate load-bearing capacity of the vault. The 3D model provided a slightly better estimate of the displacements close to the abutment (Figure 7).

## 3 MODEL STRUCTURES

Numerical modeling has been fruitfully combined with NDT and AE monitoring also in the study of scaled model structures in the laboratory.

A scaled model of a two-span masonry arch bridge has been built to investigate the effect of the central pile settlement because of riverbank erosion (Invernizzi et al., 2011).

The bridge geometry and the structural details, included the masonry bricks and mortar joints, are realized in the scale 1:2 (Figure 8). The model bridge has been subjected to incremental settlement of the pile, which was sustained on a mobile support. During the first stage of the settlement, the AE counting has been recorded. Based on the interpretation of the AE rate, it is possible to monitor the criticality of the ongoing process.

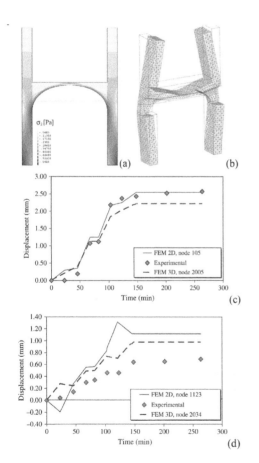

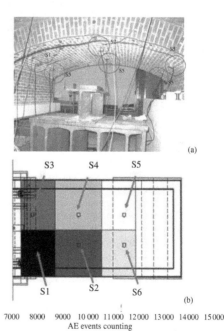

Figure 7. Vault subjected to the dead load: principal compression stress contour (a), 3D vault deformed mesh (b), time-displacement diagrams at the springing (c), and close to the keystone (d).

Figure 8. Positioning of the Acoustic Emissions transducers, and their competence areas (a), number of acoustic events in each competence area (b).

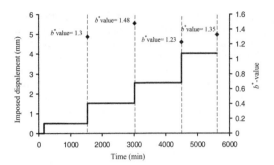

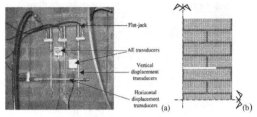

Figure 10. Double flat jack experimental set (a), and numerical discretization exploiting the double symmetry (b).

Figure 9. $b^*$-value trend during the incremental settlements of the pile.

In addition, thanks to the AE equipment, it has been possible to localize the main damaged zones. The statistical properties AE time series have been analyzed using an estimation of the $b^*$-value of the Gutenberg—Richter (GR) law permitting to determine the damage level reached in the model (Figure 9).

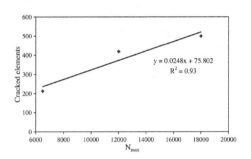

## 4 MATERIAL MESO AND MICROSTRUCTURE

Figure 11. Proportionality between $N_{max}$ and the number of cracked finite elements.

When numerical simulation is combined with AE monitoring at the meso and micro scale, detailed information about the damage evolution are obtained. In particular, the AE statistics and evolution law of the $b$-value with increasing damage can be obtained numerically.

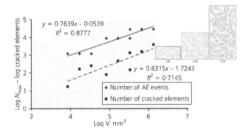

### 4.1 Smeared cracking

Finite element modeling can be performed adopting a smeared cracking constitutive law. In this way the number of crack advance at each gauss point can be put into correspondence with AE events recorded during monitoring.

Figure 12. Proportionality between $N_{max}$ and the number of cracked finite elements.

The double flat jack test has been studied in details (Carpinteri et al., 2009), combining it with AE monitoring and considering an experimental configuration where the analyzed volume of masonry is not constant.

The numerical model accounts for the detailed mesostructure of the masonry texture. Although it is not possible to obtain an easy direct relation between the acoustic emission and the amount of cracking; nevertheless, it is possible to state that the two quantities are proportional to each other when increasing sizes are considered (Figure 11).

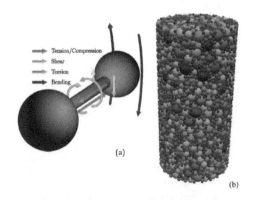

Also concrete-like materials can be studied with a similar strategy, in order to accounts for the detailed material microstructure composed of

Figure 13. Scheme of the bonded interaction between two particles (a); view of the specimen particle size distribution (b).

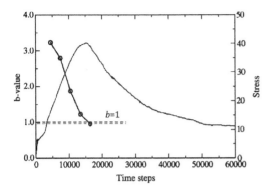

Figure 14. Evolution of the simulated $b$-value (circles) with time compared with the stress level.

matrix and aggregates (Invernizzi et al., 2013a). In this way the volumetric scaling of AE was obtained numerically (Figure 12).

## 4.2 Particle strategy

The propagation of micro cracks can be put in direct correspondence with AE events adopting a particle strategy according to the distinct element method (Invernizzi et al., 2013b). In this case, every single aggregate is modeled as a particle of appropriate size distribution, while a reciprocal lattice of beams simulates the matrix.

The method allows describing correctly the size effect on compressive strength and the scaling of AE with size. In addition, the $b$-value can be evaluated numerically, as shown in Figure 14. The evolution of the parameter obtained numerically is in good agreement with theoretical prevision and experimental results.

## 5 CONCLUSIONS

A number of numerical applications, carried out by the authors in the last fifteen years, are summarized and briefly described to emphasize the advantage of combined numerical methods and AE monitoring. It is shown how proper numerical modeling can improve the comprehension of involved phenomena and the reliability of structural assessment.

## REFERENCES

Anzani, A., Binda, L., Carpinteri, A., Invernizzi, S., Lacidogna, G. 2010, A multilevel approach for the damage assessment of historic masonry towers. *Journal of Cultural Heritage* 11(4): 459–470.

Carpinteri, A., Invernizzi, S., Lacidogna, G. 2005, In situ damage assessment and nonliner modelling of an historical masonry tower. *Engineering Structures* 27: 387–395.

Carpinteri, A., Invernizzi, S., Lacidogna, G. 2006, Numerical assessment of three medieval masonry towers subjected to different loading conditions. *Masonry International* 19: 65–76.

Carpinteri, A., Invernizzi, S., Lacidogna, G. 2007, Structural assessment of a XVIIth century masonry vault with AE and numerical techniques. *International Journal Of Architectural Heritage* 1: 214–226.

Carpinteri, A., Invernizzi, S., Lacidogna, G. 2009, Historical brick-masonry subjected to double flat-jack test: Acoustic Emissions and scale effects on cracking density. *Construction and Building Materials* 23: 2813–2820.

Carpinteri, A., Lacidogna, G., Manuello, A., Invernizzi S., Binda L. 2009, Stability of the vertical bearing structures of the Syracuse Cathedral: Experimental and numerical evaluation. *Materials & Structures* 42: 877–888.

Invernizzi, S., Lacidogna, G., Carpinteri, A. 2013a, Scaling of fracture and acoustic emission in concrete. *Magazine of Concrete Research* 65(9): 529–534.

Invernizzi, S., Lacidogna, G., Carpinteri, A. 2013b, Particle-based numerical modeling of AE statistics in disordered materials. *Meccanica* 48(1): 211–220.

Invernizzi, S., Lacidogna, G., Manuello, A., Carpinteri, A. 2011, AE monitoring and numerical simulation of a two-span model masonry arch bridge subjected to pier scour. *Strain* 47(2): 158–169.

*Emerging Technologies in Non-Destructive Testing VI – Aggelis et al. (Eds)*
*© 2016 Taylor & Francis Group, London, ISBN 978-1-138-02884-5*

# Digital cultural heritage—A challenge for the chemical engineering: Contextualizing materials in a holistic framework

M. Ioannides
*Cyprus University of Technology, Cyprus*

E. Alexakis
*Laboratory of Material Science and Engineering, School of Chemical Engineering, National Technical University of Athens, Athens, Greece*

C.M. Coughenour
*Institute for Photogrammetry, University of Stuttgart, Stuttgart, Germany*

M.L. Vincent, M.F. Gutierrez & V.M. Lopez-Menchero Bendicho
*University of Murcia, Murcia, Spain*

D. Fritsch
*Institute for Photogrammetry, University of Stuttgart, Stuttgart, Germany*

A. Moropoulou
*Laboratory of Material Science and Engineering, School of Chemical Engineering, National Technical University of Athens, Athens, Greece*

V. Rajcic
*Faculty of Civil Engineering, University of Zagreb, Zagreb, Croatia*

R. Zarnic
*Faculty of Civil and Geodetic Engineering, University of Ljubljana, Ljubljana, Slovenia*

ABSTRACT: Materials science and chemical engineering are imperative for the preservation and conservation of cultural heritage. The integration of novel methods in cultural heritage documentation, which yields an exponential amount of data, demands rigorous academic analysis. Often times these data are not integrated into the databases that form the core of cultural heritage recording systems, and therefore are often lost when the specialists complete their work. The Initial Training Network for Digital Cultural Heritage (ITN-DCH wqww.itn-dch.eu) proposes a holistic framework for cultural heritage that integrates all the data and systems in ways that allow for contextual data retrieval.

In the Troodos Mountains of Cyprus, the Byzantine church of Panagia Forviotissa Asinou (or simply Asinou), is a UNESCO World Heritage site (http://whc.unesco.org/en/list/351) and forms one of the ten monuments in the region known as the painted churches of Troodos Region (http://www.byzantine-cyprus.com/). The church represents a complex set of information, from ancient inscriptions dating to the founding of the monument in the 12th century, to modern video recordings from Cypriot television broadcasts. All of this information must be presented together in a single, holistic framework that allows researchers to approach the monument as a single unit, rather than searching for this information from disparate sources in archives and repositories around the world.

The authors will present the framework being developed by the ITN-DCH through the lens of one of their case studies at Asinou. This framework demonstrates the integration of all types of different information, and demonstrates how cultural heritage management can be conducted in such a way that unifies data and encourages transdisciplinary research. Future work will include applications at other project case studies, demonstrating its application at other sites in Europe.

## 1 INTRODUCTION—ASINOU INTERVENTIONS CHRONOLOGY

The church of Panagia Phorbiotissa, commonly known as Asinou, is built on the east bank of a small stream, three kilometres south of the village Nikitari located in the northern foothills of the Troodos Mountains (Figure 1 Aspect of Asinou Church from the Cross hill [2] Figure 1). Built around 1100, it is dedicated to Virgin Mary and it is considered as one of the most important Byzantine churches in Cyprus. In 1985, the Asinou church was included in the list of the UNESCO World Heritage Sites and until now most of its frescoes and murals, created in different chronologic periods, are preserved [1] [2].

Asinou's first frescoes covered the whole interior of the church and included at least two over doors on the exterior dated by an inscription to 1105/6. During the 12th century, narthex was added to the original church with some mural painting to be added until the 1332/33 when a comprehensive program draped the entire narthex in a coat of colour as all-enveloping as that of the 1105/6 program in the naos. The varied iconographic sources observed in Asinous's programs suggest a steady interchange not only with Byzantine but also with broader Mediterranean traditions.

Take as example the "maniera cypria" style that displaced around the 1340s by a more exuberant manner that, while still strongly affected by the Mediterranean interchange of the Crusader era, was responsive to Palaiologan developments, as well. A restoration work cycle started in the 1950s resulted in reinforcing the side walls, grouting of the cavities in the rubble fill, ringing the 12th century portion of the building with a reinforced collar to withstand the thrust of the roof beams, which were also repaired [2]. The phases of construction, wall paintings and restoration have been illustrated by Andreas Nicolaides in Figure 2a.

Figure 1. Aspect of Asinou Church from the Cross hill [2].

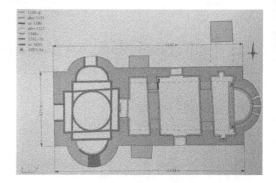

Figure 2. a) Ground Plan of the Church Panagia Phorviotissa where the colours correspond to the various eras-phases of the church construction, wall paintings and restorations [2].

Figure 2. b) A part of the interior fresco at the ASINOU monument.

## 2 EXPERIMENTAL PROCEDURE

The authors in an effort to digitally document the monument in an integrated and holistic approach: the tangible and intangible parts or the "memory" of the church: such as geometrically (Three Dimensional Representation), Material Documentation (Materials Decay), Monument's History (storytelling), Frescos, liturgy, untold story, etc. will apply their techniques and methods in order to serve the general objectives of the project; Digital Cultural Heritage.

The Laboratory of Material Sciences and Engineering (LMSE) from the School of Chemical Engineering of the National Technical University of Athens (NTUA) conducted an on-site inspection and non-destructive characterization of monument materials and of their conservation state aiming for 1) A diagnostic study-survey regarding materials decay on Architectural Surfaces (Frescoes) and Historic Structures (Masonries, Arches etc.) and 2) Assessing and evaluating incompatible materials and conservation interventions. The Non Destructive Techniques employed to make the measurements were the Infra-Red Thermography (IRT) and the Fiber Optic Microscope (FOM). Both techniques are widely used to study the murals' surface for the detection/identification of initial wear or decay and potential incompatibility concerning intervention materials.

Infrared Thermography (IRT) is an established NDT with a wide array of applications ranging from enhancing night vision to detecting defects on architectural surfaces. In the field of built cultural heritage protection, IRT is increasingly used for the investigation of structures, their materials and their preservation state. IRT investigations are carried out in either of two approaches: passive or active [3]. Passive approach (the one followed in the present study) is usually employed when surveying large areas, e.g. masonries, whereas active approach is employed when surveying small areas of interest (e.g. investigation of specimens in lab, or small surface areas) where the controlled thermal excitation is feasible. The active thermography approach (not applied in the present study) is adopted when the features of interest are in thermal equilibrium with the surroundings and therefore difficult to differentiate them from their background. Particular applications of IRT on built cultural heritage concern: 1) Materials and Decay mapping [4], 2) Evaluation of restoration and consolidation materials, and interventions on the monument scale [3] and 3) Study of moisture transfer phenomena [5].

Portable imaging systems, like FOM, provide images of high quality and adequate magnifications of the examined surfaces, which aid in the "interpretation" of data from other non-destructive (e.g. Infra-Red Thermography) or analytical techniques. These imaging systems are used to evaluate the performance and compatibility of rehabilitation interventions (e.g. consolidation, cleaning) relative to the historic materials and structures, or to study the preservation state of mosaics and other non-loading architectural elements. More particular, portable digital microscopy is employed to identify differences in the texture and composition of surfaces, for materials classification (e.g. classifications of mortars), for the study of the decay phenomena (alveolation, hard carbonate crust etc.) [3], to investigate materials' surface morphology to identify defects in historic building materials [4], for materials characterization [5], to classify decay typologies for porous stones [6], to evaluate cleaning interventions [7], consolidation interventions [8] [9] and incompatible interventions, and to study the preservation state of mosaics [10] [11]. Furthermore, images from digital microscope can be further analyzed with Digital Image Processing and morphological fusion algorithms to detect and quantify decay patterns on building materials [12].

In the present work, the particular surfaces on which IRT and FOM measurements were conducted were 1) the Churches' arches, 2) the fresco outside the church (Figure 3) next to the north entrance 3) the outer surface of the east masonry of the church (Figure 4) and 4) the frescoes inside the church.

Figure 3. a) North aspect of Asinou Church b) Fresco next to the door of the north side c) Regular photo of the fresco d) Fresco Thermograph.

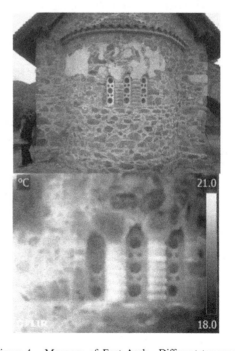

Figure 4. Masonry of East Arch—Different temperatures in the Thermograph suggest Incompatible Intervention Materials.

## 3 MATERIALS AND THE ONTOLOGICAL EXPRESSION OF HERITAGE

The study of materials is not, in of itself, a challenge in the cultural heritage world. In fact, the integrated approach to cultural heritage study should include material analysis, as it is fundamental for describing heritage within its context, whether it be an archaeological context, artwork, a monument, or another aspect of cultural heritage. The challenge, then, is the integration of these analytical processes into the data systems for heritage management and research. It is very rare to find digital libraries that integrate materials analysis into the typical object catalogues if we are to achieve a holistic approach to heritage management, study, and dissemination, it is imperative to link materials analyses with heritage, rather than separate them into silos of data that are exclusive of one another.

The CIDOC-CRM (http://www.cidoc-crm. org) is an international standard for the ontological expression of cultural heritage data. However, it has the capacity to move beyond just the basic expression of heritage data, and can include scientific observations through the CRM extension, CRMinf (http://www.ics.forth.gr/isl/index_main. php?l=e&c=663), therefore enabling the integration of material characterization information together with the heritage documentation.

## 4 GEOMETRIC DOCUMENTATION

Although previous photogrammetric work has been accomplished on the exterior and interior of the Asinou church [14] as well as at other Byzantine churches in Cyprus [15], the latest close range photogrammetric acquistion was achieved for two religious icons which are connected to the church's historical record. The first, and oldest known, icon from the Asinou church is of John the Baptist. This cultural object is currently located at the Byzantine Museum in Nicosia. The second icon, dating from the 19th century, was documented within the Asinou church.

A Canon EOS 1000D 10.10-megapixel camera with a pre-calibrated 18–55 mm f/3.5–5.6 zoom lens and CMOS sensor size of 3888 × 2592 pixels was used for the acquisition of both objects. The images were captured in RAW format and the lens was fixed at 55 mm f/5.6 in order to ensure the best image quality could be achieved. Coded targets and scales were utilized to ensure the correct scale of the object was recorded as well as proper image alignment during post processing. A total of 246 images were taken of the John the Baptist icon (Figure 5) and 903 images of the larger, and more

Figure 5. Both sides of 3D model of the John the Baptist Icon.

Figure 6. Both sides of 3D model of the Current Asinou Icon.

geometrically unique, icon dating from the 19th century (Figure 6).

Photogrammetric processing was accomplished with two software packages. Agisoft Photoscan Professional release 1.1.1.2009(http://www.agisoft. com) was used for the principle processing stages of image alignment, dense image matching, and textured mesh interpolation. The SURE software package version 1.0.9.38 (http://www.nframes. com), which is provided free for academic purposes, was utilized to produce an optimal 3D geometric record of each cultural object [16] (Figure 7). This

Figure 7. Close-up view of 3D point cloud of John the Baptist Icon produced by SURE.

Figure 8. Masonry mortar next to the front door on the South (a, b) – Grey-green crust observed when mortar is magnified (x50) with FOM (c).

processing workflow was successfully completed for both objects at both high and low resolutions as requested.

Due to the lack of optimal data acquisition conditions and time constraints, the acquisition accomplished would best be repeated with an improved planning strategy which would enhance the geometric documentation results of both icons. It would be recommended to use a more superior camera as well as planning for a better acquisition environment with improved lighting control. Since both objects are movable, this recommendation should be easily fulfilled.

## 5 RESULTS AND DISCUSSION

Regarding the fresco on the north outer masonry illustrated in Figure 3, the present study indicates the incompatible integration of outside plasters; most probably with the use of plasticizers that trigger the exfoliation of the original plaster. Possible mixture of the cement (concrete) in the repair mortgage is accelerating the decay of the porous stone. Based on the thermographic inspection of the fresco (Figure 3c), there are evident diversified temperatures on the fresco area (different colors represents different temperatures according to the scale on the right in Figure 3d) that could be attributed to the conservation varnish, which does not allow the fresco surface to respire causing irreversible damage on it. The exfoliation-flaking mechanism of the outer layer of the fresco has as a consequence the detachment of the pigments (scheduled under the secondary crystallite calcite of the outer surface) revealing the substrate without painting.

In Figure 4, the Thermograph from the east outer masonry (arch) reveals the different temperatures among the various materials have been used in maintenance interventions and the historic materials of the monument. Like before, this measurement suggests that the intervention took place with incompatible materials which comes in agreement with the general conclusions made about the monument in Dumbarton Oaks Studies book [2].

In Figure 8, a magnified picture (x50) of the yellow mortar (c) next to the central church entrance (a, b) has been measured by FOM. A grey/green crust is observed in certain areas(c), suggesting biochemical activities (biogenic decay). For validation, sample has to be distracted and measured with analytical techniques (in-lab).

To summarize, Non Destructive Techniques is an effective tool for the evaluation of past intervention took place over a monument. Infrared Thermographs revealed the differences in the materials temperatures, which subsequently mean utilization of materials with different properties suggesting incompatible interventions that have to be further investigated before the interventions themselves, exacerbate the decay process. Porous, thermal coefficients and other properties related to both original and intervention materials have to be determined in order to understand the various transport phenomena mechanisms (mass and heat flux); how latter one interact with the original materials and the impact-effect they have on them. Portable Optical Microscope (FOM) revealed decay that it not visible with the eye giving the possibility to further investigate and validate the particular type of decay with analytical techniques. FOM providing the possibility combined with Digital Image Processing, for decay mapping and have a more holistic overview of the decay proportions (over the healthy surface) and classify the different types of decay over the whole monument. With concern the interior of the church, active IRT could provide with more information about the frescoes and assess their preservation state more efficiently.

## REFERENCES

[1] D.M. Christodoulos Hadjichristodoulou, Byzantine Monuments Guides of Cyprus: The Church of Panagia Phorbiotissa of Asinou, Lefkosia: Cultural Foundation—Bank of Cyprus, Metropolis of Morphou, 2009.

[2] Annemarie Weyl Carr, Andréas Nicolaïdès, Gilles Grivaud (Contributor), Ioanna Kakoulli (Contributor), Sophia Kalopissi-Verti (Contributor) and Athanasios Papageorgiou (Contributor), Asinou across Time: Studies in the Architecture and Murals of the Panagia Phorbiotissa, Cyprus, Washington D.C.: Dumbarton Oaks Research Library and Collection, 2013.

[3] Avdelidis N.P. and Moropoulou A., "Applications of infrared thermography for the investigation of historic structures," Cultural Heritage, vol. 5(1), p. 119–127, 2004.

[4] Moropoulou A., Bakolas A., Spyrakos C., Mouzakis H., Karoglou M., Labropoulos K.C., Delegou E.T., Diamandidou D. and Katsiotis N., "NDT investigation of Holy Sepulchre Complex structures," in 14th International Conference Structural Faults & Repair, Edinburgh, UK, 2012.

[5] Moropoulou A., "Scientific Project: Planning and Programming of Conservation Materials and Interventions on the Facades of the Historic Buildings of National Bank of Greece.," NTUA, Athens, 2002.

[6] Moropoulou A. and et al., "Techniques and methodology for the preservation and environmental management of historic complexes—The case of the Medieval City of Rhodes," in In. A. Moropoulou, F. Zezza, E. Kollias & I. Papachristodoulou (Ed.), 4th International Symposium on the Conservation of the Monuments in the Mediterranean Basin, 1997.

[7] Moropoulou A., "Quality control of building materials of National Technical University of Athens Zografou Campus Athens 2004 Olympic Games Building," Scientific responsible: A. Moropoulou, National Technical University of Athens, Athens, 2004.

[8] Moropoulou A., Tsiourva Th., Theoulakis P., Christaras B. and Koui M., "Non—destructive evaluation of pilot scale treatments for porous stone consolidation in the Medieval City of Rhodes," Journal of the European Study Group on Physical, Chemical, Biological and Mathematical Techniques Applied to Archaeology, vol. 56, pp. 259–278, 1998.

[9] Moropoulou A., Koui M., Tsiourva Th., Kourteli Ch. and Papasotiriou D., "Macro—and micro non destructive tests for environmental impact assessment on architectural surfaces," in P.B. Vandiver, J.R. Druzik, J.F. Merkel, J. Stewart (Ed.), Materials Issues in Art and Archaeology, Pittsburgh: Materials Research Society, 1997.

[10] Moropoulou A., Delegou E.T., Avdelidis N.P. and Koui M., "Assessment of cleaning conservation interventions on architectural surfaces using an intergrated methodology.," in P. Vandiver, M. Goodway, J.R. Druzik, J.L. Mass (Ed.), Materials Issues in Art and Archaeology VI, Vol. 712 (pp.69–76)., Pittsburgh: Materials Research Society, 2002.

[11] Moropoulou A., Haralampopoulos G., Tsiourva Th., Theoulakis P. and Koui M., "Long term performance evaluation of consolidation treatments in situ," in In: G. Biscontin, G. Driussi (Ed.), Scienza e Beni Culturali XVI (pp. 239–255), Arcadia Ricerche S, 2000.

[12] Moropoulou A., Theoulakis P., Tsiourva Th. and Haralampopoulos G., "Compatibility evaluation of consolidation treatments in monuments scale," PACT, J. European Study Group on Physical, Chemical, Biological and Mathematical Techniques Applied to Archaeology, vol. 59, pp. 209–30, 2000.

[13] Moropoulou A., Avdelidis N.P., Delegou E.T., Gill C.H. and Smith J., "Study of deterioration mechanisms of vitreous tesserae mosaics," in G. Biscontin, G. Driussi (Ed.), Scienza e Beni Culturali XVIII (pp. 843–851), Arcadia Ricerche S.r.l., 2002.

[14] Bariami, G., M. Faka, A. Georgopoulos, M. Ioannides and D. Skarlatos. "Documenting a UNESCO WH Site in Cyprus with Complementary Techniques." International Journal of Heritage in the Digital Era 1 (2012): 27–32.

[15] Georgopoulos, A., and C. Ioannidis. "3D Visualization by Integration of Multisource Data for Monument Geometric Recording." Recording, Modeling and Visualization of Cultural Heritage, ETH Zurich, Switzerland, 2005, 375–84.

[16] Rothermel, M., Wenzel, K., Fritsch, D. and Haala, N. (2012): SURE: Photogrammetric Surface Reconstruction from Imagery. Proceedings LC3D Workshop Berlin.

[17] Moropoulou A., Karoglou M., Labropoulos K.C., Delegou E.T., Katsiotis N.K. and B.A., "Application of non-destructive techniques to assess the state of Hagia Sophia's mosaics," in T. Matikas (Ed.), Proc. SPIE 8346, Smart Sens, 2012.

[18] Kapsalas P., Maravelaki-Kalaitzaki P., Zervakis M., Delegou E.T. and Moropoulou A., "A morphological fusion algorithm for optical detection and quantification of decay patterns on stone surfaces.," Construction and Building Materials, vol. 22(3), pp. 228–238, 2008.

*Emerging Technologies in Non-Destructive Testing VI – Aggelis et al. (Eds)*
*© 2016 Taylor & Francis Group, London, ISBN 978-1-138-02884-5*

# Raman spectroscopy. A non-destructive tool for on-site chemical analysis of artifacts and monuments

E.Th. Kamilari, S.N. Kouvaritaki & M.G. Orkoula
*University of Patras, Patras, Greece*

C.G. Kontoyannis
*University of Patras, Patras, Greece*
*FORTH/ICE-HT., Patras, Greece*

ABSTRACT: Raman spectroscopy, a vibrational technique, based on a non-elastic photon scattering, extracts information that are related to vibrations of chemical bonds making possible the simultaneous qualitative and quantitative chemical analysis of the material. The non-destructive capability of this technique was explored through its application on a sulfated marble column and the subsequent quantitative determination of calcite conversion to gypsum; on a detailed chemical analysis through application micro Raman on core extractions from Nafpaktos castle and on excavated human mandibles. The changes on the mineral and collagen matrix were investigated. In all cases Raman was used in a non-destructive manner and the quality of the obtained information found equal or surpassed the information obtained through other more conventional techniques.

## 1 INTRODUCTION

Numerous analytical techniques have been used for the characterization of objects of cultural heritage such as artwork, artifacts and monuments. Commonly used methods include trace element analysis with neutron activation (Levine, 2007), determination of C and O isotopes (Herz, 1987), Scanning Electron Microscopy (SEM) (Camuffo et al., 1982, Vanco et al., 2013), Petrographic Microscopy (PM) (Vergesbelmin, 1994), X-Ray Fluorescence Spectroscopy (XRF) (Papadopoulou et al., 2006, Wehling et al., 1999), X-Ray Powder Diffraction (XRD) (Gauri and Holdren, 1981, Vanco et al., 2013), Electron Spin Resonance Spectroscopy (ESR) (Cordischi et al., 1983) and Atomic Absorption Spectrometry (AAS) (Gauri and Holdren, 1981). Despite some advantages, these techniques present certain disadvantages. The isotopic method is rather expensive and time consuming, while the trace element method cannot be applied if the artifact is polluted (Cordischi et al., 1983). The latter along with AAS and XRF are suitable for elemental analysis. XRD is suitable for crystalline substances, while SEM and PM are used for identification. All of them, with the exception of XRF, are destructive for the sample. Non-destructive technique is also the Environmental SEM, provided that a sample from the examined specimen is available (Schleicher et al., 2008). Other techniques can determine properties of materials, such as porosity, but cannot be used for the chemical analysis of the sample. FT-IR has also applied for the investigation of chemical composition of historical and cultural artifacts mostly on paintings and decoration of pottery (Barilaro et al., 2005, Akyuz et al., 2009).

Raman spectroscopy has proven to be a powerful technique for studying materials of various kinds. Raman spectroscopy is a spectroscopic technique based on the Raman effect which is observed when electromagnetic radiation is inelastically scattered while passing through matter. This effect is characterized by a change in the energy between the incident beam and scattered beam. Light is scattered at a slightly longer wavelength than the excitation laser wavelength. The wavelength differences correspond to molecular vibrations and provide information regarding the chemical bonds and thus the structure of the molecule (Das and Agrawal, 2011). Consequently, Raman spectroscopy permits the simultaneous qualitative analysis of individual components of a sample. It is also suitable for quantitative analysis.

The technique has major benefits in various research areas due to its high efficiency, speed of analysis and ease of use. Raman spectra can be acquired rapidly from solid, liquid or aqueous materials with little or no sample preparation. In addition, a small quantity of the sample is required for the analysis. It is worth noting that it is a non-destructive and noninvasive technique facilitating the direct analysis of bulk materials or products in the packaging, such as plastic containers, glass bottles and blister packs. There are some drawbacks, however, and particularly the weakness of the Raman scattering phenomenon and high levels of fluorescence from some samples which overlay the weak Raman signal. High quantum Charge-Coupled Device (CCD) detectors have significantly improved the sensitivity of Raman instruments. Regarding the fluorescent background problems, the development of Fourier- Transform (FT) spectrometers using near-infrared laser solved the problem of fluorescence interference (Zhu et al., 2014).

In archaeometry, precious objects of cultural heritage are examined and the purpose is the development of an analytical method which yields maximum information while causing minimum damage to the examined specimen. Therefore, Raman spectroscopy, due to its non-destructive nature, would be suitable for the study of artwork, archaeological objects or monuments, as the sample remains available for further investigation (Edwards and Munshi, 2005).

Nowadays, portable spectrometers have been designed and are widely used for the *in situ* characterization and identification of materials in art and archaeometry (Colomban et al., 2004, Bersani et al., 2006). The elimination of the need to transfer the samples to the laboratory is useful especially in the cases of bulk objects that are difficult to be moved or monuments of cultural heritage. The use of fiber optic Raman instrumentation has been adapted also for the *in situ* study of objects of historical importance. Fiber optic cables provide a flexible solution for an adequate interface between the device and the spot of the sample to be examined *in situ* (Vandenabeele et al., 2007). The coupling of an optical microscope to a Raman spectrometer has further increased the versatility of the method. The selective analysis of individual components of samples on a micrometer scale by focusing the laser beam on selected spots is possible and thus a 'mapping' of the sample's surface, while the amount of the sample required is even more reduced (Jimenez-Sandoval, 2000).

In this research, Raman spectroscopy was applied to test marble exposed to environment for gypsum; human mandibles were examined for alterations in collagen and mineral part of bone and for non-destructive analysis of core extraction from Nafpaktos castle.

2 EXPERIMENTAL

2.1 *Samples*

Carefully weighed mixtures of calcite-aragonite, aragonite-gypsum and gypsum-calcite in various mol% proportions were prepared from the solids and mixed mechanically.

A marble column located in Athens National Garden was examined for gypsum. Chemical analysis was performed on various core extractions from the castle of Nafpaktos which were collected by the local ephorate of antiquities. Human mandibles from local excavation were also analyzed.

2.2 *Instrumentation*

The following Raman systems were employed:

For the *in situ* measurements a portable Raman spectrometer (i-Raman plus®, B&W Tek) equipped with a 785 nm excitation diode laser and a high quantum efficiency Charge-Coupled Device (CCD) air-cooled detector. The fiber optic Raman system was also equipped with a tripod accessory to mount the head probe of the portable Raman instrument.

For the detailed mapping a Renishaw spectrometer coupled with an optic microscope (Renishaw, InVia, Raman microscope) was used (Fig. 1). A laser line at 785 nm (diode laser) was used for excitation. The system was equipped with a Charge-Coupled Device (CCD) air-cooled detector. Raman spectra were acquired from various points of the specimen under the same conditions, by focusing the laser line through the objective lens (50x) of the optic microscope onto the sample's surface.

For samples exhibiting some fluorescence a FRA-106S FT-Raman instrument (Bruker) operating with a Nd:YAG laser at 1064 nm was used. The scattered light was collected at an angle of 180° (backscattering). The system was equipped with a liquid N2-cooled Ge detector. The power of the incident laser beam was about 225 mW on the sample surface. Similarly, Raman spectra were collected from various spots of the specimen.

Figure 1. Collecting Raman spectra from a core extraction using micro-Raman.

# 3 RESULTS

## 3.1 Sulfated marble sample

Raman spectroscopy was applied for the determination of gypsum in a marble as a result of its exposure to atmospheric pollution which is responsible for the conversion of $CaCO_3$ to $CaSO_4 \cdot 2H_2O$ (Lipfert, 1989). The Raman spectrum of the marble is presented in figure 2C. The stronger in intensity Raman band at 1085 cm$^{-1}$ is assigned to C-O symmetrical stretching of calcite (Fig. 2B), whereas the $v_1$ symmetric stretch modes of the sulfate anion ($SO_4^{2-}$) can be observed at 1006 cm$^{-1}$ (Fig. 2 A) (Kontoyannis et al., 1997). Peak heights of the 1085 and 1006 cm$^{-1}$ bands were chosen to be used for quantitative analysis because the purpose was the development of an easy and reliable method for estimating each ingredient's amount. Calibration curves for calcite-gypsum and aragonite-gypsum mixtures were constructed using the measured intensity of Raman bands at 1085 cm$^{-1}$ for calcite and aragonite, and 1006 cm$^{-1}$ for gypsum. The same methodology was applied for the study of a calcite-aragonite-gypsum mixture. Typical spectra of the various mixtures are shown in figures 2D, E, F. The amount of gypsum concentrations at the examined marble was estimated between 0 and 8.5 mol%.

The results were compared with those obtained by application of quantitative X-Ray powder Diffraction (XRD) analysis. Calibration curves were constructed using the relative intensities corresponding to the 113, the 111 and the 121 reflections of the calcite, aragonite and gypsum, respectively. Both techniques were used for the determination of the amount of gypsum on the marble.

Raman spectroscopy offers the possibility of a point-by-point analysis of the sample surface, while XRD provides the average percentage of the bulk, ground powder sample. Raman spectroscopy calibration curves exhibited lower standard deviation and lower detection limits than the XRD calibration curves. In particular, the detections limits for calcite, aragonite and gypsum were calculated to be 0.1, 0.1 and 0.05 mol% respectively using the former technique, whereas XRD yielded the following detection limits for calcite, aragonite and gypsum: 4, 5 and 1–2 mol% respectively. Additionally, Raman spectroscopy is non-destructive for the sample and more rapid.

## 3.2 Samples from castle of Nafpaktos

Core extractions (a mosaic of different building materials including stones, mortars and clays) from the outer wall of castle of Nafpaktos were analyzed by means of Raman spectroscopy. Two

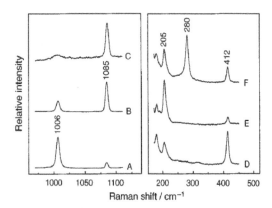

Figure 2. Raman spectra of: A, 20 mol% calcite-80 mol% gypsum; B, 80 mol% calcite-20 mol% gypsum; C, sulfated marble sample; D, 20 mol% aragonite-80 mol% gypsum; E, 80 mol% aragonite-20 mol% gypsum; and F, 40 mol% calcite-30 mol% aragonite-30 mol% gypsum.

randomly selected samples are in figure 3. Before data acquisition, specimens were placed under a stereoscope (DFC295, Leica) for a better and more in-depth observation of the variations from point to point of the cores.

As shown in figure 4, the presence of calcite can be easily identified in the core No 2 due to the bands at 280 cm$^{-1}$, 710 cm$^{-1}$ and 1085 cm$^{-1}$, which represent the lattice vibration mode, the $v_4$ internal mode and the $v_1$ internal mode (symmetric stretching) respectively (Behrens et al., 1995).

Characteristic Raman bands of calcite are distinguished also in the spectrum of core No 4 (Fig. 5). There is a band at 465 cm$^{-1}$ (Figs. 5B, C) which is attributed to the vibration of quartz (Kobayashi et al., 2008).

## 3.3 Human mandible

Raman spectroscopy was used for the study of the quality of a human mandible (Fig. 6). Figure 7 shows a Raman spectrum of the specimen which is dominated by the $v_1$ stretch modes of $PO_4^{3-}$ of bioapatite at 960 cm$^{-1}$ (Penel et al., 2005). Characteristic bands at 1244 cm$^{-1}$ and 1271 cm$^{-1}$, as well as at 1667 cm$^{-1}$, represent the amide III and amide I modes of collagen (Dehring et al., 2006, Tarnowski et al., 2004). Bands at 855 and 878 cm$^{-1}$ are assigned to the $v$(C-C) ring modes of proline and hydroxyproline respectively (Mandair et al., 2009).

Raman spectral changes induced to mandible collagen due to different storage conditions, such as dehydration, were studied. For this purpose, Raman spectra were recorded from a sample of human mandible as received and after hydration in PBS (Phosphate Bovine Solution).

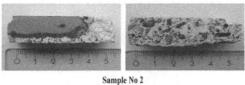

Figure 3. Core extractions from the castle of Nafpaktos.

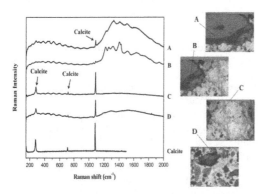

Figure 4. Raman spectra of core No 2. At the right side a photo from the stereoscope showing the exact location where the laser beam of micro-Raman spectrometer was focused.

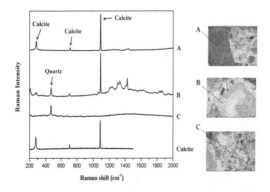

Figure 5. Raman spectra of core No 4. At the right side a photo from the stereoscope showing the exact location where the laser beam of micro-Raman spectrometer was focused.

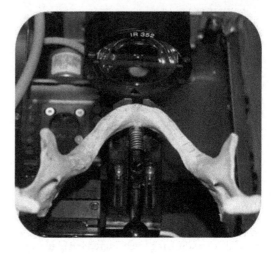

Figure 6. Human mandible.

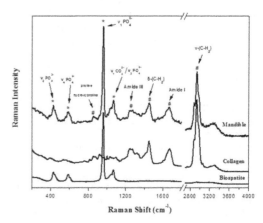

Figure 7. Raman spectrum of mandible, collagen and bioapatite. Peaks marked with (*) stem from bioapatite, while those marked with (#) from collagen.

Due to peak overlapping, deconvolution and band fitting using the appropriate software (Peakfit© v4.0, Jandel Scientific, San Rafael, CA) was considered necessary.

After deconvolution of bands in the amide I region (Figs. 8a, b), smaller in intensity bands at 1658 cm$^{-1}$ and 1680 cm$^{-1}$ were observed repeatedly and were attributed to the dehydrated form of the major bands at 1668 cm$^{-1}$ and 1690 cm$^{-1}$ respectively. The intensity of the band at 1668 cm$^{-1}$ was enhanced compared to the intensities of the other sub-bands in this region.

Intensities of sub-bands of amide I envelope were used and the following ratios were determined: 1660 cm$^{-1}$/1690 cm$^{-1}$, 1658 cm$^{-1}$/1660 cm$^{-1}$

and (1658 cm$^{-1}$ + 1668 cm$^{-1}$)/1690 cm$^{-1}$. The results showed that the presence of the band at 1658 cm$^{-1}$ was due to dehydration and that immersion in PBS can rehydrate the collagen, partially at least.

The ratio of bioapatite to collagen (Mineral to Matrix ratio) and the ratio of non-reducible to reducible cross links were determined in order to monitor the differentiation of bone quality point to point of the mandible (Morris and Mandair, 2011). In the first case, the Mineral to Matrix ratio was calculated using the intensities of the bands at 960 cm$^{-1}$ for bioapatite (Fig. 9a) and at 855 cm$^{-1}$, 875 cm$^{-1}$ and 920 cm$^{-1}$ for proline and hydroxyproline of collagen [I (960 cm$^{-1}$)/(855 cm$^{-1}$ + 875 cm$^{-1}$ +920 cm$^{-1}$)]. For the calculation of the ratio of collagen cross-links, the amide I envelope was used (Fig. 9b) and particularly the intensities of the bands at 1658 cm$^{-1}$ and 1668 cm$^{-1}$ for non-reducible cross-links and at 1680 cm$^{-1}$ and 1690 cm$^{-1}$ for reducible cross-links [I (1658 cm$^{-1}$ + 1658 cm$^{-1}$)/ (1680 cm$^{-1}$ + 1690 cm$^{-1}$) (Kohn et al., 2009). Diagrams were constructed showing the changes of bone quality within the same mandible. Variations from side to side or point to point of the bone were observed, as it is expected, which may be related to various reasons, such as chewing habits.

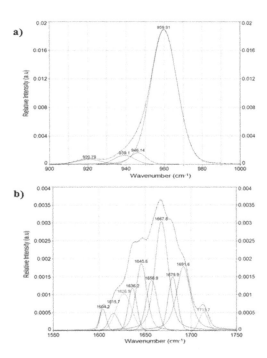

Figure 9. Deconvolution and band fitting under the: a) 902–990 cm$^{-1}$ region (bioapatite) and b) the 1590–1720 cm$^{-1}$ region (amide I) on Raman spectra of the mandible.

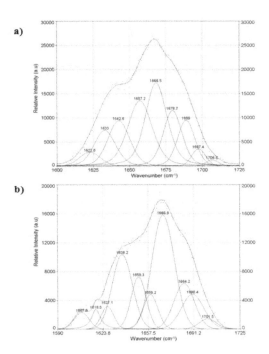

Figure 8. Deconvolution and band fitting under the amide I envelope (1590–1720 cm$^{-1}$) on Raman spectra of the mandible a) before and b) after the immersion in PBS solution.

## 4 CONCLUSIONS

Raman spectroscopy was proved to be an excellent non-destructive tool that can be used for qualitative and quantitative chemical analysis, as well as for surface "mapping" of artifacts and monuments. Among others, it was possible to monitor deterioration of monuments (marble sulfation), detailed analysis of mineral and organic matrix of bone tissues obtained from excavations and "mapping" of core extractions from a castle.

## ACKNOWLEDGEMENTS

Part of the study has been carried out under the support of the research program: THALIS-NTUA-Sustainability and Compatibility of advanced materials and technologies for the protection of cultural heritage monuments: Development of investigation criteria and methodologies – COMASUCH. This research has been co-financed by the European Union (European Social Fund – ESF) and Greek national funds through the Operational Program "Education and Lifelong Learning" of the National Strategic Reference Framework (NSRF) – Research Funding Program: THALIS Investing in knowledge society through the European Social Fund.

# REFERENCES

Akyuz, S., Akyuz, T., Basaran, S., Kocabas, I., Gulec, A., Cesmeli, H. & Ucar, B. 2009. FT-IR and EDXRF analysis of wall paintings of ancient Ainos Hagia Sophia Church. *Journal of Molecular Structure,* 924–26, 400–403.

Barilaro, D., Barone, G., Crupi, V., Donato, M.G., Majolino, D., Messina, G. & Ponterio, R. 2005. Spectroscopic techniques applied to the characterization of decorated potteries from Caltagirone (Sicily, Italy). *Journal of Molecular Structure,* 744, 827–831.

Behrens, G., Kuhn, L.T., Ubic, R. & Heuer, A.H. 1995. Raman Spectra of Vateritic Calcium Carbonate. *Spectroscopy Letters,* 28, 983–995.

Bersani, D., Lottici, P.P., Vignalil, F. & Zanichelli, G. 2006. A study of medieval illuminated manuscripts by means of portable Raman equipments. *Journal of Raman Spectroscopy,* 37, 1012–1018.

Camuffo, D., Delmonte, M., Sabbioni, C. & Vittori, O. 1982. Wetting, Deterioration and Visual Features of Stone Surfaces ina an Urban Area. *Atmospheric Environment,* 16, 2253–2259.

Colomban, P., Milande, V. & Lucas, H. 2004. On-site Raman analysis of Medici porcelain. *Journal of Raman Spectroscopy,* 35, 68–72.

Cordischi, D., Monna, D. & Segre, A.L. 1983. ESR Analysis of Marble Samples from Mediterranean Quarries of Archaeological Interest. *Archaeometry,* 25, 68–76.

Das, R.S. & Agrawal, Y.K. 2011. Raman spectroscopy: Recent advancements, techniques and applications. *Vibrational Spectroscopy,* 57, 163–176.

Dehring, K.A., Crane, N.J., Smukler, A.R., Mchugh, J.B., Roessler, B.J. & Morris, M.D. 2006. Identifying chemical changes in subchondral bone taken from murine knee joints using Raman spectroscopy. *Applied Spectroscopy,* 60, 1134–1141.

Edwards, H.G.M. & Munshi, T. 2005. Diagnostic Raman spectroscopy for the forensic detection of biomaterials and the preservation of cultural heritage. *Analytical and Bioanalytical Chemistry,* 382, 1398–1406.

Gauri, K.L. & Holdren, G.C. 1981. Pollutant effects on stone monuments. *Environmental Science & Technology,* 15, 386–390.

Herz, N. 1987. Carbon and Oxygen Isotopic Ratios - A Database For Classical Greek and Roman Marble. *Archaeometry,* 29, 35–43.

Jimenez-Sandoval, S. 2000. Micro-Raman spectroscopy: a powerful technique for materials research. *Microelectronics Journal,* 31, 419–427.

Kobayashi, T., Hirajima, T., Hiroi, Y. & Svojtka, M. 2008. Determination of $SiO_2$ Raman spectrum indicating the transformation from coesite to quartz in Gfohl migmatitic gneisses in the Moldanubian Zone, Czech Republic. *Journal of Mineralogical and Petrological Sciences,* 103, 105–111.

Kohn, D.H., Sahar, N.D., Wallace, J.M., Golcuk, K. & Morris, M.D. 2009. Exercise Alters Mineral and Matrix Composition in the Absence of Adding New Bone. *Cells Tissues Organs,* 189, 33–37.

Kontoyannis, C.G., Orkoula, M.G. & Koutsoukos, P.G. 1997. Quantitative analysis of sulfated calcium carbonates using Raman spectroscopy and X-ray powder diffraction. *Analyst,* 122, 33–38.

Levine, M.A. 2007. Determining the Provenance of native copper artifacts from Northeastern North America: evidence from instrumental neutron activation analysis. *Journal of Archaeological Science,* 34, 572–587.

Lipfert, F.W. 1989. Atmospheric Damage to Calcareous Stones: Comparison and Reconciliation of Recent Experimental Findings. *Atmospheric Environment,* 23, 415–429.

Mandair, G.S., Bateman, T.A. & Morris, M.D. 2009. Raman Spectroscopy of Murine Bone in Response to Simulated Spaceflight Conditions. *Optics in Bone Biology and Diagnostics,* 7166, 7166077-1-716607-10.

Morris, M.D. & Mandair, G.S. 2011. Raman Assessment of Bone Quality. *Clinical Orthopaedics and Related Research,* 469, 2160–2169.

Papadopoulou, D.N., Zachariadis, G.A., Anthemidis, A.N., Tsirliganis, N.C. & Stratis, J.A. 2006. Development and optimisation of a portable micro-XRF method for in situ multi-element analysis of ancient ceramics. *Talanta,* 68, 1692–1699.

Penel, G., Delfosse, C., Descamps, M. & Leroy, G. 2005. Composition of bone and apatitic biomaterials as revealed by intravital Raman microspectroscopy. *Bone,* 36, 893–901.

Schleicher, L.S., Miller, J.W., Watkins-Kenney, S.C., Carnes-Mcnaughton, L.F. & Wilde-Ramsing, M.U. 2008. Non-destructive chemical characterization of ceramic sherds from Shipwreck 31CR314 and Brunswick Town, North Carolina. *Journal of Archaeological Science,* 35, 2824–2838.

Tarnowski, C.P., Ignelzi, M.A., Wang, W., Taboas, J.M., Goldstein, S.A. & Morris, M.D. 2004. Earliest mineral and matrix changes in force-induced musculoskeletal disease as revealed by Raman microspectroscopic imaging. *Journal of Bone and Mineral Research,* 19, 64–71.

Vanco, L., Kadlecikova, M., Breza, J., Caplovic, L. & Gregor, M. 2013. Examining the ground layer of St. Anthony from Padua 19th century oil painting by Raman spectroscopy, scanning electron microscopy and X-ray diffraction. *Applied Surface Science,* 264, 692–698.

Vandenabeele, P., Castro, K., Hargreaves, M., Moens, L., Madariaga, J.M. & Edwards, H.G.M. 2007. Comparative study of mobile Raman instrumentation for art analysis. *Analytica Chimica Acta,* 588, 108–116.

Vergesbelmin, V. 1994. Pseudomorphism of gypsum after calcite, a new textural feature accounting for the marble sulfation mechanism. *Atmospheric Environment,* 28, 295–304.

Wehling, B., Vandenabeele, P., Moens, L., Klockenkamper, R., Von Bohlen, A., Van Hooydonk, G. & De Reu, M. 1999. Investigation of pigments in medieval manuscripts by micro Raman spectroscopy and total reflection X-ray fluorescence spectrometry. *Mikrochimica Acta,* 130, 253–260.

Zhu, X.Q., Xu, T., Lin, Q.Y. & Duan, Y.X. 2014. Technical Development of Raman Spectroscopy: From Instrumental to Advanced Combined Technologies. *Applied Spectroscopy Reviews,* 49, 64–82.

*Emerging Technologies in Non-Destructive Testing VI – Aggelis et al. (Eds)*
*© 2016 Taylor & Francis Group, London, ISBN 978-1-138-02884-5*

# Evaluation of seismic risk in regional areas by AE monitoring of historical buildings

G. Lacidogna, P. Cutugno, F. Accornero, S. Invernizzi & A. Carpinteri
*Department of Structural, Geotechnical and Building Engineering, Politecnico di Torino, Torino, Italy*

ABSTRACT: In this contribution a new method for evaluating seismic risk in regional areas based on Acoustic Emission (AE) technique is proposed. Most earthquakes have precursors, i.e. phenomena of changes in the Earth physical-chemical properties, that take place prior to an earthquake. Acoustic Emission in materials, and earthquakes in the Earth Crust, despite they take place on very different scales, are very similar phenomena, both are caused by a release of elastic energy from a source located in a medium. For the AE monitoring two important constructions of the Italian cultural heritage are considered: a Chapel of the "Sacred Mountain of Varallo" and the "Asinelli Tower" of Bologna. They were monitored during earthquakes sequences in the relative areas. By using the Grassberger-Procaccia algorithm a statistical method of analysis was developed that detects AEs as earthquake precursors or aftershocks. Under certain conditions it was observed that AEs precede earthquakes. These considerations reinforce the idea that the AE monitoring can be considered as an effective tool for earthquake risk evaluation.

## 1 INTRODUCTION

A new method to evaluate seismic risk in regional area is proposed as an attempt to preserve the Italian historical and architectural cultural heritage. To this purpose, the spatial and temporal correlations between the Acoustic Emission (AE) data, obtained on the monitoring sites, and the earthquakes which have occurred in specific ranges of time and space are examined. The two considered regions are those around the Cities of Varallo and Bologna.

In these cities, the authors have recently carried out the Acoustic Emission monitoring of important historical monuments to evaluate their structural stability. In particular, in the Italian Renaissance Architectural Complex of "The Sacred Mountain of Varallo" the structure of the Chapel XVII was analyzed. While in the City of Bologna the stability of the "Asinelli Tower", known as the highest leaning tower in Italy, was evaluated (Carpinteri et al., 2011; 2013).

Several studies investigated spatial and temporal correlations of epicenters, involving for example the concepts of Omori law and fractal dimension (Back et al., 2002; Parson, 2002; Corral, 2004). Other authors tried to study the complex phenomenon of seismicity using an approach that is able to analyze the spatial location and time occurrence in a combined way, without subjective a priori choices (Tosi et al. 2004). This approach leads to a self-consistent analysis and visualization of both spatial and temporal correlations based on the definition of correlation integral (Grassberger & Procaccia 1983).

Based on these considerations, we have tried not only to analyze the seismic activity in the considered regions, but also to correlate it with the AE activity detected during the structural monitoring. By adopting a modified Grassberger-Procaccia algorithm we give the cumulative probability of the events occurrence in a specific area, considering the AE records and the seismic events during the same period of time. In the modified integral the cumulative probabilities $C (r, \tau)$ are function of the regional radius of interest, $r$, and of the time interval, $\tau$, both considering the peak of AE as earthquake "precursors" or "aftershocks" (Carpinteri et al., 2007).

## 2 CHAPEL XVII AND ASINELLI TOWER'S DESCRIPTION

In Varallo, placed in Piedmont in the province of Vercelli, there is the relevant Italian Renaissance Architectural Complex named "The Sacred Mountain of Varallo". It was built in the XV century on a cliff above the town of Varallo and composed by a Basilica and forty-five chapels, some of which are isolated while others are part of monumental groups.

Because of its high level of damage due to regional earthquakes that have occurred

(Carpinteri *et al.*, 2013), we choose to consider for the AE analysis the Chapel XVII, known as the chapel of the "Transfiguration of the Christ on the Mount Tabor". This structure having a circular plant was built with stone masonry and mortar. The interior walls of the chapel are also equipped with some valuable frescoes (Fig. 1).

In Bologna, Emilia-Romagna, there is another important masterpiece of the Italian architectural and historical heritage, the "Asinelli Tower" that is

Figure 1. Sacred Mountain of Varallo, Mount Tabor Chapel XVII.

Figure 2. The Asinelli and the adjacent Garisenda Towers in the City centre of Bologna. The Asinelli tower is the tallest one on the right.

also the symbol of the City. To study the damage evolution of the structure, we have recently analyzed by the AE technique the influence of repetitive and impulsive events of natural or anthropic origin such as earthquakes, wind, or vehicle traffic (Carpinteri *et al.*, 2011).

The Asinelli Tower was built in the early XIIth century, and it rises to a height of 97.30 m above the ground. At the base, up to a height of 8.00 m, the tower is surrounded by an arcade built at the end of the fifteenth century. Studies conducted in the early twentieth century revealed that the Asinelli Tower leaned westward by 2.25 m, other recent studies have confirmed that its leaning is of 2.38 m: this is the reason why it is known as the tallest leaning tower in Italy (Fig. 2).

## 3 ACOUSTIC EMISSION MONITORING

The AE monitoring is performed by analyzing the signals received from the transducers through a threshold detection device that counts the burst signals which are greater than a certain voltage.

The USAM acquisition system used by the authors consists of 6 pre-amplified sensors, 6 units of data storage, a central unit for the synchronization operations, an internal clock, and a trigger threshold. The PieZoelectric Transducers (PZT) are calibrated over a range of frequency between 50 kHz and 800 kHz. The obtained data from this device are the cumulative counting of each mechanical wave, the acquisition time, the measured amplitude in Volt, the duration, and the number of oscillations over the threshold value for each AE wave (Carpinteri *et al.*, 2011; 2013).

### 3.1 *Chapel XVII*

Firstly we consider the analysis carried out in Chapel XVII of Varallo. The AE monitoring was led on the frescoed masonry in the North wall of the Chapel. The wall showed some damages: a vertical crack of about 3.00 m in length and frescoes detachment. Four AE sensors were located around the vertical crack, while two were placed near the frescos detachment (Fig. 3). The AE monitoring was conducted in two phases for a total duration of about 14 weeks. The first phase started on May 9, 2011 and finished on June 16, 2011; the second one from July 5, 2011 to September 5, 2011. The monitoring results related to the chapel structural integrity are reported in (Carpinteri *et al.*, 2013). We interpreted the AE data considering the amplitude and time distribution of AE signals during the cracking phenomena. From this analysis, we found that the vertical crack, monitored on the North wall of the chapel, evolved during the acqui-

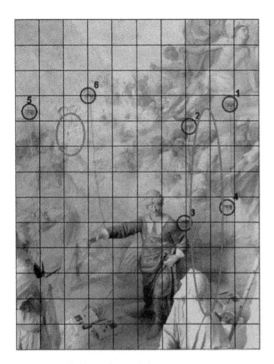

Figure 3. Chapel XVII. View of the monitored areas. Left side: sensors 5, 6, and the frescos detachment. Right side: sensors 1–4 and the vertical crack.

sition period, while the process of detachment of the frescos was mainly related to the diffusion of moisture in the mortar substrate (Carpinteri et al., 2013).

During the monitoring period, among all regional seismic events we considered 21 earthquakes with Richter magnitude ≥1.2, having epicenter within a radius from 60 to 100 km from Varallo. The strongest earthquake was a 4.3 magnitude event occurred on July 25, 2011 at 12:31 PM in the Giaveno area (epicenter about 80 km far from Varallo). Figure 4 displays the AE instantaneous rate (averaged over 1 h) and the sequence of the earthquakes as functions of time, obtained during the monitoring period. Looking at the temporal distribution of earthquakes in relation to AE rate, a quite good correspondence between AE peaks and earthquake events can be observed. Each earthquake is always anticipated or followed by AE events (Fig. 4).

### 3.2 Asinelli tower

The AE activity on the Asinelli Tower was examined in a significant zone for monitoring purposes. This was developed by attaching six piezoelectric sensors to the north-east corner of the tower at an

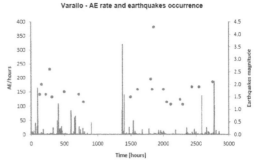

Figure 4. Chapel XVII. AE rate and nearby earthquakes occurrence as functions of time. Seismic events are indicated by red dots.

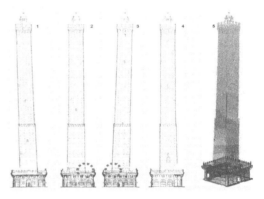

Figure 5. Front views and axonometric view of the Asinelli Tower. Faces (1) South; (2) East; (3) North; (4) West; (5) Axonometric view. The AE transducers were applied to the north-east corner of the tower, in the zones marked with a circle.

average height of ca 9.00 m above ground level, immediately above the terrace atop the arcade. In this area, the double-wall masonry has an average thickness of ca 2.45 m (Fig. 5). AE monitoring began on 23 September 2010 and ended on 10 January 2011, for a total duration of about 16 weeks. From the monitoring process carried out on a significant part of this tower, it was possible to evaluate the incidence of vehicle traffic, seismic activity and wind action on the progress of fracture and damage phenomena within the structure (Carpinteri et al., 2011). In this case among all regional seismic events we considered 43 earthquakes with Richter magnitude ≥1.2, having epicenter within a radius from 25 to 100 km from Bologna, as the most likely to affect the tower stability. The strongest earthquakes were the 4.1 magnitude event occurred on October 13, 2010 at 11:43 PM in the Rimini area (epicenter about 100 km far from

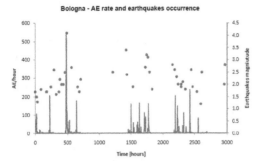

Figure 6. Asinelli Tower. AE rate and nearby earthquakes occurrence as functions of time. Seismic events are indicated by red dots.

Bologna) and the 3.4 magnitude event on November 21, 2010 at 4:10 PM in the Modena Apennines (epicenter about 50 km far from Bologna). The AE instantaneous rate (averaged over 1 h) and the sequence of the earthquakes obtained during the monitoring are displayed in Figure 6. Also in this case a correlation between peaks of AE activity in the structure and regional seismicity can be observed (Fig. 6). The tower, in fact, as in the case of the medieval towers of Alba, is very sensitive to earthquake motions (Carpinteri & Lacidogna, 2006; Carpinteri et al. 2007). The two silent windows for the AE rate evidenced in Fig. 6 (around 1000 and 2000 h), are probably due to the attenuation phenomena of the ultrasonic signals, that do not always reach the sensors during their propagation in the monitored structure.

## 4 GRASSBERGER-PROCACCIA ALGORITHM

AE in materials and earthquakes in the crust are very similar from many aspects and correlated in time, even though they occur at very different scales. In both cases, there is a release of elastic energy from a source located in the medium: respectively the tip of opening micro cracks and the seismic hypocenter (Scholz, 1968). This similarity suggests that the seismic events and the AE events can be related in space and time. Moreover, as highlighted in Carpinteri et al. (2007, 2013), before an earthquake, as deformation increases through tectonic movements, it appears that the Earth's crust surrounding the fault begins to crack transmitting their energy also to structures, as manifested by the increased level of observed AE. This is the reason why AEs can be identified as seismic precursor. Therefore, it is possible to search for a correlation between the AE parameters and the regional seismicity. In our opinion, this approach can be used to identify some warning signals that anticipate a catastrophic collapse of a structure.

Among various studies on the earthquake space-time correlation, there is a statistical method that allows to calculate the degree of correlation both in space and time between a series of AE and the local seismic recordings, collected in the same period. This analysis is based on the generalization of the space-time correlation known as the integral of Grassberger-Procaccia (1983), defined as follows:

$$C(r,\tau) = \frac{1}{N_{EQ}N_{AE}} \sum_{k=1}^{N_{EQ}} \sum_{j=1}^{N_{AE}} \Theta(r - |x_k - x_j|) \times \Theta(\tau - |t_k - t_j|), \quad (1)$$

where $N_{AE}$ is the number of peaks of AE activity registered in site, in a defined time window, $N_{EQ}$ is the number of earthquakes recorded in the surrounding area during the same time window, and $\Theta$ is the step function of Heaviside $\Theta(x) = 0$ if $x \leq 0$, $\Theta(x) = 1$ if $(x > 0)$. The index $k$ refers to the recorded seismic events $\{x_k, t_k\}$ while the index $j$ refers to the recorded AE events $\{x_j, t_j\}$.

Therefore, between all possible pairs of recorded AE and seismic events, the sum expressed by the integral of Grassberger-Procaccia can be calculated for those having the epicentral distance $|x_k - x_j| \leq r$ and the temporal distance $|t_k - t_j| \leq \tau$. Hence, $C(r,\tau)$ is the probability of occurrence of two events, an earthquake and an AE event, whose mutual spatial distance is smaller than $r$ and mutual temporal distance is smaller than $\tau$.

Anyway, this approach does not consider the chronological order of the two types of event. Since the AE time series and the earthquake sequences are closely intertwined in the time domain, the problem of the predictive ability of the AE peaks is still open. The records of AE could be both the consequences of the progressive development of micro damage, or the effect of widespread micro-seismicity. Therefore, a probabilistic analysis can be carried out discriminating between the AE events prior to the earthquake, which are precursors, and the AE following the earthquake, which are aftershocks. This analysis can be performed adopting a modified correlation integral (Carpinteri et al., 2007):

$$C_\pm(r,\tau) = \frac{1}{N_{EQ}N_{AE}} \sum_{k=1}^{N_{EQ}} \sum_{j=1}^{N_{AE}} \Theta(r - |x_k - x_j|) \times \Theta(\tau - |t_k - t_j|) \theta(|t_k - t_j|), \quad (2)$$

where "+" and "−" in the Heaviside function are used to take into account that the AE events could be respectively seismic precursors and aftershocks.

In this way, the function $C_+(r, \tau)$ gives the probability that a peak of AE, detected at a certain time, will be followed by an earthquake in the subsequent days within a radius of $r$ kilometers from the AE monitoring site. Varying the thresholds $r$ and $\tau$ in Eq. (2), two cumulative probability distributions can be constructed, one for each condition (sign "+" or "−") and then represented.

Moreover from the space-time combined correlation integral, the time correlation dimension, $Dt$, and the space correlation dimension, $Ds$, for sets of events within space-time distances $r$ and $\tau$, can be defined:

$$D_t(r,\tau) = \frac{\partial \log C(r,\tau)}{\partial \log \tau}, \quad (3)$$

and

$$D_s(r,\tau) = \frac{\partial \log C(r,\tau)}{\partial \log r}. \quad (4)$$

If $C(r,\tau)$ is a power-law in both variables, then $Dt$ and $Ds$ correspond to the temporal and spatial fractal dimensions, respectively. More in general, the behavior of $Dt$ and $Ds$ as a function of $r$ and $\tau$ characterize the clustering features of the considered events in space and time.

## 5 SPACE-TIME CORRELATION BETWEEN AE AND SEISMIC EVENTS

In this section, a correlation between seismic and acoustic events through the application of the modified integral of Grassberger-Procaccia is obtained. The data series of the analyzed AEs are shown in Figures 4 and 6, and related to the above-mentioned time intervals.

The seismic events were taken from the web-site http://iside.rm.ingv.it/iside/standard/result.jsp?rst = 1&page = EVENTS#result (Seismic Catalog of ISIDE, Italian Seismological Instrumental and Parametric Data-Base). During the defined monitoring periods the seismic events, selected with Richter magnitudes ≥1.2, were found having epicenters within a radius from 60 to 100 km around the site of Varallo, while were found in a radius from 25 to 100 km around the City of Bologna.

First of all we distinguished between the environmental contributions due to crustal trembling (external source), and the structural damage contributions (inner source). Therefore, considering the seismic precursors, we discarded all the AE signals originated from a damage source definitely localized in the buildings masonry. Then, by applying the modified Grassberger-Procaccia correlation integral to the data series, we obtained the cumulative probabilities $C_\pm(r, t)$, as a function of the radius of interest $r$ and of the interval of occurrence $\tau$, both considering the peak of AEs as earthquake "precursors" $C_+$ or as "aftershocks" $C_-$ (Fig. 7a and b).

As shown by the figures, everywhere we find $C_+ > C_-$. Therefore, it can be easily recognized that the filtered signals always have a greater probability to be seen like seismic precursors. In practice we see, as already found in Carpinteri et al. (2007a), that the monitored structures behave as a sensitive seismic receptors.

## 6 AE CLUSTERING IN TIME AS SEISMIC PRECURSORS OR AFTERSHOCKS

In the diagrams of Figure 7a and b, the cumulative probabilities $C_\pm(r, \tau)$ were not represented in the first week of the monitoring periods. Therefore,

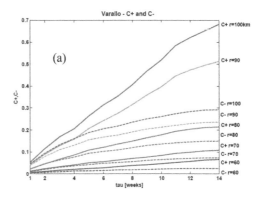

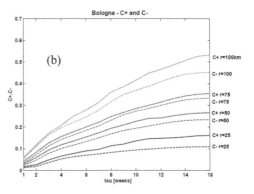

Figure 7. Evolution of the modified correlation integrals, both considering AEs as earthquake "precursors" $C_+$ or as "aftershocks" $C_-$, as a function of the radius of interest and for different time windows, during the monitoring period. (a) Chapel XVII and area around the City of Varallo; (b) Asinelli Tower and area around the City of Bologna.

to better define the properties of the AE events as earthquake precursors or aftershocks, the AE events that precede or follow the considered earthquakes, were further studied in very short periods of time.

In particular, both for the Chapel XVII and for the Asinelli Tower, through the parameter $Dt$ of Eq. (3), the AE distribution in time were analyzed. We considered a range from 3 to $1.440 \times 10^3$ minutes (equivalent to 24 hours) after the earthquakes, "aftershocks", and a range from 10 minutes to 24 hours before the earthquake, "precursors".

As mentioned in Section 4, the $Dt$ parameter of Eq. (3) characterizes the clustering features of the events in time. Studying by this parameter the time coupling of AE activity to regional seismicity, the AE events as "seismic aftershocks" ($C_-$) are found to have a non-uniform distribution over time. This is evidenced by the three regression lines of the plotted data in bi-logarithmic scale (see Fig. 8a and b).

For short time delays, $\tau = 3-10$ minutes, AEs are short-term correlated to earthquakes, as described by the higher value of $Dt = 1.38$ and $1.24$. This means that AEs following an earthquake are more likely to happen within $\tau = 3$ minutes after the seismic event diagrams. These short time delays clearly suggest that such AE activity is due to growing damage in Chapel and in the Tower, provoked

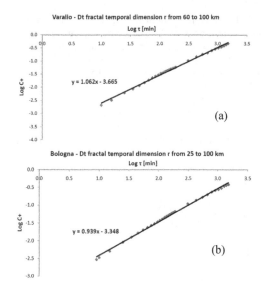

Figure 9. Analysis of the distribution over time by the parameter $Dt$ described in Eq. (3), of the AE events considered as "seismic precursors" ($C_+$). The time window amplitude $\tau$ is comprised between 10 min and 24 hours. The selected seismic events are all those considered during the defined monitoring periods. (a) Chapel XVII (Varallo); (b) Asinelli Tower (Bologna).

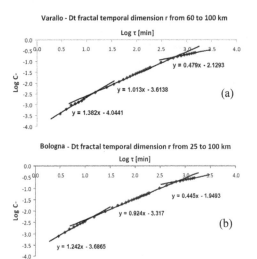

Figure 8. Analysis of the distribution over time by the parameter $Dt$, described in Eq. (3), of the AE events considered as "seismic aftershocks" ($C_-$). The time window amplitude $\tau$ is comprised between 3 min and 24 hours. The selected seismic events are all those considered during the defined monitoring periods. (a) Chapel XVII (Varallo); (b) Asinelli Tower (Bologna).

by seismic activity. While for $\tau = 1.0 \times 10^3$ to $1.440 \times 10^3$ minutes lower values of $Dt = 0.48$ and $0.45$ are obtained, this means that about one day after the earthquakes seismic shocks have almost no effect on the structural damage.

As regards AE events preceding earthquakes, i.e. considered as "seismic precursors" ($C_+$), the linear coefficients ($Dt = 1.06$ and $0.94$) in both the diagrams yield to a uniform probability density of finding AE events prior to an earthquake for a wide range of time delays, up to $\tau = 1.440 \times 10^3$ minutes (see Fig. 9a and b). Thus, observing the absence of acceleration in the AE activity before the earthquakes, it is confirmed that the precursory feature of AE events could be investigated, because it could be analyzed considering larger time windows.

## 8 CONCLUSIONS

In order to evaluate the seismic risk of different areas, it is suggested to observe a spatial and temporal correlation of these phenomena considering AE as earthquake precursors. Based on the Grassberger-Procaccia integral, a statistical method of analysis is employed. It gives the cumulative probability of the events occurrence in a specific area, considering the AE records and the seismic events during the same period of time. Two important constructions of the Italian cultural heritage were considered: the "Sacred Mountain of Varallo" and the "Asinelli Tower". Given the interesting premises that emerge from this study, the Grassberger-Procaccia integral could also be applied more extensively in several monitoring sites belonging to different seismic regions, not only for defining the seismic hazard, but also to handle it as an effective tool for earthquake risk mitigation.

## ACKNOWLEDGEMENTS

Dr E. De Filippis, Director of the Piedmont Sacred Mountains Institute, is gratefully acknowledged. The authors thank the Municipality of Bologna and Eng. R. Pisani for having allowed the study on the Asinelli Tower. Special thanks are also due to Dr. G. Niccolini for the constructive discussions on this issue.

## REFERENCES

Bak P., Christensen K., Danon L. & Scanlon T. 2002. Unified scaling laws for earthquakes. *Phys. Rev. Lett.* 88: 178501–178504.

Carpinteri A. & Lacidogna G. 2006. Structural monitoring and integrity assessment of medieval towers. *Journal of Structural Engineering* 132: 1681–1690.

Carpinteri A., Lacidogna G., Invernizzi S. & Accornero F. 2013. The Sacred Mountain of Varallo in Italy: Seismic risk assessment by Acoustic Emission and structural numerical models. *The Scientific World Journal*, 1–10. http://dx.doi.org/10.1155/2013/170291.

Carpinteri A., Lacidogna G., Manuello A. & Niccolini G. 2011. A study on the structural stability of the Asinelli Tower in Bologna. *Proc. of the 5th Conference in Emerging Technologies in NDT (5th ETNDT), Ioannina, Greece, September 19–21, 2011, 123–129.*

Carpinteri A., Lacidogna G. & Niccolini G. 2007. Acoustic emission monitoring of medieval towers considered as sensitive earthquake receptors. *Natural Hazards and Earth System Science* 7: 251–261.

Corral A. 2004. Long-term clustering, scaling and universality in the temporal occurrence of earthquakes. *Phys. Rev. Lett.* 92: 108501.

Grassberger P. & Procaccia I. 1983. Characterization of strange attractors. *Physical Review Letters* 50: 346–349.

Parson T. 2002. Global Omori law decay of triggered earthquakes: Large aftershocks outside the classical aftershock zone. *J. Geophys. Res.* 107: 2199–2218.

Scholz C. H. 1968. The frequency-magnitude rela-tion of microfracturing in rock and its relation to earthquakes. *Bull. Seismol. Soc. Am.*58: 399–415.

Tosi P., De Rubeis V., Loreto V, & Pietronero L. 2004. Space-time combined correlation integral and earthquake interactions. *Annals of Geophysics* 47: 1–6.

*Emerging Technologies in Non-Destructive Testing VI – Aggelis et al. (Eds)*
*© 2016 Taylor & Francis Group, London, ISBN 978-1-138-02884-5*

# Non-destructive evaluation of historic natural stone masonry with GPR

F. Lehmann & M. Krüger
*MPA., Universität Stuttgart, Stuttgart, Germany*

ABSTRACT: In this paper, we present the use of ground penetrating radar for the evaluation of natural stone masonry on historic structures. In the examples, the buildings had to be assessed regarding the wall's type of construction and with respect to voids behind or cracks within the outermost layer of stone. The boundary conditions for the measurement are described along with the applied radar system. The data analysis and results are presented. Account is given on the limits and possibilities of radar for the use on historic masonry structures made of natural stone.

## 1 INTRODUCTION

The internal structure of load-bearing walls in historic buildings is in many cases unknown. A frequently encountered construction includes one or two outer layers of worked stone on both sides, enclosing a more or less non-uniform filling. When the structural capacity of such a wall is to be evaluated, information on the actual outer stone thickness has to be gathered. Although this can generally be very precisely determined by the extraction of cores, this is often undesirable, if not impossible at a large amount of points, especially on historic structures.

A commonly employed method for the non-destructive evaluation of natural stone thickness is ground penetrating radar. Although the application of this method in the given context is nothing "new" per se, it still seems to lead a rather marginal existence related to its benefits. One of the reasons might be the additional challenges linked to the use of radar on sandstone and historic structures. We therefore want to present some of the merits of radar measurements for the use on historic structures, especially with respect to the evaluation of stone thickness.

## 2 RADAR MEASUREMENTS

### 2.1 *Measurement principle*

Radar measurements are a well-established tool in civil engineering for a wide range of applications (Patitz 2012, Moropoulou 2013, Jol 2009). The probably most common task is the localization of reinforcement, which certainly derives from the very distinctive radar signals evoked by conductive materials. This allows for additional conclusions like concrete cover and even an approximate estimation of the reinforcement diameter. But the method may also be used for thickness evaluation of structural members or soil layers, for void detection or comparative moisture and salinization analysis.

Radar is an electromagnetic wave in the frequency domain from about 3 MHz to 110 GHz. The range used in non-destructive structural testing is usually between 200 MHz to 2.5 GHz. Generally, antennas with a pronounced center frequency are used, although recent research now allows the application of broad-band antennas between 2 and 8 GHz as well (Kurz 2015).

Non-destructive testing with radar exploits the construction material's different electrical and magnetic properties. Radar reflections from inside the structure are produced at sudden changes of the electromagnetic wave speed, which depends on the mathematical product of the prevailing permittivity and permeability. The larger the difference between two adjacent materials, the larger is the amplitude of the reflected signal. The material thickness up to the first layer can be calculated with the triggered time from sending to receiving the signal onset, assuming that the wave speed is known. As this is generally not the case, the wave speed has to be determined at a part of the structure with known thickness or a reasonable value has to be assumed. In the latter case, precise depth calculations are consequently not possible. A further option is the calibration on point-like targets such as rebars within the material.

### 2.2 *Radar measurements on sandstone*

The use of radar on structures made of natural stone, such as sand—or limestone, inhabits some specific characteristics, which distinguish the measurements from the standard application on concrete (Maierhofer 2001, Colla 2010, Binda 1998). The most

441

evident difference is certainly the discrete construction from separate blocks of limited dimension in all three spatial directions. This leads to an increased number of side reflections during the measurements, which results in a more complex radargram. Concrete on the other hand is generally cast in slabs or beams of larger dimension, which eliminates four, respectively two boundaries in most parts.

Natural stone masonry, especially in historic structures, is often unworked on the invisible rear face. The depth information in a radar measurement is thus non-uniform to a certain degree, which can make the data interpretation more challenging. Additionally, the structured surface is often realized in a quarried or at least rough manner for optical reasons. While scanning on the surface, the radar antenna therefore not only constantly changes the distance to the reflective horizon, but is also likely to be tilted in the process. Consequently, the well-known reflection shapes in the radargram are altered and the data assessment might become inconclusive.

Most natural stones possess a good degree of native inhomogeneity, with the details most likely being unknown to the radar operator. The effects on radar measurements are a variable wave speed and especially reflections from internal surfaces, such as beddings and inclusions of foreign material. Distinct radar reflections therefore most likely point towards some noticeable feature in the structure. This is different in the evaluation of reinforced concrete, where reflections from the rebars may hide other objects.

Moisture, if present in significant amounts in the material, is a further factor to be considered (RILEM 2005, Binda 1994). In case water is inhomogeneously distributed inside the material, the resulting water front leads to a significant change in electromagnetic wave speed and relative permittivity, which yields an apparent increase of the stone thickness. Although this effect may be used for the detection of water, there is a risk that the existence of water is unknown and hence erroneous conclusions might be drawn. In return, real changes in the available structural thickness can be misinterpreted as water. Hence, special care has to be taken towards this consideration, particularly at lower wall parts near the foundation.

## 3 APPLICATION ON HISTORIC STRUCTURES

### 3.1 *Stiftskirche, Stuttgart*

The Stiftskirche (collegiate church) is the oldest church structure in Stuttgart, Germany. It was heavily damaged during the Second World War, but was reconstructed in 1958. Its southern tower was originally built in the 13th century (up to the third floor) and extended with three additional storeys in the 14th and 15th century (Figure 1). The tower is built from reed sandstone. However, information on the structure is scarce.

To examine the internal structure and condition of the tower's wall, an exemplary façade on the 4th floor (Figure 2) was assessed with radar. Based on the results, two cores were subsequently extracted at locations of former patch work for calibration and verification. The cores were additionally used for a salt analysis. For the radar assessment, each row was separately scanned in horizontal direction. Although this would have already been sufficient for an initial interpretation, the wall was also scanned vertically to gather supplementary data. A GSSI 1.6 GHz antenna was selected for the measurements.

The radargrams (Figure 3) show the horizontal radar scans on stone rows 4 and 2 (top and bottom, respectively). A background removal filter was applied to the data. The vertical joints are very clearly visible and marked with vertical dashed

Figure 1. Stiftskirche Stuttgart. The examined southern tower and façade is shown in the picture's center. The exemplary radar measurements were carried out on the 4th floor (height of the adjacent cullis).

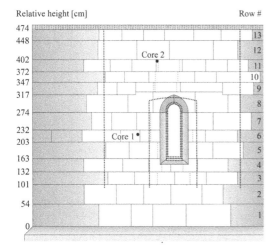

Figure 2. Measurement area on the 4th floor. The stone rows are consecutively numbered starting from the bottom. The light area marks the measured stone blocks. The dotted lines correspond to the window niche with a reduced wall thickness and the approximate extension of the interior room.

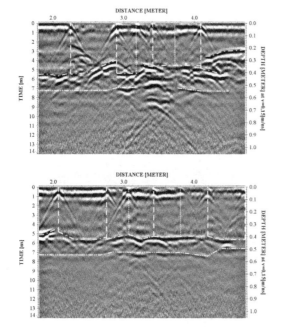

Figure 3. Radargrams from stone rows 4 (top) and 2 (bottom). The vertical dashed lines correspond to the joints, the dotted lines to obvious cracks on the surface. The horizontal dash-dotted lines indicate reflection horizons. The wave speed in the sandstone was assumed with 0.15 m/ns according to the angle of the diffraction hyperbolas.

lines in both figures. This visibility is one of the key elements for measurements on structures made of natural stone. Although the reflections from the joints superpose the radargrams, they are also used to relate any found anomaly to the specific stone within the collective. The data interpretation can usually be based on the assumption that the stone dimensions act as geometric boundaries. That is, a significant change in the measured thickness can naturally only occur in the joints. If this is not the case, some particularity may be present.

Some obvious cracks in the stone surface were recorded during the measurements and are visible in the radargrams, similar to joints. They are marked with vertical dotted lines. It is generally not possible to draw conclusions about the crack depth from radargrams, although it could be assumed that cracks went through the entire stone. For one thing they extend into the stones parallel to the main direction of wave propagation, therefore only producing recordable reflections at salient steps. Moreover, the crack face distances are most likely rather narrow, which results in a poor visibility in the graph.

The thicknesses of the outer stones appear as characteristic line parallel to the surface in both graphs in Figure 3, marked with dash-dotted lines. The average thickness is about 40 cm, calculated with a wave speed of 0.15 m/ns. Row 2 (lower graph) shows a rather undisturbed picture. A second horizon is in large parts visible at around 55 cm depth. Beyond that the signal vanishes in the noise, due to the energy loss at the previous interfaces. The radargram of row 4 (upper graph) on the other hand exhibits some smaller thicknesses. The reflection amplitude is rather strong and features no phase change in most parts, here indicated by the white-black coloring. This implies a relative permittivity of the second material greater than the outer sandstone. This means most importantly that the façade has some kind of backfill and is not separated by an air-filled gap. The exception to this is the outermost stone on the left, where a phase change can be noticed by the black-white reflection, indicating a gap.

The decreased thickness that is measured in the center corresponds to the window niche on the tower's inside. Here the wall was determined to have an outer layer of 37 cm sandstone, followed by 13 cm brick. Both values can approximately be recognized in the figure. Note that the brick most likely has a different wave speed. The given depth axis is therefore only valid for the outer sandstone.

A considerable reflection is found in the very right stone in the upper graph, which suggests a strongly different thickness compared to the other stones. However, the stone is most likely not separated from the rest of the building and thus has been placed this way.

The cores extracted from the wall draw a consistent picture. They show an outer layer of approximately 40 cm sandstone, followed by an infill of historic lime mortar with a thickness of about 10 cm and more solid sandstone thereafter.

## 3.2 Basilica St. Martin and Oswald, Weingarten

The Basilica St. Martin and Oswald in Weingarten, Germany, is a baroque abbey church from the early 18th century. The western front of its central structure is framed by two towers (Figure 4). It is constructed as multilayered stone masonry wall with a load-bearing center and a curtain wall façade made from sandstone. In the course of heritage preservation work it was questioned whether the curtain and structural walls have a connection over the entire area, or if larger parts may buckle under unfavorable conditions. It was therefore decided to use georadar measurements to identify stones without close contact to the structure behind.

For the radar assessment each stone row of the central building's façade was individually scanned. In contrast to vertical measurements, this approach allows an easy and explicit matching of the data to the individual stone blocks, because the scanning proceeds parallel to the layered arrangement. With an anticipated stone thickness of about 25 cm, a GSSI radar antenna with a center frequency of 1.6 GHz was chosen.

The knowledge of actual stone thickness was not explicitly required for the data evaluation regarding their embedment conditions. In order to check the data for plausibility it is nevertheless desirable to obtain the wave speed of the tested material. Two sandstones of the same variety already on site for substitutions (Figure 5) were used to calibrate the measurements. Although having been freshly cut and thus most probably exhibiting slightly different properties compared to the installed ashlars, they can be used for a fair estimation of the velocity. Furthermore it was possible to use the stones as model for a block without connection, showing the same reflection characteristics as equivalent built-in ones.

The velocity was determined to be 0.137 m/ns for the fresh stones, which corresponds to typical values for dry sandstone. For an assumed thickness of 25 cm, a deviation of 0.02 m/ns would result in an inaccuracy of about ± 3.6 cm, which is acceptable for the given task.

The radar measurements were evaluated regarding peculiarities up a depth of 10 ns, corresponding to a thickness of about 70 cm. The main focus of attention was set to the first 30 cm according to the question of missing backfill mortar after

Figure 4. Western façade of the Basilica St. Martin and Oswald Weingarten. Photo: "Weingarten Basilica", Clemens v. Vogelsang, flickr, CC-BY-2.0 / desaturated and cropped from original, brightened background.

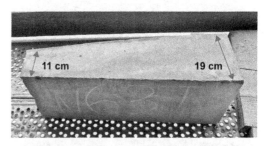

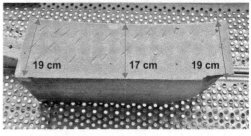

Figure 5. Sandstone ashlars used to calibrate the radar measurements and as model for stone blocks without mortar backfill.

the outermost stone layer, i.e. the curtain wall. In the data review, it das differentiated between three categories of reflection and hence three levels for possible voids. Figure 6 shows the unfiltered radargram of three intact stones with a close connection to the mortar in the vertical joint towards the structural wall. The signals are very clear up to the depth of the joint, which results from the notably plan stone surface and homogeneous stone. An ashlar with some peculiarity, in detail a greatly decreased thickness of only about 12 cm, is shown in Figure 7. However, the reflection amplitude is not as strong as it would be expected from a void. A probable explanation is a comparatively thin stone with a good contact to the mortar. Figure 8 is an example for a radargram including a stone without backfill mortar. It is easily recognized by

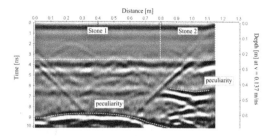

Figure 9. Example for a radargram with a detected peculiarity in a larger depth.

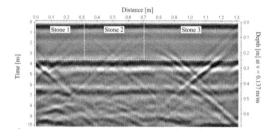

Figure 6. Example for a radargram with all three stones marked as intact.

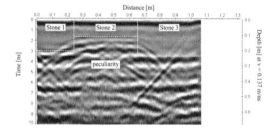

Figure 7. Example for a radargram featuring a peculiar stone.

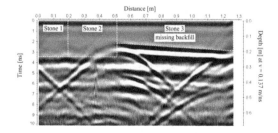

Figure 8. Example for a radargram with the typical reflection produced by a stone with missing backfill.

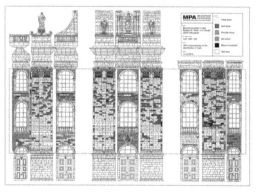

Figure 10. Overview charts of the measurement results, pinpointing the individual stones which require further attention.

the strong reflection between 14 to 18 cm. One complementary classification was introduced to mark peculiarities which were detected well behind the curtain wall (Figure 9).

The evaluation using categories allowed the creation of a readily understandable, color-coded overview chart, which maps the stone blocks according to their determined condition as "intact", "peculiar" or "void" (Figure 10). It could be used to plan and to carry out the further constructional measures.

3.3 *Blauer Turm, Bad Wimpfen*

The tower Blauer Turm (Figure 11) in Bad Wimpfen, Germany, was built around 1200 as castle keep. Since then it has often changed its purpose of use including as a fire lookout and as a prison. At least three large modifications of the tower's appearance are known from historic documents, often being carried out after incidents of partial destruction by fire.

The tower's structure is a stonework cavity wall with an outer and inner layer of each 30 to 60 cm

Figure 11. East façade of the Blauer Turm in Bad Wimpfen. Photo: "Blauer Turm in Bad Wimpfen", Peter Schmelzle, Wikimedia Commons, CC-BY-SA-2.5 / desaturated and cropped from original, brightened background.

worked limestone. The infill consists of rock and lime mortar with a thickness of about 2.2 m, decreasing towards the top. The outside stone blocks are quarry-faced with an embossing of several centimeters. The tower as a whole is well preserved. However, there are structural problems, mainly in the manifestation of an extensive number of large vertical cracks throughout the stonework. The first documented accounts of this issue were reported after the last constructive change in 1852, which brought along the neo-Gothic spire. Since then several attempts to stabilize the structure were made, all of them including some kind of circumferential bracing to prevent a baggy collapse of the tower. Yet, the reports on spreading or new gaping cracks were not stopped and thus a new investigation of the tower and the reasons for its instability was initiated.

In the course of work several ashlars with laminar separations parallel to the surface in a depth of about 10 cm were noticed. To assess whether radar or ultrasound can successfully be utilized to detect these shallow delaminations under the given conditions, it was decided to carry out a set of trial measurements. If successful, the most reasonable method would be used on all four façades to judge if there is a general problem or if the separations are only a local phenomenon. The challenges for the measurements lie in the differential head of the vertical surface and the unknown degree of contact between the separation faces.

With respect to the expected separation being in a depth of only 10 cm, it was decided to use a radar antenna with a higher frequency. Taking into account the very rough surface, it was also important to select a small antenna size. The specifications were met by a GSSI 2 GHz antenna. The radargram shown in Figure 12 is the record of a horizontal scan across five stone blocks. Again a background removal filter was applied to the data. By use of the reflection hyperbolas, the wave velocity was determined to be around 0.1 m/ns, which corresponds to typical values of rather wet limestone. A complementary determination of the actual moisture confirmed water contents close to saturation.

A picture of the measurement area is mapped below the radargram for easy comparison. The vertical lashed lines mark the joints between successive stones in both parts of the figure. Dotted lines indicate cracks that were visible on the surface. Some of the reflections from the cracks or joints seem to deviate from the straight vertical. The reason can either be an actual incline, which seems reasonable, or a tilt of the radar antenna, which is likewise a probable cause with regard to the quarried surface.

Most importantly for the evaluation, a shallow separation can be noticed in approximately 10 to

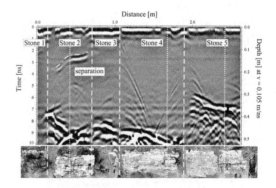

Figure 12. Radargram of a wall section at the tower Blauer Turm. Stone 2 shows a separation in about 10 to 18 cm depth.

18 cm depth in stone number 2. None of the other ashlars show a hint for a similar defect. For stones two to five, the thickness can be read from the collected data to be about 45 cm. Stone five has a depth of 10 cm less. The reflection amplitudes are comparatively strong, which is a bit surprising, since limestone and lime mortar should have a similar electromagnetic wave speed. An explanation would be the existence of an air filled gap after the outer shell of the double wall. The phase information, however, is not well-defined in most parts. Only stone 5 features a black-white wiggle trace, which here stands for a reflection at an interface towards air.

## 4 CONCLUSIONS

Using the example of three historic structures we have presented the application of georadar on natural stone cavity walls. The radar measurement technique has long proven to be a valuable tool in the assessment of civil engineering structures. On historic structures built from natural stone masonry, e.g. ashlars of sand—or limestone, it may be used to non-destructively evaluate the structure with regard to its stone block's inbound dimensions. Further information, like the existence of direct backfill, may be extracted to a certain degree under favorable conditions.

## REFERENCES

Binda, L., Colla, C., Forde, M.C. 1994. Identification of moisture capillarity in masonry using digital impulse radar. *Journal of Construction and Building Materials* 8, 101–107.

Binda, L., Lenzi, G., Saisi, A. 1998. NDE of masonry structures: use of radar tests for the characterisation of stone masonries. *NDT&E International* 31, 411–419.

Colla, C. et al. 2010. *Diagnostic by imaging: 3D GPR investigation of brick masonry and post-tensioned concrete*. 13th International Conference on Ground Penetrating Radar (GPR), Lecce, Italy. DOI 10.1109/ICGPR.2010.5550175.

Jol, H.M. (ed.) 2009. *Ground penetrating radar. Theory and applications*. Elsevier Science. ISBN 978-0-444–53348–7.

Kurz, F. 2015. *Evaluation of different measurement methods for moisture in screed samples*. 8th Competence Center for Material Moisture Conference, Karlsruhe, 7.-8. October 2015. *Accepted for publication*.

Maierhofer, C., Leipold, S. 2001. Radar investigation of masonry structures. *NDT&E International* 34, 139–147.

Moropoulou, A. et al. 2013. Non-destructive techniques as a tool for the protection of built cultural heritage. *Construction and Building Materials* 48, 1222–1239.

Patitz, G. 2012. Altes Mauerwerk zerstörungsarm mit Radar und Ultraschall bewerten. In Fouad, N.A. (ed.), *Bauphysik Kalender 2012*: 203–245. Berlin: Ernst & Sohn Verlag.

RILEM 2005. Recommendations of RILEM TC 177-MDT: Masonry durability and on-site testing. *Materials and Structures* 38, 283–288.

*Emerging Technologies in Non-Destructive Testing VI – Aggelis et al. (Eds)*
*© 2016 Taylor & Francis Group, London, ISBN 978-1-138-02884-5*

# Application of the mortar static penetration test to historical buildings

D. Liberatore, L. Sorrentino & L. Frezza
*Department of Structural and Geotechnical Engineering, "Sapienza" University of Rome, Rome, Italy*

N. Masini & M. Sileo
*CNR-IBAM (Italian Research Council, Institute for Archaeological and Monumental Heritage), Tito Scalo (PZ), Italy*

V. Racina
*MonitorING s.r.l., Potenza, Italy*

ABSTRACT: A new version of the penetration test for mortars of historical masonry is based on the idea of "static" penetration. The pin is driven at constant velocity by a stepper motor controlled by a computer. The result of the test is the penetration load as function of the penetration depth. The penetrometer has been tested on masonry walls characterized by decayed mortar with poor cohesion. The results are correlated with those of a previous percussion penetration test.

## 1 INTRODUCTION

The response of masonry constructions is noticeably affected by mortar quality, as shown, for instance, by recent seismic events in Emilia Region, Northern Italy (Sorrentino et al. 2014a,b). In this respect, penetration test, thanks to its moderate destructiveness, can provide useful information in the assessment of existing structures.

Penetration tests were originally introduced in geotechnics around 1930 and standardized by Mohr (1940). The procedure consists to insert a metal drill into the soil and to measure the corresponding resistance to penetration. The penetrometers can be either static (pressure drills) or dynamic (percussion drills). The static penetration test, known as Cone Penetration Test (CPT), consists of inserting a conic drill into the soil at controlled speed, usually 20 mm/s (ASTM 2005). The dynamic penetration test, or Standard Penetration Test (SPT), uses a sampler tube, with external diameter 50 mm, internal diameter 35 mm and length approximately 650 mm (BS 1990, ASTM 1999). On the history and applications of SPT see Rogers (2004).

In structural engineering, one of the most common tests for the *in situ* determination of the compressive strength of concrete is the Windsor probe. A metal probe is driven at high speed into the concrete by a calibrated explosive charge. The penetration depth is generally below 10 mm. The measure is the mean value of penetration over 3 blows. Given the resistance to penetration and the Mohs'

hardness of the aggregate, the test estimates the compressive strength of concrete through empiric relations (ASTM 2003, BS 1992).

A penetration test specifically devised for masonry mortars is the PNT-G, which measures the energy spent to make a hole with a normalized drill (Gucci & Barsotti 1995, Gucci & Sassu 2002). Experimental investigations show that the penetration energy is correlated with the compressive strength, for sand mortars with strength less than or equal to 4 MPa. For this test, the penetration depth is about 5 mm. Another test aimed at estimating the compressive strength of mortar is based on the measurement of the penetration depth increment of a steel probe for each hammer blow (Felicetti & Gattesco 1998).

A similar penetration test, i.e. based on the penetration of a metal pin driven by multiple blows, is aimed at estimating the friction coefficient of mortar of historical masonry, which is often decayed and without cohesion (Liberatore et al. 2001, Liberatore et al. 2004). The penetration depth is as high as 40–50 mm. The test provides the average number of blows required to drive the pin 1 mm. This test overcomes the drawbacks of many foregoing penetrometers, i.e. the high energy of the Windsor penetrometer, which is unsuited to decayed mortars, and the shallow penetration depth of the PNT-G, which only enables to obtain information about a limited depth. Through empirical relations, the test provides the friction coefficient of the mortar, given the joint thickness and the compressive stress along the vertical direction. An experimental database over more 100 buildings has been collected in recent years by means of this test.

A new version of this penetration test has been presented, based on the principle of "static" penetration (Liberatore et al. 2014). The pin is driven at constant velocity into the mortar, and the test provides the penetration load as function of the penetration depth. The penetrometer has been tested on decayed mortars, providing preliminary information on their quality. Finally, the new penetration test is compared with the previous percussion test.

## 2 DESCRIPTION OF THE PENETROMETER

The penetrometer is based on driving a metal pin into the mortar at controlled speed, and on continuously acquiring the applied load. The result of the test is the diagram of the load as function of the penetration depth. The penetration depth can reach 70 mm, so as to overcome the surface layer, which is mostly decayed because of exposition to atmospheric agents.

The apparatus consists of:

1. penetrometer;
2. control unit;
3. lock plate.

The penetrometer consists of: pin, drum, cylinder, worm screw, gearmotor, stepper motor, load cell, encoder, connecting cable (Figure 1). The pin is made of steel, has a diameter equal to 3 mm and a conic tip with angle equal to 27.5°. It is lodged into the drum, which slides inside the cylinder. The drum is driven, through the worm, by the gearmotor, which in turn is operated by the stepper motor. The cylinder head guarantees the correct alignment of the pin and enables to clamp the penetrometer to the plate, which in turn is anchored to the wall to be tested. The load is measured by a button load cell located between the end of the worm and the drum. The displacement of the pin is measured by the encoder. During the test, the penetrometer is linked to the control unit by the connecting cable.

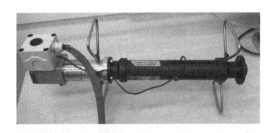

Figure 1. Penetrometer.

The control unit consists of: computer, batteries, multifunction plug (Figure 2). The control unit, featuring a touch screen, allows specifying the settings of the test (speed, acquisition step, initial and final displacements, maximum load), acquires the data during the test (displacement and applied load), pulls back the pin to the home position at the end of the test, records the data on a flash USB unit, enables to manually operate the pin, either at high or low speed, both configurable. During the test, the control unit displays in real-time the displacement of the pin and the applied load. The batteries allow executing the test without resorting to an alternating current supply. The batteries are charged through a proper cable connected to the multifunction plug.

The lock plate (Figure 3) is made of steel and is anchored to the masonry through expansion screws. The penetrometer is bolted to the plate. The

Figure 2. Control unit.

Figure 3. Lock plate.

penetrometer can be mounted on the plate in five different positions, in order to perform as many penetrations on an individual joint. At the end of the test, the penetrometer is removed from the plate, which in turn is removed from the masonry, together with the expansion screws which can be reused.

## 3 EXPERIMENTAL INVESTIGATION

The *in situ* penetration test is affected by several factors of scatter at different scales. A first group of factors consists of:

– heterogeneity of materials and constructive techniques, depending on the construction of different storeys and/or different portions of the building;
– different degree of masonry decay, depending, e.g., on the different exposure to weathering and maintenance quality;
– different compressive stress of masonry.

Besides these factors, which can be classified as "global", there are "local" factors within each wall, depending on the behaviour at the "microscale", such as the mechanism of stress transfer between the units through the mortar joints, and the degree of confinement of the mortar. Finally, scatter is present between the different penetrations on an individual joint.

The quantities involved in the test are:
$u$ = displacement of the pin;
$F$ = applied load;
$Fm$ = mean applied load;
$u_1$ = displacement corresponding to the contact between pin and mortar;
$u_2$ = displacement at the end of the test.
The displacement of the pin inside the mortar joint is $\Delta u = u_2 - u_1$.
The work spent for the displacement of the pin from $u_1$ to $u_2$ is:

$$W = \int_{u_1}^{u_2} F(u) \, du = F_m \, \Delta u \qquad (1)$$

The experimental investigation has been performed on three buildings in Southern Italy: a residential building at Avigliano (province of Potenza), the Church of Madonna del Carmine at Pomarico (province of Matera) and the Church of San Biagio at Calimera (province of Lecce). The first two buildings have not undergone past strengthening interventions, whereas the third one underwent a partial masonry repointing.

For each wall, five penetrations have been performed on a sub-horizontal mortar joint. The test has been done after the removal of the plaster.

The penetration speed has been set to 0.20 mm/s, the load has been limited to 1500 N and has been acquired at displacement steps of 0.10 mm; the maximum displacement has been varied during the campaign as specified in the following.

The preparation of the test on an individual joint (visual inspection of the panel, determination of the joint, anchoring of the plate) required about 30–45 minutes, and each penetration (clamping of the penetrometer to the plate, penetration, unclamping) about 5 minutes, for a total amount of about 55–70 minutes.

### 3.1 *Residential building, Avigliano*

The building, located in the Municipality of Avigliano, is a private house, with three storeys, uninhabited and to be renovated. The masonry has irregular fabric, with undressed limestone units and lime mortar. The mortar is decayed and has poor cohesion.

The penetration test has been carried out on two masonry panels, denoted as panel A and panel B.

For the test on panel A (Figures 4–5) the end displacement has been limited to 60 mm. The load has been acquired at displacement step 0.10 mm.

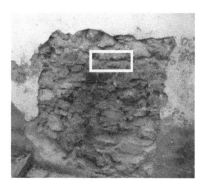

Figure 4. Residential building, Avigliano, panel A.

Figure 5. Residential building, Avigliano, panel A, detail after the test.

In Figure 6 are reported the diagrams of $F$ as function of $u$, and in Table 1 the values of $u1$, $u2$, $\Delta u$, $W$, $Fm$.

It can be noticed that the load generally increases with the displacement because of the increase of the contact surface between pin and mortar and the penetration of the pin into zones where the mortar is less decayed. Slightly decreasing branches can be noticed as well, which could be caused by local voids and by the displacements/rotations of larger aggregates induced by the displacement of the pin. Further insight into this phenomenon could come in the future from correlating the penetrometric test with porosity measures in laboratory, as already performed for the dynamic penetration test (Calia et al. 2013).

As for panel B, the end displacement has been set equal to 66 mm. The results are reported in Table 2. It can be noticed that panel B is characterized by $F_m$ values which are significantly lower than those of panel A, confirming the influence of "global" factors of scatter. For both panels, the high scatter within the joints can be noticed.

### 3.2 Madonna del Carmine, Pomarico

The Church of Madonna del Carmine (Figure 7), dating back to the end of the 16th century, is located 6 km out from the centre of Pomarico. The masonry has irregular fabric, with undressed limestone units and lime mortar. Also in this case, the mortar is decayed and has poor cohesion.

Penetration tests have been performed on two masonry panels of the façade, from outside (panel A) and from inside (panel B). The end displacement has been set equal to 60 mm. The results on panel A are reported in Table 3.

Table 1. Displacements, work and mean load, residential building, Avigliano, panel A.

| Test | $u_1$ (mm) | $u_2$ (mm) | $\Delta u$ (mm) | $W$ (J) | $F_m$ (N) |
|---|---|---|---|---|---|
| 1 | 22.0 | 60.0 | 38.0 | 17.11 | 450.18 |
| 2 | 22.0 | 60.0 | 38.0 | 10.58 | 278.51 |
| 3 | 28.5 | 60.0 | 31.5 | 10.51 | 333.79 |
| 4 | 33.0 | 60.0 | 27.0 | 4.84 | 179.30 |
| 5 | 28.0 | 60.0 | 32.0 | 6.12 | 191.19 |
| | | | | Mean = | 286.59 |
| | | | | Std dev = | 111.43 |

Table 2. Displacements, work and mean load, residential building, Avigliano, panel B.

| Test | $u_1$ (mm) | $u_2$ (mm) | $\Delta u$ (mm) | $W$ (J) | $F_m$ (N) |
|---|---|---|---|---|---|
| 1 | 27.5 | 66.0 | 38.5 | 6.17 | 160.14 |
| 2 | 23.5 | 66.0 | 42.5 | 11.53 | 271.36 |
| 3 | 40.0 | 66.0 | 26.0 | 7.85 | 302.02 |
| 4 | 48.0 | 66.0 | 18.0 | 1.24 | 68.63 |
| 5 | 7.7 | 66.0 | 58.3 | 6.79 | 116.48 |
| | | | | Mean = | 183.73 |
| | | | | Std dev = | 100.00 |

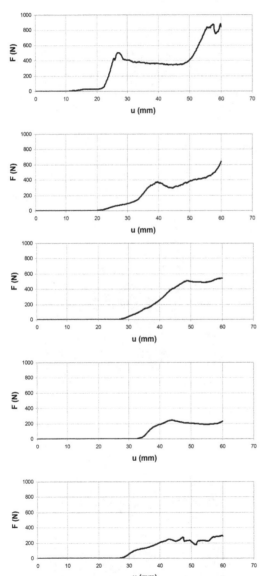

Figure 6. Load-displacement diagrams, residential building, Avigliano, panel A.

Figure 7. Madonna del Carmine, Pomarico.

Figure 8. San Biagio, Calimera.

Table 3. Displacements, work and mean load, Madonna del Carmine, Pomarico, panel A.

| Test | $u_1$ (mm) | $u_2$ (mm) | $\Delta u$ (mm) | $W$ (J) | $F_m$ (N) |
|---|---|---|---|---|---|
| 1 | 6.0 | 60.0 | 54.0 | 11.83 | 219.09 |
| 2 | 6.0 | 60.0 | 54.0 | 24.59 | 455.34 |
| 3 | 9.0 | 60.0 | 51.0 | 16.34 | 320.32 |
| 4 | 7.0 | 60.0 | 53.0 | 14.54 | 274.33 |
| 5 | 0.0 | 60.0 | 60.0 | 11.08 | 184.61 |
| | | | | Mean = | 290.74 |
| | | | | Std dev = | 105.63 |

Table 4. Displacements, work and mean load, Madonna del Carmine, Pomarico, panel B.

| Test | $u_1$ (mm) | $u_2$ (mm) | $\Delta u$ (mm) | $W$ (J) | $F_m$ (N) |
|---|---|---|---|---|---|
| 1 | 0.0 | 60.0 | 60.0 | 24.14 | 402.31 |
| 2 | 0.0 | 60.0 | 60.0 | 18.63 | 310.45 |
| 3 | 11.0 | 60.0 | 49.0 | 13.08 | 266.92 |
| 4 | 8.0 | 60.0 | 52.0 | 17.03 | 327.58 |
| 5 | 6.0 | 33.5 | 27.5 | 9.32 | 338.81 |
| | | | | Mean = | 329.21 |
| | | | | Std dev = | 49.17 |

The same settings have been used for the test on panel B. The last penetration has been interrupted at displacement $u_2 = 33.5$ mm, when the load was approaching the maximum load (1500 N). The results are reported in Table 4. It can be noticed that the mean values of $Fm$ are similar for the two panels, indicating their substantial homogeneity. Panel B shows a small scatter within the tested joint.

Table 5. Displacements, work and mean load, San Biagio, Calimera, panel SB1.

| Test | $u_1$ (mm) | $u_2$ (mm) | $\Delta u$ (mm) | $W$ (J) | $F_m$ (N) |
|---|---|---|---|---|---|
| 1 | 4.0 | 58.0 | 54.0 | 13.49 | 249.83 |
| 2 | 2.5 | 58.0 | 55.5 | 3.65 | 65.70 |
| 3 | 12.1 | 58.0 | 45.9 | 5.69 | 124.04 |
| 4 | 3.9 | 52.0 | 48.1 | 6.37 | 132.45 |
| 5 | 3.8 | 58.0 | 54.2 | 6.53 | 120.47 |
| | | | | Mean = | 138.50 |
| | | | | Std dev = | 67.57 |

Table 6. Displacements, work and mean load, San Biagio, Calimera, panel SB2.

| Test | $u_1$ (mm) | $u_2$ (mm) | $\Delta u$ (mm) | $W$ (J) | $F_m$ (N) |
|---|---|---|---|---|---|
| 1 | 0.4 | 60.0 | 59.6 | 5.30 | 89.00 |
| 2 | 0.1 | 60.0 | 59.9 | 12.99 | 216.82 |
| 3 | 0.1 | 60.0 | 59.9 | 18.04 | 301.17 |
| | | | | Mean = | 202.33 |
| | | | | Std dev = | 106.82 |

### 3.3 San Biagio, Calimera

The Church of San Biagio (Figure 8) is located 2 km east from the centre of Calimera. It is composed by a hypogeum dating back to the 11th century and a cant of a nave built in the 18th century. The masonry is made by squared calcarenite ashlars with lime mortar and some addition of cement mortar in the above walls. The building lies in a serious state of decay and at present is propped.

Penetration tests have been performed on two masonry walls: in the hypogeum (panel SB1, Table 5) and in one of the lateral façades of the nave, from outside (panel SB2, Table 6).

453

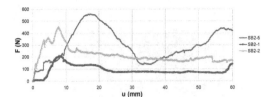

Figure 9. Load-displacement diagrams, San Biagio, Calimera, panel SB2.

Figure 9 shows the load-displacement diagrams of panel SB2. These put in evidence a different behaviour in the first 10–15 mm, corresponding to the repointing layer made by cement mortar, which is characterized by high stiffness and applied load, compared to the lower values found at deeper penetration, corresponding to lime mortar (15–60 mm).

## 4 CORRELATION OF THE STATIC TEST WITH THE PERCUSSION TEST

The static penetration test can be correlated with the percussion penetration test (Liberatore et al. 2001, Liberatore et al. 2004). In order to perform such correlation, the same walls previously mentioned have been tested with the percussion penetrometer.

The relation between the static test and the percussion test is based on the equivalence of work. The theoretical work of an individual blow of the percussion penetrometer is:

$$w = \frac{1}{2} K \delta^2 \qquad (2)$$

where:
 $K = 4200$ N/m (elastic stiffness of the spring);
 $\delta = 20$ mm (initial relative displacement of the spring);
 yielding:
 $w = 0.84$ J.

The effective work spent to displace to pin, with a single blow, is less than the theoretical work because of a number of dissipative phenomena, as:

– the impact between the hammer and the drum of the penetrometer;
– the overcoming of the static friction between pin and mortar;
– the inner friction of the apparatus.

The effective work can be written as:

$$w' = \eta w \qquad (3)$$

where $\eta$ is an efficiency coefficient ($\eta < 1$), which shall be calibrated on experimental basis.

Assuming to perform a percussion test in the same point of the static test, the equivalence of work between the two tests yields:

$$n\, \eta\, w = F_m\, \Delta u \qquad (4)$$

where $n$ is the number of blows of the percussion test. Equation 4 can be rewritten:

$$\frac{n}{\Delta u} = \frac{F_m}{\eta w} \qquad (5)$$

where the ratio $n/\Delta u$ is the average number of blows per penetration unit of the percussion test. Given the mean load $Fm$ of the static test and an estimation of the efficiency coefficient $\eta$, the average number of blows per penetration unit $n/\Delta u$ of the percussion test can be calculated according to Equation 5.

The efficiency coefficient can be estimated according to the following equation, when both the percussion and the static tests are available on the same panel:

$$\eta = \frac{F_m}{w} \frac{\Delta u}{n} \qquad (6)$$

In Table 7 are reported the mean values of $n/\Delta u$ resulting from percussion tests on five panels of the investigated buildings, together with the efficiency coefficients calculated according to Equation 6.

The values of $n/\Delta u$ are typical of poor mortars. Since the percussion test has been performed on joints different from those of the static tests, the differences between the two tests are also affected by "local" factors of scatter. The standard error on the mean efficiency coefficient is equal to $0.134/\sqrt{5} = 0.060$. Accordingly, the interval mean ± std err is equal to [0.395, 0.515]. The efficiency coefficient is rather scattered (CoV = 0.296), indi-

Table 7. Mean $n/\Delta u$ from percussion tests and efficiency coefficient.

| Building | Panel | Mean $n/\Delta u$ (mm−1) | $\eta$ |
|---|---|---|---|
| Residential building | A | 0.683 | 0.499 |
| Residential building | B | 0.694 | 0.315 |
| Madonna del Carmine | A | 0.660 | 0.524 |
| San Biagio | SB1 | 0.266 | 0.619 |
| San Biagio | SB1 | 0.760 | 0.317 |
| | | Mean = | 0.455 |
| | | Std dev = | 0.134 |

cating that further data are necessary to establish a reliable relation between the two types of test.

## 5 CONCLUSIONS

A new penetration test has been described, based on the penetration of a pin inside the masonry mortar at constant speed. The test provides the applied load as function of the displacement of the pin. Based on the force-displacement diagram, the work and the mean applied load can be calculated. The test is moderately destructive for the mortar, and non destructive for the units. Therefore, it is suitable for a widespread use.

*In situ* tests have been carried out on a residential building and two churches. Two sub-horizontal joints have been tested per building. The test on an individual joint requires 55–70 minutes. The test proves to be suitable for historical masonry.

The static penetration test has been put in relation with the percussion test, developed in previous studies, by means of the equivalence of the work spent for the displacement of the pin. Whereas the static penetration test directly provides the work, the percussion test requires the estimation of its efficiency coefficient, due to the presence of dissipative phenomena, such as the impact between the hammer and the drum of the penetrometer. The efficiency coefficient has been evaluated for five panels where both types of test have been carried out. Because the values obtained for the efficiency coefficient are rather scattered, further investigations are necessary to obtain a reliable estimation.

## ACKNOWLEDGEMENTS

The development of the penetrometer and the research have been funded under the projects:

- "AITECH" and "PRO_CULT" of IBAM-CNR (Basilicata Region ERDF 2007–13);
- "Large Equipments 2011" of "Sapienza" University of Rome; the authors are grateful to the coordinator of the project, Prof. Giorgio Monti;
- "IN-CUL.TU.RE." of IBAM-CNR, coordinated by Dr. Giovanni Quarta;
- "Dipartimento di Protezione Civile—Consorzio ReLUIS", signed on 2013–12–27, Research Line Masonry Structures. The opinions expressed in this publication are those of the authors and are not necessarily endorsed by the Dipartimento della Protezione Civile.

The penetrometer, conceived and designed by the first and the fourth author, has been made in the mechanical workshop of Mr. Felice Villano and the electromechanical workshop of Mr. Carlo Ligrani.

The authors are grateful to Mr. Raffaele Carlucci for the availability of the residential building at Avigliano, and to Prof. Michelangelo Laterza, Ente Diocesi di Matera-Irsina and Municipality of Pomarico for the availability of Madonna del Carmine.

Finally, the authors thanks Dr. Omar Al Shawa for his assistance during the preliminary laboratory tests and the drafting of the user's manual.

## REFERENCES

ASTM Standard D 1586. 1999. Standard Test Method for Penetration and Split-Barrel Sampling of Soils. *ASTM International.* West Conshohocken, PA. http://www.astm.org.

ASTM Standard C 803/C 803-M. 2003. Standard Test Method for Penetration Resistance of Hardened Concrete. *ASTM International.* West Conshohocken, PA. http://www.astm.org.

ASTM Standard D 3441. 2005. Standard Test Method for Mechanical Cone Penetration Tests of Soils. *ASTM International.* West Conshohocken, PA. http://www.astm.org.

BS 1377–9. 1990. Methods for Test for Soils for Civil Engineering Purposes. In Situ Tests. *BSI British Standards.* London. http://www.bsi-global.com.

BS 1881–207. 1992. Testing Concrete. Recommendations for the Assessment of Concrete Strength by Near-to-Surface Tests. *BSI British Standards.* London. http://www.bsi-global.com.

Calia, A., Liberatore, D. & Masini, N. 2013. Approach to the study of conservation of historical masonry mortars by means of the correlation between porosimetry and penetrometric test. First results. *Proc. of "Built Heritage 2013 Monitoring Conservation Management", Milan, Italy, 18–20 November 2013*, 1133–1140.

Felicetti, R. & Gattesco, N. 1998. A penetration test to study the mechanical response of mortar in ancient masonry building. *Materials and Structures* 31: 350–356.

Gucci, N. & Barsotti, R. 1995. A non-destructive technique for the determination of mortar load capacity in situ. *Materials and Structures* 28: 276–283.

Gucci, N. & Sassu, M. 2002. Resistenza delle murature: valutazione con metodi non distruttivi, il penetrometro PNT-G. *L'Edilizia*, XVI(2): 36–40.

Liberatore, D., Spera, G. & Cotugno, M. 2001. A new penetration test on mortar joints. Proc. of *the RILEM TC177MDT Workshop "On site control and non destructive evaluation of masonry structures and materials"; Mantua, Italy, 13–14 November 2001*, 191–202.

Liberatore, D., Spera, G. & Racina, V. 2004. Una prova penetrometrica per valutare le caratteristiche della malta: prime calibrazioni. *Proc. of the 11th Conference "L'Ingegneria Sismica in Italia"; Genoa, Italy, 25–29 January 2004.*

Liberatore, D., Masini, N., Sorrentino, L., Racina, V., Frezza, L. & Sileo, M. 2014. A static penetration test for masonry mortar. *Proc. of the SAHC 2014, 9th International Conference on Structural Analysis of Historical Constructions; Mexico City, Mexico, 14–17 October 2014.*

Mohr, H.A. 1940. Exploration of Soil Conditions and Sampling Operations. *Soil mechanics series*, Bull. 269. Graduate School of Engineering, Harvard University.

Rogers, J.D. 2004. Notes on the Standard Penetration Test. *Advanced Engineering Geology & Geotechnics*, GE 441. University of Missouri.

Sorrentino, L., Liberatore, L., Decanini, L.D. & Liberatore, D. 2014. The performance of churches in the 2012 Emilia earthquakes. *Bull Earthquake Eng* 12(5): 2299–2331. http://dx.doi.org/10.1007/s10518-013-9519-3.

Sorrentino, L., Liberatore, L., Liberatore, D. & Masiani, R. 2014. The behaviour of vernacular buildings in the 2012 Emilia earthquakes. *Bull Earthquake Eng* 12(5): 2367–2382. http://dx.doi.org/10.1007/s10518-013-9455-2.

*Emerging Technologies in Non-Destructive Testing VI – Aggelis et al. (Eds)*
*© 2016 Taylor & Francis Group, London, ISBN 978-1-138-02884-5*

# Non destructive testing to perform service of the evaluation of conservation works

R. Manganelli Del Fà, C. Riminesi, S. Rescic & P. Tiano
*CNR-ICVBC., Florence, Italy*

A. Sansonetti
*CNR-ICVBC., Milano, Italy*

ABSTRACT: The best practices of a conservation work pass through a pilot yard. In this preliminary step conservation works are tested in a small representative area: cleaning, desalination, application of consolidants and/or water-repellent treatment. Most of the tests are carried out on site using a small number of non-destructive testing—rugged, reliable and at low-cost—aimed at measuring some physical/chemical features of the surface under conservation, both before and after the application of the product. In this paper a survey about of the most diffused protocols currently used in Italy, is presented. A critical review of benefits and drawbacks, possibilities and problems offered by specific testing systems in collecting evidences, will be carried out.

## 1 INTRODUCTION

The experience of the practice in applying methods and products on building surfaces, inform us that the same product provide different performances when applied on different substrates. When a Conservator or the designer of the conservation plan is called to choose products to be used in a specific case he addresses his attention to experience and bibliographic research as a the starting point. Then when a set of different products and/or methods is selected, they should be applied in the real case to be evaluated as to their performances. Choosing a product means to evaluate it in terms of effectiveness and possible harmfulness. A pilot yard in which this choice finds the final response is the right answer. For these reasons we think the best practice in choosing products and/or methods for the conservation works, passes through the organization of a pilot yard. In this preliminary step conservation works are tested in small representative areas: cleaning, desalination, application of consolidants and/or water-repellent treatment. In this approach it is possible to outline some advantages and drawbacks; the former include that the products are tested in the real case using the real state of conservation of the substrate. For instance when a cleaning method is applied all the problems in simulating soiling are avoided (Mecchi, 2008). The latter involve the difficulty to locate a surface with enough homogeneity in order to apply the different products in the same conservation point.

## 2 DESIGNING AN ANALYTICAL PLAN

The performance of a product and/or a method is not always something involving a direct and specific "parametrization". On the contrary is something similar to the analysis of circumstantial evidences. It is of crucial importance to gather as much evidences and then to locate the equilibrium point in order to consciously choose the best product. For example in comparing the performances of two different water-repellent products, we are called to gather wettability, colour differences, water vapour permeability. May be that product A is good in colour (it does not induce any change in stone colour), but not good in permeability; how to weight the different parameters? These few considerations allow to outline that designing the correct analytical plan is of crucial importance, and only at the end of the plan the conservation scientist will be able to weight correctly the different evidences and operate the final choice. Examining the standard useful in this field (EN 16581: 2014), it is possible to observe that for the most part they include laboratory tests. Some of the tests are carried out even on site using a small number of non-destructive testing—rugged, reliable and at low-cost—aimed at measuring some physical/chemical features of the surface under conservation, both before and after the application of the product. In this paper a survey about of the most diffused protocols currently used in Italy, is presented.

Table 1. Measures of water absorption by contact sponge method on Brick and Terracotta. Ca' Granda Building. Milan.

|  | Product | % reduction respect $T_0$ $T_1$ | $T_3$ |
|---|---|---|---|
| Brick | RC80 | 99,8 | 99,4 |
|  | W290 | 99,7 | 99,4 |
|  | FLUOLINE | 99,5 | 97,5 |
| Terracotta | RC80 | 98,7 | 97,3 |
|  | W290 | 98,5 | 97,4 |
|  | FLUOLINE | 98,6 | 97,4 |

## 3 CONTACT SPONGE

The Contact sponge method is a water absorption test useful in evaluating differences induced by water repellents treatments (Vandevoorde 2009). The method consists in applying a soaked sponge with a controlled pressure on the surface to be tested. The water needed to soak the sponge is measured in advance (ml); the soaked sponge is weighed, applied and kept in contact for 1–2 minutes; then it is weighed again. The weight difference is a measure of the water absorbed by capillarity. The system provides reliable data when used on porous substrates such as brick, mortar, tuff or calcarenite. The measures are carried out before ($T_0$) and after ($T_1$) the application of the water repellent treatments. For example the data acquired on bricks in the so-called Ca' Granda building (a terracotta and brick facade) are presented. After the application of the product the absorption is decreased dramatically; the% reduction is calculated (Table 1). A good product such as siloxanes (RC80, W290) or fluorinated resins (Fluoline) decrease the absorption over the threshold of 98%, depending on the physical features of the substrate (porosity, microstructure and hydrophilicity) ($T_1$). This data is a good marker to be used in monitoring: in fact, after 3 years ($T_3$) the% reduction decrease respect $T_0$ is still good. If the product looses its functionality the test provide an immediate results, easy to be interpreted.

## 4 COLOUR MEASUREMENTS

Surface colour measurements in evaluation of cleaning operations is still to be used carefully in order not to mislead results (Mecchi, 2008). Much easier is its use in evaluation of conservation treatments, such as water-repellent products. CIE L*a*b* 1931 System results as the most used in this kind of application. The correct approach requires to measure surface colour both before and after the application of the conservation treatment. An evaluation of the difference between the colour parameters (before and after the treatment) can be obtained by considering each color as a point in the three-dimensional L*a*b* space and calculating the Euclidean distance between them (ΔE). Hence ΔE value measures the "colour difference" due to the application of the treatment. It implies that the lesser is ΔE, the better is the product, at least by chromatic point of view. It is crucial to assure the perfect repositioning of the instrument head, in order to avoid to introduce any difference which is not to ascribe to colour changes due to the product presence. This last consideration regards in particular polychrome surfaces. The use of a jig is helpful in repositioning the head of the instrument. Depending from the colour homogeneity of the surface to be treated, at least 5 colour measurements should be done, but in our experience it is better to increase the data set to almost 25, if it is allowed by dimension and morphology of the substrate to be measured. Therefore to get 25 measures is necessary to have a flat surface of almost 20/25 $cm^2$. The instrument is user-friendly, portable, light; hence it is simple to plan a monitoring of treated surfaces. In Table 2 data regarding the comparison of 3 water repellent products are reported (1 = RC80; 2 = W290; 3 = Fluoline). Colour parameters was measured before ($T_0$), after

Table 2. Colour Measures of bricks treated with water-repellent products. $T_0$ = before the treatment; $T_1$ = after the treatment; $T_2$ monitoring after 3 years natural ageing. Ca' Granda Building. Milan.

| | $T_0$ Aug. '09 | | | $T_1$ Oct. '09 | | | $T_2$ Nov. '12 | | |
|---|---|---|---|---|---|---|---|---|---|
| | L* | a* | b* | L* | a* | b* | L* | a* | b* |
| 1 | 48,2 | 10,4 | 11,4 | 38,1 | 13,6 | 14,6 | 41,4 | 12,7 | 13,4 |
| 2 | 49,7 | 8,7 | 9,3 | 41,9 | 12,7 | 15,1 | 43,5 | 10,8 | 12,0 |
| 3 | 53,1 | 15,3 | 17,1 | 51,3 | 14,9 | 17,8 | 52,0 | 14,6 | 17,2 |

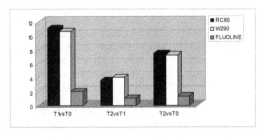

Figure 1. ΔE calculated by the values presented in Table 2 Ca' Granda Building. Milan.

the treatment ($T_1$) and monitored after 3 years of natural ageing ($T_2$). In fig. 1, the values of ΔE are reported: $T_1vsT_0$ represents the difference induced by the application of the product; $T_2vsT_1$ represents the difference induced by the ageing; $T_2vsT_0$ represents the actual difference induced by the application, at the time of monitoring.

## 5 DIELECTROMETRY BY MICROWAVE EVANESCENT METHOD (SUSI)

The SUSI instrument (acronym of Integrated Instrument for measuring Moisture and Salt content in Italian) is based on the measurement of the dielectric contrast between water (about 80) and the host material in the microwave frequency range. The dielectric constant of typical materials used in masonry (brick, plaster, mortar, stone) is about 3–5. Thanks to the high dielectric contrast, it is possible to perform early diagnostic of the moisture content and about the dissolved electrolytes (soluble salts) present within the masonry. The measure is carried out by non invasive method and in real time, needing some seconds to be completed

The dielectric constant (permittivity) determines the ability of the material to polarize itself, or to form and to orient electric dipoles when an external electric field is applied. This behavior is quantified by the real part of permittivity; on the contrary the imaginary part is related to the dissipative effects due to friction acting between the electric dipoles while chasing the variations of the electric field. The imaginary part also depends on the electrical conductivity of the material, or rather from ions mobility. Therefore, a dry material containing salts is characterized by a low electrical conductivity, even if the salts are present in high quantity.

The proposed system measures both the real part and the imaginary part of permittivity in the microwave frequency range by evanescent field method (Evanescent-Field Dielectrometry) (Olmi, 2006). The support is investigated by the electric field that spills out from the aperture of the opened resonant cavity (a microstrip line short-circuited on one side and closed on a truncated coaxial line on the other side). In this case, the sensor is a truncated coaxial line closed on one side on the resonant circuit and on the other side opened on the material under investigation. Fig. 2 shows the setup of the measurement system. It consists of a vector network analyzer (on the left in the picture), a probe (on the right in the picture), and a notebook on which is installed the software to control the instrument, as well as for the real time elaboration.

The measurement is performed in a non-destructive way keeping in contact the probe with

Figure 2. Measurement setup of SUSI instrument.

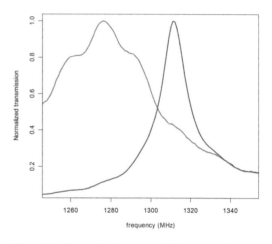

Figure 3. Normalized response of the resonant sensor on plaster with low moisture content and low salts content (blue line), and with high moisture content and high salts content (red line).

the surface of the material under test, avoiding areas with risk of detachment or with others criticality. The measurement is averaged on a hemisphere 2 cm depth of radius into the material.

On Fig. 3 is shown the sensor response, and it is characterized by the resonant frequency (the peak of the curve), and by the width of the bell curve (bandwidth). The permittivity is determined starting from these electrical parameters, and by a dedicated algorithm are derived the diagnostic parameters: moisture content (MC) and salts index (SI).

The instrument allows the following applications:

1. sub-surface diagnostic of moisture content and of soluble salts presence (Proietti, 2013); 2. evaluation of the dynamic of water diffusion in the support (Olmi, 2008);3. monitoring of wall painting related to seasonal variations (Olmi,

2006), and assessment of the effectiveness of extractive poultices (Bandini, 2008).

As regards the latter application the results of preliminary tests on the wall paintings at the Allori's *loggia* at the Pitti Palace in Florence (F. Bandini, 2008) are summarized here following. The effectiveness of several kinds of poultices were compared (cellulose pulp, mixing of cellulose pulp and sand in different ratio, and Cocoon by Westox) by SUSI measurements. Fig. 4 shows the portion of wall painting where the poultices were applied.

The used protocol was as follows: a. choice of the measurement points; b. preliminary measurement by SUSI before the application of extractive poultices; c. new measurements were performed by SUSI after 7 days from the removal of extractive poultices.

Fig. 5 presents the results of SUSI measurements regarding two kinds of poultice: cellulose pulp and Cocoon.

The Cocoon poultice demonstrates a better effectiveness after 7 days respect to the others extractive poultices. In particular the poultice made by mixing cellulose pulp and sand didn't provide good results.

## 6 DRMS

The Drilling Resistance Measurements System (DRMS) allows an assessment of the "hardness / cohesion" of natural or artificial stone materials both in laboratory and in situ, by performing real time cohesion profiles (drilling resistance versus depth). The system measures the force (N) needed for drilling and the position of the bit during drilling; it is equipped with a software program that allows continuous recording and monitoring of that force in relation to the advancement of the bit. The drilling resistance depends on mineralogical and structural characteristics of stone materials and it was correlated with the traditional mechanical parameters (Exadaktylos et al., 2000; Fratini et al., 2006). The technique is micro-destructive using bits of various diameter and consequently producing holes of 3/5 mm. The system has a drawback in presence of stone materials with abrasive constituents (such as granites) that quickly wear out the bit, producing a fictitious increase of drilling resistance. Due to this problem data corrections following specific step were proposed (Singer 2000, Delgado Rodrigues 2000); some inconveniences can be found in presence of very hard (some kind of marbles) or very soft (calcareous tuffs) materials; in such cases we could have drilling resistance values exceeding the load cell range (>100 N or <1 N). Furthermore when the system is used to verify the consolidation action due to inorganic products, such as ammonium oxalate or barium hydroxide) the results could not be meaningful because the very little increase in drilling resistance.

The advantages of the system are: it is the unique portable device that allows the direct measurement of the cohesion profile of stone materials, especially for on site tests, so it is useful in the evaluation of consolidation treatments; the instrument and the data management are user-friendly, the speed of execution of one test (1–2 minute) and the versatility of the system allows to carry out measures in almost all worksite conditions. Furthermore, last but not least, the system could be useful for the

Figure 4. The wall paintings of the Allori's small loggia at the Pitti Palace in Firenze.

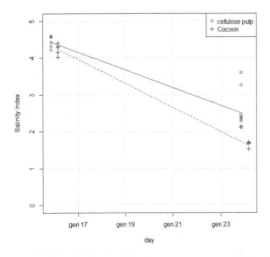

Figure 5. Plots of Salinity Index for two kinds of extractive poultices.

preliminary identification of the test area where apply the conservation treatments. This last option leads to select the test areas that are critical; in fact if these are not suitable the data collected for the comparison of different treatments and for the following monitoring, result too scattered for being used, both at short and at long time. The drilling resistance measures allow to assess the heterogeneity of the material at depth (due to the presence of grains of different hardness) and on the outer surface (in case of natural or artificial crusts); moreover it quantifies the surface cohesion for the identification of the areas which effectively need the conservative treatment; therefore a optimization of the application method and of the product concentration is reached. The study case of the wall of the Etruscan tombs in the Sovana archaeological area (Grosseto, Italy) is reported as example; the tombs are carved in a red volcanic tuff with black scoria. This kind of stone materials is particularly subject to decay (mainly loss of material and powdering) due to environmental factors (temperature variations, water infiltrations, humidity, diffusion of soluble salts with efflorescence) being a porous material, with a high water absorption coefficient. For these reasons it needs conservative treatments both consolidant and protective. Three areas were chosen for the drilling resistance tests (Fig. 6).

The results for all areas show a great heterogeneity, because of the presence of lithic fragments and/or scoria behind the surface (highlighted by peaks of cohesion profile with values of about 12–15 N), of superficial crusts and of different state of conservation (highlighted by different value of drilling resistance). The following specific situations were identified (Fig. 7):

Area 1: poor state of conservation of the external part with mean value of drilling resistance ranging from about 2–3 N in the outer layer to about 6 N in the inner part; Area 2: good state of conservation of the external part with the presence of a crust (maximum value 22 N) even if not homogeneously present. It is no clear if this crust is natural, due to

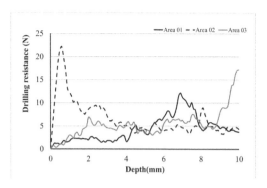

Figure 7. Drilling resistance profiles of the different area in Sovana archaeological site (Grosseto-Italy).

dissolution/re-precipitation phenomena, or artificial due to previous consolidant treatments; Area 3: medium state of conservation with average value of the drilling resistance of about 6–7 N, but with a surface layer showing lower cohesion features (about 3 N). On the basis of the drilling resistance data and by observing the different cohesion profiles the following maintenance suggestion can be drawn for a consolidating treatments and its monitoring: 1. the treatment is indicated only for Area 1 and 3 being the Area 2 in good state of conservation and does not needing a consolidation; 2. the concentration of consolidating treatment applied for the two areas must be different due to the different state of conservation, being more concentrated for Area 1. To monitor the treatment effectiveness during time a drilling monitoring every second year is suggested. Anyway it should be taken into account the high number of drilling tests to overcome the great heterogeneity of the material.

## 7 ULTRA-CLOSE RANGE PHOTOGRAMMETRY

The ultra-close range photogrammetry system is totally non-invasive, and it can be used to control the effectiveness of cleaning treatment, as well as, to monitor the durability of both cleaning treatments and application of consolidants (Barbetti, 2013, Manganelli Del Fà, 2014). The ultra-close range photogrammetry is based on the same principles of classic photogrammetry, but is applied to a different scale: this technique allows to generate a RGB points cloud of a surface acquiring only three images (or more if necessary), defined triplet, shooting from different angles, of the same area. The size of the acquired area ranges from 2 to 20 cm$^2$ with regard to the distance between the shots and the lens of the camera. The system

Figure 6. The three area in Sovana archaeological site(Grosseto-Italy).

Figure 8. Comparison of the horizontal profiles at $T_0$ and $T_1$ of the untreated surface.

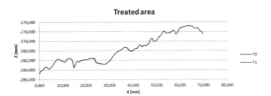

Figure 9. Comparison of the horizontal profiles at $T_0$ and $T_1$ of the treated surface.

is composed of an hardware part that consists of a digital camera Canon EOS 7D equipped with calibrated Canon EFS 60 mm macro lens, which runs on a motorized bar 260 mm long, and of three dedicated software that allow to choose the right acquisition parameters, generate the point cloud, and elaborate the 3D model. By using commercial devices it is possible to choose different focal lengths, e.g. the use of a 28 mm lens allows the reconstruction of larger areas after the acquisition of multiple shots, and the single models are mosaicked in order to obtain a single one. Choosing the area of interest the dedicated software automatically estimates the distance between the area of interest and camera sensor, indicating the best step (the distance between the snapshot); as soon as the shooting parameters were fixed, the system produces three shots of the same area, by moving the digital camera from right to left along the motorized bar. The acquired images were processed by dedicated software that generates the RGB point cloud through a specific algorithm. Once the acquisition and 3D model reconstruction have been done, it is possible to generate the Digital Elevation Model (DEM), a representation of the quotes of the surface, choosing a suitable reference plane (UCS—Users Coordinate System) through the seeding points, and finally the metric information (i.e. the xyz data) can be extracted. In the case of the Etruscan tombs of Sovana, the system was used to evaluate the effectiveness„ of consolidating products applied on the tuff substrate. A first acquisition campaign aimed at the generation of the three-dimensional model of the entire surface has been carried out on the area to be treated. After one month from the application of the product, the same area was investigated proposing the same shooting conditions: distance between the camera and the surface, acquisition step (the distance between the snapshot). The two models were subsequently superimposed: a special procedure in MathLab, developed specifically to improve the alignment of the models, allowed to extract the profiles of surface roughness to verify the modifications of the surface. The same procedure was also repeated on an untreated area close to the treated area, in order to use it as a control. The comparison of the profiles shows that at one month after the application the untreated area presents a light loss of material (Fig. 8), on the contrary for the treated area there were no changes on the surface (Fig. 9).

## 8 CONCLUSIONS

In this paper some NDT or microDT techniques were critically explored; they are engineered to be used on site for the study of the state of conservation and for the evaluation and monitoring of the conservation treatments performances (consolidants and/or water-repellent). By the description of the activities on some case studies, the pro & cons of each techniques were illustrated. The need and the usefulness of these practices are evident examining the obtained results. Nevertheless it should be emphasized that protocols are still lacking of completeness; some useful parameters are measurable only in lab; especially as regards the evaluation of the consolidant performances much research is needed. As regards protective treatments a good evaluation is possible, even if some crucial parameter as water vapour permeability, is still measurable only in lab. Anyway the proposed techniques can be used as preliminary screening in order to plan a deeper investigations by sampling and laboratory analysis.

## REFERENCES

Bandini F., Felici A., Mariotti P.I., Olmi R., Riminesi C., 2008, I dipinti murali della loggia Allori a Palazzo Pitti: una sperimentazione per l'estrazione dei sali e per il monitoraggio con dielettrometria a μonde, OPD Restauro, 20, 121–130.

Barbetti, I., Felici, A., Magrini, D., Manganelli Del Fà, R., Riminesi, C., 2013. Ultra close-range photogrammetry to assess the roughness of the wall painting surfaces after cleaning treatments. Int. J. of Cons. Science, 4, 525–534.

Delgado Rodrigues, J. Costa D., 2004. A new method for data correction in drilling resistance. Tests for the effect of drill bit wear. Int. J. Restor. of Buildings and Mon., 10, 1–18.

Exadaktylos, G. Tiano, P. and Filareto, C., 2000. Validation of a model of rotary drilling of rocks with the drilling force measurement. Int. J. Restor. of Buildings and Monuments, 3, 307–340.

EN 16581: 2014. Conserv. of Cultural Heritage. Surface protection for porous inorganic materials. Lab. test methods for the evaluation of the performance of water repellent product.

Fratini, F., Rescic, S., Tiano, P., 2006. A new portable system for determining the state of conservation of monumental stones. Mat. and Structures, 39 (2), 139–147.

Mecchi A.M., Sansonetti A., Realini M., Poli T., 2008, A Proposal for a common approach in choosing tests for the protocol evaluation of cleaning methods, Proc. 11th Int. Cong. On Deter. and Conserv. of Stone. Torun. Poland, 425–433.

Manganelli Del Fà, R., Riminesi, C., Tiano, P., 2014. Monitoring of surface pattern of artistic and architectural artifacts by means of ultra close range photogrammetry. Proc. of 6th Europ. Symp. on Religious Art, Rest. and Conserv. (ESRARC2014). Ed. Nardini, Firenze:164–167.

Olmi R., Bini M., Ignesti A., Priori S., Riminesi C. Felici A., 2006, Diagnostics and monitoring of frescoes using evanescent-field dielectrometry, Measurement Science and Technology, 17, 8, 2281–2288.

Olmi R., Riminesi C., 2008. Study of water mass transfer dynamics in frescoes by dielectric spectroscopy, Il Nuovo Cimento, 31 C, 3, 389–402.

Proietti, N., Capitani, D., Di Tullio V., Olmi R., Priori S., Riminesi C., Sansonetti A, Tasso F., Rosina E., 2015, MOdihMA at Sforza Castle in Milan: Innovative Techniques for Moisture Detection in Historical Masonry, Proc. Inter. Conf. "Built Heritage 2013" (Milano, 18–20 XI 2013), 1–6.

Fant R., Sansonetti A., Colombo C., 2013, L'apparato decorativo in terracotta della Ca' Granda: indagini conoscitive e intervento di restauro. Proc. Terrecotte del Ducato di Milano. Ed. ET Milan, 253–270.

Singer, B., Hornschild, I. & Snethlage, R., 2000, Strenght profiles correction functions for abrasive stones. Proc. of the workshop DRILLMORE, 35–42, Firenze, Centro Stampa Toscana Nuova.

Vandevoorde D. mail to: vandevoordedelphine@yahoo.co.uk, Pamplona M. mail to: pamplona@ist.utl.pt, Schalm O., Vanhellemont Y., Cnudde V., Verhaeven E. 2009, Contact sponge method: Performance of a promising tool for measuring the initial water absorption, J. of Cultural Heritage, 10, 1, 41–47.

*Emerging Technologies in Non-Destructive Testing VI – Aggelis et al. (Eds)*
*© 2016 Taylor & Francis Group, London, ISBN 978-1-138-02884-5*

# Tube-jack and sonic testing for the evaluation of the state of stress in historical masonry

E.C. Manning, L.F. Ramos & P.B. Lourenço
*ISISE, Civil Engineering Department, University of Minho, Guimarães, Portugal*

F.M. Fernandes
*ISISE, Lusiada University, Vila Nova Famalicão, Portugal*

ABSTRACT: In the investigation and diagnosis of damages to historical masonry structures, determination of the state of stress of the masonry is important. An enhanced technique, called tube-jack testing, is being developed at the University of Minho to reduce the damage caused during testing and improve accuracy. This method uses multiple cylindrical jacks inserted in a line of holes drilled in the masonry mortar joints, avoiding damage to the masonry units. Concurrently with tube-jack testing development, the effect of stress state on sonic testing is being studied. In this paper the results of tube-jack and sonic testing on masonry walls loaded in compression is presented. The tube-jack testing is used to estimate the state of stress in the masonry and the sonic test results are evaluated based on the effect of the applied load on the wall.

## 1 INTRODUCTION

In-situ non-destructive testing is often the best way to determine the characteristics of historical unreinforced masonry, reducing or eliminating the need to remove or significantly destroy material from the historical structure. However, since the in-situ tests are not performed in a controlled laboratory environment, there is always room for improvement in their methods to reduce damage to the historical masonry and increase accuracy.

This paper presents a study of two non-destructive test methods, tube-jack testing and sonic pulse velocity testing, and their use in evaluating the state of stress in unreinforced masonry. The following sections of the paper describe the test specimens built for the purpose of this study and their material properties; background, theory, test set-up and results of the single tube-jack test performed in the masonry wall; background, test set-up and results of the direct and indirect sonic pulse velocity tests performed in the same masonry wall but under different compressive stress states; and a brief discussion about the combination of non-destructive testing techniques, especially the tube-jack and sonic test methods.

## 2 TEST SPECIMENS AND MATERIAL PROPERTIES

A masonry wall composed of granite blocks and cement-lime mortar was constructed in the Laboratory of Structural Engineering (LEST) at the University of Minho in order to test the tube-jack test method and sonic tests in a controlled environment. The wall was constructed with a regular typology and single leaf. The design is shown in Figure 1.

Three small masonry wallets, several mortar cylinders, and granite cylinders were made to test the mechanical characteristics of the constituent materials. The material properties are presented in Table 1 (Manning et al., 2014a).

## 3 TUBE-JACK TESTING METHOD

### 3.1 Background

The tube-jack testing technique was first proposed by Ramos & Sharafi (2010). In these first studies several limitations of the traditional flat-jack test method were identified including the difficulty in using the saw to cut slots in the masonry and unwanted damage of masonry units in irregularly constructed masonry when joints are not aligned. It was shown that through the proposed tube-jack testing these limitations could be overcome by positioning the tube-jacks in locations along the mortar joint regardless of whether that joint is linear and by using a simple hand drill to create the holes for the tube-jacks. A numerical validation of the method was shown through finite element modeling of the tube-jack test in a homogeneous wall.

Development of the tube-jack method was continued by testing several prototype tube-jacks in a

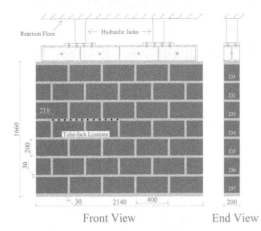

Figure 1. Design, loading configuration of the regular masonry wall, and location of tube-jacks, dimensions in mm (Manning et al., 2014b).

Table 1. Masonry mechanical characteristics.

| Specimens | Young's Modulus GPa | Poisson Ratio | Compressive Strength MPa |
|---|---|---|---|
| Masonry Wallet | 2.50 | 0.22 | 4.17 |
| Granite Cylinders | 29.82 | – | 67.90 |
| Mortar Cylinders | 0.28 | – | 0.32 |

small two block specimen by Ramos et al. (2013). The results of the tests showed that the tube-jack prototype made of rubber was the most successful but that it needed confinement. Knitted fibrous socks were tested as a way of confining the tubing material and were added to the tube-jack design. Finally, the two block specimen tests and numerical modeling showed that further testing should be performed in larger masonry specimens, hence the tests presented in this paper.

### 3.2 Theory

The tube-jack test method is very similar to that presented in existing standards and recommendations for the flat-jack method (RILEM, 2004; ASTM, 2004). However, instead of creating a slot in the masonry and inserting a flat-jack, several holes are drilled in the line of the mortar joint and cylindrically shaped jack made of rubber, tube-jacks are inserted into these holes. The line of tube-jacks creates an equivalent flat-jack. The pressure in the tubes is increased to exert pressure on the surrounding masonry. Measuring devices are placed perpendicular to and over the line of the equivalent flat-jack before the holes are drilled. The displacement of the masonry is measured using these measuring devices throughout the test. Assuming a compressive stress within the masonry, when the holes are drilled the measuring devices will record a negative displacement as the holes close. As the tube-jacks pressurize the masonry, the measuring devices will record a positive displacement. When the displacement of the masonry is reduced to zero, or the masonry returns to its original position before the drilling of the holes, the pressure in the tube-jacks can be used to determine the state of stress in the masonry using Equation 1:

$$\sigma_m = K_m K_a P_{Applied} - P_{TI} \qquad (1)$$

where: $\sigma_m$ is the local state of stress in the masonry, $K_m$ is the jack calibration factor, $K_a$ is the area correction factor, $P_{Applied}$ is the applied jack pressure required to return the average differential displacements to zero, i.e., the state before the holes were drilled, and $P_{TI}$ is the tube inflation pressure. The jack calibration factor takes into account the resistance of the jack to pressurization and depends on the tube-jack materials. The tube inflation pressure is the pressure required to inflate the tube-jack to the point where it begins to apply pressure to the masonry. The area correction factor for single tube-jack tests using only one line of tube-jacks can be determined using Equation 2:

$$K_a = \frac{A_T}{A_H} \qquad (2)$$

where: $A_T$ = total cross-sectional area of tube-jacks in contact with and applying pressure to the masonry, and $A_H$ = total cross-sectional area of the holes drilled in the masonry. A complete explanation of Equations 1 and 2 can be found elsewhere (Manning et al., unpubl.).

### 3.3 Test set-up

The masonry wall used for the tube-jack test was loaded using two hydraulic jacks centered on a steel profile above the wall, as shown in Figure 1. Before, during and after the test the load was checked and maintained. The stress level at the joint used for the test, the 4th joint up from the base of the wall, was calculated to be 0.22 MPa based on the pressure in the hydraulic jacks and the weight of the masonry above the joint.

The holes for the tube-jacks were spaced approximately 7.5 cm from center to center. Eight linear variable displacement transducers (LVDTs) were used, four on the front of the wall and four on the back. The placements of the LVDTs were between

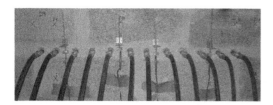

Figure 2. Tube-jacks inserted into holes drilled in the 4th horizontal mortar joint and spacing of LVDTs.

the 2nd and 3rd holes, between the 5th and the 6th holes, between the 7th and the 8th holes, and between the 10th and 11th holes, on both sides of the wall. They were labelled LVDT 1 through 4, respectively, on the front of the wall and LVDT 5 through 8, respectively, on the back of the wall.

During the test, the holes for the tube-jacks were drilled starting from the center holes and working toward the outer holes, drilling through the complete thickness of the wall. The displacements of the LVDTs were recorded both during the drilling process and after the drilling was completed. The area correction factor, $K_a$, for the single tube-jack tests was calculated to be 0.825 based on the assumption that the tube-jacks expanded to fill the entire circumference of the holes during pressurization. The jack calibration factor, $K_m$, was considered to be one. Further study is required to determine this factor.

### 3.4 Results

The results presented here are for the single tube-jack test performed in the 4th horizontal joint up from the base of the wall. Tube-jack tests were also performed in other joints of the wall and their results are presented elsewhere (Manning et al., 2014b).

The results of the hole drilling are presented in Figure 3. Since the difference between the average LVDT displacement on the front and the back of the wall is relatively small, the hole drilling did not cause bending or uneven downward displacement of the wall.

After the movement of the LVDTs had reduced to nearly zero, the rubber tube-jacks were inserted into the holes and pressurized. The results are shown in Figure 4. The calculated pressure takes into account the jack calibration factor and the area correction factor. Two pressurization cycles were conducted; the first one to 0.33 MPa calculated pressure (0.50 MPa water pressure) and the second to 0.39 MPa calculated pressure (0.60 MPa water pressure). The tube-jacks were consistent in loading the wall in both pressurization cycles as shown by the two curves overlapping during

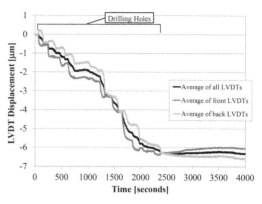

Figure 3. Average LVDT results during drilling of the holes (Manning et al., 2014b).

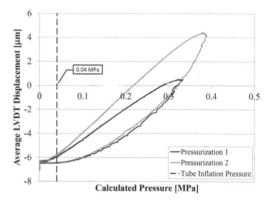

Figure 4. Tube-jack test results: calculated pressure versus average LVDT displacements. (Manning et al., 2014b).

the loading phase. The average LVDT displacements return to zero at a calculated pressure of 0.32 MPa. Given the tube inflation pressure of 0.04 MPa, the estimated local state of stress for this test was 0.28 MPa. Further research is currently being done to better understand the difference between this state of stress result from the tube-jack test, 0.28 MPa, and the calculated state of stress, 0.22 MPa (difference of 0.06 MPa).

## 4 SONIC TEST METHOD

### 4.1 Background

Sonic pulse velocity testing is a widely used non-destructive testing technique used for investigation of historic masonry structures. The technique was first applied to concrete and masonry in 1938 (Binda, 1995 as cited in Miranda, 2011). Sonic pulse velocity tests are performed by hitting the surface of the masonry with an instrumented hammer and then receiving the wave signal with a transducer placed at another point on the masonry. As the wave passes through the masonry its velocity, frequency and amplitude are changed by the properties of the masonry. By analyzing the received wave some of the properties of the masonry can be investigated. The sonic pulse velocity test was first used for detecting voids in 1942 (Carino, 2004 as cited in Miranda, 2011).

In this study, the aim was to determine if the state of compressive stress of the masonry affects the sonic wave velocity. If the level of stress of the masonry has an effect on the sonic velocity, it will be important for the state of stress to be determined in locations where sonic tests are performed in order to compare the results of the sonic tests.

### 4.2 Test set-up

The sonic tests were performed on the same masonry wall used for the tube-jack test. Transmission and reception points were marked on each end of the wall at the center of each unit (Figure 1). The points were labeled D1 through D7 with D1 at the top of the wall.

Direct and indirect sonic tests were performed using these points. In the direct tests, the sonic wave was transmitted by hitting the point on one end of the wall and receiving the wave at the same level on the other end of the wall with an accelerometer (Figure 5). To distinguish the direction of travel of the waves, the point combination was labeled with point name, the side of transmission, and then the side of reception, "L" for left and "R" for right. For example, a wave transmitted from point D1 on the left end of the wall to point D1 on the right end of the wall would be labeled, "D1 LR".

Indirect sonic tests were also performed on the ends of the masonry wall from the top points to the bottom points and vice versa. These point combinations were labeled with the transmission point first followed by the reception point second, each point including the side of the wall, left or right. For example, a wave transmitted from point D1 to point D7, both on the left end of the wall, would be labeled "D1 LD7 L".

For each point combination, ten or more sonic waves were transmitted through the masonry wall. The transmitted and received waves were recorded using a 100 Hz data acquisition system and a program created in the Labview software for the purpose of sonic wave signals (Labview, 2012). The signals were analyzed using an automatic data analysis program created using Matlab to determine the velocity of each wave (Matlab, 2009). The average velocity of the ten waves transmitted and received for each point combination was considered the average sonic wave velocity for that point combination.

### 4.3 Results

Sonic tests were performed 7 and 9 weeks after the construction of the wall. On each occasion, sonic tests were performed with zero load applied to the top of the wall and with a load producing an average stress level of 0.2 MPa within the wall.

The vertical indirect sonic tests performed on the ends of the wall at 7 weeks showed an increase in sonic velocity when the stress level was increased from zero to 0.2 MPa. The amount of increase varied greatly between the four point combinations (Figure 6).

For the direct tests, where waves passed from one end of the wall to the other, performed at 7 weeks, the results of the tests conducted from left to right and from right to left were averaged. It was assumed that since the wave was passing through the same material when it traveled from left to right

Figure 5. Direct sonic testing on the regular masonry wall.

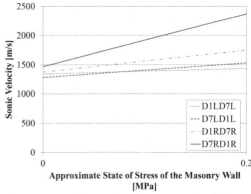

Figure 6. Results of the indirect sonic velocity tests performed at two different stress levels, 0 MPa and 0.2 MPa, 7 weeks after the construction of the wall.

and from right to left, the reason for the difference in results was the contact of the accelerometers with the surface of the granite units. Thus, the values could be averaged for one velocity between the two points.

The increase in stress in the masonry wall affected the results of the direct sonic velocity tests in a different way from the indirect tests. The increase in stress resulted in a large increase in the wave velocity for waves passing through the top course of the masonry and a slight increase in wave velocity for waves passing through the second course of masonry. All of the other direct tests performed through the courses below were not discernably affected. This is shown in a chart of the percent difference of the sonic velocity when the stress was increased from zero to 0.2 MPa (Figure 7). The percent difference, $P_{Diff}$, was calculated using Equation 3.

$$P_{Diff} = \frac{(V_{0.2Mpa} - V_{0MPa})}{V_{0MPa}} 100\% \quad (3)$$

where $V_{0.2MPa}$ = sonic velocity at 0.2 MPa and $V_{0MPa}$ = sonic velocity at 0 MPa.

One possibility for the difference in sonic velocity at the top of the wall is that the waves passing through the top levels of the masonry were affected by the steel profile above the wall. However, the steel profile was positioned on top of the wall before any of the tests were conducted. Another possibility for the increase in sonic velocity at both the top and bottom of the wall is the confinement of the masonry at these levels. The steel beam and the concrete floor beneath the wall both confine the masonry when load is applied to the wall.

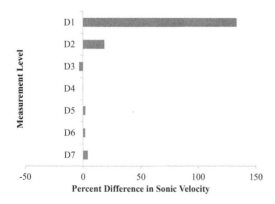

Figure 7. Results of the direct sonic velocity tests performed 7 weeks after the wall was constructed: percent difference in sonic velocity when the stress was increased from zero to 0.2 MPa.

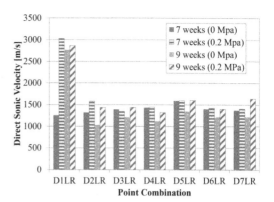

Figure 8. Comparison of direct sonic test results for stress levels of zero and 0.2 MPa at 7 and 9 weeks after construction.

Finally, the results of the direct tests performed from the left side of the wall to the right side of the wall were compared for both stress levels at 7 weeks and 9 weeks (Figure 8). The results show that for almost all point combinations the direct sonic velocity decreased between the 7th week and the 9th week. This could be due to cracks or voids forming as the mortar between the joints continued to harden. The results also show that the direct sonic velocity increased for all point combinations when the wall was loaded to a stress level of 0.2 MPa at 9 weeks.

## 5 COMBINATION OF NDT TESTS

Non-destructive test methods are often combined in the investigation of historical constructions. Using more than one test can provide additional information that a single test could not provide on its own. Bayesian analysis has been shown to increase the accuracy of the results by combining several different types of tests and reference information (Ramos et al., 2015).

Flat-jack testing and sonic testing can be used together to provide information about the characteristics and state of the existing masonry. Since tube-jack testing is being developed as an alternative to flat-jack testing, it also can be used in conjunction with sonic testing. The sonic tests can be used to determine the consistency of the masonry, if it has many voids or cracks, or is very dense. This information can indicate if the masonry is under high stresses and should be tested to determine the state of stress using the flat-jack or tube-jack methods and the areas of the masonry best suited for those tests.

Now, with the results of the sonic tests presented in this paper, it seems that the compressive stress state of the masonry also could have some impact on the sonic velocity. Thus, these two tests can be combined in a new way. Where sonic tests are performed in masonry of the same typology but the resulting velocity results are different, the state of stress of the masonry may be a factor influencing the results. Thus, differing sonic values could be an indication of higher stress levels, which can be estimated using the tube-jack test method.

## 6 CONCLUSION

In this study we used both tube-jack testing and sonic testing to investigate a masonry wall constructed in the laboratory. The purpose of the tube-jack testing was to determine the state of stress in the wall and compare that value to the calculated value based on the load applied to the top of the wall. The tube-jack testing was successful in loading the wall consistently through two loading cycles and providing a stress state estimation within a reasonable range of the calculated value. Further testing will need to be performed to study the method further and apply it to other masonry typologies.

Sonic tests were also performed on the masonry wall to determine if the indirect and direct sonic wave velocity was influenced by the state of stress in the wall. The results indicated that there was a positive correlation between the state of stress of the wall and the sonic velocity. These results will need to be verified by further testing.

Finally, a discussion was presented concerning use and combination of different non-destructive testing techniques. In particular, the new tube-jack method will be able to be used to estimate the state of stress in the masonry where sonic test results indicate there may be an increase in stress state in comparison with sonic tests in other locations on similar typology masonry.

## ACKNOWLEDGEMENTS

The authors would like to acknowledge the Fundação para a Ciência e Tecnologia, which supported this research work as a part of the Project "Improved and innovative techniques for the diagnosis and monitoring of historical masonry", PTDC/ECM/104045/2008. The authors would also like to acknowledge individuals who aided in performing the tube-jack and sonic tests presented in this paper: Kevin Vázquez, Chrysl Aranha, and Cláudia Almeida.

## REFERENCES

Binda, L. 1995. Role of Rilem Committee: Calibration of Proposed Test Methods. *RILEM* Padua.
Carino, N. 2004. Handbook on Nondestructive Testing of Concrete. 2nd Edition. Barr Harbor Drive: *ASTM—International—CRC Press LLC*.
Labview. 2012. SP1. December 2012. National Instruments.
Manning, E., Ramos, L.F. & Fernandes, F. 2014. Direct Sonic and Ultrasonic Wave Velocity in Masonry under Compressive Stress. In *9th International Masonry Conference, Guimaraes, Portugal, 7–9 July 2014*.
Manning, E., Ramos, L.F. & Fernandes, F. unpubl. Tube-jack testing for irregular masonry walls: regular masonry wall testing.
Manning, E., Ramos, L.F. & Fernandes, F. 2014. Tube-jack testing: regular masonry wall testing. In F. Peña & M. Chávez (eds), *SAHC2014–9th International Conference on Structural Analysis of Historical Constructions, Mexico City, Mexico, 14–17 October 2014*.
MATLAB R2009b. August 12, 2009. The MathWorks Inc. Natick, Massachusetts, United States.
Miranda, L. 2011. Ensaios acústicos e de macacos planos em alvenarias resistentes. PhD Thesis. Porto, Portugal: Faculdade de Engenharia, Universidade do Porto.
Ramos, L.F. & Sharafi, Z. 2010. Tube-jack testing for irregular masonry walls: first studies. *Advanced Materials Research* 133–134: 229–234.
Ramos, L.F., Manning, E., Fernandes, F., Fangueiro, R., Azenha, M., Cruz, J., Sousa, C. 2013. Tube-jack testing for irregular masonry walls: prototype development and testing. *NDT&E International* 58: 24–35.
Ramos, L.F., Miranda, T., Mishra, M., Fernandes, F.M., Manning, E. 2015. A Bayesian approach for NDT data fusion: The Saint Torcato church case study. *Engineering Structures* 84: 120–129.
RILEM Recommendation MDT.D.4: In-situ stress tests based on the flat-jack, 2004.
Standard Test Method for In Situ Compressive Stress within Solid Unit Masonry, Estimated Using Flat-jack Measurements, ASTM Standard C 1196–04, 2004.

*Emerging Technologies in Non-Destructive Testing VI – Aggelis et al. (Eds)*
*© 2016 Taylor & Francis Group, London, ISBN 978-1-138-02884-5*

# Integration of EFD, MRM and IRT for moisture mapping on historic masonry: Study cases in northern Italy

R. Olmi
*CNR-IFAC, Florence, Italy*

C. Riminesi
*CNR-ICVBC, Florence, Italy*

E. Rosina
*Politecnico di Milano, Milano, Italy*

ABSTRACT: Among the innovative techniques for detecting moisture on the historical masonry, EFD (Evanescent Field Dielectrometry) presents a high potential for providing the absolute measure of the water content, to map the salts distribution on the surfaces, without requiring to sample the materials under investigation. The integration with IRT (Infrared Thermography) results particularly useful on stone/cobbles solid masonry, because of the least reliability of the gravimetric tests on this type of wall, especially on samples collected from the layers beneath the surface. The paper presents the results on two historical buildings located in Lombardy, and the discussion regarding advantages and limits of the application: the Royal Villa in Monza and the church of Saint Ignacious in Ponte (SO).

## 1 INTRODUCTION

The presence of moisture in ancient masonry is very common, especially where neglect and lack of maintenance occurred for a long time. Water intrusion in the building structures alters the mechanical and thermal characteristics of materials, causes the loss of mortar components, and the alternant cycles of damp/drying causes damages of the surfaces up to losing the materials themselves. Diagnostics for moisture mapping and measure the water content is a mandatory requirement before intervene with any restoration or refurbishment, because the presence of water and soluble salts in the masonry will soon or later damage the restored surfaces.

Some innovative techniques are available to map the presence of water on the surface (few cm of depth), but very few have reliable application at more than 5 cm inside the structure. Gravimetric test is the only one ensuring to collect sample up to 20–30 cm inside, and to directly measure the water content (UNI, 2003). On the other hand, the test is destructive, and the accuracy is low, especially for intermediate values (for example, in mortar samples between 4 and 7% of Water content%).

At present the EU standards regard only the gravimetric tests, although the standard for the comparison and use of some steady and innovative techniques is at the last step of validation.

Among the innovative tests, Evanescent-Field Dielectrometry (EFD) is one of the techniques that had major applications in these last years. EFD is physically based on the dielectric contrast between water and dry porous materials. The technique allows to measure the sub-superficial moisture content and to detect the presence of salts in a wall by means of a non-invasive, direct contact resonant probe (Olmi 2006, Olmi 2008, Di Tullio 2010). The Moisture Content (MC) measured by means of the EFD technique refers to an average hemispherical volume of about 2 cm radius. MC and Wc (Water content) are different acronym of the percentage of water content in the sample.

A new microwave technique, developed for the detection of moisture and material discontinuities at medium depth (up to 20–25 cm) has also been applied to complement the sub-superficial investigation conducted by the EFD method. The microwave reflectometric method (MRM) (Capineri 2012, Macchioni 2013) is a non-destructive diagnostic tool based on a Continuous-Wave (CW) wide-band source. The system obtains information about the material composition (dielectric constant) by measuring the energy reflected from the investigated material.

## 2 EXPERIMENTAL TESTS ON THE STUDY CASES

The authors applied a IRT scanning and gravimetric tests (Rosina 1999, Rosina 2001, Rosina 2003) as preliminary assessment for mapping the presence of water in the structure. The repetition at different climatic conditions resulted very useful to hypothesize the cause of rising damp.

The paper presents the results on two different buildings location in Lombardy, Northern Italy.

### 2.1 Saint Ignacious Church in Ponte

The church is located in the mountains (elevation 500 m o.s.l.), it dates back to the 17th century; the structure is solid stone masonry and lime mortar for finishing and stones's joints. The interior has precious decoration: plasterworks of art and frescoes, statues and marble banisters are present.

Together with the preliminary tests, a microclimatic monitoring lasted 3 years. The hypothesis formulated after the preliminary tests was that the cause of the rising damp was water infiltration in the northern side along the road. The research team of CGT Spin Off (Siena State University) performed the additional geoelectric measures in the soil along the perimeter of the church to validate the hypothesis. Because of the local stone inside the structure, it was impossible to apply gravimetric tests in depth (the stone has low porosity and low water content., it is also very hard to drill and drilling heats the materials, causing the evaporation of water before the extraction of the sample).

Therefore is was necessary an integration and validation with non invasive techniques, as EFD is.

Table 1 shows the most significant results of gravimetric tests and Figures 2 and 3 show the photo and IR Thermogram of the northern side of the Sacristy (along the road), where the investigation showed the highest water content in the masonry.

Table 1. Highest Water content (Wc) of the samples and their location in Saint Ignacious Church.

| n sample | Wc% | Location |
|---|---|---|
| 1.1 | 9,7% | Northern side, interior, 20 cm height |
| 1.2 | 6,7% | Northern side, interior, 80 cm height |
| 3.1 | 5.7% | Sacristy (Northern side) interior, 20 cm height |
| 4.1 | 12.3% | Sacristy (Northern side) interior, 20 cm height |
| 4.2 | 9,1% | Sacristy (Northern side) interior, 80 cm height |

Figure 2. The northern side of the Sacristy.

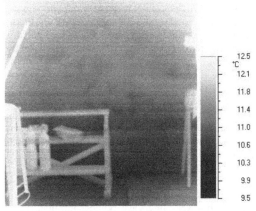

Figure 3. IRT of the northern side of the Sacristy, the colder area at the bottom of the wall indicates the presence of moisture, as the results of the gravimetric tests confirmed.

The samples 1.1, 4.1 and 4.2 have very high water content.

In addition, the results of the microclimatic monitoring show that the surfaces reach the balance with the air for almost all the seasons, at high

Figure 1. The northern side of the church, along the road.

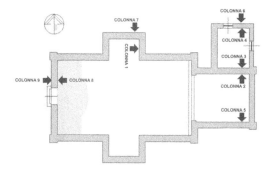

Figure 4. Location of the samples of the gravimetric tests.

Figure 5. The Royal Villa, the eastern side.

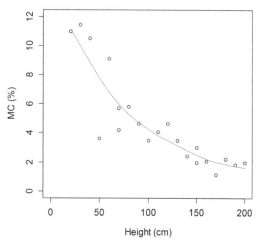

Figure 6. Vertical behaviour of MC on the Sacristy wall.

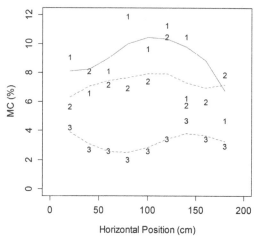

Figure 7. Horizontal behaviour of MC on the Sacristy wall.

percentage of RH inside the building. Only in few weeks during the winter and the summer sudden decreases of the exterior RH cause fast and large span drops of the RH, causing a high risk for the conservation of the surfaces.

EFD measurements in the Sacristy have been conducted on a vertical axis passing through the sample locations and on three horizontal axis located at 25, 60 and 130 cm from ground. The results are shown in Figures 6 and 7. The moisture content, as expected, decreases with the increase of the height from the floor.

Values as high as 12% are observed, in accordance with the gravimetric results (Table 1).

### 2.2 Royal Villa in Monza

The Royal Villa is located in Monza Park (in the Po plain, along the Lambro river). It is a 18th century, our storey building, with solid brick masonry as vertical structure, and stone slabs at the bottom of the exterior side of the masonry. The soil is partially sand and clay. Due to the slabs, it is impossible to collect samples by drilling the bottom of the exterior elevations. Also the use of IRT was limited to the interior, because of the low porosity and of the stone.

EFD integrated IRT and gravimetric tests, with the aim to validate the hypothesis of rising damp from the soil (where clay is a component) and the lack of an effective drainage of rain along the perimeter of the Villa. EFD and MRM where applied first on the same points of application of the other techniques, for calibrating the MRM response; then the application of the reflectometric method was also on the exterior side of the masonry. EFD measurements used a "factory" calibration, performed in laboratory on plaster samples.

Inside the building, the results of IRT showed an advanced damage of the finishing, often masking the effects of the water evaporation and the consequent decrease of the surface temperatures.

Gravimetric tests permitted to distinguish the areas where really rising damp occurred, especially in the rooms n 1, 2, 5 and 13 at half of the northern wing, at level –1.

The highest values resulted in the northern and northeastern walls. In this area the soil underneath the level –1, which was sand in the past, was recently removed due to excavation and location of the technical room for the heater. After the digging and building phases the excavation areas was covered with a new soil, that has a high clay content. At the time of the investigation (2011), without

Figure 9. Royal Villa, the interior of room n 1.

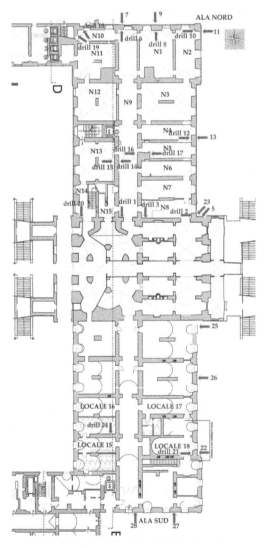

Figure 8. Royal Villa, location of the samples of the gravimetric tests at the underground level.

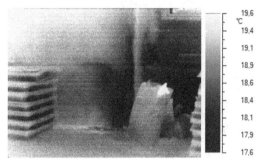

Figure 10. Royal Villa, composite of thermograms the interior of room n 1.

having set the drainage pipe, the rain remained for a long time in the soil at contact with the masonry. In the recent refurbishment of the Villa, the update of the piping system included also the new pipes for collecting and drain the rain along this side.

## 3 DISCUSSION OF RESULTS

The comparison among the gravimetric measurements and the non invasive methods allows to draw some considerations:

1. EFD measurements on quasi-dry materials, in the specific case, give lower MC values with respect to the gravimetric findings, possibly due to the different investigated volume and to the very hot and dry period when the measurement were conducted. Moreover, a small bias can be present due to differences in composition between the investigated plaster and the "standard" one used for lab calibration.
2. Measurements in presence of a considerable humidity are more in accordance. For example, the superficial gravimetric MC of the samples 10.1 and 10.2 are both about 4%, while the MC measured by EFD is between 4% and 5.6%.

Concerning the comparison among the deep-gravimetric measurements and those obtained by the MRM method, the agreement is usually good. For example, referring to the same location where the agreement between EFD and the gravimetric method is good, the MC measured by MRM close to the positions of the samples 10.1 and 10.2 is respectively 12.2% and 11.4%, in very good agreement with the results in Table 2.

Figures 11 and 12 show the deep MC values measured by means of the MRM technique on the internal and external walls respectively.

Moisture measurements have been repeated at the end of the summer period, to verify the changes in the MC distribution in the internal and external walls. From MRM measurements the statistical distribution of internal MC's appears unchanged from July to September, while the external distribution is clearly shifted to higher MC values (see Figure 13 and 14).

The integration of EFD, IRT, gravimetric tests, and MRM shows a good agreement of results in

Table 2. Highest of Water content (Wc) of the samples and their location in the Royal Villa.

| n sample | Wc% | Location |
|---|---|---|
| 6.1 | 9.2% | Corridor of the northern wing, 20 cm height |
| 8.1 | 7.3% | Room n 1, northern side, 20 cm height |
| 8.2 | 7,1% | Room n 1, northern side 60 cm height |
| 10.1 | 11.6% | Room n 2, eastern side, 20 cm height |
| 10.2 | 10.3% | Room n 2, eastern side, 60 cm height |

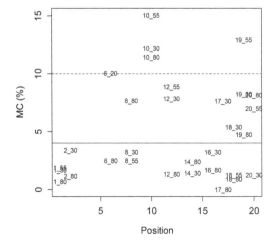

Figure 11. Deep MC on internal walls measured by MRM.

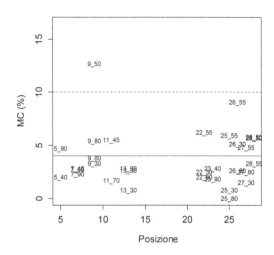

Figure 12. Deep MC on external walls measured by MRM.

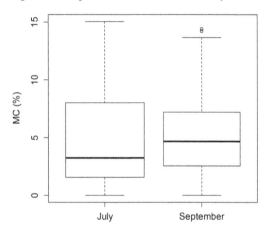

Figure 13. Season statistical distribution of deep MC on internal walls.

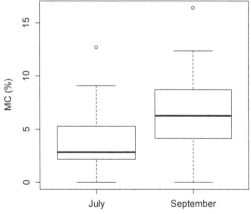

Figure 14. Season statistical distribution of deep MC on external walls.

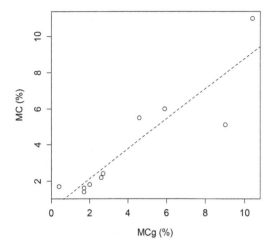

Figure 15. Correlation between dielectric (MC) and gravimetric (MCg) moisture content.

the case of high values of Wc (above 8%) and in the case of very low Wc (below 3%) for porous materials as lime mortars, bricks and highly porous stones. The agreement for sub-superficial measurements (depths up to 2 cm) is very good also for intermediate values, in particular when in presence of an homogeneous material having a flat surface, as in the first case study. Figure 15 shows the correlation between gravimetric and dielectric MC in that case: the $R^2$ is higher than 0.8.

The disagreement among gravimetric and dielectric results depends on many factors, and mainly deals with:

1. the sampling method of the gravimetric tests
2. the influence of soluble salts in the materials
3. the size of the sample in the case of the gravimetric tests and in the case of the EFD and MRM measurements
4. the irregular distribution of decay in the masonry materials
5. the flatness of the surface, mainly influencing the accuracy of the EFD method
6. the possible presence of iron and metallic plate/brakes/nails underneath the exterior slabs
7. the irregular distribution of moisture inside the thickness of the wall.

4 CONCLUSIONS

The integration with non destructive techniques improves the quality of the information, and its reliability.

Non destructiveness and real rime results permit to apply diagnostics for many months, and study the moisture diffusion in the field, without sampling the precious original materials of the historic buildings.

Long term measures permit to determine the thermal/hygrometric behavior of the buildings in their environments, determining their historic relationships and reactions to the changes of climate by means of the repeated measures of water content, temperature of the surfaces and of the air, RH of the air, distribution of salts on the surfaces. These data are the fundamental for determining the threshold of microclimatic parameter that should not be trespassed during the refurbishment and change of use of the building to ensure the conservation of the structures and their precious surfaces.

REFERENCES

Capineri, L., Falorni, P., Frosinini, C., Mannucci, M., Macchioni, N., Olmi, R., Palanti, S., Penoni, S., Pieri, S., Riminesi, C., Santacesaria, A. & Todaro, C. 2012. Microwave Reflectometry for the Diagnostic of Cultural Heritage Assets. *Proc. of PIERS 2012, Progress in Electromagnetics Research Symposium*, pp. 788–781.

Di Tullio, Proietti, N., Gobbino, M., Capitani, D., Olmi, R., Priori, S., Riminesi, C. & Giani, E. 2010. Non-destructive mapping of dampness and salts in degraded wall paintings in hypogeous buildings: the case of St. Clement at mass fresco in St. Clement Basilica, Rome. *Anal. Bioanal. Chem.*, 396, pp. 1885–1896.

Ludwig, N. & Rosina, E. 2001. Detection of the Moisted Surfaces by Active and Passive T/IR. *Atti della Fondazione Giorgio Ronchi*, No 3, May–June 2001.

Macchioni, N., Mannucci, M., Olmi, R., Palanti, S. & Riminesi, C. 2013. Microwave reflectometric tool for non-destructive assessment of decay on timber structures. *Advanced Materials Research*, 778, pp. 281–288.

Olmi, R., Bini, M., Ignesti, A., Priori, S., Riminesi, C. & Felici, A. 2006. Diagnostics and monitoring of frescoes using evanescent-field dielectrometry. *Measurement Science and Technology*, 17 (8), pp. 2281–2288.

Olmi, R. & Riminesi, C. 2008. Study of water mass transfer dynamics in frescoes by dielectric spectroscopy. *Il Nuovo Cimento*, 31 C (3), pp. 389–402.

Rosina, E. & Ludwig, N. 1999. Optimal thermographic procedures for moisture analysis in buildings materials. In *Diagnostic Imaging Technologies and Industrial Applications* Munich, Germania, SPIE procedings vol. 3827, pp. 22–33.

Rosina, E. & Grinzato, E. 2001. Infrared and Thermal Testing for Conservation of Historic Buildings. *Materials Evaluation*, ASNT Journal, 59 (8), pp. 942–954.

Rosina, E., Ludwig, N. & Redaelli, V. 2003. Metodi per la misura dell'umidità nei materiali dell'edilizia storica: legno e intonaci. *Proc. of X Congresso Nazionale AIPnD*, Ravenna 2003, ed. pp. 165–173.

UNI 11085 "Beni Culturali". Materiali lapidei naturali e artificiali. Determinazione del contenuto d'acqua: metodo ponderale, 2003.

# Pre-diagnostic prompt investigation of a historic Bell Tower by visual inspection and microwave remote sensing

A. Saisi & C. Gentile
*Department of ABC, Politecnico di Milano, Milano, Italy*

L. Valsasnini
*Architect, Monza, Italy*

ABSTRACT: The paper presents the results of a prompt investigation programme carried out on the Bell Tower of the Church Santa Maria del Carrobiolo in Monza. A preliminary documentary research and a series of accurate and systematic visual inspection on site highlighted a weak structural layout of the bell-tower and diffuse damage. The recent construction of an underground car park has triggered the movement of the several cracks. In order to assess the state of preservation of the structure a prompt dynamic investigation was carried out recording the response to ambient vibrations induced by wind, micro-tremors and swinging of bells by microwave interferometers. The preliminary results of the investigation suggest the installation of a static monitoring of the main cracks.

## 1 INTRODUCTION

Within a wide program of cataloguing of the main religious buildings in Monza, the authors carried out a pre-diagnostic survey of the Santa Maria del Carrobiolo complex. The survey, as well as the whole program, was supported by the CARIPLO Foundation, the Parish of Monza Cathedral (owner of the buildings) and the Italian Ministry of Cultural Heritage.

The complex of Santa Maria del Carrobiolo owns to the Barnabite order since 1574 and includes a church, a bell tower and several other buildings, which were built at different times and with different functions. Based on historic documents, the construction of the church and monastery dates back to XIII century, and the bell-tower construction was completed in 1339 (Magnani Pucci et al. 1997).

Direct survey of the masonry discontinuities confirmed that the tower was built after the apse and the right aisle of the church, since the tower is supported by the load-bearing walls of the apse and the right aisle. The adopted construction sequence, not identified before and apparently weak, clearly gives rise to concern about the foundation of the bell tower, its structural characteristics at the base and the interaction between the unlinked wall portions. In addition, the recent construction of an underground car park adjacent to the tower conceivably activated movement of the pre-existing cracks and of the structural discontinuities related to construction phases.

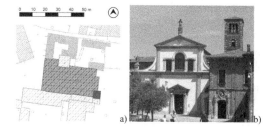

Figure 1. Plan (a) of the view of the Church of Santa Maria del Carrobiolo and of the Bell-Tower.

In order to assess the state of preservation of the bell tower, the authors set up a prompt investigation program including direct survey, visual inspection and dynamic testing using microwave remote sensing (Gentile & Saisi 2012). This first inspection suggests the installation of a static monitoring in order to control the evolution of the crack opening.

The paper presents the results of the investigation program and the criteria which led to the static monitoring of the main cracks.

## 2 THE TOWER

The complex of Santa Maria del Carrobiolo consists of several buildings, which arose at different times and with different functions. The complex was built by the Humiliati, an Italian religious order formed probably in the 12th century and suppressed in 1571, and passes to the Barnabite order in 1574.

The historic documents trace the construction of the church (probably from 1232–34) and of the monastery from XIII century, incorporating probably former buildings (Magnani Pucci et al. 1997). A date engraved on a stone placed on the bell tower dates back to 1339 the likely conclusion of the construction of the tower (Magnani Pucci et al. 1997). The tower was clearly built after the building of the apse and of the last part of the right aisle being supported by their walls (Figs. 2–3).

The direct survey led to this conclusion, observing the masonry discontinuities and the traces of brick decoration on the pilaster of both the church apse and embedded in the tower walls (Figs. 5–7).

In more details, the analysis of the discontinuities highlights that the bell tower is composed, from the base up to about the height of the church, from portions of walls not connected: during the construction of the tower (Fig. 3 and 7), its load-bearing walls were just leant against the existing walls of the church (but not toothed or linked) in the south, east and north part. The discontinuities due to the building phases look very sharp,

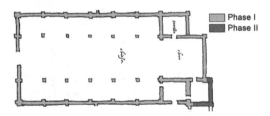

Figure 2. Building phases in a plan dated back to 1572.

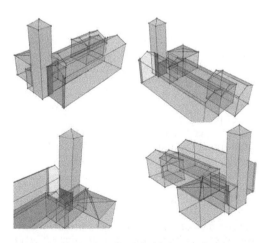

Figure 3. Reconstruction of the interaction between the Church apse and the Bell-Tower.

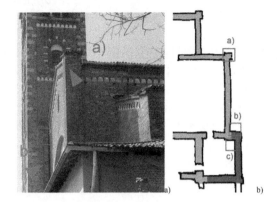

Figure 4. Localisation of the decoration of the first building phase.

Figure 5. Detail of the decoration of the first building phase called a) in Figure 4a, b.

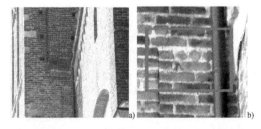

Figure 6. Localisation (a) and detail (b) of the decoration of the first building phase in the position b) in Figure 4a, b.

Figure 7. Localisation (a) and detail (b) of the decoration of the first building phase in the position c) in Figure 4a,b in the indoor and of the sharp discontinuity due to the building phase.

demonstrating the lack of connection between the portions (Fig. 7). Above the church height, the masonry texture seems homogeneous from the structural continuity point of view, even if several cracks are present.

Significant cracks are visible indoor in the center of both East and West fronts; pilasters in the center of the outdoor fronts partially hide the cracks (Fig. 8).

Other meaningful damage is concentrated at the level below the belfry and diffused on all the sides, across the large windows, most of them cutting the entire wall thickness.

The recent construction of an underground car park adjacent to the complex of Santa Maria del Carrobiolo and particularly to the East front of the bell tower, may have contributed to damage the structure and to increase the movement of the pre-existing cracks and of the sharp discontinuities related to the construction phases.

At the base of the tower, several cracks appeared on the plaster even after a recent wall painting.

The spread of cracks and the evident reopening of cracks already repaired, suggest the need of a analysis of the problem. In order to study the structural behavior of the tower a prompt dynamic investigation was carried out. The effectiveness of dynamic investigation in the evaluation of the

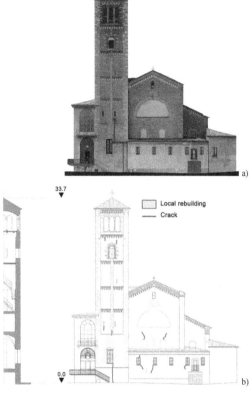

Figure 8. Orthoplane (a) and geometric survey of the East front of the Bell-Tower and of the Church apse. Cracks and local rebuilding are mapped.

state of preservation has been fully demonstrated by several authors (see e.g. Gentile & Saisi 2007, Ramos 2007, Oliveira et al. 2012, Gentile & Saisi 2013). Furthermore, a static monitoring of the main cracks for a minimum time of 18 months was planned.

## 3 THE DYNAMIC INVESTIGATION

Since the first inspection the need of a structural control was clear. In order to accelerate the survey time, the application of microwave remote sensing (Pieraccini et al. 2004, Gentile & Saisi 2012) to ambient vibration testing of the tower seemed the most suitable technique.

Microwave remote sensing might be extensively applied for identifying the dynamic characteristics of masonry towers when: (1) only the fundamental mode is mainly needed (as it happens in wide programs of preventive conservation or in post-earthquake assessment); (2) special issues have to be investigated (such as the effects of bell swinging). In those cases, the application microwave remote sensing is very quick and accurate and does not expose the test crew to hazardous conditions.

According to previous issues, ambient vibrations induced by wind, micro-tremors and swinging of bells were acquired by radar.

## 3.1 *The radar measurement system*

The radar sensor used in this work (Fig. 9) is an industrially engineered microwave interferometer (IDS, IBIS-S system) (Pieraccini et al. 2004, Gentile & Bernardini 2008, Gentile 2010, Gentile & Saisi 2012) and consists of a sensor module, a control PC and a power supply unit.

The sensor module (Fig. 9) is a coherent radar (i.e., a radar preserving the phase information of the received signal) generating, transmitting and receiving the electromagnetic signals to be processed in order to provide the deflection measurements. The sensor, including two horn antennas, has a weight of 12 kg and is installed on a tripod equipped with a rotating head, so that it can be aligned in any desired direction (Fig. 9).

Simultaneously measuring the displacement of several points on a large structure, by using a radar, involves two key steps (Gentile & Bernardini 2008, Gentile 2010): (a) acquiring consecutive radar "images" of the structure (Fig. 10) at an appropriate sampling rate, with different points of the structure being individually observable in each image; (b) using the phase variation of the back-scattered microwaves coming from each detected target point at different times to evaluate the displacement. Hence, the main information provided by the radar technique are the synthetic image of the scenario and the time histories of the points in the scenario that characterized by a good electromagnetic reflectivity.

The synthetic image of the scenario, or range profile, is simply a 1-D map of the intensity of the

Figure 10. Test layout.

received radar echoes in function of the distance of the target points generating the echoes themselves: the peaks in the range profile identify the position/range of the targets detected in the scenario at each time sample. It is worth underlining that the microwave interferometer has only 1-D imaging capabilities, i.e. different targets can be unambiguously detected if they are placed at different distances from the radar; hence, measurement errors may arise from the multiplicity of contributions coming from different points placed at the same distance from the radar (Gentile & Bernardini 2010, Gentile 2010).

Once the interesting targets have been identified from the range profile, the sensor management software provides the time histories of the deflections corresponding to the peaks in the range profile. It is worth noting that the microwave sensor measures displacement along the radar line of sight only; hence, the evaluation of actual deflections requires the prior knowledge of the direction of motion (Gentile & Bernardini 2010, Gentile 2010).

## 3.2 *The test results*

The acquisition of the tower dynamic behaviour by microwave radar interferometry was carried out in the plane East–West, perpendicular to the apse

Figure 9. View of the microwave interferometer (IDS, model IBIS-S).

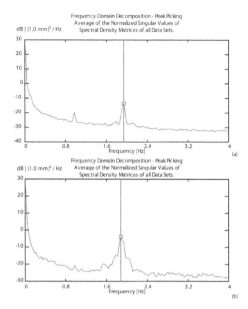

Figure 11. Singular Value Line and identification of the fundamental frequency from radar data: (a) microtremors; (b) random swinging of bells.

and to the tower front toward the underground car park (Fig. 10). The investigation identified the first dominant vibration mode of the tower in the East-West direction equal to 1.93 Hz.

Furthermore, the structure showed a slightly non linear behaviour even at the low level of ambient vibrations that existed during the tests (Gentile & Saisi 2013). The remarks derives from the comparison of the results identified from the two different data-sets acquired under different level of ambient excitation (micro-tremors and swinging of bells): the comparison, exemplified in Fig. 11, reveals slight but clear decreases of the fundamental frequency associated to the higher level of excitation caused by the bell swinging. Hence, the dynamic characteristics of the tower are possibly dependent on the amplitude of excitation/response.

## 4 THE STATIC MONITORING

The damage observed on the repaired cracks suggested the monitoring of the main cracks. In order to control the opening variations of the macroscopic cracks, 10 displacement transducers were installed at different level of the tower and connected to a wireless data logger for the automatic data acquisition, storage and transmission by modem GSM-GPRS. Figure 12 shows the general layout of the monitoring system in the tower, installed on June 2014 and still active.

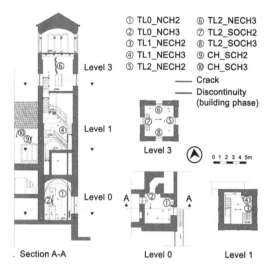

Figure 12. General layout of the monitoring system.

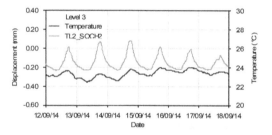

Figure 13. Temperature and TL_SOCH2 displacement variation measured between 12.9.2014 and 18.9.2014 at Level 3.

The devices are rectilinear displacement transducers (potentiometers) with a maximum stroke of 25 mm and a maximum error on the linearity of 0.2%. The system includes 5 thermocouples to measure the masonry surface temperature. The automatic acquisition has been set so as to perform the reading of the displacements and the temperature every 10 minutes.

Figure 13 shows the great influence of daily variation of temperature on the width variation of some cracks (Fig. 14). Dominant temperature behaviour affects the cracks on the South side of the Church, with nearly linear correlation.

After 10 months, the most considerable opening of the cracks seems affecting the ground level (Fig. 15). Despite the temperature influence, with a slightly closing with temperature increasing (Fig. 16), an incremental trend of the measured displacements is recognizable by plotting the displacement vs temperature. The trend of the displacement measured by the two sensors shows a

Figure 14. Detail of the localisation of the TL_SOCH2 transducer at Level 3 (Fig. 12).

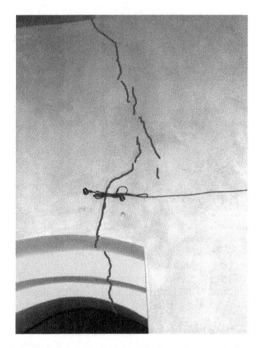

Figure 15. Detail of the localisation of the TL0_CH2 transducer at Level 0.

good agreement between the two set of data. This seems to confirm the movement at the foundation level and the effect induced by the underground structures.

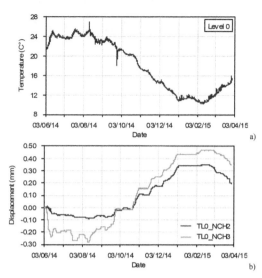

Figure 16. Temperature (a) and displacement measured at Level 0 in 10 months.

## 5 CONCLUSIONS

The research has shown the importance of a global approach in the assessment of the safety of historic buildings. The available information on historic evolution, masonry inspecton and on-site survey could help in the damage interpretation but also to give information for the further investigation.

In the case of the Bell-Tower of Santa Maria del Carrobiolo, the detailed knowledge of the building allowed to recognize the weakness of the structural layout and suggested the following dynamic testing and monitoring.

In particular, the cladding detachment is a serious problem for the Tower safety. It involves local dam-age that can be evaluated only by direct inspection.

Microwave remote sensing is probably the most recent experimental technique suitable to the non-contact measurement of deflections on heritage structures, as historic towers, in prompt investigation programs.

Notwithstanding the low level of ambient excitation that existed during the test, the identification of the fundamental frequency and of its dependence on the level of excitation provided an important information about the structural conditions of the building.

This preliminary investigation suggested the installation of static continuous monitoring of the main cracks and structural discontinuity due to the building evolution.

The research shows the importance of the combination of several investigation techniques within an accurate designed programme. The harmonization of the selected techniques, even if very far conceptually ranging from documentary research to advanced microwave interferometry, were fully successful in the prompt investigation of the structure and in its monitoring.

## ACKNOWLEDGEMENT

The research was partially supported by Fondazione CARIPLO, project "Concio d'argilla". The authors would like to thank IDS (Ingegneria Dei Sistemi, Pisa, Italy) for supplying the radar system employed in the tests. The Community of the Barnabite Order in Monza and Eng. L. Sorteni are gratefully acknowledged for the precious collaboration during the field tests.

## REFERENCES

Gentile, C. & Bernardini, G. 2008, Output-only modal identification of a reinforced concrete bridge from radar-based measurements, *NDT&E Int.*, **41**(7), 544–553.

Gentile, C. 2010, Deflection measurement on vibrating stay cables by non-contact microwave interferometer. *NDT&E Int.*, 43(3), 231–240.

Gentile, C. & Saisi, A. 2007, Ambient vibration testing of historic masonry towers for structural identification and damage assessment. *Constr. Build. Mater.* 21, *1311–1321*.

Gentile, C. & Saisi, A. 2012, Radar-based vibration measurement on historic masonry towers, In: *Emerging Technologies in Non-Destructive Testing V*, CRC Press/Balkema, 51–56.

Gentile, C. & Saisi, A. 2013, Operational modal testing of historic structures at different levels of excitation. *Constr. Build. Mater.* 48, 1273–1285.

Magnani Pucci, P., Colombo, M., Marsili, G., 1997, *La Chiesa di Santa Maria di Carrobiolo*, SPA Tipografica Sociale (in Italian).

Oliveira, C.S., Çakti, E., Stenge, D., Branco, M. 2012, Minaret behavior under earthquake loading: the case of historical Istanbul, *Earthquake Engineering & Structural Dynamics*, 41, 19–39.

Pieraccini, M., Fratini, M., Parrini, F., Macaluso, G. & Atzeni, C. 2004, Highspeed CW step-frequency coherent radar for dynamic monitoring of civil engineering structures. *Electron. Lett.*, 40(14), 907–908.

Ramos, L. 2007, *Damage identification on masonry structures based on vibration signatures*, Ph.D. Thesis, Minho, Portugal: University of Minho.

# On site investigation and continuous dynamic monitoring of a historic tower in Mantua, Italy

A. Saisi, M. Guidobaldi & C. Gentile
*Department of ABC, Politecnico di Milano, Milan, Italy*

ABSTRACT: The paper describes the strategy applied to assess the structural condition and the seismic vulnerability of the Gabbia tower in Mantua, after the seismic sequence of Spring 2012. The post-earthquake investigation highlighted the poor state of preservation of the upper part of the building and suggested the installation of a simple dynamic monitoring system, as a part of the health monitoring process helping the preservation of the historic structure. The results of the continuous dynamic monitoring for a period of about 15 months highlighted the effect of temperature on automatically identified natural frequencies as well as the non-reversible structural effects determined by a far-field earthquake.

## 1 INTRODUCTION

The Spring 2012 earthquakes highlighted the high vulnerability of the historic architectures especially in the South part of the province of Mantua and in the neighbouring Emilia-Romagna region, where several brittle collapses of towers, fortification walls and castles occurred, despite the supposed low seismicity of the area.

After the earthquake, an extensive research program was carried out to assess the structural condition of the tallest historic tower in Mantua: the *Gabbia* tower (Zuccoli 1988, Saisi & Gentile 2015) (Figs. 1 and 2). The post-earthquake survey (including extensive visual inspection, historic and documentary research, NDT, material testing and ambient vibration tests) highlighted the potential weakness of the top of the tower, due to the complex evolution of the structure (Fig. 3): the coupling of ineffective links between the building phases and the extension of the masonry decay produces local vulnerability to be better investigated.

Furthermore, a preliminary ambient vibration test turned out to detect both the global dynamic characteristics of the tower and one local mode involving the upper part of the building, so that the installation of a simple dynamic monitoring system in the tower, with seismic and structural health monitoring purposes, is suggested.

After the description of the tower and the post-earthquake survey, the paper presents the results of the first 15 months of monitoring, highlighting the effect of temperature on the automatically identified natural frequencies, the practical feasibility of damage detection methods based on natural frequency shifts and the key role of permanent

Figure 1. View of the Gabbia Tower from East.

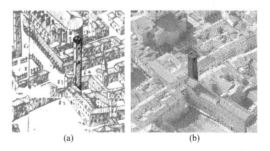

Figure 2. The Gabbia tower and the surroundings: (a) view dating back to XVII century (Bertazzolo, 1628) and (b) recent view from East.

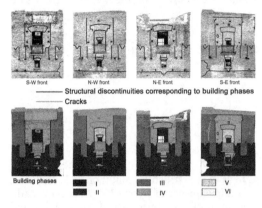

Figure 3. Map of the structural discontinuities and of the supposed building phases: (i) main building up to about 46.0 m; (ii) subsequent addition; (iii) adding of 4 corner piers; (iv) opening infilling and construction of the windows, crowning and the new roof; (v); repair of the South corner.

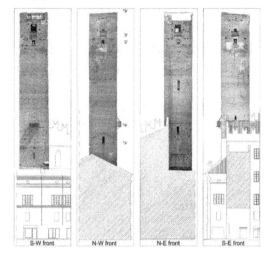

Figure 4. Fronts of the *Gabbia* Tower.

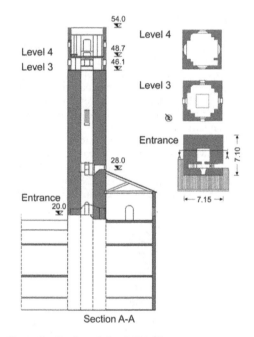

Figure 5. Section of the *Gabbia* Tower.

dynamic monitoring in the diagnosis of the investigated structure.

## 2 THE GABBIA TOWER IN MANTUA, ITALY

The Gabbia tower (Zuccoli 1988) (Figs 1, 3 and 4), about 54.0 m high and with almost square plan, is the tallest tower in Mantua and is named after the hanged dock built in XVI century on the S-W front and originally used as open-air jail. The tower was part of the defensive system of the Bonacolsi family (governing Mantua in XIII century) and some recent research dates the end of construction to 1227.

The tower is nowadays part of an important palace (Figs 1, 2, 4 and 5) whose load bearing walls seem to be not effectively linked but just drawn to the tower masonry walls, whereas the tower itself directly supports several floors and vaults. The structure is built in solid brick masonry and the load-bearing walls are about 2.4 m thick except for the upper levels, where the section decreases to about 0.7 m and a two level lodge is hosted (Fig. 5). The access to the inside of the tower, impracticable since the 1990s due to lack of maintenance, was re-established in October 2012, when provisional scaffoldings and a light wooden roof were installed to allow the visual inspection of the inner load-bearing walls.

No extensive information is available on past interventions on the structure but the passing-through discontinuities detected at the upper levels by the stratigraphic survey can be conceivably referred to the tower evolution phases. Indeed, the results of the survey allowed to recognize at least six main building phases (Fig. 3) and a clear change of the brick surface workmanship at about 8.0 m from the top. Visual inspection performed by using a movable platform after the Italian seismic sequence of May 2012 (Saisi & Gentile 2015) highlighted two different structural conditions for the tower: (a) in the main part of the building, up to about 46.0 m from the ground level, no evident structural damage is observed and the materials are mainly affected by superficial decay (b) the upper portion of the tower (i.e. the top 8.0 m) is

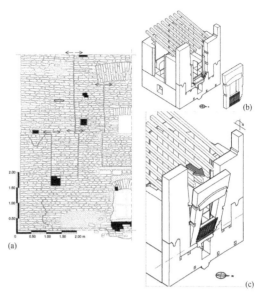

Figure 6. Details of the typical damage and discontinuities on the top of the tower: (a) discontinuities due to the merlons embedded in the masonry texture; (b) weakness of the connections; (c)-(d) infillings between the supposed merlons on the N-W front.

Figure 7. (a) Example of map of the structural discontinuities; (b) identification of the weakly restrained masonry portions and (c) of the possible overturning mechanisms due to seismic actions.

characterized by extensive masonry decay, traces of past structures on all fronts and merlon-shaped discontinuities (Fig. 3). The poor state of preservation and the vulnerability of the top levels of the structure were confirmed by the results of pulse sonic tests.

Moreover, the stratigraphic survey and the accurate mapping of the structural discontinuities (Fig. 6) allowed to support some concerns on the seismic behaviour of such portions. Because of the lack of effective links except the friction, some unrestrained portions could overturn for the combined action of the earthquake and of the roof thrust (Fig. 7). Low intensity actions, like after-shocks or far-field earthquakes could slightly move such weakly restrained elements decreasing the adhesion and accentuating the boundaries.

Figure 7a shows an example of the detailed survey of the masonry textures aimed at recognizing the structural discontinuities and the boundaries of the weakly restrained portions. Based on this investigation, the first evaluation of out-of-plane seismic behavior for each identified masonry portion not effectively linked was carried out (Fig. 7c).

## 3 PRELIMINARY DYNAMIC TESTS

Ambient Vibration Tests (AVTs) were performed on 27/11/2012. The acceleration responses were measured by high sensitivity accelerometers with a frequency range of 0.05 Hz to 500 Hz.

The response of the structure was recorded in 12 points, belonging to 4 pre-selected cross-sections (Fig. 8a). The mentioned presence of a metallic scaffolding and a wooden roof allowed to install all the accelerometers on the inside of the load bearing walls.

The modal identification was performed using time windows of 3600 s and applying the data-driven Stochastic Subspace Identification method (SSI-Data, van Overschee & de Moor 1996) available in the commercial software ARTeMIS (SVS 2012). Notwithstanding the very low level of ambient excitation, 5 vibration modes could be identified within the frequency range 0–10 Hz (Fig. 8b):

1. two closely spaced modes were detected around 1.0 Hz. These modes are dominant bending (modes $B_1$-$B_2$) and involve flexure in the two main planes of the tower, respectively;
2. the third mode (mode $B_3$) involves dominant bending in the N-E/S-W plane with slight components also in the orthogonal N-W/S-E plane;
3. the fourth mode (mode $T_1$) is characterized by dominant torsion until the height of 46.0 m and coupled torsion-bending of the upper level. The local behavior of the top region of the tower, which alters what would otherwise be classified as a pure torsion mode, can be conceivably associated to the effect exerted by the wooden roof, laying directly on the weakest part of the structure;
4. a local mode ($L_1$) was identified at 9.89 Hz and involved torsion of the top levels of the tower. The presence of a local mode provides

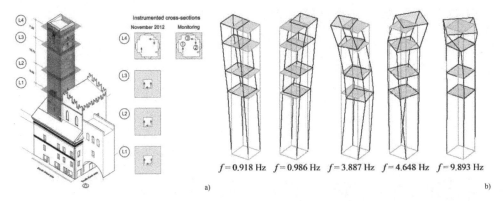

Figure 8. (a) Sensors layout (dimensions in m) adopted in the preliminary AVTs and in the continuous dynamic monitoring of the *Gabbia* tower; (b) Vibration modes identified in the preliminary AVT (SSI-Data, 27/11/2012).

further evidence of the structural change in the masonry quality and morphology highlighted in the upper part of the building by the visual inspection.

It should be further noted that the wooden roof, even if very light, is not only directly supported by the weakest structural part of the tower but is also slightly inclined in the N-E/S-W direction (as schematically shown in Fig. 8a). The roof slope, the redundant connection with the masonry walls and the thermal effects might induce not negligible thrusts on the structural walls, already very vulnerable due to the extensive decay and the lack of connection between the different addings.

## 4 DYNAMIC MONITORING AND RESULTS

A few weeks after the execution of the preliminary AVTs, a simple dynamic monitoring system was installed in the tower. The main objectives of the continuous dynamic monitoring are:

1. evaluating the dynamic response of the structure to the expected sequence of far-field earthquakes;
2. evaluating and modeling the effects of environmental factors on the natural frequencies of the tower;
3. detecting any possible anomaly or change in the structural behavior.

The system (Fig. 8a) is composed by: (a) one 4-channels data acquisition system (24-bit resolution, 102 dB dynamic range and anti-aliasing filters); (b) 3 piezoelectric accelerometers (WR model 731 A, 10 V/g sensitivity and ±0.50 g peak), mounted on the cross-section at the crowning level of the tower; (c) one temperature sensor, installed on the S-W front and measuring the outdoor temperature; (d) one industrial PC on site, for the system management and data storage.

A binary file, containing 3 acceleration time series and the temperature data, is created every hour, stored in the local PC and transmitted to Politecnico di Milano for being processed.

The continuous dynamic monitoring system has been active since late December 2012. As in the preliminary tests, modal identification was performed using time windows of 3600 s, in order to comply with the widely agreed recommendation of using an appropriate duration of the acquired time window (ranging between 1000 and 2000 times the fundamental period of the structure) to obtain accurate estimates of the modal parameters from Operational Modal Analysis (OMA). In fact, OMA methods assume that the excitation input is a zero mean Gaussian white noise and the longer is the acquired time window, the more closely this assumption is verified.

The data files received from the monitoring system are managed by a software (Busatta 2012) developed in LabVIEW and including the following tasks:

1. creation of a database with the original data (in compact format) for later developments;
2. preliminary pre-processing (i.e. de-trending, automatic recognition and extraction of possible seismic events, creation of 1 dataset per hour);
3. statistical analysis of data, including the evaluation of averaged acceleration amplitudes and temperature trends;
4. low-pass filtering and decimation of each dataset;
5. creation of a second database, with essential data records, to be used in the modal identification phase.

This section summarizes the results obtained in a period of 15 months, from 17/12/2012 to

17/03/2013, using the automated Covariance-driven SSI technique (SSI-Cov) described in (Cabboi 2013) for modal identification (10945 datasets).

Figure 9 shows the evolution in time of the outdoor temperature on the S-W front and of the automatically identified modal frequencies. The temperature tracking reveals large fluctuations, between −2°C and 45°C, with significant daily variations in sunny days. A closer inspection also highlights that the natural frequencies of the global modes (B1-B3 and T1, Fig. 8b) seem to vary accordingly with the outdoor temperature. This relationship can be better investigated by plotting each frequency with respect to the recorded temperature. Figure 10 shows the results obtained for modes B1 and B2 along with the best fit lines.

The two plots, referred to the period between 07/01/2013 and 14/06/2013, confirm that the frequency of global modes tends to increase with increased temperature. This behavior, observed also in past experiences on masonry towers (Ramos et al. 2010, Gentile et al. 2012), can be explained through the closure of superficial cracks, minor masonry discontinuities or mortar gaps induced by the thermal expansion of materials. Hence, the temporary "compacting" of the materials induces a provisional increase of stiffness and modal frequencies, as well.

The statistics of the natural frequencies identified from 17/12/2012 to 17/03/2014 are summarized in Table 1. This table includes the mean value ($f_{ave}$), the standard deviation ($\sigma_f$) and the extreme values ($f_{min}$, $f_{max}$) of each modal frequency. It should be noticed that standard deviations are larger than 0.03 Hz for all global modes and especially significant for the local mode. Indeed, the frequency evolution of mode $L_1$ looks very different from the others (Fig. 9) and characterized by significant variations, as the natural frequency ranges between 8.33 Hz and 10.33 Hz over the investigated period. This behavior deserves more explanations.

A close inspection of the frequency tracking of mode $L_1$ allows to recognize three different time windows: in a first phase (between 17/12/2012 and 14/08/2013) the frequency clearly decreases in time, from an initial value of about 10.0 Hz to a final value of about 8.5 Hz; after the summer period (from 19/09/2013 to 19/01/2014) the natural frequency increases again, even if the new maximum values do not reach those identified one year before in similar environmental conditions; finally, the modal frequency decreases again from 02/02/2014 to the end of the examined time period.

More in details, each of the mentioned three phases is characterized by the presence of some discontinuities (drops or increases), which define the general trend of the frequency tracking. Before September 2013, 3 clear drops can be detected between: (a) 03/02/2013 and 04/02/2013; (b) 14/03/2013 and 15/03/2013 and (c) 13/04/2013 and 15/04/2013. Two more discontinuities can be observed after summer: (a) between 12/10/2013 and 15/10/2013 and (b) between 05/12/2013 and 06/12/2013. This time the natural frequency increases at each step and follows backward the course charted during the previous phase. Finally, a last frequency drop is detected between 24/02/2014 and 26/02/2014, where the behavior looks similar to the one observed in the first time window.

These discontinuities divide the examined 15 months in different parts, easily observable by plotting the modal frequency versus the measured outdoor temperature (Fig. 11). The obtained

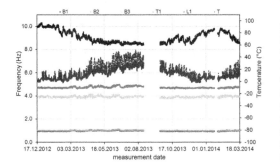

Figure 9. Variation of temperature and identified natural frequencies between 17/12/2012 and 17/03/2014.

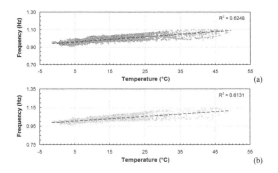

Figure 10. Natural frequency of modes $B_1$ (a) and $B_2$ (b) versus outdoor temperature between 07/01/2013 and 14/06/2013.

Table 1. Statistics of the natural frequencies identified (SSI-Cov) from 17/12/2012 to 17/03/2014.

| Mode | $f_{ave}$(Hz) | $\sigma_f$(Hz) | $f_{min}$(Hz) | $f_{max}$(Hz) |
|---|---|---|---|---|
| 1 ($B_1$) | 0.977 | 0.035 | 0.903 | 1.102 |
| 2 ($B_2$) | 1.017 | 0.028 | 0.960 | 1.148 |
| 3 ($B_3$) | 3.935 | 0.069 | 3.742 | 4.194 |
| 4 ($T_1$) | 4.741 | 0.073 | 4.600 | 5.010 |
| 5 ($L_1$) | 9.062 | 0.511 | 8.332 | 10.327 |

clouds of temperature-frequency points, corresponding to each time window defined by two consecutive discontinuities, are all characterized by similar slope of the best fit lines, whether or not the average frequency decreases (Fig. 11a, c) or increases (Fig. 11b).

The observed behavior suggests the progress of a possible damage mechanism, conceivably related to the effect exerted by the wooden roof with increased temperature, and confirms the poor structural condition and the high seismic vulnerability of the upper part of the tower, highlighting the urgent need for preservation actions. These conclusions seem to be confirmed by the frequency loss of the local mode detected after one year of monitoring, with the natural frequency unable to reach the maximum values identified one year before with almost unchanged temperatures.

It is worth mentioning that between January and June 2013, the monitoring system acquired the tower's response to various far-field earthquakes. The last recorded event, occurred on 21/06/2013 in the Garfagnana region, was the strongest of the recorded earthquakes and produced significant effects on the *Gabbia* tower, as it will be discussed in the following section.

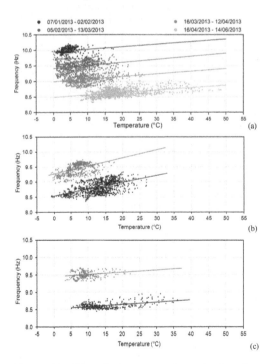

Figure 11. Natural frequency of mode $L_1$ plotted versus outdoor temperature in the time intervals: (a) 07/01/2013–14/06/2013; (b) 15/10/2013–15/01/2014; (c) 10/02/2014–17/03/2014.

## 5 TEMPERATURE AND DAMAGE EFFECTS

The continuous dynamic monitoring of the investigated tower revealed a clear dependence of the natural frequencies on temperature. This might conceal the damage effect on data, as any variation due to structural changes would be masked by the fluctuations caused by environmental factors. Therefore, the effects of temperature have been accounted for and removed by using a "multiple linear regression" mode (Cabboi et al. 2015). The Multiple Linear Regression is a statistical technique based on the relation between a single dependent variable and one or more independent variables (Hair et al. 1998). This method is most likely the simplest available technique to correlate the observed external factors (predictors) and the identified natural frequencies (dependent variables).

Regression models can be classified as static or dynamic. In the first case the relationship is defined only between simultaneously measured data. However, this characteristic can turn into a limitation when trying to model dynamic processes. Therefore, static relationships can be enhanced by taking into account also the influence of external factors measured at previous time instants, obtaining dynamic models. The very simple linear regression model, adopted in the first step of the research (Saisi et al. 2015), was further generalized so that predictions are computed considering also previous values of the dependent variables.

Therefore, the effects of temperature have been accounted for and removed by using dynamic regression models; in other words, the relationship between each natural frequency and temperature was established by taking into account not only the current temperature, but also a certain number of values recorded at previous time instants.

As previously mentioned, the earthquake occurred in Garfagnana on 21/06/2013 was the strongest of the recorded seismic events and determined accelerations on top of the tower more than 40 times larger than the usual ambient vibration responses. Therefore, in order to detect possible structural effects induced on the *Gabbia* tower, the regression model has been calibrated over the 6 months preceding the earthquake.

The comparison between identified natural frequencies and corresponding numerical predictions (Fig. 12) highlights a very good accordance, providing a sound validation of the selected model.

The dynamic regression model has been subsequently used to predict future estimates of the natural frequencies: if the model is able to predict the frequency change, its prediction error will not be affected by environmental factors and will work as an indicator of the structural condition, for any abnormal variation will be related to structural

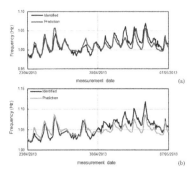

Figure 12. Typical prediction of natural frequencies obtained from a dynamic regression model: (a) Mode B1; (b) Mode B2.

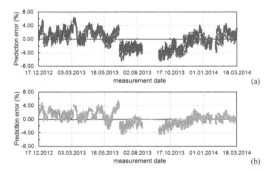

Figure 13. Change in the prediction error induced by the seismic event of 21/06/2013: (a) Mode B1; (b) Mode B2 (Dynamic regression model calibrated using data collected from 17/12/2012 to 20/06/2013).

changes. In the present application, natural frequencies have been predicted from 21/06/2013 to 17/03/2014, in order to detect possible structural effects due to the earthquake. Typical results are presented in Figure 13 for modes B1 and B2.

The following considerations can be drawn:

1. before the seismic event, the prediction error is characterized by fluctuations between −2 and 5% and does not exhibit sudden drops;
2. a clear drop of the prediction error is detected corresponding to the seismic event; in particular, the error increases (exceeding the −2% limit and reaching −4%) and departs from the previous trend. Furthermore, the error no longer oscillates around 0% but is now characterized by a negative offset;
3. during the following weeks, the prediction error neither decreases nor gets back to the original trend. At the opposite, it hovers around the new offset for few weeks and further increases after the summer period (from −4% to −6%);
4. from mid-November 2013 the difference between experimental and predicted natural frequency

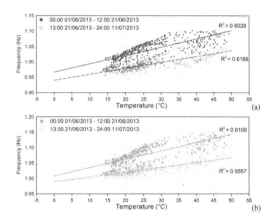

Figure 14. Change in the frequency-temperature correlation induced by the seismic event of 21/06/2013: (a) Mode B1; (b) Mode B2.

decreases, as the outdoor temperature decreases too. This behavior can be better understood by inspecting the frequency-temperature relationship before and after the earthquake. The results of this check are summarized in Figure 14, where 3 weeks before and 3 weeks after the earthquake have been considered: the regression lines exhibit a remarkable variation after the seismic event, with the range of temperature variation being almost unchanged. It can be also observed that the difference between the experimental frequencies identified before and after the earthquake decrease with a decreased temperature. As a consequence, for low temperatures, the regression model calibrated over the first 6 months of monitoring still provides quite accurate predictions.

The trend of the prediction error, along with the variation of the regression lines after the earthquake, suggests that the structural effects produced by the seismic event of 21/06/2013 are not reversible.

In order to carry on the damage detection process, the regression model should be modified to take into account the different structural behavior. In particular, the parameters should be calibrated again over the months following the seismic event.

Indeed, a new dynamic regression model has been calibrated between 21/06/2013 and 06/01/2014 and used to predict the natural frequencies until 18/03/2014. The prediction error obtained for modes B1 and B2 (Fig. 15) allows the following comments:

1. the previously observed negative offset is no longer detected as the error oscillates around 0%;
2. the prediction error now exhibits smaller fluctuations, between −4 and 4% for mode B1 and between −3 and 3% for mode B2;
3. after the summer period, a minor drop can still be detected. This behavior might suggest possible

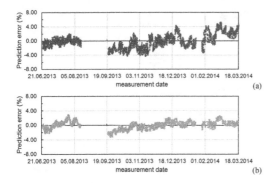

Figure 15. Change in the prediction error induced by the seismic event of 21/06/2013: (a) Mode B1; (b) Mode B2 (Dynamic regression model calibrated using data collected from 21/06/2013 to 06/01/2014).

subsequent developments of the structural effects induced by the earthquake of June 2013.

## 6 CONCLUSIONS

The paper focuses on the post-earthquake assessment of a historic masonry tower and summarizes the results of visual inspection, ambient vibration tests and long-term dynamic monitoring of the building.

Visual inspection and the stratigraphic survey of all main bearing walls clearly indicated that the upper part of the tower is characterized by the presence of several discontinuities due to the historic evolution of the building, local lack of connection and extensive masonry decay. The poor state of preservation of the same region was confirmed by the observed dynamic characteristics and one local mode, involving the upper part of the tower, was clearly identified by applying the SSI-data technique to the ambient response data collected for more than 24 hours on the historic structure. Furthermore, the natural frequency of this local mode tends to decrease as temperature increases, suggesting that the thermal expansion of materials in a very inhomogeneous area of the tower, causes a general decrease of the connection between the masonry portions.

Furthermore, the paper summarizes the main results of 15 months of continuous dynamic monitoring of the *Gabbia* tower in Mantua. The monitoring system, consisting of a limited number of accelerometers and temperature sensors installed at the top level of the structure, allowed to:

a. accurately track the evolution of natural frequencies;
b. evaluate the dependence between frequencies of the global modes and outdoor temperature;
c. highlight the progress of a damage mechanism, involving the upper part of the building,

remarked by the significant fluctuations of the natural frequency of a local mode;
d. distinguish between temperature and damage effects on the natural frequencies, revealing the sudden drop of the frequency of the fundamental modes induced by a seismic event.

## ACKNOWLEDGEMENTS

The investigation was supported by the Mantua Municipality.

M. Antico, M. Cucchi (VibLab, Laboratory of Vibrations and Dynamic Monitoring of Structures, Politecnico di Milano) and Dr. L. Cantini are gratefully acknowledged for the assistance during the visual inspection, the field tests and the installation/maintenance of the monitoring system.

## REFERENCES

Bertazzolo, G. 1628. *Urbis Mantuae Descriptio*. Biblioteca Teresiana, Mantova.
Busatta, F. 2012. *Dynamic monitoring and automated modal analysis of large structures: methodological aspects and application to a historic iron bridge*. Ph.D. Thesis, Politecnico di Milano.
Cabboi, A. 2013. *Automatic operational modal analysis: challenges and application to historic structures and infrastructures*. Ph.D. Thesis. Università di Cagliari.
Cabboi, A., Gentile, C, Guidobaldi, M., Saisi, A., 2015. Continuous dynamic monitoring of historic masonry towers using few accelerometers: methodological aspects and typical results, In: *Proc. of SMHII*. Turin. Italy
Gentile, C., Saisi, A., Cabboi, A. 2012. Dynamic monitoring of a masonry tower. In: *Proc. of the International Conference on Structural Analysis of Historical Construction (SAHC 2012)*. Wroclaw. Poland.
Hair, J., Anderson, R., Tatham, R., Black, W., 1998, *Multivariate Data Analysis*. Prentice Hall.
Ramos, L.F., Marques, L., Lourenço, P.B., De Roeck, G., Campos-Costa, A., Roque, J. 2010. Monitoring historical masonry structures with operational modal analysis: two case studies. *Mechanical System and Signal Processing* 24: 1291–1305.
Saisi, A., Gentile, C. 2015. Post-earthquake diagnostic investigation of a historic masonry tower. Journal of Cultural Heritage, http://dx.doi.org/10.1016/j.culher.2014.09.002.
Saisi, A., Gentile, C., Guidobaldi, M. 2015. Post-earthquake continuous dynamic monitoring of the Gabbia Tower in Mantua, Italy. *Construction and Building Materials* 81: 101–112.
SVS. 2012. *ARTeMIS Extractor 2011* http://www.svibs.com/.
Van Overschee, P. & De Moor, B. 1996. *Subspace identification for linear systems: theory, implementation, applications*. Kluwer Academic Publishers.
Zuccoli, N. 1988. Historic research on the Gabbia tower (in Italian). *Municipality of Mantua internal report*.

# Focus on soluble salts transport phenomena: The study cases of Leonardo monochrome at Sala delle Asse (Milan)

A. Sansonetti & M. Realini
*CNR-ICVBC, Milano, Italy*

S. Erba & E. Rosina
*Politecnico di Milano, Milano, Italy*

ABSTRACT: The program of investigations on "Sala delle Asse", which hosts a monochrome landscape attributed to Leonardo, in the Sforza Castle in Milan recently concluded the first step.
Results of the analytical tests for the characterization of materials and their damages showed the high diffusion and concentration of nitrates and sulphates on the surface of the monochrome at the edge with the restoration mortars, on the right side of the north-western wall. On the base of the scientific literature and laboratory tests, the researchers identified a threshold of RH above which deliquescence of salts could easily occur.
Microclimatic monitoring results informed that during the most humid days in spring, summer and fall, RH trespasses this threshold, with a frequency of about 30 events/year. After an accurate analysis of air Temperature (T°C) and Relative Humidity (RH) resulted that the exterior changes especially affect the interior climate at some summer conditions.

## 1 INTRODUCTION

Documentation regarding the assessment of materials and their state of conservation (Fiorio, 2007) permitted to hypothesize that the paintings require a complex, urgent, articulated intervention for the safeguard of the remaining Leonardo monochrome fragments. The verification of the first hypothesis, exposed in the preliminary phase of 2013, addressed the further analysis on the most critical areas.

On the base of the damage mapping, the first tests had the aim to ascertain and localize if anomalous water content are present in the masonry.

Despite of the masonry thickness and the elevation (first floor), the damage at the bottom of the masonry showed clear sign of water phase transition (liquid/vapor and vice versa). Moreover, due to the extension of the areas where is a diffused salt crystallization, on both the monochrome and the vault, on May 2012 the authors started the first step of a microclimatic monitoring to detect dangerous variations of Temperature (T°C) and Relative Humidity (RH) of air that could accelerate the diffusion of soluble salts solutions and increase their damage.

## 2 PRELIMINARY TESTS ON THE STUDY CASES

NDT method showed that structures and surfaces are dry: the authors applied an IRT scanning and

Figure 1. Sample 1 C. White efflorescences are concentrated alongside the border of a patch which is composed by a different cement mix respect to the surrounding matrix.

gravimetric tests (Rosina, 2008, Ludwig, 2002) as preliminary assessment for mapping the presence of water in the structure.

In the areas where the thermal gradient could show the presence of evaporative flux on going, the authors applied the gravimetric tests on a set of few samples composed by very small fragments of restoration mortars (Rosina, 2009) and they evaluated the presence of soluble salts.

The application of NMR and EDF on the masonry, confirmed the results of the IRT and gravimetric data regarding the presence and quantity of water (Capitani, 2012, Olmi, 2006, Proietti, 2015).

## 3 ANALYSIS FOR THE RESTORATION

The restoration of the monochrome is currently on going, under the direction of the Regional Office of the Ministry of Culture. The following results come from the analytical plan supporting the intervention.

The soluble salts formed whisker-like arrangements mostly localized in the outer plaster surface close to recent cement plaster. The authors sampled the surfaces and analysed the samples by means of X Ray Diffraction (X PanalyticalX'Pert PRO). Nitratine (sodium nitrate $NaNO_3$) resulted the main component of efflorescences (fig. 3 shows the diffractogram). Gypsum ($CaSO_4$ $2H_2O$) and magnesium sulphates (both Hexahydrite $MgSO_4$ $6H_2O$ and Epsomite $MgSO_4$ $7H_2O$) are also present

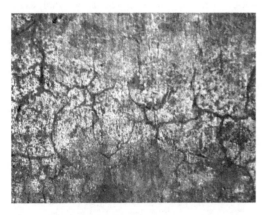

Figure 2. Sample 1. White efflorescences analysed by X Ray Diffraction (see Fig. 3).

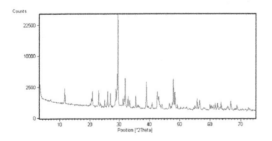

Figure 3. Sample 1, X Ray Diffractogram of powder coming from efflorescence. The pattern shows the presence of Nitratine and gypsum as soluble salts. Other insoluble mineral components are present coming form the plaster (barite and a clay).

as minor components. Nitratine efflorescences are mostly present in the proximity of the north-eastern window, on the monochrome surface. On the contrary, the sulphate salts result diffused in a lower amount in each sample coming from the hall, witnessing a diffuse sulphation of the plaster. Fragments of the original plaster have been analysed by means of Ionic Chromatography (Dionex DX 100). The obtained results corroborated XRD outcomes, quantifying the concentration of ions in the plaster; for what concern nitrates they reach 1% w/w, and their amount increase in the depth of the plaster. On the contrary, sulphates reach their maximum around 0.3% w/w at the surface of the plaster.

## 4 FIRST RESULTS AND PROSECUTION OF THE MONITORING

The results of the first part of the investigation excluded that water infiltration are the cause of the monochrome damages; on the contrary, the main causes probably are the microclimatic unbalances and diffusion of soluble salts.

The salts provenance could be the cement matrix of the restoration mortars, and the cleaning products applied in previous intervention.

The location of the most damaged areas (close to the north-eastern window) shows that further tests should investigate the microclimate variation, and the rate of air mixing close the windows, because of the lack of tightness of the window frames.

Up to now, monitoring results partially explained the dynamic of the natural ventilation inside the hall, therefore the new investigation areas included also the rooms nearby, together with an increase of sensors number to apply in the main hall and the repetition of the psychrometric survey in the second and third years.

The aim of the psychrometric investigation is to evaluate the Temperature and Humidity unbalances of the microclimate and define the thresholds for the optimal conservation of the precious surface in the room (UNI, 1999).

### 4.1 *Psychrometry*

The verification of the existing unbalance by means of the repetition of seasonal and daily psychrometric measurements allowed the authors to locate the most significant places to install the additional probes. As an example, look at the maps of December 2013 (in Figures 4, 5).

### 4.2 *Microclimatic monitoring*

In the second phase of the investigation, as shown in the previous paragraph, the authors added further 8 probes at different heights and in the nearby rooms (Sala del Gonfalone, Salette Nere, Figure 6).

The psychrometric maps showed that the highest unbalances are close to the windows and doors leading towards other rooms that have major air exchange with the exterior.

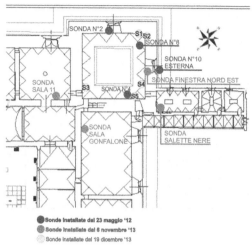

Figure 6. Map of the probes' localization.

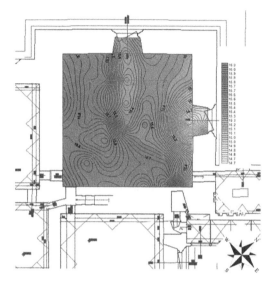

Figure 4. Map of air Temperature distribution, December 19, 2013, h 4,30 pm.

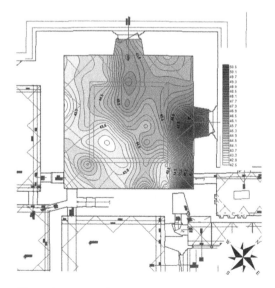

Figure 5. Map of Relative Humidity distribution, December 19, 2013, h 4,30 pm.

The probe installed in the north-western window records variations very similar to the exterior air (with a delay of 2–3 hours), but with a reduced span (daily excursion of about 3°C in late spring, 7–8°C in summer). On the contrary, the thermal difference of the values recorded along the walls, far from the windows, is about 1°C, with very smooth and gradual variations.

The probe installed close to the monochrome and the one in the north-eastern window measure values very similar to the ones measured by means of the other probes in the hall, far from the windows.

Therefore, the two probes in the two windows (north eastern and north western) measure very different values, despite of their proximity to the opening, with the same kind of window frames.

During the solar year, the values of RH vary between 30 and 70% inside the hall, the highest variations are close the north-western window, where the unbalance can be of 20% in few hours.

The other probes measure variations less frequent, small, and slow, up to 10%. Low values are frequent during the winter 2012–2013. Since April 2013 up to September, the daily average values are higher, between 40 and 60%. On the contrary, in 2014 the values are very high also during the winter (up to 70% and often above). During the summer 2014, often the values are higher than 70% due to the prolonged rainfalls, together with thunderstorm and high speed winds.

At last, the measures of probes at 4 and 8 m high show a substantially homogeneous trend, temperatures are higher about 0.5–1.5°C than the others.

In the small rooms beside the hall the temperature are the closest to the exterior ones (during the winter there are only 3–5°C of difference between the interior and exterior values) and obviously, there are the highest variations of RH.

Therefore, the corridor between the small black rooms (Salette Nere) and the hall does not prevent the entrance of airflows from the exterior, with risk of conservation for the painting of the hall.

Apart from this side and the north-western window, inside the hall the conditions are almost homogeneous, at any heights.

## 5 DISCUSSION OF RESULTS

The results of the microclimatic monitoring show risky conditions for conservation due to prolonged rainfalls (summer and late fall 2014), if considering the museum conservation standards for RH (Camuffo, 2012)

The next step of data processing is the extrapolation of the number of variations that are extremely risky for the cycles of salts solubilisation/crystallization on the precious surfaces.

Particularly, the authors processed the monitoring data in relationship to the curve of crystallization of the salts most spread on the monochrome (Nitratine $NaNO_3$). Using the diagram of state of the highlighted compound (9), the authors obtained the definition of the values of the critical line that delimits the passage into solution and vice versa the precipitation of solute. Therefore, with the precipitation of solute, we can observe the appearance of salt efflorescences on the surfaces.

The calculation of the frequency of these events is 10 in the monitoring year. Nevertheless, despite the most spread salt is Nitratine, many other salts are present, in a mixture, and therefore their conditions of crystallization change (Sawdy, 2005). Generally, the threshold of RH to obtain their deliquescence is lower, and the span of risk for occurring crystallization/solubilisation is larger.

Moreover, the curve useful to determine the number of risk events comes from the model of the materials behavior during the change of the microclimatic conditions. Although the model bases on the experimental data, they remain laboratory data, and not data collected in Sala delle Asse.

Therefore, it is preferable to consider some additional possible events, as a wider span of the one previously defined, ranging ±5% of the curve values, as the common sense of expertise suggests.

At the first step of the data processing, the authors filtered the picks and minimum values of T°C and RH in the first year of monitoring. Fig. 7 shows the colored area of the values, which are between 8 and 29°C, 21 < RH < 91.5%.

Major interest is in the phase transition between two values of T and RH, acquired in one hour, which belong to different sets (A solid, B liquid).

Moreover, it is visible also the "buffer zone" (across the dotted lines) of ±5% of the known term of the crystallization line, that includes the points that should be events of the transition phenomena.

The graphic of Figure 8 shows that the majority of the acquired points of probe n 2 are below the line of phase passage, (and below the security strip).

There are 261 points associated to the phase transition, localized in the critical area, often close to the critical events.

All 10 transition events are located between May and July, while the data of the critical strip are included between April and October.

Because of the HVAC plant is never on, and the management and visiting procedures of the hall did not vary during the year under study, the authors concluded that the variations were affected by the exterior climate.

The comparison of the exterior and interior data shows that almost all the events occur for increasing of RH > 80% lasting for some hours.

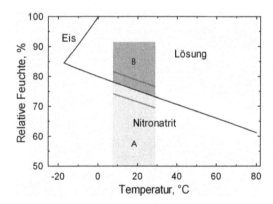

Figure 7. Crystallization curve of Nitratine (reference www.saltwiki.net/Hans-JuegenSchwarz).

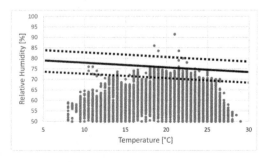

Figure 8. Scatter graph of the data acquired from 23rd May 2012 to 10th October 2013 by probe n 2.

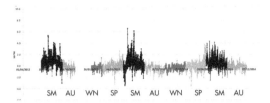

Figure 9. Differences of mixing ratio versus time between exterior (probe n 10) and interior (probe n 2).

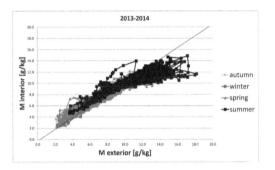

Figure 10. Exterior mixing ratio (probe n 10) and interior one (probe n 2), year 2013–2014.

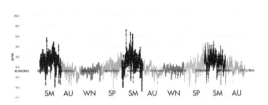

Figure 11. Differences of mixing ratio versus time between exterior (probe n 10) and interior (probe n 8).

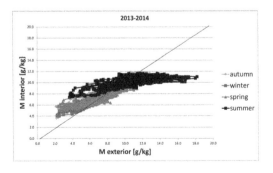

Figure 12. Exterior mixing ratio (probe n 10) and interior one (probe n 8), year 2013–2014.

On the contrary, there are not very apparent connections when the variations are fast. The calculations regarding the air mixing ratio (Camuffo, 2014) are in Figures 9, 10, 11 and 12.

The graphics show the values of mixing ratio with a similar trend: the peaks and major differences are in summer (SM black color) and at the end of the spring (SP).

In fall (AU) and especially in winter (WN), the differences are lower.

These results agree with the localization of the critical events of salts crystallization, and confirm that the time of most frequent occurrence is between May and October.

## 7 CONCLUSIONS

The investigations up to now allowed ascertaining that the prevention of further damages (after the restoration) is due to the microclimatic control.

The ongoing monitoring shows unbalances caused by the exterior variations, especially during the summer when the main wind direction and speed affect the interior microclimate of the hall and nearby rooms.

Therefore, the policy to prevent further damage regards the management of the hall, and the improvement of the window frames, to limit the exchange of air with the exterior, and obtain more even conditions inside the hall, close to the painted surfaces. Because of the lack of a certain source of water inside the masonry, the restorers plan to remove the salts only from the surfaces by means of gel systems. The use of the gel is justified with the aim to confine the release of water in the very outer layer of the plaster. On the contrary, using a soaked pad to extract the soluble salts from the interior of the structure, most probably a dangerous dynamic of salts transport between the exterior and interior layers of plaster and masonry could be activated.

The conservation plan for the future maintenance of the precious surfaces will take into consideration the restorers approach, and the further inspection will check firstly the critical areas where the salts content was higher, with the aim to early detect any whitening of colors before that salt efflorescence will visually appear.

## ACKNOWLEDGEMENTS

Authors found particularly useful the archive research of Carolina Di Biase (Politecnico di Milano) and Laura Basso (Comune di Milano), together with the report and discussion with the scientific committee of restorers that are currently working on the paintings in the hall, to whom their acknowledgment goes.

# REFERENCES

Camuffo D. Bertolin C., Bonazzi A., Campana F., & Merlo C. 2014, Past, present and future effects of climate change on a wooden inlay bookcase cabinet: a new methodology inspired by the novel European Standard EN, Journal of Cultural Heritage, 15, 26–35.

Capitani D., Di Tullio V., & Proietti N., 2012, Nuclear Magnetic Resonance to Characterize and Monitor Cultural Heritage, Prog. Nucl. Magn. Reson. Spectrosc., 64, 29–69.

Ludwig, N. 2002, Tecniche termografiche per la diagnostica sull'edilizia storica, in "Elementi di archeometria - Metodi fisici per i beni culturali "*A. Castellano, M. Martini, E. Sibilia* ed., Egea Ed., Milano 272–287.

Fiorio M.T., & Lucchini A., 2007, Nella Sala delle Asse, sulle tracce di Leonardo, Raccolta Vinciana, XXXII, 100–140.

Olmi R., Bini M., Ignesti A., Priori S., Riminesi C., & Felici A., 2006, Diagnostics and monitoring of frescoes using evanescent-field dielectrometry, Measurement Science and Technology, 17, 8, 2281–2288.

Proietti N., Capitani D., Di Tullio V., Olmi R., Priori S., Riminesi C., Sansonetti A. Tasso F., & Rosina E., 2015, MOdihMA at Sforza Castle in Milano: Innovative Techniques for Detection of MOisture in Historical Masonry, in Built Heritage: MOnitoring, Conservation, Management, L. Toniolo, M. Boriani, G. Guidi editors, Research for Development, ed. Fondazione Politecnico di Milano, Springer International 2015, Switzerland, 187–197.

Rosina E., 2008, Indicazioni metodologiche per la valutazione degli scambi termoigrometrici tra murature e ambiente, in "Contributi di ricerca e didattica per la conservazione del Castello di Malpaga", E. Rosina ed., Silvana ed, Cinisello B. 87–97.

Rosina E., 2009, Le analisi microclimatiche e del regime termo igrometrico delle murature, in "Il Canopoleno di Sassari, da casa professa a Pinacoteca. Storia e restauri", S. Gizzi, S. Della Torre, A. Casula, E. Rosina ed., Silvana ed. Cinisello B., 59–75.

Sawdy A., & Price C. 2005, Salt damage at Cleeve Abbey, England. Part I: a comparison of theoretical prediction and practical observation, Journal of Cultural Heritage, 6, 2, 125–135.

Sawdy A., & Price C, 2005, Salt damage at Cleeve Abbey, England. Part II: seasonal variability of salt distribution and implications for sampling strategies, Journal of Cultural Heritage, 6, 4, 361–367.

www.saltwiki.net/Hans-Juegen Schwarz (webpage at November 2014).

UNI 10829:1999. Beni di interesse storico e artistico—Condizioni ambientali di conservazione—Misurazione ed analisi.

*Emerging Technologies in Non-Destructive Testing VI – Aggelis et al. (Eds)*
*© 2016 Taylor & Francis Group, London, ISBN 978-1-138-02884-5*

# Novel applications of micro-destructive cutting techniques in cultural heritage

## M. Theodoridou & I. Ioannou
*Department of Civil and Environmental Engineering, University of Cyprus, Nicosia, Cyprus*

ABSTRACT: This study presents novel applications of two micro-destructive cutting techniques (DRMS and scratch tool) in cultural heritage. Both techniques have been successfully used for the characterization of natural stone and traditional composites (mortars and adobes), confirming their usefulness in the event of limited sampling. Moreover, because of their sensitivity to pore clogging, they have been used to detect salt crystallization, including subflorescence, in building stone. The DRMS was further successfully applied on stone masonry in-situ in order to assess salt weathering. Finally, both techniques have been used to assess the effectiveness of consolidation treatments. The results of the scratch tool have been utilized to develop a tomography image of the samples under investigation.

## 1 INTRODUCTION

The characterization of material properties and the diagnosis of their state of weathering and conservation are among the most important steps in the field of cultural heritage preservation. Several standardized experimental methods exist, especially for determining the properties of materials and their durability (e.g. EN 12370 & 13161). However, these are often limited in their application by the required number and size of test specimens and the controlled laboratory conditions needed to undertake the tests; this is especially true when the materials under study constitute immovable parts of heritage structures. The use of other advanced techniques, such as Positron Emission Tomography, micro-CT and synchrotron radiation ED-XRD analysis (e.g. Goethals et al. 2009, Derluyn et al. 2008, Ioannou et al. 2005), in the aforementioned field of research, offers invaluable results. However, these techniques may not always be accessible to the wider research community due to their complex nature and relatively high cost of application.

This study presents novel applications of two micro-destructive cutting techniques (the portable Drilling Resistance Measuring System (DRMS) and the scratch tool) in the field of cultural heritage. The aforementioned techniques have been successfully used for the characterization of traditional materials, the detection of salt crystallization and the evaluation of the effectiveness of consolidation treatments.

## 2 MICRO-DESTRUCTIVE TECHNIQUES

The principle of operation of both micro-destructive techniques used in this study lies with the fact that the forces needed to cut a material, either by linear (scratch tool) or rotational (DRMS) action, are related to its mechanical properties and the technological parameters applied on the cutting tool (Dagrain et al. 2010). Therefore, for a given testing configuration, the only parameter influencing the applied forces is the strength of the material.

### 2.1 Scratch tool

The scratch tool (WOMBAT) provided by Epslog Engineering was originally developed for use in the petroleum industry. The determination of the scratching resistance is based on the incremental formation of a shallow groove on the surface of the materials under study. The equipment is connected with a computer, where the force components ($F_n$ and $F_s$) required to scratch the material are automatically recorded. The analysis of these force components leads to the calculation of the intrinsic specific energy, i.e. the energy required for scratching a unit volume of the material, which can be related to uniaxial compressive strength (e.g. Dagrain et al. 2010, Detournay et al. 1995, Almenara & Detournay 1992). In this study, a sharp diamond cutter (10 mm wide) with a negative back rake angle of 15° was used for the determination of the scratching resistance in all tests.

## 2.2 DRMS

The DRMS provided by Sint Technology is a portable device that was primarily developed for laboratory and in-situ studies in the field of heritage conservation (e.g. Cnudde et al. 2008, Fernandes & Lourenço 2007, Delgado Rodrigues et al. 2002, Tiano 2001, Exadaktylos et al. 2000). During the DRMS tests, the penetration rate and rotational speed of the drill bit are kept constant, while the force required to drill is continuously recorded. The outputs of the test are given in x-y plots of the drilling force along the depth profile, while the data is registered in numeric values as well. In this study, a twist diamond drill bit (5 mm diameter) with a flat point was used. The penetration rate was 10 mm/min, while the rotational speed of the drill bit was 600 rpm.

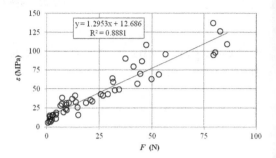

Figure 1. Correlation of the micro-drilling test results, expressed by drilling resistance force ($F$), with the scratching test results, expressed by the intrinsic specific energy ($\varepsilon$) (after Theodoridou et al. 2015).

## 3 CULTURAL HERITAGE APPLICATIONS

### 3.1 Characterization of materials

Both the DRMS and the scratch tool have recently been used, alongside a series of standardized laboratory tests, for the characterization of 50 different (weak to very strong) natural limestones (Theodoridou et al. 2015). The results showed, for the very first time, a strong ($R^2 = 0.89$) linear correlation between the drilling resistance force and the intrinsic specific energy (Fig. 1). This can be well-explained by the principle of operation of the two micro-destructive techniques, which in both cases relies on the materials' resistance to cutting, despite the fact that the mode of cutting differs in each test (i.e. scratching vs. drilling).

In addition to the aforementioned finding, the results of the same study suggested a strong interdependence between the resistance to cutting, either by scratching or drilling, and the porosity, dynamic Young's modulus and uniaxial compressive strength of the test specimens (e.g. Fig. 2). This confirms that both micro-destructive techniques may prove useful in the physico-mechanical characterization of stone, when very limited sampling is allowed (or, in the case of the DRMS, when in-situ measurements are demanded).

The scratch tool and the DRMS have also been used in the characterization of relatively weak traditional composite materials, such as mortars and adobes. In a recent study (Kyriakou 2015), 39 specimens of laboratory designed and produced mortars (aggregate size 0–4 mm), with either aerial or hydraulic lime as binder, were scratched and drilled. Additional standardized tests were performed on the same mortars for the determination of their flexural (three point bending test) and uniaxial compressive strengths. The results

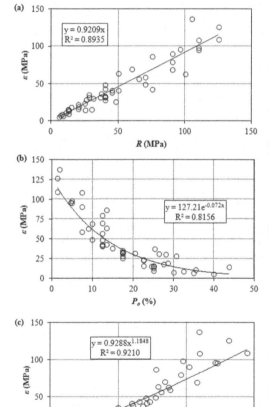

Figure 2. Correlation of the scratching test results of stones, expressed by the intrinsic specific energy ($\varepsilon$), with (a) uniaxial compressive strength ($R$), (b) open porosity ($P_o$) and (c) dynamic Young's modulus ($E_d$) (after Theodoridou et al. 2015).

suggested linear relationships between the intrinsic specific energy and the flexural and uniaxial compressive strengths (Fig 3), with high $R^2$ values (equal to 0.73 and 0.83 respectively). The same type of regression line (linear) correlated the drilling resistance with the flexural and uniaxial compressive strengths ($R^2 = 0.70$ in both cases). As with the study on natural stones, the relationship between the intrinsic specific energy and the drilling resistance force was again found linear ($R^2 = 0.66$), confirming the significant interdependence of the two aforementioned parameters.

Adobe bricks of various compositions, produced in the laboratory, were also tested using the scratch tool and the DRMS (Eftychiou 2013). Linear correlations were observed between the intrinsic specific energy, the drilling resistance force and the uniaxial compressive strength, determined using conventional laboratory testing procedures (Fig. 4). Despite the fact that the number of samples in this case was rather small, bearing in mind the inherent inhomogeneity of adobes, the results are promising and further prove the potential of the two techniques in the characterization of composite heritage materials. Nevertheless, additional research is needed to expand the database of results on adobes.

### 3.2 Salt crystallization detection

Another innovative application of the scratch tool and the DRMS aimed at detecting salts crystallizing in building stone (especially as subflorescence), before they became damaging. Laboratory tests were performed on Lympia limestone, a massive chalk with homogenous mineralogical composition (almost 100% calcitic), impregnated with $MgSO_4$ and $Na_2SO_4$ (Modestou 2012). Drilling and scratching were carried out on duplicate samples treated with a water repellent, in order to show the sensitivity of both techniques in detecting changes to the salt front location induced by surface treatments.

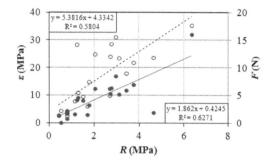

Figure 4. Correlation of uniaxial compressive strength of adobes ($R$) with the scratching test results, expressed by the intrinsic specific energy ($\varepsilon$, dashed line) and microdrilling resistance, expressed by drilling resistance force ($F$, solid line) (after Eftychiou 2013).

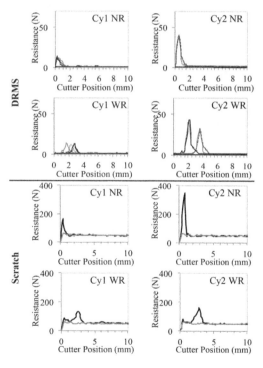

Figure 5. DRMS and scratch tool patterns for samples impregnated with $Na_2SO_4$ and dried at 70°C (2 crystallization cycles). DRMS: — reference pattern; all other lines correspond to individual drilled holes. Scratch: — average salt contaminated pattern; — average reference pattern. Cy: cycle; NR: no water repellent; WR: with water repellent. Resistance refers to recorded resistance experienced on tool. X-axis origin corresponds to the upper surface of samples (after Modestou 2012).

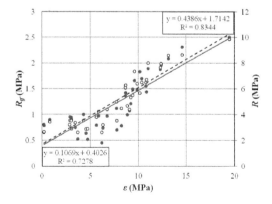

Figure 3. Correlation of the scratching test results of mortars, expressed by the intrinsic specific energy ($\varepsilon$), with flexural strength ($R_{tf}$, solid line) and uniaxial compressive strength ($R$, dashed line) (after Kyriakou 2015).

Both techniques successfully highlighted the difference in the crystallization mode and location of MgSO$_4$ and Na$_2$SO$_4$. Na$_2$SO$_4$ crystallized either at the evaporative surface of non-treated samples, or right behind this surface of samples treated with water repellent (Fig. 5). In MgSO$_4$ contaminated samples, cutting resistance peaks were detected well beneath the evaporative surface, irrespectively of the presence of water repellent (Fig. 6); this was attributed to the higher density/viscosity of the MgSO$_4$ solution, which resulted in its slower motion towards the evaporative surface upon drying.

In-situ DRMS measurements were also performed on a 110-year old masonry church (Fig. 7), built with stone of the same provenance (i.e. Lympia stone). Micro-drilling measurements performed on the discoloured area (holes 1, 6, 7, 9, and 10) corresponded to higher peak force values (Fig. 8a).

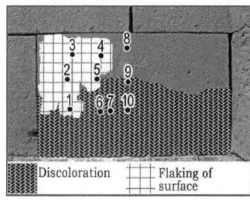

Figure 7. In-situ DRMS testing on Lympia stone masonry block. Drilling holes and weathering forms; at hole 1 both types of weathering exist (after Modestou 2012).

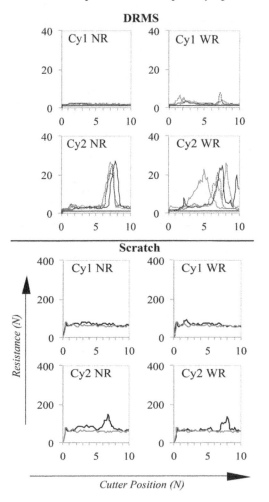

Figure 6. DRMS and scratch tool patterns for samples impregnated with MgSO$_4$ and dried at 70°C (2 crystallization cycles). DRMS: — reference pattern; all other lines correspond to individual drilled holes. Scratch: — average salt contaminated pattern; — average reference pattern. Cy: cycle; NR: no water repellent; WR: with water repellent. Resistance refers to recorded resistance experienced on tool. X-axis origin corresponds to the upper surface of samples (after Modestou 2012).

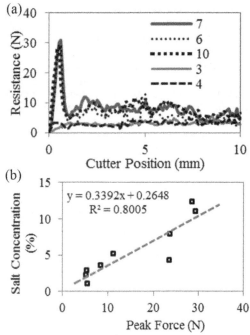

Figure 8. Results from in-situ DRMS testing. (a) Selected DRMS patterns (see Fig. 7 for hole location). (b) Relationship between peak drilling resistance and concentration of gypsum for each drilling location—no sample collected from drill hole nr. 7 (after Modestou 2012).

This was attributed to the presence of salt crystals blocking the pores of the stone. Holes on the section suffering from material loss (holes 2, 3, 4, 5, and 8) showed little or no cutting resistance peaks, likely reflecting the tendency of surface scale loss to also remove salts. Qualitative and semi-quantitative XRD analysis of the rock powder from each individual drill hole indicated a good correlation between the salt (gypsum) concentration, and the recorded drilling resistance (Fig. 8b).

The results of this study provide strong evidence that both the scratch tool and the DRMS may successfully be used in the detection of salt crystallization, because of their sensitivity to pore clogging. In-situ application of the DRMS may therefore be used to detect salts crystallizing as subflorescence before they become damaging. In this context, the DRMS may be utilized in preventive conservation. However, the results of any in-situ tests should be interpreted with due caution, since the surface of the material under investigation may either have high or low drilling resistance (compared to an unaltered material) due to weathering.

### 3.3 Evaluation of consolidation treatments

The DRMS and the scratch tool have also been successfully applied in the evaluation of the effectiveness of consolidation treatments. The two aforementioned micro-destructive techniques have been used in the laboratory to assess the performance of $Ca(OH)_2$ nanoparticles (CaLoSiL) applied on limestone (Campbell et al. 2011). The results showed that CaLoSiL was not adding resistance to scratching/drilling to the original material, thus suggesting that the treatment was only lining and not blocking the pores.

Scratch tool results were also used in the framework of another study (Dagrain et al. 2013) to develop a tomography image of samples treated with consolidants, in order to assess their effectiveness. A colour code was applied to the tomography data to visualize the mechanical strength of the test specimens in two dimensions, thus mapping and characterizing the extent of altered regions inside them (Fig. 9). The results provided strong evidence that scratching based tomography may indeed be used to characterize the effectiveness of consolidation treatments used in heritage conservation. This novel technology enables the assessment of consolidant penetration depth and of the increase in material strength; both parameters are fundamental to the evaluation of consolidation treatments.

## 4 CONCLUSIONS

This paper has provided an overview of novel applications of two micro-destructive techniques (i.e. the scratch tool and the portable DRMS) in the field of cultural heritage. Both techniques have been successfully used to characterize traditional materials: stone, lime mortars and adobes. The results highlighted the significant interdependence of the intrinsic specific energy and the drilling resistance force. Furthermore, they suggested that both aforementioned parameters are well correlated to a number of other physico-mechanical properties. As such, the DRMS and the scratch tool may prove useful in the characterization of natural stone and composite traditional materials, especially in cases where sampling is limited.

The scratch tool and the DRMS also showed clear potential in detecting salt crystallization within the porous network of stones, because of their inherent sensitivity to pore clogging. Both techniques are considered rather accessible, low cost and less time-consuming, compared to other equivalent methods, and may therefore be used for

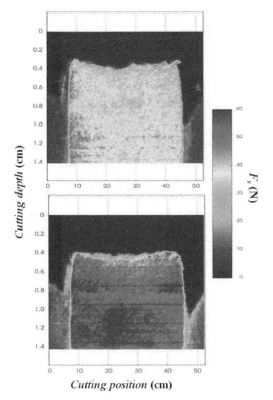

Figure 9. Tomographic visualization of Lympia stone before (up) and after (down) consolidation treatment. The colour code corresponds to the tangential force component ($F_s \approx 40$ N for non-treated specimen and $>50$ N for treated specimen) required to scratch the material (after Dagrain et al. 2013).

pro-active monitoring and prevention of salt crystallization damage in heritage structures.

Last but not least, scratching and drilling have been successfully used in the evaluation of consolidation treatments. The scratch tool results were further utilized to develop a tomography image of samples treated with consolidants. Such an image may be used to assess the penetration depth of the consolidant and the increase in the material strength induced by the treatment; both parameters are fundamental to the evaluation of consolidants.

# REFERENCES

Almenara, R. & Detournay, E. 1992. Cutting Experiments in Sandstones with Blunt PDC Cutters. In: *EuRock 1992 Conference*: London.

Campbell, A., Hamilton, A., Stratford, T., Modestou, S. & Ioannou, I. 2011. Calcium hydroxide nanoparticles for limestone conservation: Imbibition and adhesion. In: *Proceedings of CCI Symposium ICC*: Ottawa.

Cnudde, V., Silversmit, G., Boone, M., Dewanckele, J., de Samber, B., Schoonjans, T., Van Loo, D., de Witte, Y., Elburg, M., Vincze, L., van Hoorebeke, L. & Jacobs, P. 2008. Multidisciplinary characterisation of a sandstone surface crust. *Science of the Total Environment* 407: 5417–5427.

Dagrain, F., Descamps, T. & Benoit, P. 2010. Less-destructive testing of masonry materials: A comparison between scratching and drilling approaches. *Proceedings of the 8th International Masonry Conference*: Dresden.

Dagrain, F., Scaillet, J.C., Modestou, S. & Ioannou I. 2013. Evaluation of the effectiveness of masonry consolidation treatments based on scratching tomography. *Journal of Civil Engineering and Architecture* 7(5): 566–74.

Delgado Rodrigues, J., Ferreira Pinto, A. & Rodrigues da Costa, D. 2002. Tracing of decay profiles and evaluation of stone treatments by means of microdrilling techniques. *Journal of Cultural Heritage* 3: 117–125.

Derluyn, H., Poupeleer, A.S., Van Gemert, D. & Carmeliet, J. 2008. Salt crystallization in hydrophobic porous materials. In: De Clercq, H. & Charola, E. (Eds.), *Hydrophobe V: 5th International Conference on Water Repellent Treatment of Building Materials*, Brussels: 1–10.

Detournay, E., Defourny, P. & Fourmaintraux, D. 1995. Assessment of Rock Strength Properties from Cutting Tests: Preliminary Experimental Evidence. In: *Colloquium Mandanum on Chalk and Shales*: Brussels.

Eftychiou, M. 2013. Optimization of traditional adobes. M.Sc. Thesis, University of Cyprus.

EN 12370. 1999. Natural stone test methods. Determination of resistance to salt crystallization. Brussels: CEN.

EN 13161. 2008. Natural stone test methods. Determination of flexural strength under constant moment. Brussels: CEN.

Exadaktylos, G., Tiano, P. & Filareto, C. 2000. Validation of a model of rotary drilling of rocks with the drilling force measurement. *International Journal for Restoration of Buildings and Monuments* 3: 307–340.

Fernandes, F. & Lourenço, P. 2007. Evaluation of the Compressive Strength of Ancient Clay Bricks Using Microdrilling. *Journal of Materials in Civil Engineering* 19: 791–800.

Goethals, P., Volkaert, J.P., Roels, S. & Carmeliet, J. 2009. Comparison of Positron Emission Tomography and X-ray radiography for studies of physical processes in sandstone. *Engineering Geology* 103: 134–138.

Ioannou, I., Hall, C., Hoff, W.D., Pugsley, V.A. & Jacques, S.D.M. 2005. Synchrotron radiation energy-dispersive X-ray diffraction analysis of salt distribution in Lépine limestone. *The Analyst* 130: 1006–1008.

Kyriakou, L. 2015. Optimization of lime-based mortars with nano-additives, M.Sc. Thesis, University of Cyprus.

Modestou S. 2012. Salt weathering in natural stone and micro-destructive mapping of the crystallization front. M.Sc. Thesis, University of Cyprus.

Theodoridou, M., Ioannou, I. & Dagrain, F. 2015. Micro-destructive cutting techniques for the characterization of natural limestone. *International Journal of Rock Mechanics & Mining Sciences* 76: 98–103.

Tiano, P. 2001. The use of microdrilling techniques for the characterisation of stone materials, in Site control and non destructive evaluation of masonry structures and materials. In: *Proceedings of the RILEM TC177MDT International Workshop*: Mantova.

*Emerging Technologies in Non-Destructive Testing VI – Aggelis et al. (Eds)*
*© 2016 Taylor & Francis Group, London, ISBN 978-1-138-02884-5*

# Integrated ND methodologies for the evaluation of the adhesion of frescoes on stone masonry walls

M.R. Valluzzi, G. Salemi, R. Deiana & E. Faresin
*DBC—Department of Cultural Heritage, University of Padova, Padova, Italy*

G. Giacomello, M. Giaretton, M. Panizza & M. Pasetto
*DICEA—Department of Civil, Environmental and Architectural Engineering, University of Padova, Padova, Italy*

S. Calò, M. Battistella & A. Frestazzi
*Venetian Cluster of Cultural and Environmental Heritage, Mestre Venezia, Italy*

ABSTRACT: The paper presents the results of experimental laboratory tests performed on a three-leaf stone masonry wall coated by a plaster simulating various compositions and finishing works, aimed at pointing out the influence of execution techniques on the adhesion of frescoes connected to structural elements under loading. In particular, the wall was subjected to cyclic uniaxial compression test and, contextually, Non-Destructive (ND) and visual procedures, i.e., Infrared Thermography (IRT) and laser scanning system were used to monitor the crack pattern of the masonry. Results showed the good response of frescoed portions compliant with proper practice in comparison with the ones executed with lower quality. ND techniques showed also a good consistency in detecting the cracking development during loading until failure occurred.

## 1 INTRODUCTION

The condition of artistic assets in historic buildings may be affected by intrinsic defects or deterioration and by vulnerabilities of the structural substrate and the efficiency of its connection with the other components. Therefore, the preservation of frescoes or paintings is strictly connected to the identification of combined solutions, able to take into account the mutual interaction with their bearing components, e.g., vaults or walls. In this connection, multidisciplinary approaches based on Non-Destructive Techniques (NDT) are suggested (Binda & Saisi 2009) (Bosiljkov et al. 2010). Nevertheless, due to the large variability of masonry typologies, the complexity of the problem and the consequent approximation of ND evaluations, experimental calibrations with destructive tests are still in need, in order to increase reliability of investigation procedures, also in the field of cultural heritage preservation.

Limited comprehensive works are available on this subject (Cotic et al. 2013) (Calderini et al. 2014) (D'Ayala & Lagomarsino 2015). They refer to the application of complementary NDTs in laboratory (e.g., ground penetrating radar and IR thermography), on plastered multi-leaf stone masonry subjected to shear loads. Although limits in penetration depth were highlighted, NDTs were able to detect the texture of the walls and the gradual plaster delamination and crack propagation under in-plane cyclic tests.

In particular, IRT inspections are particularly effective as full non-contact method to qualify substrate morphology and deterioration but also quantify defects and weaknesses below the finishing layer (Grinzato et al. 2002) (Grinzato 2012).

The paper presents an experimental contribution aimed at evaluating the effect of preparation and finishing works on the behavior of plasters applied on masonry walls. In particular, the plaster was applied on the irregular surface of a full scale rubble multi-leaf stone masonry panel, to observe the performance at the plaster-to-wall interface and measure the effect of increasing compressive loads. The wall was previously injected to improve a monolithic behavior in its thickness (Valluzzi et al. 2004) (Silva et al. 2014). Nevertheless, in comparison with more solid substrates, the selected wall typology remain sensitive to severe actions (e.g., earthquakes), and macroscopic damage effects as detachment and crack propagation are expected, even under compressive loads. The plaster was applied in various strips covering three portions of the wall façades, simulating the optimal selection of materials, preparation and workmanship of frescoes (proper practice) or the effect of some weaknesses (low binder content, different care in execution). Cyclic uniaxial compression test were performed, in combination with visual inspections and no-contact investigation techniques, i.e., laser scanning and IR thermography. These methods were applied simultaneously on a central portion at one side of the wall. The main results, in terms of mechanical performance and comparison with ND procedures, are discussed in the paper.

## 2 EXPERIMENTAL TESTS

### 2.1 Wall preparation

The masonry wall had base section of $150 \times 50$ cm$^2$ and height of 220 cm. The thickness was composed by two outer leaves of 15 cm and internal core of 20 cm. Calcareous stone, and mortar and grout injection based on hydraulic lime were used.

The wall was plastered on both main faces, applying a fresco preparation on side A and a *marmorino* (marble plaster or stucco) on side B. Each side was composed by three vertical strips finished with different degrees, in order to check the influence of defects in execution and composition on mechanical damage. In particular, one of the end strips was executed in compliance with the best working practice, the central one included defects or limits in composition or finishing, and the last one was a poorly executed plaster, particularly poor in binder. The two qualities of strip for the two types of plaster and the worst layer were positioned on corresponding sides of the wall (Figure 1).

The surface of the wall was previously wet with slaked lime to improve the adhesion of the finishing layers. Then, to smooth the surface, as common base for both fresco and *marmorino*

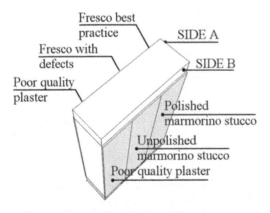

Figure 1. Scheme of position of finishing strips on wall.

Figure 2. Plastering on side B: rough cast coat (a) and *intonachino* layers (b).

Table 1. Description of plaster strips applied to wall (coats are ordered from lowest to most superficial layers) (a/b = aggregate/binder ratio in weight).

| Side | Strip | Description |
|---|---|---|
| A | left: fresco best practice | - two rough cast coats (a/b = 3:1)<br>- two fresco plaster coats (a/b = 1:1)<br>- fresco chromatic decoration |
|  | center: fresco with defects | - two rough cast coats (a/b = 3:1)<br>- fresco chromatic decoration |
|  | right: poor quality plaster | - two rough cast coats (a/b = 5:1)<br>- fresco chromatic decoration |
| B | left: poor quality plaster | - four rough cast coats (a/b = 5:1) |
|  | center: unpolished *marmorino* stucco | - two rough cast coats (a/b = 3:1)<br>- two *intonachino* (fine plaster) finishes (a/b 1:1) |
|  | right: polished *marmorino* stucco | - two rough cast coats (a/b = 3:1)<br>- two *intonachino* (fine plaster) (a/b 1:1)<br>- three slaked lime finishes (a/b = 1:5, 1:7 and 0:1)<br>- polishing |

strips, a rough cast in *cocciopesto* (crushed clay-based mortar) was applied. The ancient finishing techniques were reproduced with materials that most closely reflected optimal standards of workmanship, in order to provide samples similar to historical fine surface finishes (Figure 2). The samples were made of slaked lime as binder, water as solvent, and sand and powdered stone as aggregates in different proportions, grain sizes and thicknesses.

Details of the composition and finishing layers of the various strips covering the two sides of the wall are given in Table 1.

## 2.2 Compression tests

Uniaxial compression tests were performed using a universal Amsler machine of 10 MN. Potentiometers were used to measure the displacements: two positioned vertically at the edges in the façades (V1 to V4, in Figure 3), three on each side of the thickness to monitor the possible opening of the section (H1 to H6), and one horizontally at the midlevel (H7 and H8) (Figure 3).

The load was applied at increasing steps from 0 to a maximum of 2290 kN, which corresponded to the failure of the wall. At each step the panel was unloaded and moved out the machine to perform the visual inspections. Figure 4 shows the stress-strain curves for the main load steps applied to the wall.

No evidence of damage was detected in the plaster layers u ntil about 1500 kN. Then out-of-

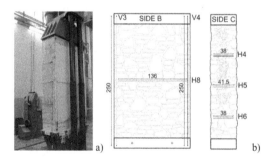

Figure 3. Test setup (a) and example of layout of potentiometers on two sides of panel (b) (dimensions in cm).

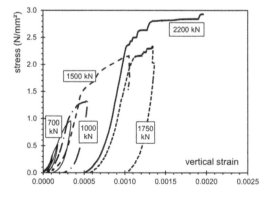

Figure 4. Cyclic behavior of panel (mean vertical deformations).

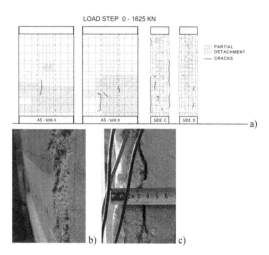

Figure 5. Distribution of damage at the first significant phase of loading: general view (a), spalling at side A (b), detachment at side B (c).

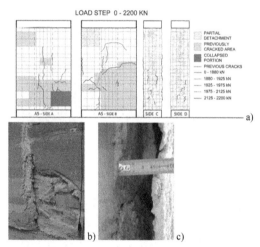

Figure 6. Distribution of damage at the last significant phase of loading: general view (a), rupture at side A (b), large separation of interface at side B (c).

plane behavior started. For increasing load steps, in side A (fresco) some cracks appeared; in side B the *marmorino* started to detach from the substrate (Figure 5 and Figure 6). The most cracked zone was the one covered by the poorest finishing, i.e., the poor quality plaster. In general, the detachment occurred in the rough cast layer, thus avoiding the disaggregation of the upper plaster strata. At the very last load capacity, although the wall was severely crushed, the two well-executed strips of *marmorino* were able to stand compact until

Figure 7. View of sides of panel at masonry failure.

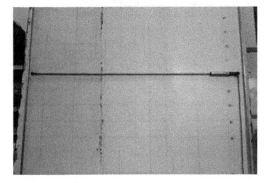

Figure 8. Portion of wall inspected with NDTs.

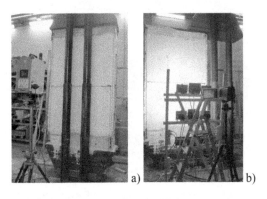

Figure 9. Preparation of tests with laser scanning (a) and IR thermography (b).

the end; the fall down of a large portion occurred because of the high out-of-plane expulsion of the wall (Figure 7). This confirmed the importance of care in the phases of selection of material, preparation and execution of the layers, in order to avoid the damage of valuable finishes.

2.3 *Laser scanning and IR thermography*

The ND methods were applied on a central portion (about 100 cm high) of the side B of panel (Figure 8 and Figure 9).

The laser acquisition was carried out with a Konica Minolta Vivid 910 digitizer, a triangulation stripe scanner with a maximum geometrical resolution of 0.17 mm. A wide lens with focal length of 8 mm and maximum resolution of 0.09 mm was used.

This instrument employs laser-beam light sectioning technology to scan objects using a slit beam. A CCD camera acquires light reflected from the object, then 3D data are created by triangulation, to have distance information. The laser beam works with an high-precision galvanometric mirror ($640 \times 480$ pixels). The triangulation is made by the camera and projector pointing to the surface, and the point of the surface itself. As result, a cloud of points is given, each point having coordinates x, y and z, where x and y indicate the position on the plane image, and z is the depth of the point in the same plane image.

The data were processed with Polygon Editing Tool (PET). A preliminary surface model was obtained, then the scans were cleared from data not belonging to the geometry of the wall portion, in order to obtain a 3D high resolution model. The model can be also visualized as a point grid with photographic information. In this case, the texture can be applied using a single digital shot or a stitched mosaic. This method gives very rich surface details and, in this specific application, information on crack distribution were obtained. The localization of damage was confirmed by the shell-shell deviation, in which the distance between corresponding points of two scans (close scans on the temporal scale) was analyzed. The frequency for each distance value was computed, so that histograms can represent the differences (Figure 10).

The IR thermography was carried out with a NEC AVIO H2640 camera, equipped with a high resolution sensor ($640 \times 480$ pixels), able to read temperature differences up to a maximum of 0.03 °C (temperature range from −40 °C to +120 °C). For this study, the emissivity of plaster was set to 0.9, the temperature and the relative humidity of the air were 21°C and 65%, respectively.

To detect a minimum temperature gradient, before each IR acquisition the surface of the wall was heated with a set of lamps placed at a suitable distance from the wall. Figure 11 shows the IR frames referred, for the various load steps, to temperatures ranging from about 16.5 °C to 19.5 °C. The comparison of results at the same temperature showed anomalies highlighted in a different way depending on the type of plaster. Evaluating the wall from left to right is in fact known as the anomalies related to the irregularities of the blocks forming the wall will be more clearly evident on the left side, less refined than the right side, where instead it is possible to detect only macroscopic variations of the temperature, probably related to a

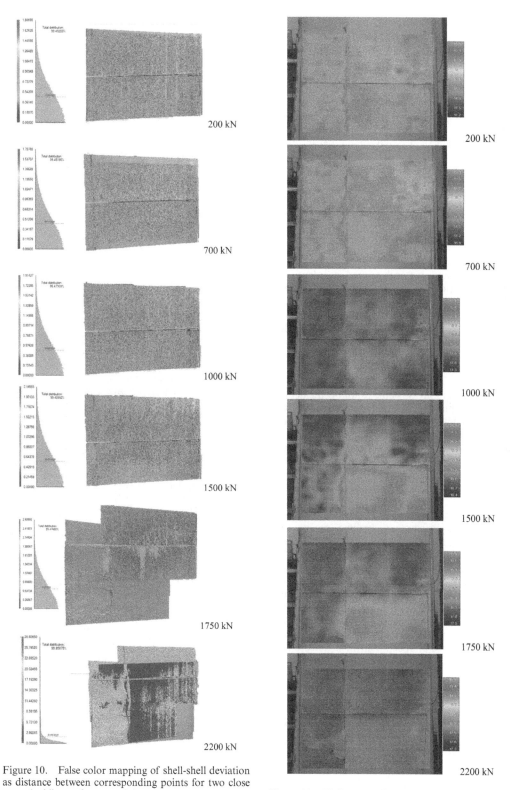

Figure 10. False color mapping of shell-shell deviation as distance between corresponding points for two close scans (red lines refer to cracks).

Figure 11. IR frames collected at various load steps.

gradual and smooth detachment of the plaster. On the right side of the wall are also not easily detectable micro-fractures or damage underlying the plaster. Furthermore, in case of a plaster performed in a workmanlike manner, able to mask the failures of the masonry, the IR measurements can appreciate only the detachment of the plaster, without information about the degree of damage to the underlying masonry and therefore the risk associated with it. The detachment of the plaster in the block on the right side of the wall, which occurred at the maximum load, is in fact a clear example of the importance of the measurement of the temperature variations on plastered surfaces, especially in the case of polished *marmorino* stucco.

## 3 CONCLUSIONS

An experimental research aimed at identifying the effect of preparation and finishing of plasters simulating frescoes and painting substrates in cultural heritage was performed. Uniaxial cyclic tests showed the compact behavior of well-executed layers and the crumbling of the poorest ones.

Non-contact methods able to provide images at distance from the surface to investigate were also adopted.

IR thermography was able to follow the detachment of the plaster during loading, thus confirming its reliability as ND method to be applied for the assessment of the current conditions of valuable artistic assets. Nevertheless, as this study showed, variation in temperature may be associated with the unevenness of the surface, therefore the crosscheck with precision survey methods, as laser scanning, is suggested.

Laser scanning was able to detail variations in the geometric surface (e.g., spalling, cracks) during increasing loads. This method can be also suggested to monitor the movements of components decorated with artistic surfaces and their safety conditions.

Further developments of the research include the mechanical characterization of the plaster-to-masonry adhesion taking into account both regular or irregular surfaces (stone and brickwork) and the influence of specific intrinsic defect in the plaster.

## ACKNOLEDGEMENTS

This research was supported by the contract PON-Provaci "Tecnologie per la protezione sismica e la valorizzazione di complessi di interesse culturale" (2011–2014), Italy. The authors wish to thank M. Maso for the plastering of the wall and C. Cerato

for the contribution to the experimental investigation during her MSc. thesis. Tests were carried out at the Laboratory of Structural Materials Testing of the University of Padova, Italy.

## REFERENCES

Binda L, Lualdi M., Saisi A., Zanzi L., Gianinetto M., Roche G. 2003. NDT applied to the diagnosis of historic buildings: a case history. Proc. of the Int. Conf. Structural Faults & Repair—2003, July 1–3, London (10 pp).

Binda L. & Saisi A. 2009. Application of NDTs to the diagnosis of Historic Structures. Proc. of the Int. Conf. NDTCE'09, Non-Destructive Testing in Civil Engineering, Nantes, June 30th–July 3rd, 2009 (28 pp).

Bosiljkov V., Uranjek M., Zarnic R., Bokan-Bosiljkov v. 2010. An integrated diagnostic approach for the assessment of historic masonry structures. Journal of Cultural Heritage, 11(3):239–249.

Cotic P., Jaglicic Z., Bosiljkov V. 2013. Validation of non-destructive characterization of the structure and seismic damage propagation of plaster and texture in multi-leaf stone masonry walls of cultural-artistic value. Journal of Cultural Heritage, 15(5)490–498.

Calderini C., Degli Abbati S., Cotic P., Krzan M., Bosiljkov V. 2014. In-plane shear tests on masonry panels with plaster: correlation of structural damage and damage on artistic assets. Bulletin of Earthquake Engineering, 13(1)237–256.

D'Ayala D. & Lagomarsino S. 2015. Performance-based assessment of cultural heritage assets: outcomes of the European FP7 PERPETUATE project. Bulletin of Earthquake Engineering, 13:5–12

Grinzato E., Bison P.G., Marinetti S. 2002. Monitoring of ancient buildings by the thermal method. Journal of Cultural Heritage, 3(1):21–9.

Grinzato E. 2012. IR Thermography Applied to the Cultural Heritage Conservation. 18th World Conference on Nondestructive Testing, 16–20 April 2012, Durban, South Africa.

Silva B., Dalla Benetta M., da Porto F., Valluzzi M.R. 2014. Compression and sonic tests to assess effectiveness of grout injection on three-leaf stone masonry walls. International Journal of Architectural Heritage: Conservation, Analysis, and Restoration. 8(3): 408–435.

Valluzzi M.R., da Porto F., Modena C. 2004. Behavior and modeling of strengthened three-leaf stone masonry walls. RILEM Materials and Structures, 37:184–192.

*Emerging Technologies in Non-Destructive Testing VI – Aggelis et al. (Eds)*
*© 2016 Taylor & Francis Group, London, ISBN 978-1-138-02884-5*

# Detection and localization of debonding damage in composite-masonry strengthening systems with the Acoustic Emission technique

E. Verstrynge & K. Van Balen
*Department of Civil Engineering, Building Materials and Building Technology Division, KU Leuven, Heverlee, Belgium*

M. Wevers
*Department of Metallurgy and Materials Engineering, KU Leuven, Heverlee, Belgium*

B. Ghiassi & D.V. Oliveira
*Department of Civil Engineering, ISISE, University of Minho, Guimarães, Portugal*

ABSTRACT: Different types of strengthening systems, based on fiber reinforced materials, are under investigation for external strengthening of historical masonry structures. A full characterization of the bond behavior and of the short—and long-term failure mechanisms is crucial to ensure effective design, compatibility and durability of the strengthening solution. In this paper, the effectiveness of the Acoustic Emission (AE) technique for debonding characterization and localization on Fiber Reinforced Polymer (FRP)- and Steel Reinforced Grout (SRG)-strengthened clay bricks is investigated. The AE technique proofs to be efficient for damage detection during accelerated ageing tests under thermal cycles and during experimental shear bond tests. AE data demonstrated the thermal incompatibility between brick and epoxy-bonded FRP composites during the accelerated ageing tests and debonding damage was successfully detected, characterized and located during the shear bond tests.

## 1 INTRODUCTION

Fiber reinforced materials are frequently used as externally bonded reinforcement for structural enhancement of concrete and masonry structures. They have well known advantages such as low weight to strength ratio and versatility in application. In recent years, composite materials such as fiber reinforced polymers (FRP) and steel reinforced grouts (SRG) have been under investigation for strengthening of (historical) masonry structures. Thereby, fully characterizing the bond behavior and failure mechanisms and studying the compatibility with the masonry substrate are crucial to ensure effective design and durability of the strengthening solution (Valluzzi et al., 2012). Aspects such as failure initiation, interfacial damage propagation, damage localization and long-term bond quality are still under investigation.

To ensure the development of efficient and durable FRP strengthening solutions, non-destructive techniques are essential for the following tasks:

- Characterization of the debonding mechanisms, to evaluate the efficiency of the applied strengthening technique and to support numerical modeling (parameter estimation and validation);

- Durability and compatibility assessment of the strengthening system (long-term behavior);
- Detection, localization and quantification of interfacial defects or progressive delamination for performance assessment, maintenance and early-warning systems (on-site monitoring).

In this paper, the effectiveness of the acoustic emission (AE) technique for debonding characterization and localization on FRP—and SRG-strengthened clay bricks is investigated. The bond degradation will be analyzed with the AE technique during an accelerated ageing test under thermal cycles and during experimental shear bond tests. The different damage mechanisms that occur during a debonding process will be characterized and subsequent debonding areas will be located.

## 2 DETECTION OF FRP DEBONDING

### 2.1 *Characterization of debonding mechanisms*

Failure in FRP-strengthened masonry elements typically occurs due to FRP rupture or FRP debonding from the masonry substrate. Debonding in the masonry substrate, denoted as cohesive

511

failure, occurs due to the lower mechanical properties of masonry compared to the repair material and the adhesive. Interfacial debonding, denoted as adhesive failure, normally occurs in case of poor surface preparation, e.g. when the surface is too smooth or wet upon application of the adhesive. It has been observed that environmental conditions, especially moist environments, can change the cohesive failure to adhesive failure. The tests described in this paper also indicated that specimens subjected to accelerated ageing tests are more likely to show adhesive failure. Also a combination of cohesive and adhesive failure surfaces, denoted mixed failure mode, can occur.

In case of strengthening with steel reinforced grout (SRG), in addition to masonry cohesive failure and adhesive debonding at the mortar-brick interface, debonding at the fiber-mortar interface can occur. The latter, being the most observed failure mode in the tests described in this paper, is followed by slipping of the fibers in the matrix.

## 2.2 NDT for debonding detection

Visual inspection and hammer tapping are the most widely used in-situ non-destructive testing methods for bond monitoring in FRP-strengthened elements, while several other methods are being applied such as digital image correlation (DIC), infrared (IR) thermography (Lai et al., 2012), ultrasonic testing (Mahmoud et al., 2010), shearography (Taillade et al., 2011) and acoustic emission (AE) testing.

DIC and IR thermography were applied during previous bond tests on similar specimens and setups. DIC has been used during shear bond tests on GFRP—and SRG-strengthened bricks to obtain the evolution of strains on the FRP surface (Ghiassi, Xavier, et al., 2013). The use of active IR thermography for detection of interfacial flaws and FRP delamination induced by environmental ageing, with specimens similar to the ones used in this study, is reported in (Ghiassi, Silva, et al., 2013).

## 2.3 FRP debonding detection with AE

For an introduction into the principles of AE testing in civil engineering (Grosse & Ohtsu, 2008) and research on the application of this technique in masonry (De Santis & Tomor, 2013; Verstrynge et al., 2009), the reader is referred to literature.

Limited results are reported in the literature regarding the analysis of debonding phenomena in externally strengthened masonry and concrete components by means of the acoustic emission technique. AE monitoring during FRP debonding from concrete beams and slabs was studied by (Carpinteri et al., 2007), who detected the propagation of flexural cracks in an FRP-strengthened beam, and by (Degala et al., 2009), who observed the progressive debonding of CFRP strips from concrete slabs and differentiated between CFRP debonding and concrete failure (flexural, compressive or shear failure) by looking at the relative intensity of the AE signals. Shear behavior of strengthened masonry walls was analyzed with the acoustic emission technique by (Masera et al., 2011) who observed decrease of the signal peak frequency upon failure of the masonry specimens. In the presented study, the debonding mechanism itself will be the object of investigation.

## 3 EXPERIMENTAL PROGRAM

The experimental study focuses on the detection of debonding with the AE technique during an accelerated ageing test under thermal fluctuations and during laboratory shear bond tests on two types of strengthening systems. Clay bricks were strengthened with Glass Fiber Reinforced Polymer (GFRP) and with Steel Reinforced Grout (SRG). Twelve single-lap shear bond tests were performed with AE detection, being three reference specimens and three aged specimens for each strengthening type.

### 3.1 Materials and test specimens

Test specimens consisted of single bricks strengthened with GFRP and SRG composites. Solid clay bricks with dimensions of $200 \times 100 \times 50$ mm were used as substrate. The composite materials were cut in 50 mm width and applied to the bricks' surface along 150 mm length of the brick with a 40 mm unbonded part near the loaded end. GFRP strips were applied to the bricks' surfaces following the wet lay-up procedure. A two-part epoxy primer was applied for preparation of the substrate and a two-part epoxy resin was used as matrix for the GFRP.

For SRG-strengthened brick specimens, a 1-directional medium density steel fiber net was used as reinforcement. The steel fibers were placed on a 3 mm thick layer of a lime-based mortar that was applied on the sand-blasted brick's surface. Then, another 3 mm mortar layer was applied to cover the steel fibers. Mechanical properties of the bricks and strengthening materials are presented in Table 1 as the mean value of five tests and the coefficients of variation (CoV).

### 3.2 Experimental setup

#### 3.2.1 Accelerated ageing tests
To investigate the effect of environmental exposure, the specimens were exposed to 180 temperature cycles in a climatic chamber. In each cycle, the temperature was kept constant at +10 °C for 2 h.

Table 1. Mechanical properties of strengthening material and bricks.

| Masonry brick | | Mean value | CoV (%) |
|---|---|---|---|
| Compressive strength | $f_{cb}$ (MPa) | 14.2 | 15.7 |
| Flex. tensile strength | $f_{tb}$ (MPa) | 1.6 | 24.6 |
| GFRP strips | | | |
| Tensile strength | $f_{tf}$ (MPa) | 1250 | 15.0 |
| Elastic modulus | $E_f$ (GPa) | 75.0 | 8.2 |
| Ultimate deformation | $\varepsilon$ (%) | 3.0 | 20.2 |
| Epoxy resin | | | |
| Tensile strength | $f_{tm}$ (MPa) | 53.8 | 9.7 |
| Elastic modulus | $E_m$ (GPa) | 2.5 | 9.5 |
| Primer | | | |
| Tensile strength | $f_{tm}$ (MPa) | 51.4 | 11.1 |
| Elastic modulus | $E_m$ (GPa) | 2.4 | 6.1 |
| Mortar | | | |
| Compressive strength | $f_{cm}$ (MPa) | 12.7 | 10.1 |
| Steel fibers | | | |
| Tensile strength | $f_{ts}$ (MPa) | 2980 | 2.9 |

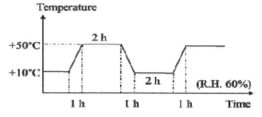

Figure 1. Hygrothermal exposure cycle.

It was then increased to +50 °C in 1 h, followed by 2 h constant temperature at +50 °C. Then, the temperature was decreased again to +10 °C in 1 h, resulting in 6 h cycles of exposure, see Figure 1.

During the accelerated ageing process, AE hits were monitored on four specimens using a 4-channel Vallen AMSY-5 system with 150–500 kHz operation frequency and 5 MHz sampling rate. Four 150 kHz resonance sensors were attached to the middle of the side of a brick by means of hot melt glue, which was chosen to resist temperatures of 50 °C without softening. The preamplifier gain was set to 34 dB with a fixed threshold level of 50 dB. To calculate the AE energy, the AE signal is squared and integrated and the energy unit (eu) is given by 1 eu = $10^{-14}$ V²s.

### 3.2.2 Shear bond tests

Single-lap shear bond tests were performed using a closed-loop servo-controlled testing machine with maximum load capacity of 50 kN. A rigid supporting steel frame was used to support the specimens and avoid misalignments in the load application. The specimens were pulled monotonically with a speed rate of 5 μm/sec under displacement control and the resulting load was measured by means of a load cell. The relative slip between the composite material and the brick was measured with two LVDTs glued at the loaded end and one LVDT glued at the free end, (Figure 2). Four 150 kHz resonance AE sensors were attached two by two on opposite sides of the bricks.

To locate the AE sources in real time, a standard planar location algorithm is applied (Vallen Systeme GmbH, 2004), which assumes isotropic and homogeneous velocity of wave propagation and iterates until a minimum location error is obtained, based on the wave velocity, the sensor locations and the arrival time difference of the AE event at the different sensors. Setting the correct wave velocity is particularly difficult for the setup at hand, since the limited size of the specimens causes reflections and boundary effects and the mechanical properties of the involved materials, and thus the wave velocities, are not fully isotropic. In addition, the setup is in fact 3D (AE sensors are placed on the side of the specimen while cracks occur towards the front surface), while a planar sensor setup and location algorithm are applied; the wavelength is equal to the velocity/frequency ratio (approximately 1000 m/s/150–500 kHz = 2–7 mm) and poses a lower limit for the location accuracy; Crack formation during the test will increase the heterogeneity of the specimen and hinder source location towards the end of the test.

Some of these issues can be solved by applying more advanced location algorithms and more

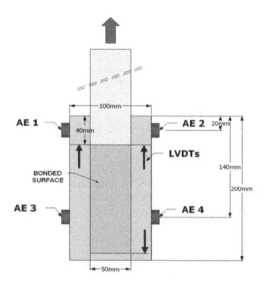

Figure 2. Single-lap shear bond test: test instrumentation and specimen dimensions.

accurate arrival time determination. Since the location accuracy is not the main focus of the present research, a pragmatic approach was followed for the calibration. A grid (20 × 30 mm) is drawn on the back of the bricks and the wave velocity in each specimen is determined by searching for the minimum average source location error, by means of pencil lead breaks before the test. This resulted in an average location error limited to 12 mm in the middle area of the bricks and a location error between 5–17 mm for the middle point of the grid. This latter point has equal distance to all sensors, a large error at this location thus indicates a non-homogeneous specimen or a non-exact positioning of the AE sensors.

## 4 RESULTS AND DISCUSSION

### 4.1 Damage detection during environmental ageing

Damage progress was monitored by means of acoustic emission detection on two SRG—and two GFRP-strengthened bricks. The average number of AE hits per day recorded for each specimen within a period of 45 days is presented in Figure 3. Limited AE activity is recorded for each type of specimen and, as a first observation, it can be mentioned that not much difference is observed between the SRG—and GFRP-strengthened specimens.

When the moments of AE energy emission are compared for both types of specimen, an important difference is observed. AE emissions in the SRG-strengthened specimens occur randomly, while for the GFRP-strengthened bricks, the majority of AE energy is emitted during temperature decrease (Figure 4). This is an indication of the different damage sources. The AE hits which are detected from SRG-strengthened specimens probably originate from the further hardening, shrinking or cracking of mortar. In the GFRP-strengthened bricks, the AE output during temperature decrease is a manifestation of the thermal

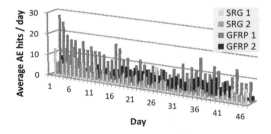

Figure 3. Average number of AE hits per day, recorded on 2 SRG-strengthened and 2 GFRP-strengthened specimens during environmental ageing test.

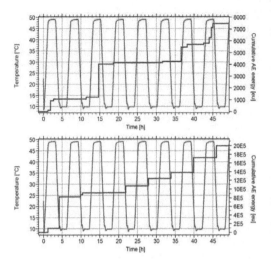

Figure 4. SRG-strengthened specimen: typical temperature fluctuation and random emission of AE energy (above); GFRP-strengthened specimen: typical temperature fluctuation and periodically emitted AE energy (below).

incompatibility between the epoxy glue and the brick. Since the thermal expansion coefficient of epoxy can be up to 10 times larger compared to brick, the temperature cycling causes stress concentrations which might lead to damage propagation at the brick-GFRP interface.

### 4.2 Damage detection during shear bond tests

Typical AE results obtained from the debonding tests on GFRP-strengthened brick specimens are presented in Figures 5–6. The results in Figure 5 are presented for a specimen with mixed cohesive/adhesive failure mode, in terms of cumulative AE energy and slip development during the test. Generally, the debonding phenomenon can be divided into three main regions: elastic range, micro-cracking range, macro-cracking and progressive fracture. In the elastic range, the system deforms without any crack generation or AE activity. The small displacement measured at this stage is due to the elastic deformation of the FRP composite. As the applied force increases, micro-cracks appear in the interfacial region and they can be distinguished by initiation of AE activity with low emitted energies. As the debonding progresses, macro-cracks are formed and propagate along the interface with higher fracture energy being released. The cumulative AE energy increases with a stepwise pattern in which each sudden jump can be attributed to macro-fracture events. A sudden release of AE energy is also observed at the moment of full debonding at the end of the test. The cumulative AE energy

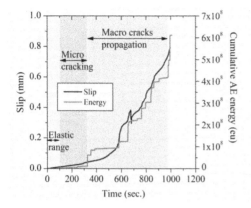

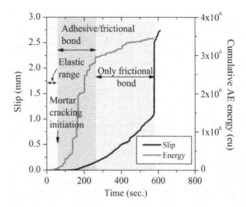

Figure 5. Typical AE results in a GFRP-strengthened brick specimen with cohesive/ adhesive failure mode: evolution of slip and cumulative AE energy.

Figure 7. Typical AE results in an SRG-strengthened brick specimens: evolution of slip and cumulative AE energy.

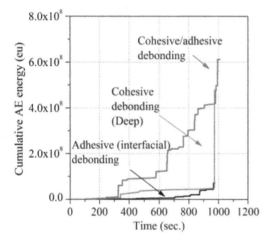

Figure 6. Comparison of AE output for different failure modes in GFRP-strengthened bricks.

could thus be applied to define the subsequent regions of fracture progress.

The effect of failure mode on the AE outputs is investigated in Figure 6. A clear distinction is found between AE outputs of specimens with different failure modes. In the specimen with cohesive debonding, the AE energy release remains relatively low throughout the test, accompanied by a sudden and large amount of AE energy release when debonding occurs at the end of the test. The observed behavior confirms the brittle and sudden nature of the cohesive debonding. In the specimen with cohesive/adhesive failure, a progressive release of energy is observed during the test. In the specimen with adhesive debonding mode, progressive detection of AE energies is observed until complete debonding. However, the magnitude of the detected energies is much lower than the ones detected in the specimens with cohesive failure mode, due to the different nature and fracture properties of brick and FRP/brick interface.

Figure 7 presents a typical result obtained from an SRG-strengthened brick specimen, which failed with slipping of the steel fibers and mortar cover separation. Mainly, three regions representing different mechanisms can be observed during the debonding process. AE activities before the first cracking of the mortar are negligible. During the mortar cracking, the rate of AE activities increases and high AE energy is detected as the force increases. The resisting mechanisms in this region are adhesive bond and friction between the steel fiber and mortar. Detachment of the bond is accompanied by releasing relatively high fracture energies, observed as sudden jumps in the AE cumulative energy curve. As the debonding progresses, the bond diminishes and friction governs the failure mechanism resulting in a reduction of the detected AE energy rate. The debonding occurs with a sudden force reduction and slip increase. In contrary to the GFRP-strengthened specimens, no direct relation can be observed between the measured slip and AE cumulative energy.

In Figure 8, a comparison is made between two SRG-strengthened specimens with different failure modes, namely brick/mortar detachment and fibers slipping. A brittle behavior is observed in the specimen with brick/mortar detachment failure. The detected AE energy level in this specimen is very low during the test followed by a sudden release of energy at the moment of debonding. On the other hand, the fibers slipping failure mode produces a progressive release of energy during the test while the adhesive bond diminishes, followed by a reduction of the AE energy rate in the stage

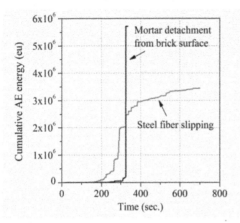

Figure 8. Comparison of AE output for different failure modes in SRG-strengthened bricks.

governed by frictional resistance. A more detailed analysis and characterization of debonding mechanisms based on AE results was presented in (Ghiassi et al., 2014).

### 4.3 Location of debonding

During the single-lap shear bond tests, AE sources were located in real time. For the SRG-strengthened specimens, very few AE sources are located due to heterogeneity of the propagation path. For the GFRP-strengthened specimens, AE source location starts with the onset of the macro-fracture range at the loaded end of the laminate and progressively moves down during the test. This is in accordance with the expected downward movement of the debonded area.

This downshift of located AE events as a function of time is illustrated in Figure 9, which presents the total energy of all AE events located in zones of 1 cm perpendicular to the loading direction. The energy plots are made at specific time intervals, as indicated on the force-slip curves. Progressive debonding can more distinctively be observed for the specimen with predominantly cohesive debonding (= progressive cohesive/adhesive debonding) and the cohesive debonding phenomenon typically produces more and higher-energy AE sources.

## 5 CONCLUSIONS

Acoustic emission data obtained during the accelerated ageing test demonstrated the thermal incompatibility between the clay bricks and epoxy-bonded FRP composite, since AE hits were predominantly detected during temperature decrease for the GFRP-strengthened brick specimens.

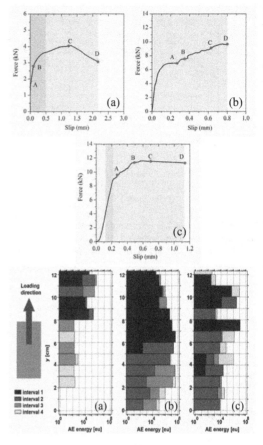

Figure 9. Cumulated energy of located AE events on SRG-strengthened specimen (a), GFRP-strengthened specimen with progressive cohesive/adhesive debonding (b) and with predominantly adhesive debonding (c). Time intervals 1 to 4 refer to the moments indicated on the force-slip curves (e.g. from the start to A is interval 1, from A to B is interval 2).

This conclusion strengthens the general consensus that besides epoxy-based systems, new strengthening techniques should be developed with better mechanical and thermal compatibility with the masonry substrate.

During the experimental shear bond tests, debonding damage was successfully detected, characterized and located, although location accuracy is limited due to the relative dimensions of the applied setup.

## ACKNOWLEDGEMENTS

The authors acknowledge the financial support of the Research Foundation—Flanders (FWO) for the mobility grant offered to Els Verstrynge.

## REFERENCES

Carpinteri, A., Lacidogna, G., & Paggi, M. (2007). Acoustic emission monitoring and numerical modeling of FRP delamination in RC beams with non-rectangular cross-section. *Materials and Structures, 40*(6), 553–566.

De Santis, S., & Tomor, A.K. (2013). Laboratory and field studies on the use of acoustic emission for masonry bridges. *Ndt & E International, 55*, 64–74.

Degala, S., Rizzo, P., Ramanathan, K., & Harries, K.A. (2009). Acoustic emission monitoring of CFRP reinforced concrete slabs. *Construction and Building Materials, 23*(5), 2016–2026.

Ghiassi, B., Silva, S.M., Oliveira, D., Lourenço, P.B., & Bragança, L. (2013). *Assessment of the bond quality degradation in FRP-strengthened masonry using IR thermography technique* Paper presented at the FRPRCS11, Guimaraes, Portugal.

Ghiassi, B., Verstrynge, E., Lourenco, P.B., & Oliveira, D.V. (2014). Characterization of debonding in FRP-strengthened masonry using the acoustic emission technique. *Engineering Structures, 66*, 24–34.

Ghiassi, B., Xavier, J., Oliveira, D.V., & Lourenço, P.B. (2013). Application of digital image correlation in investigating the bond between FRP and masonry. *Compos struct, 106*, 340–349.

Grosse, C.U., & Ohtsu, M. (Eds.). (2008). Acoustic emission testing—basics for research—applications in civil engineering: Springer.

Lai, W.L., Lee, K.K., Kou, S.C., Poon, C.S., & Tsang, W.F. (2012). A study of full-field debond behaviour and durability of CFRP-concrete composite beams by pulsed infrared thermography (IRT). *NDT & E International, 52*(0), 112–121.

Mahmoud, A.M., Ammar, H.H., Mukdadi, O.M., Ray, I., Imani, F.S., Chen, A., & Davalos, J.F. (2010). Nondestructive ultrasonic evaluation of CFRP–concrete specimens subjected to accelerated aging conditions. *NDT & E International, 43*(7), 635–641.

Masera, D., Bocca, P., & Grazzini, A. (2011). Frequency Analysis of Acoustic Emission Signal to Monitor Damage Evolution in Masonry Structures. *9th International Conference on Damage Assessment of Structures (Damas 2011)*, 305.

Taillade, F., Quiertant, M., Benzarti, K., & Aubagnac, C. (2011). Shearography and pulsed stimulated infrared thermography applied to a nondestructive evaluation of FRP strengthening systems bonded on concrete structures. *Constr Build Mater, 25*(2), 568–574.

Valluzzi, M.R., Oliveira, D.V., Caratelli, A., & et al. (2012). Round robin test for composite-to-brick shear bond characterization. *J Mater Struct, 45*, 1761–1791.

Verstrynge, E., Schueremans, L., Van Gemert, D., & Wevers, M. (2009). Monitoring and predicting masonry's creep failure with the acoustic emission technique. *NDT & E International, 42*(6), 518–523.

*Emerging Technologies in Non-Destructive Testing VI – Aggelis et al. (Eds)*
*© 2016 Taylor & Francis Group, London, ISBN 978-1-138-02884-5*

# NDTs in the monitoring and preservation of historical architectural surfaces

L. Falchi & E. Zendri
*Department of Environmental Science, Informatics and Statistics, Ca'Foscari University of Venice, Venice, Italy*

G. Driussi
*Arcadia Ricerche Srl, Marghera-Venice, Italy*

ABSTRACT: Diagnosis and post intervention monitoring are central action in a new approach related to the conservation and maintenance of architectural surfaces, which takes into account not only the restoration moment, but also the behavior of the materials over time. In this regard, ND techniques could give a significant contribution in the definition of the conservation state. This paper collects part of the authors experience in regard to NDTs used in the monitoring of historical surfaces and in the study of the transformation during CH maintenance. Furthermore, it deals with the problem of the definition of suitable indicators, taking into account the know-how of our research groups in connection to specific case studies.

## 1 INTRODUCTION

The present contribution deals with the use of Non Destructive Techniques (NDTs) in the assessment of the conservation state of historical architectural surfaces and with the monitoring of these surfaces over time: before, during and after maintenance and restoration interventions (Zeid 2012; Moropoulou 2013). The aim of the paper is to offer a brief review of ND methods suitable for the conservation of historical buildings tailored on the authors experience. Therefore, some NDT applications on real case studies, in relation to selected types of decay commonly found is reported. Furthermore, a discussion on two important themes regarding NDT and Cultural Heritage conservation is proposed: i) the role and the good practices for ND monitoring; ii) the problem of the individuation of indicators and threshold values suggestive of the state of conservation of the materials and of the efficacy/durability of conservative interventions.

In this contribution the term "historical architectural surfaces" is used to indicate the external visible surface and also the first layers inside the building, e.g. the renders, plasters, stone-panels, paint layers and the interface with the underlying masonry/ wall or few cm inside stone walls. These surfaces, in direct contact with the surrounding environment underwent several degradation phenomena defined as the changing over time of the materials' properties and characteristics, leading to their failure as building components (Moropoulou, 2013).

The conservation of historical architectonical surfaces is a particularly sensitive task, because it involves the necessity to preserve materials and structures bearer of cultural, social, economic values, often difficult to estimate.

In view of past practices based only on prior experience, or on inadequate and non-systematic identification of the prevailing problems, in the last years the consciousness of the benefits of a wider approach, involving also advanced NDTs, is increased. Cultural Agencies, institutions and boards for Cultural Heritage and Preservation request more often to dedicate a part of the restoration budget to specific diagnostic phases (Cadignani 2009), and stress the importance of choosing ND or micro-invasive methods to assess the conservation state. In fact, extensive use of analytical techniques requiring destructive or invasive sampling cannot be admissible anymore by the conservators professional ethic.

### 1.1 Pre-intervention diagnosis and post-intervention monitoring

Regarding the use of NDT in the assessment of the conservation state, a distinction could be made between the diagnostic phase prior to a restoration intervention and the monitoring phase post intervention and over time (Price 1996).

The diagnostic phase should be tailored on the specific problems of the building and of the intervention planned, in order to avoid unnecessary analysis. The NDTs can play a central role in the diagnostic phase by addressing invasive sampling, or allowing a more extensive evaluation of some properties, such as colour, water absorption, surface texture (Menezes 2015, Flores-Colen 2008). However, in the first diagnostic and knowledge phase the collection of samples and the use of destructive techniques is often unavoidable and should not be *a priori* disapproved.

The monitoring phase should also be addressed to answer specific conservation questions, but in this case the importance of a non-invasive inspection could be enhanced. NDTs in the monitoring phase should be as far as possible stand-alone techniques in order to avoid repeated and extended sample collection.

The central point when choosing which analyses are necessary is the balance cost-benefit, even in qualitative terms. Performing an analysis implies economic costs, availability of analytic techniques and of skilled professionals, furthermore in the field of cultural heritage it implies also a "cultural-social" cost (e.g. the collection of samples might be a in itself a cost since it affect the artifact integrity). Non or micro invasive, low-cost, available, easy to do, effective investigation should be carried out before more expensive and invasive investigation (Menezes 2015, Balayssac, 2012, Zendri 2011). Specific protocols and best practices can be developed in order to reduce the subjectivity and to obtain complementary information from several analytical methods.

However, it is not only the cost-benefit balance that often prevent the carrying out of a planned monitoring over time after restoration intervention, but often the absence of well-specified and explicit request by the institutions to destine special founds of the intervention budget, to which is added a general carelessness regarding the long-term intervention investigation and the evaluation of treatment durability. Only of late some authors addresses the lack of information regarding past intervention practices reporting the in-situ evaluations of previous conservations in their publications. But this is complicated by the difficulty of knowing all the parameters involved, the absence of information regarding the products used for the restoration and the lack of untreated reference surfaces (Svahn 2006, Haake & Simon 2004, Favaro 2005, Tabasso 2004, Zendri 2010).

## 1.2 *NDTs vs DTs*

NDTs are valuable resources in order to limit the diagnostic and monitoring costs, but it is not easy to automatically substitute a DTs with a NDTs. First of all, the definition of specific indicators and parameters of decay is necessary to individuate suitable NDTs for specific situations (Tabasso 2004), then a correlation between DTs or traditional methodologies and NDTs is needed.

Regarding the possible indicators of decay, they might be environmental indicators (i.e. assessment of microclimate variation and definition of threshold values); physical (water—absorption, material coherence, etc.) and chemical (material composition and degradation products) properties of the original or superimposed materials.

The establishment of known (quantitative or at least semi-quantitative) correlation between the results of NDTs and DTs is still an open challenge, in particular when the reading of the specific property is strictly dependant to a selected methodology. While Camuffo & Bertolin 2012, speaking on methods to assess the moisture content of a building material, says that it is theoretically incorrect to perform calibrations by comparing methodologies based on different physical principles, different works have been published correlating the results of different techniques (Grinzato 1998). To complicate the theme, in the field of historical surfaces preservation it is still difficult to find sufficient information about some NDT and some methods may be labour intensive or inaccurate and difficult to use in-situ (Svahn 2006). The authors opinion is that the possibility of a non-invasive and promptly individuation of physical and chemical variation is decisive for timely maintenance intervention even if the correlation between NDT and DT might be theoretically incorrect and practically complex.

A more systematic elaboration of the results of methods applied in the past might help in the extraction of further information and to individuate correlation between techniques. Furthermore, the help of expertise from other research fields and the constitution of multidisciplinary research teams is requested for the management of the great quantity of data obtainable by the more recent NDTs.

## 2 CONSIDERATION ABOUT COMMON DECAY IN HISTORICAL SURFACES: THE ROLE OF NDT FOR DIAGNOSIS AND POST-INTERVENTION MONITORING

The decay forms of historical architectural surfaces presented in this contribution, chosen because of their frequency and seriousness, are: detachments; disaggregation/loss of material coherence, discolouration and deposits, rising damp and moisture.

They might regard both the original materials and the restoration materials.

## 2.1 Detachments and scaling

Detachments of natural or artificial stone surfaces are divided and defined by the Illustrated Glossary on Stone Deterioration set up by ICOMOS as blistering, bursting and delaminations (Vergès-Belmin 2008). These decay forms are ascribable to several causes, such as the action of water, of in water soluble salts, free-thawing cycles, thermo-hygrometrical variations, alone or as a concurrence of them. Their detection in the severe cases could be easily obtained through a visual and tactile investigation by using traditional techniques (e.g. hammer tapering the surface), but, in order to obtain a timely individuation, a less subjective observation and to evaluate the extent of the phenomenon, other ND method could support the investigation.

IR-termography can be used to evidence non visible in-homogeneities, sub-surface or structural defects of the surfaces, suggesting underlying detachments, which can be revealed because of the induced thermal anomalies on the surface (Avdelidis 2003, Daffara 2011). In our research group experience IR-termography has been successfully used to detect surface air voids behind thin paint or mortar layers by using a Nikon Laird S-270thermocamera working in the middle IR range (PtSi sensor sensible between 3 and 5 μm) instead of the long wave infrared instruments LWIR (8–15 μm) usually used in building material investigation (Melchiorre 2011).

The investigation method have been developed on laboratory specimens consisting of cement lime substrate cover by a layer of an acrylic copolymer binding media resembling external contemporary paints. The delamination of paint layer from the support was induced by capillary absorption of water. The methodology was then applied in the investigation of the mural paintings of Leon Tarasewicz in the Perusini's Tower of Corno di Rosazzo, Udine (Italy), painted in 1994 with a vinyl paint. Already in 2010 the mural paint was affected by extended delamination, disaggregation, fouling, salt sub-efflorescences (Melchiorre 2011, Scrascia 2010). Active thermography was performed on the models by applying a pulse heat stimulation with a controlled IR lamp to obtain enough contrast, then for the mural painting the wall temperature was set as the minimum of the acquisition scale and the investigated zone were heated at least 5°C more. The middle wavelength IR reflectography resulted to be particularly sensible to the IR reflected component (while in LWIR the emitted component cover the reflected one) and therefore, the technique was able to detect a number of subsurface defects, such as detaches of the painting layers more extended than what is visible by naked eye observation, cracks, and subsurface defects caused by interventions as visible in Figure 1 (Daffara 2011).

Also UltraSound (US) measurement of the external surfaces could indicate the presence of concealed detachments. In 2009–2010 US investigation of arches 49–50 of the "Procuratie Vecchie" of Saint Mark's Square in Venice was commissioned by the Superintendence for Architectural and Landscaping and done by Arcadia Ricerche S. r. l. The measurements were carried out with a Controls 58-E4800 UPV with indirect configuration of the measurements. Figure 2 show the surface tomography of the upper part of arch 50; it is possible to notice the lower US wave speed in correspondence of a discontinuity/ detachment of the stone.

In the detection of detachments, it is advisable that US investigation and IR termography

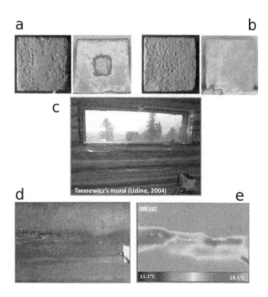

Figure 1. a, b: raking light and thermal images of lab models. C, d, e: pictures and thermal images of Leon Tarasewicz painting in the Perusini's Tower of Corno di Rosazzo, Udine.

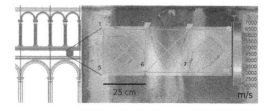

Figure 2. 2D US tomography of Arch nr. 50, "Procuratie Vecchie" of Saint Mark's Square in Venice.

or reflectography is used together with other techniques such as hammer impact/rebound methods and visual and tactile investigation. In fact, comparisons between DT tests on concrete samples and ND elastic wave velocity, measured in situ, shows how US-NDT could be affected by moisture, heterogeneity, porosity, etc., and IR thermal methods relate to a shallow layer of the material (Grinzato 2004). Up to now, ND evaluation of complex structures, such as masonry, is challenging, allowing often only a qualitative result. US measurements/tomography and IR thermography can be used for evaluating the restoration intervention, together with DT, and as stand alone techniques for post-intervention monitoring. In this case an adequate budget should be reserved since both the techniques requires the use of quite expensive instrumentation and trained personnel.

### 2.2 Disaggregation/loss of material coherence

Disaggregation (called disintegration by Vergès & Belmin 2008) is the detachment of single grains or aggregates of grains. It affects the surface of the stone or can occur in depth. It can be further divided into crumbling, granular disintegration, powdering, chalking, sugaring, sanding. The disaggregation are often easy to notice by naked eye observation, but their extent in depth is not so easily assessable. In general, a different behavior of the not-coherent parts in comparison to the underlying health material is expected and the individuation of the disaggregation extent could be made thanks to the different response of the interfaces to external stimuli.

US measurement can be used to evidence the presence of disaggregation, alone or in addiction to micro-invasive drilling. Tomographic visualization of US measurement is a powerful tool in visualizing the effectiveness of the consolidation intervention. In this respect, the in-lab evaluation of consolidation interventions (Jorne 2014; Sciola 2010) evidence that US can evidence density and strength gradients originated from different penetration and diffusion capacity of the consolidant and that the US velocity in heterogeneous materials depends not only on compactness but also on attenuation due to dispersion at the internal interfaces porous media particles–grout. The in-situ assessment of the effectiveness of consolidation intervention have been carried out during several investigation/restoration, e.g. on decohese and exfoliated Frabosa marble slabs of The Holy Shroud Chapel in Turin after the fire and restoration of 1997–2010. Here 3D tomography was obtained by complementing the surface measurements with endoscopic measurements in pre-existent cavities, evidencing the not always homogeneous distribution of the consolidation resins and grouts used (Sciola 2010). Other two interesting examples of consolidation assessment and monitoring are the US investigations of 'Loggia dei Viretti' at the Sacra di San Michele near Turin (Driussi 2012) and of the Ghirlandina Tower in Modena (Cadignani 2009, Driussi 2014). In both sites, 2D US tomography allowed the evaluation of the consolidation intervention and of the distribution and depth reached by the consolidation product. The intervention methodologies involved in both cases a timely assessment of the conservation state by visual observation, elaboration of decay maps, US investigation, static survey, and in the case of Ghirlandina Tower thermographic laser scanner survey. An experimental/applicative phase followed, with individuation of suitable consolidation and protection products thanks to in-lab and in-situ tests. The restoration phase was followed by the assessment of the effectiveness of the treatment by different NDTs, among them, 2D US tomography. In the case of Ghirlandina Tower a virtuos management allowed to dedicate a part of the budget for the post-intervention monitoring and all documented and observation materials have been systematically collected in the web-GIS SICaR program. In "Loggia dei Viretti" a consolidation treatment of the columns by impregnation under vacuum with acrylic water-based emulsion was necessary in order to reestablish a sufficient coherence of the stone. The 2D US tomography done on the columns of the Loggia before and after the intervention allow to evaluate the effectiveness of the treatment by measuring the higher material density reached after the consolidation under vacuum and the notable continuity of the US speed within the stone even in vertical sections (Fig. 3). Also in the case of Ghirlandina tower (Fig. 4) the 2D US tomography evidenced the effectiveness of

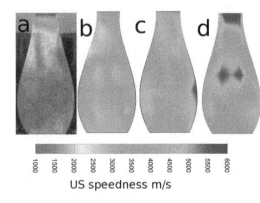

Figure 3. a, b: 'Loggia dei Viretti' at the Sacra di San Michele near Turin; c, d: 2DUS tomography of a column before and after consolidation.

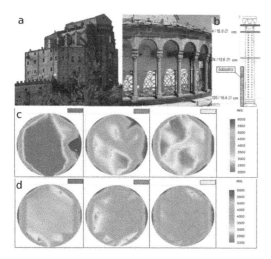

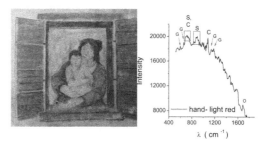

Figure 4. 2D US tomography of a column of Ghirlandina Tiwer balustrade, a: before the intervention (June 2009), b after the intervention (September 2011), c (July 2012), d (July 2013).

Figure 5. 3D US tomography of a column Palatium Vetus in Libertà Square in Alessandria (Piemonte, Italy) 2011.

the treatment on the columns of the upper balustrades, furthermore the repeated measurement allow the monitoring of the intervention and testify the enduring effectiveness of the consolidation over time, till 5 years (last monitoring: September 2014).

The possibility to mediate the single US measurement over a larger area thanks to 2D or 3D reconstruction and the evaluation of the US pattern inside the stone allow a better reconstruction of the material density (Schabowicz 2014), as can be seen also in Figure 5 showing a 3D US tomography done with direct configuration on a column Palatium Vetus in Libertà Square in Alessandria (Piemonte, Italy) in 2011. However, a difficulty in evaluating the data obtained from US investigation of consolidated materials might be encountered: is the adhesion of the consolidation product with the original material sufficient enough, or if the product is just filling the porosity without linking chemically/intimately to them? It might be often necessary to investigate a core-drilled sample to definitely evaluate the adhesion between the product and the original material.

## 2.3 *Features induced by material loss, discolouration and deposits*

Feature induced by material loss, such as alveolization, erosion, mechanical damages), discoloration and deposit (discoulouration, efflorescence encrustation, film, patina, soiling, staining) might be consider as pathologies strictly linked to the external surface (Vergès-Belmin 2008). The NDT suitable for the diagnosis and monitoring of these pathologies can be divided in techniques used to assess and characterize the material composition and techniques used to assess the physical properties of the material. The same techniques are suitable for the monitoring of preservation treatment applied to the surfaces, their effectiveness and weathering over time.

The visual observation, integrated by observation under different light wavelengths (UV-IR-light) and contact microscope, is the first useful tool for the determination of these kind of features. The observation can be systematized and documented by the elaboration of decay maps (Moropoulou 2013).

Techniques such as portable FT-IR reflectance, or Raman spectroscopy can be used for the detection of the surface composition, of original and superimposed materials, of damaging salts and in the selection of micro-samples for the laboratory analyses (Veneranda 2014). Also the decay of protective layers could be monitored over time. Unfortunately, these instruments are not always easily available at low costs and needs trained personnel. An experimentation done with a mobile Raman hand-held instrument (Xantus-1™, laser wl 785.59 nm, acquisition range 400–2000 $cm^{-1}$) in the framework of a didactic Workshop held in Cibiana (Belluno, Italy) by Ca' Foscari University of Venice (Zendri 2014, unpubl.) allowed to evidence the surface composition (Fig. 6). The presence of calcite, silicates, iron oxides, cadmium red and cobalt blue and of an organic component in correspondence of repainted areas and of gypsum due probably to a degradation process was individuated. The composition was confirmed and better assessed by IR-ATR spectroscopy on few micro-samples.

The relation of the surface, original or treated by protectives, in respect to liquid water could be assessed by simple NDTs such as the in-situ contact angle measurement (on horizontal surfaces with a portable microscope) or the water absorption at low

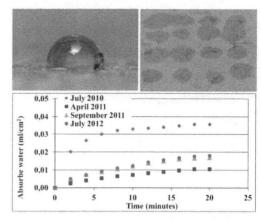

Figure 6. La Mare" mural pinting in Cbiana di Cadore and raman spectra of point 3. S = silicates; G = gypsum; C = calcite.

Figure 7. Contact angle and capillary water absorption at low pressure measurement at Ghirlandina Tower, Modena.

pressure with the pipette mehod (CNR-ICR Normal 44/93), Karsten tube (Menezes 2015.), or with the contact sponge method (Vandevoorde 2009).

Recalling the case of the upper balustrades of the Ghirlandina Tower (Driussi 2014, Cadignani 2009), the wettability of the surfaces before and after the application of protection products have been assessed on horizontal surfaces by in-situ measurement of contact angle and water absorption at low pressure, obtaining both the external water repellency and a more massive determination of water uptake capacity. The measurements (Fig. 7) highlighted the effectiveness of the protective in reducing the stone wettability and absorption after the intervention and the permanence of a sufficient protection and compatibility over the last five years (restoration intervention in 2010), even if in the last monitoring campaign (September 2014) the water-repellency of the more exposed surfaces was reduced.

The colour measurement and the determination of the tristimulus coordinates in the L*a*b* space complements the analysis and monitoring of surfaces and are necessary in evaluating the compatibility and weathering of protective layers, by recording quantitatively the colour variation of the exposed surface and taking into account that variation (expressed in terms of ΔE%) minor of ΔE 3 are consider as acceptable. (Zendri 2011). However, the colour of a surface could be an ambiguous indicator, since it might be influenced by the environmental condition (dampness, sunlight) or might not correspond directly to degradation phenomena. The results obtained must be interpreted: a case by case evaluation should be done in order to assess whether a colour variation is significant or not and, when protection products are present, if the colour variation is linked to a decrease of the protection performances (Zendri 2011).

## 2.4 Rising damp and moisture within historical masonries

Rising damp, i.e. capillary rise of water from the ground to the walls of building, and moisture, i.e. the entrance and retention of water inside the structures, are well-known phenomena in ancient buildings, which might lead to serious failure of renders and masonries surfaces. Furthermore, water act as carrier for soluble salts and pollutants, causing physicochemical decay of the material. The individuation and the promptly use of mitigation measures against these phenomena is therefore a central point in the surfaces conservation. However, the core problem in the evaluation of the mitigation measures against rising damp is how to measure materials moisture, hence how to assess the drying occurrence after the treatment (Franzoni 2014). Moisture measurement should be accurate, quantitative and sensitive of small moisture changes, measure deep in the wall, unaffected by the great heterogeneity of historical masonries and allow unambiguous interpretation (Franzoni 2014). A large literature on and reviewing ND methods for assessing rising damp and moisture is available (Camuffo & Bertolin 2012, Zeid 2012, Franzoni 2014, Pinchin 2008, Grinzato 2002), but, when dealing with ND determination of moisture is still difficult/impossible to satisfy the above-mentioned requirements. NDT such IR thermography can give only surface measurements; the determination of electrical resistance, impedance or capacitance (Zeid 2012) is often affected by material heterogeneity and absence of a good electrical contact and in general, most of the usable NDT are far from give quantitative results in this kind of application, their accuracy and reliability is still lower than invasive detection based on gravimetric methods. In the diagnostic phase the sample collection is still required, but in a post-intervention, monitoring phase also a qualitative surface analysis could be indicative of variation in the humidity presence and distribution at least on the surfaces. In the authors experience, the use of ND monitoring by capacitance measurements and thermograms of the humidity gave interesting results in the case of the monitoring of Ghirlandina Tower (Driussi 2014).

## 3 CONCLUSIONS

The post intervention monitoring is a central action in the conservation and it alone permits timely interventions, aware, farsighted and well-planned monitoring. An increased contribution of NDTs in the post intervention monitoring and safeguarding of historical architectural surfaces could be achieved, upon reflection on the NDTs role in specific case studies. In particular, the reported literature and the practical experience of the Authors suggest that the following issues should be further addressed:

- development of specific ND methods for the post-intervention monitoring;
- development of best practices suitable and adaptable to different practical situations;
- correlation between the outcomes of different techniques.

## REFERENCES

Abu-Zeid, N. et al. 2006. Non-invasive characterisation of ancient foundations in Venice using the electrical resistivity imaging technique. *NDT&E International* 39(1): pp.67–75.

Abu-Zeid, N. Cocco, G. & Santarato, G. 2012. *Guardare all'interno delle murature, La tomografia della resistività elettrica per la caratterizzazione di murature storiche*. Padova: Il prato publishing house srl.

Avdelidis, N.P., Moropoulou, A. & Theoulakis, P. 2003. Detection of water deposits and movement in porous materials by infrared imaging. *Infrared Physics and Technology* 44: pp.183–190.

Balayssac, J.-P. et al. 2012. Description of the general outlines of the French project SENSO—Quality assessment and limits of different NDT methods. *Construction and Building Materials* 35: pp.131–138.

Cadignani, R. editor 2009. *The Ghirlandina Tower, Conservation project*. Rome: luca sossella editore srl.

Camuffo, D. & Bertolin, C. 2012. Towards standardisation of moisture content measurements in cultural heritage materials. *E-Preservation Science* 9: pp.23–35.

CNR-ICR NorMaL 44/93 1993. *Assorbimento d'acqua a bassa pressione* (Italian normative on stone material-capillary water absorption at low pressure).

Daffara, C., Fontana, R., Melchiorre, M., Scrascia, S. & Zendri, E. 2011. Optical techniques for the characterization of surface-subsurface defects in painted layers. *SPIE Proc. 8084*.

Driussi, G. Longega, G., Morabito, Z. & Tonon M. 2012. Restauro, conoscenza scientifica e risorse economiche, un processo virtuoso nel Restauro della Loggia dei Viretti alla Sacra di San Michele. *Science and Cultural Heritage; Proc. intern. Conf.* Brixen: Ed. Arcadia Ricerche srl.

Driussi G.& Morabito Z. 2014. Technical report for the Superintendence for Cultural and Environmental Heritage of Bologna, Modena, Reggio Emilia (internal report, unpubl.).

Favaro, M. et al. 2005. The Four Virtues of the Porta della Carta, Ducal Palace, Venice. *Studies in Conservation* 50(1): pp.109–127.

Flores-Colen, I., de Brito, J. & de Freitas, V.P. 2008. Stains in facades' rendering—Diagnosis and maintenance techniques' classification. *Construction and Building Materials* 22(3): pp.211–221.

Franzoni, E. 2014. Rising damp removal from historical masonries: A still open challenge. *Construction and Building Materials* 54: pp.123–136.

Grinzato, E., Vavilov, V. & Kauppinen, T. 1998. Quantitative infrared thermography in buildings. *Energy and Buildings* 29(1): pp.1–9.

Grinzato, E. et al., 2004. Comparison of ultrasonic velocity and IR thermography for the characterisation of stones. *Infrared Physics & Technology* 46 (1–2): pp.63–68.

Haake, S. & Simon. S. 2004. The Bologna Cocktail—evaluation of Consolidation treatments on monuments in France and Italy after 20 years of natural ageing. *Deterioration and Conservation of Stone; proc. 10th int. con.* Stockholm: ICOMOS. pp.423–430.

Vergès-Belmin V. editor. 2008. Illustrated glossary on stone deterioration patterns. ICOMOS-ISCS.

Janssens, K.H.A. & Grieken, R.v. 2004. Comprehensive analytical chemistry. London: Elsevier.

Jorne, F., Henriques, F.M.A. & Baltazar, L.G. 2014. Evaluation of consolidation of grout injection with ultrasonic tomography. *Construction and Building Materials* 66: pp.494–506.

Melchiorre, M. & Zendri, E. & Daffara, C. 2011. Thermal Imaging for the examination and conservation of contemporary mural paintings. *10th Int. Conf., Art'11*, Florence.

Menezes, A., Glória Gomes, M. & Flores-Colen, I. 2015. In-situ assessment of physical performance and degradation analysis of rendering walls. *Construction and Building Materials* 75: pp.283–292.

Morabito, Z. & Mazzucato, A. & Tonon, M. 2006. Il consolidamento superficiale. stato dell'arte e sperimentazione per una metodologia di valutazione efficace. *Lo Stato dell'Arte 2, IGIIC*, Genoa.

Moropoulou, A. et al., 2013. Non-destructive techniques as a tool for the protection of built cultural heritage. *Construction and Building Materials* 48: pp.1222–1239.

Pinchin, S.E. 2008. Techniques for monitoring moisture in walls, *Reviews in Conservation* 53(2): pp.33–45.

Price, C.A. 1996. Stone conservation: an overview of current research, Second Edition. *Research in conservation*. Santa Monica: Getty Conservation Institute.

Schabowicz, K. 2014. Ultrasonic tomography—The latest nondestructive technique for testing concrete members—Description, test methodology, application example. *Archives of Civil and Mechanical Engineering* 14(2): pp.295–303.

Sciola, N. 2010. Sviluppo di sistemi diagnostici non distruttivi per lo studio microstrutturale di materiali lapidei. *Masther thesis in "Chemical sciences for conservation and restoration"* Supervisor G. Driussi, G. Biscontin, Ca' Foscari University of Venice.

Scrascia, S. 2010. Tecniche termografiche e di imaging IR per lo studio di film pittorici su muratura. *Masther thesis in "Chemical sciences for conservation and restoration"* Supervisor E. Zendri, Ca' Foscari University of Venice.

Svahn, H., 2006. Final Report for the Research and Development Project Non-destructive Field Tests in Stone Conservation, Literature Study. Stockholm: Riksantikvarieämbetet.

Tabasso Laurenzi, M. 2004. Products and Methods for the Conservation of Stone: Problems and Trends. In *The 10th International Congress on the Deterioration and Conservation of Stone:* 269–282. Stockholm: ICOMOS Sweden.

Vandevoorde, D. et al., 2009. Contact sponge method: Performance of a promising tool for measuring the initial water absorption. *Journal of Cultural Heritage* 10(1): pp.41–47.

Veneranda, M. et al., 2014. In-situ and laboratory Raman analysis in the field of cultural heritage: The case of a mural painting. *Journal of Raman Spectroscopy* 45(February): p.228–237.

Zendri, E., Biscontin G., Battagliarin M. & Longega G. 2000. Valutazione del comportamento nel tempo di trattamenti protettivi a base di polimeri acrilici. *La prova del tempo Proc.16th int. Con.* Venice: Arcadia Ricerche. pp.173–180.

Zendri, E. et al. 2011. The monitoring of architectural stone surface treatment: choice of parameters and their threshold values. *International conference on preventive conservation of architectural heritage*, Nanjing China.

Zendri, E., Izzo F.C., Bragagnolo, C., Toffoletto, E., De Franceschi & J., Trazzi R. 2014. I murales di Cibiana di Cadore. Studio di Affresco. La Mare. *Report of the didactic Workshop "The murales of Cibiana: Art and Science"*, Cibiana, 22–23 October 2014 (unpubl.).

*Optical sensors for structural health monitoring*

*Emerging Technologies in Non-Destructive Testing VI – Aggelis et al. (Eds)*
*© 2016 Taylor & Francis Group, London, ISBN 978-1-138-02884-5*

# The novel potential for embedded strain measurements offered by micro-structured optical fiber Bragg gratings

T. Geernaert, S. Sulejmani, C. Sonnenfeld, H. Thienpont & F. Berghmans
*Brussels Photonics Team B-PHOT, Department of Applied Physics and Photonics, Vrije Universiteit Brussel, Brussels, Belgium*

G. Luyckx & J. Degrieck
*Department of Materials Science and Engineering, Universiteit Gent, Gent, Belgium*

D. Van Hemelrijck
*Department of Mechanics of Materials and Constructions, Vrije Universiteit Brussel, Brussels, Belgium*

ABSTRACT: Optical fibre sensors have been used since many years in the field of composite material production and structural health monitoring with varying levels of success. So far these sensors have relied on conventional step-index fibres. We introduce our recently developed highly birefringent butterfly micro-structured optical fibre that allows temperature-insensitive stress and strain sensing when combined with a fibre Bragg grating sensor. We explain how the integration of this specialty fibre sensor technology within fibre-reinforced composite materials provides great opportunities for internal multidimensional strain monitoring during the production or use phase of composite materials and for shear strain measurements and disbond detection in adhesive joints. Such devices can therefore play an important role in the domain of structural health monitoring instrumentation.

## 1 INTRODUCTION

### 1.1 *Fibre Bragg gratings in micro-structured optical fibres*

One of the most well known possibilities to turn an optical fiber into a strain sensor is to use a so-called Fiber Bragg Grating (FBG) (Othonos 1999). This device is a wavelength-selective filter fabricated inside the core of an optical fiber for which the reflected wavelength changes under the influence of external perturbations. Although these sensors have many advantages such as a low weight, small size, immunity to electromagnetic interference, high linearity of the response, etc., they still show a number of shortcomings. For example, a FBG in a conventional step-index optical fiber is sensitive to both temperature changes and axial strain leading to cross-sensitivity problems. Furthermore the sensitivity of FBGs in conventional optical fiber to transverse strain remains considerably lower than to axial strain, while monitoring transverse strains is most important in many applications.

To overcome the current limitations of FBG-based sensors, we have designed and fabricated a dedicated Micro-structured Optical Fiber (MOF) to replace the conventional single-mode fibers that are most often used. Indeed Micro-structured Optical Fiber Bragg Grating (MOFBG) sensors are becoming increasingly popular owing to the peculiar characteristics of MOFs that cannot be achieved using conventional optical fiber technology (Russell 2006). More specifically, the design flexibility of such MOFs allows developing sensors that exhibit selective sensitivities to e.g. axial strain, transverse strain or even shear stress, whilst being negligibly cross-sensitive to temperature changes (Frazao 2008, Canning 2012, Pinto 2009, Frazao 2005, Sorensen 2006, Chen 2008, Wang 2009, Geernaert 2009, Luyckx 2009, Zhang 2012, Fernandes 2012). This is a great asset for Structural Health Monitoring (SHM) applications in general and for SHM of composite materials and assemblies in particular.

### 1.2 *Butterfly micro-structured optical fibre*

Because of the specific shape of the microstructure in our MOF we refer to this fiber as the butterfly MOF (Martynkien 2010). This MOF is highly birefringent (on the order of $10^{-3}$) and therefore contains two orthogonally polarized modes that can operate individually (Fig. 1). The central part

529

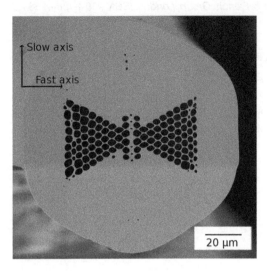

Figure 1. Scanning Electron Microscope photograph of the cross-section of the fabricated Butterfly MOF.

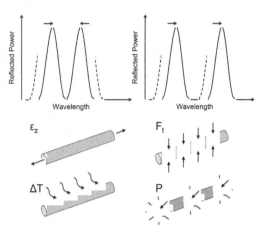

Figure 2. Sensing principle of the butterfly MOFBG. Load applied transversally to the fiber is encoded in the spacing Δλ between the two Bragg peaks. (Left) Response of the sensor to temperature changes or axial strain—both peaks move in the same way and Δλ remains unchanged. (Right) Response of the sensor to transverse load—the peaks move in opposite direction and Δλ changes.

of the core region is doped with $GeO_2$ to allow for the inscription of FBGs using conventional ultraviolet grating writing techniques (Berghmans 2014, Geernaert 2008, 2010).

A Bragg grating inscribed in this MOF then yields two Bragg reflection peaks, one corresponding to each mode. In addition, owing to the mechanical asymmetry of the fibre's cross-section, hydrostatic pressure or a transverse mechanical load induces material birefringence via the stress-optic effect and is therefore encoded in the spectral distance between those peaks Δλ, which is linked with the MOF's modal birefringence.

When the fibre experiences a transverse load, the Bragg peaks will move towards or away from each other. On the other hand Δλ is practically insensitive to temperature and to longitudinal strain (Sonnenfeld 2011). By doing so the combination of a highly birefringent micro-structure with a mechanically asymmetric side micro-structure allows bringing a selectivity mechanism to MOFBG-based sensors (Fig. 2).

Both negative (dΔλ/dP < 0 – the peak separation decreases with pressure) and positive (dΔλ/dP > 0 – the peak separation increases with pressure) pressure sensitivities can be obtained by controlling the fabrication process and the resulting micro-structure deformations and air filling factor in the core region. We obtained experimental values of −15 pm/MPa and +33 pm/MPa for different versions of the butterfly MOF, in very good agreement with the numerically simulated sensitivity values (Sulejmani 2012). The optimization of the microstructure of a MOF thus leads to record-high mechanical sensitivities, more than 20 times higher than in conventional birefringent optical fibers, In addition such pressure measurements benefit from the low sensitivity to temperature (expressed as dΔλ/dT) that is below 0.02 pm/°C.

## 2 THREE-DIMENSIONAL STRAIN MEASUREMENTS IN COMPOSITE MATERIALS

### 2.1 Three-dimensional strain measurement approach

A substantial amount of research has already been conducted on techniques for monitoring the structural integrity of fibre-reinforced polymers during their lifetime (Karbhari 2005, Housner 1997, Takeda 2008). The integration of multi-axial strain FBG sensors within such a material would allow obtaining useful information about the strain state of the host material e.g. to evaluate residual stresses during fabrication or for structural health monitoring applications. We have proposed such a sensor based on the butterfly MOFBG.

Due to the asymmetry of the micro-structure, the sensitivity of the butterfly MOFBG to transverse strain varies in a sinusoidal manner with the angular orientation of the microstructure with respect to the direction of the applied load. This dependence of the sensitivity on the angular orientation can be exploited to quantify the strain along

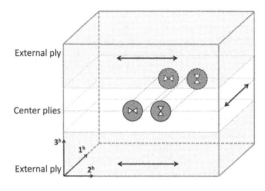

Figure 3. Illustration of 2 closely space MOFBGs embedded in the middle of a laminate composite (not drawn to scale) with the longitudinal direction of the optical fibers aligned with the reinforcement fibers of the embedding plies. The left MOFBG is oriented with α = 0°, whilst the right sensor is oriented with α = 90°. The 2 MOFBGs are stripped over a few centimeters at the grating location. The axis system $(1^h, 2^h, 3^h)$ represents the coordinate system within the host composite material. The directions of the carbon reinforcement fibers in the plies are indicated by the black double arrows.

the 3 directions of space inside a Carbon Fiber Reinforced Polymer (CFRP) composite material. To do so, one needs to relate the normal strains in the 3 directions of space in the composite material to the measured changes in the 4 Bragg wavelengths returned by two closely spaced parallel MOFBG sensors embedded in a composite coupon with different angular orientations of their microstructure with respect to the through-the-thickness direction of the composite laminate. This configuration is illustrated in Figure 3.

The relation between the strains and the temperature in the material at the location of the sensors and the Bragg wavelength shifts is defined via a Transfer Coefficient matrix [TC] that relates the temperature and strain distributions in the host composite material to those in the optical fiber sensor in the coordinate system (1h,2h,3h) of the host composite material (Luyckx 2010, Voet 2010) and a sensitivity matrix [K] that relates the temperature and strain changes in the optical fiber core to the output of the optical signal (Lawrence 1999, Nelson 1998). Those relations are represented by the following equations:

$$\begin{pmatrix} \varepsilon_1^h \\ \varepsilon_2^h \\ \varepsilon_3^h \end{pmatrix} = TC\,K^{-1} \begin{pmatrix} \Delta\lambda_{B1,1'}/\lambda_{B1,1'} \\ \Delta\lambda_{B1,2'}/\lambda_{B1,2'} \\ \Delta\lambda_{B2,1'}/\lambda_{B2,1'} \end{pmatrix} \quad (1)$$

$$\begin{pmatrix} \varepsilon_1^h \\ \varepsilon_2^h \\ \varepsilon_3^h \end{pmatrix} = \begin{pmatrix} TC_{11} & TC_{12} & TC_{13} \\ TC_{21} & TC_{22} & TC_{23} \\ TC_{31} & TC_{32} & TC_{33} \end{pmatrix} \begin{pmatrix} \varepsilon_1^{sf} \\ \varepsilon_2^{sf} \\ \varepsilon_3^{sf} \end{pmatrix} \quad (2)$$

$$\begin{pmatrix} \Delta\lambda_{B1,1'}/\lambda_{B1,1'} \\ \Delta\lambda_{B1,2'}/\lambda_{B1,2'} \\ \Delta\lambda_{B2,2'}/\lambda_{B2,1'} \end{pmatrix} = K \begin{pmatrix} \varepsilon_{1'\,MOF1} \\ \varepsilon_{2'\,MOF1} \\ \varepsilon_{3'\,MOF1} \end{pmatrix} \quad (3)$$

The optimal sensor system uses a combination of 2 parallel MOFBGs with 0° and 90° angular orientations integrated at mid-thickness within the laminate. The strain resolution in this configuration exhibits a 6-times higher mathematical stability and a 6-times improved strain resolution in the transverse directions when compared to the state-of-the-art (Voet 2010).

### 2.2 Experimental confirmation of three-dimensional strain measurements

For the experimental confirmation cross-ply laminate CFRP coupons with embedded MOFBG sensors have been prepared and tested. The composite coupons were made by stacking 8 prepreg plies of carbon fiber-reinforced polymer (CFRP, M10/T300 provided by Hexcel) and were processed in an autoclave using the vacuum bagging technique. We obtain a good agreement between our finite element modeling results and the strains measured for both the longitudinal and transverse out-of-plane laoding tests. The calculated axial strain in the composite material shows a relatively larger relative error of up to 50% when compared to the experimental data of the strain gauge (Fig. 4).

However, and since the axial strains are very low, this corresponds to a difference of only 24 με in the MOFBG configuration 0°/90°. The values of the transverse in-plane strain also exhibit some differences with respect to the measured strains, which can be explained by their low values. This stems from the fact that the expected strains along the 1h—and 2h-directions have the same magnitude as the strain measurement resolutions of about 5 με (Sonnenfeld 2015).

### 2.3 Composite cure monitoring and residual strain quantification

Since it is possible to efficiently integrate Butterfly MOFBG sensors in Carbon Fiber Reinforced Composite (CFRP) material in the lay-up phase, these sensors also allow to monitor its manufacturing process. Figure 5 shows a first drop of Δλ (part B), which corresponds to the polymerization

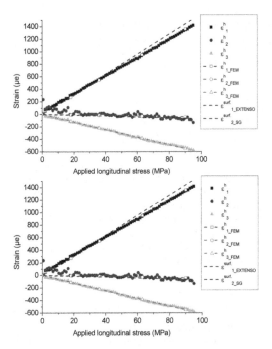

Figure 4. Measured strains in $[90/0]_{2s}$ and $[0/90]_{2s}$ laminates as a function of the applied longitudinal (top) and transverse out-of-plane (bottom) stress. $\varepsilon^h_1$, $\varepsilon^h_2$ and $\varepsilon^h_3$ have been calculated with Equation (4). $\varepsilon^h_{1\_FEM}$, $\varepsilon^h_{2\_FEM}$ and $\varepsilon^h_{3\_FEM}$ are the results of our finite element simulations. $\varepsilon^{surf.}_{1\_EXTENSO}$ is the strain measured by the extensometer and $\varepsilon^{surf.}_{2\_SG}$ is the strain measured by the strain gauge glued onto the sample.

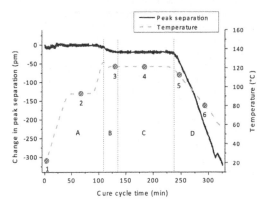

Figure 5. Changes in the temperature and in the peak separation $\Delta\lambda$ during the entire cure cycle. Adapted from (Sonnenfeld 2013).

of the sample. The cooling phase (part D) features a large decrease of $\Delta\lambda$ associated with the build-up of residual strains during the consolidation phase. Since the phase modal birefringence of the Butterfly PCF is inherently insensitive to temperature, the changes in $\Delta\lambda$ are due to thermally induced transverse strain resulting from the changing strain state in the CFRP material as it cures.

In part B, $\Delta\lambda$ decreases with 22 pm which corresponds to a compressive transverse strain of about -100 μstrain. In region C, the sensor signal remains constant meaning that the MOFBG sensor measures no transverse strain. One can therefore reasonably assume that the cure reaction has been completed and that the composite material is formed. The cooling down to room temperature (part D) is associated with a large decrease of $\Delta\lambda$ and thus with the development of substantial transverse residual strain of about −1100 μstrain in the composite material. The conversion from change in $\Delta\lambda$ to actual transverse strain values is obtained using data from previous sensor calibrations, as explained in (Sonnenfeld 2013).

## 3 SHEAR STRAIN AND DISBOND MONITORING IN SINGLE LAP ADHESIVE JOINTS

### 3.1 Shear strain measurement approach

Besides strain measurements in composite materials we can also measure shear strain in glued joints with a butterfly MOF that is embedded such that the transverse strain sensing axes of the MOF are aligned with the directions of principal stress in a shear loaded Single Lap adhesive Joint (SLJ) (Sulejmani 2013, Sulejmani 2014) as shown in Figure 6. We obtained a shear stress sensitivity of about 60 pm/MPa, which corresponds to a shear strain sensitivity of 10 fm/με (Fig. 7). Compared to conventional birefringent fibers and as discussed in (Sulejmani 2013), our dedicated MOF design has a fourfold larger sensitivity. These results also show that by changing the angular orientation of the butterfly MOFBG sensor it can be used for either shear strain sensing or for transverse strain sensing, which provides opportunities for combined normal and shear strain sensing with the same type of sensor.

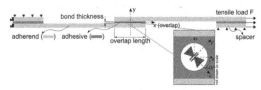

Figure 6. Configuration of the tested and modeled SLJ with an optical fiber embedded in the center of the adhesive layer.

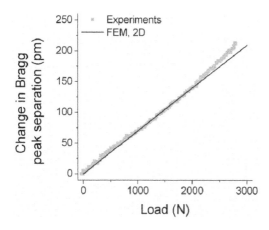

Figure 7. The Bragg peak separation increases due to tensile loading with a sensor response of 67.4 pm/kN. Results from 2D FEM modeling of the SLJ are in very good agreement with the experimental results.

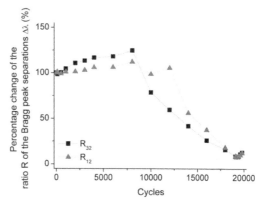

Figure 9. Evolution of the ratios $R_{12}$ and $R_{32}$, of the Bragg peak separations of the different sensors. These ratios change significantly when cracks start to propagate.

Figure 8. Photograph of the sample placed in the tensile loading test setup.

### 3.2 Disbond monitoring of adhesive joints

The combination of multiple shear stress sensors can provide insight in the existence and propagation of disbonds in SLJs. To demonstrate that ability, we have embedded three butterfly MOFBG shear stress sensors in a SLJ at the specific locations where the magnitude of the peel stress is minimal (in the center and near the edges). We subjected the sample to a cyclic tensile loading test the response of the 3 sensors and their (collective) ability to detect disbond initiation and propagation by monitoring their response throughout the fatigue test (Fig. 8).

By reconstructing the response of the sensors through 2D FE analysis, we could derive the length of the cracks on both sides of the SLJ. Indeed, as long as the bond line remains intact, the ratios $R_{12}$ and $R_{32}$ (representing the response of a MOFBG shear sensor at the edge of the SLJ compared with that of a sensor in the center of the SLJ) will nearly remain constant, independently of the loading level. However, once a significant disbond starts to appear at one or both sides, these ratios will change (Fig. 9).

As derived in (Sulejmani 2014), with the presented sensor configuration based on three butterfly MOFBG shear stress sensors, the minimum length of a detectable disbond is 100 μm, while the exact loading level does not need to be known (but should be above 17% of the failure load).

## 4 CONCLUSION

We have introduced a dedicated micro-structured optical fibre to offer new potential to FBG-based strain measurements. The butterfly MOF has been developed to provide a high transverse strain sensitivity and a low cross-sensitivity between strain measurements and temperature changes. We have shown our approach for production and 3D strain monitoring in fibre-reinforced polymers and for shear measurements in adhesive bonds. The results of the first support the potential to use the proposed butterfly MOFBGs as in-situ multi-axial strain sensors in laminate composites in view of future structural health monitoring applications. The shear strain measurements in the latter illustrate how the initiation and growth of disbonds in an adhesive joint can be quantitatively evaluated in a self-referencing manner, provided that one uses at least three shear stress MOFBG sensors.

## ACKNOWLEDGEMENTS

The authors would like to acknowledge financial support from the Agency for Innovation by Science and Technology (IWT) for funding this research with the SBO Project grant 120024 'Self-Sensing Composites'. The authors also acknowledge the Research Foundation—Flanders (FWO), the Methusalem and Hercules Foundations Flanders, the Belgian Science Policy Interuniversity Attraction Pole P7/35 and the COST action TD1001. T. Geernaert is post-doctoral research fellow with the Research Foundation Flanders (FWO).

## REFERENCES

Berghmans, F., Geernaert, T., Baghdasaryan, T., Thienpont, H. 2014. Challenges in the fabrication of fibre Bragg gratings in silica and polymer microstructured optical fibres. *Laser & Photonics Reviews* 8(1): 27–52.

Canning, J. 2012. Properties of Specialist Fibres and Bragg Gratings for Optical Fiber Sensors. Journal of Sensors: 598178.

Chen, C., Laronche, A., Bouwmans, G., Bigot, L., Quiquempois, Y., Albert, J. 2008. Sensitivity of photonic crystal fiber modes to temperature, strain and external refractive index. *Optics Express* 16(13): 9645–9653.

Fernandes, L.A., Becker, M., Frazão, O., Schuster, K., Kobelke, J., Rothhardt, M., Bartelt, H., Santos, J.L., Marques, P.V.S. 2012. Temperature and Strain Sensing With Femtosecond Laser Written Bragg Gratings in Defect and Nondefect Suspended-Silica-Core Fibers. *IEEE Photonics Technology Letters* 24(7): 554–556.

Frazao, O., Santos, J.L., Araujo, F.M., Ferreira, L.A. 2008. Optical sensing with photonic crystal fibers. Laser & Photonics Reviews 2(6): 449–459.

Frazão, O., Carvalho, J.P., Ferreira, L.A., Araújo, F.M. Santos, J.L. 2005. Discrimination of strain and temperature using Bragg gratings in microstructured and standard optical fibres. *Measurement Science and Technology* 16: 2109–2113.

Geernaert, T., Nasilowski, T., Chah, K., Szpulak, M., Olszewski, J., Statkiewicz, G., Wojcik, J., Urbanczyk, W., Poturaj, K., Becker, M., Rothhardt, M., Bartelt, H., Berghmans, F., Thienpont, H. 2008. Fiber Bragg Gratings in Germanium-Doped Highly Birefringent Microstructured Optical Fibers. *IEEE Photonics Technology Letters* 20(8): 554–556.

Geernaert, T., Luyckx, G., Voet, E., Nasilowski, T., Chah, K., Becker, M., Bartelt, H., Urbanczyk, W., Wojcik, J., De Waele, W., Degrieck, J., Terryn, H., Berghmans, F., Thienpont, H. 2009. Transversal Load Sensing With Fiber Bragg Gratings in Microstructured Optical Fibers. *IEEE Photonics Technology Letters* 21(1): 6–8.

Geernaert, T., Becker, M., Nasilowski, T., Wojcik, J., Urbanczyk, W., Rothhardt, M., Bartelt, H., Terryn, H., Berghmans, F., Thienpont, H. 2010. Bragg Grating Inscription in GeO2-Doped Microstructured Optical Fibers. *IEEE Journal of Lightwave Technology* 28(10): 1459–1467.

Housner G., Bergman L., Caughey T., Chassiakos A., Claus R., Masri S., Skelton R., Soong T., Spencer B. and Yao J. 1997. Past, Present, and Future Journal of Engineering Mechanics. *Structural Control* 123: 897–971.

Karbhari, V.M. 2005. Health Monitoring, Damage Prognosis and Service-Life Prediction — Issues Related to Implementation Sensing Issues. *Civil Structural Health Monitoring* ed F. Ansari (Springer Netherlands).

Lawrence, C.M., Nelson, D.V., Udd, E., Bennett, T. 1999. A fiber optic sensor for transverse strain measurement. *Experimental Mechanics* 39(3): 202–209.

Luyckx, G., Voet, E., Geernaert, T., Chah, K., Nasilowski, T., De Waele, W., Van Paepegem, W., Becker, M., Bartelt, H., Urbanczyk, W., Wojcik, J., Degrieck, J., Berghmans, F., Thienpont, H. 2009. Response of FBGs in Microstructured and Bow Tie Fibers Embedded in Laminated Composite. *IEEE Photonics Technology Letters* 21(18): 1290–1292.

Luyckx, G., Voet, E., Waele, W.D., Degrieck, J. 2010. Multi-axial strain transfer from laminated CFRP composites to embedded Bragg sensor: I. Parametric study. *Smart Materials and Structures* 19(10): 105017.

Martynkien, T., Statkiewicz-Barabach, G., Olszewski, J. Wojcik, J., Mergo, P., Geernaert, T., Sonnenfeld, C., Anuszkiewicz, A., Szczurowski, M.K., Tarnowski, K., Makara, M., Skorupski, K., Klimek, J., Poturaj, K., Urbanczyk, W., Nasilowski, T., Berghmans, F., Thienpont, H. 2010. Highly birefringent microstructured fibers with enhanced sensitivity to hydrostatic pressure. *Optics Express* 18(14): 15113–15121.

Nelson, D.V., Makino, A., Lawrence, C.M., Seim, J.M., Schulz, W.L., Udd, E. 1998 Determination of the K-matrix for the multi-parameter fiber grating sensor in AD072 fibercore fiber. *Proceedings of SPIE* 3489: 79–85.

Othonos, A. and Kalli, K. 1999. Fiber Bragg gratings: Fundamentals and Applications in Telecommunications and Sensing. Boston: Artech House, Boston.

Pinto, A.M.R., Lopez-Amo, M. 2009. Photonic Crystal Fibers for Sensing Applications. *Journal of Sensors*: 871580.

Russell, P. St. J. 2006. Photonic-crystal fibers. J. Lightw. Technol. 24(12): 4729–4749.

Sørensen, H.R., Canning, J., Lægsgaard, J., Hansen, K. 2006. Control of the wavelength dependent thermooptic coefficients in structured fibres. Optics Express 14(14): 6428–6433.

Sonnenfeld, C., Sulejmani, S., Geernaert, T., Eve, S., Lammens, N., Luyckx, G., Voet, E., Degrieck, J., Urbanczyk, W., Mergo, P., Becker, M., Bartelt, H., Berghmans, F., Thienpont, H. 2011. Microstructured Optical Fiber Sensors Embedded in a Laminate Composite for Smart Material Applications. *Sensors* 11: 2566–2579.

Sonnenfeld, C., Luyckx, G., Collombet, F., Grunevald, Y-H., Douchin, B., Crouzeix, L., Torres, M., Geernaert, T., Sulejmani, S., Chah, K., Mergo, P., Thienpont, H., Berghmans, F. 2013. Cure cycle monitoring of laminated carbon fiber-reinforced plastic with fiber Bragg gratings in microstructured optical fiber. *Proceedings of ICCM 19*: 3327–3335.

Sonnenfeld, C. *et al.* 2015. Microstructured optical fiber Bragg grating as an internal three-dimensional strain sensor for composite laminates. *Smart Mater. Struct.* 24: 055003.

Sulejmani, S., Sonnenfeld, C., Geernaert, T., Mergo, P., Makara, M., Poturaj, K., Skorupski, K., Martynkien, T., Statkiewicz-Barabach, G., Olzeswki, J., Urbanczyk, W., Caucheteur, C., Chah, K., Mégret, P., Terryn, H., Van Roosbroeck, J., Berghmans, F., Thienpont, H. 2012. Control Over the Pressure Sensitivity of Bragg Grating-Based Sensors in Highly Birefringent Microstructured Optical Fibers. *IEEE Photonics Technology Letters* 24(6): 527–529.

Sulejmani, S., Sonnenfeld, C., Geernaert, T., Luyckx, G., Van Hemelrijck, D., Mergo, P., Urbanczyk, W., Chah, K., Caucheteur, C., Mégret, P., Thienpont, H., Berghmans, F. 2013. Shear stress sensing with Bragg grating-based sensors in microstructured optical fibers. *Optics Express* 21(17): 20404–20416.

Sulejmani, S., Sonnenfeld, C., Geernaert, T., Van Hemelrijck, D., Luyckx, G., Mergo, P., Urbanczyk, W., Chah, K., Caucheteur, C., Mégret, P., Thienpont, H., Berghmans, F. 2014. Fiber Bragg grating-based shear strain sensors for adhesive bond monitoring. *Proceedings of SPIE* 9128:9128–12.

Sulejmani, S., Sonnenfeld, C., Geernaert, T., Luyckx, G., Mergo, P., Urbanczyk, W., Chah, K., Thienpont, H., Berghmans, F. 2014. Disbond monitoring in adhesive joints using shear stress optical fiber sensors. *Smart Mater. Struct.* 23: 075006.

Takeda, N., Minakuchi S. 2008. Recent Development of Structural Health Monitoring Technologies for Aircraft Composite Structures. *Proceedings of the 26th International Congress of the Aeronautical Sciences* (Anchorage, Alaska, USA).

Voet, E., Luyckx, G., De Waele, W., Degrieck, J. 2010. Multi-axial strain transfer from laminated CFRP composites to embedded Bragg sensor: II. Experimental validation. *Smart Materials and Structures* 19(10): 105018.

Wang, Y., Bartelt, H., Ecke, W., Willsch, R., Kobelke, J., Kautz, M., Brueckner, S., Rothhardt, M. 2009. Sensing properties of fiber Bragg gratings in small-core Ge-doped photonic crystal fibers. Optics Communications 282: 1129–1134.

Zhang, J.-H., Liu, N.-L., Wang, Y., Ji, L.-L., Lu, P.-X. 2012. Dual-Peak Bragg Gratings Inscribed in an All-Solid Photonic Bandgap Fiber for Sensing Applications. Chinese Physics Letters 29(7): 074205–1–4.

*Emerging Technologies in Non-Destructive Testing VI – Aggelis et al. (Eds)*
*© 2016 Taylor & Francis Group, London, ISBN 978-1-138-02884-5*

# Combining embedded Fibre Bragg Grating sensors and modal analysis techniques to monitor fatigue induced propagating delaminations in composite laminates

A. Lamberti
*Department of Mechanical Engineering, Vrije Universiteit Brussel (VUB), Elsene, Belgium*

G. Chiesura
*Department of Material Science and Engineering, Ghent University, Zwijnaarde (Ghent), Belgium*

B. De Pauw
*Department of Mechanical Engineering and Department of Applied Physics and Photonics*
*Vrije Universiteit Brussel (VUB), Elsene, Belgium*

S. Vanlanduit
*Department of Mechanical Engineering, Vrije Universiteit Brussel (VUB), Elsene, Belgium*
*Faculty of Applied Engineering, University of Antwerp, Antwerp, Belgium*

ABSTRACT: In the last decades, the increasing need of non-destructive testing and structural health monitoring has boosted the development of new sensing technologies. Among the emerging technologies, fibre based sensors, such as Fibre Bragg Gratings (FBG), occupy a fundamental role because of their intrinsic advantages. For instance, their reduced dimensions make them suitable to be embedded in composite materials and monitor them both during the manufacturing and during the service. FBG sensors embedded in composite materials have already been used to detect cracks and delaminations. However few studies exist in literature regarding the employment of embedded FBG sensors for monitoring in real time fatigue induced propagating damages. In this paper we present a procedure to monitor propagating delaminations in composite laminates undergoing dynamic fatigue loading. The procedure is based on the combination of embedded FBG sensors and modal parameter estimation. Using a composite laminated beam with two embedded FBG sensors and one initial delamination, we tracked the delamination length by analyzing the shifts induced by the delamination growth in the beam natural frequencies. To apply the fatigue load and induce delamination propagation, we used an electro-mechanical shaker. During the loading, we processed the strain measurements acquired via the FBGs by means of a modal analysis technique. The changes of the natural frequencies registered during the delamination propagation are reported in this paper. The natural shifts obtained from the FBG sensors are eventually compared with those measured by means of a surface mounted accelerometer.

## 1 INTRODUCTION

Structural Health Monitoring (SHM) of in-service structures is nowadays fundamental not only for safety reasons but also for minimizing downtime and for decreasing maintenance costs. Several SHM methodologies exist in literature, based for instance on acoustic emission, visual inspection, infrared thermography, holographic interferometry and modal analysis. The latter is a technique capable to detect the presence of structural damages by monitoring the changes occurring in modal parameters, such us natural frequencies, damping and mode shapes. Previous studies have demonstrated

the effectiveness of modal analysis in detecting and locating damages both in metals (Salawu 1997, Chinchakar 2001, Patil and Maiti 2003, Fan and Pizhong 2011) and in composites materials (Žak et al. 1999, Kessler et al. 2002, Ghaffari et al. 2009, Nasiri et al. 2011). In most of these studies, the modal parameters have been retrieved using either surface mounted accelerometers or Laser Doppler Vibrometers (LDV). However, if accelerometers have the disadvantage of being intrusive (especially when multiplexed), LDV sensing systems require that the structure under test is accessible. A valid alternative to solve these problems has arrived from the use of Optical Fibre (OF) sensors. Such sensors are much lighter and smaller than accelerometers and allow spatial multiplexing. Moreover, they are insensitive to electromagnetic disturbances and resistant to corrosion. When used with composites materials, OFs can be either surface mounted or embedded and they can serve as in-situ sensors for monitoring the composite structural integrity during its operating life. The combination of OF sensors and modal analysis techniques for health monitoring of composites structures has been recently investigated by different authors. Watkins et al. (2002) used surface mounted OF sensors and neural network to detect and locate delaminations in composite beams via the analysis of the beam modal frequencies shifts. Cusano et al. (2006) employed embedded Fibre Bragg Grating (FBG) to perform experimental modal analysis of an aircraft model wing, while Capoluongo et al. (2007) showed that FBGs can detect resonant modes frequency and amplitude variations up to a frequency of 1.5 kHz.

In this paper we investigate the use of embedded FBG sensors to monitor the modal frequencies shifts induced by a delamination which propagates in a composite beam subjected to fatigue load. To the best of our knowledge this has never been accomplished before, although studies regarding the monitoring of fatigue induced delamination via static and quasi-static OF strain measurements have already been conducted (Shin and Chiang 2006, De Baere et al. 2008, Epaarachchi et al. 2010, Bernasconi et al. 2011). Using a Carbon Fiber Reinforced (CFR) beam with two embedded FBG sensors and one initial delamination, we succeeded in tracking the delamination size by accurately monitoring the shifts of the modal frequencies. To make the delamination propagate, we applied a fatigue load using an electro-mechanical shaker. During the loading, we acquired the strain time histories by demodulating the FBG output signals with an in house recently developed algorithm (Lamberti et al. 2014a, Lamberti et al. 2014b). At the same time, for sake of comparison, we measured the out of plane beam accelerations using a surface mounted accelerometer. Successively, we performed a modal analysis via a polyreference least-square modal parameter estimator (Peeters et al. 2004). The obtained beam modal frequencies shifts are here reported and analyzed.

## 2 FIBRE BRAGG GRATING SENSORS

### 2.1 Fibre Bragg grating working principle

A Bragg grating is a periodic change of the refractive index in the core region of an optical fibre. When broadband light is transmitted into the optical fibre, Bragg gratings act as a wavelength

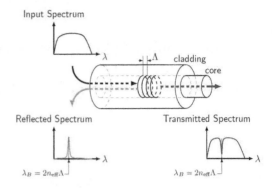

Figure 1. Schematic representation of optical Fibre Bragg Grating (FBG) principle.

selective filter that reflects a single wavelength called the Bragg wavelength, $\lambda_B$ (Figure 1). The Bragg wavelength is given by:

$$\lambda_B = 2 n_{eff} \Lambda \qquad (1)$$

where $\Lambda$ is the grating pitch (Figure 1), and $n_{eff}$ is the effective refractive index over the length L of the grating. If thermo-mechanical loads are imposed on the fibre, both $\Lambda$ and $n_{eff}$ vary, therefore the Bragg wavelength shifts in the wavelength space. In case of only mechanical loads, the *strain-sensitivity* is given by:

$$\frac{\Delta \lambda_B}{\lambda_B} = (1 - p_{eff}) \varepsilon_{zz} \qquad (2)$$

where $\varepsilon_{zz}$ is the strain induced in the fibre axial direction and $p_{eff}$ is the effective photo elastic coefficient. A typical value of $p_{eff}$ for $GeO_2$ doped (quartz) glass-fibre is 0.204.

### 2.2 Fast phase correlation algorithm for demodulating FBG signals

The strain distribution occurring in a FBG sensor can be retrieved by tracking the shifts of the Bragg wavelength. Such tracking can be performed by opportunely demodulating the FBG reflected spectrum. Several demodulation schemes are available in literature. In this paper, a Fast Phase Correlation (FPC) algorithm (Lamberti et al. 2014a, Lamberti et al. 2014b) recently developed by the authors is employed. In this algorithm, the wavelength shift between two reflected spectra $R(\lambda)$ and $R'(\lambda) = R(\lambda + \Delta\lambda)$ is computed with the following equation

$$\Delta\lambda = \underset{2 \leq k \leq M}{\text{median}} \left( (\angle \mathfrak{R}'(k) - \angle \mathfrak{R}(k)) \frac{N k \, \delta\lambda}{2\pi} \right) \qquad (3)$$

In Equation 3, $\mathfrak{R}(k)$ and $\mathfrak{R}'(k)$ are the Fourier transforms of $R(\lambda)$ and $R'(\lambda)$, respectively, $k$ is the generic Fourier spectral line, $M$ is the maximum number of Fourier spectral lines used by the algorithm, $N$ is the number of samples in each reflected spectrum. The symbol $\angle$ indicates the phase of the complex number while $\delta\lambda$ is the wavelength resolution. Compared with other demodulation schemes, the FPC is less influenced both by wavelength resolution and by distortion occurring in the reflected spectra. Moreover, the FPC requires low execution time and it is therefore suitable for real time sensing applications.

## 3 EXPERIMENTS

### 3.1 The test beam

A schematic of the test beam used in this article is reported in Figure 2. The beam was manufactured by assembling two different CFR laminated parts, indicated with L and S in Figure 2. Both parts are made of the same M10/T300 unidirectional pre-impregnated (prepreg) layers of carbon fibres, but they differ for size and stacking sequence. Part L is 50 cm long and 4 cm wide and has a stacking sequence of $[0/90/0/90/0]_s$. Part S, is 15 cm shorter than L and is formed by 16 CFR layers oriented at 0 degree. In both layups, the carbon fibres at 0 degrees are aligned with the direction of the beam maximum dimension. Two optical fibres (FBG1 and FBG2 in Figure 2) are embedded in part S, between the 2th and 3rd layer. Both fibres are Draw Tower fibre Bragg gratings provided by FBGS—Technologies GmbH. They have grating length of 8 mm, core diameter of 6 μm and cladding diameter 125 μm.

To create a delamination, a Teflon inclusion of 6.5 cm × 4 cm has been used during the assembling of L with S. Figure 2b shows the relative position of the two FBGs with respect to the delaminated area after the assembling. Both FBGs are placed 2 layers above the delamination plane while the distance from the delamination tip is about 2 cm for FBG1 and 4 cm for FBG2. Once assembled, the beam was cured at 120 degrees Celsius in autoclave. After curing, the beam lateral edge was white painted and marked with a graduating scale to allow better visual inspection of the delamination extension. The post-curing Bragg wavelengths of the two FBG sensors resulted to be 1536.35 nm and 1556.07 nm for FBG1 and FBG2, respectively. It is worth to note that, due to the different stacking sequences used during the manufacturing, part S has higher flexural rigidity than L. This facilitates the delamination growth when an external load is applied using the experimental setup described in the next session.

### 3.2 Experimental set up and procedure

The manufactured composite beam was mounted in the setup depicted in Figures 3 and 4a. The first 14 cm of the beam bottom side (see Figure 2a) were clamped using an opportunely designed steel support (Figure 3). The top side of the beam was connected to an electro-mechanical shaker. The connection was made eccentric in order to increase the chances of delamination propagation. Preliminary feasibility studies conducted by the authors on

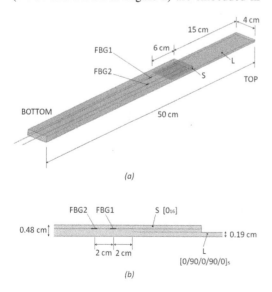

Figure 2. Schematic of the composite beam with two FBG sensors embedded and one delamination: perspective view of the entire beam (a); lateral view of the region close to the delaminated area (b).

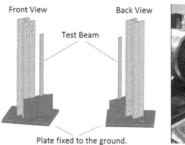

Figure 3. Schematic of the steel support (left); supported beam with electromechanical shaker attached (right). The support consists of a 4 cm thick plate fixed to the floor and welded to a 1.2 cm thick H-shaped beam. The CFR test beam is clamped between two 1 cm thick steel plates which are bolted to the H-shaped support beam.

sinilar beams, had in fact showed that pure bending loads were not suitable to induce delamination propagation, due to the limitation on the shaker maximum applicable displacement. On the contrary, the use of an eccentric connection allowed the shaker to induce at the delamination interface higher stress concentration resulting from a combination of both bending and torsion.

The shaker was voltage driven by an amplifier controlled with a computer through a NI USB-6341 data acquisition card (NI DAQ). The two optical fibres where connected to a FBG scan 700 interrogator whose acquisition sample rate was controlled via a computer generated trigger signal. To compare the results obtained with the FBG sensors, a PCB accelerometer was surface mounted on the beam in correspondence of FBG1 position. Figure 4b shows the experimental procedure adopted to make the delamination propagate under fatigue load and to monitor the beam modal frequencies. The procedure is divided in 4 steps. In the first step a 6 Hz sinusoidal signal lasting 128 seconds and having amplitude of 8 V is generated in Matlab®, sent to the NI DAQ, amplified and then applied by the shaker to the test beam. In the second step the beam is excited for 8 seconds with a Schroeder-phased multisine. During the excitation, the FBG 700 interrogator acquires and store the FBGs reflected spectra with a sampling frequency of 1 kHz using an in-house developed LabVIEW code. The same sampling rate is used by the NI DAQ to measure and save the accelerations. In the third step the FBG spectra are demodulating using the FPC algorithm described in Section 2.2 and the time evolutions of the Bragg wavelengths are retrieved. The final step deals with the modal parameters estimation. In this step both the accelerations and the demodulated FBG signals are transformed from the time to the frequency domain using a Fast Fourier Transform (FFT) and feed to a polyreference least-square modal parameter estimator (Peeters et al. 2004). Once the estimated natural frequencies are stored, the procedure is repeated from the first step. This procedure allows to monitor changes in the natural frequencies approximately every 2 minutes.

## 4 RESULTS AND DISCUSSION

The procedure presented in Figure 4b was repeated $N_s$ = 90 times, which means that the composite beam underwent a total of 69120 fatigue cycles. Figure 5 shows two views of the beam lateral side during the fatigue loading.

The beam modal frequencies were estimated every 768 cycles. Figure 6 reports three beam frequency response amplitudes corresponding to 1, 15360 and 46080 fatigue cycles and obtained by applying the polyreference least-square modal parameter estimator to the FBG dynamic strain measurements. The analysis of the synthesized response corresponding to the beam initial condition (fatigue cycle = 1) provided three stable natural frequencies: $f_1$ = 139.937 Hz, $f_2$ = 199.631 Hz and $f_3$ = 283.292 Hz. The other two synthesized responses of Figure 6 visually explain how the initial beam natural frequencies shift when the fatigue load cycles increase. Such changes are due to the delamination growth induced by the fatigue loads. Figure 7–10 show the delamination propagation and the evolution of the three beam natural frequencies as function of the fatigue load

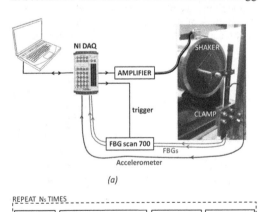

Figure 4. Experimental setup (a) and block diagram of the experimental procedure (b).

Figure 5. Lateral views of the CFR beam during the fatigue test. Open (left) and closed (right) delamination condition. The delamination tip and extension can be tracked on the graduated scale reproduced on the beam lateral side.

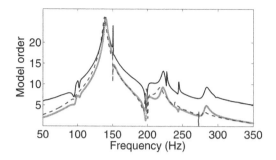

Figure 6. Synthesized beam frequency response amplitudes obtained from FBG measurements after 1 (continuous black), 15360 (dashed black) and 46080 (grey circles) fatigue cycles.

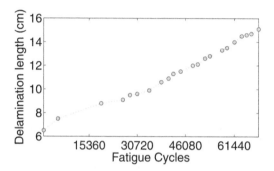

Figure 7. Evolution of the delamination size as function of the fatigue load cycle obtained by visual inspection.

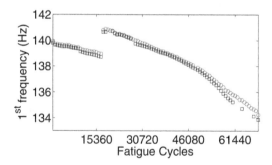

Figure 8. Shift of the 1st beam modal frequency obtained from FBGs strains (blue circles) and from accelerations (black squares) measurements.

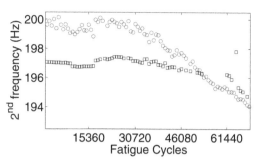

Figure 9. Shift of the 2nd beam modal frequency obtained from FBGs strains (blue circles) and from accelerations (black squares) measurements.

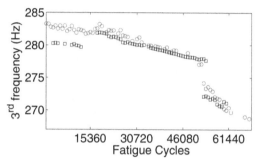

Figure 10. Shift of the 3rd beam modal frequency obtained from FBGs strains (blue circles) and from accelerations (black squares) measurements.

cycles. Figure 7 was obtained by visually inspecting the delamination length on the beam lateral side (see Figure) while Figure 8–10 were obtained by processing in real time both the FBG strains and the accelerations. From the comparative analysis of these 4 figures it is possible to observe that:

- The delamination size increases almost linearly with the fatigue cycles and reaches a final length (15.1 cm) more than double the initial one (6.5 cm)

- The beam modal frequencies estimated via the embedded FBG sensors are in good agreement with those obtained with the PCB accelerometer, especially for $f_1$ and $f_3$. For $f_2$, the dicrepancy between acceleration and FBG measurements is more evident for fatigue cycles below 46080. This might be due to the fact that the measurands are not of the same nature, therefore they produce different SNR levels which are probably not constant in function of frequency.
- All three natural frequencies decrease as the delamination propagates. In the case of FBG sensors, the percentage decrease between initial and final condition is 4.07% for $f_1$ (Figure 8), 2.79% for $f_2$ (Figure 9) and 5.15% for $f_3$ (Figure 10).
- The first modal frequency $f_1$ presents a discontinuity after 16896 fatigue cycles, corresponding to a delamination length of approximately 8.5 cm. Such behavior was not expected and does not have an evident explanation. A possible explanation, according to the authors, might be associated to changes in the input forces caused by the high coupling between the composite beam and the shaker. Additional investigations are already planned and undergoing.

– The second natural frequency $f_2$ shows higher initial variability therefore, compared to $f_1$ and $f_3$, it is less capable to indicate early stage delamination propagation. On the contrary $f_1$ estimates exhibit the lowest variability therefore $f_1$ is the best indicator of beam structural modifications.

## 5 CONCLUSIVE REMARKS AND FUTURE DEVELOPMENTS

In this paper we investigated the feasibility of using modal analysis combined with embedded fibre Bragg grating sensors to detect propagating delamination in beamlike composite structures. We performed experiments on a CFR beam carrying an initial delamination and instrumented with two embedded FBGs and one surface mounted PCB accelerometer. Using an electro-mechanical shaker, we were able to make the delamination propagate under the effect of a fatigue load. At the same time, we were able to monitor the changes induced in the beam natural frequencies. We showed that the FBG sensors provide similar or even better results than accelerometers, with the additional benefit of being much less intrusive. The proposed method appear to be appropriate to monitorstiffness changes of real life structures. It offers many advantages: it is easy to implement, it is cost effective and it can be also applied to relatively large structure in operating conditions. The main drawback of the method is that it can not be used for damage localization purposes. However, the multiplexing capabilities of FBG sensors might offer a valid solution to such limitation. Future work will focus on the use of similar structures but carrying embedded optical fibres with more Bragg gratings opportunely multiplexed. The modal parameter technique will then be extended to the estimation of the strain mode shapes which should provide more information about the damage location.

## ACKNOWLEDGEMENT

The authors appreciate the support received from the Flemish Agency for Innovation by Science and Technology (IWT) for the SBO project grants, 120024 (SSC). The authors are grateful for partial financial support of the Research Foundation, Flanders (FWO).

## REFERENCES

Bernasconi, A., M. Carboni, & L. Comolli (2011). Monitoring of fatigue crack growth in composite adhesively bonded joints using fiber bragg gratings. *Engineerign Procedia, ICCM11* 10, 207–2012.

Capoluongo, P., C. Ambrosino, S. Campopiano, A. Cutolo, M. Giordano, I. Bovio, L. Lecce, & A. Cusano (2007). Modal analysis and damage detection by Fiber Bragg grating sensors. *Sensors Actuators A Phys. 133*(2), 415–424.

Chinchakar, S. (2001). Determination of crack location in beams using natural frequencies. *J. Sound Vib 247*, 417–429.

Cusano, a., P. Capoluongo, S. Campopiano, a. Cutolo, M. Giordano, M. Caponero, F. Felli, & a. Paolozzi (2006). Dynamic measurements on a star tracker prototype of AMS using fiber optic sensors. *Smart Mater. Struct. 15*(2), 441–450.

De Baere, I., G. Luyckx, W. Van Paepegem, & J. Degrieck (2008). The use of optical fibres for fatigue testing of fibrereinforced thermoplastics. *Proceedings 4th International Conference Emerging Technologies in Non-Destructive Testing*, 65–70.

Epaarachchi, J., J. Canning, & M. Stevenson (2010). The response of embedded nir (830 nm) fiber bragg grating sensors in glass fiber composites under fatigue loading. *J. Compos. Mater. 44*(7), 809–819.

Fan, W. & Q. Pizhong (2011). Vibration-based damage identification methods: A review and comparative study. *Structural Health Monitoring 10*, 83–108.

Ghaffari, H., A. Zabihollah, E. Saeedi, & R. Ahmadi (2009). Vibration based damage detection in smart non-uniform thickness laminated composite beams. *TIC-STH, IEEE Toronto International Conference*, 176–181.

Kessler, S., S. Spearing, M. Atalla, E. Cesnik, & C. Soutis (2002). Damage detection in composite materials using frequency response methods. *Composites: Part B 33*, 87–95.

Lamberti, A., S. Vanlanduit, B. De Pauw, & F. Berghmans (2014a). A novel fast phase correlation algorithm for peak wavelength detection of Fiber Bragg Grating sensors. *Opt. Express 22*(6), 7099–7112.

Lamberti, A., S. Vanlanduit, B. De Pauw, & F. Berghmans (2014b). Influence of fiber bragg grating spectrum degradation on the performance of sensor interrogation algorithms. *Sensors 14*(12), 24258–24277.

Nasiri, M., M. Mahjoob, & A. Aghakasiri (2011). Damage detection in a composite plate using modal analysis and artificial intelligence. *Appl Compos Mater 18*, 513–520.

Patil, D. & S. Maiti (2003). Detection of multiple cracks using frequency measurements. *Engineering Fracture Mechanics 70*, 1553–1032.

Peeters, B., H. Van der Auweraer, & P. Guillaume (2004). The polymax frequency-domain method: a new standard for modal parameter estimation? *Journal of Shock and Vibration 11*(3–4), 395–409.

Salawu, O. (1997). Detection of structural damage through changes in frequency: A review. *Engineering Structures 19*, 791–723.

Shin, C. & C. Chiang (2006). Fatigue damage monitoring in polymeric com- posites using multiple fiber bragg gratings. *Int. J. Fatigue 28*(10), 1315–1321.

Watkins, S., G. Sanders, F. Akhavan, & K. Chandrashekhara (2002). Modal analysis using fiber optic sensors and neural networks for prediction of composite beam delamination. *Smart Mater. Struct. 11*, 489–495.

Zak, A., M. Krawczuk, & W. Ostachowicz (1999). Numerical and experimental investigation of free vibration of multilayer delaminated composite beams and plates. *Computational Mechanics 26*, 309–315.

# SMARTFIBER: Miniaturized optical-fiber sensor based health monitoring system

N. Lammens, G. Luyckx, E. Voet, W. Van Paepegem & J. Degrieck
*Department of Materials Science and Engineering, Ghent University, Zwijnaarde, Belgium*

ABSTRACT: SMARTFIBER is a EU funded FP7 research project, aiming to resolve technological and practical issues holding back the industrial uptake of optical fiber sensing technology. In this paper, a broad overview of the most significant achievements—from a structural health monitoring perspective—within the project are discussed. In a first section, the development of an automated, robotic lay-down system for the optical fiber system is discussed. Secondly, a finite element method is presented capable of predicting the structural behavior of a composite structure with the embedded health monitoring system. It is shown to be capable of accurately modelling the resin pocket formation surrounding the inclusion, and predicting the resulting mechanical response of the structure. Thirdly, a numerical technique for coating optimization is presented, allowing the coating properties to be tuned to reduce the impact of the sensor network on the host performance. Finally, a novel interrogation technique known as PDL is demonstrated, leading to an increased resolution in transverse strain sensing with traditional FBG sensors, enabling (amongst others) very accurate residual strain measurements.

## 1 INTRODUCTION

### 1.1 Optical fiber sensing in composite materials

Composite structures have received a large amount of interest both from the research community as well as industry. Their high strength and low weight make them ideal materials for high-performance, cutting-edge applications such as (aero)space, wind industry, sporting applications.

While the benefits of composite materials are easily understood, their fibrous and layered nature also entails a strong orthotropic material response and advanced failure mechanisms such as delaminations, fiber breakage, matrix cracks… This behavior raises the challenges in proper, safe design and exploitation of composite structures. As a result, Non-Destructive Testing and Evaluation (NDT&E) techniques have become an important area of research for composite materials.

Within the SMARTFIBER project, optical Fiber Bragg Grating (FBGs) sensors are used as a method for NDT. FBGs are the optical counterparts of traditional foil strain gauges. A Bragg grating reflects a specific wavelength (called the Bragg-wavelength) when a broadband light-source is coupled into the fiber. This is shown in Figure 1. As the optical fiber is stretched, the periodicity of the grating changes, resulting in a different reflected wavelength. By tracking the shift in Bragg-wavelength, it is possible to measure strain in the optical fiber.

This simplified explanation is valid for understanding the axial strain sensitivity of optical fibers. However, it is known that optical fibers are also sensitive to transverse strain (Voet et al., 2010, Luyckx et al., 2010). In the presence of transverse strains, an optical fiber will start reflecting two different Bragg peaks, known as "birefringence". In this case, the difference between both Bragg peaks gives an indication of the differential transverse strain on the fiber.

Amongst the many different NDT&E techniques, optical fiber sensor possess a distinct quality giving them an advantage over other sensor techniques. Their small dimensions (commercial fibers are

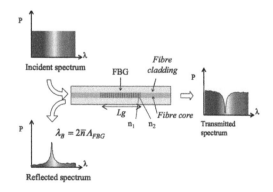

Figure 1. Basic operating principle of fiber Bragg gratings (Voet, 2011).

being manufactured at 80–125 μm diameter while research is continuously lowering this value even further), immunity to electromagnetic interference and fibrous nature make them ideally suited to be embedded inside the composite host during manufacturing. This allows the optical fiber sensor to provide information on the internal stress/strain state of the composite host, both during manufacturing as well as during the in-service lifetime. A deployed sensor-network based on these optical fiber sensors can then provide continuous health-monitoring of the application, ensuring the safe operation of the structure and timely maintenance when required.

### 1.2 Industrial challenges

Many researchers have published results on embedded optical fiber sensors in composite materials, illustrating the wide range of parameters that can be accurately sensed by the technique. Unfortunately, even with this large amount of data and knowledge available, the industrial uptake of optical fiber sensing in structural applications has been limited.

- A major issue holding back the industrial uptake is related to the fragility of the optical fibers. The careful (and fully manual) placement of the optical fiber sensors during composite manufacturing results in a very time-consuming and costly process requiring specially trained operators with experience in handling these sensors. Additionally, this manual procedure results in a low repeatability of the process when series production is envisaged.
- Secondly, even if this first hurdle is overcome or accepted, the optical fiber sensor needs to be connected to a suitable read-out device in order to measure the response of the sensor. This effectively means that the fragile optical fiber sensors need to exit the rigid composite host at some location(s). This is shown in Figure 2, where several optical fiber lines can be seen exiting the composite lay-up. While such a configuration is workable in a laboratory environment, it hardly qualifies as an industrially robust solution. Handling of such a part would require excessive care and increased risk of breaking (and thus losing) the sensor network.
- In addition to these two practical issues, the process of embedding an optical fiber sensor inside any type of host, will inevitably create some distortions in the host material. As these distortions may affect the behavior and strength of the final part, it is vital that these interactions be studied and optimized.
- Finally, the orthotropic nature of composites requires the knowledge of the full-field strain field before statements about the health-status can be made safely. It has been shown in lit-

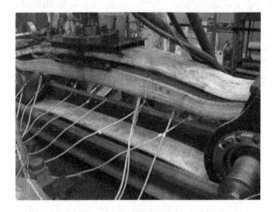

Figure 2. Multiple optical fiber lines exiting a composite host in a laboratory test (FBGS).

erature that optical fiber sensors can provide multi-axial strain data when embedded inside a composite host (Voet et al., 2010, Luyckx et al., 2010). However, this traditionally means that specialty fibers (polarization maintaining or photonic-crystal fibers) should be used at an increased cost, as traditional optical fiber sensors have an initial threshold before transverse strains can be accurately detected.

The major issues holding back the industrial uptake, mentioned in the previous section, were tackled within the SMARTFIBER project and are discussed in the following sections.

In the next section, the problem of fragility and efficient lay-down of the optical fiber sensor is discussed. This has led to the development of a novel robotic placement system capable of laying down the optical fiber sensor system in a controllable way during manufacturing.

Next, the issue of fragile egress point is tackled by the development of a state-of-the-art, fully embeddable interrogator system. Embedding any foreign structure (e.g. the embeddable interrogator system) will lead to local distortions of the composite, a finite element model, develop within the project, is presented capable of determining the resin rich zones surrounding the inclusion and the influence on structural behavior of the part.

Even the small optical fiber sensors themselves will inevitably lead to distortions and stress concentrations in the composite host. The numerical technique developed within SMARTFIBER allows the optimization of the optical fiber coating in order to reduce the impact on the surrounding host material.

Finally, the issue of multi-axial strain sensing is tackled by the demonstration of a new sensing technique (or rather, a read-out approach) known as PDL, capable of providing high resolution transverse strain sensing in traditional optical fiber sensors.

## 2 SMARTFIBER

### 2.1 Automated placement system

To overcome the issues related to the manual embedding procedure, a fully automated robot-head was built, capable of laying down the entire SMARTFIBER solution. Figure 3 shows the robot placement system, placing the optical fiber sensor on a predetermined curved path inside a laminate.

In addition to laying down an optical fiber along a predetermined path, the robotic system is capable of pre-straining the optical fiber during lay-down, thus ensuring a straight path and allowing the ability to measure compressive strains during curing. Figure 4 shows the result of a lay-down test on two different optical fiber lines, showing good repeatability of the process.

### 2.2 Miniaturized read-out system

To overcome the issue regarding the fragile egress point of the optical fiber out of the composite, SMARTFIBER developed a miniaturized and fully embeddable read-out system based on state-of-the-art photonic and electronic technologies. The necessary power and data transmission to a logging device is achieved using wireless technology.

A known issue when embedding structures (such as the SMARTFIBER read-out) is the creation of resin rich zones surrounding the inclusions. These are a consequence of the fibrous nature of the composite, and form weak(er) points in the composite which will influence the strength and stiffness properties of the host (Shivakumar and Emmanwori, 2004). Obviously, the impact of the sensor network on the structural strength should be kept to an absolute minimum.

Figure 3. Automatic SMARTFIBER placement robot (SmartFiber).

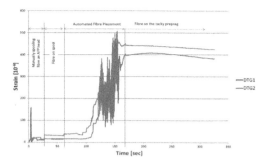

Figure 4. Optical fiber pre-straining during automated lay-down.

To study the effects of geometry and resulting resin pockets surrounding the (arbitrary) inclusions on structural performance, a novel finite element method was developed. This method was shown to allow the accurate and efficient simulation of resin pockets while requiring only a minimal amount of experimental tests to determine the material behavior. Figure 5 shows the correspondence between an experimentally measured resin pocket surrounding the SMARTFIBER read-out device, and the resin pocket geometry as predicted using the finite element approach in a quasi-isotropic Glass Fiber Reinforced Polymer (GFRP). Several other experiments were conducted, showing the general applicability and accuracy of the method.

Using the predicted resin pocket geometries, cured material response could be simulated and compared to experimental tests. Figure 6 shows the comparison between finite element and experimental data in a 3-point bending experiment.

Additional tests were performed under 4-point bending loads and for different inclusion materials and geometries. In all cases, the results show a very good correspondence between simulation and experiment. Using this tool therefore opens up the possibility of optimizing the shape of any random inclusion in order to minimize its effect on the host structure.

### 2.3 Optimization of optical fiber coating

As was already stated previously, the process of embedding an optical fiber inside a composite host will (locally) distort the composite and affect the stress—and strain-field in its vicinity. Even when the optical fiber is aligned with the reinforcements, mismatches in material properties between the optical fiber line and the host lead to the occurrence of stress-concentrations. As optical fibers are generally coated (to simplify handling), the material properties of the coating offer a degree-of-freedom which can be tuned to optimize the stress—and strain-field surrounding the sensor.

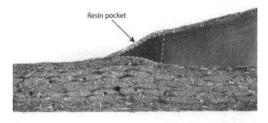

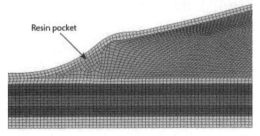

Figure 5. (top) Experimental resin pocket surrounding the SMARTFIBER read-out, (bottom) finite element prediction of resin pocket.

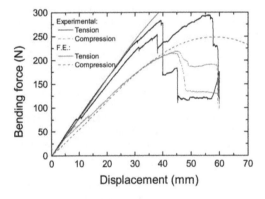

Figure 6. Comparison between experimental (black lines) and numerical (red lines) force-displacement curves in a 3-point bend test.

This possibility was already elaborated on in a couple of publications, showing the theoretical ability to optimize coating properties (E-modulus, Poisson ratio, coefficient of thermal expansion,...) (Dasgupta and Sirkis, 1992, Hadjiprocopiou et al., 1996). However, these studies focused on changing material properties (E-modulus, Poisson ratio) which are difficult to modify given the limited amount of materials commonly/commercially available for coating application. Additionally, the load cases considered in these studies are limited to mostly academic load cases such as purely axial (i.e. along the optical fiber) or purely transverse.

Within SMARTFIBER, the focus was put on optimizing the coating thickness (b) relative to the optical fiber diameter (a), as this was a parameter which could be modified as part of the SMARTFIBER project. In addition, the project set out to define a methodology which was generally valid for any kind of loading conceivable in an industrial application.

### 2.3.1 TC-matrix approach

The basic idea behind the methodology proposed, is that embedding an optical fiber sensor has a deterministic effect on the stress—and strain-fields. As a result, the relationship between the coating-composite interfacial strain ($\varepsilon^s$) and the far-field strains ($\varepsilon^\infty$) can be captured in a simple relationship, similar to the one proposed by (Luyckx et al., 2010) and (Voet et al., 2010). The equation is given as:

$$\begin{bmatrix} \varepsilon_1^s(\theta) \\ \vdots \\ \varepsilon_6^s(\theta) \end{bmatrix} = [TC(\theta)] \begin{bmatrix} \varepsilon_1^\infty \\ \vdots \\ \varepsilon_6^\infty \end{bmatrix}$$

in which TC represents the so-called transfer coefficient matrix. This TC-matrix will vary with position on the interface ($\theta$), and depend on the material properties of host and coating as well as the coating thickness ratio (b/a). This principle is illustrated in Figure 7.

Under the assumptions of small displacements and linear elasticity, the TC-matrix coefficients can be found through the superposition principle by applying all 6 far-field strains independently.

Obviously, the geometrical extent of the strain-field distortion is limited. Hence, if the optical fiber is sufficiently thin, or embedded in a sufficiently thick layer of aligned reinforcements (to avoid resin pockets), the effect of embedding an optical fiber becomes independent of laminate stacking. Further simplifications can then be made by assuming a unidirectional host material in which the reinforcements are aligned with the direction of the optical fiber sensor. In this case, the amount of independent load simulations necessary can be reduced to a total of 3 load cases (one axial, one transverse and one shear load in a plane perpendicular to the fiber axis). The result of these calculations is a set of TC-matrices over the coating-composite interface giving the relationship between the far-field strains and the interfacial strains.

By performing these TC-matrix calculations for several coating thicknesses (expressed dimensionless as the ratio of coating outer diameter to optical fiber diameter b/a), the effect of changing coating thickness for any load case (given by a set of far-field strains) can be determined. This allows the user to extract the optimal coating

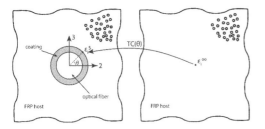

Figure 7. TC-matrix principle mapping far-field strains to coating-composite interfacial strains.

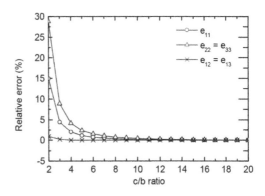

Figure 8. Relative error in TC matrix coefficients in finite samples by assuming an infinite host.

thickness for a given combination of coating and host materials and load case, without requiring independent calculations for each combination of interest.

### 2.3.2 Size and lay-up effects

The procedure described uses a model of a UD composite with finite dimensions (defined by $c$). It stands to reason that edge-effects will disappear as the dimensions of the part become larger. Using finite element analysis, it becomes possible to assess at what point the dimensions of the part no longer influence the TC-matrix coefficients and the part is equivalent to an infinitely wide sample. Figure 8 shows the relative error made by assuming an (theoretically) infinitely wide sample, while it actually has finite dimensions.

As expected, the error reduces to 0% for larger samples. This data can be used to determine the minimal layer thickness before an infinite wide host can be assumed. Below this limit, separate simulations need to be performed for all sample dimensions of interest.

Similar observations can be made for the type of lay-up considered. While the method described assumes a UD laminate, the method can equally be employed in any other type of lay-up, as long as the layer thickness in which the optical fiber is embedded is sufficiently thick. From that point on, the method is independent of precise material lay-up. Once again, F.E. simulations allow to determine from what layer thickness the method is applicable to other lay-ups.

The flexibility and performance of the TC-matrix approach is most pronounced in samples which can be considered "infinite" according to the above description. In this case, only a limited amount of simulations is required for any load case, any lay-up and any part dimensions. The only input required for the method is a set of far-field strain components. This entails that no details need to be provided about the precise material lay-up or loading conditions which might be proprietary company information.

### 2.4 High-accuracy transverse strain sensing

In a final topic the ability to perform high-accuracy transverse strain sensing with traditional optical fiber sensors was considered.

As discussed in the introduction of this work, a FBG exposed to a multi-axial strain-field will reflect two Bragg peaks, each responding differently to the applied strain fields. This phenomenon is known as birefringence. A known issue related to this birefringence, is that both reflected Bragg peaks are almost identical for small levels of differential transverse strains. Using traditional (amplitude based) read-out systems, it is impossible to discern both Bragg peaks below a certain threshold in transverse strain (or equivalently, Bragg peak separation). The traditional approach in overcoming this issue, is by pre-straining the optical fiber (known as polarization-maintaining fibers) or using advanced fibers such as photonic crystal fibers. These types of advanced fibers introduce an initial birefringence, causing the Bragg peaks to be sufficiently separated for immediate detection of transverse strains. However, these specialty fibers have their difficulties in modelling and are usually more costly than traditional fibers. Additionally, proper embedding and orientation of these fibers is challenging and time-consuming.

### 2.4.1 Polarization dependent loss measurements

While the two Bragg peaks reflected in a birefringent fiber may have almost identical wavelengths, they do possess a different polarization state (indicated as $E_x$ and $E_y$). Using a technique known as Polarization Dependent Loss (PDL), it is possible to discriminate between different polarizations of the incoming light, thereby enabling us to measure the two Bragg peaks independently (Caucheteur et al., 2007).

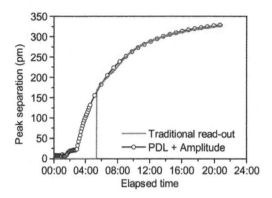

Figure 9. Bragg peak separation using traditional and PDL read-out techniques.

2.4.2 *Residual strain measurements*

The accuracy of the PDL method was illustrated in a series of cure monitoring experiments. A cross-ply ([0$_2$/90$_2$]$_{2s}$) lay-up, known to create large differential transversal strains, was cured with a standard optical fiber embedded in the mid-plane. Both traditional read-out equipment as well as the novel PDL technique were used to read out the sensor data. Figure 9 shows the evolution of Bragg peak separation over the entire curing cycle (Lammens et al., 2013) for both techniques.

The results in Figure 9 clearly illustrate the improved accuracy of the PDL method over traditional read-out techniques. Using this technique essentially allows the measurement of differential transverse strains with virtually no initial threshold, while in the example shown in Figure 9, a minimum birefringence of minimally 150 pm is necessary before peak separation can be detected with traditional amplitude measurements. The use of the PDL technique therefore enables much more accurate strain sensing in applications where only small levels of differential transverse strains are expected, hence increasing the accuracy of a potential health monitoring system based on these sensors.

3 CONCLUSIONS

This work has provided a general overview of the achievements obtained within the SMARTFIBER project. It has shown the development of a novel finite element approach for modeling the resin pocket geometries surrounding arbitrary inclusions, and described a methodology to perform numerical coating optimization. Finally, a novel read-out technique known as polarization dependent loss was illustrated as a way to increase the accuracy and sensitivity of transverse strain read-out using traditional optical fiber Bragg gratings.

ACKNOWLEDGEMENTS

The research leading to these results has received funding from the European Union Seventh Framework Programme FP7/2007–2013 under grant agreement n° 257733 (SmartFiber, http://www.smartfiber-fp7.eu).

REFERENCES

Caucheteur, C., Bette, S., Garcia-Olcina, R., Wuilpart, M., Sales, S., Capmany, J. & Megret, P. 2007. Transverse strain measurements using the birefringence effect in fiber Bragg gratings. *IEEE Photonics Technology Letters*, 19, 966–968.

Dasgupta, A. & Sirkis, J.S. 1992. Importance of Coatings to Optical Fiber Sensors Embedded in Smart Structures. *AIAA Journal*, 30, 1337–1343.

FBGS http://www.fbgs.com.

Hadjiprocopiou, M., Reed, G.T., Hollaway, L. & Thorne, A.M. 1996. Optimization of fibre coating properties for fiber optic smart structures. *Smart Materials & Structures*, 5, 441–448.

Lammens, N., Kinet, D., Chah, K., Luyckx, G., Caucheteur, C., Degrieck, J. & Megret, P. 2013. Residual strain monitoring of out-of-autoclave cured parts by use of polarization dependent loss measurements in embedded optical fiber Bragg gratings. *Composites Part A - Applied Science and Manufacturing*, 52, 38–44.

Luyckx, G., Voet, E., De Waele, W. & Degrieck, J. 2010. Multi-axial strain transfer from laminated CFRP composites to embedded Bragg sensor, I: parametric study. *Smart Materials & Structures*, 19.

SMARTFIBER SmartFiber Final Technology white paper (http://www.smartfiber-fp7.eu).

Shivakumar, K. & Emmanwori, L. 2004. Mechanics of failure of composite laminates with an embedded fiber optic sensor. *Journal of Composite Materials*, 38, 669–680.

Voet, E. 2011. *In-situ deformation monitoring of aerospace qualified composites with embedded improved draw tower fibre Bragg gratings*. Ghent University. Faculty of Engineering and Architecture.

Voet, E., Luyckx, G., De Waele, W. & Degrieck, J. 2010. Multi-axial strain transfer from laminated CFRP composites to embedded Bragg sensor, II: experimental validation. *Smart Materials & Structures*, 19.

*Electromagnetics and X-rays*

*Emerging Technologies in Non-Destructive Testing VI – Aggelis et al. (Eds)*
*© 2016 Taylor & Francis Group, London, ISBN 978-1-138-02884-5*

# Micro-CT as a well-established technique to investigate the internal damage state of a composite laminate subjected to fatigue

G. Chiesura, G. Luyckx, E. Voet, W. Van Paepegem & J. Degrieck
*Department of Material Science and Engineering, Ghent University, Ghent, Belgium*

M.N. Boone, J. Dhaene & L. Van Hoorebeke
*UGCT—Department of Physics and Astronomy, Ghent University, Ghent, Belgium*

ABSTRACT: Micro Computed Tomography (micro-CT) has become an established non-destructive technique for microstructure investigation of composite materials, where due to the intrinsic difficulty in defining suitable damage models, it can be proved useful as such in helping to interpret and validate them. In this study the evolution of damage on several cross-ply laminates having embedded fibre optics of different diameters was investigated over several millions of fatigue cycles. High-resolution 3D X-ray tomography has been performed at different intervals in the vicinity of the sensor and damage has been quantified by counting the overall number of matrix cracks in the scanned volume. A clear evolution of damage has been noticed in accordance with a measured stiffness degradation of the laminate. Nevertheless, the results reveal that for the given loading conditions damage is not evolving in the vicinity of the fibre optics.

## 1 INTRODUCTION

When compared to metals, composite materials exhibit some interesting advantages as for instance, a higher stiffness-to-weight ratio, lower density and corrosion resistance. In addition, they exhibit a gradual stiffness degradation during in-life operation and not a sudden loss of properties as it happens for a metallic component when reaching its end-life. Furthermore, composite materials allow also the integration of sensors during their production. The latter can then be used during the whole lifetime of the component in order to gain information on its structural health. A desirable sensor technology which can be employed for this purpose is represented by fibre optics, which thanks to their small dimensions ($\approx 200$ μm) and flexibility can be successfully embedded in between composite plies. This will allow for sensing the "real strain" to which the structure is subjected inside the composite laminate. However, questions arise when an optical fibre is embedded in a composite layup, e.g. what is the distortion created by the sensor in the surrounding composite layers? Will this distortion affect the structural integrity of the composite? Is the strain sensed by the fibre optic an artefact caused by the presence of the sensor itself? In this study, these questions are addressed via an experimental approach, where fibre optics were embed-

ded in carbon fibres (CFRP) and glass fibres (GFRP) prepreg laminates and cycled for several millions of cycles under tensile loading. In order to investigate the internal state of the laminates, non-destructive assessments via micro-computed tomography (micro-CT or μCT) were performed and damage in the scanned volume was traced and quantified. Some cracks were noticed at the fibre optic location directly after manufacturing in some carbon fibre samples: this was directly attributed to the relatively high (180 °C) process temperature. Apart from this, no evidence of damage evolution on the fibre optics surroundings were found over several millions of fatigue cycles. High-resolution (i.e. 2 to 4 μm voxel size) CT scans, both for carbon fibre and glass fibre composite, were obtained and the details around the fibre optic were highlighted. This technique has proven to be of relevance for internal material investigation and the damage assessment was proven by an overall stiffness decay measured on the sample. It has also to be noticed that the scans performed on the GFRP, presented more difficulties in defining the fibre optics outline. In fact, the difference in composition between the reinforcing fibres and optical fibre materials in this case was minimal (i.e. both optical fibre and reinforcing fibres have the same density and are made of glass), but this was achieved after an optimization of the scanning parameters.

## 2 EXPERIMENTAL SET-UP

### 2.1 Sample production

Two different laminates having both a $[90_2,0_2]_{2s}$ lay-up were produced by an autoclave process. The first one was a carbon-fibre UD prepreg M55 J-M18 by Hexcel®, which was cured for 2 hours at 180 °C, imposing a vacuum level of -80 kPa and an external pressure of 7 bars. The second laminate was produced with MTM28 glass-fibre UD prepreg layers by Umeco® and was cured for 2 hours at 120 °C, using vacuum level of -80 kPa and 5 bars of additional pressure. In the mid-plane of each laminate, optical fibres of different diameters (the outer diameter was ranging from 106 μm to 195 μm) were aligned with the reinforcing fibres direction and embedded between two 0° layers. For the GFRP laminate the fibre optics were carrying also a grating, which was positioned in the centre of the plate, while for the CFRP dummy fibres were used instead. After autoclave manufacturing, the laminates were cut to coupon size according to the ISO 527–4 standard. Due to the limitations imposed by the micro-CT setup (which will be explained in more detail in subparagraph 2.3) the carbon fibre samples had the following dimensions $t \times w \times l = 2.5\ mm \times 12\ mm \times 250\ mm$, while the glass fibre samples were $5\ mm \times 19\ mm \times 250\ mm$. Aluminium tabs were glued to the specimens ends, allowing an overall gauge length of approx. 150 mm.

### 2.2 Fatigue loading set-up

A preliminary set of static tests in accordance with the ASTM D3039standard have been performed on three reference samples for each material in order to evaluate the Ultimate Tensile Strength (UTS) and therefore define the load amplitude of the fatigue cycles. The test procedure consisted of a set of cumulative cycle intervals, which were interspersed with static tensile tests; these tests served to evaluate the laminate stiffness and its reduction due to occurring damages. At this stage, the samples were also dismounted from the testing rig, moved to the μCT facility and scanned. Figure 1 summarizes the test procedure in a schematic.

The fatigue tests were performed in accordance with the ASTM D3479M standard. A tension-tension sinusoidal load controlled cycle was selected in order not to buckle the specimens, thus avoiding unwanted premature damage. As a testing rig, the 100 kN 8801servo-hydraulic machine from Intron® was used. All the samples were tested at a cycle frequency of 5 Hz, as a good compromise between testing time (which was of 55 hours for $10^6$ cycles) and the risk of overheating the samples with consequent matrix degrada-

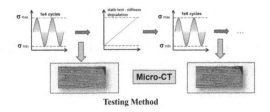

Figure 1. Schematic of the testing method: the fatigue test has been divided in several cycle intervals; between each of them a static test was performed and the sample was scanned for damage assessment.

tion. In fact, in a preliminary testing stage, the effect of the loading frequency was evaluated by measuring the temperature increase during fatigue cycling. Due to the relatively low strains levels and the cross-ply lay-up, no relevant increase of temperature was noticed. The CFRP samples were initially cycled between 50 and 450 MPa (~50% of the UTS) for a cumulative of 1,000 cycles. The cycle was then resumed at the same stress level till 1,000,000 cycles. Afterwards the load level was increased to 600 MPa (~70% of the expected UTS) and cycled for another 1,000 cycles, resulting in a cumulative of 1,001,000 cycles. The following load cycles were continued at the same stress level till 2,000,000 cycles and, as last, till 4,000,000 cycles.

For the GFRP samples instead, the amplitude was set between 30 and 130 MPa (~35% of the UTS) and kept for all the stages till 4,000,000 cycles. Tests were interrupted at 10,000, 1,000,000, 2,000,000 and 3,000,000 cycles. The mean strain on the central region of the gauge length was recorded with a dynamic extensometer during the static tests. For the GFRP samples, the longitudinal strain on the outer 90° layers was compared to the corresponding strain measured from the FBGs inside the laminate (between two 0° layers). In Figure 2 the samples prepared for testing are depicted. One can notice the aluminium strips on the central gauge length region which were used as marks for the CT scans in order to scan repetitively the same volume.

### 2.3 CT-scanners

As already mentioned in the previous subparagraph, the samples were scanned in the central region of their gauge length with a voxel size ranging from 2 to 4 μm. The obtainable 3D scan resolution is normally much lower than the voxel size, but with optimized scanning parameters, the latter can lead to a reasonable esteem of it (although there is no one to one correspondence). For the sake of simplicity, in the following the term resolution will be used. The scanned volume was of approx. $t \times w \times l = 2 \times 4 \times 4\ mm^3$ for the 2 μm voxel size scans and

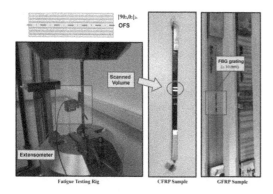

Figure 2. Testing setup: (top-left) schematic of the lay-up for the laminate and fibre optics embedding location, (bottom-left) sample mounted between the grips of the servo-hydraulic testing machine and (right) both carbon and glass fibres specimens prepared for testing.

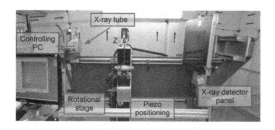

Figure 3. High-energy CT scanner optimized for research (Hector) used to scan at high resolution the composite specimens.

of $4 \times 8 \times 8$ mm$^3$ for the 4 µm ones, respectively. The scans were obtained with two different setups, property of the UGCT—Centre for X-ray Tomography of Ghent University. Both of them had an X-ray tube emitting X-rays through the sample, which was fixed vertically on a rotational stage, and a flat panel detector to record digital radiographs. In Figure 3 the HECTOR (High-Energy CT scanner Optimized for Research) scanner is depicted. The X-ray tube is a high—energy XWT 240-SE microfocus source (from X-RAY WorX®) which was set at a Voltage of 140 kV and 8 W of target power, while maintaining a minimum focal spot size of approx. 4 µm. A $40 \times 40$ cm$^2$ PerkinElmer 1620 CN3 CS flat panel detector and a total of 9 motorized axes of freedom allows for a high flexibility scanning. Nevertheless, the high-precision stage mounted on the manipulator allows for fine positioning of the sample and therefore a spatial resolution up to 4 µm is also achievable.

The second setup used is similar to the scanner just described but scaled-down in order to achieve better resolution up to 2 µm on a smaller sample size. The X-ray tube, a medium energy FXE–160.50 dual head open type source (from Feinfocus®). The detector used for these scans is a Varian 2520 V Paxscan a-Si flat panel with $1820 \times 1460$ pixels, 127 µm pixel size, covering a $20 \times 25$ cm$^2$ area. The sample manipulator is an XYZ-theta CT system (UPR-160F AIR) with ultra-precision air-bearing rotation motor from PI|miCos GmbH.

The X-ray source was set at 100 kV and 3 W of target power, resulting in a focal spot size of 2 µm. The Source to Object Distance (SOD) was set to 14.4 mm (27.5 mm for the 4 µm voxel size) and the Source to Detector Distance (SDD) to approximately 870 mm, resulting in a magnification around 60x and a corresponding voxel size of 2 µm. The stated distances, along with the conical X-ray beam, results in a geometric magnification, and the small focal spot size minimizes image smoothing. The reduction of spatial resolution ensures high image quality at this magnification. Moreover, the bigger the panel detector area is, the larger the obtainable magnification will be. On the other hand, one should also account for noise effects which could derive from the large SDD and a low energy of the outgoing beam (i.e. sample material with high attenuation). In some scans also a metal plate filter (e.g. aluminum or copper) was placed in between the source and the sample, with the aim of attenuating the higher energy beam component responsible for the so called beam-hardening artifact. In a tomographic dataset, X-ray radiographs are acquired at different viewing angles. The stability of the scanning geometry, either due to sample movement or focal spot drift, is monitored during the CT scan by returning to a reference point a number of times. For the scans discussed in this paper, no shift was observed. In total 1500 projections were recorded over 360° and the data were reconstructed with the in-house developed software package Octopus Reconstruction. The reconstructed pictures were post-processed through Matlab® and analyzed. The stack of cross-sections can be further employed to render the whole scanned volume. For this purpose, 3D renderings were made with VGStudio Max from the original reconstructed images.

## 3 SCAN RESULTS

### 3.1 *CT's cross-sections*

In this paragraph the cross-section images will be firstly introduced. Then, in the following section, the 3D renderings relative to the same scans will be presented. As already mentioned, these cross-sectional images were obtained from post-processing

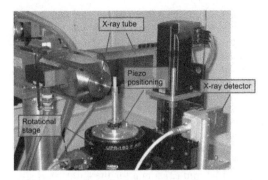

Figure 4. Modular 900 nm medium-energy micro-CT scanner used for the non-destructive evaluation of the composite samples.

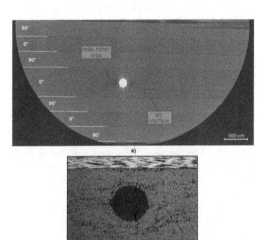

Figure 5. (a) CT cross-section of a carbon fibre sample having a 125 μm optical fibre embedded. (b) The scan is compared with the corresponding picture taken with a microscope after cutting and polishing the sample.

of a stack of 20 adjacent reconstructed slices; this allowed reducing the noise level and enhancing the contrast-to-noise-ratio, with consequent details smoothing. In Figure 5, the post-processed cross-section of a CFRP laminate having a 125 μm embedded optical fibre (195 μm diameter considering its coating) is given. Clearly recognizable in the image are the differently oriented 0° and 90° plies, which are defining the cross-ply lay-up. Each 0° or 90° layer is composed by two prepreg plies; in the centre of the laminate, the fibre was placed in the middle of a four plies layer oriented at 0°. Most relevant to be noted, is the crack which has developed at the sensor/composite interface.

Comparing the fibre close-up with an optical microscopy image taken from a polished cross-section on a similar sample (Fig. 5b), one can notice some analogies. Firstly, the crack presents the exact same geometry in both images. The samples were analysed shortly after manufacturing and were not yet subjected to any mechanical loading. In other words, probably the stress build-up during curing of the prepreg layers has been sufficiently high to induce the damage. This is explainable by the low resistance of the composite plies in the transverse direction (which is governed only by the resin) and by the thermal expansion mismatch between the sensor and the hosting composite. Secondly, even the richer resin content interfaces between two adjacent plies can be distinguished on the fibre neighbourhood. Particularly, it is noticeable how the plies are forced to redistribute around the sensor, resulting in resin richer areas sideways the fibre.

In Figure 6 the same post-processed image has been obtained for a GFRP sample having a 125 μm fibre embedded. Two important findings are worth to be reported: the contrast and the quality of the scan have been proved high enough even though the composition of the reinforcing fibre (E-glass) and of the optical fibre is very similar. Moreover, no crack appeared on the sensor surroundings and this can be explained by the lower curing temperature imposed during manufacturing of the laminate. In addition to this, also the thickness of each prepreg layer for the MTM28 was approx. double the amount of the M55 J-M18 (0.312 mm and 0.156 mm, respectively), allowing the stresses surrounding the fibre to decrease. However, a similar scan was also done on a $(90,0)_s$ cross-ply MTM28 prepreg sample (halved number of plies and total thickness of 2.5 mm) and on the surroundings of the fibre (195 μm outer coating diameter) no cracks were found.

## 3.2 3D renderings

The reconstructed stack of cross-sections was rendered with VGStudio Max in a 3D volume of approx. $t \times w \times l = 2.5 \times 5 \times 5\ mm^3$ for a 2 μm voxel size scan (or $5 \times 5 \times 10\ mm^3$ for a 4 μm voxel size). This allowed to freely examining the surroundings of the fibre optic sensor and the adjacent composite layers in order to check for damage. As suspected the crack at the sensor/composite interface for the CFRP sample was running along the whole fibre, suggesting that it was not a casual and localized damage, but more a systematic effect instead, mostly related to a wrong combination of materials/manufacturing parameters as already mentioned in the previous subsection. In Figure 7 all this information is presented for a 3D rendering

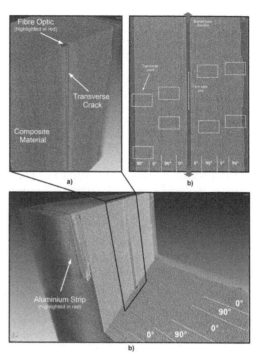

Figure 6. CT cross-section of a glass fibre sample having a 125 μm optical fibre embedded that was scanned at 4 μm voxel size. The enlargement is the result of a finer scan obtained at 2 μm voxel size.

Figure 7. (c) 3D rendering of a CT scan on the fibre optics surroundings for a carbon fibre sample, (a) detail of the crack evolving through the whole fibre coating and (b) cross-section taken along the fibre axis; the matrix cracks are indicated in white boxes.

of a scan on a CFRP sample which, in this case, was loaded in fatigue for 2,000,000 cycles. In red colour the fibre optic, as well as the aluminium strips used to mark the scanning volume, are highlighted for clarity in the images. Besides the already mentioned crack evolving through the coating of the fibre optic, no additional damage was assessed on the sensor surroundings. Instead, damage was found on the 90° plies as it is noticed on the top right side of the figure (Fig. 7b), where the transversal cracks have been highlighted in white boxes.

In order to quantitatively assess the damage evolution over 4,000,000 fatigue cycles, the overall number of cracks lying on a section taken along the fibre optic axis has been tracked for each scan. In addition to this, the stress-strain curve (i.e. sample stiffness) has been determined at each static tensile test. The results for the laminate longitudinal elastic modulus evaluated at different fatigue cycles

Table 1. Laminate stiffness degradation and overall number of transverse cracks for a M55 J-M18 sample tested over 2,000,000 fatigue cycles.

| Cycles nr. | $\sigma_{max}$ [Mpa] | $E_{11}$ [GPa] | Cracks nr. |
|---|---|---|---|
| 1,000 | 450 | 144.18 | 3 |
| 1,000,000 | 450 | 141.35 | 12 |
| 1,000,001 | 600 | 127.68 | 13 |
| 2,000,000 | 600 | 121.88 | 20 |

for a M55 J-M18 sample are presented in Table 1. Although not always evident for all the tested samples, here a clear correspondence between stiffness degradation and transverse cracks development could be noticed.

4 CONCLUSION

In this study the potential of the non-destructive technique known as High-Resolution X-ray Computed Tomography or μCT was presented for internal material investigation on CFRP and GFRP composite laminates. The scanned samples had fibre optics embedded, and the µCT proved capable of clearly visualizing the fibre outlines even for GFRP, where the attenuation difference between the materials is small (i.e. reinforcing glass fibre/silica optical fibre have comparable density, therefore similar attenuation coefficients). Moreover, the technique was exploited to assess the quality of in-composite embedded fibre optics. This has revealed in some cases internal damage on the fibre optic surroundings already after manufacturing of the laminates. The samples were then subjected to tension-tension fatigue and cycled to failure. Despite this initial damage, no significant reduction of the fatigue lifetime was encountered for the samples having fibre optics embedded. Even for the "already damaged samples", no further crack growth on the sensor surroundings was noticed after 4,000,000 cycles.

## ACKNOWLEDGMENTS

The research leading to these results has received funding from the Flemish Agency for Innovation by Science and Technology (IWT) – through the program for Strategic Basic Research (SBO) under grant agreement n° 120024 (Self Sensing Composites). The author also gratefully acknowledges the significant support of UGCT—the "Centre for X-ray Tomography" of Ghent University. The Special Research Fund of the Ghent University (BOF) is acknowledged for the post-doctoral grant of M. N. Boone.

## REFERENCES

Berthelot J., El Mahi A. & Le Corre J.F., 2001. Development of transverse cracking in cross-ply laminates during fatigue tests, *Composites Science and Technology* 61(12): 1711–1721.

Crupi V., Epasto G. & Guglielmino E., 2011. Computed Tomography analysis of damage in composites subjected to impact loading, *Frattura ed Integrità Strutturale*, 17: 32–41.

Dierick M., Van Loo D., Masschaele B., Van den Bulcke J., Van Acker J., Cnudde V., Van Hoorebeke L., 2014. Recent micro-CT scanner developments at UGCT, *Nuclear Instruments and Methods in Physics Research B*, 324: 35–40.

Hahn H.T., 1979. Fatigue behavior and life prediction of composite laminates, *Composite materials: testing and design; Proc. fifth conference ASTM STP 674*, 383–417, 20–22 March 1978, New Orleans.

Hufenbach W., Bohm R., Gude M., Berthel M., Hornig A., Ruc˘evskis S. & Andrich M., 2012. A test device for damage characterization of composites based on in situ computed tomography, *Composites Science and Technology* 72: 1361–1367.

Ketterer J., 2009. Fatigue crack initiation in cross-ply carbon fiber laminates, *Georgia Institute of Technology*.

Luyckx G., Voet E., Lammens N. & Degrieck J., 2011. Strain Measurements of Composite Laminates with Embedded Fibre Bragg Gratings: Criticism and Opportunities for Research, *Sensors*, 11: 384–408.

Masschaele B.C., Cnudde V., Dierick M., Jacobs P., Van Hoorebeke L. & Vlassenbroeck J., 2007. UGCT: New x-ray radiography and tomography facility, *Nuclear Instruments & Methods in Physics Research Section a-Accelerators Spectrometers Detectors and Associated Equipment*, 580(1): 266–269.

Masschaele B., Dierick M., Cnudde V., Van Loo D., Boone M., Brabant L., Pauwels E., Cnudde V., Van Hoorebeke L., 2013. HECTOR: A 240kV micro-CT setup optimized for research, *Journal of Physics: Conference Series 463; Proc. 11th International Conference on X-ray Microscopy*, 5–10 August 2012, Shanghai.

Shivakumar K. & Emmanwori L., 2004. Mechanics of failure of composite laminates with an embedded fiber optic sensor, *Journal of Composite Materials*, 38(8): 669–680.

Sisodia S., Kazemahvazi S., Zenkert D., Edgren F., Fatigue Testing Of Composites With In-Situ Full-Field Strain Measurement, *Proc. ICCM-18*, Jeju Island, Korea, 21–26 August, 2011.

Takeda N., 2002. Characterization of microscopic damage in composite laminates and real-time monitoring by embedded optical fiber sensors, *International Journal of Fatigue*, 24: 281–289.

Vlassenbroeck J., Dierick M., Masschaele B., Cnudde V., Hoorebeke L. & Jacobs P., 2007. Software tools for quantification of X-ray micro-tomography, *Nuclear Instruments & Methods in Physics Research Section a-Accelerators Spectrometers Detectors and Associated Equipment*, 580(1): 442–445.

*Emerging Technologies in Non-Destructive Testing VI – Aggelis et al. (Eds)*
*© 2016 Taylor & Francis Group, London, ISBN 978-1-138-02884-5*

# Damage detection and classification in composite structure after water-jet cutting using computed tomography and wavelet analysis

A. Katunin
*Institute of Fundamentals of Machinery Design, Silesian University of Technology, Gliwice, Poland*

ABSTRACT: The proposed method of processing of X-ray Computed Tomography (CT) data is based on its three-dimensional (3D) wavelet analysis, which allows for detection and localization of sudden changes in voxels of CT data array. The applied wavelet transform can be considered as filtering using sets of low-pass and high-pass filters over directions of a domain. The internal damage, which initially has different spectrum than the healthy regions of a structure, are emphasized after the wavelet analysis and depending on its intensity in the CT data array the wavelet coefficients become different for them. The tests were performed on the carbon fibre reinforced composite plate. The water-jet method was used for cutting of a circular hole inside. During the cutting process several delaminations in various layers occurred. Moreover, during the manufacturing process several air pockets in the matrix appeared. Application of the proposed processing algorithm allows not only for detection and localization of internal damage but also its classification by a type. It is possible due to analysing the magnitude of wavelet coefficients after wavelet-based decomposition and characteristic shape and dimensions of various types of damage. This allows for automation of the examination process.

## 1 INTRODUCTION

The problem of Non-Destructive Testing (NDT) of composite structures becomes more and more important in modern examination studies, since nowadays composite materials are very often applied for manufacturing of structural parts in aircraft, aerospace, automotive, naval industries and many others. From a great variety of NDT methods applied for structural damage assessment one of the most precise is the computed tomography, which allows for identification of even micro—and nanoscale damage in a tested structure. In spite of a great resolution and detection performance of this testing method several problems may occur during interpretation of results and it is difficult to process the output data due to its great capacity.

One of the main problems with CT data handling is the isotropy of detected and localized damage, i.e. damaged regions usually have different colours in comparison with healthy regions due to different absorption of radiation and such a difference is the only information concerned with detected and localized damage. Thus, the identification and classification of damage observed in reconstructed CT scans is not obvious and additional signal processing techniques should be applied in order to identify a type of damage and make it distinguishable. Additional difficulties appear due to presence of measurement noise as well as reconstruction noise,

which might be improperly classified as a structural damage.

Several studies on extraction and classification of damage features from 3D arrays of reconstructed CT scans have been performed. The earliest studies (Chu & Lee 2004, Chu et al. 2004) were based on a thresholding performed using Otsu's method extended to the 3D space. This technique solves a problem partially—the data still contain a lot of measurement noise and reconstruction artefacts. Other, more advanced algorithms of feature extraction described e.g. by Lontoc-Roy et al. (2006), are based on 3D neighbourhood analysis, however the procedure was performed on binarized input data, which results in loss of information. The post-processing algorithm presented by Perret et al. (2007) is based on selection of upper and lower thresholds by the trial-and-error method. The more advanced noise reduction algorithm with use of anisotropic diffusion filter was proposed in (Heinzl et al. 2007), while in (Huang et al. 2003) the authors used 3D Gaussian filter for noise reduction. Another feature extraction algorithm was presented in (Kaestner et al. 2006), where the authors used non-linear diffusion filter to enhance the contrast and then the adaptive thresholding procedure for overall enhancement of measurement data and availability of features extraction from them. Several studies of feature extraction from CT scans were performed using wavelet analysis. The authors of (Roeding &

Westenberg 1998) were probably the first who described the advantages of application of 3D wavelet transform (WT) to CT data. An algorithm, which was based on 3D discrete wavelet transform (DWT), was successfully applied to 3D data from CT scans by Chen & Ning (2004) for denoising and feature extraction. Another approach was proposed by Moss et al. (2005). Considering effectiveness of wavelet-based filtering and denoising it was decided to use 3D DWT for a construction of novel algorithm for identification and classification of internal damage in composite structures.

The proposed algorithm is based on 3D DWT and thresholding of wavelet coefficients and allows for extraction of damage features from reconstructed CT scans. The algorithm was tested on a composite plate, in which a circular hole was cut using a water-jet method. The cutting procedure caused delamination around a hole with various directions of propagation. The tested plate was subjected to X-ray CT scanning and obtained reconstructed 3D data array was used as an input in the proposed algorithm. By application of 3D DWT it was possible to identify the defects in the structure and filter out the most of measurement noise. The application of classification algorithm, which is based on evaluation of characteristic dimensions of extracted volumetric features and values of wavelet coefficients, allows classifying the defects into three classes: delamination, air pockets and measurement noise. The CT scanning together with a proposed algorithm allows for automation of identification and classification of defects in composite plates and could be useful in a quality control of composite elements as well as in NDT of structures being in operation.

## 2 MATERIALS AND TESTING

### 2.1 Tested structure

The tested structure was prepared in the form of an epoxy-based laminate with a stacking sequence of [±45]$_s$ reinforced by carbon fibre with dimensions of $300 \times 150 \times 5$ mm. A circular hole in this structure was cut using water-jet method on the Trumpf® Trumatic WS 2500 cutting system. The cutting process was performed using a water jet with corundum particles with a nominal pressure of 300 MPa. The diameter of a jet was in the range of 0.8÷1 mm and the velocity of cutting was 1 m/min. In order to prevent an initiation of delaminations in plates during cutting an initial pressure of a jet was set to 70 MPa. Moreover, the cutting process was initiated in the middle of a cut contour and the cutting head was moved following the circular trajectory. A picture of a tested structure with a hole is presented in Figure 1.

### 2.2 Ultrasonic scanning

Despite a composite-optimised process parameters and cutting trajectories, the ultrasonic scanning of these plates indicated the presence of significant delaminations area around cut hole observable in Figure 1.

The Ultrasonic Testing (UT) was performed on the air-coupled ultrasonic transducers system HFUS 2400 AirTech manufactured by the Ingenieurbüro Dr. Hillger. The focusing distance between 250 kHz emitter AirTech 4412 and receiver AirTech 4422 probes was set to 50 mm. The attenuation range was defined from -31 dB to 0 dB with 16 levels in between. The wave running time was set to 250 μs, which ensures the acceptable quality of the obtained C-scan (Figure 2).

Figure 1. The tested structure.

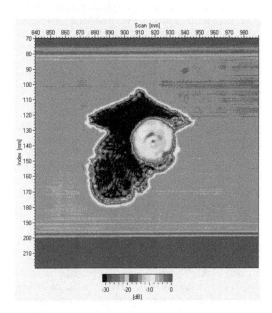

Figure 2. The C-scan of the tested structure.

The delamination, resulted from the water-jet cutting, is clearly detectable in C-scan, however this technique allows for damage detection and localization in two dimensions, which results in loss of information of damage location in the normal direction to the surface of the tested structure. Moreover, it is still impossible to determine a type of a damage. The next step of damage localization was performed using CT.

## 2.3 CT scanning

The CT scanning procedure was performed on the X-ray tomograph v|tome|x L 450 manufactured by GE® Sensing & Inspection Technologies GmbH. The GE® detector of type DXR250 with 410 × 410 mm active area was used during the scanning process. The parameters of scanning were as follows: accelerating voltage of 200 kV, current of 180 µA, Cu-filter with thickness of 0.5 mm. The RTG lamp of a microfocus type (which allows obtaining the tomography resolution up to 5 µm) has the following parameters: maximal voltage of 300 kV, maximal power of 500 W, conic-type ray with an angle of 40°. The scanning was performed on the area of 100.7 × 90 mm of the tested plate centered on the hole with a resolution of 50 µm.

The CT scans were reconstructed to a 3D data array and exported to the tomograph-dedicated software myVGL by Volume Graphics. Examples of virtual 2D slices of the tested structure are shown in Figure 3.

The damaged regions are clearly detectable in the virtual slices presented in Figure 3 but are still of isotropic character. The damage is represented now by three spatial dimensions, which allows performing the classification procedure. Obtained 3D data array with dimensions of 452 × 504 × 83 voxels was exported to Matlab® environment for further processing.

## 3 DAMAGE IDENTIFICATION AND CLASSIFICATION

### 3.1 Description of damage extraction algorithm

The exported array was subjected to 3D DWT using the B-spline wavelet of order 4. This wavelet was selected basing on previous analyses (Katunin 2011). As a result, eight sets of coefficients (one set of approximation $a$ and seven sets of directional detail $d_{1,...,7}$ coefficients) were obtained. Since the measured data was highly biased by the measurement noise and artefacts it was necessary to apply the pre-filtering procedure. It was realized by hard thresholding of the data with a threshold level of 25% from the maximal value of every set of detail coefficients. Such a procedure allows filtering out most of low-magnitude noise and the blurring surroundings, resulted by application of decomposition algorithm, of detected defects. After thresholding the labels on the CT images were removed and the boundary effect was reduced using the zero-padding method. After these operations the resulted arrays of detail coefficients (without the array of approximation coefficients) were reconstructed using 3D Inverse DWT (IDWT) algorithm. Each of the sets of detail coefficients was reconstructed separately and then their absolute values were added up, which formed single 3D arrays—the sets of $D$-coefficients. The scheme of this algorithm is presented in Figure 4.

The results of damage extraction is presented in the form of a 3D plot in Figure 5. In order to increase the visibility of particular features on a 3D plot the transparency property, depending on the magnitude of the $D$-coefficients, was used.

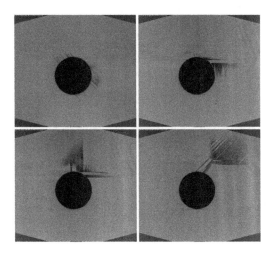

Figure 3. Exemplary virtual 2D slices of a CT scan.

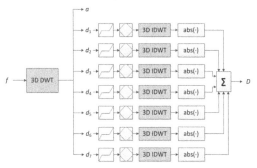

Figure 4. A scheme of wavelet-based processing algorithm.

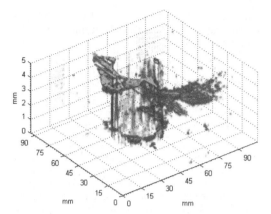

Figure 5. Results of damage extraction using proposed algorithm.

### 3.2 Description of damage classification algorithm and classification results

In order to classify the defects in the tested structure the set of $D$-coefficients was thresholded with an empirically determined threshold level of 2.7%. Such thresholding allows for filtering out low-magnitude $D$-coefficients, which have no dominant influence on the defects identification, and preparing the data for classification. The thresholding procedure is based on the wavelet-based hard-thresholding approach applied for obtained $D$-coefficients, which extracts the significant components of a signal (i.e. useful diagnostic information) based on the magnitudes of $D$-coefficients. The hard-thresholding approach was chosen due to better preservation of boundaries of locally increased magnitudes of $D$-coefficients with respect to soft one. In this approach the analysed $D$-coefficients below the threshold value $\lambda$ were set to zero.

The next step of preparing the data for classification was a 3D boundary tracking. During this operation the sets of non-zero voxels, which represent various types of defects, were identified and classified to several clusters. The classification procedure was based on the geometric properties of the resulted 3D array as well as on the magnitudes of $D$-coefficients after thresholding. It should be considered that during decomposition/reconstruction procedures the highest magnitudes of $D$-coefficients were obtained for the greatest singularities in the analysed array and thus the defects locations. Following this, the geometrical criteria were applied in order to classify three types of defects:

- delamination, which is characterized by the great spatial dimensions of the sets of non-zero voxels in the tangent plane to the surface of the plate and low dimensions of non-zero voxels in the normal direction to the surface of a structure, it was also assumed that the $D$-coefficients are high for this case of defects,
- air pockets (appeared during manufacturing of a structure), which have near-spherical geometry and the number of voxels with high-magnitude $D$-coefficients is higher than 20 close-located voxels,
- noise (resulted from CT scanning as well as processing procedures), which is characterized by low-magnitude $D$-coefficients with a small number of non-zero voxels.

Using the above-discussed measures three types of defects were effectively classified. The results of classification are presented in Figures 6–8.

In order to compare the effectiveness of the proposed method of extraction of damaged regions from CT scans the quantitative analysis of the area of the delamination was performed. Firstly, the surface area of delamination was determined from the C-scan of the tested structure (see Figure 2).

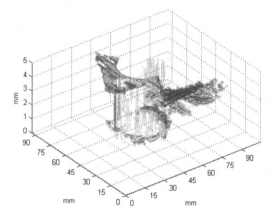

Figure 6. Classified set of delaminated regions.

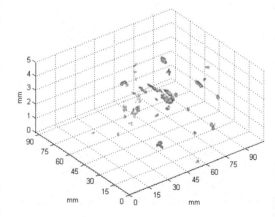

Figure 7. Classified set of air pockets.

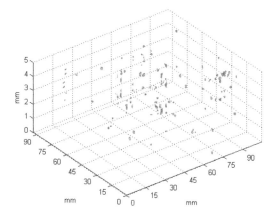

Figure 8. Classified set of noise.

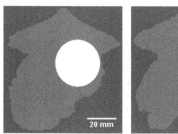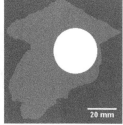

Figure 9. Delaminated region: left—UT, right—CT.

Since the obtained result after processing of the CT scan is the 3D dataset (see Figure 5) the top projection was considered during determination of an area for comparability purpose. Using binarization and some logical operations on the analysed images the boundaries of delamination were determined in both cases (see Figure 9).

At it can be noticed the area of delamination determined using UT and CT differs and equals 3024 mm$^2$ and 2892 mm$^2$, respectively. The difference between calculated amounts of delamination area equals 4.36%. This difference results from different resolutions of applied testing techniques as well as wavelet-based filtering operations, which filtered out some low-magnitude regions in the analysed dataset.

## 4 CONCLUSIONS

The study presented a novel approach of damage identification and classification based on wavelet analysis of the CT scan of the tested structure. The proposed algorithm effectively classified defined types of damage (delamination, air pockets and noise) basing on the values of $D$-coefficients obtained from isotropic 3D DWT decomposition/reconstruction procedures and geometric properties of specific types of damage. By appropriate configuring the parameters of the proposed algorithm the classification procedure can be automated.

Further studies in this area will be concentrated on development of the processing algorithms as well as development of new and more effective classification procedures with use of soft computing and artificial intelligence methods.

## ACKNOWLEDGEMENTS

The research was partially financed by the National Science Centre (Poland) granted according the decision no. DEC-2011/03/N/ST8/06205.

## REFERENCES

Chen, Z. & Ning, R. 2004. Breast volume denoising and noise characterization by 3D wavelet transform. *Computerized Medical Imaging and Graphics* 28(5): 235–246.

Chu, C.P. & Lee, D.J. 2004. Bilevel thresholding of slices image of sludge floc. *Environmental Science and Technology* 38(4): 1161–1169.

Chu, C.P., Lee, D.J. & Tay, J.H. 2004. Bilevel thresholding of floc images. *Journal of Colloid and Interface Science* 273(2): 483–489.

Heinzl, C., Kastner, J. & Gröller, E. 2007. Surface extraction from multi-material components for metrology using dual energy CT. *IEEE Transactions on Visualization and Computer Graphics* 13(6): 1520–1527.

Huang, R., Ma, K.-L., McCormick, P. & Ward, W. Visualizing industrial CT volume data for nondestructive testing applications. *Proc. 14th IEEE Conf. Visualization VIS'03, Seattle, WA, 24–26 October 2003*, 547–554.

Kaestner, A., Schneebeli, M. & Graf, F. 2006. Visualizing three-dimensional root networks using computed tomography. *Geoderma* 136(1–2): 459–469.

Katunin, A. 2011. Damage identification in composite plates using two-dimensional B-spline wavelets. *Mechanical Systems and Signal Processing* 25(8): 3153–3167.

Lontoc-Roy, M., Dutilleul, P., Prasher, S.O., Han, L., Brouillet, T. & Smith, D.L. 2006. Advances in the acquisition and analysis of CT scan data to isolate a crop root system from the soil medium and quantify root system complexity in 3D-space. *Geoderma* 137(1–2): 231–241.

Moss, W.C., Haase, S., Lyle, J.M., Agard, D.A. & Sedat, J.W. 2005. A novel 3D wavelet-based filter for visualizing features in noisy data. *Journal of Microscopy* 219(2): 43–49.

Perret, J.S., Al-Belushi, M.E. & Deadman, M. 2007. Non-destructive visualization and quantification of roots using computed tomography. *Soil Biology & Biochemistry* 39(2): 391–399.

Roerdink, J.B.T.M. & Westenberg, M.A. 1998. Wavelet-based volume visualization. *Tech. rep. 98-9-06*, University of Groningen.

*Emerging Technologies in Non-Destructive Testing VI – Aggelis et al. (Eds)*
*© 2016 Taylor & Francis Group, London, ISBN 978-1-138-02884-5*

# Post Weld Heat Treatment surface residual stress measurements using X-ray diffraction

S. Kumar
*Rolls-Royce@NTU Corporate Lab, Singapore*

M.J. Tan & B.S. Wong
*Nanyang Technological University, Singapore*

N. Weeks
*Rolls-Royce Singapore Pte Ltd., Singapore*

ABSTRACT: The main objective of this study is to investigate the use of Non-Destructive Testing (NDT) techniques to measure residual stresses induced on TIG (Tungsten Inert Gas) welded titanium specimens. Residual stress is an unavoidable problem which occurs during any manufacturing, fabrication process or during repair and becomes a limitation to the service life of a component.

In this approach, X-Ray Diffraction (XRD) is used for measurement of residual stress, and the results obtained from this work describes the residual stress distribution on the weld-bead, HAZ and the base material before and after post weld heat treatment in both the parallel and perpendicular directions. In addition, the repeatability of the X-ray diffraction method was analyzed. This analysis using X-ray diffraction will help optimise Post Weld Heat Treatment (PWHT) process with the aim to reduce residual stress significantly.

*Keywords*: residual stress; Tungsten Inert Gas welding; X-ray diffraction; Post weld heat treatment; Bragg's law.

## 1 INTRODUCTION

Residual stress in a structure or component is an important factor to be considered for service performance and structural integrity of gas turbine components. The distribution and magnitude of residual stress varies for different manufacturing and repair processes, and it is a major uncertainty in a product's service life. In the service of aero-engines, components are subjected to various repair processes i.e. welding, brazing, etc. Residual stresses developed due to these repair processes can be measured with the help of both destructive and non-destructive techniques. The most commonly used techniques are the central hole-drilling on the destructive side and X-ray diffraction on the non-destructive side. Since aero-engine components incur high cost for manufacturing, this work looks primarily at using a non-destructive method for measurement.

Residual stress is a major factor which reduces the life expectancy of a component, so relieving stress is an important consideration which can be carried out by heat treatment on the stressed region. (James et al., 2010).

## 2 THEORY

### 2.1 *Stress analysis using X-ray diffraction*

Residual stress measurement using X-ray diffraction is the most preferred method when compared to the mechanical method (destructive method) and nonlinear elastic method (ultrasonic and magnetic method), but for X-ray diffraction the depth and spatial resolution have lower order of magnitude. X-ray diffraction can be used for greater depths but as a destructive technique by removing material to reach the required depth. In order to determine residual stress, the strain in the crystal lattice must be measured with reference to a precisely known orientation from a sample surface (Suzuki et al., 2011).

### 2.2 *Principle of X-ray diffraction for residual stress measurement*

The measurement of residual stress using X-ray diffraction relies on the interaction between X-ray beam and the crystal lattice. The basics of all XRD measurements are explained by Bragg's Law.

A crystalline material consists of a periodic plane of atoms that can cause a constructive or destructive interference pattern caused by diffraction. The nature of the interference depends on the inter-planar spacing and the wavelength of the incident radiation. This is commonly described by Bragg's law as,

$$n\lambda = 2dSin\theta \qquad (1)$$

where n = an integer, $\lambda$ = the wavelength of the incident wave, $d$ = spacing between the planes in the atomic lattice, and $\theta$ = angle between the incident ray and the scattering planes

### 2.3 Strain measurement

The inter-planar spacing of a material that is free from strain will produce a characteristic diffraction pattern for that material. When that material is stilted, elongations and contractions are produced within the crystal lattice, which change the inter-planar spacing of the lattice planes. This induced change in distance between the inter-atomic spacing "d" will cause a shift in the diffraction pattern. By precise measurement of this shift, the change in the inter-planar spacing can be evaluated and thus the strain within the material is deduced using this equation (Fitzpatrick et al., 2005).

$$\varepsilon = (d_n - d_0) / d_0 \qquad (2)$$

where $\varepsilon$ = strain, $d_n$ = unknown inter-planar spacing and $d_0$ = known inter-planar spacing

Strain along any particular direction can be measured with the known inter-planar distance of the corresponding crystal lattice for the chosen material (Fig. 1).

### 2.4 Stress distribution

The strain within the material is quite important to find the stress, from Hooke's Law

$$\sigma = E\varepsilon \qquad (3)$$

where $\sigma$ = stress, E = elastic modulus and $\varepsilon$ = strain.

It is known that a tensile force producing a strain in the x-direction will produce not only a linear strain in that direction and also strains in the lateral directions. Assuming a state of plane stress exists, i.e. $\sigma z = 0$ (stress in z-direction), and the stresses are biaxial, then the ratio of the lateral to longitudinal strains is determined by Poisson's ratio. If the X-ray measurement has been made, with a known value of $d_0$ and the X-ray analyser determines $d_n$, then assuming $\sigma z = 0$ to a general case, where the sum of the principal stresses along

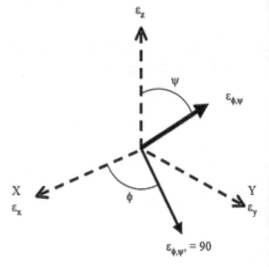

Figure 1. Principle Axis (X, Y and Z) of Strain (Fitzpatrick et al. 2005).

x-direction and y-direction can be obtained using this equation

$$(d_n - d_0) / d_0 = -(\nu / E)(\sigma x + \sigma y) \qquad (4)$$

where $\nu$ = poisson's ratio, E = elastic modulus, $\sigma x$ = principle stress in x-direction, $\sigma y$ = principle stress in y-direction, $d_n$ = unknown inter-planar spacing and $d_0$ = known inter-planar spacing (Fitzpatrick et al., 2005).

### 2.5 Penetration depth

X-rays are absorbed when passed through a material, so naturally that will affect the intensity and in-turn the intensity affects the penetration depth. Penetration depth decreases exponentially with a decrease in intensity, the rationale for this is as follows: the attenuation, loss in signal strength, is proportional to the distance travelled in a material; hence the contribution to the diffracted beam from layers, or planes, deeper down in the material will reduce (Fitzpatrick et al., 2005).

## 3 EXPERIMENTAL DETAILS

The Xstress 3000 G3 from Stresstech Group (Stresstech Group, V 1.5, 2013) is the portable analyser used for measurement.

### 3.1 Materials

The material tested is Grade 5 Ti-6 Al-4V AMS 4911. The machined specimen is a flat plate of

50 mm × 50 mm with 2 mm thickness in 'as-rolled' condition. The specimen was then welded using the following TIG (Tungsten Inert Gas) welding parameters: 40 amps and 10 volts with inert gas shielding of flow rate of 6 lit/min (Fig. 2), this is a 'bead on plate' weld with complete penetration over the entire thickness of the plate. These welding parameters represent typical values for weld repairs of aero-engine components.

### 3.2 Xstress 3000 G3 measurement procedure

Initially, sample surface preparation was carried out, followed by setting up the hardware (Fig. 3).

The Goniometer was setup up, followed by the system calibration(Fig. 4).

The accurate determination of the shift of the intensity peak is one of the most important tasks in the residual stress measurement by X-ray diffraction. There are large number of methods to localise the peak. In the software there are two global methods: Cross-correlation and Peak fit.

The Peak Fit method is based on the assumption that there exists a function that describes the intensity distribution. Based on the shape of the intensity distribution it can be assumed that the bell shape of the Gaussian fit function is close. A Pearson VII function is found to most accurately describe the intensity distribution in the back reflection region used in residual stress measurement (Stresstech Group, V 1.5, 2013).

## 4 RESULTS AND DISCUSSION

In this work, X-ray diffraction was applied to measure residual stress. Xstress 3000 G3 analyzer (Fig. 4) is calibrated using the "stress-free" samples in order to establish lattice parameter $d_0$, and a reference specimen was also used for calibration purpose. As previously discussed, the residual stress was measured at the surface to a depth of about 7 to 10 micron, and the results were obtained from points on the weld-bead, HAZ and base material (Fig.5).

The longitudinal stresses have the largest tensile values in the weld-bead due to the severe restraint condition, and small compressive values on the base material regions. It can be seen that the magnitude of the tensile stress reached about 357 MPa at the center of the weld-bead, and the HAZ on either sides of the weld-bead also has tensile stress between a range of 327 MPa to 286 MPa. And as we move away, the stress changes from tensile to very low values of compressive stress. The resid-

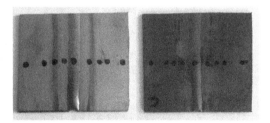

Figure 2. Specimen before and after PWHT.

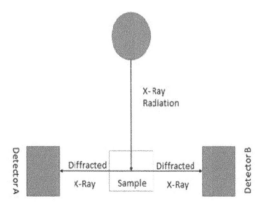

Figure 3. X-stress 3000 residual stress measurement technique (Schematic).

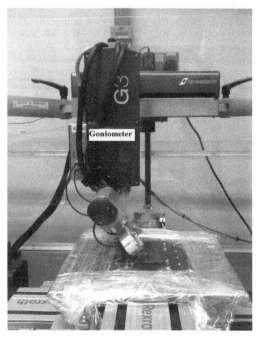

Figure 4. X-ray analyser for stress measurement with Goniometer.

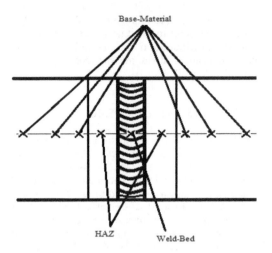

Figure 5. X-ray diffraction measurement zones (Schematic).

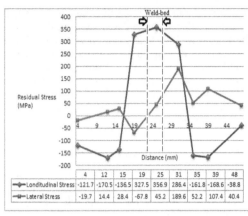

Figure 6. Residual stress before Post-weld heat treatment (PWHT).

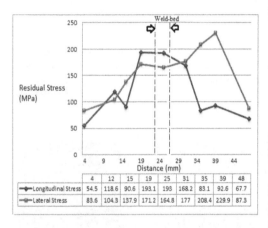

Figure 7. Residual stress after Post-weld heat treatment (PWHT).

ual stress is also measured in the lateral direction and the stress ranges from 190 MPa to −70 MPa (Fig. 6); it can be seen that the stress changes from compressive to tensile in a linear fashion along the weld-bead and HAZ(Kudrayavtsev et al., 2010). Ti-6 Al-4V has high yield strength when compared to commercial pure titanium so it will have less residual stress built up when welded (Song et al., 2014).

The residual stresses are again measured in the same areas after Post Weld Heat Treatment (PWHT) as per the Rolls-Royce specifications for PWHT by radiation heating in a furnace.

After PWHT the residual stresses for both longitudinal and lateral direction (Fig. 7) has shifted completely to tensile residual stress, the longitudinal tensile residual stress in the weld-bead and the HAZ have reduced by 42% on an average, and totally the stress in the longitudinal direction ranges between 50 MPa to 200 MPa. The stress in the lateral direction has changed on the base material region, with an increase in the tensile stress by 100 MPa approx., and between the weld bead and HAZ the stress has shifted from a linearly increasing stress value (Kudrayavtsev et al., 2010) to a similar stress value throughout that region.

The stress in both the directions were found to exist between a range of 50 MPa to 220 MPa (Fig. 7) after PWHT. Tensile residual stresses in the lateral and longitudinal directions may be used for determining the weld-life performance (Pardowska et al., 2009).

An observation was also made from the stress analysis is that instead of doing a uniform heat treatment in the weld-zone and the surroundings, a varying temperature profile with the peak setpoint temperature at the weld-bead and gradually decreasing the temperature to the base material will help reduce the increase in the tensile stress in both lateral and longitudinal directions in the base material region.

To clarify the accuracy of the measurements, a repeatability study of X-ray diffraction equipment has been performed. Repeatability quantifies the measure of accuracy and precision. The measurements were taken for three weld-beads with similar welding conditions and the variation between the test determinations for each specimen determines the repeatability.

The Repeatability percentage (Equipment Variation Percentage) is 16.5%, neglecting human errors, which is less than 20% and good, but can be made better.

Table 1. Repeatability Analysis of X-ray diffraction for Similar Welds.

| Similar Welds | 1 MPa | 2 MPa | 3 MPa |
| --- | --- | --- | --- |
| Test Determination 1 | 319.3 | 313.9 | 385.9 |
| Test Determination 2 | 364.7 | 333.8 | 337.1 |
| Test Determination 3 | 356.9 | 329.5 | 367.8 |
| Test Determination 4 | 314.6 | 316.7 | 383.1 |
| Repeat Standard Deviation | | | 20.4 |
| Repeatability | | | 56.6 |
| Repeatability Percentage (Equipment Variation Percentage) | | | 16.5% |

## 5 CONCLUSION

The XRD process has proved to be capable of determining residual stress in the surface of the test piece. Post weld heat treatment reduces the tensile residual stress in the weld-bead and HAZ, but increase the tensile residual stress in the base material, so this study highlighted the opportunity to investigate a localized PWHT method to reduce the effect on the base material

The residual stress measurement using X-ray diffraction analyzer Xstress 3000 is good but not very accurate as the equipment variability percentage is 16.5%. This study does not differ from earlier studies which have suggested that X-ray diffraction as a non-destructive technique is good for surface residual stress measurement.

Even though X-ray diffraction residual stress measurement is the best developed non-destructive method available for the characterization of the residual stress distribution, the equipment variation percentage of X-ray diffraction is less than that of destructive methods, so more measurements are usually required for X-ray diffraction for better accuracy (Rossini et al., 2012).

Further studies to compare these techniques would help evaluate this process and also the study into residual stress variation at different depths is also important to further understand the weld properties.

## ACKNOWLEDGEMENTS

This work was conducted within the Rolls-Royce@ NTU Corporate Lab with support from the National Research Foundation (NRF) Singapore under the Corp Lab@University Scheme.

## REFRENCES

Farajian-Sohi, M. Nitschke-Pagel, Th. Dilger, K. 2009. Residual stress relaxation in welded joints under static and cyclic loading, *JCPDS International Centre for Diffraction Data,* U.S.A.

Fitzpatrick, M.E. Fry, A.T. Holdway, P. Kandil F.A. Shackleton, J. & Suominen, L. Sep. 2005. Determination of Residual Stresses by X-ray Diffraction. *National Physical Laboratory*, Issue 2, United Kingdom.

Govinda Rao, P. Clvrsv Prasad, D. Sreeamulu, D. Chitti Babu, V. & Vykunta Rao, M. Apr. 2013.Determination of residual stresses of welded joints prepared under the influence of mechanical vibrations by hole drilling method and compared by finite element analysis. *International Journal for Mechanical Engineering and Technology (IJMET)*, Vol. 4, Pages: 542–553. India.

James, M.N. 2010. Residual Stress Influences in Mechanical Engineering. *Proceedings of the XVIII Congreso Nacional de Ingenieria Mecánica.* Spain.

Kudrayavtsev, Y. Kleiman, J. June 2010. Residual stress in welded elements. *3rd International CANDU In-service inspection and NDT*, Canada.

Non-destructive Testing—Test Method for Residual Stress analysis by X-ray Diffraction, *British Standards, BS EN 15305:2008*, United Kingdom.

Pardowska, A.M. Price, W.H.J. Finlayson T.R. Ibrahim, R. Evaluation. 2009. of Residual Stress Measurements Before and After Post-Weld Heat Treatment in the Weld Repairs. *International Conference on Neutron Scattering*, Journal of Physics—Conference series 201, U.S.A.

Paul, P.S. 1986. X-ray Diffraction Residual Stress Techniques, *Metals Handbook.10. Metals Park: American Society for Metals*, Pages: 380–392. U.S.A.

Rossini, N.S. Dassisti, M. Benyounis, K.Y. Olabi, A.G. Mar. 2012. Methods of Measuring Residual Stresses in Components. *Materials & Design*, Volume 35, Pages: 572–588, Elsevier, U.K.

Song, S. Paradowska, Anna, M. Dong, P. Mar. 2014. Investigation of Residual Stresses Distribution in Titanium Weldments. *Material Science Forum*, Vol.777, Pages: 171–175, Trans Tech publications, Switzerland.

Suzuki, T. Hatsuhiko, O. Sugiyama, M. Nose, T. Imafuku, M. Tomota, Y. Suzuki, H. Moriai, A. July 2011. Residual Stress Measurement of Welding Area by Neutron Diffraction Method. *Nippon Steel* Technical Report no. 100, Japan.

Stresstech Group, V1.5, Sept. 2013. Xtronic Guide, Finland.

*Emerging Technologies in Non-Destructive Testing VI – Aggelis et al. (Eds)*
*© 2016 Taylor & Francis Group, London, ISBN 978-1-138-02884-5*

# Enhancement of spatial resolution using metamaterial sensor in NonDestructive Evaluation

Adriana Savin, Nicoleta Iftimie & Rozina Steigmann
*National Institute of Research and Development for Technical Physics, Iasi, Romania*
*University Al.I. Cuza Iasi, Romania*

Alina Bruma
*CRISMAT Laboratory, National Graduate School of Engineering, University of Caen on Normandy,*
*Caen, France*

ABSTRACT: This paper presents a possibility to enhance the spatial resolution of eNDE methods that operate in radiofrequency and high frequency range using a sensor with metamaterial lens that can manipulate the evanescent waves that appear in the space between strips and respective, carbon fibers. Also, interruptions, short-cuts of MSG as well as nonalignment of carbon fibers, lack of resin or voids, and delaminations induced by impacts with low energy can be detected and emphasized.

## 1 INTRODUCTION

In the past several decades, a number of Nondestructive Evaluation (NDE) techniques have been developed for detecting effect of damages/embedded objects in homogeneous media.

Basically, the NDE of materials consists in the application of a physical field to the examined object and evaluating the interaction between the field and the eventual material discontinuities. If the physical field applied to the examined object is an electromagnetic field with frequencies ranging in interval tens of Hertz to tens of giga Hertz, the procedure is electromagnetic Nondestructive Evaluation (eNDE). This is applied to the examined having high conductivity, in which, under the action of electromagnetic incident field, eddy current are induced, according to Faraday's law (Bladel 2007). The induced currents create a secondary electromagnetic field opposing the incident one. The presence of material inhomogeneities (voids, inclusions, cracks, with lower electrical conductivity) will disturb induced eddy currents and will change the apparent impedance of sensors.

Damage detection is a primary concern in composite structures, because they are prone to multiple damages forms that can be hidden within the structure. Damages can include matrix cracking, fiber breakage and delamination which can be caused by impacts, fatigue or overloading. Further, Metallic Strip Gratings (MSG) in the form of flat periodic arrays of thin strips are used today across a wide range of Electromagnetic (EM) frequencies and applications. For example, they are used as a filter or polarizer, as conductive strips or microstrips, and also can be found in almost all rigid and flexible printed circuits. On the same note, these structures are also intensively studied from theoretical point of view, in order to give a better understanding of their behavior in common applications and to design new types of metamaterials starting from the existence of surface plasmons polaritons (Kolomenski et al. 2009). These applications impose a rigorous and rapid quality control.

This paper presents a possibility to enhance the spatial resolution of eNDE methods that operate at frequencies of tens at hundreds of MHz using metamaterial sensors. The possibility to manipulate the evanescent waves that appear in the space between strips and respective, carbon fibers, allows an improvement of the spatial resolution with at least $\lambda/2000$.

## 2 METAMATERIAL SENSOR FOR eNDE AND THEORY

The Metamaterials (MM), electromagnetic structures with distinguished properties have started to be studied especially in the last few years. Electromagnetic MM belong to the class of artificially engineered materials, can provide an engineered response to electromagnetic radiation that is not available from the class of naturally occurring materials. These are often defined as the structures of metallic and/or dielectric elements, periodically arranged in three or two dimensions (Pendry 1999). The size of the structure is typically smaller than the free space wavelength of incoming electromagnetic waves. Nowadays, a multitude of MM structural elements type are known, conferring special electromagnetic properties. Depending on the frequency of the incident electromagnetic field, the type and geometrical shape, MM may have a high relative magnetic permeability either positive or negative (Pendry 1999). These properties strongly depend on the geometry of MM rather than their composition (Cai et al. 2007) and experimentally demonstrated by Smith et al. 2004, MMs have attracted intensive research interest from engineers and physicist in recent years because of their wide application in perfect lens (Shelby 2001), slow light (Bai et al. 2010), data storage (Wuttig &Yamada 2007), etc. All researchers reported until now, capitalize, in different manners the possibility to realize perfect lens. The metamaterial lens assure the possibility to apply of electromagnetic MM in eNDE. For a MM slab characterized by effective permittivity $\varepsilon_{eff}$ and effective magnetic permeability $\mu_{eff}$, the refractive index is

$$n = \sqrt{\varepsilon_{eff} \mu_{eff}} \quad (1)$$

and the impedance is given by

$$Z = \sqrt{\frac{\mu_{eff}}{\varepsilon_{eff}}} \quad (2)$$

The connection between ε and μ for a medium as well as the wave propagation through can be categorized into the following classes (Engheta & Ziolkowski 2006, Zouhudi & Shivola 2008), as presented in Figure 1.

When $\varepsilon_{eff} = -1$ and $\mu_{eff} = -1$, the refractive index of MM slab is n = −1 (Veselago 1968) and the surface impedance Z = 1, so there is no mismatch and consequently no reflection on the interface slab-air. This metamaterial slab forms perfect lens (Pendry 2000) and is focusing the electromagnetic field and also the evanescent waves (Pendry 2000). Due to experimental difficulties in obtaining a perfect lens, the manipulation of the evanescent modes can be made with a new type of electromagnetic sensors with MM lens that have, at the operation frequency, either $\varepsilon_{eff} = -1$ and electric evanescent modes can be manipulated, either $\mu_{eff} = -1$, and the lens can focus magnetic evanescent modes (Pendry 1999). The principle of the lens transducer made from two Conical Swiss Rolls (CSR), used for manipulating the evanescent and abnormal modes generated in slits of MSG or carbon fibers of CFRP excited with electromagnetic waves, polarized TEz and respectively TMz is shown in Figure 2.

The proper detection system is made from a lens transducer consisting of two identical CSR having the large basis front to front (Grimberg & Savin 2011). The focal distance of this lens is f = l, where l represents the height of CSR (Savin & Steigmann (2014). A conductive screen with a circular aperture having the diameter d ≪ λ (λ is the wavelength in vacuum) (Fig.2) is placed near the focal object point. A detection coil is placed in the focal image point, converting the localized energy into electromagnetic force (emf). The sample is raster scanned, recording the energy image pixel or electromagnetic signature.

Using the Fourier optics method (Born & Wolf, 1975), an object O(x,y) that can represent the eigenmode $e_v$ or $h_v$ in function of the polarization of incident electromagnetic field, has, while pass-

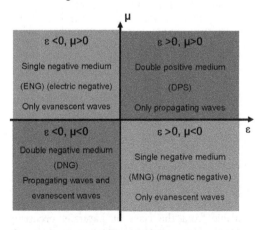

Figure 1. Schematic representation of materials depending on ε and μ.

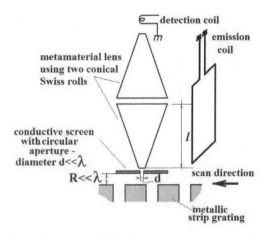

Figure 2. Detection of evanescent and abnormal modes generated in the slits of a metallic strip grating using lens transducer.

ing through the circular aperture and the lens, an imaging I(x′,y′) given by

$$I(x',y') = \frac{1}{\lambda^2 d_1 d_2} \int\int_{-\infty}^{\infty} \exp\left[i\frac{k\left((x'-x_1)^2+(y'-y_1)^2\right)}{2d_2}\right]$$
$$\times P(x,y)\exp\left[i\frac{k(x^2+y^2)}{2f}\right]$$
$$\left(\int\int_{-\infty}^{\infty} O(x,y)\exp\left[i\frac{k\left((x_1-x)^2+(y_1-y)^2\right)}{2d_1}\right]dxdy\right)dx_1 dy_1 \quad (3)$$

where P(x,y) is the pupil function defined as

$$P(x,y) = \begin{cases} 1 & x^2+y^2 \leq d^2 \\ 0 & \text{otherwise} \end{cases} \quad (4)$$

O(x,y) is the object defined as

$$O(x,y) = \begin{cases} e_y(x,y) \text{ for TEz polarized incident waves} \\ h_y(x,y) \text{ for TEz polarized incident waves} \end{cases} \quad (5)$$

$d_1 = R + l$ is the distance from the object to the center of the lens, $d_2 = l$ is the distance from the center of the lens to the detecting coil and $K = 2\pi/\lambda$ is the wave number.

## 3 STUDIED SAMPLES

The functioning of MM sensor has been verified using two types of materials:

MSG with polyimide support, one with 65 μm thickness with Ag conductive strips of 14 μm thickness, 1.2 mm width and 0.8 mm distance between conductive traces; another one with 10 μm thickness 0.6 mm width and 0.4 mm distance between conductive traces. The Ag strips were realized by successive deposition of Ag paste, using an adequate stencil by screen printed method. The adhesion of Ag on polyimide has been done with a thin film of resin. The conductive silver past made from microparticules with concentration >80%, density 10.49 g/cm³, resistivity 1÷3*10⁻⁵Ωcm was used (SPI supplies web site). Silver paste exhibit, in addition to a reasonably fast drying rate even at room temperature, good adhesion to most substrates and high conductivity. At frequencies around the value of 500 MHz, the permittivity of silver is $\varepsilon_m = -48.8 + j \cdot 3.16$ (Palik 1985).

Nowadays metallic films on flexible substrates (polyimide or plastic) have attracted more and more attention because have superior mechanical properties comparatively with the ones deposited on glass currently in many aspects. The use of polyimide implies a mechanically flexible substrate which can be easy manipulated into unique three-dimensional designs. So it provides an ideal surface for the selective attachment of various important bioactive species onto the device accomplished (Schlesinger 2010). The flexible MSG structure is presented in Figure 3a.

Plates from FRPC composite materials having 12 layers of carbon fibers woven type 5 H satin (Carbon T300 3 K 5HS) with a layout that assures the quasi-isotropic properties, with thickness of 4.2 mm and volume ratio 0.5 ± 0.1, made by CETEX The Netherlands are studied. The polymer matrix is made of PolyPhenylene Sulfide (PPS), an organic polymer consisting of aromatic rings linked with sulfides, resisting to chemical and thermal attack, and the gas released due to ignition of matrix is substantially low (TENCATE website).

The composites carbon fiber woven—PPS have increased strength to impact. The matrix of FRPC has the transverse electric conductivity between 10 S/m and 100 S/m is paramagnetic, and the carbon fibers of average conductivity are embedded into this matrix allowing eNDE. The sample were impacted with energy of 8 J using FRACTOVIS PLUS 9350-CEAST-Instron USA with a hemispherical bumper head with 20 mm diameter and 2.045 kg weight, in order to induce delaminations. Under impact, the FRPC suffer delamination that is usually accompanied by a dent. The dent causes a reduction in the spacing between fibers in the thickness direction and this causes an increase in fiber contact leading to decrease of electrical

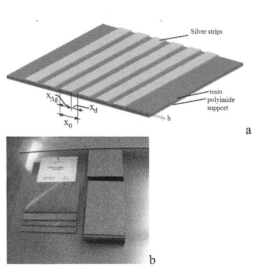

Figure 3. Studied samples: a) MSG; b) FRPC.

resistance in the thickness direction and modifies the electrical conductivity local both in the plane of the fibers and perpendicularly on fibers. The energy absorbed by the composite serves at the plastic deformation of the composite in the contact zone, being dissipated through internal friction between the matrix's molecules, carbon fibers, matrix-carbon fibers as well as at the creation of delaminations.

## 4 CONDITIONS REQUIRED FOR GENERATION OF EVANESCENT MODES IN MSG

The TMz polarization of incident field was created with a rectangular frame having one turn from 1.2 mm diameter Cu wire. The frame having 20 × 60 mm dimensions was placed at 3 mm height perpendicularly on the sample (CFRP/MSG) (Grimberg et al. 2012), the working frequency being 476 MHz. The position of the frame is with short side perpendicularly on the direction of the MSG traces/ carbon fibers.

The TMz wave illuminating the sample to be tested, generate evanescent waves (Grimberg 2013). The generation of the incident field is made within the frequency range of hundreds of MHz (adequate to eNDE) by the rectangular frame circulated by an alternative electrical current (Grimberg et al. 2008).

The calculation of the field can be made using the dyadic Green's functions method for free space as well as the method of an integral on the source volume (Grimberg et al. 2000).

## 5 EXPERIMENTAL SET-UP

The scheme of the measurement equipment is presented in Figure 4.

The sensor is connected to a Network/Spectrum/Impedance Analyzer type 4395 A Agilent USA and is maintained fixed during the measurements. The tested samples are set and displaced in front of the sensor with a motorized XY stage Newmark USA. The measurement system is commanded by PC through RS232 interface for displacement system and IEEE 488.2 for analyzer. The acquisition and storage of data are made by software developed in Matlab 2012b. The electromotive force induced in the reception coil of the measurement system is the average of 10 measurements in the same point in order to reduce the effects of the white noise, the bandwidth of the analyzer being set-up to 10 Hz, also to diminish the noise level.

The generation and detection of evanescent waves from slits/fibers has been made using and absolute send receiver sensor and the equipment presented in Figure 4.

The frequency dependency of lens effective magnetic permeability has been determined measuring the S parameters ($S_{11}$ and $S_{12}$) and applying the effective medium method (Grimberg 2013, Chen et al. 2004). The measurements of S parameters were carried out with Agilent S Parameters Test kit 87511 A coupled to Agilent 41395 Analyzer. In Figure 5 is presented the dependence by frequency of effective magnetic permeability of the lens used for manipulation of evanescent and abnormal modes.

The real component of the effective magnetic permeability reaches the maximum value at 473.8 MHz and the value of −1 at 476 MHz. The distance between screen aperture and the surface to be examined has been maintained at 20 μm ± 1 μm.

The rectangular frame used for generation of $TM_z$ polarized EM field having one turn with size of 20 × 60 mm, 1 mm diameter of Cu wire.

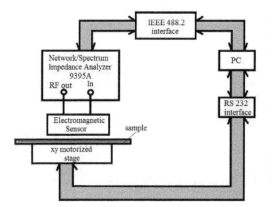

Figure 4. Experimental set-up.

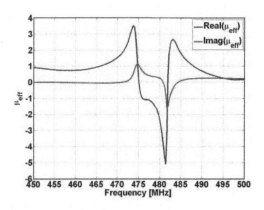

Figure 5. The dependence by frequency of effective magnetic permeability of the lens.

# 6 RESULTS AND DISCUSSIONS

The method and the developed sensor can serve for eNDE of conductive strips, in order to detect eventually interruptions of the strips or short-circuits between the traces.

According to Savin & Steigmann (2014), in the case of thick MSG with silver strips and subwavelength features, excited with $TM_z$ polarized incident electromagnetic field, it has been shown that only a single evanescent mode appears in slits; this mode disappears when water is inserted in slits ($\varepsilon_{water}$ = 81). This mode could be detected and visualised using MM lens described above. The theoretical and experimental study of eigenmodes that appear in this type of MSG, open new domains of applications in eNDE of stratified structures.

A region of 16 mm × 16 mm from MSG has been scanned with 0.1 mm steps in both directions. The scanning along x direction was made so that shall correspond to a period of grating, $x_0$. The working frequency was 476 MHz. During the measurements, the excitation frame and the lens have been maintained in fixed position, the MSG being displaced with a motorized XY stage Newmark USA.

The emf induced in reception coil represents the average of 10 successive measurements, in order to reduce the effect of white noise. For MSG with features compatible with λ of incident electromagnetic field, TMz polarized, the structure presents known selective properties of transmission and reflection (Balanis 1989). For thick MSG, according to Peterson et al. 1998, in the case TMz waves, the reflection coefficient for a strip grating with $x_0 \ll \lambda_0$ is practically 1. The presence of the low frequency plasmons, suggested by the presence of evanescent waves in slits (Grimberg 2012), having maxima of the field in the interior of the trace, opens the perspective of using the layered mesostructures in subwavelength regime as sensors and biosensors.

In Figure 6 is presented the dependency of emf amplitude induced in the reception coil of sensor at the scanning of the MSG taken into study.

It can be observed that the increasing of the sensor spatial resolution using metamaterials lens leads to a clear visibility of the traces profile. The roughness feature is emphasized by the propagation of surface polaritons (Kolomenski et al. 2009). It is known that biosensing characteristic is strongly dependent on the deposition condition that affects the physical properties of the thin film.

Extending the analysis to a layered structure having the features $x_{Ag}$ = 0.6 mm, $x_d$ = 0.4 mm and thickness of 10 μm with liquid dielectric media in the slits, excited with $TM_z$ polarized wave with λ = 0.6 m, we present the image of the evanescent and abnormal modes created in slits at the scanning of 1 × 1 mm² region with 10 μm step on both directions. When the slits are filled with air, only evanescent mode will be generated, the amplitude of the signal induced in the reception coil has the shape presented in Figure 7a.

In the case of excitation with TMz polarized wave, for large values of the liquid dielectric constants larger than 10, when $\varepsilon_d$ = 17.9, (isopropyl alcohol), in slits are generated abnormal modes (Fig 7 b). Because the real component of the propagation constant $\beta_v$ for isopropyl alcohol is smaller ($\beta_v$ = 4.2 + i·104.321) than the imaginary component, abnormal modes will be generated in slits, the electromagnetic image of these modes shows a similar behavior like in the case of air in the slits. The amplitude of the signal is smaller and has a central maximum more flat (Fig 7b).

For the second type of material taken into study, the electromagnetic behavior of composite was simulated by FDTD software, the samples being CAD designed following textiles features, and compared with eNDE tests. One cell of carbon fiber woven has been designed in TexGen—Textile Geometric modeler software (Fig. 8a) and exported as CAD format in order to be used in FDTD software—XFDTD Remcom USA. The 5 × 5 tows were represented in different colors in order to easily follow their intersection as woven. Cell dimension in FDTD simulation is 0.04 mm, the grid size being 146 × 146 × 45 cells. The perfectly matched layer boundary condition was applied at the grid boundaries.

The result of the simulation is presented in Figure 8b. The image is a snapshot from a field sequence, showing the $H_y$ field progress at a particular slice of the geometry. Because the carbon fibers act as MSG, the apparition of evanescent waves can be emphasized.

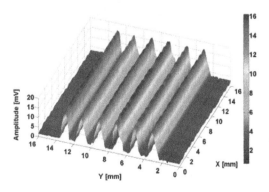

Figure 6. The results at the scanning of 16 × 16 mm² surface.

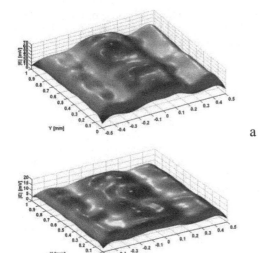

Figure 7. Image of evanescent and abnormal modes generated in slits for TMz polarized excitation at frequency of 476 MHz: (a) slits filled with air; (b) slits filled with isopropyl alcohol.

Figure 8. One cell carbon woven simulation: a) TexGen—different colors show the tows intersections in the 5 HS woven; b) XFDTD—$H_y$ propagation in a plane orthogonal at composite.

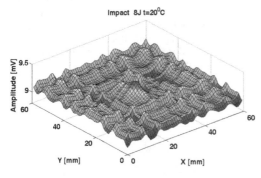

Figure 9. The answer of the sensor with metamaterials lens for 8 J impact.

A region of 60 × 60 mm from the sample has been scanned in both directions with 1 mm step, using the electromagnetic sensors described above.

Thus, we present the results obtained at testing of the FRPC samples impacted with 8 J, in Figure 9 is show, the answer of sensor with metamaterials lens in 3D presentation.

It can be shown that the metamaterials lens allow the increasing of spatial resolution, the layout of the woven being emphasized. The proposed method can be thus extended not only for the evaluation of MSG but also for the eNDE of FRPC in order to evaluate delaminations as well as the woven layout.

## 7 CONCLUSIONS

Two CSR were selected to have appropriate working frequencies and approximate the same effective magnetic permeability to construct the MM lens.

The excitation of the studied structure with a TMz polarized electromagnetic wave can be achieved with a rectangular frame coil, the plane of the frame being perpendicularly to the surface of the scanned structure.

In a thick MSG having silver strips with dimensions $x_{Ag} = 0.6$ mm, $x_d = 0.4$ mm, h = 10 µm, excited with a TMz electromagnetic wave with 476 MHz frequency, a single purely evanescent mode appears in the slits, that can be detected using MM lens and a reception coil placed in its focal plane.

In order to improve the quality of transmission through lens, (without reducing the value of $S_{21}$ parameter too much) and the intensification of the evanescent mode, the circular aperture is placed in the front of the first CSR with 20 µm lift-off.

Due to the amplitude distribution of the evanescent mode/modes in the slits, it can be considered that at the interface metal-air, on the vertical walls of the strip, SPPs appear when the structure is excited with TMz polarized wave.

Using the sensor with MM lens, interruptions, short-cuts of MSG as well as nonalignment of carbon fibers, lack of resin or voids, and delaminations induced by impacts with low energy can be detected and emphasized.

## 8 AUTHOR CONTRIBUTIONS

The manuscript was written through contributions of all authors. All authors have given approval to the final version of the manuscript and contributed equally.

## ACKNOWLEDGEMENTS

This paper is partially supported by Romanian Ministry of Education and Scientific Research under projects PN-II-ID-PCE-2012–4-0437 and Nucleus Program—Contract PN 09 43–01–04.

## REFERENCES

Bai, Q., Liu, C., Chen, J., Cheng, C. & Kang, M. 2010. Tunable slow light in semiconductor metamaterial in a broad terahertz regime. J. Appl. Phys. 107, 093104:1–093104:8.

Balanis, C. 1989. *Advanced Engineering Electromagnetics*, Wiley Eds, NY.

Bladel, J. 2007. *Electromagnetic fields*. 2nd Ed. Wiley-IEEE Press.

Born, M. & Wolf, E. 1975. *Principle of Optics*, 5th Ed. Pergamon Press: Oxford, UK.

Cai, W., Chettiar, U.K., Kildishev, A.V. & Shalaev V.M. 2007. Optical cloaking with metamaterials. Nature Photonics. 1. 4: 224–227.

Chen, X., Grzegorczyk, T.M., Wu, B.I., Pacheco, J.J. &Kong, J.A. 2004. Robust method to retrieve the constitutive effective parameters of metamaterials. Phys. Rev. E 70, 016608.

Engheta, N. & Ziolkowski R.W. 2006. *Electromagnetic Metamaterials: Physics and Engineering Explorations*. Wiley. NY.

Grimberg, R. & Savin, A. 2011. Electromagnetic transducer for evaluation of structure and integrity of the composite materials with polymer matrix reinforced with carbon fibers, Patent RO126245-A0/2011.

Grimberg, R. 2013. Electromagnetic metamaterials. Mater. Sci. Eng. B. 178: 1285–1295.

Grimberg, R., Savin, A. & Steigmann, R. 2012. Electromagnetic imaging using evanescent waves, NDT and E International., 46: 70–76.

Grimberg, R., Savin, A., Steigmann, R., Serghiac, B. & Bruma, A. 2011a. Electromagnetic non-destructive evaluation using metamaterials. *INSIGHT*. 53. 3: 132–137.

Grimberg, R., Udpa, L. & Udpa, SS. 2008. Electromagnetic transducer for the determination of soil condi-tion. *International Journal of Applied Electromagnetics and Mechanics*. 28. 2–01: 201–210.

Kolomenski, A., Kolomenskii, A., Noel, J., Peng, S. & Schuessler, H. 2009. Propagation length of surface plasmons in a metal film with roughness. *Applied Optics*, 48(30), 5683–5691.

Palik, E.D. 1985. *Handbook of Optical Constants of Solids*; Academic Press: London, UK.

Pendry, J., Holden, A.J., Robbins, D.J. & Stewart, W.J. 1999 Magnetism from conductors and enhanced non-linear phenomena, *IEEE Trans. Microw Theory Tech.* 47: 47–58.

Pendry, J.B. 2000. Negative Refraction Makes on Perfect Lens. *Phys. Rev. Lett.* 85: 3966–3969.

Peterson, A.E., Ray, S.L. & Mittro, R.1989. Computational Methods for Electromagnetics, IEE Press, NY.

Schlesinger M. 2010. Deposition on nonconductors, Ch. 15 on *Modern Electroplating*, 5th Ed., John Wiley & Sons: 413–420.

Shelby, R.A., Smith D.R. & Schultz S. 2001. Experimental Verification of a Negative Index of Refraction. *Science*. 6. 292: 77–79.

Smith, D.R., Pendry, J.B. & Wiltshire, M.C.K. 2004. Metamaterials and negative refractive index. *Science*. 305: 788–792.

Veselago, V.G. 1968. The electrodynamics of substances with simultaneously negative values of $\varepsilon$ and $\mu$. *Physics-Uspekhi*. 10: 509–514.

Wuttig, M. & Yamada, N. 2007 Phase-change materials for rewriteable data storage. Nat. Mater. 6: 824–832.

Zouhudi, S., Shivola, A. & Vinogradov, A.P. 2008. *Metamaterials and Plasmonics: Fundamentals, Modelling, Applications*. Springer. NY.

www.tencate.com

www.2 spi.com/catalog/spec_prep/plus_prop.html

# Author index

Accornero, F. 433
Aggelis, D.G. 93, 125, 235, 255, 287, 367
Alexakis, E. 421
Alver, N. 185
Anastasopoulos, A. 197
Anisimov, A.G. 205
Antikainen, J. 309
Arroud, G. 353

Baccouche, Y. 161
Badogiannis, E. 191
Bagheri, A. 153, 173
Balachandran, W. 247
Battistella, M. 505
Benedictus, R. 213
Bentahar, M. 161
Berghmans, F. 529
Bollas, K. 197
Boone, M.N. 551
Bossuyt, F. 293
Bruma, A. 319, 569

Calò, S. 505
Carpinteri, A. 413, 433
Chacon, J.L.F. 341
Chai, H.K. 49
Chang, K.C. 25, 33, 61, 85
Chatziangelou, M. 383
Cheng, L. 247
Chiesura, G. 293, 537, 551
Chong, A.Y.B. 267
Choumanidis, D. 191
Christaras, B. 383
Colla, C. 397
Concu, G. 335
Coughenour, C.M. 421
Craus, M.L. 319
Cutugno, P. 433

Dannemann, M. 375
De Baere, D. 287
De Belie, N. 227
De Nicolo, B. 335
De Pauw, B. 537

De Sutter, S. 93
Degrieck, J. 133, 141, 293, 529, 543, 551
Deiana, R. 505
Delrue, S. 117, 133, 141, 167, 179
Deraemaeker, A. 367
Devlioti, K. 383
Devriendt, C. 235
Dhaene, J. 551
Di Cara, A. 315
Diakides, I. 103
Dimitraki, L. 383
Dirckx, J.J.J. 261
Doulamis, A. 389
Driussi, G. 519

El Guerjouma, R. 161
Erba, S. 493
Ercan, E. 185

Falchi, L. 519
Faresin, E. 505
Feiteira, J. 227
Fernandes, F.M. 465
Frestazzi, A. 505
Frezza, L. 449
Fritsch, D. 421
Fujiwara, Y. 85

Gan, T.-H. 247, 267, 341
García-Fuentes, M.A. 397
Geernaert, T. 529
Geller, S. 361
Gentile, C. 477, 485
Georgopoulos, A. 405
Ghiassi, B. 511
Giacomello, G. 505
Giaretton, M. 505
Givalos, G.L. 327
Gortsas, T. 103
Gresil, M. 109
Grosse, C.U. 227
Groves, R.M. 205, 213
Grum, J. 147, 241
Gruyaert, E. 227

Gude, M. 361
Guidobaldi, M. 485
Guillaume, P. 287
Gutierrez, M.F. 421
Gwarek, W. 267

Hara, M. 33
Hashinoki, M. 33
Hatziioannidis, I. 267
Hernández, J.L. 397
Hettler, J. 117, 179
Holeczek, K. 375

Idjimarene, S. 161
Iftimie, N. 569
Iliopoulos, A.N. 235
Iliopoulos, S.N. 93, 125
Invernizzi, S. 413, 433
Ioannides, M. 421
Ioannou, I. 499

Jackson, B.K. 347
Jimenez, J. 341

Kamilari, E.Th. 427
Kappatos, V. 247, 267, 341
Karaiskos, G. 367
Katunin, A. 557
Kaufmann, M. 293
Kek, T. 147, 241
Kersemans, M. 133, 141
Kinoshita, T. 33
Kioussi, A. 389
Kobayashi, Y. 41, 55
Kogia, M. 247
Komninou, E. 191
Konstantinova, T.E. 319
Kontoyannis, C.G. 327, 427
Kopyt, P. 267
Koui, M. 267
Kourousis, D. 197
Kouvaritaki, S.N. 427
Krüger, M. 441
Kumar, S. 563
Kusić, D. 241

La Malfa Ribolla, E. 173, 315
Lacidogna, G. 413, 433
Lamberti, A. 537
Lammens, N. 543
Lee, F.W. 49
Lehmann, F. 441
Li, W. 347
Liberatore, D. 449
Lim, K.S. 49
Lopez-Menchero Bendicho, V.M. 421
Lourenço, P.B. 465
Luyckx, G. 293, 529, 543, 551

Malm, F. 227
Malo, S. 319
Manganelli Del Fà, R. 457
Manikas, E.D. 299
Manning, E.C. 465
Martens, A. 133, 141
Masini, N. 449
Matikas, T.E. 299
Matsumoto, K. 55
Mechri, C. 161
Meiss, A. 397
Minnebo, P. 255
Modler, N. 375
Mohimi, A. 247
Mojškerc, B. 147
Möller, U. 281
Momoki, S. 55
Morii, T. 79
Moropoulou, A. 383, 389, 421
Muller, A. 109
Müller, B. 205

Nahm, M. 227
Nomikos, P. 191
Novak, A. 161

Ochôa, P. 213
Ogura, N. 61
Ohtsu, M. 3
Oikonomou, P. 191
Oliveira, D.V. 511
Olmi, R. 471
Orkoula, M.G. 327, 427

Panizza, M. 505
Papasalouros, D. 197
Papaspyridakou, P.S. 327
Pasetto, M. 505
Peeters, J. 261, 353
Pfeiffer, H. 219

Pistone, E. 153
Pitropakis, I. 219
Plagge, R. 281
Polyzos, D. 103, 125
Pyl, L. 133

Racina, V. 449
Rajcic, V. 421
Ramos, L.F. 465
Realini, M. 493
Rescic, S. 457
Ribbens, B. 261
Riminesi, C. 457, 471
Rizzo, P. 153, 173, 315
Romero, A. 341
Rosina, E. 471, 493

Saisi, A. 477, 485
Salemi, G. 505
Salski, B. 267
Sansonetti, A. 457, 493
Sarris, J. 275
Savin, A. 319, 569
Scalerandi, M. 161
Schlangen, E. 227
Schoonacker, M. 219
Sekler, H. 219
Selcuk, C. 247, 267, 341
Shiotani, T. 25, 41, 55, 61, 69, 79, 85
Siakavellas, N.J. 275
Sileo, M. 449
Sinke, J. 205
Sofianos, A. 191
Sol, H. 133
Solodov, I. 13
Sonnenfeld, C. 529
Sorrentino, L. 449
Soua, S. 341
Soutis, C. 109
Starke, E. 375
Steenackers, G. 261, 353
Steigmann, R. 569
Stelzmann, M. 281
Strantza, M. 287
Sulejmani, S. 529
Sülün, Ö.Y. 185
Suzuki, T. 79

Tabatabaeipour, M. 117, 179
Tamrakar, S.B. 25, 85
Tan, M.J. 563
Tan, S.M. 267
Tanarslan, H.M. 185

Tatsis, E.N. 299
Theodorakeas, P. 267
Theodoridou, M. 499
Thienpont, H. 529
Thomas, J.-H. 161
Tiano, P. 457
Toumi, S. 161
Tournat, V. 161
Trulli, N. 335
Tsangouri, E. 227, 367
Tsinopoulos, S.T. 103
Turchenko, V. 319
Tysmans, T. 93
Tziviloglou, E. 227

Valdés, M. 335
Valluzzi, M.R. 505
Valsasnini, L. 477
Van Balen, K. 511
Van Den Abeele, K. 117, 133, 141, 167, 179
Van Hemelrijck, D. 235, 255, 287, 367, 529
Van Hoorebeke, L. 551
Van Paepegem, W. 133, 141, 543, 551
Vanfleteren, J. 293
Vanlanduit, S. 537
Vardaki, M.Z. 327
Verbruggen, S. 93
Verstrynge, E. 511
Vervust, T. 293
Vincent, M.L. 421
Voet, E. 543, 551

Wall, J.J. 347
Warwick, J.L.W. 347
Weeks, N. 563
Wevers, M. 219, 511
Wille, S. 353
Winkler, A. 361, 375
Wong, B.S. 563

Xiaolei, D. 309

Yang, Y. 293
Yatsumoto, H. 61

Zarnic, R. 421
Zastavnik, F. 133
Zendri, E. 519
Zoidis, N.V. 299